Scientific Basis for Nuclear Waste Management XXXI

MATERIALS RESEARCH SOCIETY
SYMPOSIUM PROCEEDINGS VOLUME 1107

Scientific Basis for Nuclear Waste Management XXXI

Symposium held September 16–21, 2007, Sheffield, United Kingdom

EDITORS:

William E. Lee
Imperial College London
London, United Kingdom

John W. Roberts
University of Sheffield
Sheffield, United Kingdom

Neil C. Hyatt
University of Sheffield
Sheffield, United Kingdom

Robin W. Grimes
Imperial College London
London, United Kingdom

Materials Research Society
Warrendale, Pennsylvania

CAMBRIDGE
UNIVERSITY PRESS

University Printing House, Cambridge CB2 8BS, United Kingdom

One Liberty Plaza, 20th Floor, New York, NY 10006, USA

477 Williamstown Road, Port Melbourne, VIC 3207, Australia

314-321, 3rd Floor, Plot 3, Splendor Forum, Jasola District Centre, New Delhi - 110025, India

79 Anson Road, #06-04/06, Singapore 079906

Cambridge University Press is part of the University of Cambridge.

It furthers the University's mission by disseminating knowledge in the pursuit of education, learning and research at the highest international levels of excellence.

www.cambridge.org
Information on this title: www.cambridge.org/9781605110790

Materials Research Society
506 Keystone Drive, Warrendale, PA 15086
http://www.mrs.org

First published 2008
First paperback edition 2012

Single article reprints from this publication are available through University Microfilms Inc., 300 North Zeeb Road, Ann Arbor, MI 48106

CODEN: MRSPDH

A catalogue record for this publication is available from the British Library

ISBN 978-1-605-11079-0 Hardback
ISBN 978-1-107-40850-0 Paperback

CONTENTS

ILW WASTEFORMS

GLASSES AND GLASS COMPOSITE MATERIALS

LEACHING

CERAMIC WASTEFORMS

RADIATION EFFECTS IN CERAMICS

PLUTONIUM WASTEFORMS

MOBILIZATION, MIGRATION AND RETENTION

PERFORMANCE ASSESSMENT

PREFACE

The 31st International Symposium on the Scientific Basis for Nuclear Waste Management organized and hosted by the Immobilisation Science Laboratory (ISL) of the University of Sheffield was held September 16–21, 2007 in Sheffield, England. 140 attendees from 18 countries heard over 60 talks and viewed an equal number of posters during the four days of scientific sessions, interrupted by an excellent social program including a Yorkshire themed reception in the Winter Gardens in Sheffield, a visit to the Duke of Devonshire's home Chatsworth in Derbyshire, and a banquet in the magnificent Cutlers' Hall in the city centre.

The Symposium offered an overview of the status of the UK's clean-up program and reviewed international research programs including those in Europe and the United States. Topics covered in the sessions included: Geological Disposal Concepts; ILW Wasteforms; Glasses and Glass Composite Materials; Leaching; Ceramic Wasteforms; Radiation Effects; Pu Wasteforms; Spent Fuel; Engineered Barrier Systems; Mobilization, Migration and Retention; and Performance Assessment. Invited lectures by leading experts kicked off each session providing a snapshot of the state-of-the-art in the pertinent topic.

A technical visit to Sellafield was arranged for the last day including tours of many of the plant and storage facilities.

Many people worked hard to ensure the success of this Symposium and we must take this opportunity to recognize their efforts and thank them here. As well as ourselves the International Technical Committee comprised: Bruce Begg, Boris Burakov, Neil Chapman, Sergey Dmitriev, Rod Ewing, Graham Fairhall, Bernd Grambow, Jim Marra, Neil Milestone, Denis Strachan, Pierre Van Iseghem, and Etienne Vernaz. This group was responsible for sifting abstracts and organizing and chairing the sessions, organizing two reviews for, and the final acceptance of, each paper.

However, the key person in organizing this meeting was John Roberts (the ISL Manager) who worked tirelessly over many months (through flood and famine!) to ensure its success. Most members of the ISL put in tremendous efforts, particularly in the weeks leading up to the conference, to make sure things ran smoothly.

Finally, we have to thank our sponsors (the Nuclear Decommissioning Authority, AWE, Nexia Solutions, Imperial College London, the ISL, AMEC, and SIA Radon), particularly the NDA who provided funds to enable the conference to meet its main objective of being the world's premier forum for scientific research into nuclear waste management.

William E. Lee
John W. Roberts
Neil C. Hyatt
Robin W. Grimes

London and Sheffield, UK

February 2008

MATERIALS RESEARCH SOCIETY SYMPOSIUM PROCEEDINGS

Volume 1066 — Amorphous and Polycrystalline Thin-Film Silicon Science and Technology—2008, A. Nathan, J. Yang, S. Miyazaki, H. Hou, A. Flewitt, 2008, ISBN 978-1-60511-036-3
Volume 1067E —Materials and Devices for "Beyond CMOS" Scaling, S. Ramanathan, 2008, ISBN 978-1-60511-037-0
Volume 1068 — Advances in GaN, GaAs, SiC and Related Alloys on Silicon Substrates, T. Li, J. Redwing, M. Mastro, E.L. Piner, A. Dadgar, 2008, ISBN 978-1-60511-038-7
Volume 1069 — Silicon Carbide 2008—Materials, Processing and Devices, A. Powell, M. Dudley, C.M. Johnson, S-H. Ryu, 2008, ISBN 978-1-60511-039-4
Volume 1070 — Doping Engineering for Front-End Processing, B.J. Pawlak, M. Law, K. Suguro, M.L. Pelaz, 2008, ISBN 978-1-60511-040-0
Volume 1071 — Materials Science and Technology for Nonvolatile Memories, O. Auciello, D. Wouters, S. Soss, S. Hong, 2008, ISBN 978-1-60511-041-7
Volume 1072E —Phase-Change Materials for Reconfigurable Electronics and Memory Applications, S. Raoux, A.H. Edwards, M. Wuttig, P.J. Fons, P.C. Taylor, 2008, ISBN 978-1-60511-042-4
Volume 1073E —Materials Science of High-k Dielectric Stacks—From Fundamentals to Technology, L. Pantisano, E. Gusev, M. Green, M. Niwa, 2008, ISBN 978-1-60511-043-1
Volume 1074E —Synthesis and Metrology of Nanoscale Oxides and Thin Films, V. Craciun, D. Kumar, S.J. Pennycook, K.K. Singh, 2008, ISBN 978-1-60511-044-8
Volume 1075E —Passive and Electromechanical Materials and Integration, Y.S. Cho, H.A.C. Tilmans, T. Tsurumi, G.K. Fedder, 2008, ISBN 978-1-60511-045-5
Volume 1076 — Materials and Devices for Laser Remote Sensing and Optical Communication, A. Aksnes, F. Amzajerdian, N. Peyghambarian, 2008, ISBN 978-1-60511-046-2
Volume 1077E —Functional Plasmonics and Nanophotonics, S. Maier, 2008, ISBN 978-1-60511-047-9
Volume 1078E —Materials and Technology for Flexible, Conformable and Stretchable Sensors and Transistors, 2008, ISBN 978-1-60511-048-6
Volume 1079E —Materials and Processes for Advanced Interconnects for Microelectronics, J. Gambino, S. Ogawa, C.L. Gan, Z. Tokei, 2008, ISBN 978-1-60511-049-3
Volume 1080E —Semiconductor Nanowires—Growth, Physics, Devices and Applications, H. Riel, T. Kamins, H. Fan, S. Fischer, C. Thelander, 2008, ISBN 978-1-60511-050-9
Volume 1081E —Carbon Nanotubes and Related Low-Dimensional Materials, L-C. Chen, J. Robertson, Z.L. Wang, D.B. Geohegan, 2008, ISBN 978-1-60511-051-6
Volume 1082E —Ionic Liquids in Materials Synthesis and Application, H. Yang, G.A. Baker, J.S Wilkes, 2008, ISBN 978-1-60511-052-3
Volume 1083E —Coupled Mechanical, Electrical and Thermal Behaviors of Nanomaterials, L. Shi, M. Zhou, M-F. Yu, V. Tomar, 2008, ISBN 978-1-60511-053-0
Volume 1084E —Weak Interaction Phenomena—Modeling and Simulation from First Principles, E. Schwegler, 2008, ISBN 978-1-60511-054-7
Volume 1085E —Nanoscale Tribology—Impact for Materials and Devices, Y. Ando, R.W. Carpick, R. Bennewitz, W.G. Sawyer, 2008, ISBN 978-1-60511-055-4
Volume 1086E —Mechanics of Nanoscale Materials, C. Friesen, R.C. Cammarata, A. Hodge, O.L. Warren, 2008, ISBN 978-1-60511-056-1

MATERIALS RESEARCH SOCIETY SYMPOSIUM PROCEEDINGS

Volume 1087E —Crystal-Shape Control and Shape-Dependent Properties—Methods, Mechanism, Theory and Simulation, K-S. Choi, A.S. Barnard, D.J. Srolovitz, H. Xu, 2008, ISBN 978-1-60511-057-8
Volume 1088E —Advances and Applications of Surface Electron Microscopy, D.L. Adler, E. Bauer, G.L. Kellogg, A. Scholl, 2008, ISBN 978-1-60511-058-5
Volume 1089E —Focused Ion Beams for Materials Characterization and Micromachining, L. Holzer, M.D. Uchic, C. Volkert, A. Minor, 2008, ISBN 978-1-60511-059-2
Volume 1090E —Materials Structures—The Nabarro Legacy, P. Müllner, S. Sant, 2008, ISBN 978-1-60511-060-8
Volume 1091E —Conjugated Organic Materials—Synthesis, Structure, Device and Applications, Z. Bao, J. Locklin, W. You, J. Li, 2008, ISBN 978-1-60511-061-5
Volume 1092E —Signal Transduction Across the Biology-Technology Interface, K. Plaxco, T. Tarasow, M. Berggren, A. Dodabalapur, 2008, ISBN 978-1-60511-062-2
Volume 1093E —Designer Biointerfaces, E. Chaikof, A. Chilkoti, J. Elisseeff, J. Lahann, 2008, ISBN 978-1-60511-063-9
Volume 1094E —From Biological Materials to Biomimetic Material Synthesis, N. Kröger, R. Qiu, R. Naik, D. Kaplan, 2008, ISBN 978-1-60511-064-6
Volume 1095E —Responsive Biomaterials for Biomedical Applications, J. Cheng, A. Khademhosseini, H-Q. Mao, M. Stevens, C. Wang, 2008, ISBN 978-1-60511-065-3
Volume 1096E —Molecular Motors, Nanomachines and Active Nanostructures, H. Hess, A. Flood, H. Linke, A.J. Turberfield, 2008, ISBN 978-1-60511-066-0
Volume 1097E —Mechanical Behavior of Biological Materials and Biomaterials, J. Zhou, A.G. Checa, O.O. Popoola, E.D. Rekow, 2008, ISBN 978-1-60511-067-7
Volume 1098E —The Hydrogen Economy, A. Dillon, C. Moen, B. Choudhury, J. Keller, 2008, ISBN 978-1-60511-068-4
Volume 1099E —Heterostructures, Functionalization and Nanoscale Optimization in Superconductivity, T. Aytug, V. Maroni, B. Holzapfel, T. Kiss, X. Li, 2008, ISBN 978-1-60511-069-1
Volume 1100E —Materials Research for Electrical Energy Storage, J.B. Goodenough, H.D. Abruña, M.V. Buchanan, 2008, ISBN 978-1-60511-070-7
Volume 1101E —Light Management in Photovoltaic Devices—Theory and Practice, C. Ballif, R. Ellingson, M. Topic, M. Zeman, 2008, ISBN 978-1-60511-071-4
Volume 1102E —Energy Harvesting—From Fundamentals to Devices, H. Radousky, J. Holbery, B. O'Handley, N. Kioussis, 2008, ISBN 978-1-60511-072-1
Volume 1103E —Health and Environmental Impacts of Nanoscale Materials—Safety by Design, S. Tinkle, 2008, ISBN 978-1-60511-073-8
Volume 1104 — Actinides 2008—Basic Science, Applications and Technology, B. Chung, J. Thompson, D. Shuh, T. Albrecht-Schmitt, T. Gouder, 2008, ISBN 978-1-60511-074-5
Volume 1105E —The Role of Lifelong Education in Nanoscience and Engineering, D. Palma, L. Bell, R. Chang, R. Tomellini, 2008, ISBN 978-1-60511-075-2
Volume 1106E —The Business of Nanotechnology, L. Merhari, A. Gandhi, S. Giordani, L. Tsakalakos, C. Tsamis, 2008, ISBN 978-1-60511-076-9
Volume 1107 — Scientific Basis for Nuclear Waste Management XXXI, W.E. Lee, J.W. Roberts, N.C. Hyatt, R.W. Grimes, 2008, ISBN 978-1-60511-079-0

Prior Materials Research Society Symposium Proceedings available by contacting Materials Research Society

National and International Programs

Mater. Res. Soc. Symp. Proc. Vol. 1107 © 2008 Materials Research Society

The UK National Nuclear Laboratory and Waste Management R&D

Graham A. Fairhall
Nexia Solutions,
Sellafield, Seascale,
Cumbria, CA20 1PG, UK

INTRODUCTION

In the autumn of 2006, the UK government declared its expectation to set up a National Nuclear Laboratory (NNL) as soon as possible (subject to satisfactory contractual terms). The aim would be to base the Laboratory around Nexia Solutions and its 'state of the art' facility at Sellafield in Cumbria. The initial phase of the work to recommend formation of the laboratory is well advanced, and the NNL is expected to have a remit for the following roles:

— to play a key role in supporting the UK's strategic R&D requirements
— to operate world-class facilities
— to ensure key skills are safeguarded and enhanced
— to play a key role in the development of the UK's R&D supply base.

Within this remit it is to be expected that the NNL will play a key role in the R&D for successful management of radioactive waste that the UK will need for at least the rest of this century. This paper provides an overview of the role expected to be required of the NNL and the likely scope of its waste management R&D. Examples are given from current programmes which the NNL's predecessor (Nexia Solutions) is undertaking.

WHAT IS THE NATIONAL NUCLEAR LABORATORY?

The National Nuclear Laboratory will be a customer-funded organisation which will undertake R&D for nuclear-fission related applications across the entire UK nuclear industry. Placing Nexia Solutions at the heart of the NNL will ensure that the laboratory has a strong heritage of nuclear-related skills and expertise. This unique skill base was assembled when Nexia Solutions was established out of the former Research & Technology division of BNFL. It encompasses the nuclear fuel and waste management expertise of BNFL, the reactor R&T activities of the former Magnox Electric and the nuclear science skills from AEA(T), which was acquired by BNFL in 2003. Consequently the NNL is expected to constitute the bulk of the UK's remaining civil nuclear fission research capability and all of the active research facilities.

The technical skill base is impressive in its scope; it covers the full scope of fission reactors and their associated fuel cycles. Also included is the necessary challenge of decommissioning and clean-up. The figure below illustrates the range of the skills and expertise available in Nexia Solutions.

Reactor Systems & Services	**Fuel cycle &waste management**	**Decommissioning & Clean-up**
Reactor system support Support to reactor operations Nuclear analysis Post-irradiation examination Fuel design & performance Production of experimental fuels	Reprocessing & Wastes management technical support Development & testing services Plant monitoring Inspection, asset care & life-time extension Remote engineering solutions Waste characterisation Aerial & liquid effluent treatment Cementation/encapsulation Vitrification	Waste retrieval Waste process development Decontamination Clean-out of residual materials Dismantling of radioactive plant Characterisation of contaminated land Environmental impact assessment.

Scope of skills available to the UK's National Nuclear Laboratory

WHY IS THE NNL NEEDED?

The intent to establish the NNL recognises the key element that technology, and R&D in particular, plays in the safe and effective utilisation of nuclear fission. The UK has a long term need to secure its technology base in this area, which can be summarised as follows:

- The UK has a need for civil nuclear fission technology beyond the timescales in which commercial R&D organisations work
- Facilities for the study of radioactive materials need to be used effectively for the benefit of the entire UK nuclear industry
- A stable sustainable technology and skill base will be needed
- An approach is required to address the loss of skills in nuclear technology
- R&D plays a critical role in helping to deliver programmes of work cheaper, safer and faster through innovative technologies, it also helps to train and develop the skill base of the future and allow the UK to assess future options

VISION FOR THE NNL

The stakeholders who have been advising the UK government on the formation of the NNL recommend that it should:

"Lead the application of world class science and technology, in the field of national and international nuclear energy and associated fuel cycles, to enhance the social and economic value of the UK economy, whilst maintaining and developing key strategic skills and facilities".

The scope of the NNL's mission will be 3-fold:

1. Maintenance of capability in existing nuclear power and fuel cycle systems
2. To provide the UK with the technology in order to exploit the use of power generation by nuclear fission in the future
3. To enable the UK to undertake the necessary clean-up of the historic legacy and to develop long term solutions to the disposal (or disposition) of radioactive waste.

Strategic R&D programmes

The figure below proposes a future scenario for nuclear fission-related activities in the UK from which it is possible to understand the scope and duration of strategic R&D programmes which the National Nuclear Laboratory might undertake. While this scenario is not unique (certainly the timescales are not fixed) it does illustrate the challenges that a National Nuclear Laboratory would be well placed to address:-

- There is a need to maintain national priorities in nuclear technology over many decades
- There is a need to renew the skill base over more than one generation of scientists and engineers
- There is a need to ensure co-ordination between the different parts of the scenario to gain maximum benefit from the technology.

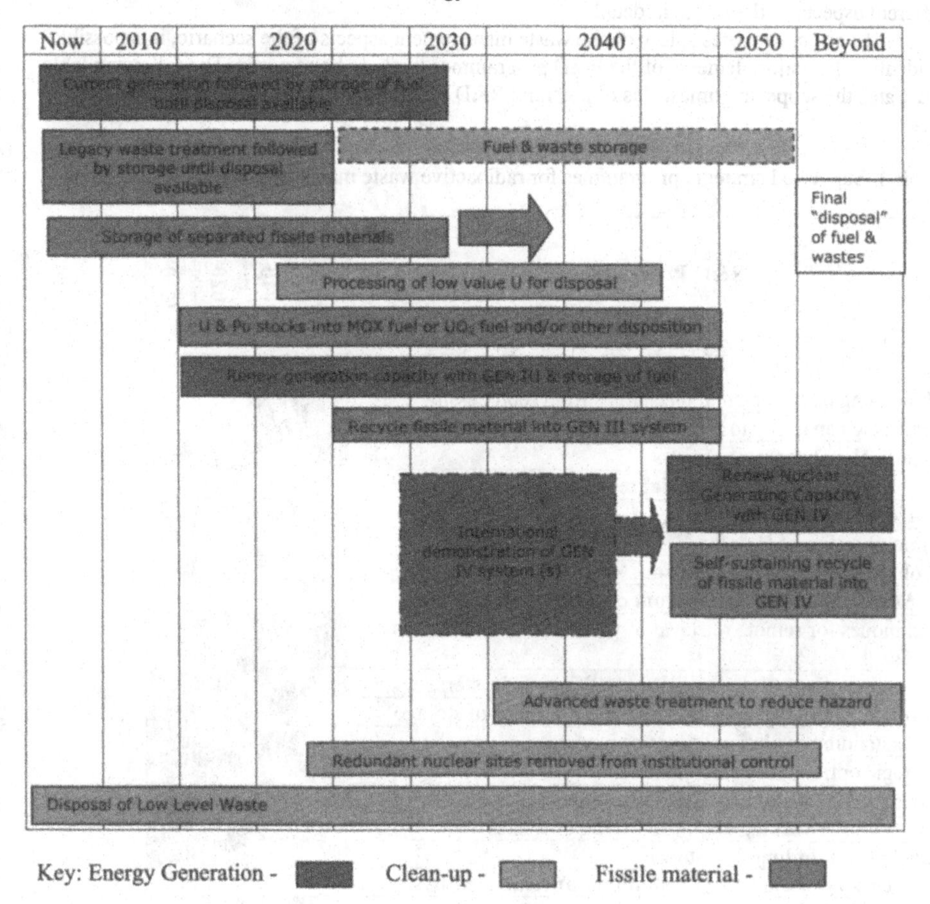

Figure 1. Scenario for future nuclear-fission related activities in the UK

NNL RESPONSE TO THE CHALLENGE OF MANAGING RADIOACTIVE WASTES

A major aspect of the National Nuclear Laboratory's programmes can be expected to include all aspects of managing radioactive wastes, both from the UK's legacy from its early decades of developing nuclear power and from its current nuclear generation systems.

In general terms the challenges are covered by simple statements, such as, to achieve the safe and effective management of radioactive wastes for the medium term and to ensure that conditioned radioactive wastes are compatible with the schemes eventually developed for final and permanent management. In reality this requires a range of R&D programmes to address the different aspects of the issues in detail.

If one concentrates solely on the waste management aspects of the scenario, it is possible to identify the major elements of the R&D programmes likely to be required. The following table illustrates the scope and timescales of potential R&D programmes in the waste management area.

Table I. Suggested strategic programmes for radioactive waste management

R&D Programmes	Now	2010	2020	2030
Operation of current reactors & chemical plant				
Reducing Pu-residues from MOX manufacture	■	■		
Minimising environmental discharges from reprocessing	■	■		
Increasing capability to reprocess orphan fuels & fuel residues	■	■		
Spent fuel management options	■	■	■	■
Fissile material management				
Re-use of separated U & Pu	■	■	■	■
Immobilisation of U & Pu	■	■	■	■
Evolution of materials in storage	■	■	■	■
Management of wastes from current nuclear systems				
Techniques for remote retrieval of wastes from vaults, silos & ponds	■	■	■	
Processing of wastes to facilitate conditioning for storage	■	■	■	■
Development of wasteforms suitable for decades of storage		■	■	■
Understanding evolution of wastes in storage		■	■	■
Strategic options for fissile materials	■	■	■	■
Compatibility with final management/disposal		■	■	■
Options for reworking and repackaging of wastes		■	■	■
Development of long-life stores			■	■
Decommissioning of redundant nuclear facilities				
Characterisation & survey of redundant facilities	■	■	■	
Retrieval & capture of residual mobile radioactivity from vessels, silos, etc	■	■	■	
Fixing of mobile radioactivity to prevent release during dismantling	■	■		

Advanced decontamination techniques	■	■	■	
Decontamination of graphite cores to permit treatment as LLW		■	■	■
Remote dismantling		■	■	■
Treatment of degradable & combustible materials	■	■	■	
Development of dose rate mapping and remote viewing techniques	■	■	■	
Management of end-points for nuclear sites				
Transparent, defensible decision making and optioneering processes	■	■		
Non-intrusive site characterisation for contaminated land	■	■		
Understanding contaminant behaviour	■	■	■	■
Modelling current and future impacts of contaminated land/water and disposal facilities on human health and the environment	■	■	■	■
Evaluating of remediation technologies	■	■	■	■
Impacts of future climate and environmental change on site end points	■	■	■	■
Monitoring and validation of site end points			■	■
Low level waste disposal				
Modelling impacts of disposal facilities on human health and environment.	■	■	■	■
Techniques for undertaking BPM studies and SEA	■	■		
Impacts of future climate and environmental change on LLW disposal facilities.		■	■	■
Development of wasteforms suitable for new disposal facilities		■	■	■
Establishment of new LLW disposal facilities		■	■	■
Improved national LLW inventory dataset	■	■		

Examples of Radioactive Waste Management R&D

Obviously the R&D programmes in the NNL have not be formed but an understanding of the nature of its work can be obtained from examples of R&D that Nexia Solutions has been undertaking to address a variety of challenges from radioactive waste management.

Characterising legacy waste and decommissioning wastes

The UK has a wide range of challenges from radioactive materials and radioactive facilities some derived from reactor and fuel processing programmes carried out in the 1950's and '60s, others from ongoing operations. In most cases successful treatment requires a more thorough understanding of the materials to be treated. Examples on R&D in this area are:

Impact of hydrided fuel

Uranium metal fuel still present in the graphite core of one of the UK's 'original atomic piles' may contain potentially hazardous quantities of uranium hydride. A programme of research using knowledge of oxidation chemistry coupled to a thermal analysis using

Computational Fluid Dynamics (CFD) has been used to predict the temperature distribution within fuel channels and surrounding graphite. It has been successfully shown that pessimistically assessed amounts of uranium hydride in the core, reacting after seismic disturbance, lead to only very slight temperature rises in the adjacent core materials. The implications are significant, demonstrating that the core is sufficiently benign to allow physical intrusion into the core.

Sampling of sludge

A different challenge is to be able to characterise waste material in-situ. One example, (reported in detail elsewhere in this conference) results from the need to retrieve large quantities of corroded fuel cladding material from a reactor fuel storage pond. There are concerns about the amount of radioactivity that will be leached from the sludge as it is agitated during hydraulic retrieval operations and this will impact on subsequent effluent treatment. The R&D response has been to develop a method of measuring the activity released from samples of the real material during simulated retrieval operations. A 'bell jar' apparatus has been developed which can be immersed in the sludge within the fuel storage pond in question. The apparatus is designed with the means to sample the sludge, before and after controlled agitation.

Improving the knowledge of the waste behaviour has a great impact downstream on the provisions to treat waste and effluent arisings.

Advanced monitoring and interrogation techniques

The difficulty of obtaining access to radioactive wastes and facilities remains a major challenge to any need to improve understanding through characterisation. The NNL will be expected to bring new approaches and techniques to bear that overcome the inherent difficulties of high radiation fields, thick biological shielding and hazardous levels of contamination.

As an example, Nexia Solutions has been leading a programme to exploit the properties of cosmic rays as a way of imaging radioactive wastes and facilities.

Naturally occurring muons, produced from cosmic rays have great potential for imaging through thick biological shielding because of their high energies (up to 10^{12} GeV) and therefore offer a superior penetrative power than X-rays or gamma rays. An initial programme of studies, working with a consortium of UK universities has already shown that there is sufficient flux of muons at ground level to offer the basis of viable measurements. Simple planar arrays of detectors have also shown the ability to track muon events both entering and leaving the object of interest. Further work on large and more sensitive detectors offers promise that a prototype device to interrogate waste packages and even whole building could be constructed.

Optimising waste treatment from current operations

Even within a fully operational industrialised process there are often opportunities to improve operations and reduce wastes. The NNL will be expected to work closely with customers and undertake R&D that can be applied to an industrial setting.

Vitrification Test Rig

One example, from the Nexia Solutions portfolio, is the construction and operation of a full size non-active vitrification facility to develop process improvements that can then be applied to the vitrification plants at Sellafield. The aims are to provide the necessary underpinning development work so that the Sellafield customer can:

- Process highly active liquor faster
- Improve understanding of process and its limitations
- Develop flowsheets for alternative waste compositions
- Provide underpinning product quality and operability data

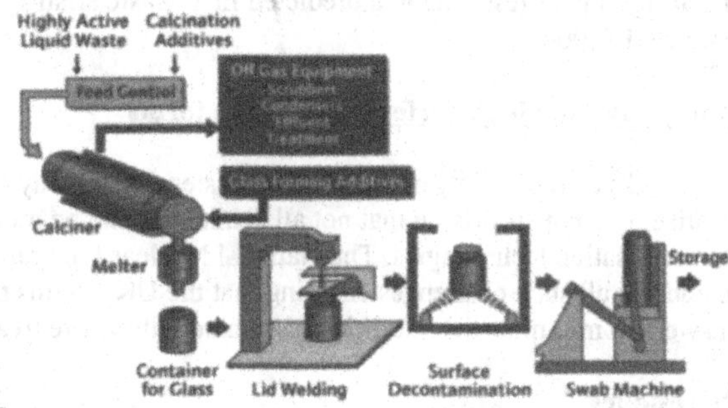

Calciner within the Vitrification Test Rig

A number of campaigns of non-active, full-scale operation have already provided the customer with the basis to increase the incorporation rate of waste in the glass, increase the rate of calcination of the highly active liquid waste and understand how to vitrify non-standard waste more efficiently.

Full Scale Cementation Test Rig

Optimisation of encapsulated sludge

Operations to treat intermediate-level waste are also amenable to optimisation and, as an example, a couple of years ago Nexia Solutions was able to make a significant contribution to radioactive waste clean-up at Sellafield involving radioactive sludge stored in concrete tanks since the 1950's. The intention was to treat the sludge by dewatering in an existing ultrafiltration facility and to cement the concentrated sludge to produce a radioactive waste suitable for long term storage and disposal. It was clearly necessary to confirm that the sludge could be treated satisfactorily through the existing ultrafilter and cementation plants and this was achieved by a programme of laboratory and pilot scale work using simulated waste. Once this was successfully achieved attention then focussed on the large volumes (12,000 x 500 litre drums) of cemented waste product that would be produced. A further research programme then addressed the potential for cementing a more concentrated sludge. It was necessary to work within the constraints of the existing cementation plant to ensure that the sludge and cement powders could be satisfactorily mixed.

Experimental trials were undertaken based on a step-wise increase in the floc concentration ensuring that the process could at all times be compatible with the active cementation plant. Testing of the products produced confirmed that they met requirements for long term storage and eventual disposal. The final encapsulated sludge concentration was double the design intent resulting in a predicted final waste arising of some 6000 drums rather than the predicted 12000.

Alternative and high performance waste forms

Given the variety of radioactive wastes that already exist in the UK and ones that are yet to arise, it is not surprising that not all challenges can be met by the currently deployed waste immobilisation technologies. The National Nuclear Laboratory is expected to have a capability to address all such challenges ensuring that the UK obtains the most appropriate technical answers to minimise the liabilities associated with waste treatment, storage and disposal.

Pu residues

Some of the more demanding wastes which will need to be treated are plutonium residues which have accumulated over the previous 5 decades of nuclear power and which are currently arising from fuel and nuclear material processing operations. A key priority is for the residues to be packaged in a stable form, using packaging materials which will not degrade, thus eliminating the need for repackaging in the future. In the longer term, the aim is to have a wasteform suitable for geological disposal or other long term treatment options. Nexia Solutions has been pioneering the use of ceramics to produce compact and stable wasteforms, suitable for many of the plutonium residues. A vital part has been to leverage knowledge from elsewhere in the world and this has led to a successful collaboration with ANSTO (Australian Nuclear Science & Technology Organisation). The work has shown that a glass-ceramic wasteform, combines the necessary processing flexibility of glasses with the chemical durability of ceramics and promises to provide an ideal solution for immobilising actinides.

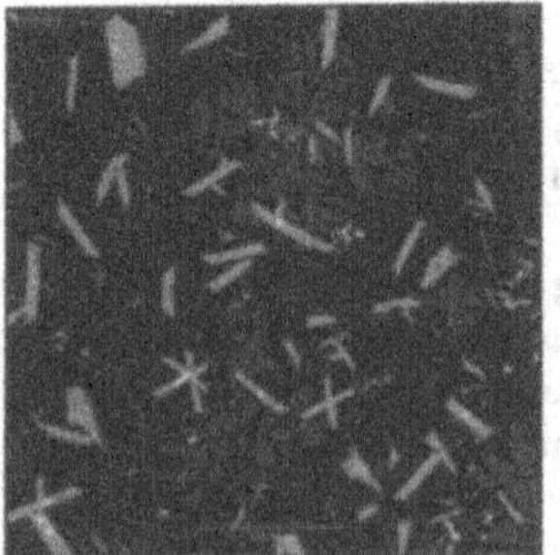

SEM micro graph of a glass-ceramic showing zirconlite phases in light, calcium fluoride in grey and vitreous phase as dark background.

The key crystalline phase in the glass-ceramic is zirconolite (ideally $CaZrTi_2O_7$), a naturally occurring titanate mineral and the main host for plutonium and other actinides in the Synroc formulation developed by ANSTO. Zirconolite has considerable chemical flexibility, allowing cationic substitutions in its structure in addition to the actinides. Extensive complementary studies on naturally occurring zirconolites have confirmed that they survive in the natural environment and are capable of locking up actinides in the crystalline matrix for millions of years.

The R&D programme has been progressing in stages and is now reaching the point of non-active full scale equipment trials.

Management of nuclear materials

The role to be played by the National Nuclear Laboratory in strategic R&D has already been indicated and this will undoubtedly cover radioactive wastes as well as strategic issues such as reactors and fuel cycles. Within the UK there are several issues which require clear strategic direction and among them is the matter of the future use of nuclear materials; plutonium and uranium.

The Nuclear Decommissioning Agency (NDA) is fully aware of the challenge of the managing the nuclear materials which the UK holds. It has commissioned several strategic R&D programmes, an example of which is one on plutonium disposition. Nexia Solutions has been contracted carry out research which will result in an evaluation of the technical feasibility of a number of disposition options, including reactor re-use and immobilization of plutonium as a waste for disposal.

Pu-recycle

In the case of the re-use options, a series of evaluations of fuelling current and future UK reactor designs have been made. These studies used Nexia Solutions' in-house capability to model reactor core performance and addressed the potential perturbations that MOX fuel would bring to the core and the overall goal of reducing or destroying the plutonium stockpile.

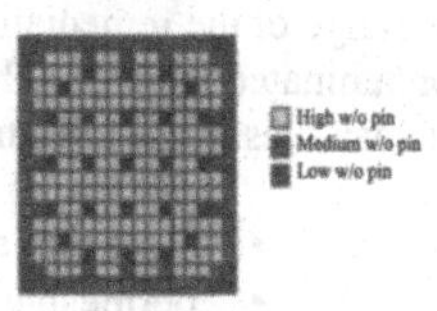

MOX assembly with Pu zoning

The work to date has already achieved some of the objectives set in the re-use evaluation work programme. In particular, the work completed to date on Sizewell 'B' and AP1000 has demonstrated how much Pu can be removed from the UK Pu stockpile and over what timescale. It has been clearly demonstrated that if an all-reactor option is to be considered to remove the existing Pu stockpile, then additional reactors are required. With the advent of new designs that can accommodate a higher MOX core fraction (50 to 100%), the mission can be achieved on potentially shorter timescales.

Pu immobilisation

The other option for plutonium disposition is to produce a suitable wasteform for storage and eventual disposal. Studies have focused initially upon the use of a glass matrix, although ceramic matrices remain an option. The strong synergies with other programmes on vitrification and ceramics illustrate one way in which the retention of skills within a national laboratory will deliver benefits across several R&D programmes.

The study on vitrification of plutonium stocks is based upon the assimilation into a durable glass waste form. A preliminary experimental survey assessed a selection of potential glass systems on the basis of Pu-surrogate (cerium) loading, durability, and ease of processing. Following this, a number of borosilicate compositions have been taken forward into a more detailed investigation in order to fully qualify their potential for Pu-immobilisation. Work thus far has used hafnium as a simulate for plutonium, in order to reduce the costs and expedite the experimental work. As options are refined, it is intended that studies will use plutonium, an example of where the extensive facilities of the NNL's technology centre for working with radioactive materials will bear fruit.

Contaminated land remediation and low level waste disposal

Radioactively contaminated land and waste disposal represent large long term liabilities for the UK and it is logical that the National Nuclear Laboratory should play a strong role in the R&D required. Some of the issues require strategic R&D programmes and cannot be solved by a limited series of research tasks. There are also synergies that can usefully be developed between contaminated land management, low level waste disposal and geological disposal.

Remediation of contaminated land

Nexia Solutions has been undertaking a strategic R&D programme on remediation of contaminated land for the Nuclear Decommissioning Authority and this provides an example of how a research programme in the NNL might be developed. The initial aim has been to broaden the range of the remediation technologies in order to inform strategic decision-making on contaminated land. It includes acquisition and detailed characterisation of soils from a number of nuclear sites in the UK and is evaluating the following 4 technologies over a 2 year period:

- electrokinetics,
- permeable reactive barriers (PRBs),
- enhanced soil washing; and
- *In situ* stabilisation (ISS).

The experimental studies are being complemented by a set of electronic tools to help site owners readily evaluate optioneering studies, and understand the effect on cost and performance of varying the factors that affect remediation efficiency.

Uranium leaching from low level waste disposal

A related project (which is described in detail later in this conference) shows how the need to understand the future behaviour of disposal sites requires an in-depth capability in a range of technical skills. Assessment of the UK Low Level Waste Repository (LLWR) has previously shown that the prevailing conditions in disposal trenches are chemically reducing and this limits the uranium release through the extreme low solubility of U(IV) solids. A recent R&D project by Nexia Solutions has enhanced the understanding by examining the additional effects that the physical and chemical nature of the uranium wastes may have on the release of uranium. The work combined an understanding of the chemical constituents in the LLWR, understanding of the geochemistry of the disposal site, a knowledge of historic waste disposals together with a 3-D modelling capability to predict potential future activity leaching into groundwater. Overall, the model correctly represented the range of fluoride and uranium concentrations that are measured in leachate from the LLWR trenches. Overall the study builds confidence in the inherent safety features that are provided by the uranium waste residues and the reducing chemical conditions of the LLWR trenches.

Summary

The National Nuclear Laboratory is expected to be established with a broad range of technical capabilities. These capabilities, coupled with an ability to undertake experimental studies on radioactive materials provide a powerful resource to tackle the UK's challenges in radioactive waste management. Precise details of the NNL's R&D portfolio have yet to be decided but insights from examples of R&D carried out by Nexia Solutions emphasise that to tackle challenges of radioactive waste management successfully, the NNL needs to:-

- Develop a significant component of strategic R&D in its portfolio
- Maintain the capability to undertake work that is immediately applicable to industrial deployment
- Facilitate the application of technology from non-nuclear areas
- Work appropriately with other nuclear and non-nuclear technology organisations

Mater. Res. Soc. Symp. Proc. Vol. 1107 © 2008 Materials Research Society

Opportunities for Research in the UK for Decommissioning and Disposal

Neil Smart
NDA,
Herdus House
Westlakes Science & Technology Park
Moor Row
Cumbria
CA24 3HU.

ABSTRACT

The NDA remit as set out within the Energy Act includes – *"promote, and where necessary fund, research relevant to nuclear clean up"*. The NDA need to underpin delivery and / or accelerate programmes to fulfil the overall mission and technical underpinning of these activities is critical. In this paper we will present consideration of the investment required in nuclear waste Research and Development.

Firstly, NDA set the requirement for nuclear sites to write down within the Life Time Plans (LTP), at a high level, the proposed technical baseline underpinning the LTP activities; furthermore we required technology gaps / opportunities in the technical baselines to be outlined in a R&D requirements section to the LTP. Criteria were established to categorise the R&D in three areas:

- "needs" - those development activities needed to underpin the proposed technical solutions
- "risks" – those activities required to reduce / eliminate key risks to the proposed technical solutions
- "opportunities" – innovations / changes to the technical baselines

The purpose of production of the technical baselines and underpinning R&D requirements is to establish an auditable trail through the LTP from programme components into how the programme will be delivered.

NDA believes the production of the technical baselines and R&D requirements will be of benefit to the Site License Companies (SLC) in terms of ensuring a focus on overall programme delivery and not just short term activities. Furthermore, we can ensure that investment in technology is targeted at priority areas, with common issues and requirements identified and solutions on a broader scale will be achievable.

INTRODUCTION

The Nuclear Decommissioning Authority (NDA) is a non-departmental public body, set up in April 2005 by the UK Government under the Energy Act 2004 to take strategic responsibility for the UK's nuclear legacy.

The NDA mission is clear: 'To deliver a world class programme of safe, cost-effective, accelerated and environmentally responsible decommissioning of the UK's civil nuclear legacy in an open and transparent manner and with due regard to the socio-economic impacts on our communities'. In line with the mission, the NDA's main objective is to decommission and clean-up the civil public sector nuclear legacy safely, securely, cost effectively and in ways that protect the environment for this and future generations. The NDA does not carry out clean-up work itself but has in place contracts with site licensee companies (SLCs), who are responsible for the day-to-day decommissioning and clean-up activity on each UK site. Individual sites develop LTPs that set out the short, medium and long-term priorities for the decommissioning and clean-up of each site.

Critical to achieving the NDA main objective and overall mission is to accelerate and deliver clean-up programmes through the application of appropriate and innovative technology. That's why the remit as stipulated in the Energy Act is to: '*promote, and where necessary fund, generic research relevant to nuclear clean-up*". The NDA have therefore considered the investment required in Research and Development (R&D) both directly and indirectly (i.e. through the Site Licensee Company clean-up programmes) to ensure appropriate delivery of Lifetime Plans and to maximise the return on the investment made.

The sites are required to state within the LTP, at a high level, the proposed technical baseline that underpins the LTP decommissioning and clean-up activities. In addition the sites are required to identify technology gaps / opportunities in the technical baselines within the R&D requirements section of the LTP. R&D is categorised in three key areas:

1. The 'needs' - providing solutions to known and common issues
2. The 'risks' - providing options to avoid or mitigate the risks
3. The 'opportunities' - delivering innovative improvements to the LTP to achieve the NDA's mission

The purpose of including technical baselines and underpinning R&D requirements within the LTP is to establish an auditable trail through the LTP and a direct link between the programme components and programme delivery. The LTPs 05 was the first programme to attempt this process. This has been updated in LTP 2006 and LTP 2007.

Historically, the short-term benefits gained from carrying risks associated with the technical underpinning of projects led to significant cost implications and delays to projects and programmes. Today, NDA believe the technical baselines and identification of R&D requirements will help the SLCs to focus on overall programme delivery and not just short-term activities. In addition, NDA can ensure that investment in technology is targeted at priority areas, with common issues and requirements identified, achieving solutions on a broader scale.

Following the production of the LTPs, NDA have completed a review of the 'NDA R&D needs, risks and opportunities'. We have considered the information submitted by the SLCs to the NDA in the LTPs , in terms of the R&D requirements. In doing so, we have compared and contrasted the plans from different sites and evaluated commonalities, differences and potential omissions, with a view to sharing the 'NDA R&D needs, risks and opportunities' across the entire technical supply chain.

KEY FINDINGS OF THE LTP 2006 REVIEW

A top down review of our overall technology needs, risks and opportunities has identified common key issues.

Key issues identified:

Balance of R&D programmes
Owing to the mature nature of the industry, the vast majority of R&D development activities are integrated directly with on-plant deployment projects and therefore solution driven in their application. Judgements are therefore made about the degree to which remaining knowledge gaps need to be filled or whether the proposed solution is sufficiently robust to deal with the remaining uncertainties. Given the need to accelerate clean-up programmes in line with the NDA's mission, NDA fully support this approach but recognise the importance of maintaining an appropriate level of underpinning scientific knowledge of the applied processes. In addition NDA will continue to monitor activities to maintain the adequate skills to support the clean-up projects.

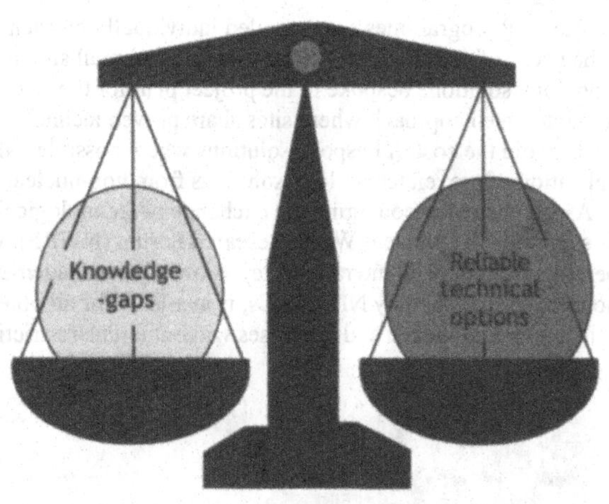

NDA R&D requirements
The analysis of the full life cycle of existing liabilities overlaid with the need to deliver the NDA's mission yields significant R&D challenges. NDA expect an increase in R&D investment from the SLCs over the next ten years, if delivery is to be assured. We will be monitoring the developing R&D programmes to ensure the activities are being undertaken in line with the delivery of the NDA strategy. We do however expect that confidence in the technology development activities will grow as the programme of clean-up activities accelerates

Underpinning science
Although most of the R&D activities are focused to support clean-up projects, NDA recognises that these rely upon a strong underlying science base. If clean-up is to be successful it is important to develop the science base as new challenges emerge. To support this, our aim is to stimulate the academic sector to help meet our science needs.

In collaboration with Nexia Solutions, NDA are supporting a series of University Research Alliances (URA) to develop and maintain a network of basic science capability and skills to achieve the short and long-term aims of our mission. Furthermore, the NDA has supported student bursaries, where additional funding has been made available to support PhD projects aligned to the NDA needs, risks and opportunities. Nine awards were made in 06 and an equivalent number is expected in 2007.

These URA's and a series of smaller University contracts provide a range of underpinning science support to the NDA mission. Also, graduates from PhD programmes are providing an influx of new talent into the decommissioning supply chain.

Good practice

The LTP programmes are compiled individually by each site from a 'grass-roots' assessment of the needs of each project, culminating in an overall site plan. This often leads to unique technology solutions bespoke to the project plan for the site. As a result, NDA are encouraging a more integrated approach where sites share proven technology solutions for everyone's benefit and so avoid the cost of bespoke solutions where possible. NDA also fully supported the application of proven technology solutions from non-nuclear fields within the nuclear industry.

As a means of encouraging the exchange of technological knowledge and processes, NDA has supported the Nuclear Waste Research Forum (NWRF), where the SLCs and other nuclear operators from the UK interact. A key knowledge management issue is that all processes and knowledge developed by NDA SLCs, is available for all other SLCs to access. This enables exchange of knowledge and processes without undue restrictions.

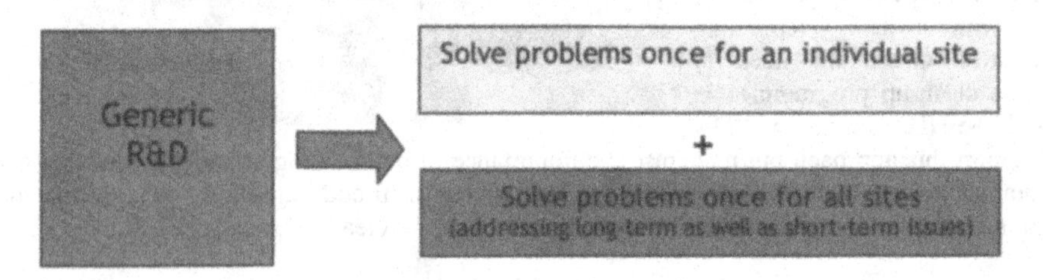

Additionally, international experience in terms of proven technology capability should be considered further. A number of nations have had substantial clean-up programmes over the previous two decades, with proven delivery capability. As improvements in the supply chain management take place within the competitive clean-up market, more proven technology options will be proposed, requiring minimal development activities.

Contingency activities

A further review of the underpinning research and development activities of high priority projects will establish whether the level of activities (basic science and contingency work) are aligned to the uncertainties and risks associated with the projects. Whilst NDA believe these high priority projects are technically underpinned against current plans, the level of risk associated with the waste streams involved may require more investment to develop contingency options in parallel.

Safety and environmentally driven research

High standards of safety and environmental performance are a fundamental requirement of the SLCs and NDA are supporting SLC R&D activities that provide improvement in these areas. As the R&D drivers are mainly the same, R&D activities specific to safety and the environment have not been separated out. NDA will actively encourage the SLCs to work together on specific technical projects offering improved safety or environmental performance and therefore mutual benefit.

Development of common technology solutions

The site LTPs indicated a range of common problems that would benefit from combining efforts. Examples include a widespread requirement for local and mobile effluent technology and sludge handling technology. Clearly, these processes will have commonality in terms of engineering design, waste disposal, IX technology etc.

Development of common waste packaging solutions

Most sites need to package and store ILW, whilst awaiting a long-term disposal option. Each waste stream requires process development and assessment for suitability for eventual disposal. There are opportunities to share pre-existing solutions to common waste materials on different sites. There are similar opportunities to establish common formulations for cement and associated fillers, across the waste spectrum, enabling a higher level of confidence in waste packages and security of supply.

Availability of facilities to meet market needs

Whilst the analysis of 'NDA R&D needs, risks and opportunities' has highlighted work to be carried out, no assessment has been made of the ability of the supply chain to meet these requirements in terms of facilities. One specific area of concern is the changing need in terms of measurement capability and the availability of laboratory facilities and laboratory standards.

Technology transfer

Accelerating clean-up programmes means accelerating the development process and transfer of new technology options into the market place. A number of interfaces need improvement (a) SLC to SLC (b) Nuclear research organisations to SLC (c) non-nuclear sector to SLC.

MAJOR TECHNICAL ISSUES

Major technical needs and opportunities were identified as worthy of highlighting due to importance or widespread interest:

Materials characterisation

The LTPs identified issues surrounding the characterisation of materials for treatment and disposal.

- Techniques to rapidly assay low levels of radiation and contamination in order to sentence materials for LLW disposal and segregate materials for release as clean or exempt materials
- Development of techniques to characterise contamination of structures
- Development of techniques to characterise site and contaminated land
- Development of techniques to characterise waste properties – radiochemical, chemical and physical, including facilities for ILW characterisation

Waste processing

- Development of proven sludge handling techniques
- Remote handling techniques for fuel debris and highly activated materials
- Methodology and techniques for waste segregation

- Recycle of materials
- Graphite management

Management of strategic nuclear materials

- Development of immobilisation technology for separated plutonium
- Process development for the conversion of uranium hexafluoride to a more stable form
- Long-term options for the UK approach to the management of spent fuel

Plant termination

- Improved decontamination technology to either enable man access or waste re-categorisation
- Improved effluent management to process decontamination reagents
- Development of local and mobile effluent treatment capability
- Technology to carry out size reduction of large items
- Remote dismantling technologies

Site restoration

- Surveying and characterisation of land contamination
- Ground remediation technology for active and non-active contaminated land
- Development of consistent protocols to underpin site end point considerations for a wide variety of sites

CONCLUSIONS

The NDA are committed to the identification of Research and Development Needs, Risks and Opportunities underpinning the NDA mission. These are published in the open forum, where ever possible, with a view to the encouragement of supply chain engagement and development of an innovation driven culture.

Mater. Res. Soc. Symp. Proc. Vol. 1107 © 2008 Materials Research Society

Overview of the CEA French Research Program on Nuclear Waste

Etienne Y. Vernaz[1] and Christophe Poinssot[2]
[1]CEA, Marcoule, Department of Waste Treatment and Conditioning,
[2]CEA, Saclay, Department of Physics & Chemistry, Service for the Studies of the Radionuclides Behaviour, F-91191 Gif-sur-Yvette cedex, France

ABSTRACT

This presentation gives an overview of the French major research program on nuclear waste and presents some perspectives in the context of the 2006 Act, while focusing on the main CEA contributions. Development of new conditioning processes combined with in-depth studies of the waste-packages long term behaviour studies allows enhancing the expected performance of waste-packages. Geological disposal performances have been demonstrated based in particular on the relevant understanding of the radionuclides chemistry and migration in complex environment, including tracing field-experiments in ANDRA underground research lab. French research and plans for waste from future reactors will be discussed. It is still a very prospective area as fuel materials that will be used in these reactors is not decided. The capacity to recycle most of the fuel component will be a major point in the choice of an integrated concept including reactor and fuel cycle.

INTRODUCTION

The French context

With an installed capacity of 62 MWe in 58 PWR, France roughly produced more than 80% of its electricity, i.e. 450 TWh, by nuclear fission. Nuclear energy is therefore a major component of the French energetic "mix". France also made the strategic choice to reprocess spent nuclear UOX fuels in order to recover all recyclable matters: recovered plutonium is currently entirely recycled in the current reactor as MOX fuel whereas 30% of recovered uranium is recycled in UOX fuel. This choice also allows to minimize the volume of produced waste and to dispose off only the "ultimate" waste conditioned in suitable matrices. Since spring 1957 when first glasses of containment were elaborated in Saclay, the CEA is supporting a long-term research effort on nuclear waste management. This policy of research aimed to permanently improve the industrial practice of recycling / treatment chosen by France since its first nuclear program.

From "Bataille Act" to the "June 28, 2006 Act"

In 1990, controversies on the issue of radioactive waste led the French government to decide a moratorium on the selection of a relevant geological disposal site and to draw up a Bill to set the framework for research work on the long term management of long-lived high-level radioactive waste. Passed on 30 December 1991, the so call "Bataille Act" (from the name of the

author of the parliamentary report), gives 15 years to the French research to study three complimentary research lines:
- Reduce long-term toxicity by partitioning and transmutation of long-lived radionuclide present in the waste;
- Investigate the possibilities of reversible or irreversible deep geological disposal, in particular through the construction of an underground research laboratory
- Investigate the processes for the conditioning and long-term storage of this waste in surface.

CEA was involved during these 15 years in these three lines of research with more than 300 researchers and a total budget of 1.6 Billion Euros.
By the end of this 15 years period, the research results were evaluated by the National Review Board (Commission Nationale d'Evaluation), the government (Ministry of research and Ministry of Industry) and by the Parliament (Office for the Assessment of Scientific and Technological Options) and debated with the citizens through public debates. All these evaluations and debates leads the French Parliament to pass on 28 June 2006 a new law entitled "Sustainable Management of Radioactive Materials and Waste". This Act gives a clear roadmap for the future, in particular:
- It confirms deep geological disposal as the reference solution for ultimate HLW or ILW waste. Corresponding studies have to be conducted in the perspective of a license application in 2015 and the commission of the repository in 2025.
- It consolidates the effort of research on actinides recycling in fourth generation reactors. For this purpose a pilot reactor have to be commissioned before 31 December 2020.
- It specifies that any owner of ILW generated before 2015 shall condition them no later than 2030.

This law gives then to France a strong guaranty that all nuclear wastes will be properly managed with a good balance between progressive industrial implementation and long-term high-quality research that ensures relevance of the technical choices.

DISMANTLING AND LEGACY WASTE

One major development, in terms of waste management, concerns operations of legacy waste retrieval and conditioning, scheduled by Areva and CEA through to 2035 at La Hague, Marcoule and other CEA sites. Moreover, the construction of a durable nuclear power requires showing that one can dismantle the old installations and the first generation of reactor under satisfactory economic and safety conditions.
Simultaneously with research on reactors of the future and the associated cycle, an operational structure was installed at CEA so that the whole of the operations of cleaning, dismantling and old waste retrieval is completed at the end of a thirty years period. This work, undertaken on about thirty projects, involves an investment of approximately 400 Million Euros per annum. To guarantee the continuity of the financing two dedicated funds were made up for CEA installations dismantling. One made up in 2001 and of an amount of 3.96 Billion Euros, is intended for the civil installations. The other, created at the end of 2004 and of an amount of 4.18 Billion Euros, is intended for the military installations [1].

Taking into account the importance of these projects CEA developed an important know-how on decontamination and tele-operation techniques, allowing minimizing at the same time the produced effluents and waste but also the radiation doses received by the personnel.
In synergy with the operational teams, R&D is performed on innovating processes of

decontamination as for instance surfactants association to optimize degreasing efficiency of rinsing operations of nuclear facilities, caesium removal from high salinity liquid waste using calixarenes selectivity, decontamination process using drying gels or aqueous foams, concrete electro-decontamination, gloves decontamination with supercritical CO_2, etc.[2]

WASTE CONDITIONING

Vitrification and CCM implantation in La Hague

The R7T7-glass specification was designed at the end of the 80's to accommodate the FP-solutions arising from the reprocessing of a nominal UOx spent fuel initially enriched at 3.5% of ^{235}U and irradiated at 33.000 MWe.d/t. About 0.75 glass canister is produced per ton of reprocessed UOX spent fuel in the current industrial Vitrification facilities. The current Vitrification technology is based on a hot metallic melter, with an upper limited melting temperature of 1150°C.

For over 20 years the CEA at Marcoule have been developing direct induction Cold Crucible Melter (CCM) to vitrify high-level nuclear wastes as well as non-nuclear toxic waste or enamels. The cold crucible is a compact water-cooled melter in which the radioactive waste and the glass additives are melted by direct high frequency induction. The cooling of the melter produces a solidified glass layer that protects the melter's inner wall from corrosion. Because the heat is transferred directly to the melt, high operating temperature can be achieved with no impact on the melter itself. A fully nuclearized CCM is under the final stage of qualification in Marcoule, before an industrial implementation in one of the La Hague Vitrification line, scheduled on 2010. This CCM allows melting corrosives glass compositions or melting glass at higher temperature. It will be used to vitrify old stream as Uranium Molybdenum waste. It could also allow increasing glass throughputs, increasing waste loading or melting new matrix necessary for other waste streams as decontamination liquids.

Development of new glasses

Glass formulation is a complex approach, combining material science, material behaviour, and constrains imposed by the technological vitrification process as well. Specific physical, thermal and chemical conditions are needed to get an homogeneous glass matrix upon cooling of the molten homogeneous liquid: the glass waste form is a compromise between the target volume reduction factor, the solubility limits of the chemical elements, the technological constraints of the process that have to operate under highly active conditions, and the glass ability to resist toward devitrification, aqueous corrosion and self-radiations[3]
Some high-level solutions generated by reprocessing legacy fuel like UMo and MoSnAl contain high molybdenum concentrations. Molybdenum is known to be sparingly soluble in conventional borosilicate glass, and basic studies have been done to find suitable glass formulations for such waste. The selected matrix is a SiO_2-B_2O_3 -P_2O_5 glass ceramic that is melted at about 1250°C; the matrix comprises unconnected micro beads of partly crystallized phosphate glass uniformly dispersed in a borosilicate glass [4]. On going research on glass crystallisation aim at evaluating how such glass ceramics could be used for future wastes.

Nuclear fuel burn-up is increasing in nuclear power plants. In the future, MOX fuels will also be reprocessed, and the corresponding fission products solutions will be vitrified. In this context, new glass formulations are studied to accept more FP-oxides in the glass network [5]. A major concern is the ability to increase the solubility limits in the glass network of some FP without decreasing the glass long term behaviour. Such material design is linked with the development of the CCM technology capable of achieving higher melting temperature than today. The main families of glasses under studies that would allow a FP-loading charge of up to 25 FP-oxides wt. % are lanthanide enriched borosilicate glass and lanthanide alumina-silicate. A major concern is the ability of the new glass compositions to accommodate a higher loading charge of noble metal. Specific studies are performed to understand the effect of these particles on the properties of molten glass. Insoluble Ruthenium oxide for instance has a strong influence on electrical conductivity that could impact the heat generation homogeneity in the molten bath [6].

A new axe of research related to the R&D program devoted to the new nuclear systems, is to evaluate the feasibility of a specific conditioning matrix for noble metals. The goal would be to condition the flux of "fines" (as currently recovered in the PUREX process during the clarification stage), in a specific metallic ingot, instead of vitrifying these fines. This would reduce the final noble metals content in the glass below the current acceptable limit of 3 wt. %.

Other researches related to ILW vitrification include studies on sulphur solubility in nuclear glass, development of new compositions for decontamination stream, and glass compositions able to confine a large range of chemical sludge.

Cement

Compared with other solidification techniques, the cementation process is relatively simple and inexpensive. Moreover, calcium silicate cements show many advantages: easy supply, good mechanical strength, compatibility with aqueous wastes, good self-shielding, and high alkalinity which allows precipitating and thus confining many radionuclides. That is the reason why cementation is a widely applied technique for the conditioning of low- and intermediate-level radioactive wastes. At the request of waste producers (CEA research centers, AREVA NC...), the Marcoule «cementation » laboratory has developed, at the lab-scale, cement-based materials to immobilize a wide range of wastes: evaporator concentrates, sludge, ion exchange resins, graphite, magnesium-containing powders. The studies are carried out by taking out both industrial requirements and near-surface disposal specifications. R & D activities include both applied investigations aiming at developing cement matrices for waste producers and more prospective ones dealing with:

- the methodology to elaborate cement-based grout formula for the encapsulation of radioactive wastes [7],
- the mechanisms involved in cement-waste interactions,
- the development of new binders showing a better compatibility with the waste or the disposal environment, like sulfo-alumineux cement or low pH cement.
- the transport of gases through cementitious materials (matrices and canisters).

Bitumen

Even if bitumen will not be processed anymore in future plant, some old sludge have still to be bituminized and some 70 000 drums are existing in La Hague and Marcoule.

For more than a decade, bitumen long term behaviour studies have been going on, in relation with storage and disposal scenario. These studies include physicochemical properties of bitumen under thermal, radiological or aqueous stress. For these studies, synthetic and inactive materials are mainly used, but comparisons with industrial or active material behaviour are also performed for validation. Study of stress behaviour under alpha and gamma irradiation allowed modelling the swelling of bitumen drums under self-irradiation.
The understanding of leaching phenomenology is achieved on one hand by simulating the storage conditions varying the chemical composition of the bituminized sludge, the nature of the leachant and the mechanical constraints, and on the other hand by monitoring the water uptake, the salts releases and the evolution of the porous structure. Then a predictive modelling, reproducing both the kinetics of leaching and the morphological changes of the altered material, is built and validated against experiments step by step [8].

Alternatives processes

Alternatives processes of incineration-vitrification coupling CCM and plasma torches are developed with the aim of drastically reduce ILW volume while confining them in a more durable glassy matrix. The proposed technology relies on the large skill developed by the CEA's vitrification teams on the cold crucible melter but also on the oxygen plasma transferred mode plasma torches that have been tested for more than 10 years in Marcoule. The SHIVA process ((Systeme Hybride d'Incineration Vitrification Avancé - figure 1) has been developed to study the feasibility these technologies for the future [9]. It allows performing in the same vessel the incineration of the burnable wastes, the vitrification of their mineral charge and the off-gases combustion. Significant advantages can be obtained by supplying the waste directly into oxygen arc plasma located above a glass bath heated by direct induction in a cold crucible [10]. The temperature is very high and so is the efficiency of the combustion in the exited free oxygen rich atmosphere that also promotes to a good oxidation of the glass. The treatments of different kind of wastes have been investigated. Burnable wastes such as ionic exchange resins, bituminous wastes, sulfate slurries, graphitic sludge, have been successfully incinerated with a good incorporation of their mineral charge in the glass [9]. In addition mineral wastes such as sludges issued from nuclear treatment have been treated too. The current studies concern the incineration – vitrification of chlorinated organic wastes. In this case, the main difficulty is to manage the volatile metallic chlorides in the process. Some studies are also in progress about the treatment of mixed wastes containing organics and metals.

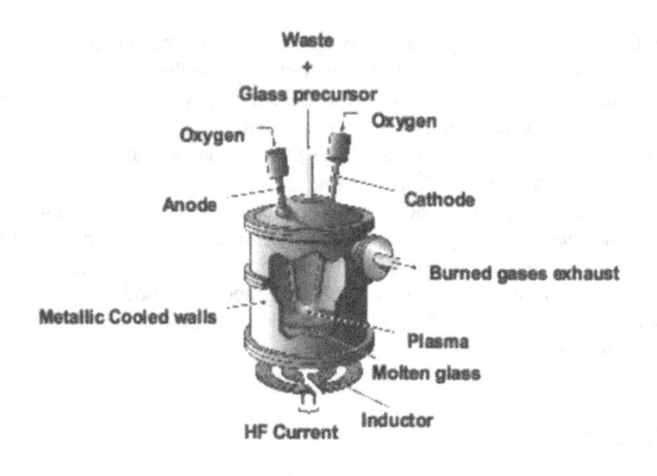

Fig.1 : The SHIVA process principle

RESEARCH RELATED TO FINAL DISPOSAL

Final disposal is since the 2006 Waste Management Act the reference solution for the management of the nuclear waste. For HLW and ILW, the French repository will be located within the 250 km² of the extrapolation zone defined around the ANDRA underground research lab (URL) of Meuse/Haute-Marne. This site, located in Eastern part of Paris sedimentary basin, is composed of an altercation of limestones and clay-rich formation. The repository will be emplaced within the Callovo-Oxfordian argilites (~150 My.old), around 500m depth.
CEA significantly supports some R&D for preparing the future implementation of a geological disposal. In particular, it has developed since the early 80's a comprehensive and reliable approach to build predictive models to assess the long term evolution of the waste packages and the long term fate of radionuclides in the geosphere.

Long-term behaviour of waste packages

The primary objective of a repository is to delay and limit the breakthrough of radionuclides to the biosphere in order to decrease the potential radiotoxic impact, considering the natural radioactive decay. The radionuclides being initially embedded within a robust wasteform, it was of major importance to exhaustively study their long term evolution. CEA hence developed in connection with ANDRA a comprehensive integrated approach of the wasteform long term behaviour [11].

- First step is to assess the evolution as a function of time of the **boundary conditions** in the repository: it implies, not only to perform a detailed characterization of the geological environment, but also to estimate the potential normal and incidental scenario of evolution. These scenarios allow identifying and studying the different elementary mechanisms that will infer the evolution of the wasteforms. In particular, we emphasis the necessity to account for the primary stage of evolution in closed system, before the canister failure, during which wasteform can significantly evolves.

- Using the guideline of these evolution scenarios, relevant **elementary processes** are therefore studied at a mechanistic level. Indeed, empirical approaches which are often used for industrial purposes can not be applied for long term since no validation will ever be possible. Extrapolating wasteform performances up to geological timescale needs to be based on a mechanistic understanding.
- Then, more integrated approaches have to be developed to assess the **couplings** between the different mechanisms which simultaneously occur within a given scenario. Such couplings may indeed modify significantly the global evolution.
- Finally, the models describing the different processes have to be integrated in a global predictive model which often requires to be simplified by selecting the most significant processes and parameters. A global predictive model would lead to very complex models and codes, with a number of parameters preventing to draw any doubtless conclusions... The reliable approach is rather to progressively simplify the phenomenology by focusing on the governing processes. Furthermore, **local conservative approaches** are often necessary to overcome the lack of knowledge and wrap the general trend. We therefore distinguish the scientific models which can be very intricate from the **operational models** which are developed on the focused aim of a long term reliable prediction.
- Long term prediction can not be strictly speaking checked towards experiments. However, **natural or human analogues** as well as field experiments give unique opportunity to check towards similar systems the relevance of the models: (i) they can allow identifying some mechanisms, which are not easily observed in laboratory experiments and may be important for the long term, (ii) they give us some very long term integrated experiments, towards which predictive models can be qualitatively validated.

This integrated approach defined the new **Science of long term behaviour** that CEA has been promoting and developed for several decades for the different type of waste forms: bituminized and concrete wasteform, nuclear glass and also spent nuclear fuel, although it is not considered in France as a waste since it still contains valuable matters (U and Pu). These researches allow CEA to give to ANDRA some reliable models predicting the lifetime and alteration rate of the various wasteforms as a function of time and boundary conditions. For example, nuclear glass wasteform was demonstrated to last more than 300 ky., even in pessimistic scenario where large amount of iron corrosion product act as silica sink [12]. Current research effort is now focused on the near-field interactions, between the wasteform interaction and near-field materials (canister, backfill material, argillites). For example, several experiments are conducted to understand the couplings between the nuclear glass dissolution, which depends among others on the aqueous silica concentration, the potential sorption of silica on canister corrosion products or on the geological material, the potential precipitation of newly formed Si-bearing minerals … These "glass/canister/clay" interactions are the key of the understanding of the long term performance of the nuclear glass confinement performance in a repository.

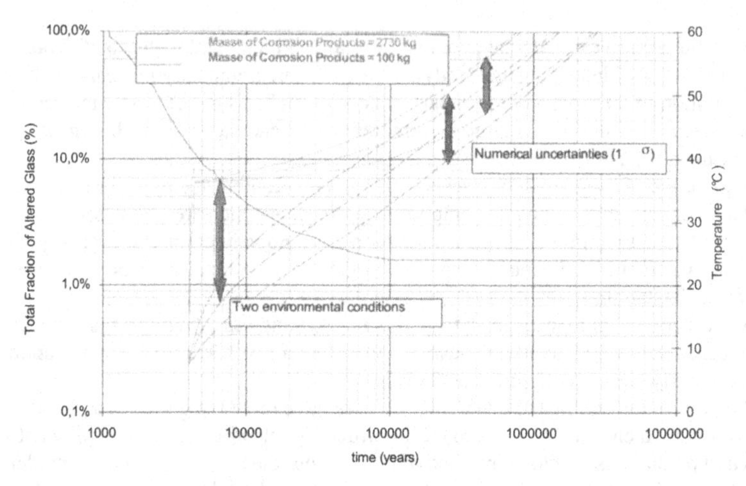

Figure 2. Operational modelling: example of calculation of glass lifetime in two environmental conditions

In order to enlighten all the potential option for the back-end of the fuel cycle, R&D was also performed on spent nuclear fuel (SNF), although SNF is not considered as a waste. We hence study both the SNF evolution in closed system (first 10 ky.) and the SNF alteration in presence of water after 10 ky. We demonstrated that SNF performance is dominated by the release of the so-called Instant Release Fraction (IRF) which has been conservatively estimated to 12-16% for a UOX fuel due to the large uncertainties concerning the long term stability of grain boundaries [13]. Matrix alteration was demonstrated to be dominated in the early stage by radiolytic dissolution, but which is rapidly overcame by the corrosion [14].

Behaviour of radionuclides in a geological repository

CEA has also been widely involved in the research concerning the chemistry of radionuclides within a repository. Indeed, once the radionuclides have been released by alteration of the wasteform, they will been involved in several type of processes that need to be modelled to predict the long term fate of radionuclides:

- complexation and redox reaction within the aqueous phase, which defines the relevant species and the speciation of the radionuclides
- interaction at the solid/solution interfaces with the materials, the sorption processes
- Couplings with the transport processes (mainly diffusion)

CEA has developed both experimental and modelling work on these three processes in order to build both a mechanistic understanding of each processes, but also of their couplings, and to derive predictive operational models. Specific focus is stressed on the development of operational models which are able to link the in-depth mechanistic understanding and the application to long term prediction. For instance:

- Development of operational thermodynamic database which complete the well-known reference database (like the NEA TDB) in order to cover the whole range of potential chemical reactions, even when data are scarce and poor. Indeed, we demonstrated that predictions are more reliable if all the significant reactions are accounted for, even if the uncertainties are large, instead of focusing on the sole reactions for which we have reliable data [15]

- Development of operational models to predict the evolution of sorption as a function of the chemical conditions. Indeed, classical complexation surface models is not thought to be fully designed to perform long term prediction due to the large number of parameters required, their intrinsic inconsistency when electrostatic terms is neglected… We therefore developed a more general approach considering that all sorption reactions can be depicted as an ion exchange (which does not mean that the actual molecular process is an ion exchange). This Ions Exchanger Thermodynamic Theory (IXT2) developed since the mid 90's by Ly et al. [16], succeeds to depict with a significant accuracy the evolution of sorption as a function of pH, ionic strength, RN concentrations …
- Development of operational understanding of the evolution of the diffusion properties of RN as a function of the argillites mineralogical composition. In particular, we focus our study on the anionic behaviour since anionic RN mostly contributes to the long term impact of the repository (^{36}Cl, ^{125}I, ^{79}Se ...) [17]. In particular, we demonstrated that the significant decrease of the effective diffusion coefficient of iodine and chlorine normalised to the diffusion in water below 400m is to be related to the apparition of clay minerals which excludes the radionuclides from a significant part of the porosity by electrostatic repelling (clay surfaces are negatively charged).

Beyond this lab approach, CEA was also involved in the development and implementation of the tracing experiments within the ANDRA URL. The aim of these experiments was to validate the models developed at the centimetric scale in the lab, on in-situ experiments at the decimetric to metric scales [18].

Status of the ANDRA research for the future repository

ANDRA is since 1991 in charge of selecting, designing, coordinating research and implementing a potential repository in France. Beyond the specific CEA contribution, ANDRA has therefore developed and performed a very wide R&D project which allow them to demonstrate in their 2006 report the feasibility and safety of a future repository within the Callovo-Oxfordian layer. They produce some early design as well as some first safety assessment of the potential repository. Their studies demonstrate that the long term safety of the repository is quite ensured within such a geological environment and that the long term impact will be dominated by the anionic RN, such as ^{36}Cl, ^{125}I, ^{79}Se … (Figure 3)

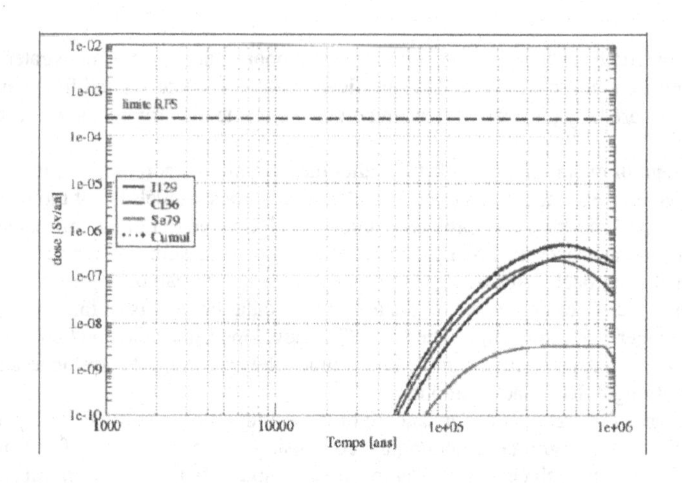

Fig. 3: Long term impact of a potential repository of nuclear glass waste packages within the Callovo-Oxfordian argillites of the Meuse/Haute-Marne.

FRENCH RESEARCH AND PLANS FOR WASTE FROM FUTURE REACTORS

French research on wastes from future reactors is still a very prospective area as fuel materials that will be used in these reactors is not yet decided.

Due to the expected increasing demand for nuclear energy in the world, the next fourth generation reactor could be more evolutionary than revolutionary, as for instance sodium fast breeders. In this case the main concern about waste treatment would come from the strong increase of burn-up that would increase the amount of noble metal and minor actinides in the fuel. A separated treatment of noble metal could be necessary along a strong recycling of minor actinides. Other improvement could be searched as for instance stainless steel hulls decontamination by melting.

On a longer term more, "revolutionary" reactors could be done as gas cooled high temperature one. Many challenges have to be passed to demonstrate the sustainability of such high temperature materials, especially for the fuel. In this case the capacity to recycle most of the fuel component will be a major point in the choice of an integrated concept including reactor and fuel cycle.

On the very long term possible reprocessing of spent fuel by an innovative pyrochemical process is studied. In this case the whole of the fission products would be found in chloride or fluoride medium. Some research has been done on containment materials that would incorporate large amount of Cl or F as for instance chloro-sodalite minerals. However such materials are probably not the best and large volume of waste are expected. A more efficient solution would be a selective precipitation of fission product to retrieve them from the melted salt and to obtain a durable waste form in the same operation. For example precipitation of lanthanides by phosphate, to get durable monazite has been demonstrated with a lanthanide recovery of 90 to 99% [19].

CONCLUSION

A significant and long-term research effort has been carried out by CEA in close relation both with the industrial, but also with the academic research labs (University, CNRS…). It allows France to progressively build a general management plan for all nuclear waste with dedicated optimised solutions for each. Furthermore, by associating many high-grade research teams, France ensures the relevance and reliability of the scientific results which supports the political decisions that have been progressively taken. Although the new Waste Management Act gives an explicit planning of industrial implementation, it also sustains a long term research effort that will allow a constant dynamics of progress. Finally, the issue of waste management is not only a technical one but also a political and societal one. Therefore, it has also been recognised that continuous effort has to be done to provide objective and understandable information towards the citizens. This effort of pedagogy in direction of public opinion led CEA to create in Marcoule a new interactive space of communication and pedagogy on nuclear energy and waste management, entitled **VISIATOME**. The sharp success of the **Visiatome**, testifies the growing interest of our fellow-citizens for societal and energetic problems.

REFERENCES

1. CLEFS CEA – No.53 Winter 2005-2006
2. Fournel, B., A. Roubaud, S.Sarrade, "Supercritical Fluids for Nuclear Wastes Decontamination. CEA Research Status", GLOBAL 2007, 9-13 Sept, Boise Idaho
3. S. Peuget, J.-N. Cachia, C. Jegou, X. Deschanels, D. Roudil,V. Broudic, J.M. Delaye, J.-M. Bart; Irradiation stability of R7T7-type borosilicate glass, Journal of Nuclear Materials, Volume 354 (2006) 1-13
4. O. Pinet, J.L. Dussossoy, C. David, C. Fillet "Glass matrices for immobilizing nuclear waste containing molybdenum and phosphorus" Journal of Nuclear Materials (accepted November 2007)
5. J.L. Dussossoy, S. Schuller, T. Advocat. "The formulation of new high level waste glasses with an increased waste loading charges" International Congress on Glass, Strasbourg 2007
6. C. Simonnet and A. Grandjean. Mixed ionic and electronic conductivity of RuO2–glass composites from molten state to glassy state. Journal of Non-Crystalline Solids, Volume 351, Issues 19-20, 1 July 2005, Pages 1611-1618.
7. C. Cau Dit Coumes, S. Courtois, « Investigating Main Effects and All Two Factor Interactions when One of the Factors has more than Two Levels – Application to the Choice of a Cement for Nuclear Waste Encapsulation », Chemometrics and Intelligent Laboratory Systems 80 (2006) 167-175
8. J. Sercombe, B. Gwinner, C. Tiffreau, B. Simondi-Teisseire, F. Adenot "Modelling of bituminized radioactive waste leaching. Part I: Constitutive equations" Journal of Nuclear Materials 349 (2006) 96–106
9. Girold C., Lemort F.and O. Pinet O. "The vitrification as pathway for longlife organic waste treatment" Waste Management Conference , 26/02/2006 - 28/02/2006, Tucson , Etats-unis (Waste Management Symposium 2006 proceedings)

10. Girold C et Al. « Intermediate level mixed waste vitrification in a refractory free furnace using oxygen plasma ». Int. Conf. On Thermal Treatment Tech. Proceedings. Orlando, FL – USA (2003).
11. Poinssot C., Toulhoat P. (2007) : From nanoseconds to million years : predicting the long term fate of radionuclides in a deep geological disposal, SFC07 congress, chemistry of the future, the future of chemistry, Paris, Jul.07.
12. Gin,S. et al. Assessment of nuclear glass behavior in a geological disposal conditions: Current state of knowledge and recent advances. European Nuclear Conference ENC 2005. Versailles. 12-12-2005.
13. L. Johnson, C. Ferry, C. Poinssot, P. Lovera (2005) Spent fuel radionuclide source term model for assessing spent fuel performance in geological disposal. Part I- assessment of the instant release fraction, Journal of Nuclear Material, **346**, *56-65*. (Poinssot et al., 2005, 2006)
14. Bion L. (2003), BASSIST, an applied thermodynamic database for radionuclide chemistry, Radiochim.Acta, **91**, 633-637.
15. Gorgeon, L (1994), Contribution à la modélisation physico-chimique de la rétention de radioéléments à vie longue par des matériaux argileux, PhD thesis, University Pierre & Marie Curie, Paris.
16. Jacquier P., Ly J., Beaucaire C. (2004), the ion-exchange properties of the Tournemire argillite. Study of the H, Na, K, Cs, Ca and Mg behaviour, Applied Clay Science, **26**, 163-170.
17. Bazer-Bachi F., Tevissen E., Descostes M., Grenut B., Meier P., Simmonot O., Sardin M. (2006),Characterization of iodine retention on Callovo-Oxfordian argillites and its influence on iodide migration, Physics and Chemistry of the Earth, **31**, 517-522.
18. Dewonck S, Blin V ;, Descostes M., Radwan J., Poinssot C., Eikenberg J. (2007), In-situ tracer diffusion experiments at various depths in the Callovo-Oxfordian rock formation, Migration conference, Munich, Aug.07.
19. D. Hudry. presented at the 7èmes Journées Scientifiques de Marcoule - La grande Motte 2007(unpublished).

Disposal Concepts

Mater. Res. Soc. Symp. Proc. Vol. 1107 © 2008 Materials Research Society

Disposability of the UK's Intermediate Level Wastes

Paul K. Abraitis[1] and Glyn Davies[2]
[1] Environment Agency, Nuclear Waste Assessment Team, Ghyll Mount, Gillan Way, Penrith, Cumbria, CA11 9BP
[2] Nuclear Installations Inspectorate, Nuclear Directorate, Redgrave Court, Merton Road, Bootle, Merseyside, L20 7HS

ABSTRACT

The UK's intermediate level waste (ILW) comprises a range of waste types, which vary widely in terms of physical and chemical properties, radionuclide content (including fissile) and half-life. ILW needs to be conditioned in such a way as to provide confidence that a future disposal facility for such wastes will be able to meet or exceed the basic long-term safety and environmental protection standards. The reactive properties and heterogeneous nature of certain ILW can present waste immobilisation challenges and, depending on the waste conditioning process that is adopted, may ultimately preclude the production of truly passive wasteforms. Of considerable interest is the extent of long-term containment likely to be afforded by the waste container, the wasteform and any engineered barriers within the disposal system. A regulatory view of the current uncertainties is provided and areas where further understanding would be beneficial are identified.

BACKGROUND

In the UK the upper activity boundaries for low level waste (LLW) are 4 GBq/te alpha and 12 GBq/te beta–gamma [1]. By definition, intermediate level waste (ILW) has an activity concentration higher than this, but does not generate so much radiogenic heat that this needs be taken into account in the design of storage or disposal facilities. The ILW category encompasses a wide range of wastes, of differing radionuclide content, half-life, and physical and chemical properties. These include: cladding and other waste materials from Magnox and Advanced Gas Cooled Reactor (AGR) fuel elements; unirradiated fuel and certain experimental fuel declared as waste; sludges; ion exchange resins; desiccants and catalysts; miscellaneous activated and contaminated materials; and plutonium contaminated materials. Large quantities of ILW are expected to arise over the next century, as operational nuclear facilities reach the end of their lives, decommissioning proceeds and stored operational wastes are retrieved.

The Government's policy states that there should be a 'systematic and progressive reduction of the hazard' [2] represented by the UK's public sector civil nuclear liabilities. These liabilities include a considerable legacy of potentially mobile unconditioned wastes, some of which are currently held in older storage facilities that do not meet modern standards. There is currently no disposal facility for ILW in the UK, however the need for a progressive reduction in hazard requires an environmentally sustainable solution that provides for occupational and public safety. These wastes therefore require treatment/conditioning to a safe passive form for storage and eventual disposal. The Nuclear Decommissioning Authority (NDA) has been given responsibility for planning and implementing geological disposal and has now incorporated the skills and technology from United Kingdom Nirex Ltd (Nirex). To ensure the safe management and future disposability of these wastes, the nuclear safety and environmental protection regulators have developed joint regulatory arrangements relating to the conditioning of ILW [3].

THE NATURE OF ILW IN THE UK

A total ILW volume of 217,000 m^3 (mass 250,000 tonnes) is reported in the 2004 UK Radioactive Waste Inventory [4], and it is projected that a disposal volume of up to 290,000 m^3 of conditioned waste may arise. Pending the availability of a disposal facility (or facilities) for such waste, ILW is currently stored on the nuclear licensed sites. The only exception is a small amount disposed of at sea prior to the moratorium on sea dumping adopted in the early 1980s by the parties to the London Convention.

Approximately 16,000 m^3 of ILW has been conditioned to date, producing about 32,000 individual waste packages. ILW is generally encapsulated using cementitious materials, typically based on Ordinary Portland Cement (OPC) mixed with pulverised fly ash (PFA) or blast furnace slag (BFS), although a number of alternative encapsulants (e.g. polymers) have been used for particular wastes, for the purpose of achieving specific wasteform properties. Typically waste conditioning has involved immobilisation within vented, single skinned, 500 litre stainless steel drums. In addition to direct encapsulation, certain wastes have been supercompacted and the resulting pucks cemented into 500 litre drums. Future conditioning of a number of major legacy waste streams is expected to involve the use of larger (3m^3), double skinned containers.

Much of the UK's ILW has been produced at Sellafield, where there are a number of legacy waste storage facilities and operational ILW conditioning. ILW conditioning plants are also operating at Dounreay, Harwell, Windscale and Trawsfynydd. Further waste conditioning plants will be needed and several are in the design phase at Sellafield.

The major components of ILW include metals (stainless steel, other steel, Magnox, aluminium, Zircaloy, plus others, such as lead), organics (cellulosic materials, plastics, rubbers, ion exchange resins), inorganics (concrete, cement, sand and rubble; graphite, glass and ceramics, sludges, flocculants and zeolites), and contaminated soils.

DISPOSABILITY CHALLENGES

Disposal concept and safety case

The UK approach has been to develop a generic disposal concept for conditioned ILW. Proposals for the conditioning of ILW are assessed against this developing concept, so as to provide confidence in future disposability. The current Phased Geological Repository Concept (PGRC) envisages a period of interim surface storage followed by transport of the wastes to the repository and emplacement in disposal vaults deep (hundreds of metres) underground [5]. In order to take account of the concerns of some stakeholders, the concept includes the possibility of an extended period (possibly as long as hundreds of years) prior to closure during which wastes would be monitored and remain retrievable. As a result, a target of 500 years for waste container integrity has been introduced [6].

The wide variation in ILW composition has implications for the design of the disposal concept, as does the potentially reactive and heterogeneous nature of many UK wastes. In particular, the disposal system must be able to safely manage the large quantities of gas that are anticipated to be produced, to accommodate significant quantities of fissile material and to contain a significant inventory of long-lived radionuclides, so as to provide environmental protection in the long-term.

The conceptual design envisages stacking waste packages in vaults prior to backfilling and closure. A range of package designs are available though the material of fabrication and choice of waste matrix is not prescribed. ILW packages have vents to allow generated gases to escape. A porous, cementitious backfill is envisaged in the reference design and is intended to prevent over-pressurisation and to provide a chemical buffer. It is envisaged that the reducing, hyperalkine environment afforded by the presence of corroding metal and a high surface area, cementious backfill will provide a chemical barrier in which the mobility of many radionuclides is very low. In addition, it is argued that the cement will enhance the lifetime of stainless steel containers and will react with carbon dioxide that is produced by organic matter degradation.

The design of the disposal system is yet to be optimised and a safety case has yet to be fully developed. This will require site-specific information and further knowledge of the conditioned waste inventory. Provided that a publicly and technically suitable site becomes available, the Environment Agency believes that deep disposal is a viable strategy for the UK's ILW and that an acceptable safety case for a deep disposal facility (or facilities) could be generated in the medium term [7].

Reactive metals

A number of metallic waste components (magnesium, aluminium and uranium) can react with water within cementitious grouts to yield corrosion products and hydrogen [8]. For example, magnesium (a major component of Magnox fuel cladding) can react with water yielding gaseous hydrogen and magnesium hydroxide as the solid corrosion product:

$$Mg + 2H_2O \rightarrow Mg(OH)_2 + H_2 \qquad (1).$$

This reaction is accompanied by a volume increase in the solid phase of approximately 77%. The reaction typically proceeds via an acute phase of rapid reaction, declining rapidly (over a period that may be as long as 18 months) to a very low reaction rate, which may decrease further with time.

Similarly metallic uranium (e.g. the major component of spent Magnox fuel) and aluminium yield hydrogen and higher molar volume corrosion products on reaction with water in cement. Example reactions include:

$$U + 2H_2O \rightarrow UO_2 + 2H_2 \qquad (2).$$
$$2Al + 6H_2O \rightarrow 2Al(OH)_3 + 3H_2 \qquad (3).$$

The exact nature of the reactions and the solid products will depend on such factors as the pH and Eh of the system, temperature, the presence of solutes that can react to form salts and the availability of water. In the case of the reaction of aluminium in wet, hyperalkaline systems, the hydrogen generation rates during the acute phase can be such as to approach explosive limits and suitable control measures would be required to avoid such situations. In the case of uranium, a recent study has demonstrated that, under certain conditions, corrosion can continue at an acute rate with no decrease in corrosion rate with time. In addition, a number of experiments have shown that uranium can form self-heating reaction products in a cement matrix under certain environmental conditions. Inspections of uranium-bearing waste packages that have been in storage for more than a decade have identified minor protrusions on the container walls in a limited number of cases. One theory is that these are the result of expansive uranium–grout reactions near the container wall. Further work is ongoing to understand these observations and to consider the implications in terms of package longevity and disposability.

It is clear then that some metal wastes are not chemically passive following encapsulation in OPC-based grouts. Pressurisation caused by hydrogen evolution in the acute phase, followed by gradual expansion due to corrosion product formation could potentially challenge waste package integrity in the long-term. While waste remains in dry storage, the extent of the reaction may be limited by water availability and limits on the mass transport of any available water within the grout matrix. When the waste is subsequently in a closed repository system saturated with groundwater it would seem likely that further reaction will occur. For example, the ingress of water through package gas vents could restart corrosion of any available reactive metal. Ultimately this could lead to extensive degradation of the physical barriers and could ultimately enable rapid radionuclide release from the repository near field if the other barriers to migration prove ineffective. A loss of containment at the package scale is inevitable in the long-term, but slow, gradual release from a monolithic wasteform would be preferable to relatively rapid release from a highly degraded package.

Fissile wastes

The UK's ILW disposal inventory is likely to contain many thousands of times the theoretical minimum critical mass for the predominant fissile radionuclides, plutonium-239 (^{239}Pu) and uranium-235 (^{235}U). In addition, large stocks of fissile material are not currently classified as waste, but some of this material could eventually be declared as waste to be disposed. Ideally, the conditioning of fissile wastes will ensure that the fissile radionuclides are suitably immobilised to minimise the potential for migration with groundwater. However, given the half-lives of ^{239}Pu (24,000 years) and ^{235}U (7×10^8 years) complete immobilisation is impossible to guarantee. Mobility may be increased, for example, by wasteform degradation and the presence of solubility-enhancing organic materials within the disposal inventory (in some cases directly associated with fissile material).

Limits on fissile material in waste packages can be defined with some degree of confidence for the earlier waste management stages (e.g. for operations and transport) such that criticality is unlikely to be induced by identified fault sequences. However, over longer timescales, e.g. in a closed geological repository scenario, there is a possibility of mobilisation and subsequent re-accumulation of fissile material once engineered containment is breached or significantly degraded.

Theoretically, even relatively small amounts of fissile material derived from a small number of waste packages could be redistributed into a reactive configuration resulting in a criticality event. Limiting fissile package loadings may reduce the likelihood of criticality events following disposal but cannot preclude such possibilities. Furthermore, striving for reduction in the already low likelihood of criticality must be balanced against the fact that reduced fissile loadings can only be achieved by increasing the number of packages, with associated increases in costs and in the risks associated with transport and handling [9]. A key safety case argument is that any criticality events of the types envisaged within a deep geological disposal system would have low consequences at the surface, based on current understanding [10, 11].

Organics

The ILW inventory includes a wide variety of organic materials and substances rich in organic matter. These include cellulose (e.g. paper and tissues), bird guano, algae, complexing agents (e.g. citric acid, EDTA), organic preparations and polymers. Recognition and characterisation of organic materials in waste streams can be challenging. In addition, organic materials may be added to waste or engineered repository system components, for example in the form of organic-based superplasticisers, which could be used in encapsulating grouts and for structural cement formulations. The presence of organic materials in ILW raises a number of issues including:

- the potential for radionuclide solubility to be enhanced through formation of chemical complexes (e.g. the complexation of actinide cations with the organics or with the products of organic matter degradation);
- gas generation by degradation of organic matter, resulting in effects such as wasteform cracking or release of radioactive material via gas pathways (e.g. carbon-14 labelled gases);
- direct interactions that might adversely affect the long-term behaviour of the waste package (e.g. generation of acidic degradation products that may adversely affect the encapsulating matrix or waste container).

In some cases an improved understanding of the long term behaviour of organics may be important to provide assurance that their effects pose no significant problems for disposal. Otherwise improved segregation of the waste, or treatment to mitigate the effects of organic materials (e.g. incineration, pyrolysis) might need to be considered to avoid undesirable interactions.

Carbon-14

The ILW inventory contains significant amounts of carbon-14 (^{14}C) associated with irradiated graphite, irradiated metals and ^{14}C-labelled organic wastes. As much as 10,000 TBq ^{14}C could be associated with the irradiated graphite in reactor cores, which will ultimately need to be managed as ILW. A significant inventory of carbides may also be associated with irradiated metals (including Magnox cladding, metallic uranium and steel). For example, some ^{14}C arises as elemental carbon or metal carbides, formed by neutron irradiation of ^{14}N present as an impurity in Magnox cladding and fuel. Carbides yield methane and other short-chained hydrocarbons on reaction with water and, unlike carbon dioxide, these are not anticipated to react with the repository backfill to form insoluble carbonates.

Current assessments, based on models which assume no dissolution of ^{14}C-bearing methane in groundwater, instantaneous migration to the biosphere and efficient assimilation of ^{14}C by plants, suggest that the post-closure regulatory risk target [12] may be challenged. Further work is ongoing to review the validity of these assumptions [13]. A major uncertainty is the rate and form of ^{14}C release from different waste materials on contact with groundwater.

Waste package longevity

The waste package comprises the wasteform and the waste container, which have important containment functions, both individually and in combination. Disposal packages will need to retain integrity throughout the period in which they need to be safely handled and moved. In the UK context there will be a period of storage, the duration of which is not currently known, prior to emplacement within a repository, and potentially a further extended period of retrievable storage prior to repository closure. Thus package lifetimes may need to extend to centuries. Potential challenges to waste package longevity include the following [14]:

- External corrosion of the container – generalised and localised corrosion of metallic containers and any package handling features, including atmospheric stress corrosion cracking. The extent and progress of such processes will depend on factors such as the storage environment (e.g. temperature, humidity) and the presence of surface contaminants (e.g. salt deposits);
- Internal corrosion of the container – processes occurring within the wasteform may generate acidic or other potentially aggressive species (e.g. chlorides). Such species could promote "inside out" corrosion. Galvanic coupling, radiolytic and microbially influenced corrosion have also been postulated as potentially significant;
- Evolution of the wasteform – this may involve: interactions between metals and water to produce gases and expansive corrosion products; continuation of cement hydration reactions to produce mineral phases with different physical characteristics (e.g. molar volume, sorption capacity, porosity); degradation of the encapsulant (e.g. via radiolysis); and interactions of organic materials within the wasteform to produce gases, voids and solubility- or corrosion-enhancing products.

Ultimately such processes could lead to package failure (e.g. through a significant breach of the container wall or a loss of mechanical integrity of the package overall such that it cannot be safely handled or stacked). The following strategies could be adopted to promote or enhance package longevity:

- Ensuring that package storage conditions and practices are optimised, so as to minimise or prevent the corrosion of metallic containers;
- Using "enhanced containers". These could be fabricated from highly corrosion resistant alloys (e.g. duplex stainless steels; nickel, copper and titanium alloys) or novel materials (e.g. carbon fibre reinforced concrete). Packages may also be designed to accommodate any wasteform evolution (e.g. an inner vessel to accommodate expansion) or with features that allow package finishing (e.g. an inner annulus that can be filled at a later date, or ports to allow addition of further encapsulant prior to transfer to the repository).
- Using alternative waste conditioning matrices. These would be designed both to be inert with respect to the container material and to produce a passive wasteform. In the case of reactive metal wastes, for example, this might include low water content superplasticised grouts or polymeric encapsulants.
- Removing waste reactivity by pretreatment, e.g. the dissolution or incineration of reactive waste, followed by conditioning of the residue;
- Effective sorting and segregation of incompatible wastes for appropriate conditioning;
- Remediation or reworking of failed packages or those tending to failure (e.g. injection grouting of cracked wasteforms, overpacking of failed containers).

There are likely to be considerable worker safety and environmental impacts should reworking or remediation of degraded packages be required during any period of extended storage [15]. Package degradation during the operational phase of the repository could negatively affect long-term safety and increase environmental impacts following repository closure. Caution is needed in projecting package lifetimes to several centuries or more given the current state of knowledge and the uncertainties relating to waste package ageing. There is a need to better understand package longevity and the corresponding degradation mechanisms over a long period of storage. This would inform any requirements to produce improved packages for certain waste streams, or to make provision for remediation or reworking of any packages which may fail.

Waste characterisation

Adequate characterisation is essential to enable optimisation of wasteforms and prediction of the short and long-term performance of conditioned waste. In addition, there will need to be sufficient knowledge of the conditioned waste to demonstrate compliance with future repository waste acceptance criteria. The characterisation challenge for a given waste stream may vary depending on the waste conditioning method that is selected (e.g. analysis of a homogeneous liquor resulting from waste dissolution compared to assay of raw, heterogeneous wastes for direct encapsulation). Examples of waste characterisation challenges include:

- For certain large volume historic ILW, extensive characterisation may be problematic for reasons of physical/chemical heterogeneity, volume and accessibility. Here reliance on historical waste records data and confirmatory measurements (e.g. gross gamma dose rate) may be necessary, together with statistical arguments regarding the distribution of waste components across a large number of packages.
- The limitations in non-destructive assay may preclude the reliable measurement of the fissile content of certain wastes. For example, it is difficult to measure fissile contents using neutron techniques in wet waste systems (e.g. where the water content is greater than about 5%). In other cases, where fissile loadings are so low so as to challenge detection limits, the uncertainties due to low signal-to-noise ratios may mean that highly pessimistic biases must be applied in assigning fissile loadings to a waste package.

It must be noted that while adequate characterisation is essential, unnecessary over characterisation resulting in additional cost and unjustifiable occupational radiation exposure must be avoided if systematic and progressive reduction in the hazard is to be achieved.

CONCLUSIONS

This paper has summarised the nature of the UK's ILW and some of the more significant disposability challenges that it presents. There is a need to progressively reduce the hazards associated with the storage of raw wastes. These include certain high volume, legacy waste streams currently held in older storage facilities that do not meet modern standards. In the short-term, prior to the availability of a disposal facility, the major challenge is ensuring that waste conditioning produces sustainable, passively safe waste packages that are suitably robust for any period of interim storage. Provided that a publicly and technically acceptable site becomes available, the Environment Agency believes that deep disposal is a viable strategy for the UK's

ILW and that an acceptable safety case for a deep disposal facility (or facilities) could be generated in the medium term.

REFERENCES

[1] Defra, DTI and the Devolved Administrations (2007). Policy for the Long Term Management of Solid Low Level Radioactive Waste in the United Kingdom.

[2] "Review of Radioactive Waste Management Policy: Final Conclusions", Cm2919, HMSO, July 1995.

[3] Health and Safety Executive, Environment Agency and Scottish Environment Protection Agency, Conditioning of intermediate level radioactive waste on nuclear licensed sites, Guidance to Industry, 2005.

[4] Defra and Nirex (2006). Radioactive Waste in the UK: Main Report. DEFRA Report DEFRA/RAS/05.002, Nirex Report N/090.

[5] Nirex (2005). The Viability of a Phased Geological Repository Concept for the Long-term management of the UK's Radioactive Waste. Nirex Report N/122.

[6] Nirex (2005). Waste Package Specification and Guidance Documentation. Volume 1 - Specification. Nirex Report N/104.

[7] Environment Agency (2005). Review of Nirex Report: 'The Viability of a Phased Geological Repository Concept for the Long-term Management of the UK's Radioactive Waste'. NWAT/Nirex/05/003 (Version 3.1).

[8] G. A. Fairhall and J. D. Palmer, 'The Encapsulation of Magnox Swarf in Cement in the United Kingdom', Cem. Concr. Res. (1992), 22, p. 293–298.

[9] P. K. Abraitis and A. J. Baker, 'Regulatory Oversight of the Conditioning of UK Intermediate Level Radioactive Wastes: Post-Closure Criticality Safety Aspects', Proceedings of ICEM'05, Glasgow, 2005.

[10] T. W. Hicks and T. H. Green, (1999). A Review of the Treatment of Criticality in Post-Closure Safety Assessment of Radioactive Waste Disposal. The Environment Agency Research and Development Technical Report No. P222.

[11] Environment Agency (2005). NWAT Review of Nirex's Approach to the Setting of Fissile Limits for Waste Packages and Assessment of Possible Post-closure Criticality Events. NWAT/Nirex/04/002 (Version 1.3).

[12] Environment Agency, Scottish Environmental Protection Agency and Department of the Environment for Northern Ireland (1997). Radioactive Substances Act 1993 - Disposal Facilities on Land for Low and Intermediate Level Radioactive Wastes: Guidance on Requirements for Authorisation. Bristol: Environment Agency.

[13] Nirex (2006). C-14: How we are addressing the issues. Nirex Technical Note.

[14] Nirex (2006). Package Longevity: Current Status and Route Map. Nirex Technical Note.

[15] Environment Agency (2005). Reworking Intermediate Level Waste. Science Report SC040047.

Mater. Res. Soc. Symp. Proc. Vol. 1107 © 2008 Materials Research Society

Experience in Progressing the Planning Application for New LLW Disposal Facilities for Dounreay

David Broughton and Michael S Tait
UKAEA, Dounreay
Caithness, UK, KW14 7TZ

ABSTRACT

An integral part of decommissioning the Nuclear Decommissioning Agency's (NDA) Dounreay site is the management of the solid low level radioactive waste (LLW). The United Kingdom Atomic Energy Authority (UKAEA) has developed and progressed a technical and stakeholder programme that has enabled it to submit a robust Planning Application to Highland Council (HC) for New LLW Disposal Facilities at Dounreay and to submit substantive preliminary safety and environmental cases to the Nuclear Installations Inspectorate (NII) and the Scottish Environment Protection Agency (SEPA). To UKAEA's knowledge this is the most advanced project in the UK for new LLW disposal facilities. Experience has been gained in progressing the Best Practicable Environmental Option (BPEO) Study, working with regulators in unfamiliar areas, and undertaking groundbreaking stakeholder consultation. Key lessons learnt are that stakeholders should be engaged in dialogue on the project as early as possible, documentation must be high quality and tailored to its audience, and internationally respected and credible consultants must be involved.

INTRODUCTION

The project to develop and implement a long term strategy for managing all Dounreay's existing and future LLW was initiated in 1999 and is novel in the UK. Between 38,000m^3 and 64,000m^3 (packaged volume) of LLW and between 26,000m^3 and 45,000m^3 of the high volume but low activity (HVLA) category[1] of LLW is estimated to be produced. This is in addition to the LLW that has already been disposed of in existing authorised facilities, which are now closed and have interim caps on them. Current LLW arisings are stored temporarily on site in new and suitably converted facilities.

A BPEO Study [1] was completed in 2004 and an Overall Strategy [2] was developed and published in 2005. A Planning Application [3] was submitted to HC in June 2006 for a key element of the strategy, the construction at Dounreay of New LLW Disposal Facilities.
The estimated cost of the New LLW Disposal Facilities project including operations is around £140M.

UKAEA's experience on this project is presented for the interest of others who may need to develop similar facilities in the UK as part of their decommissioning programmes.

UK BACKGROUND

There are current UK and Scottish initiatives and policy developments that are closely associated with this project's technical work and its assessment methods. The project has been developing during the same period as the reviews of both ILW by CoRWM and LLW by

[1] building rubble and soil from demolition

DEFRA and the Devolved Administrations [4]. The project has contributed to both these programmes. There have been instances where the project has been forcing the pace.

The authors have also been involved with developing stakeholder engagement ideas at European level through the EU funded COWAM and RISCOM II - EU 5th Framework programmes.

While this project has been progressing a post closure safety case of the UK national LLW disposal facility near Drigg, Cumbria has been completed and presented to the Environment Agency. Coupled to this there has been examination by NDA of the future capacity to receive LLW at this national facility.

POLICY FOR PROGRESSING THE PROJECT

At the beginning of the project a long term view was taken and it was appreciated that a successful solution would only be achieved through gaining public and regulatory acceptance. It was also known that the technical work required to assess different options would be complex.

UKAEA knew that no approvals would be given by regulators without a robust justification for a proposal that had been derived from a credible assessment of feasible options. Stakeholders would demand an input into any options selection process and the planning application process itself requires an assessment of options considered. It was clear to UKAEA that the process to be used to assess the options from all aspects had to be robust. UKAEA followed UK established practice and followed a BPEO approach which was introduced by the Royal Commission on Environmental Pollution [5].

The stages of the project up to obtaining planning permission are steps in a formal legal process. There is the possibility that the planning application might go to a public Planning Inquiry (PI). The administrative systems introduced for all activities in the project have been designed and maintained to enable fast access to information which would be essential during any PI. During the selection of consultants to assist UKAEA their future credibility in a PI courtroom environment was assessed.

BEST PRACTICABLE ENVIRONMENTAL OPTION STUDY

The BPEO identified for the management of Dounreay's LLW is to dispose of it in new, shallow, below-surface disposal vaults constructed at Dounreay; and to retrieve the already disposed LLW from the existing facilities, and re-dispose of it in the new facilities. The concept design for these new facilities is given in Figure 1.

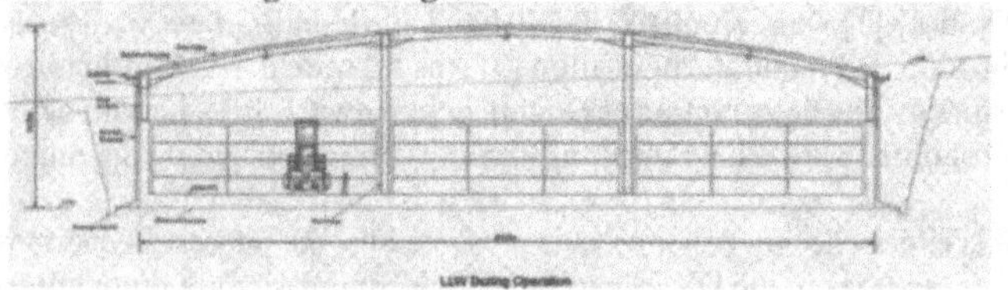

Figure 1. Concept design for New LLW Disposal Facilities

The BPEO Study has been extensively reported in references [1] and [6]. This paper provides comments on the stakeholder consultation part of the BPEO study and lessons learnt that will be of interest to other future developers.

Stakeholder Consultation

UKAEA developed a way of involving stakeholders in the BPEO Study and published its plans in a brochure "Restoring our Environment". The stakeholder consultation programme has been a success and has contributed strongly to the credibility of the study and its outcome.

Three stakeholder panels were held:

- *Internal Stakeholder Panel* consisting of eight people who worked on the site, though not necessarily for UKAEA, and who were not associated with the project
- *Youth Stakeholder Panel* consisting of three sixth-year students from Wick and Farr High Schools
- *External Stakeholder Panel* consisting of fourteen Highland councillors, officials and representatives of community organisations

All attendees were provided in advance, specifically prepared summaries of the technical work and options under consideration which had calibrated scorings. Using a range of methods, including facilitated discussions, "post-it" note displays, and interactive computer presentations, panel members were able to share their thoughts on different aspects of the issues raised, and UKAEA was able to understand and appreciate these views. A noticeable feature of the discussions at the panels was that participants did not stick to preconceived ideas when they were exposed to the wider experiences and views of others.

Each stakeholder panel carried out a number of "weightings" on the attribute scores and so each panel arrived at a number of option rankings. In total the stakeholder panels carried out seven attribute weightings coming from different perspectives. This was done in real time with the rankings being seen to change on a display screen as weightings were debated and entered into the computer model. There was a consistency of view across all the stakeholder panels' weightings with the near surface below-ground option always ranking first and the indefinite above-ground storage always ranking last. Considerable more detail on sensitivities etc. can be found in the previously quoted references.

Following the stakeholder panel meetings, a period of three months was set aside for wider public participation. A consultation document was issued to stakeholders who had registered previously an interest in Dounreay decommissioning. The consultation document, technical information and the reports of the stakeholder panels were also placed on the UKAEA's website for unlimited access.

Twenty three responses were received, eleven from individuals and twelve from community or local government organisations. This number of responses is in line with other public consultations on similar nuclear issues. The content of the responses was wide with very detailed ones from local councils and simpler ones from individuals. However, all respondents had given serious thought to the issue and their responses were well presented. UKAEA made detailed individual responses to each person or organisation. It is important to ensure respondents are informed as to how their views have been taken into account.

UKAEA's observations on the BPEO study process

Undertaking the BPEO study and submitting the final report to NII, SEPA and HC raised the awareness of this process and encouraged discussion on the role of the BPEO study process. It was indeed a trail blazing initiative, particularly with the involvement of stakeholders. UKAEA had assumed that the BPEO study process was uncontroversial but the regulators expressed some reservations. Reflecting on this UKAEA now understands some of the reasons. UKAEA probably underestimated the problem the regulators had when no formal procedures or guidance on how to deal with a BPEO study were available , how it should be assessed, or indeed whether they were required to assess it at all. Other future developers should note that there is now available a guidance document that the regulators jointly commissioned [7].

This BPEO study is an integral part of what will be a later formal authorisation process and the regulators were cautious in case early involvement could be viewed as prejudicing their later determinations.

Owing to the political sensitivities involved with site identification, UKAEA naturally was able to go into more detail on the Dounreay options than the local or distant generic options so the regulators may have felt there was an imbalance in assessment of the options. This was coupled with some regulators considering that BPEO studies should have been undertaken at a higher level of broader policy assessment than the specific issue of Dounreay LLW.

NII places strong emphasis on current policy and what is required of a licensed site operator. Thus NII considered that those options which fell within current government policy should be leading candidates in the BPEO study. Similarly, long term storage of LLW is not government policy and should not be an option to be considered.

SEPA had some concern that the range of attributes considered might not be highlighting the environmental issues sufficiently.

If UK policy on an issue is generally clear then UKAEA senses that the regulators feel more comfortable with Best Practical Means (BPM) studies focussing on the particular project.

On revisiting the Final BPEO Study Report and its sixty supporting technical studies, UKAEA now observes that many detailed BPM issues had actually been addressed during the BPEO process and would recommend that others embarking on a similar BPEO study that will have a subsequent associated BPM, should consider carefully what issues, and to what depth, are included in which studies.

As neither a BPEO study report nor an overall strategy document is a formal document in the regulators' approval process, a developer should not expect to receive a formal endorsement from them. The developer should strive to achieve the position where the regulators are content for the documents to become public knowledge.

BEST PRACTICABLE MEANS STUDY

The BPM process and report can be considered as one of the links between the identified option from the BPEO Study and the submissions put to the regulators. The BPM assessment supports the ESC and underpins the site justification process.

A modification was made to the design during the BPM development through consultation with the near neighbours to the Dounreay site. A number of "walk-in" information and separate discussion meetings was held with the near neighbours[2] and the residents of the nearest village of Reay. UKAEA's initial proposals for the location of the new LLW disposal facilities were displayed and explained at these meetings. In general, the residents of Reay had little concern over the proposals but thought the facilities should be as near to the existing Licensed Site as possible and visually unobtrusive. However, the nearer neighbours were

---

[2] the hamlets of Buldoo, Achreamie and Upper Dounreay at the southern edge of the UKAEA owned land at Dounreay

vociferous in their objection to the facilities being too near to their homes and properties. The initial location had been chosen as it was the highest ground farthest from the sea to cater, as far as practicable, for future sea level rise and coastal erosion. UKAEA took note of the neighbours' concerns and found it was feasible to move the location nearer to the Licensed Site and further away from their homes. This exercise immediately highlighted the constraints in siting the facilities.

A difficult concept for the neighbours to accept was the requirement of UK policy and guidance to consider sea level rise in the siting of the facilities. The performance assessment work considered climate change and sea level rise models to determine a line above which land is unlikely to be eroded and inundated by the sea. This is shown in Figure 2 as the red line. UKAEA is required to justify its siting of the facilities with regard to potential future shorelines. This of course means that the facilities are not as far away from the neighbours' properties as they would like. Their argument is that UKAEA and the regulators are concerned more about safety in 10,000 years time, rather than the effect on them now. They also question why there is concern about the integrity of the facilities and inundation by the sea in 10,000 years time if LLW reduces to radioactivity levels close to that of the local soils in only 300 years. This debate has highlighted that well meaning conservatism in government guidance is not automatically accepted as a good thing by people directly and immediately affected by its consequences.

More accurate location of the main geological faults identified in the site characterisation programme also placed a tighter restriction on the area in which the facilities could be sited. UKAEA has now optimised the location for the new facilities but has not succeeded in gaining the support of the nearest neighbours in Buldoo. Resolution of these differences of opinion will rightly be determined by the planning permission process.

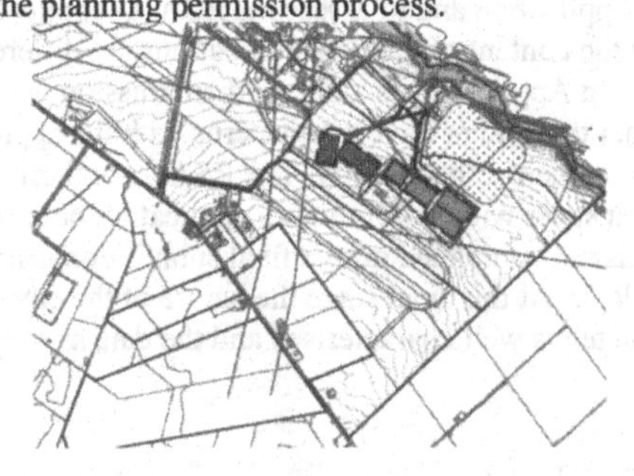

Figure 2. Position of New LLW Disposal Facilities in the Planning Application

Local Consultation

UKAEA has considered whether it could have involved these near neighbours as a specific stakeholder group earlier in the stakeholder consultation programme[3]. It would have been useful for both parties, but future UKAEA projects and other developers must be careful how this is accomplished.

---

[3] One of the nearest neighbours did in fact respond to the general public consultation through correspondence and all could have if they had wished to

At the stakeholder panels individuals were nominated by the formal bodies that they represented e.g. NFU, HC. For consistency, the nearest neighbours should have to form a group[4] and select representatives. However, they are without doubt a special group and in a future situation UKAEA should probably constitute a near neighbours' stakeholder panel to follow the more widely representative one.

Local councillors or community representatives are what their titles suggest, and exert their right to represent their constituents. Going past them direct to their constituents needs prior discussion, care and organisation. Whatever the situation, fast and accurate feed back is essential in local networking.

KEY AUTHORISATIONS

In order to progress the realisation of the New LLW Disposal Facilities UKAEA needs to achieve two key authorisations:

- Planning Permission from HC under the Town and Country Planning and Environmental Impact Assessment legislation
- Radioactive Waste Disposal Authorisation from SEPA under the Radioactive Substances Act 1993 (RSA 93)

The Planning Application is sent by HC to its statutory consultees for comment. SEPA is both a statutory consultee and an authorisation body in its own right. A new situation has arisen for SEPA, as this is the first time it has been involved in a request to authorise new LLW disposal facilities. This requires SEPA to respond to HC on the Environmental Statement contained in the Planning Application and to assess the RSA 93 application.

The timescales and the content of the two applications are different, though complementary. The Planning Application is a once-off submission that requires statutory consultees to inform HC that they either have no concerns with the application or that it does not conform to regulations within their responsibility. The RSA 93 authorisation process on the other hand is a staged process that starts with a preliminary application based on general data and theoretical modelling, and is not completed until a final submission is made when the facilties are completed some years later. At this final stage the nature of the environment in which the facilities have been constructed is well characterised and the data as robust as possible.

Planning Application

Through the OJEU[5] process Jacobs Babtie was engaged in September 2005 with the remit to produce the Planning Application and Preliminary RSA 93 Application by the end March 2006. This was a very tight timescale but was achieved. Jacobs Babtie sub-contracted Galson Sciences Ltd. for the specialist performance assessment and RSA 93 work. Production of the Planning Application progressed in parallel with the BPM study and local stakeholder liaison. Integration of developments was achieved easily and quickly with UKAEA and its contractors acting in a coordinated manner.

The completed Planning Application submission, submitted to HC at the end June 2006, comprised of four A4 and one A3 size folders containing high quality presentations of designs,

---

[4] they did this in early 2007 in order to join the DSG

[5] the Official Journal of the European Union in which government bodies must advertise upcoming projects of specific value

layouts, graphics for environmental data and visualisation of the facilities from different locations in the neighbourhood.

SEPA decided that it could not respond to HC that it was satisfied that the development could go ahead on the basis of the planning application documents without undertaking a review of the preliminary RSA 93 documentation and the Environmental Statement (ES) Because of the detail and time required for this second review, a period of eighteen months has had to be allowed over and above the normal planning timescales. SEPA is now only halfway through its review.

RSA 93 Application

As far as UKAEA is aware, no RSA 93 Application has yet been made in the UK to SEPA or the Environment Agency for new LLW disposal facilities. Guidance is available in the "Guidance on requirements for authorisation" (GRA) [8]. However, this does not differentiate between ILW and LLW and is under revision at present. Differentiation will be made between near surface and deep disposal. It is unlikely that any major changes in criteria will be made for LLW so UKAEA is confident it is following a justifiable approach.

The GRA recommends an early application which recognises the staged submissions approach explained earlier. UKAEA is still unable to submit an early application to SEPA for legal reasons connected with NDA's site management competition programme. However in order for SEPA to to be aware of the future application and respond to HC on the Planning Application, UKAEA submitted the first draft of the Environmental Safety Case (ESC) to SEPA in May 2006. The significant point for other future developers to note is that more detailed performance assessment information is required in the ESC and RSA 93 submissions than the ES of the Planning Application.

EXPERIENCE IN WORKING WITH OFFICIAL BODIES

Although this project is concerned solely with a solution for Dounreay's LLW, the Scottish Executive (SE) is influential in coordinating the regulators' and other official bodies' input to the project. The SE is working closely with the UK Government and other Devolved Administrations to implement their joint policy on LLW management and recognises that Dounreay's experience should facilitate other future similar developments.

NII and SEPA wish to be kept aware of developments with prior warning of submissions. However it is UKAEA's experience on this project that while valuable informal discussion and help can be obtained from the regulators' staff, if a definitive response is needed on an issue, then a formal submission has to be made.

UKAEA and other developers have to take into account in their project timescales that adequate time is needed for NII and SEPA to interact on issues of policy and regulations to achieve a sensible outcome that is unlikely to be challenged.

UKAEA and HC planning officials have brought together all the proposed developments at Dounreay into a comprehensive framework document [9]. This has been accepted as supplementary information to the Local Plan and allows HC officials and elected members to set an individual planning application in context with the overall long term plan.

During the course of this project the UKAEA has changed from being "the authority" working with the Department of Trade and Industry (DTI) to being a contractor to the NDA. The course of this project and the funding for it is now determined by NDA and UKAEA recognises that the next project stage of selecting contractors for detail design and construction stage could be influenced by the site competition timetable and the proposals for this project contained in the different tenderers' submissions.

CONCLUSIONS

UKAEA has successfully submitted a planning application for New LLW Disposal Facilities at Dounreay to HC after exhaustive BPEO and BPM studies which have been groundbreaking in their technical and stakeholder consultation aspects. One key lesson learnt is that stakeholder engagement should be initiated as early as possible and maintained throughout the planning application process. Another is the necessity to produce high quality documentation that is specifically tailored to its audience. Although UKAEA regrets that there is still opposition to the project from some of its nearest neighbours, it believes that its open approach in planning, stakeholder engagement and regulatory dialogue has helped to gain support for the project and was the appropriate way to proceed.

ACKNOWLEDGEMENTS

UKAEA acknowledges that all the work described in this paper has been funded by the DTI and NDA. UKAEA also acknowledges the work and advice given by the consultants Jacobs Babtie, Galson Sciences Ltd., Quintessa and Morton Fraser which was often over and above their contractual commitments.

REFERENCES

1 Dounreay Solid Low Level Waste Long Term Strategy Development – Best Practicable Environmental Option Study – Final Report – UKAEA Report no. GNGL(04)TR75, April 2004
2 Dounreay Solid Low Level Waste Strategy Development – Overall Strategy, UKAEA Report No. GNGL(05)P51 Issue 1, March 2005
3 Planning Application, UKAEA Report No. LLW(06)S2/68, June 2006
4 Policy for the Long Term Management of Solid Low Level Radioactive Waste in the United Kingdom – Summary of comments and Government response – 26 March 2007, defra PB12522A
5 RCEP (1988). Royal Commission on Environmental Pollution Twelfth Report: Best Practicable Environmental Option. Cm310. London, HMSO
6 ICEM03- 4514 Sept. 03 – ASME
7 Guidance for the Environment Agencies' Assessment of Best Practicable Environmental Option Studies at Nuclear Sites – SEPA and EA, February 2004
8 Disposal Facilities on land for low and intermediate level radioactive wastes: Guidance on requirements for authorisation. Environment Agency, SEPA, Dept. for the Environment for Northern Ireland
9 Dounreay Planning Framework (January 2006) – Highland Council

Mater. Res. Soc. Symp. Proc. Vol. 1107 © 2008 Materials Research Society

Deep Borehole Disposal of Plutonium

Fergus G.F. Gibb[1], Kathleen J. Taylor[2] and Boris E. Burakov[3]
[1]Immobilisation Science Laboratory, Department of Engineering Materials, University of Sheffield, Sheffield S1 3JD, UK
[2]Department of Geography, University of Sheffield, Sheffield S10 2TN, UK
[3]Laboratory of Applied Mineralogy and Radiochemistry, V.G. Khlopin Radium Institute, St Petersburg 194021, Russia.

ABSTRACT

Excess plutonium not destined for burning as MOX or in Generation IV reactors is both a long-term waste management problem and a security threat. Immobilisation in mineral and ceramic-based waste forms for interim safe storage and eventual disposal is a widely proposed first step. The safest and most secure form of geological disposal for Pu yet suggested is in very deep boreholes and we propose here that the key to successful combination of these immobilisation and disposal concepts is the encapsulation of the waste form in small cylinders of recrystallized granite. The underlying science is discussed and the results of high pressure and temperature experiments on zircon, depleted UO_2 and Ce-doped cubic zirconia enclosed in granitic melts are presented. The outcomes of these experiments demonstrate the viability of the proposed solution and that Pu could be successfully isolated from its environment for many millions of years.

INTRODUCTION

Plutonium, along with the other actinides produced in nuclear reactors, presents a long-term management problem and could constitute a serious security threat. Much of the world Pu inventory of around 1800 tonnes [1] exists in the form of spent fuel from which the Pu has not been (and may never be) separated. Pu not destined for burning as MOX fuel or in Generation IV reactors or required for other purposes must eventually be disposed of safely. For unseparated Pu this could be through direct disposal of the spent fuel or by separation and subsequent disposal.

IMMOBILISATION AND STORAGE

The borosilicate glasses currently used for vitrification of most of the fission products from spent fuel have a limited capacity for actinides and are generally not appropriate for their immobilisation. Lanthanum borosilicate glasses, which can accommodate up to 10% Pu, offer an improvement but require melting temperatures around 1500°C. Calcium borosilicate and phosphate glasses are other possibilities still under investigation. There is, however, a growing majority view that Pu is best immobilised in mineral and ceramic-based waste forms [1-6] and minerals and ceramics that can accommodate substantial amounts of Pu and other actinides in their crystal lattices are being extensively investigated. Among the more promising candidates are the zircon [$(Zr>Hf)SiO_4$] - hafnon [$(Hf>Zr)SiO_4$] series, baddeleyite [$(ZrHf)O_2$], cubic zirconia [ZrO_2], zirconolite [$Ca(ZrHf)Ti_2O_7$], monazite [$(CeLaTh)PO_4$], uranium dioxide and various perovskite [$(CaNaFeCe)(TiNb)O_3$] and pyrochlore [$NaCaNb_2O_6F$] based compositions. An alternative currently gaining popularity is immobilisation in a "low specification MOX" not destined for burning, possibly utilising depleted U. This has the advantage that it could be manufactured using existing fuel fabrication procedures and facilities.

No regulatory approval has yet been sought for geological disposal of such Pu-bearing waste forms, nor has any mechanism or route been identified for it, and the assumption is that the waste form would be placed in safe interim storage. Such storage needs to be dry, requiring a surface facility protected against flooding etc., and secure - ideally deep underground. The necessary compromise could prove difficult and expensive. For storage, sub-critical amounts of the waste form would need to be packaged, possibly in small metal containers. To reduce their volume these packages could be hot isostatically pressed but doing so would place constraints on the wall thickness of the container, reducing its long-term durability.

DISPOSAL

The most widespread view is that any such packages containing the Pu-bearing waste form would eventually be disposed of in some form of repository designed for high-level wastes as and when such a facility becomes available.

Mined repositories

Proposed mined and engineered repositories, such as those based on the Swedish KBS-3 generic concept, are at depths of 300-1000 m and are therefore geologically shallow. They may be in hard rocks (granite, gneiss etc.) as favoured in Finland, Sweden and Canada or in soft rocks (clay, shale) as preferred in France, Belgium and Switzerland. The main concern with any mined repository is that eventually groundwater will gain access to the waste, leaching out the radionuclides and transporting them back to the biosphere before decay has rendered them radiologically harmless. In the case of Pu-containing wastes there is also a concern that some, unspecified, geochemical process might concentrate the Pu leading to potential criticallity.

As a result, much attention is being focussed on the durability and leaching behaviour of proposed Pu-bearing waste forms [1,2]. The situation is further complicated by concerns over the effects of damage to the crystalline structure (metamictisation) from self-irradiation resulting in expansion and potentially enhanced leachability of the Pu [2,4,6-8]. These effects have been evaluated for both external radiation [7,8] and internal self-irradiation of waste forms doped with Pu and other actinides [4,7-9]. Although the results are somewhat inconclusive [4,10] there is little doubt that some of the proposed waste forms, such as zircon, are susceptible to metamictisation [2,4,11]. Although this need not rule them out as potentially useful waste forms, especially if aqueous leaching is not an issue (see below), the effects of radiation damage are often used to question their suitability for Pu containment in geological disposals [12].

Deep boreholes

The concept of using deep boreholes for disposal, sometimes known as very deep disposal (VDD) or deep borehole disposal (DBD), is emerging as a potentially superior form of geological disposal for certain types of higher activity radioactive wastes [13,15] and a particularly strong case can be made for DBD of Pu and other fissile materials [16]. DBD offers many advantages over mined repositories [13,14,16] as it relies more on the geological barrier and less on engineered barriers, the performances of which are uncertain on the timescale of 10^5-10^6 years required. In particular, at over 4 km compared with a few hundred metres, the greater depth and less dynamic hydrogeological conditions increase confidence in the ability of the geological barrier to prevent return of the redionuclides to the biosphere. In addition to greater safety, DBD offers maximum security against terrorist and accidental intervention, wider availability of geologically suitable sites, minimal environmental disruption and potentially a more cost effective solution.

In a study on the future of nuclear power in the USA [15] it was recommended that DBD "merited a significant R & D program" as a disposal option for spent fuel. In the UK the Committee on Radioactive Waste Management recommended geological disposal for all high-level wastes to the Government, stating [17] that any decision on the exact form of such disposal "should leave open the possibility that other long term management options" (*i.e., than mined repositories*) "for example borehole disposal, could emerge as practical alternatives".

The two arguments commonly put forward against DBD are that sinking the large diameter (0.6-0.8 m) holes required to depths of 4 km or more are at the limits of current deep drilling technology and that retrieval of the waste packages would be extremely difficult and expensive. For the version of DBD we are proposing for Pu the former is not relevant as the holes need be no wider than 0.27 m. Fully cased and cemented boreholes this size and larger are routinely sunk to more than 6 km in the geothermal energy industry for around US$8 M each [18] and commercial drilling rigs with this capability are currently available. For most high-level wastes retrievability is a very debateable issue [19] but for Pu, where security is paramount, it is undesirable and the difficulties of retrieval, which could not be done easily or covertly, are a positive advantage.

GRANITE ENCAPSULATION

We propose here a scheme for the disposal of Pu in deep boreholes that eliminates the possibility of aqueous leaching and hence the uncertainties about the long-term performance of mineral and ceramic-based Pu waste forms. The key to this scheme is the prior encapsulation of the Pu-bearing waste form (including low-specification MOX) in rock identical to the granitic host rock of the borehole deployment zone. This could be achieved by disseminating small (1 – 10 mm) pieces of the preferred waste form in crushed granite that is then partially melted and completely recrystallized under conditions of suitably slow cooling.

Until relatively recently it was widely believed that medium to coarse grained granites could only be formed by extremely slow natural cooling over hundreds, if not many thousands, of years. However, within the last few years Attrill and Gibb have successfully demonstrated that, under the conditions of DBD, a typical S-type granite like E93/7 [20,21] can be extensively melted and completely recrystallized [22] in a few months. In their experiments, which were designed to investigate high-temperature DBD of HLW, Attrill and Gibb found [20] that, at a pressure of 150 MPa (the pressure at depths of ~ 4.5 km in continental crust), granite E93/7 begins to melt just below 700°C in the presence of small amounts of water. The amount of melting increases with temperature and water content up to saturation (= 4-5% H_2O depending on temperature). For example, with only 1.5% of total H_2O a temperature of 800°C is required to produce 40% of melting but with 5% H_2O 80% of melt can be generated at only 750°C. The granitic silicate liquids produced after more than 30 days under the conditions for over ~50% of melting are thought to be very close to equilibrium partial melts [20]. Linear cooling experiments [22] over the range 800°C to 560°C required for solidification showed that the amount of quenched liquid (glass) decreases with a reduction in the cooling rate, becoming negligible at rates around 0.1°C/hour. Hence, even from high degrees of partial melting, granite can be recrystallized completely to a medium grained, holocrystalline rock when cooled more slowly than 0.1°C/hour.

On the basis of these studies the Pu-bearing waste form could be encapsulated in cylinders of granite by various methods. The simplest would be to seal the mixture of crushed granite and waste form with a suitable amount of H_2O (4-5%) into an appropriately sized, thin-walled metal container that is capable of deforming to equalise interior and exterior pressures. The container is then placed in a large pressure vessel and heated (possibly by RF heating) and cooled as required. Ideally, the mixture should be held at over 760°C for around 30 days at a pressure of 150 MPa and an oxygen fugacity close to the Ni/NiO buffer before cooling to 550°C at a rate of 0.1°C/hour or less. Below 550°C the now solid granite cylinder need only be cooled slowly enough to allow annealing and avoid contraction cracking. The entire process would take about 120 days but several cylinders could be processed simultaneously. After cooling the metal container could be removed, or not, depending on the disposal strategy. Alternatively, the granite cylinders could be produced in expendable refractory moulds. More sophisticated variants could aim to produce a granite cylinder in which the Pu-bearing waste forms are absent from the outer margins.

Crucial to the proposed scheme for the disposal of Pu-bearing waste forms in deep boreholes is that during the encapsulation process the waste form does not dissolve in, or react with, the silicate melt and there is no diffusion of Pu out of the waste form. Although the behaviour of natural analogues (zircon, uraninite, monazite etc.) crystallized from various parent magmas is very encouraging in this respect, a series of experiments was undertaken to investigate this and the results are reported below.

EXPERIMENTAL RESULTS

Using the same experimental procedures as detailed in Attrill and Gibb [20] pieces of zircon, UO2 and Ce-doped cubic zirconia were enclosed in partially melted granite E93/7 at high temperatures for several months. A full account of these experiments will be given elsewhere but we summarise here two of the most important with their results and discuss their significance for the DBD of Pu.

A 0.187 g cylindrical pellet of depleted UO_2 (manufactured in the same way as fuel pellets and kindly supplied by BNFL) was sandwiched between the flat faces of two crystals of natural zircon (containing 1.3 wt.% Hf). The "sandwich" was placed in an 8 mm diameter gold capsule surrounded by 0.788 g of powdered granite and 0.023 g of H_2O. After welding closed the capsule was placed in a cold seal pressure vessel and held at 760°C and 150 MPa for 6.6 months to generate ~60 vol.% of melting [20] before quenching.

Polished thin sections through the sample (figure 1a and b) show the UO_2 pellet and zircon crystals still in mutual contact and enclosed in partially melted granite. It appears that the contacts of both the UO_2 and zircon with the silicate liquid (now quenched to glass) are perfectly sharp with no evidence of reaction or corrosion (figure 1 c and d). The granite/UO_2, granite/zircon and zircon/UO_2 contacts were all investigated by EPMA (although only the first two are relevant to this paper). Sequences of spot analyses across the granite/UO_2 interfaces revealed no detectable U in the silicate liquid adjacent to the contact and no detectable Si, Al, Na or K in the UO_2 pellet close to the granite. Similarly for the granite/zircon contacts, EPMA detected no Si, Al, Na or K in the margins of the zircon crystals and no Zr or Hf in the granitic glass adjacent to the zircon. [Granite E93/7 contains 49 ppm of Zr, 2 ppm of Hf and 4 ppm of U, which even if concentrated in the melt phase would remain below the detection limit of the electron microprobe. However, as these elements are almost certainly present in refractory accessory minerals such as zircon and monazite, they are unlikely to have entered the melt phase during the experiment.]

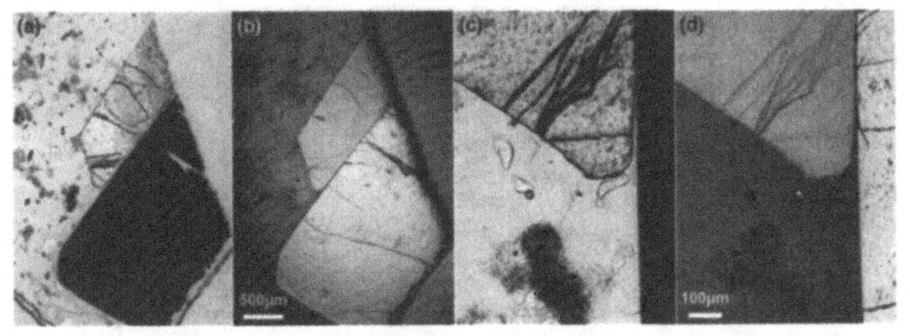

Figure 1. Photomicrographs of the zircon/UO_2 "sandwich" after the experiment in (a) transmitted light and (b) reflected light. The zircons are upper centre and lower right; the UO_2 is lower centre. Details of the zircon/UO_2/granite contacts from (a) and (b) are shown in transmitted light (c) and reflected light (d).

It is evident from the optical examination of the interfaces and the electron microprobe analyses that there has been no dissolution of either the zircon or UO_2 in the liquid phase of the partially melted granite. Nor has there been any significant diffusion of elements across the interfaces in either direction despite the zircon and UO_2 having been in contact with granitic melt at 760°C for over 6 months.

To study the behaviour of a ceramic-based waste form, a similar experiment was carried out using a gem quality single crystal of 20% yttria stabilised cubic zirconia doped with 0.3% CeO_2 to simulate tetravalent actinides like Pu. The crystal (figure 2a), weighing 0.102 g, was enclosed in 0.735 g of powdered granite to which 0.022 g of H_2O was added to give a total H_2O content of 3.43 wt.%. The sample was held at 780°C and 150 MPa for 4 months during which ~70% of melt was generated [20] before quenching. Optical examination of sections through the sample (figures 2b and c) and SEM imaging (figure 3) revealed a perfectly sharp interface between the zirconia and the silicate liquid (quenched to glass).

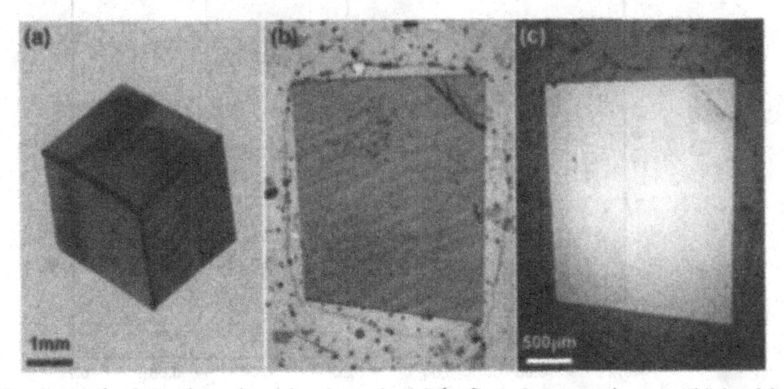

Figure 2. Crystal of Ce-doped cubic zirconia (a) before the experiment; (b) in thin section enclosed in quenched partially melted granite in transmitted light and (c) in reflected light.

Figure 3. Secondary electron image of the interface between the cubic zirconia crystal (left) and quenched, partially melted granite (right).

Electron microprobe analyses of the glass immediately adjacent to the zirconia crystal (figure 4) found no Zr, Y or Ce above the detection limits (about 150 ppm) indicating that if any material diffused out of the crystal during the experiment it must have been in insignificant amounts. Similarly, analyses of the margins of the zirconia crystal revealed that no Si, Al, Na or K had migrated in from the granitic melt. Laser ablation ICP-MS analyses along traverses across the zirconia/liquid interface confirmed that no reaction or diffusion of elements had occurred between the zirconia and the silicate melt during the experiment.

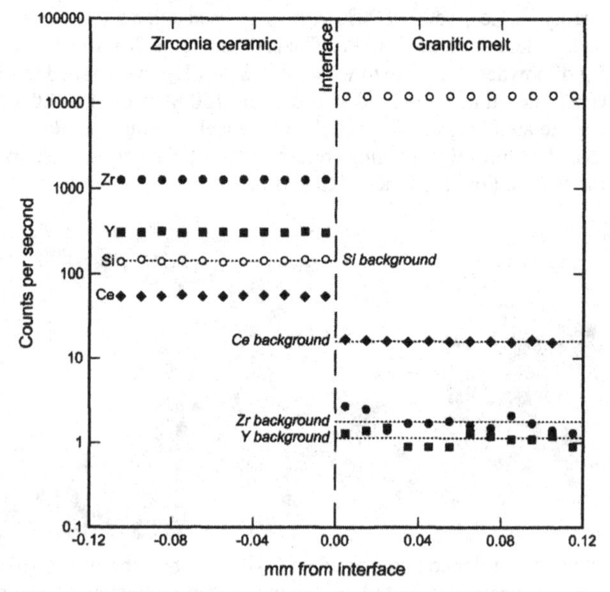

Figure 4. EPMA count rates for Si, Zr, Ce and Y at intervals on a traverse across the zirconia/granitic liquid interface.

DISCUSSION AND CONCLUSIONS

Following the manufacture of granite cylinders containing the Pu-bearing waste form by partial melting and recrystallization, they would be disposed of by insertion into a fully cased borehole sunk into granite to a depth of around 6 km. When deployment of the cylinders is complete (e.g., over the bottom 2 km of the borehole), the casing can be withdrawn from part, or all, of the hole (although this may not be necessary) and the borehole is sealed at intervals above the deployment zone. The purpose of the sealing is to deny fluids originating in and around the disposal zone access to the higher reaches of the borehole and possibly the surface. Sealing could be accomplished by a number of different methods and use a variety of materials but undoubtedly the safest and most permanent would be rock welding [14], which itself could be implemented by different means, including down-hole electrical heating.

In the fullness of (geological) time the intra-rock fluids present in the host granite will seep slowly into the borehole and invade the spaces around the granite cylinders. These fluids will most likely be dense saline brines [23] that have equilibrated with the granite in which they have resided for many millions of years. Consequently, they will also be in thermodynamic equilibrium with the recrystallized granite cylinders and so there will be no tendency for any reaction or mineralogical alteration of the cylinders that might allow the fluids access to the Pu-bearing waste forms. It is highly significant in this context that natural zircons, monazites and uraninites in granites and similar rocks survive for thousands of millions of years under just such conditions without any loss of their actinides (U and Th), even when completely metamict as a result of self-irradiation.

The outcomes of the experiments described above confirm that waste forms such as zircon, zirconia and UO_2 (analogous to low specification MOX) being proposed for Pu immobilisation will not react with, or release their Pu to, the silicate liquid during the partial melting and recrystallization process required for their encapsulation in granite. Under the conditions of the proposed encapsulation either the kinetics of any reaction or diffusion are too slow for any effects to be detected or, more probably, the waste forms are in stable or metastable equilibrium with the granite. Under the much lower temperatures involved in the eventual disposal these phases, like their natural analogues, will survive and retain their Pu for as long as they are enclosed in the granite and protected from aqueous leaching. In effect, the Pu will remain isolated from its immediate environment, let alone the biosphere, until the physical destruction of the enclosing continental crust by geological processes. At a depth greater than 4 km, by even the most unfavourable estimate, this would require many millions (possibly hundreds of millions) of years.

The amounts of Pu that could be disposed of using the proposed DBD scheme will depend principally on the exact waste form, the Pu loading of the waste form, the ratio of waste form to granite and the size and configuration of the borehole. Evaluating the limits of these parameters and the effects of various combinations is beyond the scope of this paper but a useful insight can be gained from a conservative, but realistic, example. Using an yttria-stabilised cubic zirconia containing 5 wt.% Hf and 14 wt.% Pu as the waste form and encapsulating it in a 0.25 m diameter granite cylinder at a volume ratio of 10% would give a disposal of 4.18 kg of Pu per metre of borehole. Hence a 6 km deep borehole with waste cylinders deployed over the lowermost 2 km would dispose of ~ 8 tonnes of Pu and keep it isolated from the human environment effectively for ever.

ACKNOWLEDGEMENTS

We are grateful to Neil Chapman for discussions on the disposal concept, Alan Cox for the LA-ICP-MS analyses, Martin Stennett for discussion of waste forms and Mike Cooper for help with the figures. We acknowledge the support of the NDA/Nexia Solutions (FGFG) and an EPSRC research studentship (KJT).

REFERENCES

1. R.C. Ewing, Proc. Nat. Acad. Sci. USA. 96, 3432 (1999).
2. G.R. Lumpkin, Elements 2, 365 (2006).
3. I. Muller and W.J. Weber, Mat. Res. Soc. Bull. 26, 698 (2001).
4. I. Farnan, H, Cho and W.J. Weber, Nature 445, 190 (2007).
5. B.E. Burakov and E.B. Anderson, Crystalline ceramics developed for the immobilisation of actinide wastes in Russia, Proc. 8th Int. Conf. on Environmental Management ICEM'01, Bruges, Belgium (2001).
6. W.J. Weber, R.C. Ewing, C.R.A. Catlow, T. Diaz de la Rubia, L.W. Hobbs, C. Kinoshita, H. Matzke, A.T. Motta, M. Nastasi, E.K.H. Salje, E.R. Vance and S.J. Zinkle, J. Mat. Res. 13, 1434 (1998)
7. W.J. Weber, R.C. Ewing and L.M. Wang, J. Mat. Res. 9, 588 (1994).
8. K.E. Sickafus, H. Matzke, T. Hartmann, K. Yasuda, J.A. Valdez, P. Chodak, M. Nastasi and R.A. Verrall, J. Nucl. Mat. 274, 66 (1999).
9. B.E. Burakov, M. Yagovkina, M. Zamoryanskaya, A.A. Kitsay, V.M. Garbuzov, E.B. Anderson and A.S. Pankov, Mat. Res. Symp. Proc. 807 (2004) pp213-217.
10. T. Geisler, B. Burakov, M. Yagovkina, V. Garbuzov, M. Zamoryanskaya, V. Zirlin and L. Nicolaeva, J. Nucl. Mat. 336, 22 (2005).
11. R.C. Ewing, Nature 445, 161 (2007).
12. R. Edwards, New Scientist 2586, 26 (2007).
13. F. Gibb, Imperial Engineer (Royal School of Mines, London) 3, 11 (2005).
14. F.G.F. Gibb, K.P. Travis, N.A. McTaggart, D. Burley and K.W. Hesketh, Nucl.Technology (in press).
15. M.I.T., The Future of Nuclear Power: An Interdisciplinary MIT Study (Massachusetts Institute of Technology, Cambridge, 2003)
16. N. Chapman and F. Gibb, Radwaste Solutions 10 (4) 26-37 (2003).
17. CoRWM, Managing Our Radioactive Waste Safely: CoRWM's Recommendations to Government, (Committee on Radioactive Waste management, London, 2006).
18. M.I.T., The Future of Geothermal Energy: Impact of Enhanced Geothermal Systems [EGS] on the United States in the 21st Century, (Massachusetts Institute of Technology, Cambridge, 2006)
19. F. Gibb, Geoscientist 16 (1), 14-16 (2006).
20. P.G. Attrill and F.G.F. Gibb, Lithos 67, 103 (2003).
21. B.W. Chappell and A.J.R. White, Pacific Geology 8, 173 (1974).
22. P.G. Attrill and F.G.F. Gibb, Lithos 67, 119 (2003).
23. P. Moller, S.M. Weise, E. Althaus, W. Bach, H.J. Behr, R. Borchardt, K. Brauer, J. Drescher, J. Erzinger, E. Faber, B.T. Hansen, E.E. Horn, E. Huenges, H. Kampf, W. Kessels, T. Kirsten, D. Landwehr, M. Lodemann, L. Machon, A. Pekdeger, H-U. Pielow, C. Reutel, K. Simon, J. Walther, F.H. Weinlich and M. Zimmer, J. Geophys. Res. 102, 18233 (1997).

Mater. Res. Soc. Symp. Proc. Vol. 1107 © 2008 Materials Research Society

Stability of Cubic Zirconia in a Granitic System Under High Pressure & Temperature

Fergus G.F. Gibb[1], Boris E. Burakov[2], Kathleen J. Taylor[3] and Yana Domracheva[2]
[1]Immobilisation Science Laboratory, Department of Engineering Materials, University of Sheffield, Mappin Street, Sheffield S1 3JD, UK.
[2]Laboratory of Applied Mineralogy and Radiogeochemistry, The V.G. Khlopin Radium Institute, 28, 2-nd Murinskiy Ave., St. Petersburg, 194021, Russia.
[3]Department of Geography, University of Sheffield, Winter Street, Sheffield S10 2TN, UK.

ABSTRACT

Cubic zirconia is a well known, highly durable material with potential uses as an actinide host phase in ceramic waste forms and inert matrix fuels and in containers for very deep borehole disposal of some highly radioactive wastes. To investigate the behaviour of this material under the conditions of possible use, a cube of ~ 2.5 mm edge was made from a single crystal of yttria-stabilized cubic zirconia doped with 0.3 wt.% CeO_2. The cube was enclosed in powdered granite within a gold capsule and a small amount of H_2O added before sealing. The sealed capsule was held for 4 months in a cold-seal pressure vessel at a temperature of 780°C and a pressure 150 MPa, simulating both the conditions of a deep borehole disposal involving partial melting of the host rock and the conditions under which the actinide waste form might be encapsulated in granite prior to disposal. At the end of the experiment the quenched, largely glassy, sample was cut into thin slices and studied by optical microscopy, EMPA, SEM and cathodoluminescence methods. The results show that no corrosion of the zirconia crystal or reaction with the granite melt occurred and that no detectable diffusion of elements, including Ce, in or out of the zirconia took place on the timescale of the experiment. Consequently, it appears that cubic zirconia could perform most satisfactorily as both an actinide host waste form for encapsulation in solid granite for very deep disposal and as a container material for deep borehole disposal of highly radioactive wastes (HLW), including spent fuel.

INTRODUCTION

Cubic zirconia based crystalline ceramics have considerable potential for the immobilisation and storage of actinides, for use in inert matrix fuels, as waste forms for geological disposal of actinides and for use in containers for the high-temperature very deep borehole disposal of HLW.

Immobilisation and storage

Because of their ability to accommodate actinide ions in the crystal structure and their durability under a wide range of physical and chemical conditions, cubic zirconia ceramics have been proposed as suitable host matrices for the immobilization and safe interim storage of Pu and other actinides [1-14]. Ceramics based on gadolinia-stabilized cubic zirconia or hafnia-zirconia solid solutions could also be used for the safe storage of significant amounts of excess weapons-grade Pu [15] without the risk of criticallity. A study of cubic zirconia irradiated by Xe^{2+} and I^{+} ion implantation confirmed the high resistance of its crystalline structure to external radiation damage [7,10,12,13,16] and gadolinia-stabilized cubic zirconia doped with ^{238}Pu has demonstrated an extremely high resistance to damage by self-irradiation [17,21,24]. Also, the chemical durability of zirconia-based ceramics was confirmed using samples doped with rare-earth elements [2], ^{237}Np [9, 19], ^{239}Pu [18,20] and ^{238}Pu [17,24]. Other important and relevant features of zirconia doped with ^{239}Pu [5,8,15,20,22,23], ^{238}Pu [17,21,24], ^{237}Np [9,19,23], ^{243}Am and ^{248}Cm [25] were investigated in previous studies. It is expected that all such actinide-bearing materials used for immobilisation and storage would eventually go to some form of geological disposal.

Inert matrix (ceramic) fuels

Zirconia-plutonia solid solutions, $(Zr,Pu)O_2$, with Pu^{4+} substituting for Zr^{4+} in the zirconia structure, are prospective inert matrix ceramic fuels [1] that could be competitive with mixed oxide fuel, MOX (= UO_2+PuO_2). It has been reported that burning such fuels in a LWR could lead to transmutation of more than 85 % of the total Pu and 98 % of the ^{239}Pu in the fuel [5]. It was also proposed [8] that, in contrast to spent MOX fuel, spent inert matrix fuel could be disposed of directly into geological formations without reprocessing.

Waste forms

Deep borehole disposal (DBD) of highly radioactive materials, including HLW, spent fuel and fissile materials like Pu, is emerging as a safe, secure, environmentally sound and potentially cost-effective alternative to conventional mined and engineered repositories [26-29]. At depths of 3-6 km (as against 300-800 m for mined repositories) the hydrogeological conditions are much less dynamic. They are stable over very long periods and the geological barrier provides much greater security against any return of radionuclides to the biosphere. Such geological disposals, deep into the granitic basement of the continental crust are sometimes referred to as very deep disposal (VDD) to distinguish them from mined repositories, which are deep in engineering terms but geologically shallow. Among the different forms of borehole disposal currently under investigation [27,29-31] are low temperature versions (LTVDD) aimed specifically at disposal of Pu and other actinides in packages containing sub-critical amounts of the actinide-bearing waste form which are deposited at the bottom of the borehole and sealed in.

The exceptional physical and chemical durabilities of cubic zirconia suggest it could be ideal as a waste form for DBD of Pu. The waste form could be placed in some kind of simple container purely for physical protection during transport and deployment. The high resistance to aqueous leaching of the waste form, combined with the extremely low hydraulic flow rates of the intra-rock fluids through the borehole at the depths involved [26,27,30], might be sufficient to ensure safe containment of the radionuclides on the necessary timescale.

Another material that has been demonstrated as a suitable host for actinides is zircon ($ZrSiO_4$). The exceptional stability and durability of zircon, especially in granite, is well known but both natural (U, Th-bearing) and Pu-doped synthetic zircons suffer structural damage from α-radiation [32,33]. There is some concern that the consequent metamictisation and swelling could lead to enhanced susceptibility to aqueous leaching [34] but this should not rule out zircon as a potentially useful waste form. There is, however, evidence [24,32,35] that cubic zirconia is more resistant to radiation damage and has even lower leachability than zircon.

To minimise any potential leaching, we propose encapsulating the waste form in a cylinder of the same granitic rock as that into which the borehole is drilled. This cylinder would then be in chemical equilibrium with the intra-rock fluids at the bottom of the borehole, which have equilibrated with their host rock over geological time, ensuring immunity to reaction or corrosion. One method of encapsulation would be to mix the actinide-bearing zirconia ceramic with crushed granite and then partially melt and recrystallize the granite under the same pressure as would prevail in the disposal [27,29-31,36,37].

Containers

One of the potentially safest forms of DBD being proposed for reprocessing waste and spent nuclear fuel is a high temperature variant (HTVDD) [27,29-31] in which heat from the waste packages is sufficient to partially melt the host granite [36]. As the heat output decays slowly, the silicate melt recrystallizes [37] to seal the packages into a sarcophagus of solid granite tens of cm thick. A basic requirement of this method is a container that will survive not only the transport and deployment in the borehole, but the high temperatures associated with the host rock melting and recrystallization, i.e., it must withstand temperatures up to, and possibly above, 900°C for periods of several months [29,30,37]. Among the materials being investigated for HTVDD containers, possibly as liners or surface coatings, is cubic zirconia [41].

OBJECTIVE

Successful encapsulation for DBD of actinide-bearing zirconia in granite by melting and recrystallization requires that no dissolution of the actinides by the silicate melt occurs during the process. Equally, if cubic zirconia is to be a suitable material for the manufacture or coating of HTVDD containers, it is essential that dissolution of the zirconia in the granitic melt over the melting and recrystallization period is insignificant. The main objective of this paper is to describe the first results of experimental investigations of the behaviour of cubic zirconia in granitic liquids at the high temperatures and pressures appropriate to both the HTVDD of HLW/SNF and the encapsulation of zirconia waste forms in crystalline granite prior to LTVDD.

EXPERIMENTAL

For the investigation we used a gem-quality single crystal of yttria-stabilized cubic zirconia doped with a small quantity of Ce to simulate tetravalent actinides such as Pu and Np. This has the chemical composition (wt.% by EPMA); $ZrO_2 = 79.9$; $Y_2O_3 = 20.0$; $HfO_2 = 1.3$; $CeO_2 = 0.3$. The crystal was cut and polished to a cube of approximately 2.5 mm edge weighing 0.1 g (figure 1a), and placed in an 8 mm diameter gold capsule (figure 1b) containing 0.73 g of powdered granite E93/7 [36] to which 0.02 g of H_2O had been added. Since the granite contains 0.58% of H_2O^+, the rock-water system in which the crystal was embedded contained a total of 3.23% H_2O. E93/7 is a typical S-type granite [36,38] of Caledonian age consisting mainly of perthitic feldspar and quartz with minor plagioclase, biotite and muscovite.

After welding closed and weighing, the capsule was placed in a vertical cold-seal pressure vessel (Fig. 2) and held at a temperature of 780°C and pressure of 150 MPa for 4 months before quenching and returning to atmospheric pressure (for details of the experimental method, see [36]). The capsule had not changed weight during the experiment indicating that the system had remained closed. The sample, now in the form of a solid, largely glassy, slug (figure

1b) was cut into transverse slices and made into polished thin sections for examination by optical microscopy, EMPA, SEM and CL imaging and spectroscopy.

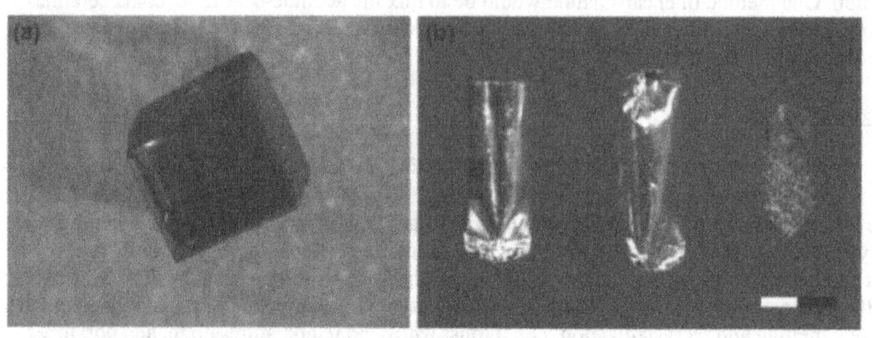

Figure 1. (a) Cube cut and polished from a single crystal of cubic zirconia: (b) gold capsules used in experiments; left – empty, centre – filled and sealed, right – sample removed from capsule after experiment (scale bar = 1 cm).

Figure 2. Cold seal pressure vessels and furnaces used in experiments.

RESULTS

Optical examination

Examination of polished thin sections in both transmitted and reflected light (figure 3) revealed that the interfaces between the zirconia crystal and the enclosing granitic liquid (now quenched to glass) are sharp with no signs of any corrosion or reaction. Further, there is no suggestion of changes in the optical properties of either the granitic glass or the zirconia as the interface is approached, which might indicate diffusion of elements from one material into the other.

Figure 3. Photomicrographs of the cubic zirconia crystal enclosed in partially melted granite taken in transmitted (a & c) and reflected (b & d) plane polarised light.

It is noticeable in figure 3 that a crack surrounds the zirconia crystal a short distance into the granitic glass. This is almost certainly a result of the rapid quenching of the silicate melt and such cracks would not occur under the conditions of slow cooling (< 0.1°C per hour) required for complete recrystallization of the granite [37].

EPMA

Points along traverses from within the crystal, across the interface and out to near the outer edge of the glass (avoiding obvious relic crystals, mainly of quartz and feldspar) were analysed by EPMA. Particular attention was paid to elements abundant in the granite but absent from the original crystal, such as Na, K, Al and Si. No evidence was found of any migration of these elements into the crystal. Despite the fact that trace quantities of Zr are present in the original granite, no Zr was detected (at a detection limit of ~ 150 ppm) in the glass, even within a few microns of the crystal. Further, no Y or Ce was detected in the glass but, given the low initial concentration of Ce in the crystal, the latter may be of debateable significance. Nevertheless, the unambiguous conclusion of the EPMA analyses is that no diffusion of elements occurred between the zirconia crystal and the silicate melt during the experiment (at least at the level of EPMA detection limits).

SEM and cathodoluminescence

Figure 4. Secondary electron image of the boundary between the crystal of cubic zirconia (left) and granite melt (right) after the experiment.

No evidence of interaction between the zirconia crystal and the granite melt was observed by SEM (figure 4). The edge of the crystal is sharp, without obvious cracks or damage and no new phases formed along the boundary of the zirconia crystal were observed.
During the experiment the oxygen fugacity (fO_2) was close to the Ni/NiO buffer, i.e. relatively reducing conditions. We therefore considered the possibility that Ce might change valence state from Ce^{4+} to Ce^{3+} and be released into the silicate melt from the zirconia with the formation of $Ce_2Si_2O_7$ or some Ce^{3+}-doped silicates. Trivalent Ce gives rise to intensive cathodoluminescence (CL) emission [39] that allows the detection of Ce admixture at very low contents (10^{-5} – 10^{-6} wt.%). Cerium admixture in 4+ valence state does not, in practice, give rise to any luminescence and the valence state of cerium can thus be determined explicitly from CL spectra. Consequently, the cathodoluminescence of the zirconia crystal and enclosing granitic glass were investigated. The spectra were acquired using an accelerating voltage of 15kV and a beam current of 10-50nA. The CL spectrometer [40] was installed in the optical microscope port of a Camebax electron microprobe. The CL spectra from different areas of the zirconia crystal after the experiment are similar to the CL spectra of the initial sample (figure 5a) and it is concluded that the cerium in the zirconia crystal remained in 4+ valence state.

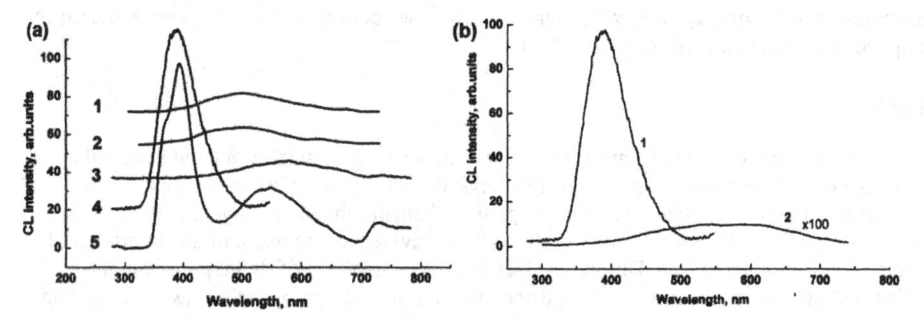

Figure 5. Cathodoluminescence (CL) spectra of : (a) undoped crystal of yttria-stabilised cubic zirconia (1), Ce-doped crystal of yttria-stabilised cubic zirconia after experiment (2), crystal of monoclinic zirconia doped with Ce^{4+} (3), zircon $ZrSiO_4$ doped with Ce^{3+} (4), $Ce_2Si_7O_7$ (5); (b) polycrystalline zircon $ZrSiO_4$ doped with Ce^{3+} (1) and silicate melt adjacent to the cubic zirconia crystal after the experiment (with CL intensity increased x 100) (2).

The intensity of the spectrum of the granite melt is very weak (figure 5b) and no traces of Ce^{3+} were observed in the silicate melt around crystal boundaries. Thus, cathodoluminescence investigations are consistent with the conclusion of the EPMA that no outward diffusion of cerium had occurred.

DISCUSSION AND CONCLUSIONS

The experiment on the Ce-doped cubic zirconia single crystal was carried out under conditions which closely simulate those expected in both the HTVDD of HLW/SNF and in the encapsulation of Pu-bearing zirconia wasteforms in recrystallized granite prior to disposal in a deep borehole. Examination of the products of the experiment by optical, SEM, EPMA and CL methods failed to reveal any evidence of corrosion of the crystal, reaction with the silicate melt or diffusion of elements across the interface after 4 months at 780°C. Either the zirconia is in thermodynamic equilibrium with the granitic liquid under these conditions or the kinetics of any reaction or diffusion are extremely slow. In either event, it is clear that the stability of cubic zirconia in granitic melts at high temperature and pressure is such that it would be ideal for the two purposes proposed above.

As a container construction material or protective covering zirconia would emerge essentially unaffected from the high-temperature phase of the HTVDD, lasting several months, during which a sarcophagus of melted and recrystallized granite is formed around the waste packages.

As a Pu/actinide-containing waste form, crystalline zirconia could be encapsulated in solid cylinders of recrystallized granite for disposal in deep boreholes without any risk of the radionuclide(s) finding their way into the granite, thus protecting them from any subsequent exposure to aqueous leaching.

ACKNOWLEDGEMENTS

This work presented in this paper was supported in part by the Immobilization Science Laboratory, EPSRC and the V.G. Khlopin Radium Institute.

REFERENCES

1. D.F. Carrol, *J. Am. Ceram. Soc.*, **46**, 195-196 (1963)
2. R.B. Heimann and T.T. Vandergraaf, *J. Mater. Science Letters*, **7**, 583-585 (1988).
3. Anderson, B. E. Burakov, and V.G. Vasiliev, *Proc. Intern. Conf. SAFE WASTE'93*, 13-18/06/1993, Avignon, France, Vol. **2**, 29-33 (1993).
4. K. Kuramoto, Y. Makino, T. Yanagi, S. Muraoka and Y. Ito, *Proc. Intern. Conf. GLOBAL'95*, Versailles, France, 11-14/09/1995, Vol. **2**, 1838-1845 (1995).
5. H. Furuya, S. Muraoka and T. Muromura, *Disposal of Weapon Plutonium* ed. E.R. Merz and C.E. Walter, Kluwer Academic Publishers, Dordrecht, 107-121 (1996).
6. B. E. Burakov, E. E. Anderson, B.Ya. Galkin, V.A. Starchenko and V.G. Vasiliev, *Disposal of Weapon Plutonium* ed. E.R. Merz and C.E. Walter, Kluwer Academic Publishers, Dordrecht, 85-89 (1996).
7. C. Degueldre, P. Heimgartner, G. Ledergerber, N. Sasajima, K. Hojou, T. Muromura, L. Wang, W. Gong and R Ewing, *Mat. Res. Soc. Symp. Proc.*, Vol. **439**, 625- (1997).
8. B.E. Burakov and E.B. Anderson, *Proc. 2nd Intern. Symp. NUCEF'98*, JAERI-Conf.99-004 (**Part I**), 295-306 (1998).
9. H. Kinoshita, K. Kuramoto, M. Uno, S. Yamanaka, H. Mitamura and T. Banba, *Proc. 2nd Intern. Symp. NUCEF'98*, JAERI-Conf.99-004 (**Part I**), 307-326 (1998).
10. K.E. Sickafus, C. Wetterland, N.P. Baker, N. Yu, R. Devanathan, M. Nastasi and N. Bordes, *J. Mater. Sci. Engin.*, **A 253**, 78-85 (1998).
11. W.L. Gong, W. Lutze and R.C. Ewing, *Mat. Res. Soc. Symp. Proc.*, Vol. **556**, 63-70 (1999).
12. K.E. Sickafus, R.J. Hanrahan, *Am. Ceram., Soc. Bull.*, Vol. **78**, No 1 (1999).
13. K.E. Sickafus, Hj. Matzke, Th. Hartman, K. Yasuda, J.A. Valdez, P. Chadak, M. Nastasi and V.A. Verrall, *J. Nucl. Mat.*, **274**, 66-77 (1999).
14. W.L. Gong, W. Lutze and R.C. Ewing, *J. Nucl. Mat.*, **277**, 239-249 (2000).
15. L.F. Timofeeva, B.A. Nadytko, V.K. Orlov, E.E. Malyukov, V.I. Molomin, V.A. Zhmak, E.A. Semova and N.V. Shishkov, *J. Nucl. Science and Tech.*, Suppl. **3**, 729-732 (2002).

16. L.M. Wang, S.X. Wang and R.C. Ewing, *Philos. Mag. Lett.*, Vol. **80**, 341-347 (2000).
17. B. Burakov, E. Anderson, M. Yagovkina, M. Zamoryanskaya and E. Nikolaeva, *J. Nucl. Science and Tech.*, Suppl. **3**, 733-736 (2002).
18. B.E. Burakov and E.B. Anderson, CD-ROM *Proc. Intern. Conf. Waste Management'02*, 24- 28/02/2002, Tucson, AZ, USA (2002).
19. H. Kinoshita, K. Kuramoto, M. Uno, S. Yamanaka, H. Mitamura and T. Banba, *Mat. Res. Soc. Symp. Proc.*, Vol. **608**, 393-398 (2000).
20. B.E. Burakov and E.B. Anderson, *Excess Weapons Plutonium Immobilization in Russia* ed. J.L. Jardine and G.B. Borisov, UCRL-ID-138361, *Proc. Meeting for Coordination and Review of Work, St. Petersburg, Russia*, 1-4/11/1999, 167-179 and 251-252 (2000).
21. B.E. Burakov, E.B. Anderson, M.V. Zamoryanskaya, M.A. Yagovkina and E.V. Nikolaeva, *Mat. Res. Soc. Symp. Proc.*, Vol. **713**, 333-336 (2002).
22. B. Burakov and E. Anderson, *CD-ROM Proc. 8th Intern. Conf. ICEM'01*, 30/09-04/10/2001, Bruges, Belgium, sess. 39 (2001).
23. T. Yamashita, K. Kuramoto, M. Nakada, S. Yamazaki, T. Sato and T.Matsui, *J. Nucl. Science and Tech.*, Suppl. **3**, 585-591 (2002).
24. B.E. Burakov, M.A. Yagovkina, M.V. Zamoryanskaya, A.A. Kitsay, V.M. Garbuzov, E.B. Anderson and A.S. Pankov, *Mat. Res. Soc. Symp. Proc.*, Vol. **807**, 213-217 (2004).
25. P.E. Raison, R.G. Haire and Z. Assefa, *J. Nucl. Science and Tech.*, Suppl. **3**, 725-728 (2002).
26. N. Chapman and F. Gibb, *Radwaste Solutions,* **10/4,** 26-37 (2003).
27. F. Gibb, *Imperial Engineer* (Royal School of Mines, London), **3,** 11-13 (2005)
28. M.I.T., *The Future of Nuclear Power: An Interdisciplinary MIT Study,* Massachusetts Institute of Technology , Cambridge (2003).
29. F.G.F. Gibb, K.P. Travis, N.A. McTaggart, D. Burley and K.W. Hesketh, *Nuclear Technology,* [In Press (2007)].
30. F.G.F. Gibb, *J. Geol. Soc.*, **157**, 27-37 (2000).
31. F.G.F. Gibb and P.G. Atrill, *Geology*, **31**, 657-660 (2003).
32. G.R. Lumpkin, *Elements,* **2,** 365-372 (2006).
33. I. Farnan, H. Cho and W.J. Weber, *Nature,* **445,** 190-193 (2007).
34. R.C. Ewing, R.F. Haaker and W Lutze, *Leachability of Zircon as a Function of Alpha Dose,* in W.Lutze (Ed.) Scientific basis for Nuclear Waste Management V, 389-397. Elsevier, New York (1982).
35. T. Geisler, B. Burakov, M. Yagovkina, V. Garbuzov, M. Zamoryanskaya, V.Zirlin and L. Nikolaeva. *J. Nucl. Mat.,* **336,** 22-30 (2005).
36. P.G. Attrill and F.G.F. Gibb, *Lithos,* **67,** 103-117 (2003).
37. P.G. Attrill and F.G.F. Gibb, *Lithos,* **67,** 119-133 (2003)
38. B.W. Chappel and A.J.R. White, *Pacific Geology,* **8,** 173-174 (1974).
39. M.V. Zamoryanskaya and B.E. Burakov, *Neorg. Mater.*, Vol. **36**, No. 8, 1011-1015, *in Russian* (2000).
40. M.V. Zamoryanskaya, S.G. Konnikov and A.N. Zamoryanskii, *Instruments and Experimental Techniques*, Vol. **47**, No. 4, 477-483 (2004).
41. K. Taylor and F. Gibb, *Proc. University Research Alliance Conf.*, Sellafield, p.5 (2004).

Mater. Res. Soc. Symp. Proc. Vol. 1107 © 2008 Materials Research Society

Inner Material Requirements and Candidates Screening for Spent Fuel Disposal Canister

Francesc Puig[1], Javier Dies[1], Manuel Sevilla[1], Joan de Pablo[2], Juan José Pueyo[3], Lourdes Miralles[3] and Aurora Martínez-Esparza[4]

[1]Dept. de Física i Enginyeria Nuclear. ETSEIB-UPC, Diagonal 647 PC, 08028 Barcelona. Spain
[2]Dept. d'Enginyeria Química. ETSEIB-UPC, Diagonal 647 H-4, 08028 Barcelona. Spain
[3]Dept. de Geoquímica, Facultat de Geologia (UB), Martí i Franquès, s/n, 08028 Barcelona, Spain
[4]ENRESA, C/ Emilio Vargas 7, 28043 Madrid, Spain

ABSTRACT

In the context of the present Spanish 'once-through' nuclear fuel cycle, the need arises to complete the geological repository reference concept with a spent fuel canister final design. One of the main issues in its design is selecting the inner material to be placed inside the canister, between the steel walls and the spent fuel assemblies. The primary purpose of this material will be to avoid the possibility of a criticality event once the canister walls have been finally breached by corrosion and the spent fuel is flooded with groundwater. That is an important role because the increase in heat generation from such an event would act against spent fuel stability and compromise bentonite barrier functions, negatively affecting overall repository performance. To prevent this possibility a detailed set of requirements for a material to fulfil this role in the repository environment have been devised and presented in this paper. With these requirements in view, eight potentially interesting candidates were selected and evaluated: cast iron or steel, borosilicate glass, spinel, depleted uranium, dehydrated zeolites, haematite, phosphates, and olivine. Among these, the first four materials or material families are found promising for this application. In addition, other relevant non-performance-related aspects of candidate materials, which could help on decision making, are also considered and evaluated.

INTRODUCTION

The present Spanish radioactive waste management policy has adopted a 'once-through' nuclear fuel cycle followed by a geological repository for spent fuel, either in granite or clay. Present design includes steel canisters where corrosion is allowed provide it can withstand mechanical loads and corrosion effects for a confinement period of over 10^3 years. Beyond that time the *bentonite barrier*, located between the canister and the host rock (Figure 1), takes over the role of radionuclide confinement.

The carbon steel canisters hold either 4 PWR or 12 BWR spent fuel assemblies (Figure 2) in configurations that inevitably leave significant void space, i.e. 1.343 m^3 for PWR, and 1.296 m^3 for BWR canisters; (81.1% and 78.3% of their total inner volume). This void space will ultimately be filled by groundwater once the canister is breached by corrosion, which acts as a neutron moderator that may trigger a self-sustained nuclear chain reaction given a high water-to-fissile material ratio. Although there is no risk of explosion, the increase in heat generation may adversely affect the bentonite barrier (Figure 1) and fuel corrosion rates, compromising long term performance [1,2]. Consequently, the canister needs to be designed with an appropriate inner material, and the main purpose of the work summarized in this paper is to provide a short list of suitable candidates. This list could help to identify the most promising alternatives for final selection or point out specific research needs.

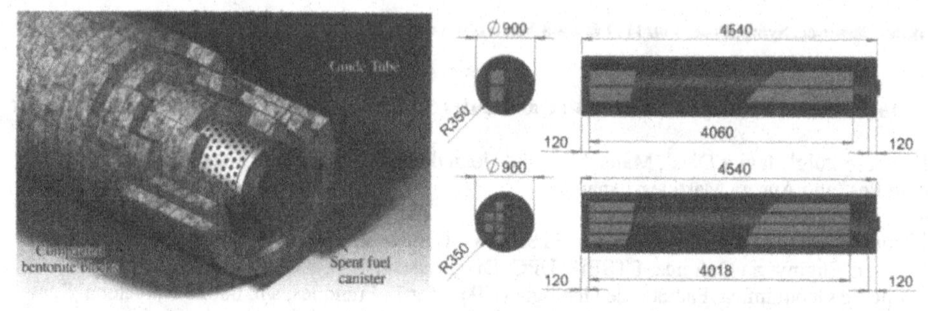

Figure 1. Spent fuel canister emplacement into a repository gallery.

Figure 2. Canister broken-out section views (in mm). Above: holding 4 PWR 17×17 fuel assemblies. Below: holding 12 BWR 10×10 FAs.

MATERIAL REQUIREMENTS

A complete list of evaluation criteria covering all the material properties needed to fulfil the repository demands has been developed and is divided into five groups.

Criticality

From criticality studies on similar canisters containing 12 BWR fresh fuel assemblies enriched to 3.6% [3], and where 0.95 is considered as the usual limit of the effective neutron multiplication factor (k_{eff}) for preventing criticality, it is inferred that at least 60% of the free volume should be filled to prevent k_{eff} reaching 0.95 if flooded. This provides a conservative case to account for PWR higher enrichment and possible burnup variability. The specific criticality requirements considered for the fill material are as follows:

1. Capacity to fill over 60% of the inner free volume of the canister. Additionally, the particulate fill should not compact under its own weight by more than 10% after canister sealing, according to the results of Agrenius [4] and Mennerdahl [5].
2. A significant neutron absorption capability of the material.
3. Prevent the presence of hydrogen, or any other effective neutron moderator element.
4. Ensure that previous criticality avoidance demands are fulfilled throughout repository life. These include:
   - Radiation resistance: essential material properties should not be affected by radiation doses.
   - Thermal stability: appropriate thermal conductivity is required to avoid temperature rises that could harm materials and fuel assemblies, and obviously sufficient thermal stability to sustain whatever temperature prevails as a result of the material thermal conductivity.
   - Chemical stability: stability towards corrosive substances in groundwater or introduced during repository construction, in addition to low solubility and lixiviation rates.

General requirements

These are demands on materials to ensure long term fulfilment of criticality criteria:

1. Thermodynamic equilibrium with the system conditions and repository materials.
2. Homogeneity between batches to allow a valid performance assessment for all canisters.
3. Good rheological properties for materials in particulate form to ensure proper canister filling.

4. Capability to fill the canister without risk of damaging it or the fuel assemblies.
5. Absence of any serious interference on fabrication, encapsulation or any other process.
6. Disassembly should be possible if quality control results after sealing are not satisfactory.
7. The canister should allow retrievability, if needed, for at least a few decades after its disposal.

General requirements to avoid

Some of the issues that must be avoided for inferior performance or viability:
1. Limited availability of the fill material.
2. Interactions with the potential to increase canister, cladding or fuel corrosion rates.
3. Capability to increase transport of ions or water through the bentonite or to chemically alter its properties.
4. Propensity to retain significant amounts of air, leading to the formation of nitric acid by radiolysis contributing to possible Stress Corrosion Cracking (SCC).

Performance improvement properties

These properties include those designed to improve the canister mechanical resistance or increase the time needed to breach it and those aimed to improve radionuclide confinement.
1. Moderate to high mechanical strength and/or incompressibility when properly packed.
2. Sorption of some key long-lived radionuclides or any means to reduce hydraulic conductivity.

Other interesting properties

These are features not related to the performance of the material, though still relevant:
1. Well-documented long-term durability, either from natural or archaeological data.
2. Low material density to reduce the additional weight.
3. Overall low cost of material, including raw material acquisition and processing to final form.
4. Radiation shielding to reduce the dose received by the bentonite and ease of manipulation.
5. Inherent simplicity of all processes, from its treatment, and manufacturing to its transport, assembly and insertion into the canister, including facilities and required equipment.

Hydrogen gas production due to corrosion and its effects on performance are considered but not included as criteria because of a lack of consensus on its implications for disposal.

CANDIDATE MATERIALS AND EVALUATION

The initial selection of candidate materials has been based on research team experience, similar reports in literature [6–8], current canister designs [9–17], Spanish canisters requirements, and some key desirable properties. The materials evaluated in detail are:

Cast iron or steel

Nodular cast iron has been selected and tested in similar designs [15,18,19], allowing a solid frame to be built as an independent component and easily assembled into the canister. However, it does not achieve high occupation rates, (typically around 50% for equivalent Spanish canisters variants), according to our calculations. If that was insufficient, steel shot ($\text{Ø} \approx 0.5$ mm) could be used, enabling higher and more homogeneous volume occupation ratios (up to 70% have been found feasible [16]). It also offers good radiation resistance [20] and thermal conductivity (~ 34 W/m·K), but poor neutron absorption capability. Iron can maintain reducing conditions in the near field, reducing fuel corrosion rates and stimulating actinide precipitation, and its corrosion products could also contribute to radionuclide sorption [21,22].

Borosilicate glass

Borosilicate glass (BSG) is much lighter (2,250 kg/m^3) and has relatively good mechanical properties. Volume occupation ratios with glass beads could be rather high, which coupled with its high neutron absorption capability would guarantee criticality avoidance. Good leaching resistance and high thermal and radiation stability are also useful [23,24], but its low thermal conductivity (~ 1.2 W/m·K) may be a problem. The higher process complexity, probably requiring vibration during canister filling [16], should be taken into account as well.

Spinel

Spinel ($MgAl_2O_4$) can be obtained in large sintered pieces and seamless welded from newly developed techniques [25,26] and also in particulate form. Its high mechanical strength (E = 190 GPa), outstanding corrosion resistance [27], good radiation tolerance [28], high thermal stability (fusion at 2135°C) and exceptional thermal conductivity for a ceramic material (~ 15 W/m·K) [28] has allowed it to be considered a candidate in either form.

Depleted uranium (DU)

Recently the U.S. NRC has decided to consider DU a form of low-level waste [29], which may involve significant disposal costs. However, DU is still a potential energy source alike used fuel, so their joint disposal, with DU as the canister fill, may reduce disposal and possible future recovery costs without compromising safety or security. DU is a fair neutron absorber and could also help to maintain reducing conditions in the near field [30], contributing to the stability of the spent fuel and preventing its lixiviation by saturating the the groundwater with U. DU could be used as spheres of DUO_2 (with a thermal conductivity around 7.5 W/m·K), which would possess essentially the same chemical properties as spent fuel itself, preventing its separation by any chemical means at any stage of disposal.

Dehydrated zeolites

Zeolites are a low density, readily availability materials with very high stability that could be used as particulate fill. They are very resistant to corrosion, can withstand high temperatures and show high sorption capabilities toward some key radioactive elements, such as Cs, I, Th, U, Ra or Pu. The water contained in natural zeolites can be released by heating, but their highly porous nature (from 40 to 70%) is a serious problem for effective volume filling. They also show very low thermal conductivity (0.2 to 0.3 W/m·K) [31], which may compromise heat evacuation.

Haematite

Haematite (α–Fe_2O_3) has high sorption capabilities for heavy elements, like U, Pu, Np, Am or Sr; and under reducing conditions it can fix H_2 by reduction, transforming to magnetite

(Fe_3O_4). It could be processed to obtain near spherical particles, to provide high canister occupation ratios, and neither its thermal stability nor conductivity (~ 13 W/m·K) would pose any problem. Magnetite may also help to buffer redox conditions, but its long term behaviour is unknown.

Phosphates

Among natural phosphates, those of the apatite group, $Ca_5(PO_4)_3(OH,F,Cl)$, are the most abundant and chosen here as reference materials. They are known for their ability to incorporate various radionuclides (U, Th, Am, I, Pu or Np). Apatite has high thermal stability and very high dissolution resistance. It contains only small amounts of H atoms, but its highly porous nature (48% to 58% [8]) poses a much greater difficulty, in addition to its low thermal conductivity.

Olivine

The attractiveness of olivine $(Mg,Fe)_2SiO_4$ lies in its sorption capacity, its ability to maintain reducing conditions [32], its low density and cost, and it is expected to reach moderate packing efficiency in a sand-like form [16]. Serpentine, appearing on its alteration in the presence of Si, also possesses similar capabilities with regard to sorption and reducing conditions fostering. Its thermal stability is not a problem, while its modest conductivity (~ 4.5 W/m·K) and the fact that usually contains impurities [6] could be of some concern. Finally, although dissolution of olivine or serpentine is not expected under repository conditions, irradiation effects should be clarified.

RESULTS AND DISCUSSION

A set of matrices (Figures 4 to 6), which summarise previous evaluations and many others, is presented here, serving as easy reference and basic decision making tools. Evaluation parameters are divided into **(1)** criticality avoidance (Figure 4), **(2)** summary of criticality requirements plus remaining general requirements (Figure 5) and finally, **(3)** additional properties and implications (Figure 6). Figure 3 serves as legend to Figures 4 to 6. Since parameters evaluation may not be apparent or entirely straightforward from the earlier data, the most relevant issues are marked (numbered) and briefly discussed for each summary matrix.

Fulfillment Degree for Proposed Requirements or Objectives *(Table 1 and Table 2)*	Colour / Pattern	Evaluation of Additional Properties *(Table 3)*
Fulfils with certainty		Highly positive
Probably fulfils		Positive
Significant uncertainty exists		Medium
Probably doesn't fulfil		Quite negative
Doesn't fulfil with certainty		Negative
Pending evaluation		Pending evaluation

Figure 3. Legend of Figures 4 to 6.

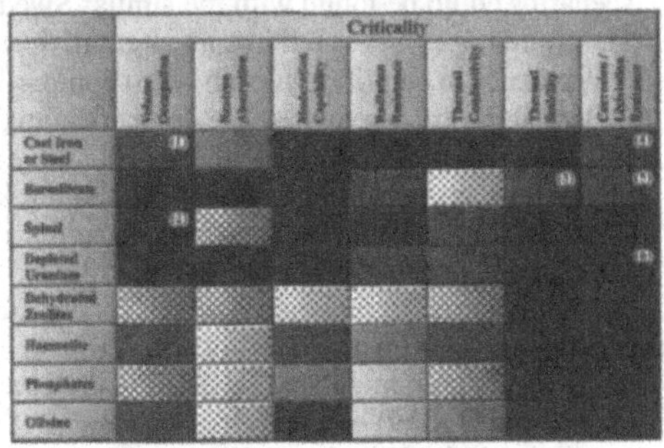

Figure 4. Summary matrix for criticality requirements evaluation.

For Figure 4, the issues that require detailed explanation are as follows:

(1) Cast iron evaluation is shown as first option, as it seems to involve some additional benefits. Volume occupation ratio of steel shot would be better, and may be used if needed.

(2) Spinel is evaluated in particulate form.

(3) Possible devitrification at high temperatures is discussed under Figure 5 (comment #3).

(4) Although iron is not corrosion resistant, the corrosion process should be slow enough not to endanger any short-term demands. Long-term requirements are mainly focused on material steadiness, not allowing the formation of gaps, and iron corrosion products are foreseen to perform well in this regard.

(5) BSG dissolution rate is anticipated to be slow, and even in the event of boron selective lixiviation and removal, volume occupation should be high enough to avoid criticality.

(6) DUO_2 shows very high stability under reducing conditions, and would behave in the same way as used fuel, ensuring it remained around the fissile material as long as the latter remained in place.

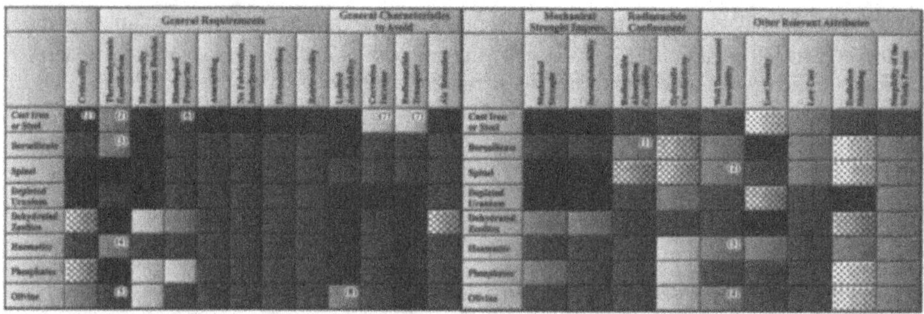

Figure 5. Evaluation summary matrix of overall criticality criteria, general requirements and other characteristics to avoid.

Figure 6. Summary matrix for mechanical and confinement improvements, and other non-performance-related attributes.

Issues related to Figure 5 are listed below:

(1) Present experience indicates that criticality avoidance requirements are not fulfilled (for PWR assemblies) if burnup credit is not taken into account, but substantial safety margins are achieved if that is done [33–36]. Oversby [37] also concluded that no credible criticality scenarios could be found with the similar Swedish repository concept. If greater safety margins are desired, steel shot could be an alternative canister fill material.

(2) While not in equilibrium, iron corrosion products are expected to fulfil their long term roles.

(3) BSG stability, due to possible devitrification, is a complex, not yet well established issue that linked to radiation resistance, thermal stability and corrosion resistance. Despite the existing uncertainty, its performance is considered unlikely to be seriously compromised.

(4) Haematite is also not in equilibrium, but its alteration products should perform similar roles.

(5) Olivine is not in equilibrium in a granitic environment [6], where quartz and soluble Si are present, altering to serpentine, which, as mentioned earlier, is expected to perform well enough.

(6) Rheological properties are not applicable to cast iron, but what can be evaluated alternatively is the ability of liquid metal to fill the mould, and fabrication tests carried out by SKB [15,38] and Posiva Oy [19] have given good results. If steel shot is chosen instead, both its intrinsic properties and fulfilled tests [16,39] point to good rheological properties.

(7) A debate exists on the effects of H_2 generation upon materials corrosion [40], so it is not clear if canister corrosion or radionuclide transport increases when using iron or steel [41–45].
(8) Olivine may be difficult to obtain in pure form, as previously noted by Oversby [6].

Relevant comments for Figure 6 are the following:
(1) The alteration layer formed on the surface of HLW BSG has important retention capabilities, but sorption for external radionuclides present in groundwater will have to be evaluated.
(2) Spinel has a well-documented long-term durability as a mineral, but there is far less experience with the new fabrication techniques mentioned [25] and nuclear applications.
(3) Long-term performance of haematite would need to be tested under reducing conditions with bentonite and steel, despite well proven durability under other natural conditions.
(4) Olivine has well-proven durability, but not in granitic environment, as it cannot occur in rocks containing quartz [6].

CONCLUSIONS

Among the candidates examined, cast iron or steel, borosilicate glass, spinel, and DU are judged adequate to fulfil the requirements established in this study, for canister use. However some uncertainties still exist, and testing will be required before final selection and employment. For cast iron, the main problems are related to criticality margins without burnup credit, and possible undesirable effects of H_2 generation. BSG beads may present some thermal conductivity problems, but they could probably be solved by using thermal shunts. DU may be subject to some regulatory issues about its management and presents the handicap of its weight. Finally, for spinel, the main foreseen difficulties are related to available experience and fabrication techniques.

REFERENCES

[1] J. De Pablo, I. Casas, J. Gimenez, M. Molera, M. Rovira, L. Duro, and J. Bruno, Geochim. Cosmochim. Acta **63** (19-20), 3097-3103 (1999).
[2] M. Amme, T. Wiss, H. Thiele, P. Boulet, and H. Lang, J. Nucl. Mater. **341** (2-3), 209-223 (2005).
[3] S. Risenmark, BR 93-150 (ABB Atom AB, Västerås, 1993).
[4] L. Agrenius, SKB Rapport 20.152 (SKB, Stockholm, 1993).
[5] D. Mennerdahl, SKI Report 02:51 (Statens kärnkraftinspektion, Stockholm, 1998).
[6] V.M. Oversby, SKB Report Number 94-3420-01 (SKB, Stockholm, 1994).
[7] R.D. McCright, UCRL-ID-119442 (LLNL, Livermore, CA, 1995).
[8] B. de la Cruz , P. Rivas, A. Hernández, C. Marín, M.V. Villar, and A. de la Iglesia, Publicación Técnica PT-01/99 (ENRESA, Madrid, 1999).
[9] L. Werme, SKB Technical Report TR-98-08 (SKB, Stockholm, 1998).
[10] J. Dies, F. Puig, M. Sevilla, J. de Pablo, J.J. Pueyo, L. Miralles, and A. Martínez-Esparza, Publicación Técnica PT-03/2006 (ENRESA, Madrid, 2006).
[11] W.H. Bowyer, SKI Report 99:28 (Statens kärnkraftinspektion, Stockholm, 1999).
[12] H. Raiko, POSIVA 2005-02 (Posiva Oy, Helsinki, 2005).
[13] DOE (U.S. Department of Energy), DOE/RW-0539-1, Rev. 1 (Office of Civilian Radioactive Waste Management, 2002).

[14] CRWMS M&O (Civilian Radioactive Waste Management System Management and Operating Contractor), SDD-UDC-SE-000001, Rev. 1 (Las Vegas, NV, 2000).
[15] C.-G. Andersson, P. Eriksson, M. Westman, and G. Emilsson, SKB Technical Report TR-04-23 (SKB, Stockholm, 2004).
[16] C.W. Forsberg, ORNL/TM-13502 (ORNL, Oak Ridge, TN, 1997).
[17] L. Johnson, J. Schneider, P. Zuidema, P. Gribi, G. Mayer, and P. Smith, Technical Report NTB 02-05 (Nagra, Wettingen, 2002).
[18] C.-G. Andersson, SKB Technical Report TR-02-07 (SKB, Stockholm, 2002).
[19] H. Raiko, Working Report 2005-53 (Posiva Oy, Helsinki, 2005).
[20] M.W. Guinan, SKB Technical Report TR-01-32 (SKB, Stockholm, 2001).
[21] T. Missana, M. García-Gutiérrez, and V. Fernández, Geochim. Cosmochim. Acta **67** (14), 2543-2550 (2003).
[22] M. Rovira, I. Casas, J. Giménez, F. Clarens, and J. de Pablo, Publicación Técnica PT-03/2004 (ENRESA, Madrid, 2004).
[23] L.W. Gray, UCRL-ID-118819 (LLNL, Livermore, CA, 1996).
[24] I.W. Donald, B.L. Metcalfe, and R.N.J. Taylor, J. Mater. Sci. **32** (22), 5851-5887 (1997).
[25] A.E. Rokhvarger and A.B. Khizh, U.S. Patent No. 5 911 941 (15 June 1999).
[26] A.E. Rokhvarger and A.B. Khizh, U.S. Patent No. 6 054 700 (25 April 2000).
[27] K.R. Wilfinger, UCRL-ID--130734 (LLNL, Livermore, CA, 1998).
[28] J. Adams, M. Cowgill, P. Moskowitz, and A. Rokhvarger, "Effect of radiation on spinel ceramics for permanent containers for nuclear waste transportation and storage", in Proc. of the Amer. Ceramic Soc. 102nd Ann. Mtg. and Exposition (St. Louis, MO, 2000).
[29] U.S. Nuclear Regulatory Commission, In the Matter of Louisiana Energy Services, L.P. (National Enrichment Services), CLI-05-05, 70-3103-ML, January 18, 2005.
[30] C.W. Forsberg and L.R. Dole in Scientific Basis for Nuclear Waste Management XXVI, edited by R. Finch and D. Bullen (Mater. Res. Soc. Symp. Proc. 757, Pittsburg, PA, 2003), pp. 677-684.
[31] E. Hahne, K. Spindler, and A. Griesinger, SFB 270 B4 (Univ. Stuttgart, 1996).
[32] L. Duro, F. El Aamrani, M. Rovira, J. Giménez, I. Casas, J. de Pablo, and J. Bruno, Appl. Geochem. **20** (7), 1284-1291 (2005).
[33] L. Agrenius, SKB Technical Report TR-02-17 (SKB, Stockholm, 2002).
[34] T. Hicks, A. Prescott, SKI Report 00:13 (Statens kärnkraftinspektion, Stockholm, 2000).
[35] M. Anttila, Working Report 99-03 (Posiva Oy, Helsinki, 1999).
[36] M. Anttila, Working Report 2005-13 (Posiva Oy, Helsinki, 2005).
[37] V.M. Oversby in Scientific Basis for Nuclear Waste Management XXI, edited by I.G. McKinley and C. McCombie (Mater. Res. Soc. Symp. Proc. 506, Pittsburg, PA, 1998), pp. 781-788.
[38] C.-G. Andersson, SKB Technical Report TR-98-09 (SKB, Stockholm, 1998).
[39] C.W. Forsberg, S.N. Storch, and K.W. Childs, "Depleted uranium dioxide as SNF Waste Package Particulate Fill: Engineering Properties" 9th International High-Level Radioactive Waste Management Conference (Las Vegas, NV, 2001).
[40] M.E. Broczkowski, J.J. Noel, and D.W. Shoesmith, J. Nucl. Mater. **346** (1), 16-23 (2005).
[41] K. Spahiu, J. Devoy, D.Q. Cui, and M. Lundstrom, Radiochim. Acta **92** (9-11), 597-601 (2004).
[42] A. Zielinski and P. Domzalicki , J. Mater. Process. Technol. **133** (1-2), 230-235 (2003).
[43] B. Bonin, M. Colin, and A. Dutfoy, J. Nucl. Mater. **281** (1), 1-14 (2000).
[44] S.T. Horseman, J.F. Harrington, and P. Sellin, Eng. Geol. **54** (1-2), 139-149 (1999).
[45] L. Ortiz, G. Volckaert, and D. Mallants, Eng. Geol. **64** (2-3), 287-296 (2002).

Mater. Res. Soc. Symp. Proc. Vol. 1107 © 2008 Materials Research Society

Characterisation of Partial Melting and Solidification of Granite E93/7 by the Acoustic Emission Technique

Lyubka M. Spasova, Fergus G.F. Gibb and Michael I. Ojovan
Immobilisation Science Laboratory, Department of Engineering Materials, University of Sheffield, Mappin Street, Sheffield, S1 3JD, UK

ABSTRACT

The acoustic emission (AE) technique was used to detect and characterise the processes associated with generation of stress waves during melting and solidification of granite E93/7 at a pressure of 0.15 GPa. The AE signals recorded as a result of partial melting of the granite at a temperature of 780 °C and subsequent solidification during cooling were distinguished from the equipment noise and their parameters used to characterise the AE sources associated with the phase transformations during melting and solidification of the granite. The mechanisms generating AE during granite melting were differentiated by AE signals with their highest peaks in the frequency spectrum at 170 and 268 kHz. The transformation of the liquid into glass during solidification of the partially melted granite generated AE waves in an essentially broad range of frequencies between 100 and 300 kHz. This preliminary work demonstrates the potential of the AE technique for use in applications related to deep borehole disposal of radioactive wastes.

INTRODUCTION

Over the last 40 years non-destructive methods for testing and evaluation, such as acoustic emission (AE), have been extensively developed and applied in industry for assessment of the mechanical performance of various structures under conditions varying from compression or tensile loading to thermal shock and hydraulic tests [1,2]. In the field of materials science AE has been shown to be a valuable tool, not only for characterising the mechanical performance of materials under stress in relation to their microstructure, but also for detecting microstructural processes active during phase transformations in metals, alloys, ceramics and glasses [3-7].

AE is a naturally occurring phenomenon associated with the release of stored elastic energy within a material in the form of transient waves with frequencies typically in the range 20 kHz to 1.2 MHz [1]. The AE technique has been successfully used to study phase transformations due to its high sensitivity to micro-scale changes in the shape or volume of a material and its capability for application *in-situ* and in real time under extreme experimental conditions such as high temperature (>1000 °C) or pressure. The latter is usually achieved by attachment of waveguides to the samples or with advanced approaches such as the development of a non-contact laser AE technique used during processing of ceramic materials [8].

In this work a new application of the AE technique to monitor and characterise the processes of granite melting and solidification during cooling has been explored. The study has focused on granites due to their potential role as a host environment for the disposal of high level radioactive waste (HLW) packages [9]. The processes of melting and recrystallisation of granite predicted to occur as a result of the heat generated by containers of HLW have been studied in detail by Attrill and Gibb [10,11] under conditions simulating those in a very deep geological disposal.

EXPERIMENTAL DETAILS

The initial material used for this AE study was prepared from a solid piece of granite E93/7 crushed to powder with a particle size <500 μm [10]. The granite powder was placed in a gold capsule with a diameter of 8 mm and a wall thickness of 0.3 mm. The capsule contained a small amount of water (=4.64 wt% of the total granite-water system, giving water saturation [10]) inserted at the bottom of the capsule with a micropipette before the granite powder was added. Immediately after filling, the capsule was weighed, crimped and welded with a carbon arc microwelder. To prevent possible evaporation of the water from the capsule during welding, the lower part of it was immersed in cold water. After welding the capsule was cleaned in acetone, dried for 5 minutes and weighed. To ensure that the capsule was sealed properly it was placed in an oven at 110 °C for a minimum of 12 hours. After removal from the oven the capsule was left to cool for 15 minutes and weighed again to ensure that no water had been lost.

A cold-seal pressure vessel and furnace (figure 1a) were used to maintain the gold capsule at a temperature of 780 °C and pressure of 0.15 GPa. The pressure vessel had a diameter of 4.2 cm and a length of 40 cm. The gold capsule was placed into the bore of the pressure vessel (diameter = 9.5 mm). The bore below the capsule was packed with stainless steel rods to keep it at the top of the vertically mounted pressure vessel (figure 1a). The furnace used and the temperature control were as described by Dalton and Gibb [12]. Two thermocouples: one Pt-Rh mounted just "above" the gold capsule in a well in the pressure vessel and one internal Cr-Al just "below" the capsule were used to monitor the temperature and record it on a PC hard drive.

Water was used as the pressure medium and the vessel pressurised at 0.15 GPa (or 22 000 psi) at least 24 hours before the start of the heating cycle. The sample was heated to the target temperature of 780 °C±0.5 °C without significant "overshoot" and held at that temperature for 337.5 hours (~14 days). According to Attrill and Gibb [10] under these conditions (780 °C, 0.15 GPa and 4.64 wt% water), about 80% by volume of the granite powder would be expected to melt. For cooling of the sample the power supply to the furnace was simply switched off and it was allowed to cool down to room temperature in less than 24 hours. Even though this cooling rate does not simulate real conditions in a deep borehole repository for nuclear wastes [9], it was used for two reasons. First, for the purpose of this initial AE study, a short duration experiment was considered sufficient to demonstrate the feasibility of the method for detection and characterisation of the processes occurring during granite melting and solidification. Second, the results obtained under this cooling rate extended the work performed by Attrill and Gibb [11] and confirmed their finding that it is too fast to allow recrystallisation of the granite melt. In addition, fast cooling rates like those applied in the current experiment are also thought likely to occur during a "self-burial" of sealed radioactive sources [13].

The physical changes during melting and cooling of the granite generated AE signals detected by a piezoelectric transducer (or sensor) attached to the cold end of the pressure vessel (figure 1b). A G-clamp was used to secure the transducer in position during the experiment as the pressure vessel itself functioned as a waveguide because (a) it was made of NiCo alloy which has low acoustic signal attenuation, (b) it is not practically possible to attach a waveguide directly to the gold capsule inside the pressure vessel in our experimental setup and (c) the transducer needs to be prevented from overheating. A wideband transducer type WD, calibrated and supplied by Physical Acoustics Corporation (PAC), was used due to its high sensitivity over large operational ranges of frequencies (from 100 to 1000 kHz) and temperatures (up to 177 °C). The temperature at the cold end of the pressure vessel during the experiments did not exceed 100 °C.

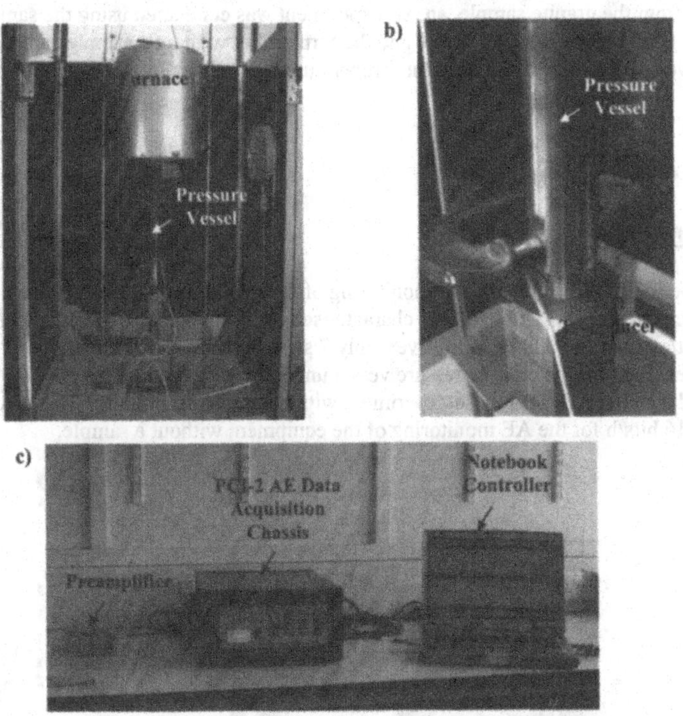

Figure 1. (a) Equipment for melting the granite sample at a temperature of 780 °C under pressure of 0.15 GPa (furnace raised), (b) a WD transducer attached to the cold end of the pressure vessel and (c) computer-based instrumentation for recording AE signals.

The output electrical signal (voltage) generated by the transducer was amplified by a low-noise preamplifier type 2/4/6 from PAC set at 40 dB and then transmitted to the main AE recording instrument - a PCI-2 based AE system (calibrated and supplied by PAC) used for AE data recording and analysis (figure 1c).

The electrical signals generated by the WD sensor were collected when the amplitude of the signal exceeded the pre-determined detection level (threshold) of 40 dB (or 10 mV). Each of the generated AE signals was additionally filtered by a bandpass filter operating between 20 kHz and 3 MHz. For each AE signal (or hit) processed and recorded, a set of parameters was stored on the PC and used for post-test analysis. The main acoustic parameters used to characterise the AE sources are amplitude, duration, rise time, number of counts per signal and absolute (or ABS) energy, as defined in [1,14]. In this study the duration of the AE signals was found to be the most distinguishing feature of the detected AE waves. Duration of a hit is defined as the time interval between the first and the last crossing of the threshold level for the AE signal [14].

During our experiment, extraneous noise due to the thermal expansion and shrinkage of the pressure vessel and the furnace was also present. In order to filter this noise from the AE data collected from the granite sample, an AE experiment was conducted using the same heating and cooling regime for the pressure vessel and the furnace but without a sample capsule in the pressure vessel. For this experiment the temperature was held at 780 °C±0.5 °C for less than 48 hours.

RESULTS AND DISCUSSION

AE data filtering

The period (337.5 hours) of AE monitoring of the granite sample with the furnace held at 780 °C, denoted as 2 in figure 2a, was characterised by a large number (3758) of AE hits with a long duration (501 to 8000 μs). However, only 7 signals with a duration between 500 and 8000 μs were recorded for the empty pressure vessel under the same conditions over 43 hours. The calculated average AE rate for the experiment with the granite sample is 12.76 hits/h whereas it is only 0.16 hits/h for the AE monitoring of the equipment without a sample.

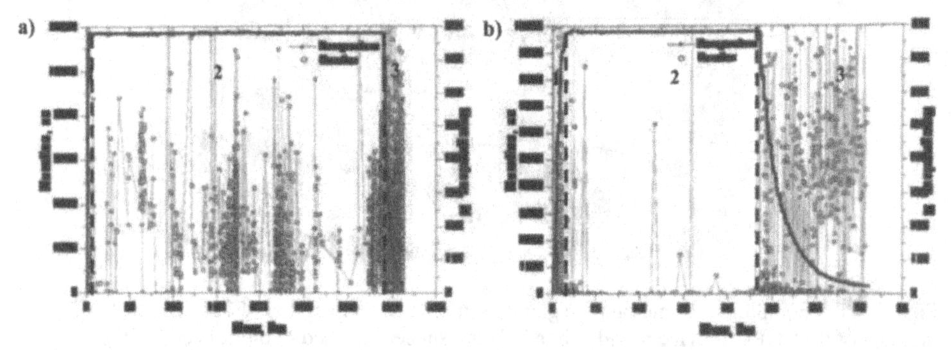

Figure 2. Duration of the AE hits versus time recorded from (a) the experiment with the granite sample and (b) the equipment without a sample.

The number of hits with a long duration (501 to 8000 μs) recorded during the cooling of the empty equipment (period 3 from figure 2b) was significantly larger (288) than that from the period 2. However, this group of signals (distinguished by a duration between 501 and 8000 μs) was found to exhibit the main difference between the detected AE waves from both experiments. Subsequently, these long duration signals were selected from the total number of hits recorded during the periods 2 and 3 from figure 2a and used for further analysis focused on their frequency and frequency-time characteristics. In this study the signal parameters characterising the AE sources during heating to the target temperature of 780 °C (period 1 in figures 2a and 2b) were not analysed due to fact that the temperature was increased manually with varying heating rate making it difficult to identify and filter the noise generated by the equipment.

Frequency and frequency-time analysis of the AE signals recorded during constant temperature melting of the granite sample

The AE signal waveforms with a duration between 501 and 8000 μs recorded during melting of the granite sample were characterised by primary frequencies (defined as the highest peak in the frequency spectrum obtained by fast Fourier transformation (FFT) [15]) distributed in two main bands - between 150 and 200 kHz and 250 and 300 kHz respectively (figure 3a). More precisely, the two largest populations in the 3758 recorded hits were 1035 at a primary frequency of 170 kHz and 887 at a primary frequency of 268 kHz.

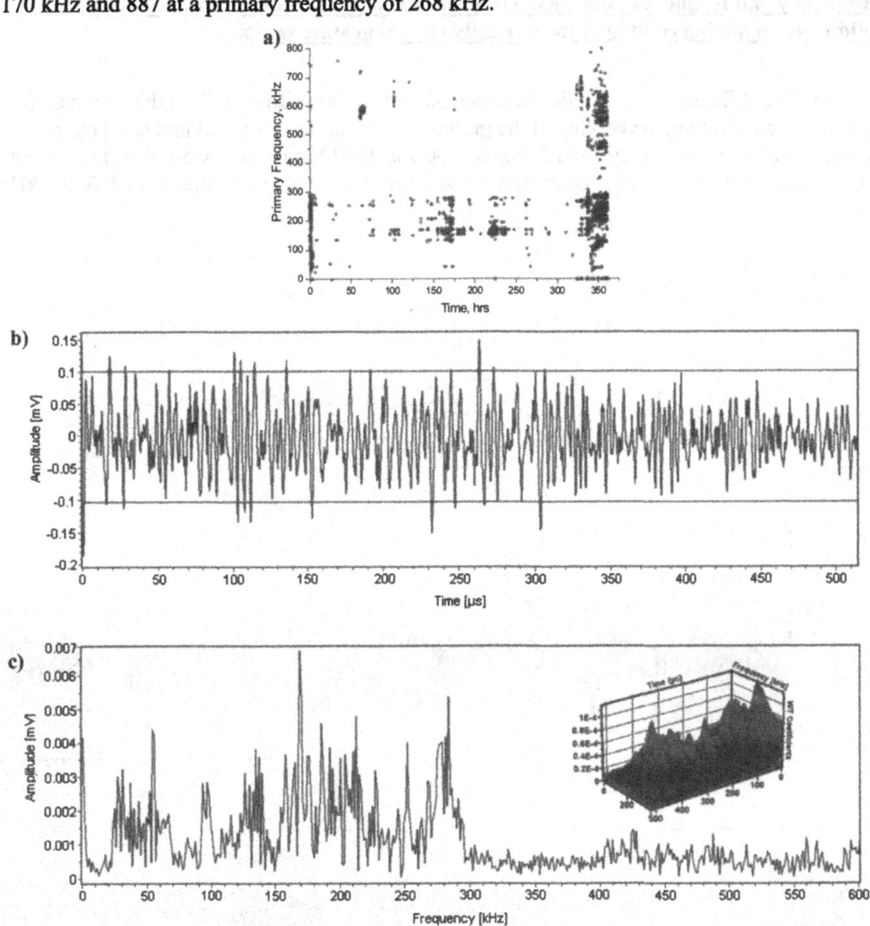

Figure 3. (a) Primary frequency of the AE hits versus time, (b) typical AE signal waveform, (c) its power spectrum by FFT and (d) insert 3D plot of WT with a primary frequency at 170 kHz recorded during melting of the granite sample.

A typical signal waveform recorded during period 2 of the AE monitoring of the granite sample with a primary frequency at 170 kHz and its frequency spectrum are presented in figures 3b and 3c. From the 3D plot of wavelet transformation (WT) [16] shown in figure 3d, it can be seen that this acoustic wave consists of a sequence of closely spaced (in time) short events.

This continuous type of AE [17] has been reported as characteristic of phase transformations in different materials [4-7] and therefore is likely to be associated with the melting of the phases present in the granite [10].

Frequency and frequency-time analysis of the AE signals recorded as a result of solidification during cooling of the partially melted granite sample

The 1793 AE hits with a duration between 501 and 8000 μs filtered from the total number of signals recorded during the cooling of the granite (figure 4a) are clustered into three main groups. These are between 200 and 300 kHz, 100 and 150 kHz and below 50 kHz. The largest population (1062) of AE hits is characterised by a primary frequency in the range 200-300 kHz.

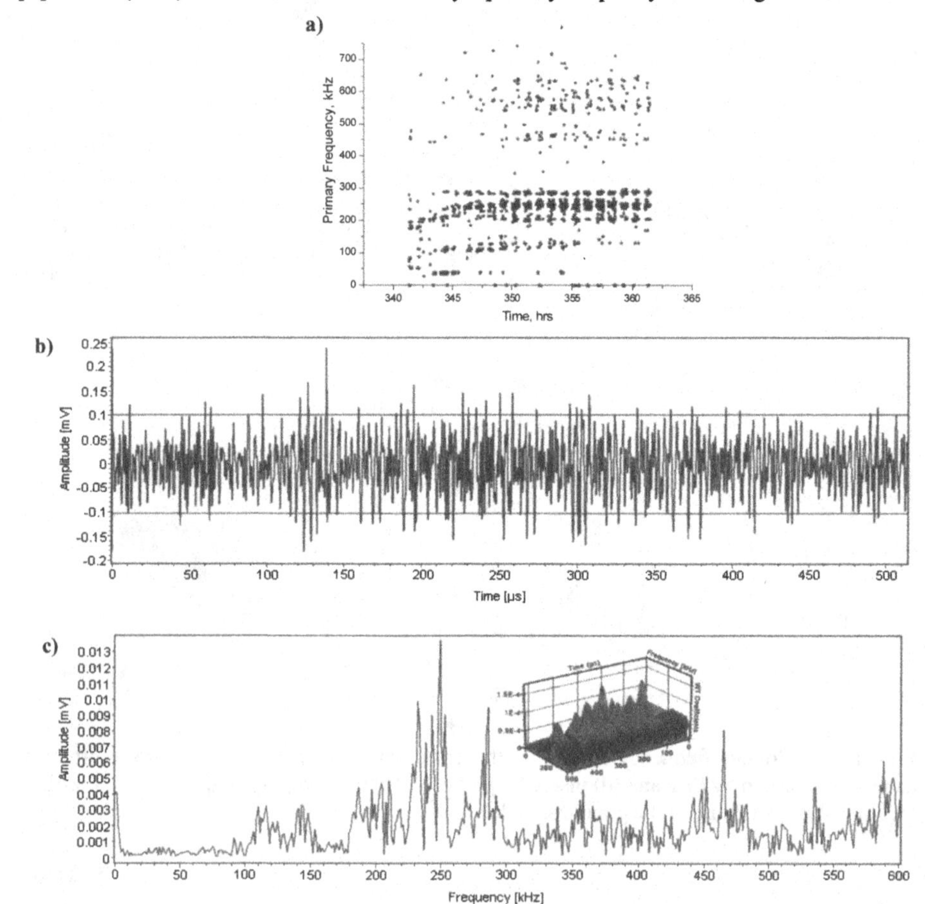

Figure 4. (a) Primary frequency of the AE hits versus time, (b) typical AE signal waveform, (c) its power spectrum by FFT and (d) insert 3D plot of WT with a primary frequency at 249 kHz recorded during fast cooling of the partially melted granite sample.

A typical AE signal waveform detected during cooling of the granite sample with a primary frequency at 249 kHz and its Fourier spectrum and a 3D plot of WT are shown in figures 4b, 4c and 4d respectively. Again this type of AE signal can be regarded as composed of a sequence of short events (figure 4d) as found for the signal waveform shown in figure 3b.

CONCLUSIONS

- AE waves from the processes of melting granite E93/7 and its subsequent solidification during cooling at a pressure of 0.15 GPa can be detected when the noise produced by the equipment is filtered out.
- The melting of the granite sample for 337.5 hours at a temperature of 780 °C is associated with the recording of thousands of AE signals with a duration between 501 and 8000 μs and primary frequencies mainly at 170 kHz and 268 kHz.
- The solidification of the partially melted granite is also characterised by long duration AE signals between 501 and 8000 μs but with primary frequencies mainly distributed between 200 and 300 kHz.
- The wavelet analysis of the signal waveforms acquired in the highest populations during this melting and solidification showed that these long duration AE signals consist of a series of short events, closely spaced in time.
- Based on these preliminary results it appears that the AE technique can be used as a tool to monitor melting and solidification during heating and cooling of granite and therefore could have applications for following changes in a deep geological repository for HLW.

REFERENCES

[1] A.A. Pollock, "Acoustic emission inspection", Technical Report TR–103–96–12/98, Physical Acoustics Corporation, 1989.
[2] K. Ono, *J. Strain Anal.*, **40** (1), Special Issue Paper N1 (2005).
[3] M.M. Lyakhovitskii, M.A. Pokrasin, V.V. Roshchupkin, N.L. Sobol and A.I. Chernov, *High Temperature* **44** (2), 221-225 (2006).
[4] S.M.C. Van Bohemen, M.J.M. Hermans and G. Den Ouden, *Mater. Sci. Tech.* **18**, 1524-1529 (2002).
[5] V.P. Melekhin, M.M. Mit'ko, V.F. Balakirev, M.A. Spiridonov and D.V. Chernykh, *Instrum. Exp. Tech.* **47** (5), 698-700 (2004).
[6] E. Dul'kin, *J. Superconductivity* **15** (6), 527-529 (2002).
[7] J. Steinberg and A.E. Lord, *J. Amer. Ceram. Soc.* **63** (3-4), 234-235 (1980).
[8] M. Enoki and S. Nishinoiri, *Adv. Mater. Res.* **13-14**, 283-290 (2006).
[9] F.G.F. Gibb, *J. Geol. Soc.* (London) **157**, 27-36 (2000).
[10] P.G. Attrill and F.G.F. Gibb, *Lithos* **67**, 103-117 (2003).

[11] P.G. Attrill and F.G.F. Gibb, *Lithos* **67**, 119-133 (2003).
[12] J.A. Dalton and F.G.F Gibb, *Mineral. Mag.* **60**, 337-345 (1996).
[13] M.I. Ojovan and F.G.F. Gibb, *Proc. WM'05*, Tucson, Arizona, WM 5072 (2005).
[14] Standard Terminology for Nondestructive Examination, ASTM E1613, ASTM International, Vol. 03.03, West Conshohocken, Pa, 1989.
[15] C.D. McGillem and G.R. Cooper, "Continuous and discrete signal and system analysis", 1984, CBS College Publishing, pp. 174-180.
[16] H. Suzuki, T. Kinjo, Y. Hayashi, M. Takemoto and K. Ono, *J. Acoustic Emission* **14** (2), 69-84 (1996).
[17] K. Ono, "Encyclopedia of Acoustics", ed. Malcolm J. Crocker, John Wiley & Sons, 1997, pp. 797-809.

Mater. Res. Soc. Symp. Proc. Vol. 1107 © 2008 Materials Research Society

A Model for Predicting the Temperature Distribution Around Radioactive Waste Containers in Very Deep Geological Boreholes

Karl P. Travis, Neil A. McTaggart, Fergus G. F. Gibb and David Burley
Immobilisation Science Laboratory, Department of Engineering Materials,
University of Sheffield, Mappin Street, Sheffield S1 3JD, U. K.

ABSTRACT

We present a mathematical model for determining the temperature field around radioactive waste containers in very deep geological boreholes. The model is first used to predict the temperature rise for some simple, but well-established cases with known solutions in order to verify the numerical work. The temperature distribution is then determined for two variants of the deep bore hole concept; a low temperature variant and a high temperature variant. The results from these studies are discussed in terms of their utility in establishing deep borehole disposal as a workable concept.

INTRODUCTION

Disposal of high level waste (HLW) including spent nuclear fuel and fissile materials in very deep geological boreholes is starting to gain serious attention as a possible safe, long term solution for the management of these wastes [1]. Very deep disposal in the present context refers to large diameter boreholes drilled 4-5 km into the granitic basement of the continental crust. This form of disposal offers many advantages over mined and engineered repositories (usually 300 -800 m in depth). There is, for example, an order of magnitude increase in the geological barrier, providing protection from catastrophic events [1-2]. An additional safety feature arises from the high confining pressures of around 150 MPa at depths of 4.5 km, which ensures that volatile radionuclides, such as I^{129}, remain in a condensed phase. The cost of drilling the boreholes (around US$1.5 million/km [3]) also makes them economically more attractive than the alternative mined repositories; According to a 2002 DOE estimate, Yucca Mountain is projected to cost US$58 billion to build and operate, while a 2005 NIREX study concluded that a HLW repository built and operated in the UK, and based on the Swedish SKB-3 concept, would cost £4.9 billion [4]. Mined repositories have a head start in that they are at an advanced stage of development in several countries. For the case of deep boreholes, predictive computer modelling will have a significant role to play in verifying some of its claims. For an example of a related thermal modelling study, see the SKB report by Marsic *et al.* [5].

In the very deep borehole concept we are developing, there are two main variants: a low temperature scheme and a high temperature scheme. The latter form relies on the heat output from the waste being of sufficient magnitude to partially melt the surrounding granite, which will then cool and recrystallise to permanently seal the waste into its own granite sarcophagus [6-7]. In both variants it is important to determine the temporal and spatial distribution of the temperature in the borehole and its environment, but the success of the high temperature variant depends crucially on attaining a temperature above the solidus of granite (for details on experimental studies of the partial melting of granite see Attrill and Gibb [8-9]). In this paper we describe the approach we have taken to modelling the flow of heat in this disposal system. We describe the underlying physical model and the numerical methods employed to obtain a solution, together with the computer code.

THEORY

The principle mechanism for the transport of heat in a solid mass of rock is conduction. In this case, the temperature rise at a point defined by position vector $\boldsymbol{r}$ (in a Cartesian coordinate system) and time t, $T(\boldsymbol{r},t)$, is given by the solution of the 3-dimensional heat conduction equation,

$$\rho c \frac{\partial T}{\partial t} = \frac{\partial}{\partial \boldsymbol{r}} \cdot \left(K \frac{\partial T}{\partial \boldsymbol{r}} \right) + S, \tag{1}$$

where ρ is the density of the material, c its specific heat, K its thermal conductivity and S is the heat produced per unit time per unit volume (source term). We have assumed that any thermal gradients are not too large to render Fourier's law invalid. In a typical borehole disposal system, heat transfer will occur across several different materials including the waste, the container, borehole fluid, borehole casing, surrounding rock, and any backfill or spacers. In some variants, a refractory plug emplaced at the bottom of the borehole constitutes an additional material the heat transfer properties of which need to be accounted for. In practice, the borehole and its surroundings are partitioned into a number of zones, each one containing a different material taken to have spatially uniform thermophysical properties. Solution of equation 1 for a realistic disposal scenario requires a computational approach.

FINITE DIFFERENCE CALCULATIONS

It is convenient to solve equation 1 using an Eulerian numerical method i.e. the coordinate space is discretized by superimposing a fixed spatial grid together with a discretization of time. The set of partial differential equations are then transformed into a sparse, linear system of finite difference equations that are conveniently expressed in matrix form (in the fully implicit formulation) by

$$\boldsymbol{A}\boldsymbol{T}^{t+\Delta t} = \boldsymbol{B}\boldsymbol{T}^{t+\Delta t} + \boldsymbol{S}, \tag{2}$$

where $\boldsymbol{A}$ and $\boldsymbol{B}$ are square matrices that contain all the physical and geometric parameters, $\boldsymbol{T}$ is a column vector containing all the temperatures at the mesh points, $\boldsymbol{S}$ is a column vector containing contributions from the source term and some boundary conditions, while Δt is the discrete time step size.

In our modelling work we have employed a number of simplifications to make the solution more tractable. We first assume that the boreholes and waste containers are perfect cylinders giving an axis-symmetric system. The mathematical problem can then be reformulated in cylindrical polar coordinates, with only the radial and axial coordinates retained. Thermal conductivities and diffusivities are assumed to be piecewise constant across the different zones.

We have constructed a dedicated code, GRANITE , written in Fortran 90, for solving the heat conduction problem in a wide range of borehole scenarios using the method of finite differences. It has been designed to be flexible enough to handle a number of different mesh structures, emplacement of waste containers in batches, or sequential emplacement *etc*. The advantage of a computer model is that a broad envelope of operating conditions can be explored to quickly test the feasibility of the concept prior to any expensive (but necessary) programme of experimental work.

The numerical solution of equation 1 is evaluated on a spatially non-uniform 2-dimensional mesh arranged such that the origin is placed at the midpoint of the container emplaced at the bottom of the borehole (i.e. on the borehole axis). A finer mesh spacing is typically employed in the region of the waste, borehole, casing and first few metres of the surrounding rock than is applied elsewhere. Figure 1 shows a cross section of a single container emplaced in a single borehole along with the axial and radial coordinate system employed.

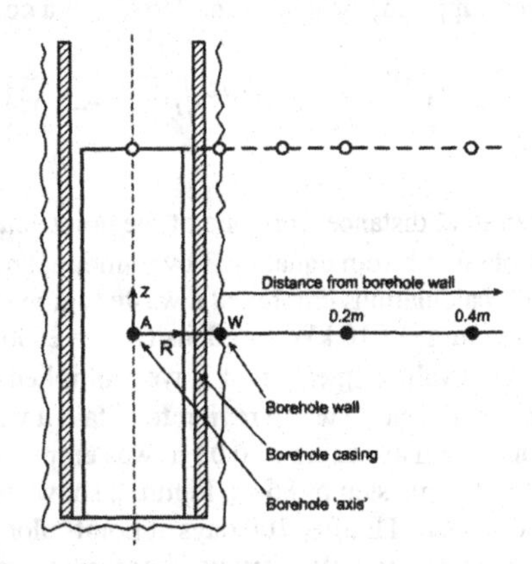

Figure 1.Cross section of a single borehole showing coordinate system employed.

The mathematical model is completed by specification of the initial and boundary conditions. These take the form: (i) the conditions at the axis are that the temperature remains finite and that there is zero flux across this boundary; (ii) at large distances from the source the temperature is set to its ambient value; (iii) at inter-regional boundaries it is ensured that the temperature and flux are the same on either side of the boundary; (iv) the initial condition takes an ambient temperature over all spatial regions. (In dealing with problems involving multiple containers in which additional containers are added sequentially, the current temperature distribution at the time a new container is fed in is taken to be the new initial condition).

Heat transport may also occur by free and forced convection (convection resulting from groundwater flow). The depth of the boreholes we propose to use rules out forced convection. This leaves free convection (resulting from buoyancy) to consider. This form of convection will arise due to the presence of fluids in the cracks and fissures of the surrounding granite. Separate calculations employing a Boussinesq approximation indicate that any upward transport of heat via this mechanism is likely to be minimal [7].

RESULTS

Validation against simple models

Verification of the GRANITE code was achieved by modelling a hierarchy of simple cases involving point and line sources of heat, each of which has a known analytical or semi-analytical solution. We now give brief details here of two of these verification exercises.

An analytic solution of the (3D) heat conduction equation (in the absence of any sources or sinks of heat) gives the temperature rise as a Gaussian function, and may be regarded as being due to a point source of heat [10]. From this solution we can obtain the temperature rise for a continuous point source of heat by integration. If we arrange a series of these continuous point sources uniformly along the *z*-axis from $z = a$ to $z = b$ say, and then integrate over *z*, we obtain an expression for the temperature rise at a point, *P*, say, with cylindrical coordinates, *R* and *z* due to a continuous finite line source supplying heat (per unit length) at a constant rate, $q_L \rho c$,

$$T = \frac{q}{4\pi\kappa} \int_a^b dz' \left(R^2 + (z - z')^2\right)^{-1/2} erfc\left[(4\kappa t)^{-1/2} \sqrt{R^2 + (z - z')^2}\right] \tag{3}$$

where $\sqrt{R^2 + (z - z')^2}$ is the radial distance from one of the point sources to the point, *P*. Values of the temperature rise were obtained from equation 3 by numerical quadrature for a line source of length 2 m. The GRANITE calculations employed a waste region of diameter 0.5 m, length 2 m with a heat output per unit volume of 10 kW m^{-3}, density, $\rho = 2600$ kg m^{-3} and thermal diffusivity, $\kappa = 1.1 \times 10^{-6}$ m^2. In evaluating equation 3, we multiplied the heat source density by the cross sectional area of the (cylindrical) waste region to obtain a value for the heat output per unit length, q_L. A uniform mesh with a spacing of 0.03 m was employed in the GRANITE calculations along with a discrete time step of 864 s. Figure 2 shows the temperature rise obtained from equation 3 and GRANITE after 100 days at points along a radius extending out from the midpoint of the waste region (z = 0). Very good agreement is obtained between GRANITE and the finite line source calculations.

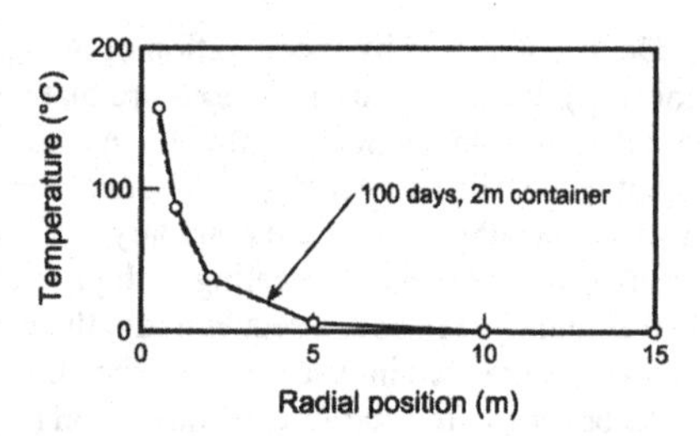

Figure 2. Radial temperature variation. Straight lines are finite line source solutions obtained using equation 3, broken lines are GRANITE calculations for a 2 m waste region.

Other useful models may be constructed from the previous result. Hodgkinson constructed a model for the conductive flow of heat around a cylindrical waste region in which the heat output diminished exponentially with time [11]. This model is an excellent first approximation to a geological disposal scenario and can be solved to yield a semi-analytical expression for the temperature rise. The solution obtained by Hodgkinson is,

$$T(R,z,t) = \frac{q_0 e^{-\lambda t}}{4\rho c\kappa} \int_0^t d\mu \frac{e^{-\lambda\mu}}{\mu} \left\{ erf\left(\frac{z+b}{2\sqrt{\kappa\mu}}\right) - erf\left(\frac{z-b}{2\sqrt{\kappa\mu}}\right) \right\} \int_0^r R' dR' I_0\left(\frac{RR'}{2\kappa\mu}\right) \exp\left\{\frac{-\left(R^2 + R'^2\right)}{4\kappa\mu}\right\}, \tag{4}$$

where b is the cylinder half length, r is the cylinder radius, I_0 is the modified Bessel function, λ is the decay constant of the heat source, while R and z are the radial and axial coordinates respectively. The GRANITE code was used to model the above case and compared with the semi-analytical solution of equation 4. The parameters used were: $c = 879$ J kg^{-1} K^{-1}, $\rho = 2600$ kg m^{-3}, $q_0 = 1$ kW m^{-3}, $\kappa = 1.1 \times 10^{-6}$ m^2 s^{-1}, $t_{1/2}$ ($= \ln 2/\lambda$) = 30 years, $b = 1$ m and $r = 0.25$ m. For these calculations a mesh spacing in the radial direction of 0.015 m was employed together with a time step of 1800 s. Solutions were obtained at $z = 0$ for 3 different values of R/m = {0, 0.25, 1, 10} for a time span of 6 decades on a log scale. The temperatures generated by GRANITE are plotted along with those obtained from equation 4 (using numerical quadrature) in figure 3. Qualitative agreement is obtained between the two sets of results. Quantitative agreement is not quite as good - GRANITE temperatures are always greater – but is probably an artefact stemming from the treatment of the rock-waste boundary in our finite difference method. The magnitude of the maximum temperature reached by the rock diminishes with increasing distance from the source while the time taken to attain this maximum increases. At the borehole axis (R = 0), a maximum temperature rise of 77 °C is reached after 179 days, for R = 0.25 m, this maximum is 62 °C established after 291 days, for R = 1 m, the maximum rise is 25 °C after a period of about 1.5 years, while at R = 10 m, the maximum rise is only 2 °C, achieved after 8 years.

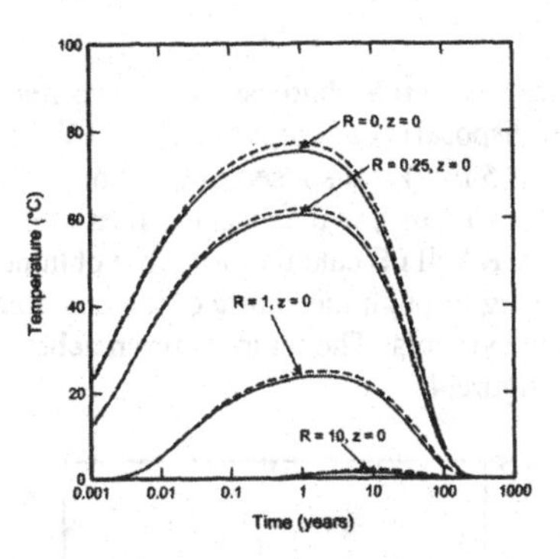

Figure 3. Temperature rise versus time plot. Straight lines are semi-analytical solutions obtained from equation 4, broken lines are GRANITE calculations.

Modelling Realistic Very Deep Disposal Scenarios

We now consider two realistic (realistic in the sense that actual nuclear industry data is used for the heat source and account is taken of the thermal properties of typical materials likely to be involved in a future deep borehole disposal site) case studies in which the number and type of engineered barriers/materials differs: a low temperature disposal and a high temperature disposal system. Any heat flow modelling must take into account the different physical and thermal properties of these materials. In both cases we include in the model a mild steel casing of thickness 15 mm inserted into the borehole. In the low temperature case, grout (we assume this to be similar to those employed in the Geothermal Energy industry e.g. a Halliburton ThermoLock™ cement) is poured into the boreholes to fill any voids and finally the holes are capped with a backfill material. In the high temperature case, the backfill is employed in place of the grout. Material properties used in the calculations are given in table I.

Table I. Parameters used for the numerical models.

Material	K (W/m/kg)	c (kg/m^3)	ρ (kg/m^3)
Granite	2.20	775	2630
Backfill	1.84	1738	2149
Stainless Steel	16.60	515	7900
Mild Steel	54.94	510.66	7861
Grout	1.15	1005	2350

Real waste packages contain a complex mixture of radioisotopes, all decaying at different rates. We have therefore used a code developed by the nuclear industry, called FISPIN, which estimates the heat decay curve for a given waste package [12]. The actual heat decay curves are complex and cannot be fitted by simple analytical forms. Consequently, we used splines to fit each heat decay curve and used these to calculate heat values in the code.

Low temperature disposal

In this example we consider a *single* stainless steel container of vitrified reprocessing waste with an initial (at time of disposal) heat output of 11.57 kW m^{-3}. This container (diameter is 0.43 m, height 1.4 m, thickness 5 mm) is disposed of in a bore hole of diameter 0.60 m. The waste density was taken to be 2721 kg m^{-3}, with a specific heat of 403.8 J kg^{-1} m^{-3} and a thermal conductivity of 4.33 W m^{-1} K^{-1}. GRANITE calculations were obtained for two sets of points lying along horizontal radii passing through the centre of the container and through the top of the container (see figure 1 for key to symbols) The temperature reached at each of these points is plotted as a function of time in figure 4.

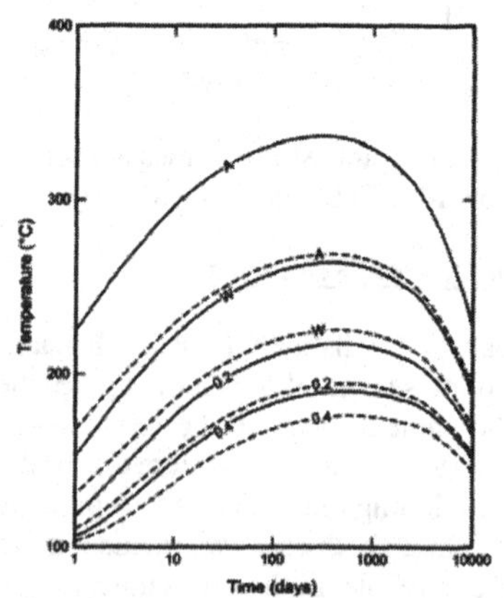

Figure 4. Temperature versus time for the low temperature disposal case study. Solid lines refer to positions along a horizontal radius passing through the centre of the container while broken lines refer to positions along a horizontal radius passing through the top of the container.

Figure 4 looks qualitatively similar to the results obtained using the Hodgkinson model described in the last section in which the temperature rises to a maximum and then falls off with increasing time. At the centre of the container this maximum value is 337°C (237 °C above ambient) after 309 days, while along the central axis the maximum at the borehole wall is 264 °C after 389 days and at a point 0.4 m out from the borehole wall it reaches 190 °C after 559 days. Along the top axis, these maxima are consistently lower and take a little longer to be achieved. The data displayed in figure 4 show that at no point will the temperature exceed the glass transition temperature of the vitrified waste (~505 °C), and further that it will not be high enough to cause failure of the container.

High temperature disposal

In this example we consider a *stack* of 5 stainless steel containers of 45 GWd/t PWR UO_2 spent fuel with a heat output at disposal of 15.31 kW m^{-3}. These containers are disposed of in a borehole of diameter 0.8 m. The container diameter is 0.63 m while its height is taken to be 1.5 m and its wall thickness is 3 cm. The waste density was taken to be 8601.5 kg m^{-3}, with a specific heat of 402.6 J kg^{-1} m^{-3} and a thermal conductivity of 4.33 W m^{-1} K^{-1}. GRANITE calculations were obtained for the same 8 points used in the low temperature case. The absolute temperature reached at each of these points is plotted as a function of time in figure 5.

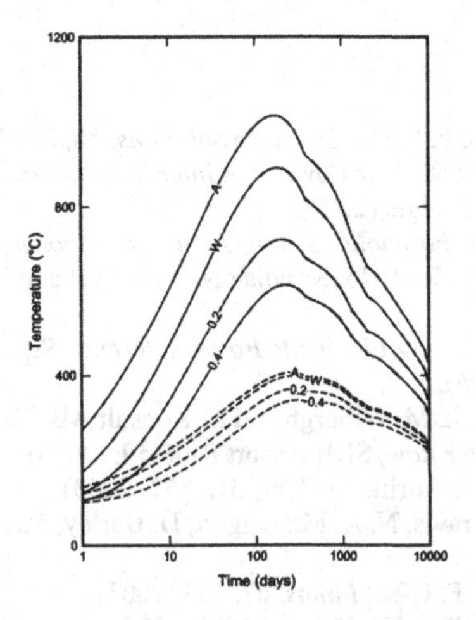

Figure 5. Plot of temperature versus time for the high temperature disposal case study. Solid lines refer to positions along a horizontal radius passing through the midpoint of the stack while broken lines refer to positions along a horizontal radius passing through the top of the stack.

Figure 5 shows that the maximum temperatures reached at the borehole wall range from 890 °C after 180 days midway up the stack to 397 °C after 258 days at the top. At pressures of approx. 150 MPa, temperatures in excess of 700 °C are required to cause partial melting of the granite [9]. Our model data show that by a suitable choice of waste loading, the high temperature variant of the deep disposal is able to generate temperatures sufficiently high for a long enough period of time to effect this melting, making the concept evidently viable. An estimate of the thickness of the granite sarcophagus can be obtained from figure 5. Mid-way up the stack it would be ~0.25 m but no melting would occur at the top edge and thus in this case, the waste would not be completely sealed inside a core of recrystallised granite. The utility of a computational model means that one can work backwards to determine the necessary heat source/ container arrangement required to ensure this condition is met.

CONCLUSIONS

We have constructed a mathematical model describing the conductive flow of heat around containers of nuclear waste buried in deep geological boreholes. The resulting system of differential equations has been solved using a numerical method. A dedicated code, GRANITE, has been written to carry out the necessary calculations. The code was verified against a hierarchy of simple models of increasing complexity, each of which has a known solution. Two such verification case studies have been described in this paper. The GRANITE code has been designed to be sufficiently flexible to handle a large number of different disposal scenarios, including boreholes containing single or multiple waste containers with any number of engineered barriers. Two disposal systems have been described: a low temperature variant, and a high temperature version. In both cases the model has provided temperature data that, in conjunction with appropriate experimental data may be used to establish the feasibility of this type of disposal.

REFERENCES

1. N. Chapman and F. G. F. Gibb, *Radwaste Solutions*, **10**, 26 (2003).
2. M. I. T., *The Future of Nuclear Power: An Interdisciplinary MIT study*, Massachusetts Institute of Technology, Cambridge, (2003).
3. T. Harrison, *Very deep borehole: Deutag's opinion on boring, container emplacement and retrievability*, SKB Report R-00-35, Swedish Nuclear Fuel and Waste Management Co., Stockholm, Sweden (2000).
4. NIREX Technical Note, *Cost Estimate For A Reference Repository Concept For UK HLW/SNF*, No. 484281, (2005).
5. N. Marsic, B. Grundfelt, M. Wiborgh and K. Konsult AB, *Very deep hole concept: thermal effects on groundwater flow*, SKB Report R-06-59, (2006).
6. F. G. F. Gibb and P. G. Attrill, *Geology,* **31,** 657 (2003).
7. F. G. F. Gibb, K. P. Travis, N. A. McTaggart, D. Burley, *Nuclear Technology,* in press (2007).
8. P. G. Attrill and F. G. F. Gibb, *Lithos,* **67,** 103 (2003).
9. P. G. Attrill and F. G. F. Gibb, *Lithos,* **67,** 119 (2003).
10. H. S. Carslaw and J. C. Jaeger *Conduction of heat in solids,* (Clarendon press, Oxford. 1959), pp 510.
11. D. P. Hodgkinson, *Deep rock disposal of high-level radioactive waste: transient heat conduction from dispersed blocks*, Report R8783, Atomic Energy Research Establishment, Harwell, UK, (1977).
12. R. F. Burstall, *FISPIN: A Computer code for Nuclide Inventory Calculations*, ND-R-328(R) (1979).

ILW Wasteforms

Mater. Res. Soc. Symp. Proc. Vol. 1107 © 2008 Materials Research Society

The Use of Activated Slags as Immobilisation Matrices for ILW

NB Milestone[1], Y Bai[1], C.H. Yang [2], Y.J. Shi [2], X.C. Li [3]

1. *Immobilisation Science Laboratory, University of Sheffield, U.K.*
2. *Department of Building Materials & Engineering, Chongqing University, P.R. China*
3. *Yunnan Construction Concrete Co. Ltd, P.R. China*

ABSTRACT

Composite cements where large amounts of blast furnace slag (BFS) replace Portland cement are currently used for immobilisation of ILW. Hydration of BFS is activated by the small amount of OPC present but the amount of reaction that occurs is limited at ambient temperatures. Increasing the temperature increases the hydration of the BFS but large amounts still remain unreacted, leaving a porous matrix where the capillary pores remain filled with a highly alkaline solution. This solution causes corrosion of reactive metals giving rise to expansive reactions and hydrogen release, and it can destroy the structure of zeolites releasing the adsorbed species.

Apart from OPC, BFS hydration can be activated by other compounds such as hydroxides, sulphates, silicates, and calcium aluminate cements. The use of these alternatives gives rise to binders such as ettringite and strätlingite which have a different chemistry where the pore solution has a lower pH. Corrosion of metals does not readily occur in these binders. This may be due to the reduced pH but could also arise from the lack of pore water, as these binders bind more water in their structure so that it is not available for transport of ionic species. This extra water binding also has potential for immobilisation of sludges where high w/s ratios are necessitated by the need to transport the sludge.

This paper will review some of the alternative activators for slag hydration and present experimental results on several systems where slag has been activated with compounds other than OPC.

INTRODUCTION

The conventional method for encapsulating low and intermediate level nuclear waste (LLW and ILW) in the UK is to use composite cement blends based on high levels of replacement of ordinary Portland cement (OPC) with supplementary cementing materials (SCM's) such as pulverised fuel ash (PFA) and blast furnace slag (BFS). Typically levels up 9:1 BFS:OPC or 4:1 PFA:OPC are used. The use of cement for encapsulation/immobilisation of nuclear waste offers a number of advantages. These include low cost, ready availability and a known consistent performance, good workability and ease of remote working, high alkalinity which helps immobilise many radionucleides as hydroxides or carbonates, and formation of a strong, dense, low permeability matrix with high adsorption properties and known durability. The composite cement formulations have been chosen, not for their strength development, but rather for their ability to reduce the exothermic temperature rise associated with the large volumes of cement used in containers which can be as large as 3m^3, although for ILW only 500L drums are routinely used. Despite the use of coarse ground cement and BFS, temperatures as high as 80°C can be reached at the centre of drums that contain these composite cements.

To obtain grouts with sufficient fluidity to allow encapsulation of complex shapes such as fuel rod assemblies cladding etc., relatively high water/solids ratios are used, typically ~0.37 – 0.42. The very low cement content of these composites, together with the presence of coarse particles of slag particularly, results in hardened matrices where large amounts of slag or PFA remain unreacted, even after 2 years of hydration, along with significant amounts of capillary water [1]. Porosities are typically in excess of 20% and change little with time after the first few weeks of hydration. While many of the operational wastes can be satisfactorily encapsulated in these systems, the high pH generated by the OPC creates problems with wastes containing reactive metals such as magnesium, aluminium and uranium which corrode generating hydrogen and expansive corrosion products. The typical reactions that occur for aluminium are:

Dissolution of the protective oxide layer

$$Al_2O_3 + 2OH^- + 3H_2O \rightarrow 2Al(OH)_4^- \quad (1)$$

Attack on metal

$$Al + 2OH^- + 2H_2O \rightarrow Al(OH)_4^- + H_2\uparrow \quad (2)$$

Formation of corrosion product as pH falls

$$Al(OH)_4^- \rightarrow Al(OH)_3 + OH^- \quad (3)$$

The soluble $Al(OH)_4^-$ species attacks the C-S-H forming strätlingite, C_2ASH_8*, at the edge of the hydrated cement matrix. Reduction in pH around the metal due to corrosion results in the formation of expansive $Al(OH)_3$.[2]

A number of factors affect the rate at which these metals corrode, including pH, available water and temperature. Addition of SCM's in composite cements reduces the pore solution pH by ~1 unit which slows the reaction but does not prevent it and laboratory samples ultimately crack [2]. Small volumes of some legacy wastes that are stored in less than satisfactory conditions cannot be cemented using conventional cements, so that a 'toolbox' of different cement types with different chemistries is needed [3].

Calcium sulfoaluminate cements have been shown to have potential for encapsulation of reactive metals [4] but further work is needed to demonstrate their long term durability as a wasteform. These cements can show large exotherms upon hydration so temperature rise would be problem. The binder is ettringite, $C_3A.3CaSO_4.32H_2O$, which accommodates large amounts of bound water in its structure but its decomposition temperature could limit its use. Aluminium corrosion in a CSA matrix is very limited but it is not clear whether it is pH or lack of water that stops the corrosion.

We have been investigating the use of activated slags as potential encapsulating cements with the aim of developing a low pH system. Alkali activated slags were first developed by Glukhovsky [5] and have been subsequently developed throughout the world [6, 7]. Ground granulated blast furnace slag (GGBS) is typically activated with alkaline activators such as sodium hydroxide or sodium silicate but the internal pH of these systems remains high at around 11.5–12. While GGBS is normally activated with alkaline material, it is possible to use gypsum or sodium sulphate although the strengths reached are often lower and take longer to develop. These systems are similar to the super-sulfated cements which use limited amounts of OPC and anhydrite for activation [8]. Preliminary data on a sodium sulphate activated system was presented at the last symposium [9]. It is these systems on which we have concentrated our efforts and are described in this paper.

---

* Cement nomenclature: C = CaO, S = SiO_2. A = Al_2O_3, H = H_2O

EXPERIMENTAL

Ordinary Portland cement (OPC), PFA and BFS (fineness 286 m^2/kg) supplied by BNFL plc. were used to prepare comparison samples. A GGBS with fineness 520 m^2/kg and glass content 95% supplied by North East Slag Cement and Frodingham Cement Company Ltd was used to manufacture the $CaSO_4$ and Na_2SO_4 activated GGBS/PFA matrices. The chemical compositions of raw materials are given in Table 1. Analytical grade Na_2SO_4 (Fisher) and anhydrite (Chance and Hunt Ltd UK) were used as activators. Mixing was conducted with distilled water.

Table 1. Chemical composition of OPC, BFS, GGBS and anhydrite

Material	CaO	SiO_2	Al_2O_3	Fe_2O_3	MgO	SO_3	Na_2O	K_2O	LOI
OPC	64.58	20.96	5.24	2.61	2.09	2.46	0.28	0.59	0.73
BFS	42.1	34.5	13.74	0.97	7.29	-*	0.22	0.49	-1.05
GGBS	40.00	35.01	13.16	0.60	8.02	1.7*	0.20	0.72	0.75
PFA	1.62	49.53	26.45	8.70	1.56	0.88	< 0.01	4.58	4.10

* sample contains reduced sulfur

The first series of samples examined the behaviour of GGBS activated with Na_2SO_4, $CaSO_4$ and Na_2CO_3 with a second series evaluating the potential of further reducing the pH of the matrices by adding low calcium PFA. In the first series, the water/solid was fixed at 0.35. When Na_2SO_4 and Na_2CO_3 were used to activate GGBS, they were added at 2% Na_2O equivalent by mass of the GGBS. When anhydrite was used, it was added at 20% by mass of the total solid (i.e. GGBS and anhydrite). For the second series, Na_2SO_4 was added at 2% Na_2O equivalent by mass of the GGBS with part of the GGBS replaced with low calcium PFA at 0%, 10%, 20% and 30%.

Na_2SO_4 and Na_2CO_3 were dissolved in distilled water and mixed with GGBS. When PFA was used, the GGBS and PFA were preblended, and then mixed with the solution. For the anhydrite activated GGBS, the GGBS and anhydrite powders were preblended, and then mixed with the required amount of distilled water. The resulting grouts were poured into polypropylene pots and vibrated until air bubbles stopped appearing on the surface. The samples were sealed and cured at 40 °C and 95% relative humidity in an environmental chamber until examination.

The initial pH of the grout was determined by using a close range pH paper in the matrix before casting the samples. The heat of hydration of the 9:1 BFS:OPC composite system and the sodium sulfate activated GGBS/PFA systems were tested in a Wexham JAF Calorimeter using a silicone oil bath at a constant temperature of 40°C. Standard compression tests were measured on three 50 x 50 x 50 mm cubes at 3, 7, 28, 91 and 180 days. The phases formed in the hardened matrices at 28 days were examined by XRD (Siemens, CuKα radiation, scanning speed of 0.5°/min, step size of 0.02° from 5 to 65°2θ). Hardened pastes were soaked in acetone to stop the hydration and vacuum dried. The dried samples were then ground in an agate mortar to a fine powder of below 63 µm for further analysis. Scanning electron microscopy (SEM) was carried out on fracture surfaces. The vacuum dried samples were coated with gold and examined under a JEOL JSM6400 scanning electron microscope. Porosity was measured using a Micromeritics Mercury Intrusion Porosimeter.

Cs leaching was carried out by dissolving 3% CsCl by weight of total solid in the mixing water used to prepare the blends. Leaching was carried out following BS EN 12457-1:2002 except that a particle size of 2-4 mm was used. Analysis was by atomic emission spectroscopy and Cs loss calculated as a percentage of total Cs added.

RESULTS AND DISCUSSION

The initial pHs for sodium salt activated slags are higher than expected from their aqueous solutions (Table 2) indicating that the reaction between the activator and slag releases additional OH^- ions. Only for the anhydrite activated slag is the pH lowered below that expected for OPC systems. Table 3 shows that the measured pH of the Na_2SO_4 activated GGBS could be reduced when part of the GGBS was replaced with a low calcium PFA. However, it must be noted that increasing the PFA content prolonged the setting time. Beyond 30% addition, the system did not set, even after two days reaction, although the initial fluidity was improved. This is a feature of activated slags where strength development is slow.

Table 2. Initial pH of $CaSO_4$, Na_2SO_4 and Na_2CO_3 activated GGBS

$CaSO_4$/GGBS	Na_2SO_4/GGBS	Na_2CO_3/GGBS
11.5	12.5	13.0

Table 3. Effect of PFA on the pH of Na_2SO_4 activated GGBS

0% PFA	10% PFA	20% PFA	30% PFA
12.5	12.0	11.6	10.6

Figure 1 shows the heat evolution rate of 9:1 BFS:OPC system and the sodium sulfate activated GGBS/PFA systems. As it can be seen, the peaks for the maximum rate of heat output (Qmax) of the sodium sulfate activated systems appear later than that of the 9:1 BFS/OPC system. The increase of the PFA content, with the exception of 20%, caused further delay of the appearance of the Qmax. This indicates the prolonged setting and hydration nature of the sodium sulfate activated system, particularly with the addition of PFA which can cause a further delay in setting and reaction. From Figure 1 it also can be seen that with the increase of PFA, the Qmax was reduced with the only exception being at the 20% level. Figure 2 shows the total heat output of both systems. It can be seen that the sodium sulfate activated systems generate less heat than the 9:1 BFS:OPC system, and with an increase of the PFA content, the general trend is to reduce further the total heat output. Thus, it can be concluded that compared to the current 9:1 BFS:OPC system, the sodium sulfate activated GGBS/PFA can be categorised as a low heat output cement system, which should be beneficial for the current plant operations as the formulations used were selected primarily due to their lower heat output. Furthermore, as it can be seen from Figure 3, the strength of this sulfate activated GGBS/PFA system is more than sufficient for waste immobilisation purpose.

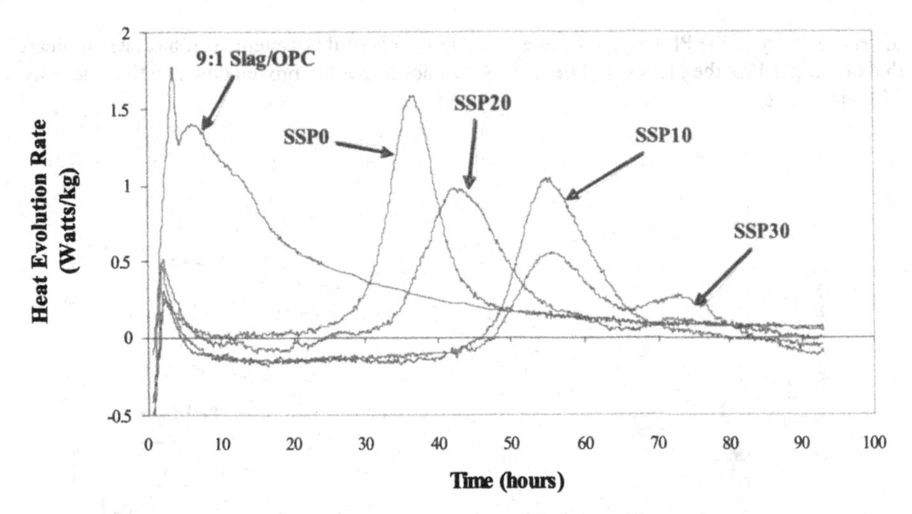

Figure 1: Heat Evolution Profiles of 9:1 BFS:OPC and Na_2SO_4 Activated GGBS/PFA;
(SSP0 2% Na_2SO_4 activated GGBS 0% PFA; SSP10 2% Na_2SO_4 activated GGBS 10% PFA; SSP20 2% Na_2SO_4 activated GGBS 20% PFA; SSP30 2% Na_2SO_4 activated GGBS 30%PFA)

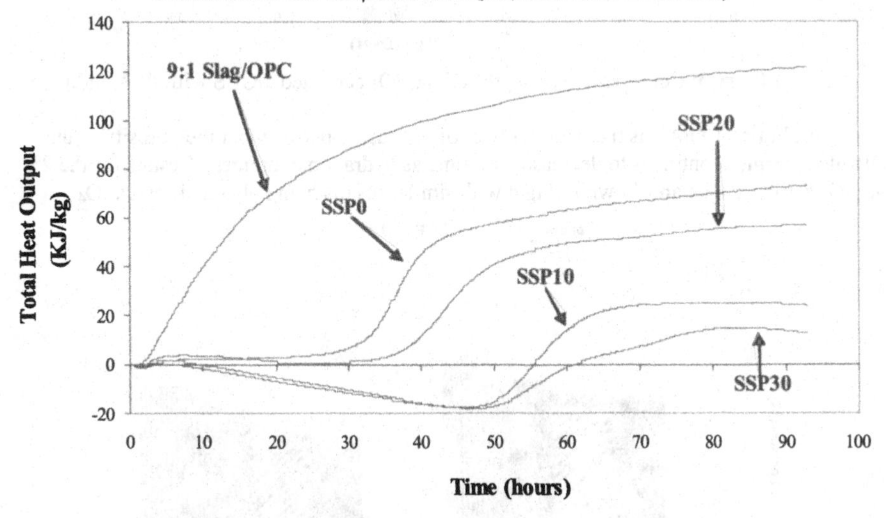

Figure 2: Total Heat Output of 9:1 BFS:OPC and Na_2SO_4 Activated GGBS/PFA

Figure 3 shows the compressive strength development of the sodium sulfate activated GGBS/PFA system. It can be seen that even at 3 days, 18 MPa compressive strength can be achieved with 30% PFA addition, which continues to increase. This is more than enough for waste immobilisation purposes. However, it should be noted that with increasing PFA addition, the compressive strength was reduced, indicating a reduced reaction from PFA addition. The lower reactivity of the PFA might be due to the lower pH of this system, as it has been indicated that below pH 13.2 the glass structure of PFA can not be readily broken down so the reactivity of PFA is reduced.

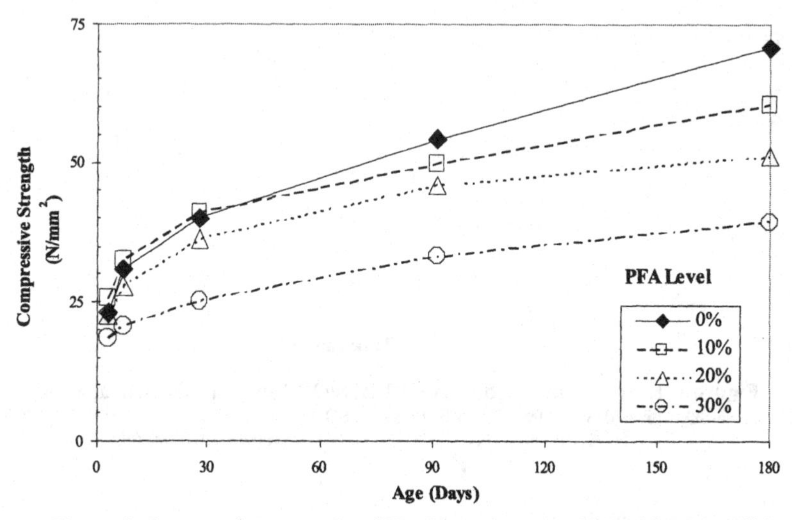

Figure 3: Compressive strengths of Na_2SO_4 activated GGBS with PFA additions

Addition of PFA has the overall effect of increasing porosity but the porosity in the activated samples continues to decrease with time as hydration continues. Results for the 2% Na_2SO_4 activated slag are shown in Fig 4 with similar results being obtained for $CaSO_4$.

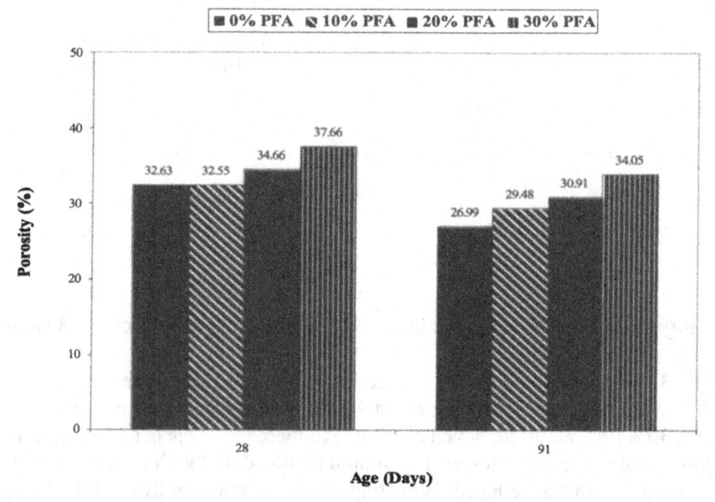

Figure 4: Porosity of 2% Na_2SO_4 activated GGBS with PFA additions

This changing porosity has an effect on the manner in which ions are leached from the matrix. One of the most difficult ions to immobilise is Cs^+. While at early ages the amount of Cs leached from the sulfate activated slags is comparable with that for the conventional BFS:OPC blends, with increasing time less Cs is released as the system continues to hydrate. (Fig 5).

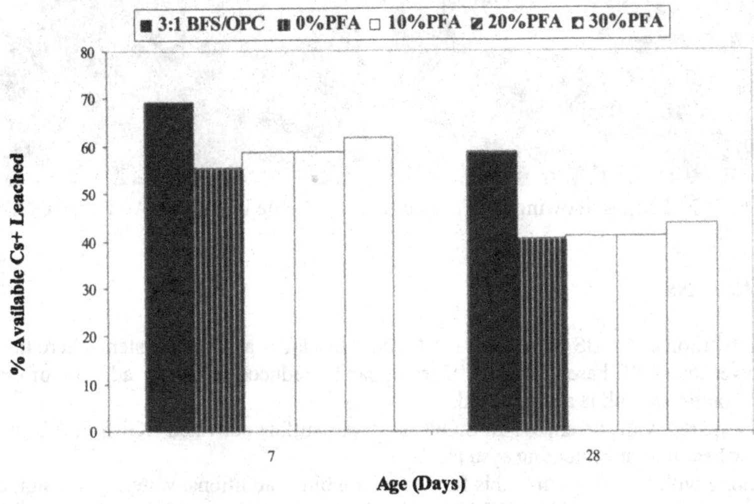

Figure 5: Leaching of Cs^+ from slag mixes

Aluminium corrosion, or lack of it in the Na_2CO_3, Na_2SO_4 and anhydrite activated systems, was reported at the last symposium [9]. Corrosion and expansion occurred in the Na_2CO_3 activated GGBS system but was reduced for the Na_2SO_4 and anhydrite activated systems. For the Na_2SO_4 and $CaSO_4$ activated systems, while the major reaction product is C-S-H, significant amounts of ettringite are present (Fig.6), even as early as 3 days. In the calcium sulfoaluminate system the same phenomenon was seen [4]. Ettringite can chemically bind large amounts of water (32 molecular of H_2O in the unit cell), which might contribute to the reduced corrosion and generation of hydrogen gas. This is seen as needle like crystals in both $CaSO_4$ and Na_2SO_4 activated slags (Fig. 6), although it is more pronounced in the $CaSO_4$ activated material where more sulphate is present. These results suggest that reducing the pH and allowing some form of self desiccation will lower the risk of Al corrosion in a similar way to that observed with CSA cements [4]. However, in the Na_2SO_4 activated slags the pH is still relatively high at low PFA additions yet corrosion is limited. Hence, the difference in the corrosion behaviour among the different activated slag systems may be due to the second factor – free water content. However, it is not yet clear why aluminium corrosion was reduced and further work is needed in order to promote this system for future nuclear waste immobilisation operations.

Figure 6: SEM Images Showing the Formation of Ettringite in Na_2SO_4 Activated GGBS/PFA

CONCLUSIONS

- Activation of GGBS with $CaSO_4$ or Na_2SO_4 produces a cement system where the pH is lower than OPC based cements. The pH can be reduced further by addition of PFA although strength is also reduced.
- Compared with current 9:1 BFS:OPC systems, sulfate activated GGBS/PFA system is a low heat output cementing system.
- Along with C-S-H, ettringite is formed which binds additional water. Corrosion of aluminium is reduced in a GGBS system that has been activated with $CaSO_4$ or Na_2SO_4. This could be due to lower pH or to reduced amount of available water.

REFERENCES

1. J-P Gorce, and NB Milestone, *Probing the Microstructure and Water Phases in Composite Cement Blends,* Cem. Concr. Res. 37 (2007) 310–318
2. A Setiadi, NB Milestone, J Hill, and M Hayes, *Corrosion of aluminium and magnesium in BFS composite cements*, Adv. App. Ceram. 2006 105 (4) 191 -196
3. NB Milestone, "*Reactions in cement encapsulated nuclear wastes: need for toolbox of different cement types*", Adv. Appl. Ceram. 105 (1) 13-18, 2006
4. Q Zhou, NB Milestone, and M Hayes, *An alternative to Portland cement for waste immobilisation – the calcium sulfoaluminate cement system,* J. Hazard. Mat. 136 (2006) 120–129
5. VD Glukhovsky, et al., *Binder*, US Patent 4410365, 1983
6. CJ Shi, X Wu, and MS Tang, *Research on alkali-activated cementitious systems in China: a review,* Adv. Cem. Res., 5, (17), 1-7, (1993)
7. DM Roy, and MR Silsbee, *Alkali activated cementitious materials: an overview*, Mater. Res. Soc. Symp. Proc., 1992, 245, pp. 153-164.
8. M Moranville-Regourd in *Lea's Chemistry of Cement and Concrete* pp 664-667. pub Elsevier 2004.
9. Y Bai, NB Milestone, and C Yang, *Sodium sulphate activated GGBS/PFA and its potential as a Nuclear Waste Immobilisation matrix,* Mat. Res. Soc. Symp. Proc. 2005 932 759- 766

Mater. Res. Soc. Symp. Proc. Vol. 1107 © 2008 Materials Research Society

Durability of a Cementitious Wasteform for Intermediate Level Waste

Peter J. McGlinn, Daniel R.M. Brew, Laurence P. Aldridge, Timothy E. Payne, Kylie P. Olufson, Kathryn E. Prince and Ian J. Kelly
Australian Nuclear Science and Technology Organisation (ANSTO),
New Illawarra Road, Lucas Heights, NSW, Australia 2234

ABSTRACT

Cementitious material is the most commonly used encapsulation medium for low and intermediate level radioactive waste. This paper focuses on the aqueous durability of a Materials Testing Reactor (MTR) cementitious wasteform - a possible candidate for the proposed intermediate level waste management facility in Australia. A series of medium term (up to 92 months) durability tests, without leachate replacement, were conducted on samples of this wasteform.

The wasteform was made from cement, ground granulated blast furnace slag and a simulated waste liquor. The compressive strength (39 MPa) was typical of MTR cement wasteforms and well above that required for handling or storage. The wasteform was an inhomogeneous mixture containing calcite, a calcium silicate hydrate phase, hydrotalcite and unreacted slag particles. After leaching for 92 months the crystallinity of the calcium silicate hydrate phase increased.

The majority of the releases of Ca, Si, Al, Sr, S, Na and K was reached within 4 days of leaching, with the maxima ie. the highest concentrations in the leachates, occurring at 3 months for Ca, Al, Sr, S, Na and K, and at 1 month for Si. For the longer leach periods (6 months and 3 months respectively) there was a slight reduction in concentration in the leachates, and these levels were similar to those for the longest period of 92 months, suggesting steady-state conditions prevailing after 3 to 6 months of leaching. The highest releases of matrix elements were for Na (37%), K (40%) and S (16%). Releases for elements such as Ca, Na, Al and Sr were similar in magnitude to those reported by the UKAEA in earlier MTR studies.

After leaching for 92 months there was an alteration layer about 80 μm deep where calcium has been depleted. Na, K and Sr showed signs of diffusion towards the outer part of the cement samples.

INTRODUCTION

Current concepts for the disposal of radioactive wastes will employ multiple barriers to contain the radionuclides which are held in a wasteform. Cementitious material is the most commonly used encapsulation medium and its physical and aqueous durability is important. The aim of this study was to evaluate the short- to medium-term aqueous durability of a possible candidate wasteform for Australia's intermediate level waste and gain an understanding of the factors that control its degradation process. The samples tested simulated the Materials Testing Reactor (MTR) wasteform and were prepared by AEA Technology in the United Kingdom. The samples were acquired by ANSTO in 1998. The wasteform has a formulation of 9:1 wt% ground granulated blast furnace slag to ordinary Portland cement. The real waste contains a number of fission products [1] but the wasteform studied in this work contained simulated inactive waste liquor. The waste liquor contained salts dissolved in H_2SO_4 and neutralised in caustic solution, as described in [2]. Elemental concentrations in the cement are given below under Results and Discussion.

Some wasteform samples were leach tested over a series of shorter periods (up to 6 months), whilst another batch was leached for 92 months, at temperature for the first 12 months and then further leached at room temperature for the remaining 80 months. Overall, the experimental plan is to examine the leaching behaviour and evaluate the results to determine a realistic testing scenario. The tests include leaching using a series of different surface area: volume ratios (SA:V) at two different temperatures, 40 and 85°C (to accelerate leaching and to simulate the range of conditions possible in a repository). The SA:V ratios selected were $0.003mm^{-1}$, $0.03mm^{-1}$ and $0.3mm^{-1}$. This report gives the preliminary results from the study.

EXPERIMENTAL METHODOLOGY

The bulk elemental composition of the wasteform (leached and unleached) was carried out by X-ray Fluorescence Spectroscopy (XRF) on duplicate specimens. Powder X-Ray Diffraction (XRD) was used to analyse the mineralogy of the wasteform. Scanning Electron Microscopy (SEM), incorporating an Energy Dispersive X-ray Spectrometer (EDS), was used to characterize the unleached specimens and those leached for 92 months. Secondary Ion Mass Spectrometry (SIMS) was used for elemental depth profiling to compare the unleached specimen with that leached for 92 months. Leachates from the durability tests were analysed using a Perkin Elmer PE-SCIEX Elan 6000 Inductively Coupled Plasma-Mass Spectrometer (ICP-MS) and a Perkin Elmer Optima 5300DV ICP-AES Inductively Coupled Plasma – Atomic Emission Spectrometer (ICP-AES).

The XRF spectrometer was a Philips PW2400, a wavelength dispersive instrument with a rhodium anode tube. XRD was carried out on a Siemens D500 diffractometer using Co K_α-radiation (0.17903 nm) and fitted with a graphite monochromator and proportional counter. SEM was carried out with a JEOL JSM-6300 instrument operated at 15 kV, and fitted with a NORAN Voyager IV X-ray microanalysis system. SIMS was performed using a Cameca ims 5f secondary ion mass spectrometer. Variations in major element ion yields were examined using a beam of 12.5 keV O-primary ions focussed to a spot about 20 μm in diameter.

The leach test was based on the ISO 6961 standard [3] with the leach protocol modified so that tests were of the non-replacement leachant type. The leach tests at 1, 3 and 6 months, were carried out at temperature and from 12 months all further leaching took place at room temperature (25 ± 5°C). Leach tests were carried out in triplicate, in deionised water under three different Surface Area: Volume ratios (SA:V) and initially at two temperatures, 40 and 85°C. The results presented here are for the tests carried out at 40°C and a SA:V ratio of $0.03mm^{-1}$, and include the 92-month tests at room temperature. Results and findings of the tests at other temperatures and SA:V ratios are not reported here due to space limitations and will form part of subsequent papers.

As discussed later the leaching results indicated that the major releases took place at early durations and a second set of leach tests was carried out to determine the duration over which this initial reaction took place. These subsequent tests were carried out with a SA:V ratio of $0.03mm^{-1}$ at 40°C and terminated at 1, 4, 19, 28 and 47 days respectively.

RESULTS AND DISCUSSION

The unleached samples were white in colour on the surface, but dark grey and mottled on the interior. Optical microscopy showed the white surface layer on the cement cylinders was approximately 800μm deep. The surface of the samples was smooth with no signs of cracking. After leaching, irrespective of the leaching period (1, 3, 6 or 92 months), the specimens exhibited cracking on the surface (see Figure 1). The cracking is most likely due to swelling from long-term immersion in water. Further investigation into this phenomenon is planned.

The density of the unleached wasteform was 1990 kg m^{-3}, and the compressive strength about 39 MPa. The compressive strength of MTR cement from previous studies [2] ranged from 30-50 MPa, depending on the waste: cement ratio and the aging of the waste liquor, but were more typically between 30 and 40 MPa. Such compressive strengths are well above the strength required for wasteform handling and storage.

Figure 1. Photographs of unleached specimen (left) and that leached for 92 months (right) showing the development of cracking (samples 8 cm high and 4 cm in diameter). The top of the leached specimen was sectioned to provide the specimen for SEM/EDS examination.

The concentrations of all detectable elements in the unleached and leached wasteforms are given in Table 1. Elements P, Rb, Y, Mo, Ru, Te, Cs, La, Pr, Nd, Sm and Hg that are present in MTR waste liquor [2] were not detected in our samples at the levels usually encountered. Some elements show a slight increase in concentration after leaching as all the values are relative abundances and increases after leaching reflect more significant decreases in the concentrations of other elements such as Na and K.

Table 1. Average elemental concentrations (wt%) of unleached and leached (92 months) wasteforms by XRF.

Element	Na	Mg	Al	Si	S	K	Ca	Ti	Mn	Fe	Sr	Zr	Ba
Unleached	5.4	3.3	6.9	13	0.76	0.41	30	0.37	0.20	0.51	0.06	0.03	0.10
Leached	3.1	3.3	7.1	14	0.72	0.31	31	0.37	0.21	0.53	0.06	0.03	0.10

XRD spectra were obtained of the unleached wasteform and those leached at 1, 3, 6 and 92 months. Figure 2 shows the spectra of the unleached and 92-month leached specimens. Calcite, hydrotalcite and a calcium silicate hydrate are shown to be present in both the unleached and leached wasteforms. The white exterior layer was composed of calcite. As the length of leaching time increased from 1 to 92 months the crystallinity of the calcium silicate hydrate phase also increased and the XRD pattern displays some reflections similar to that of okenite.

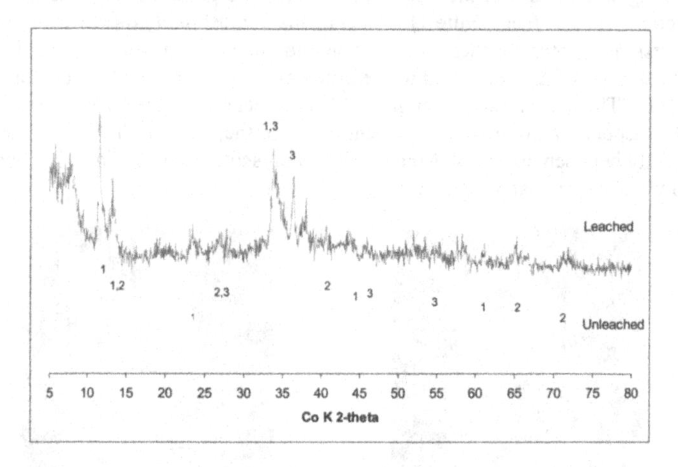

Figure 2. XRD spectra of the wasteforms - unleached and leached for 92 months (*1 – calcium silicate hydrate phase; 2 – hydrotalcite; 3 – calcite)*.

Leach testing shows the pH of the leachates were 12.9 at 1 month; 13.4 at 3 months; and 13.2 at both 6 and 92 months. Leachate data for K, Na, Si, Al, Ca, S and Sr are plotted as fractional releases against time in Figure 2. A feature of the 92-month leachates was the presence of a pale-coloured, translucent precipitate in the bottom of the leach vessels. It is not known if the precipitate formed after removal of the experiments from the 40°C oven and subsequent cooling of the saturated leachate, or if it has formed at temperature during the first 12 months of leaching. No analysis has been made of this precipitate but it forms part of our ongoing studies.

Figure 3 shows that fractional releases generally decrease in the order Na ~ K > S > Al ~ Sr > Si ~ Ca. Releases for Al, Ca, Sr, K, Na and S peak at 3 months, with releases of Al, Ca, Sr and S decreasing between 3 and 6 months. The releases thereafter for all elements are similar to those at 92 months. These results suggest that for leaching times greater than 3 months, precipitation and/ or sorption occurs for Al, Ca, Sr and S, removing them from solution. Steady-state conditions then prevail for longer leaching times. There is some evidence for precipitation in the 92 month leachates. Except for 1 month, Ca and Si releases are similar for all leach periods, which is unusual as generally Ca has higher releases than Si during the leaching of cement (ordinary Portland cement, for example).

The results of the second series of leach tests (1, 4, 19, 28 and 47 days) are shown in Figure 4 and unequivocally show that the major releases of the matrix elements occur in less than 4 days. Further tests are now being carried out at different leaching temperatures to further study this phenomenon.

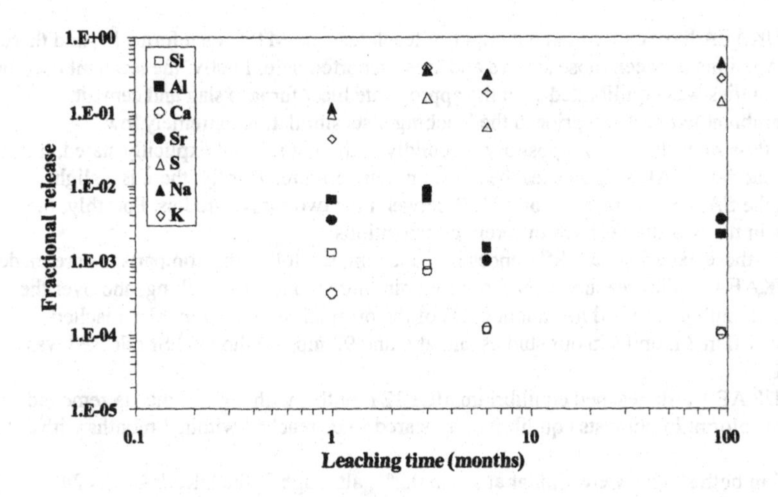

Figure 3. Fractional release of major matrix elements and Sr for each leach period at 40°C and SA:V ratio of 0.03 mm^{-1}. Note that the 92-month tests were conducted at ambient temperature for the last 85% of the leaching period.

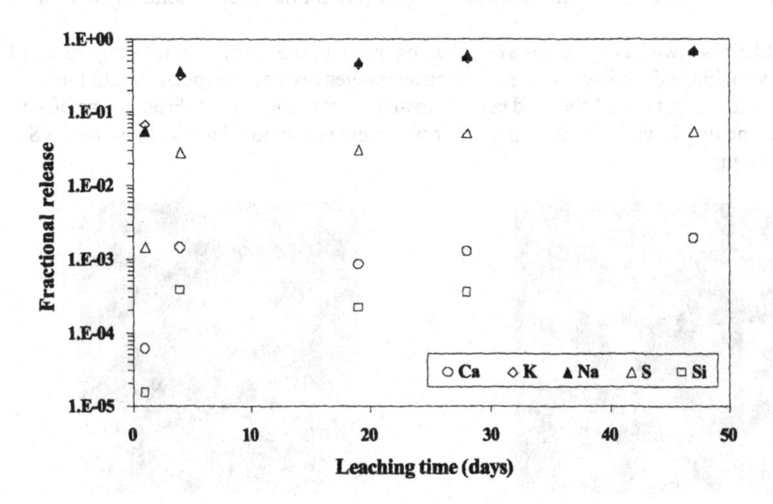

Figure 4. Fractional releases of major matrix elements at 40°C and SA:V ratio of 0.03 mm^{-1}, as determined from leachate analyses. Si was not detected by ICP-AES for the 47 day period and analysis by ICP-MS (with lower detection limit) was not available at the time of writing.

The UKAEA have carried out a number of leach tests on MTR wasteforms [4] and there are several variations between those studies and these reported here. Firstly, the leachant used in the UKAEA studies was equilibrated with the appropriate blast furnace slag and cement materials at ambient temperature prior to the leaching tests simulating extremely low groundwater flow rates through a repository. Secondly, although it is not explicitly stated, it can be assumed that the UKAEA studies leached at room temperature. Thirdly, there is a slight difference in the SA:V ratio (by a factor of 1.7) between the two sets of studies. Fourthly, the waste liquors in the two studies have different compositions.

Taking these experimental differences into account, the following comparisons are made:

- Sr in the UKAEA studies reached equilibrium within three months of leaching, and over the complete 24 month period had lost about 0.2% of the original Sr. Strontium also reached equilibrium within 3 months in our studies and at 6 and 92 months the total Sr released was about 0.1%.
- Na in the UKAEA tests reached equilibrium after 12 months, with 26% of the Na removed from the wasteform. In our tests equilibrium appeared to be reached within 3 months with 32% removed.
- Al released in both studies were similar at about 0.2%, although in the UKAEA tests 24 months was required to attain steady-state.
- Ca releases were similar for both studies in that Ca concentrations decreased with time rather than maintaining equilibrium, although the rate of decrease in concentration was greater in the UKAEA tests compared to ours. At the end of leaching period - 24 months for the UKAEA studies and 92 months in this work - the total Ca loss was similar at approximately 0.01%.

SEM/ EDS showed a high degree of inhomogeneity in the wasteform (see Figure 5). EDS analysis was variable and the back-scattered electron images divided components into three groups - unreacted slag, reacted slag and cement matrix (including calcite). From a wasteform point of view, the matrix is the primary phase for waste encapsulation thus the focus of the SEM/ EDS investigations.

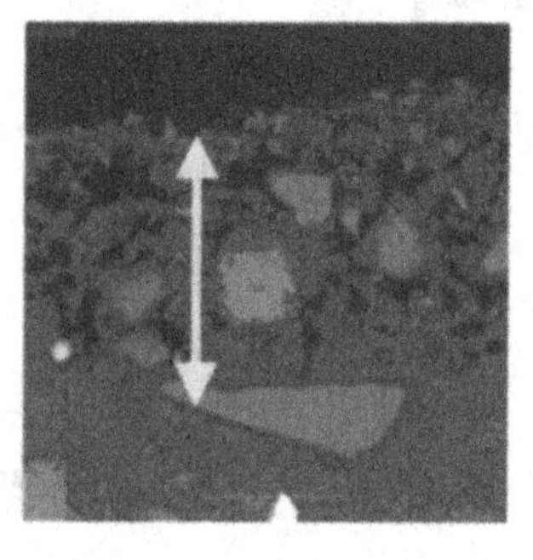

Figure 5. SEM photomicrographs of unleached (left image) and leached (right image) specimens (both x500 magnification). Composition is primarily slag particles in a cement matrix. Porosity caused by leaching evident to a depth of about 80 μm (shown by white arrow).

The cement specimen leached for 92 months exhibited corrosion to a variable depth, on average about 80 μm (see Figure 5). Analysis by EDS was carried out on a few areas within the altered zone, and distal to the zone for comparison of elemental composition. The most notable feature was the decrease in Ca content within the corroded zone, dropping on average by about 10 wt% compared to that in the unaltered zone. The Ca content within the unaltered zone was similar to that of the matrix in the unleached wasteform. Concentrations of Al and Si in the leached cement matrix (both within and outside the altered zone) were not significantly different from that of the unleached wasteform. These observations reflect a dissolution front which leaves behind a zone of decalcified CSH surrounding an uncorroded core [5].

EDS analysis on selected areas showed that Na and K levels were lower in the leached specimen than the unleached. These results correlate with the analysis of the wasteform before and after leaching by XRF. EDS also showed that Na and K were in higher concentrations closer to the surface of the 92-month leached specimen than in the underlying cement, suggesting that diffusion is an important transport mechanism for these two elements through the matrix.

A number of SIMS analyses of Ca, Si, Al, Mg, Fe, Na, Na and Sr were carried out on selected areas of the unleached sample and the sample leached for 92 months. The spot size was between 20 and 30 μm and some care had to be taken to ensure that the areas chosen comprised mainly the cement matrix. Areas were analysed from the surface into the interior of the samples at about 0.5mm intervals, with an additional five analyses made at about 50μm apart near the surface. While the Ca: Si ratio varied by about 20%, the relative abundance of Al, Fe and Mg varied in a random fashion across the surface suggesting that inhomogeneous slag fragments were being included in the spot analysis. Sodium and Sr show evidence of a diffusion profile in the leached samples, as evident by Figure 6.

We suggest that the difference between the SIMS and SEM/ EDS results for Na reflect the inhomogeneity of the sample and are being investigated further.

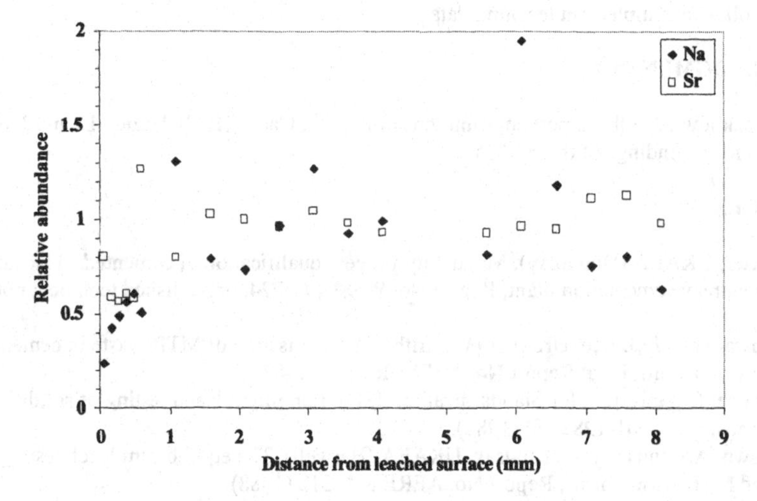

Figure 6. The distribution profile of Sr and Na as determined by SIMS. Results were normalised to the Si analysis and show significant anomalies from a straight diffusion profile suggesting that the Sr and Na have an affinity to parts of the cement matrix.

SIMS and SEM/ EDS results imply that the degradation process observed in our studies appears to result from the combined action of diffusion and dissolution at the water-cement interface [6].

CONCLUSIONS

The wasteform was an inhomogeneous mixture of calcite, hydrotalcite and a calcium silicate hydrate. The crystallinity of the calcium silicate hydrate phase increased with leaching time. After leaching for 92 months there was an alteration layer about 80 μm deep depleted in Ca whilst Na, K and Sr showed signs of diffusion towards the outer part of the cement samples.

The compressive strength (39 MPa) of the wasteform was typical of that reported in earlier studies and well above that required for handling or storage.

The majority of the releases of Ca, Si, Al, Sr, S, Na and K in the non-replacement tests was reached within 4 days of leaching, with the maxima occurring at 3 months for Ca, Al, Sr, S, Na and K, and at 1 month for Si. For the longer leach periods (6 months and 3 months respectively) there was a slight reduction in concentration in the leachates, and these levels were similar to those for 92 months, suggesting steady-state conditions prevailing after 3 to 6 months of leaching. The highest releases of matrix elements were for Na (37%), K (40%) and S (16%), reflecting their presence, typically, in the pore fluids of cements. Releases for Ca, Na, Al and Sr were similar to those reported by the UKAEA in earlier studies on MTR cement.

This study has been an excellent opportunity to study the long term behaviour of this material and enables the results from shorter studies to be set in context. It suggests that, for this particular wasteform, shorter term data can be extrapolated to the longer term, although longer term tests are nonetheless important to provide the detailed surface and bulk characterisation of the leached solids to complement leaching data.

ACKNOWLEDGMENTS

The authors acknowledge the expert contributions of Patricia Gadd (XRF), Huijun Li and Joel Davis (SEM) to the findings of the studies.

REFERENCES

1. P.B. Carter, UKAEA (Dounreay), Manual for process qualification of cemented MTR raffinate at the Dounreay cementation plant, Report No. WSSD(99)P24, unpublished technical note (2000).
2. C.G. Howard and D.J. Lee, UKAEA (Winfrith), Immobilisation of MTR waste in cement (product evaluation), Final Report No. AEEW-R 2312 (1987).
3. International Organization for Standardization (ISO), Long-term leach testing of solidified waste forms, ISO 6961-1982 (E) (1982).
4. P.E. Brown, A.J. Inns and M. Lindsay, UKAEA (Harwell), The equilibrium leach testing of cemented MTR waste forms, Report No. AERE R 13312 (1988).
5. E. Revertegat, C. Richet and P. Gégout, *Cement and Concrete Research*, **22**, pp. 259-272 (1992).
6. P. Faucon, J.F. Jacquinot, F. Adenot, J. Virlet and J.C. Petit, Materials Research Symposium Proceedings, **465**, pp. 295-302 (1997).

Mater. Res. Soc. Symp. Proc. Vol. 1107 © 2008 Materials Research Society

Thermodynamic Simulation and Experimental Study of Irradiated Reactor Graphite Waste Processing with REE Oxides

Olga K. Karlina, Vsevolod L. Klimov, and Galina Yu. Pavlova
Moscow SIA "Radon", 2/14, 7-th Rostovsky per., Moscow, 119121, Russian Federation
Michael I. Ojovan
Immobilisation Science Laboratory, Department of Engineering Materials, University of Sheffield, S1 3JD, UK

ABSTRACT

Thermochemical processing of reactor graphite waste is based on self-sustaining reaction $4Al + 3TiO_2 + 3C = 3TiC + 2Al_2O_3$ which chemically binds ^{14}C from the irradiated graphite in the titanium carbide. Thermochemical processing was investigated to analyse the behaviour of rare earth elements (REE), where REE = Y, La, Ce, Nd, Sm, Eu and Gd. Both thermodynamic simulations and laboratory scale experiments were used. The REEs in the irradiated reactor graphite are formed as activation products of impurities and spread over the graphite bricks surfaces as well as arise from fission of nuclear fuel. REEs can be used also to substitute for waste actinides as well as to increase the durability of carbide-corundum ceramics relative to waste actinides.

Thermodynamic calculations and X-ray diffraction analysis of ceramic specimens synthesized revealed that durable REE's aluminates with perovskite, β-alumina and garnet structures are formed by interaction of REE oxides with the Al_2O_3 melt during the self-propagating reaction of ceramic formation.

The porous carbide-corundum ceramics synthesized have a high hydrolytic durability, e.g. the normalised leaching rates of ^{137}Cs, ^{90}Sr and Nd are of the order of $10^{-7} - 10^{-8}$ g/(cm^2·day).

INTRODUCTION

Operation of uranium-graphite reactors resulted in production of significant amounts of irradiated graphite waste in the form of dust, powder, chips and lumps. This waste resulted both from technological operations and incidents and contains inclusions of metallic and ceramic nuclear fuel and other reactor components. This waste contains activation products such as ^{10}Be, ^{14}C, ^{36}Cl, ^{41}Ca and ^{59}Ni, actinides in fuel particles such as $^{234,236,238}U$, ^{237}Np, $^{239,240,242}Pu$, ^{243}Am and $^{242,243,244}Cm$, decay products such as $^{134,135,137}Cs$, ^{90}Sr, ^{151}Sm and $^{154,155}Eu$. Similar waste can be produced when cleaning-decontaminating surfaces of graphite blocks heavily contaminated by radionuclides.

Among the radionuclides present in irradiated graphite the long-lived radionuclide ^{14}C with half-life of 5730 years is particularly hazardous. ^{14}C is readily incorporated into organic matter molecular structure of living species including humans, moreover it enters into RNA and DNA molecules [1].

Disperse radioactive wastes intended for long-term storage and disposal in Russia must be immobilised (consolidated) accordingly to Russian regulatory requirements, which were developed accounting for IAEA recommendations [2]. To solidify disperse irradiated graphite waste the self-sustaining high-temperature synthesis (SHS) based on exothermic chemical reaction

$$3C(graphite) + 4Al + 3TiO_2 = 3TiC + 2Al_2O_3$$

was suggested in Russia [3–5]. The product of this reaction is a sintered corundum-titanium carbide composite ceramic retaining waste radionuclides including chemically-bound biologically-hazardous radionuclide ^{14}C. As a result, graphite, including carbon ^{14}C, is held in stable titanium carbide. Monolithification of disperse irradiated graphite waste simultaneously with isolation of ^{14}C from environment was previously considered in [6–11]. The irradiated reactor graphite waste typically contains REEs which can also be used as substitutes for actinides when modelling reactor graphite waste processing. This paper is devoted to both thermodynamic simulation and experimental investigation of the behaviour of REEs in irradiated reactor graphite waste during SHS processing.

THERMODYNAMIC ANALYSIS

A detailed thermodynamic simulation of the abovementioned reaction and C–Al–TiO_2 system was previously performed [7–10]. In the present investigation the software TERRA [12] involving the IVTANTHERMO database [13] on thermodynamic properties of chemical substances was used for thermodynamic simulation.

As is usual, in all thermodynamic calculations the gaseous phase was treated as an ideal gas. Formation of the following condensed species was considered: C, Al, Al_2O_3, Al_2OC, Al_4O_4C, Al_4C_3, Ti, TiO, TiO_2, Ti_2O_3, Ti_3O_5, Ti_4O_7, TiC, Y, Y_2O_3, $Y_3Al_5O_{12}$, Ce, Ce_2O_3, CeO_2, $CeAlO_3$, Ce_2O_3, CeC_2, Nd, Nd_2O_3, $NdAlO_3$, NdC_2, Sm, Sm_2O_3, $SmAlO_3$, SmC_2, Eu, EuO, Eu_2O_3, $EuAlO_3$, EuC_2, Gd, Gd_2O_3, $GdAlO_3$, GdC_2. Thermodynamic functions of Al_2OC and Al_4O_4C have been calculated in [14]. Thermodynamic functions of TiC were taken from the NIST-JANAF tables [15]. Thermodynamic functions of yttrium-aluminum garnet $Y_3Al_5O_{12}$, $LnAlO_3$ aluminates (Ln = La, Ce, Nd, Sm, Eu, Gd), and LnC_2 carbides have been calculated over a wide range of temperature especially for the purposes of this simulation [16–18].

The introduction of Ln_2O_3 additives in amounts up to 15–20 wt.% into the $3C+4Al+3TiO_2$ stoichiometric mixture does not diminish the equilibrium reaction temperature which remains 2327 K so that the resulting Al_2O_3 reaction product is liquid.

Thermodynamic calculations have shown the $LnAlO_3$ aluminates-perovskites and yttrium-aluminum garnet ($Y_3Al_5O_{12}$) are formed by reaction of the $3C+4Al+3TiO_2$ mixture with lanthanide and yttrium oxides (Figure 1). These results are in agreement with known state diagrams of the Al_2O_3–Ln_2O_3 binary system [19]. In contrast the formation of the LnC_2 lanthanide carbides lacks support from thermodynamic calculations.

As follows from the thermodynamic calculations a correction of starting blend is required when adding REE oxides in order to prevent aluminum oxycarbide formation (Figure 1).

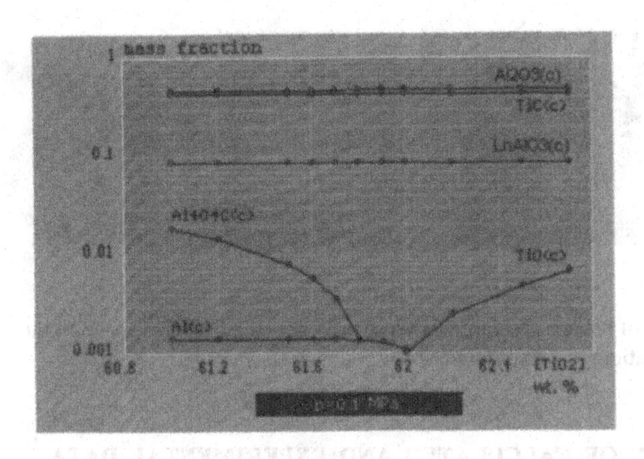

Figure 1. Correction of composition of final product of graphite processing in the $3C+4Al+3TiO_2$ stoichiometric mixture with 7 wt.% of the Ln_2O_3 additives by variation of TiO_2 content.

CERAMIC SYNTHESIS

The SHS experiments were planned so that the most difficult stages, such as preliminary pressing of initial batch and hot pressing of reaction products, were avoided. Simulant mixtures were prepared from fine-dispersed powders. Aluminum and titanium dioxide powders were available industrial products, the graphite was obtained by crushing graphite blocks from uranium-graphite reactors, and the REEs oxides were reagent grade. Irradiated graphite from the AM research reactor sleeves with specific activity of ^{137}Cs $1.15 \cdot 10^6$ Bq/kg was used for leach tests of carbide-corundum matrices synthesized. Mixture components were thoroughly mixed, the mixture obtained was poured into corundum crucibles with the volume of 0.2 – 1 L. The REE oxides were introduced into the starting blend in amounts of 7 wt.% (above 100 % of main mixture). Combustion reaction in the case of SHS in air atmosphere was ignited by the use of a small amount of titanium powder on the surface. When the SHS was carried out in an inert atmosphere (argon) we have used the thermite mixture. After ignition a combustion front was formed which brightly glowed so that the downward motion of reaction front was seen through the crucible walls. Several stages of SHS process are shown in Figure 2. The final SHS product is a hard porous ceramic material.

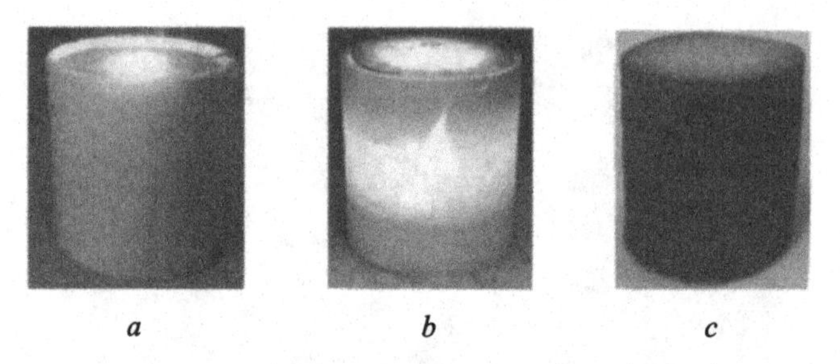

a *b* *c*

Figure 2. View of reactor graphite waste processing using SHS: *a* – reaction initiation; *b* – reaction propagation; *c* –final product.

COMPARISON OF CALCULATED AND EXPERIMENTAL DATA

The ceramics produced were analysed on X-ray diffractometer DRON–4 and X'Pert PRO (Philips) with radiation CuK_{α} (Figure 3). The results of laboratory tests were compared with data from thermodynamic modelling for the final product phase composition. We found a satisfactory compliance of calculated and experimental data on final product phase composition (Table I).

Formation of the $LnAlO_3$ aluminate with perovskite structure in the final product is characteristic of the mixtures investigated, except for the mixture with europium oxide. In mixtures with lanthanum and cerium oxides the $La_2O_3 \cdot 12.2Al_2O_3$ and $CeAl_{11}O_{18}$ aluminates with β-alumina structure form, consistent with phase diagrams of these systems [19]. The distinguishing feature of the mixture with europium oxide is a change of europium to a valence of two with formation of the $EuAl_{12}O_{19}$ and $EuO \cdot 10.5Al_2O_3$ aluminates. In the mixture with yttria the yttrium-aluminum garnet $Y_3Al_5O_{12}$ and yttrium-aluminum oxide $YAlO_3$ are produced.

REEs involved in corundum-titanium ceramic matrix in the form of perovskite, β-alumina and garnet provide high chemical durability to the final product. Data on the neodymium leaching from the matrix produced are given in the Table II. As mentioned above, the corundum-titanium ceramic matrix synthesized has a high porosity (around 50 %). Because of this, determination of the contact surface of the pores with leachate is problematic. Leach tests were conducted according to the Standard of Russian Federation GOST R 52126–2003 using powdered specimens [20] which is an analog of ASTM C 1285–94 test method used to determine the chemical durability of nuclear waste glasses. Table III gives data on the leach rate of ^{137}Cs and ^{90}Sr from the corundum-carbide matrix synthesized by processing reactor graphite powder with the $BaTiO_3$ modifier-additive [5].

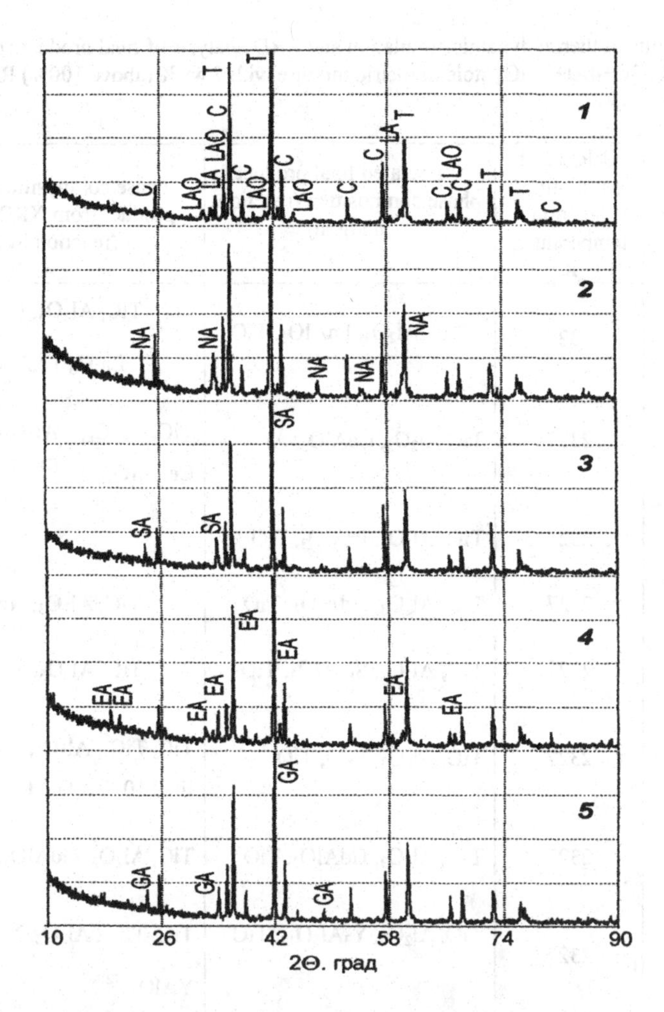

Figure 3. X-ray diffractograms of final products of SHS reaction of the $3C + 4Al + 3TiO_2$ stoichiometric mixture with REEs oxide additives: ***1*** – La_2O_3; ***2*** – Nd_2O_3; ***3*** – Sm_2O_3; ***4*** – Eu_2O_3; ***5*** – Gd_2O_3; peaks designation: **C** – corundum; **T** – TiC; **LAO** –$La_2O_3 \cdot 12.2Al_2O_3$; **LA** – $LaAlO_3$; **NA** – $NdAlO_3$; **SA** – $SmAlO_3$; **EA** – $EuAl_{12}O_{19}$, $EuO \cdot 10.5Al_2O_3$; **GA** – $GdAlO_3$, $AlGdO_3$.

Table I. Results of thermodynamic simulation and XRD analysis of final product composition of reaction of the $3C+4Al+3TiO_2$ stoichiometric mixture with 7 wt.% (above 100%) REEs oxide additives.

REE oxide	Calculated equilibrium reaction temperature, K	Calculated final product phase composition (reaction in argon)	Phase composition of final product from XRD analysis (reaction in air)
La_2O_3	2327	TiC, Al_2O_3, $LaAlO_3$, TiO	TiC, Al_2O_3, $LaAlO_3$, $La_2O_3 \cdot 12.2Al_2O_3$
CeO_2 [1]	2327	TiC Al_2O_3, $CeAlO_3$, Al	TiC, TiC_x, Al_2O_3, $CeAlO_3$, $CeAl_{11}O_{18}$, C [2]
Pr_2O_3	2327	TiC, Al_2O_3, $PrAlO_3$, TiO	–
Nd_2O_3	2327	TiC, Al_2O_3, $NdAlO_3$, TiO	TiC, Al_2O_3, $NdAlO_3$
Sm_2O_3	2327	TiC, Al_2O_3, $SmAlO_3$, TiO	TiC, Al_2O_3, $SmAlO_3$
Eu_2O_3	2327	TiC, Al_2O_3, EuO, TiO	TiC, TiC_x, Al_2O_3, $EuAl_{12}O_{19}$, $EuO \cdot 10.5Al_2O_3$, Ti_9O_{17}, C [2]
Gd_2O_3	2327	TiC, Al_2O_3, $GdAlO_3$, TiO	TiC, Al_2O_3, $GdAlO_3$, $AlGdO_3$
Y_2O_3	2327	TiC, Al_2O_3, $Y_3Al_5O_{12}$, TiO	TiC, TiC_x, Al_2O_3, $Y_3Al_5O_{12}$, $YAlO_3$ [2]

[1] The starting blend, wt.%: $3C+4Al+3TiO_2$ mixture (80) + $2Al+3CeO_2$ mixture (20).

[2] Analysis was taken on X'Pert PRO (Philips) diffractometer using CuK_α-radiation.

Table II. Normalised leaching rate of neodymium from carbide-corundum matrix, g/(cm^2·day)

Leaching time, days	Specimens	
	No. 1	*No. 2*
2	$1.1 \cdot 10^{-5}$	$1.0 \cdot 10^{-5}$
4	$5.0 \cdot 10^{-6}$	$4.2 \cdot 10^{-6}$
6	$4.4 \cdot 10^{-6}$	$3.9 \cdot 10^{-6}$
8	$3.7 \cdot 10^{-6}$	$3.3 \cdot 10^{-6}$
10	$2.2 \cdot 10^{-6}$	$2.0 \cdot 10^{-6}$
12	$2.6 \cdot 10^{-6}$	$2.3 \cdot 10^{-6}$

Table III. The leach rate of ^{137}Cs и ^{90}Sr from the carbide-corundum matrix with the modifying additive, g/(cm^2·day)

Leaching time, days	Normalised leaching rate			
	of ^{137}Cs		of ^{90}Sr	
	Sp. No. 11-c	*Sp. No. 12-c*	*Sp. No. 11-s*	*Sp. No. 12-s*
1	$4.8 \cdot 10^{-5}$	$6.2 \cdot 10^{-5}$	$1.9 \cdot 10^{-5}$	$1.8 \cdot 10^{-5}$
3	$1.2 \cdot 10^{-5}$	$7.2 \cdot 10^{-6}$	$1.1 \cdot 10^{-5}$	$1.1 \cdot 10^{-5}$
7	$2.4 \cdot 10^{-6}$	$3.1 \cdot 10^{-6}$	$4.1 \cdot 10^{-6}$	$4.6 \cdot 10^{-6}$
10	$1.4 \cdot 10^{-6}$	$1.6 \cdot 10^{-6}$	$5.5 \cdot 10^{-7}$	$2.3 \cdot 10^{-6}$
14	$1.1 \cdot 10^{-6}$	$7.2 \cdot 10^{-7}$	$7.0 \cdot 10^{-7}$	$7.9 \cdot 10^{-7}$
21	$3.8 \cdot 10^{-7}$	$2.6 \cdot 10^{-7}$	$2.4 \cdot 10^{-7}$	$2.6 \cdot 10^{-7}$
44	$1.6 \cdot 10^{-7}$	$1.2 \cdot 10^{-7}$	$1.1 \cdot 10^{-7}$	$9.6 \cdot 10^{-8}$
65	$4.6 \cdot 10^{-8}$	–	$2.2 \cdot 10^{-8}$	$6.7 \cdot 10^{-8}$

CONCLUSIONS

Chemically durable carbide-corundum ceramics with REE aluminates of perovskite structure form as a result of reactor graphite waste with REE oxides processing using the SHS on the base of the exothermic reaction $3C(graphite) + 4Al + 3TiO_2 = 3TiC + 2Al_2O_3$. The normalized leaching rates of ^{137}Cs and ^{90}Sr from the carbide-corundum ceramics synthesized is ~10^{-7} g/(cm^2·day) on 44-th day.

ACKNOWLEDGMENTS

The authors are grateful to Dr. Nadezhda Penionzhkevich for XRD analysis and Ms Natalie Manyukova for computer assistance.

REFERENCES

1. Rublevsky V.P., Golenetsky S.P., Kirdin T.S. "Radiocarbon in Biosphere", Atomizdat, Moscow, 1979.
2. "Safety Rules in Managing Radioactive Wastes from Nuclear Power Stations NP-002-04", Moscow: Federal Service on Ecological, Technological and Atomic Survey, 2004.
3. Merzhanov A.G., Borovinskaya I.P., Makhonin N.S., et al., *Patent RU 2065220* (18 March 1994).
4. Klimov V.L., Karlina O.K., Pavlova G.Yu., et al., *Patent RU 2192057* (28 June 2001).
5. Dmitriev S.A., Karlina O.K., Klimov V.L., et al., *Patent RU 2242814* (01 April 2003).
6. Ojovan M.I., Karlina O.K., Klimov V.L., and Pavlova G.Yu. in *Scientific Basis for Nuclear Waste Management* XXIII, edited by R.W. Smith and D.W. Shoesmith (Materials Research Society, Warrendale, PA, 2000) pp. 565–570.
7. Ojovan M.I., Karlina O.K., Klimov V.L., et al. in *Scientific Basis for Nuclear Waste Management* XXVI, edited by R.J. Finch and D.B. Bullen (Materials Research Society, Warrendale, PA, 2003) pp. 615–620.
8. Karlina O.K., Klimov V.L., Pavlova G.Yu., et al., *Atomic Energy* **94** (6), 405–410 (2003).
9. Karlina O.K., Klimov V.L., Pavlova G.Yu., et al., *Atomic Energy* **101** (5), 830–837 (2006).
10. Karlina O.K., Klimov V.L., Pavlova G.Yu., et al., *Intern. J. of SHS* **14** (1), 77–86 (2005).
11. Karlina O.K., Klimov V.L., Ojovan M.I., et al., *J. of Nuclear Materials* **345**, 84–85 (2005).
12. Trusov B.G. in *III Intern. Symposium «Combustion and Plasmochemistry»*. Aug. 24 – 26, 2005. Almaty, Kazakhstan. (Kazakh University, Almaty 2005)–pp. 52 – 57.
13. Belov G.V., Iorish V.S., and Yungman V.S., *CALPHAD* **23** (2), 173 – 180 (1999).
14. Klimov V.L., Bergman G.A., and Karlina O.K., *Russian J. of Physical Chemistry* **80** (11), 1816–1818 (2006).
15. NIST–JANAF Thermochemical Tables, 4th ed., edited by Chase M.W., Jr., *J. Phys. Chem. Ref. Data*, 1998. – Monograph No. 9.
16. Report of Moscow SIA "Radon" No. 553, 2002.
17. Report of Moscow SIA "Radon" No. 612, 2003.
18. Report of Moscow SIA "Radon" No. 663, 2004.
19. Toropov N.A., Barzakovsky V.P., Lapin V.V., and Kurtseva N.N., "Phase Diagrams of Silicate Systems, Handbook, Issue 1, Binary Systems", Nauka, Leningrad, 1969.
20. Standard of Russian Federation "GOST R 52126–2003. Radioactive Waste. Determining Chemical Durability of Solidified High-Level Waste by Long-Time Leaching Test", Moscow, 2003.

Mater. Res. Soc. Symp. Proc. Vol. 1107 © 2008 Materials Research Society

The Role and Management of Free Water in the Production of Durable Radioactive Waste Products Using Hydraulic Cements

Michael J. Angus, Ed Butcher, Ian H. Godfrey and Neil B. Milestone[1]
Nexia Solutions, Havelock Road, Derwent Howe, Workington, CA14 3YQ, UK
[1]Sheffield University, Mappin St., Sheffield, S1 3JD, UK

ABSTRACT

Water is a necessary component in the production of encapsulated wastes based on hydraulic cements which are widely used for immobilization of intermediate and low level waste (ILW) and (LLW). Apart from providing the fluidity required to readily transport slurry wastes, it plays an essential role in hydrating the cement. Too low a water content prevents homogeneous mixing of the cement binder and waste and does not provide the fluidity needed for effective infilling of solid wastes. The water left after hydration creates a porous network that allows egress of gaseous corrosion/radiolytic degradation products such as hydrogen. A broad envelope (i.e. range) of acceptable water/binder ratios is essential for effective process control, particularly for the encapsulation of slurry wastes which have widely varying water contents.

Nevertheless, the presence of large amounts of free water in the pore system of the hardened matrix allows easy transport of soluble ions such as hydroxide, which can lead to metal corrosion, and the increased permeability of the system increases the leachability. Therefore effective management of the 'free' water content of a waste product will allow optimisation of both the encapsulation process and the product quality and durability.

This paper describes a range of innovative approaches to 'water management', including the use of alternative hydraulic cements, modification of powder characteristics and use of superplasticised composite OPC grouts and examines the contribution of ^{1}H NMR relaxometry in providing improved understanding of the distribution of water within the pores of the hardened cement matrix.

INTRODUCTION

The encapsulation of ILW at UK nuclear sites is typically achieved using ordinary Portland cement (OPC) with large additions of either blast furnace slag (BFS) or pulverised fuel ash (PFA) to control the heat of hydration and product temperatures during early curing. Typically, a cement grout used for infilling a solid waste using a vibrogrouting process will be required to achieve the following performance;

- sufficient fluidity for up to 2.5 hours from mixing to enable infilling,
- capable of being pumped and if necessary, vibrated without segregation,
- able to displace residual water from the solid waste,
- setting within 24 hours with minimum bleed.

To compensate for variability in cement powder performance and waste composition, the encapsulation plants currently operating within the UK nuclear industry use a relatively high water/binder ratio to achieve these requirements consistently. However, the consequences can be bleed water production requiring treatment as secondary waste, increased porosity resulting in poor leach performance and greater availability of free water to participate in waste-matrix interactions such as the corrosion of reactive metal wastes. The UK nuclear industry is therefore investigating a range of alternative methods to introduce additional flexibility to mix design.

This paper highlights some of the results from investigations into the following options;

- management of powder particle size distribution (in this case, BFS) to improve fluidity at a given water content or to maintain fluidity at a lower water content,
- use of polycarboxylate superplasticisers to improve fluidity at a given water content,
- use of alternative cements such as magnesium phosphate and calcium sulphoaluminate which combine higher water binding with pore water chemistries likely to be more compatible than OPC with some waste streams.

While each of these methods shows promise, there is also a need to fully understand the role played by water in potentially deleterious ageing reactions. To this end, ^{1}H NMR relaxometry has been used to follow the changing distribution of water between capillary pores, gel pores and hydrate phases during ageing of OPC-based and CSA-based cements over periods of up to 2 years. This technique has the following advantages over alternative methods:

- it is non-invasive and non-destructive so the same sample can be examined with time,
- only ^{1}H nuclei in the liquid phase of the sample are detected,
- NMR distinguishes between water filled pores and air voids, can probe closed porosity and pore diameters below 5nm, does not suffer from an "ink bottle" effect and does not damage the microstructure being probed (unlike mercury porosimetry for instance).

EXPERIMENTAL DETAILS AND RESULTS

The effect of water content on flow properties of standard grouts

The Colflow technique is commonly used in the UK as a key performance and quality assurance standard and measures the distance that 1136ml (1 quart imperial) of grout/slurry will flow in a confined channel under gravity. Typically a flow of 200 to 250mm is required for vibrogrouting and potentially much higher fluidities if vibration is not used. To demonstrate the effect of water-solids mass ratio (w/s) on fluidity, a series of 3 litre scale mixes was prepared at a range of w/s ratios, in which powders were added to the water with low shear mixing over a 5-minute period then mixed for a further 5 minutes, followed by high shear mixing over 10 minutes. These mixes were with a 9:1 wt% BFS/OPC blend, the highest BFS content normally used for waste encapsulation. The fluidities measured immediately after mixing are given in Figure 1. These results show that the w/s ratio is an important variable in determining fluidity. However, the grout fluidity increase with w/s was at the expense of an increase in bleed volume.

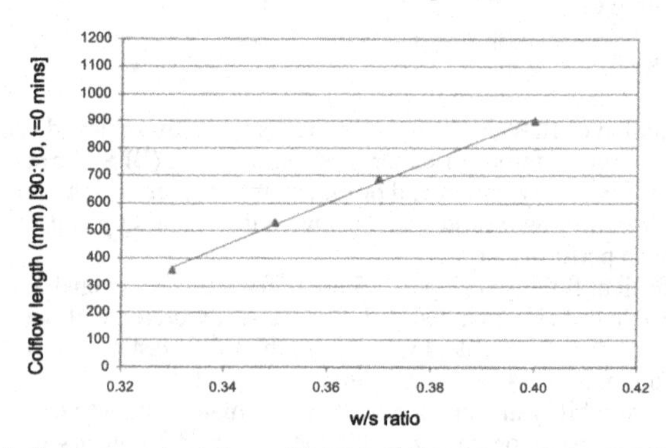

Figure 1 Relationship between w/s ratio and initial Colflow length for 9:1 BFS/OPC Blends

The effect of water content on leaching and porosity

^{134}Cs-loaded clinoptilolite, a zeolitic ion exchanger, was encapsulated in OPC, at a clinoptilolite/cement ratio 1:9, and at a range of w/s ratios. Samples (45mm dia x 80mm high) were cured for 28 days and subjected to leach tests in distilled water (350ml) with complete replacement of the leachate at intervals over 231 days according to the method of Sambell et al [1]. The leachate was filtered and analysed by gamma-spectrometry. Separate samples were also characterized by mercury intrusion porosimetry. Results are shown in Table I.

Table I Average leach rates and porosity for clinoptilolite-cement at various w/s

w/s	Average leach rate cm.d^{-1}	Mercury Porosity cm^3.g^{-1}
0.25	2.1x10^{-4}	0.065
0.40	5.3 x10^{-4}	0.132
0.50	9.3 x10^{-4}	0.159
0.60	12.0 x10^{-4}	0.19

Because of the low waste loading and high OPC content, the pozzolanic and ion exchange reactions between clinoptilolite and cement were maximized and a large fraction of the Cs was leached. For these reasons, as well as the high heat of hydration, these blends would not be suitable for full-scale application. Nevertheless, they provide a clear demonstration of the effect of water content on leach performance.

The management of water content by powder particle size distribution or superplasticisers

To investigate the effect of powder particle size distribution on grout fluidity, BFS from various sources was modified by grinding and "scalping" (sieving to remove the largest particle size distribution). This process produced powders with a broad particle size distribution and a controlled balance of coarse and fine material. Acceptance tests using the mixing procedure outlined above, including Colflow fluidity testing, were carried out to compare the modified BFS with "BNFL specification" BFS not subjected to these additional processes. Colflow results over 2.5 hours for a ground BFS, scalped to 300µm and further modified in terms of particle size distribution, are compared with BNFL specification BFS, for 9:1 wt% BFS/OPC at w/s=0.33 in Figure 2.

Although it is not current practice for organic admixtures (e.g. superplasticisers) to be used for waste encapsulation by the UK nuclear industry, due to the requirement to demonstrate that they would not have a detrimental effect on repository performance as a result of radionuclide complexation and leaching, the effect of superplasticisers has also been investigated. A wide range of trials has been undertaken using BFS, PFA and alternative cements. For example, 9:1 BFS/OPC, w/s=0.33, was prepared using Advacast 550, a polycarboxylated polyether comb polymer superplasticiser (at a 0.1% v/w loading). The fluidities are also shown in Figure 2. It is clear that superplasticisers and modification of the particle size distribution of the BFS give increased fluidity without production of excess bleed.

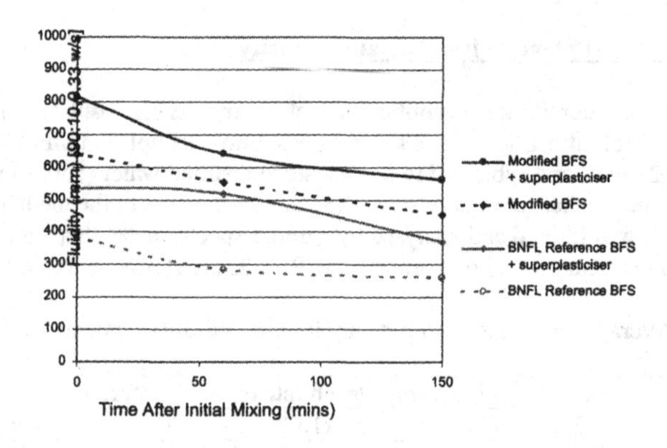

Figure 2 Grout fluidity of standard and superplasticised mixes at 0.33w/s

Management of water content using magnesium phosphate cement

In magnesium phosphate (MP) cement, a greater proportion of the mix water is chemically bound within hydration products than in OPC at the typical water contents required to formulate these cements. This reduces the availability of water to participate in waste-matrix interactions. In the case of reactive metal corrosion of aluminium, further benefit is gained as a result of the low internal pH of MP cement [2]. The principal reaction in commercial MP cements involves magnesium oxide reacting with ammonium dihydrogen phosphate in the presence of water. This reaction initially yields schertelite $Mg(NH_4)_2(HPO_4)_2.4H_2O$ and subsequently, depending on water availability, struvite $MgNH_4PO_4.6H_2O$ as the main reaction products. Potassium or sodium dihydrogen phosphate salts can also be used as a direct replacement for the ammonium salt. The overall reaction for the ammonium analogue is:

$$MgO + NH_4H_2PO_4 + 5H_2O \rightarrow MgNH_4PO_4.6H_2O$$

Trials have been carried out to develop MP formulations with consistent flow, exotherm and setting behaviour. These have been based on proprietary cements, as well as blends of the individual components. Small scale (3 litre) high shear mixes were carried out in the same way as for conventional cements, with flow measured at various times up to 2.5 hours after completion of high shear mixing (during which time low shear mixing was continued). Results for a blend based on MgO (dead burned), potassium dihydrogen phosphate, PFA and boric acid and a proprietary blend (Devlab 1846) are compared with a conventional 3:1 BFS/OPC mix in Table II. High fluidity and low pH were obtained.

Management of water content using calcium sulphoaluminate cements

In calcium sulfoaluminate (CSA) cements, a high proportion of the mix water is also chemically bound. The pore solution pH is less alkaline than OPC as lime is not produced during hydration, so CSA is also considered to have potential for reducing certain waste-matrix interactions. In CSA, the principal reactive component is calcium sulfoaluminate, or yeelimite,

($3CaO.3Al_2O_3.CaSO_4$,denoted as $C_4A_3\hat{S}$[1]). Hydration occurs along a number of reaction paths depending on the components present [3]. Principally $C_4A_3\hat{S}$ reacts with gypsum or anhydrous calcium sulfate in the presence of water to form ettringite (C6AŜ3H32) and alumina gel;
:

$$C_4A_3\hat{S} + 2C\hat{S}H_2 + 34H \longrightarrow C_6A\hat{S}_3H_{32} + 2AH_3$$

If reduced quantities of sulphate are present, monosulphate ($C_4A\hat{S}H_{12}$) is also produced:

$$2C_4A_3\hat{S} + 2C\hat{S}H_2 + 52H \longrightarrow C_6A\hat{S}_3H_{32} + C_4A\hat{S}H_{12} + 4AH_3$$

As ettringite is the primary reaction product, rather than a secondary degradation product, as with sulphate attack of OPC, it would not lead to wasteform expansion and failure.

A range of trials was carried out using a commercial CSA cement, Rockfast 450 (fineness of 450 m^2/kg), supplied by Lafarge UK, blended with different proportions of gypsum and PFA. Measurement methods were the same as for MP cements. Results are included in Table III for typical successful blends. It should be noted that while these blends have a high w/s relative to BFS/OPC, the amount of water (and gypsum) added is close to the stoichiometric value required for production of ettringite at 100% hydration, and therefore, in theory, free water is minimised. Preliminary corrosion experiments for these cements incorporating aluminium have also shown excellent performance compared to OPC blends [2].

Table II. Comparison of results for Nexia Solutions blend, proprietary MP and 3:1 BFS/OPC

Components		Colflow					pH	Temp (°C)	Set (hrs)	
Powder	w/s	0 min	30 min	60 min	120 min	150 min			Initial	Final
KH_2PO_4=1532g MgO=681g PFA=1500g	0.30	1000	1000	880	820	820	5.4-6.4	10-31	3	<24
Devlab 1846, Borax 0.66%	0.2	1000	1000	950	860	830	5.5-6.4	15-20	4	<24
3:1 BFS/OPC	0.35	440				270	12.8		3.5	<24

Zero bleed was measured for all mixes.

Table III Mix results for 3 litre CSA cement studies

Components % by mass				Colflow					pH	Temp (°C)	Set (hrs)	
CSA	Gypsum	PFA	w/s	0 min	30 min	60 min	90 min	120 min			Initial	Final
60	40		0.55	1230	1260	880	790	n.m	10.8-11.2	12-14	6	8
60	40		0.58	1360	1200	990	780	700	10.6-11	11-18	n.m	n.m
45	30	25	0.50	1360	1080	890	770	600	9.5-10.8	11-17	n.m	n.m

n.m. = not measured. Zero bleed was measured for all mixes.

Investigation of water content by NMR Relaxometry

[1] Cement Nomenclature used: C=CaO, S=SiO_2, A=Al_2O_3, Ŝ=SO_3, H=H_2O

This technique exploits the property of nuclear spin possessed by ^{1}H where in a magnetic field, the thermal equilibrium of magnetic moments of the nuclear spin tend to be aligned parallel to the applied magnetic field. The nuclear spins can be disturbed from equilibrium by applying a radiofrequency pulse, and the processes by which the nuclei return to their original thermal equilibrium state is referred to as relaxation. Longitudinal relaxation (spin-lattice relaxation) and transverse relaxation (spin-spin relaxation) can be measured using specific pulse sequences such as 'inversion recovery' for longitudinal relaxation and the Carr-Purcell-Meiboom-Gill (CPMG) sequence for transverse relaxation. The result of the CPMG experiment is a signal intensity that decays over time as the nuclei return to thermal equilibrium, and for a bulk liquid the decay of the signal $s(t)$ is generally exponential:

$$s(t) = s(0)e^{-t/T_2}$$

where $s(0)$ is the signal intensity at time t=0 and T_2 is the transverse relaxation time constant. The rate of relaxation of nuclei is related to molecular motion and the local environment of the nuclei. In the case of liquids in a porous material it may be shown that the transverse relaxation time is inversely proportional to the ratio of the pore surface area to the pore volume. This means that the decay of the signal from ^{1}H nuclei of molecules confined in small pores is more rapid than for ^{1}H nuclei of molecules confined in large pores.

Two cement systems have been investigated using this method, BFS/ OPC blends at two ratios 3:1 and 9:1 with w/s of 0.37 and hydrated at 20 and 60°C and some preliminary work on CSA cement hydrated at 20°C and w/s 0.67. [4]. Relaxometry measurements were performed using the CPMG sequence at a ^{1}H resonance frequency of 20 MHz. For each sample relaxometry measurements were performed following 1, 3, 7, 14, 90, 360 and 720 days curing. Inverse Laplace transformation of CPMG signal was performed to obtain the distribution of T_2 relaxation times present in the sample.

The amount of evaporable water in the BFS:OPC blends was estimated by freeze-drying as a function of curing time. Water is extracted from the cement by sublimation and the samples then weighed and the mass difference calculated as a percentage of the initial weight and attributed to the amount of evaporable water present in the blends. TGA was performed on the freeze-dried samples to estimate the amount of chemically bound water from the mass loss between 105°C and 950°C. This method provides a good indication of the degree of hydration of the blend, although the contribution of the cement and BFS fractions cannot be separated.

NMR Results for BFS/OPC

A general shift of the water from capillary pores towards gel pores is observed for each sample over the first few days of hydration, with little change in the water distribution after about 90 days (as measured for 20°C specimens) or 28 days (as measured for 60°C specimens) (see Figure 3). There are differences between the four trials for the distributions of relaxation times (and hence the distribution of water between different pore sizes) at long curing times. A higher gel water content is seen in the fully cured 3:1 BFS/OPC sample cured at 20°C compared to that cured at 60°C, despite the latter having the higher gel water content after 3 days. The 9:1 BFS/OPC samples show that water is found largely in meso-pores, with a measurable population of water in capillary pores remaining after full curing. These results were consistent with MIP measurements. Using the TGA and relaxometry measurements it is possible to estimate the distribution of evaporable water between gel, meso and capillary pore water (see Table IV). It is clear that much of the water remains in the capillary pores for 9:1 BFS/OPC cement and that the capillary pore water population is higher for samples cured at higher temperatures. At high BFS content (9:1BFS/OPC) it appears there is insufficient alkali to hydrate the BFS which is largely acting as a filler. A higher OPC content may therefore reduce the long term availability of water to support corrosion.

Table IV Calculated distribution of water between pore populations at 90 days

Trial	Gel pores (vol. %)	Meso-pores (vol. %)	Capillary pores (vol. %)	Evaporable (vol. %)
3:1 BFS/OPC, 20°C	22.7	14.3	0.3	37.3
3:1 BFS/OPC, 60°C	14.5	20.3	2.1	36.9
9:1 BFS/OPC, 20°C	8.6	28.8	5.5	42.9
9:1 BFS/OPC, 60°C	5.7	22.7	13.9	42.3

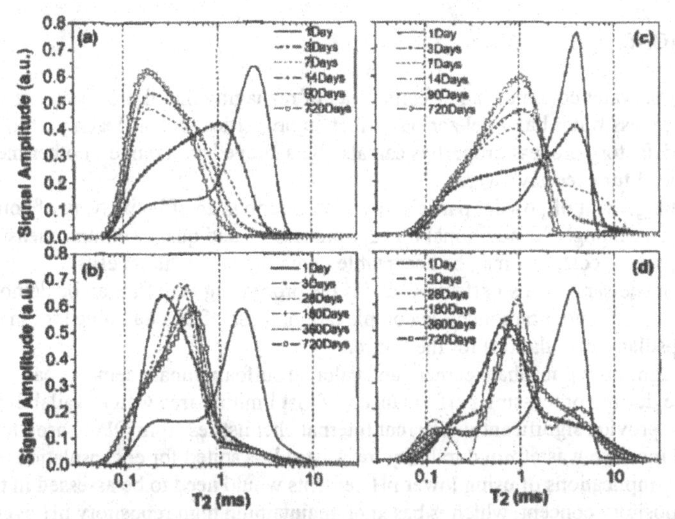

Figure 3. **^{1}H T_2 relaxation time distributions using an echo time of 25μs. (a) 3:1 BFS:OPC blend hydrated at 20°C, (b) 3:1 BFS:OPC blend hydrated at 60°C, (c) 9:1 BFS:OPC blend hydrated at 20°C, (d) 9:1 BFS:OPC blend hydrated at 60°C**

NMR Results for CSA/$CaSO_4$

Initially, due to a high w/s ratio, a large proportion of the water is present in large capillaries, giving the peak at 10 ms after 1 day (Fig 4). During curing there is gradual reduction in the large capillary pore water to water in pores with a considerably lower diameter. This is unlike OPC systems where a continuum of pore sizes exists. Some very small pore sizes are present at later times although capillary water remains significant. It should be noted that the hydration mechanism and products of CSA cements are different to those of conventional cements, with the main product being ettringite with a long needle-like structure, rather than C-

S-H fibrous growth. This differing geometry may contribute to the differences in the relaxation time distributions measured for CSA blends compared with OPC blends.

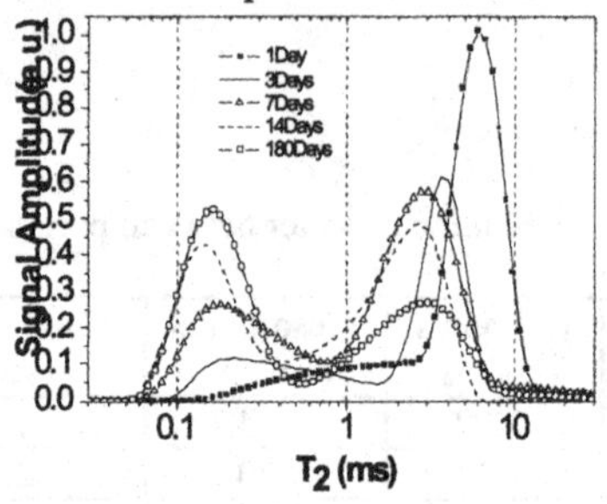

Figure 4: T_2 relaxation time distributions recorded ^{1}H NMR relaxometry using an echo time of 25 μs for a 3:1 CSA:$CaSO_4$ blend hydrated at 20°C.

CONCLUSIONS

Control of water content is an effective way of managing initial processing properties, but unnecessary excess water has a deleterious effect on properties such as leach performance. However, satisfactory process properties can also be achieved by a range of other methods without the need for excessive water.

The management of powder particle size distribution is an effective way of controlling water demand. Although the additional powder processing and quality control measures will increase the process cost, this may be acceptable given the potential benefits.

Superplasticisers are also effective, and if the processing benefits can be demonstrated not to be offset by preferential radionuclide complexation, these offer a valuable tool in design of cement encapsulant formulations for the future.

Both magnesium phosphate cement and calcium sulfoaluminate cements can be formulated to achieve the desired processing performances, whilst limiting free water availability. In addition, they provide significantly different internal chemistries from OPC-based formulations, which would increase wasteform durability for wastes less suited for encapsulation using OPC. However, the implications of using lower pH cements would need to be assessed in the context of the UK repository concept, which is based on maintaining high repository pH over a long timescale.

Proton NMR relaxometry has provided useful insight into the changing distribution of water between pore populations, as OPC- and CSA-based cements hydrate.

ACKNOWLEDGEMENTS

This paper includes results from studies carried out on Nexia Solutions' behalf by Prof N Miles, Nottingham University (powder properties), JP Gorce, Sheffield University (NMR) and M Hayes, previously Nexia Solutions, now AMEC (alternative cements). The authors are grateful to the NDA for financial support for the NMR relaxometry work.

REFERENCES

1. R Sambell, C Smitton and A Elsden, Nucl. *Chem. Waste Manage.*, **3**, 125-129 (1982)
2. M Hayes and IH Godfrey, *Waste Management 07, Feb 25 – March 1 2007, Tucson, AZ*
3. Q Zhou, NB Milestone and M Hayes, *J. Hazardous Materials*, 136, 120-129 (2006)
4. JP Gorce and NB Milestone, *Cement and Concrete Research*, 37 (2007) 310–318

Mater. Res. Soc. Symp. Proc. Vol. 1107 © 2008 Materials Research Society

Evaluation of Durability of Mortars and Concretes Used in Ancient Structures

A. S. Aloy[1] (Khlopin Radium Institute, St. Petersburg, Russia, 28, 2-nd Murinsky pr.)
J.R.Harbour[2], E.W.Holtzscheiter[2], C.A.Langton[2] (Savannah River National Laboratory, Aiken, 29808, USA)

ABSTRACT

The data on historic mortars and concretes provide qualitative and quantitative information to evaluate long-term behavior of cement materials in repositories and to understand processes that may occur in repositories (e.g., interaction with other materials and radionuclide transfer). Beyond that, such information is important to demonstrate safety aspects of the repositories to the public and stakeholders.

A number of reports have been devoted to study of historical mortars and concretes used in the Western countries. The purpose of this paper is to review studies on compositions and structures of analogs, located mainly over the former Soviet Union's territory.

INTRODUCTION

The current safest and technically feasible method for long-term containment of radioactive waste is considered to be its disposal in geological repositories, i.e., subsurface repositories for low-activity and intermediate-activity waste, and deep geological repositories for high-activity and some types of intermediate-activity waste.

The durability of waste containment in repositories depends on the conditions of the entire containment system. This system includes engineering barriers (a matrix that reliably retains radionuclides, a container for packaging waste, and a buffer material), as well as natural barriers (rock materials of the repository). This system should be selected (or designed) to assure the required containment of hazardous radionuclides for the required period of time stipulated in the major radiation safety requirements.

In their laboratory and full-scale studies, scientists evaluate natural analogs to assess stability and durability of cement materials intended for use in the waste repositories.

Studies of natural analogs allow evaluation of available natural conditions and materials, thereby providing information applicable for the waste repositories. Another aspect of natural analog studies is related to studies of man-made materials existing under natural conditions during the time comparable with the time planned for the period of waste disposal in repositories. Such analogs are defined as archeological or industrial analogs [1]. Regarding cements and concretes, historic mortars and concretes can be considered to be archeological analogs.

The data on historic mortars and concretes which can provide qualitative and quantitative information to evaluate long-term behavior of cement materials in repositories and to understand processes that may occur in repositories (e.g., interaction with other materials and radionuclide transfer). Beyond that, such information is important to demonstrate safety aspects of the repositories to the public and stakeholders.

Historic mortars and concretes include ancient (dating back to antique times) and medieval mortars and concretes, as well as old "modern" concretes and cements.

BINDING MATERIALS USED IN CONSTRUCTION IN ARMENIA IN THE FIFTH-EIGHTH CENTURIES

Binding materials used for medieval construction in Armenia were described in [2], specifically, the masonry mortar used in structures from the fifth-eighth centuries (Table I). These were lime-based mortars with various locally found additives, such as tuffs, pumice, quartz, limestone, and basalt, used as filler or aggregate.

The porosity of the samples was found to be in the range of 27.6-52.8%, density ranged from 1.1 to 1.7 g/cm^3, and the water saturation from 17.2 to 45.1%. Therefore, long-lasting and durable lime mortars appeared to be relatively light and porous materials with the macro-hardness of 3 by Mohs' scale. Micro-hardness of the binding material, filler, and the interface between them varied. In many cases, micro-hardness of the interface either exceeds micro-hardness of the filler (for example, in sample # 1, micro-hardness in the interface, filler and the binder was 119,254, and 309 kg/mm^2, respectively), or it was in between the hardness of the binder and filler, which meant better durability of the reaction layer that had been formed with time on the interface between the filler and the binder.

The results of the differential thermal analysis (DTA) of all samples indicated hygroscopic moisture and hydrates of amorphous silica (endothermic peak 150-170 °C), small amounts of either carboaluminates or garnet hydrates (endothermic peak 300-400 °C), silicate hydrates (endothermic peak~720 °C), and a large amount of carbonates (endothermic peak 750-840 °C).

The pH values of water taken from the samples (7.9-8.9) demonstrate a fairly deep carbonation of the mortars. The chemical analysis of the samples showed that the processes that occurred in those binding materials resulted in generation of significant quantities of carbonates and compounds such as calcium silicate hydrates (Table II).

The authors concluded that the major reason for durability of the analyzed mortars had been an adequate selection of formulation for preparation of the mortars that, due to a very long carbonation and reactions in the filler-binder interface, assured a high weather resistance.

Table I. Locations of Sampling and Age of Structures [2]

Sample #	Location of Sampling	Time of mortar placement	Notes
1	Masonry mortar in Pogos Petros church in Erevan	5th-6th century	
2	Masonry mortar in Ererujskaya three-aisle church, 4 km away from Ani near the village of Anipezma (former Bagratide capital)	5th century	Currently ruins
3	Masonry mortar of Karnautskaya (Diraklarskaya) single-aisle church, 8 km away from Leninakan, in the village of Karnaut	5th century	Ruins
4	Masonry mortar of Avansky single-aisle cathedral, 6 km away from Erevan	590-609	

5	Masonry mortar of Katukhki church in Erevan	13th century	
6	Masonry mortar in Dvinsky cathedral in Dvin (ancient capital of Armenia)	608-615	
7	Masonry mortar in a cathedral in the village of Ptgni of Kotajsky region, 18 km away from Erevan	6th century	Dilapidated
8	Masonry mortar of Grande Talinsky domed cathedral	7th century	The design is similar to Dvinsky cathedral, but preserved slightly

Table II. Results of Chemical Analysis of Samples

Sample#	Chemical composition, %										
	SiO_2 soluble	Fe_2O_3	Al_2O_3	Insoluble residue	CaO	MgO	Na_2O	K_2O	SO_2	Loss by calcinations (850 °C)	Total
1	9.85	1.48	6.22	39.34	21.27	1.14	0.67	0.11	1.61	17.98	99.67
2	10.80	1.08	6.77	31.42	23.32	1.63	0.61	0.53	0.89	22.22	99.27
3	9.25	1.64	5.86	47.26	16.81	1.05	0.86	0.42	0.31	15.98	99.44
4	6.30	2.40	5.86	40.50	22.09	2.42	0.55	0.25	0.20	18.82	99.33
5	8.40	2.36	6.29	35.56	23.29	1.47	-	-	0.27	20.66	98.30
6	4.40	1.68	3.82	18.44	27.57	2.68	0.40	0.32	29.57	11.16	100.04
7	17.85	3.12	7.33	22.0	23.05	2.82	0.47	0.27	0.86	21.18	99.85
8	2.25	0.92	2.43	77.92	6.86	0.61	0.16	0.05	0.17	8.56	99.83

LIME-BASED MORTARS OF THE SIXTH-SEVENTH CENTURIES IN ARMENIA, GEORGIA, AND UKRAINE

Reference [3] discusses physical and chemical properties contributed to durability of some lime-based mortars of the medieval period. Table III provides brief review of compositions and physicochemical properties of the analyzed lime-based mortars.

Table III shows that, although many various recipes preparation of durable mortars were used at that time, the phase composition of their hardening (regardless of climatic zones and time of construction) is fairly uniform. All of them contained calcium and magnesium carbonates, calcium silicate hydrates, calcium aluminate hydrates, calcium aluminosilicate hydrates, and amorphous silica, resulting from various reactions among mortar and cement components and between mortar and cement components and the environment.

Table III. Compositions and Physicochemical Properties of Lime-Based Mortars and Concretes [3]

Name of structure	Date of construction (century)	Ratio		Volume weight, g/cm^2	Water absorption, %	Porosity, %	Compression strength, kg/cm^2	Hardness on scale of 10
		binder	filler					
Masonry mortar of Kelasurskaya Wall (near Sukhumi)	2th	1	3.5	1.73	18.0	31.3	108	3
Masonry mortar of Karnautskaya church (near Leninakan, Armenia)	4th-5th	1	3.8	1.6	21.0	33.5	-	3
Masonry mortar of Kelasurskaya Wall (Sukhumi)	6th	1	2.17	1.49	25.9	38.7	102	3
Stucco of Golden Gates in Kiev	9th	1	3.5	1.58	24.4	38.6	100	2.5
Masonry mortar of Tamara`s Fortress (Military Georgian Road, 12 km from Tbilisi)	12th	1	2	1.85	14.73	27.3	108	-
Masonry mortar of minaret Juma (Baku)	14th	1	1.5	1.92	13.2	25.5	-	3.5
Stucco of monastery (Kharkiv, Ukraine)	17th	1	3.5	1.37	33.9	46.6	32	2.5
Bell tower stucco (Kharkiv, Ukraine)	Early 20th	1	2.75	1.81	17.2	31.0	311	7.5

Reaction rims (or fringes) consisting of newly formed compounds were observed in the binder-filler interface. Those fringes primarily consisted of calcite, calcium silicate hydrates, and calcium aluminate hydrates and hydrated silica. In many cases, the reaction rims are connected with one another and, also, with the binder and the filler by "tongues," which significantly improves their bonding. The concentric layered structure of the fringe confirms its gradual wavelike formation. The micro-hardness of the interface was higher than that of the binder by, on the average, 50%. The interface layer had a tendency of hardening with time. The quantity and size of the reaction fringes are proportional to the age of the mortars, amount of soluble silica, and activity of the silicon-containing filler. The reaction fringes in lime-based mortars usually form around grains of the fillers, with the grain size of 0.1-0.2 mm.

The size of the reaction fringe as a function of the mortar age was also described in [4] and [5] (Figure 1).

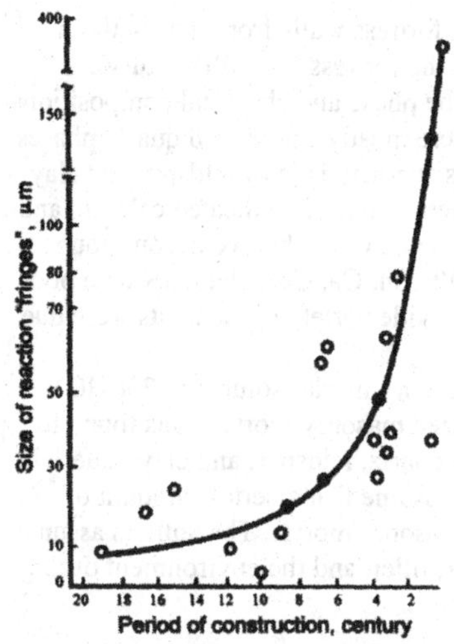

Figure 1. Sizes of Reaction Fringes as a Function of Lime-Based Mortar Age [4]

The macrostructure of long-lasting mortars and concretes and their durability vary significantly. Their open porosity ranges from 8 to 50%, with the average value of about 30%. A wide range of these parameters shows that they should not be considered criteria for durability and that increased porosity does not prevent durability (under normal weather conditions).

The authors observed some connections between durability of the lime mortars and their pH and electric conductivity: pH tends to decrease and electric conductivity tends to slightly increase.

The data provided are in a good agreement with theoretical assumptions on chemistry and thermodynamics of long-term hardening of lime-based mortars [6]. According to the theory, calcium and magnesium carbonates [7] are considered to be the most stable products of the reactions associated with long-term hardening. Calcium silicate hydrates may react with carbonic acid and form calcite and amorphous silica can also be considered stable [8].

External and internal factors determine kinetics of long-term hardening of lime-based mortars. Carbonation can be considered the main external factor, while interactions among components of the mortars can be considered the major internal factor. A very insignificant speed of interaction between the binder and silica or other fillers resulted in the term "inert fillers." This term is occasionally used even today, although, D.I. Mendeleev [9] mentioned long ago that completely inert fillers did not exist.

MASONRY MORTARS IN AZERBAIJAN, TWELFTH-FOURTEENTH CENTURIES

Chemical interactions in the lime-based mortars associated with hardening under natural conditions during a number of centuries are described in [10]. For this purpose, the authors analyzed samples of the masonry mortars of medieval structures in the city of Gyanji

(Azerbaijan): walls of a house (twelfth-fourteenth centuries), fortress walls from the southern part of the fortress and fortress walls from the eastern part of the fortress (twelfth century).

XRD, DTA, and spectral analysis were used to evaluate the phase and chemical compositions of the masonry mortars. According to the XRD data, there were mostly calcite and quartz phases in the samples. The X-ray diffraction patterns contained lines characteristic of feldspar and clay materials (d=13.77 A, 11.84 A, 9.18 A, and 7.00 A). The spectral analysis indicated calcium and silicon as the major elements. Beyond that, the following elements were observed: iron group elements (Cr, Mn, Ti, Ni, Co, V, Se), chalcophile elements (Pb, Sn, Ga, Cu), elements with poor volatility (Be, Y, Yb, Mo, Zr), and alkaline earth elements. A wide variety of elements were due to the presence of significant quantities of clay materials.

The dry residue obtained by evaporation of part of the masonry mortar soluble in 3% HC1 was subject to spectral analysis. The soluble part of all analyzed masonry mortars was found to contain 9.3-11 % of silicon. Taking into account the fact that quartz, feldspar, and clay materials do not dissolve in 3% HC1 under normal conditions, we can assume that a certain amount of silicon is present in the soluble compounds of the hardened masonry mortar. The authors assume that these compounds resulted from interactions of the binder, filler, and the environment during a long period of time.

MORTARS IN EARLY MONUMENTAL CONSTRUCTION IN ST. PETERSBURG, RUSSIA

In St. Petersburg, Russia, early cathedrals and public buildings date back to earlier than Middle Ages, however, lime-based binding materials were used for their construction, rather than Portland cement. For this reason, we decided to include discussions on those types of mortars in this section of the report.

Mortar samples from walls, ceilings, roofs, and foundations from Peter and Paul Cathedral of Peter and Paul Fortress (1712-1733), Gostiny Dvor Department Store (1761-1775), and St. Isaac Cathedral (1818-1858.) were analyzed (a total of 13 samples [11]).

Out of 13 analyzed mortar samples, 12 mortars were made of hydraulic lime, $Ca(OH)_2$ (9 from a strongly hydraulic lime), and only one mortar was made of thin air hardening lime. The almost exclusive application of hydraulic limes for monumental ground-based construction, in spite of the fact that air hardening lime was also known at that time, can be explained by considerations of strength and durability. This approach appeared more advanced than the Western European approach where, at that time and later, until Parker's invention, the use of air hardening limes with hydraulic additives prevailed.

The fillers in St. Petersburg's mortars mostly contained grains of quartz and various feldspars. Sometimes grains of various minerals, clays, chunks of granite and other rocks were used. Apart from natural fillers, some mortars contained pozzolans and carbonates. However, only one type of mortar, with the most hydraulic lime, contained only natural fillers.

A significant amount of non-carbonated calcium silicate hydrates (aluminate hydrates) were found in all thirteen mortars. Carbonation in these relatively young mortars (137-245 years old at the time of analysis) is very far from being complete. At first sight, distribution of free lime seems unexpected in mortars of different ages. Calcium hydroxide as found in three mortars of the aboveground parts of Peter and Paul Cathedral, the earliest structure, and only in one mortar of the underground foundation of Gostiny Dvor. No calcium hydroxide was found in mortars used for construction of St. Isaac cathedral, the latest structure. Thinness of seams and homogeneity of the mortar used for construction of Peter and Paul Fortress and the fact that the Gostiny Dvor mortar is located 1.5 meters underground, makes it difficult for CO_2 of air access the mortars, thereby causing retardation of $Ca(OH)_2$ carbonation.

To a certain extent, the binding material was found to interact with fillers in all analyzed mortars. Grains of quartz, feldspars, and, especially, pozzolans are surrounded by reaction fringes that can be clearly seen under magnification (Figure 2). Pozzolans have the widest fringes with a zoned structure, from a gradual transition from the periphery part of the filler grains to the binder. The reaction products are not only present in the fringes on the surface of the grains, but also in pores where calcium hydroxide enters during mixing.

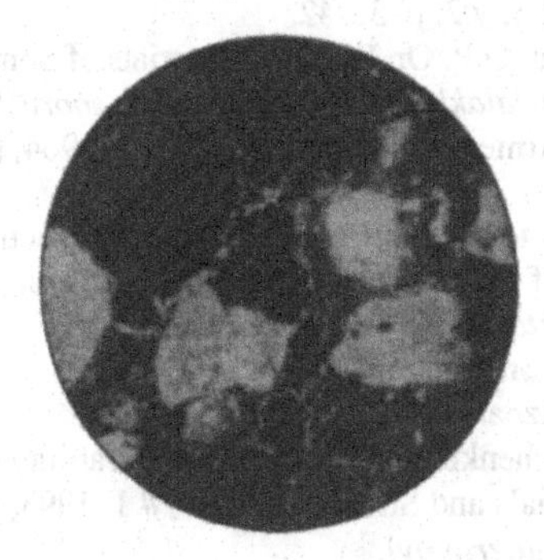

Figure 2. Microstructure of Mortar from Ceiling of Peter and Paul Cathedral, ×60, Nicoli+ [11]

CONCLUSION

Studies of the composition of ancient mortars and concretes show that, with a wide variety of compositions for long lasting lime-based mortars, the phase composition of long hardening products is similar. These products contain calcium (and magnesium) carbonate and silicate hydrates in much smaller quantities. Calcium aluminate hydrates and calcium aluminosilicate hydrates can be occasionally present, as well as amorphous silica. All these compounds resulted from reactions of the components of mortars and concretes with one another and with the environment. The mineralogical composition of historical lime mortars is similar to modern pozzolan cements. This important observation makes it possible to draw a parallel between ancient lime mortars and modern cements.

The microstructure of the artificial stone that hardened for a long time shows new-layered formations on the interface between the binder and the filler (pozzolan or inert filler). In the course of centuries, these new formations improve micro-hardness of the interface by 50% above average, with the tendency to increase with time. The size and quantity of layers are proportional to the age of the mortar and the amount of active silicon dioxide in the fillers.

Based on the study for ancient and medieval lime-based mortars and cements from Western countries [12] and results of this review it is possible to say that a role of the initial compositions for durability of the materials was not of primary importance. The major physicochemical reasons for their durability were associated with their phase compositions, as well as macro- and microstructure that depended on interactions of the mortar components with one another and with the environment.

Regarding modern cements, the processes inside the cement block during a long-lasting hardening make them more durable and impermeable due to pozzolanic reactions. On their surfaces, low permeability is explained by generation of a protective carbonate layer.

REFERENCES

1. Miller B., Chapman N., Postcards from the Past: Archeological and Industrial Analogs for Deep Repository Materials/Radwaste Magazine, 1995, v.2, p. 32-42.
2. Papkova L.P., Gevorkyan K.O., Mchedlov-Petrosyan O.P. On Binding Materials of Some Ancient Structures in Armenia. *(O vyzhutscikh materialakh nekotorykh drevnikh sooruzhenij Armenii. I* Publications of Academy of Science of Armenian Republic, XVII, #2, 1964, pp. 61-65.
3. Papkova L.P. Studies of Physico-Chemical Reasons for Durability of Certain Construction Materials. Abstract of Thesis, Charkiv, University of Charkiv, 1966, p.22 *{Issledovanie phisiko-khimicheskikh prichin dologevechnosti nekotorykx stroitel;nykh materialov).*
4. Mchedlov-Petrosyan O.P. Chemistry of Inorganic Construction Materials *{Chimiya neorgnicheskikh stroitel;nykh materialov)* M.: Strojizdat, 1988, p. 250.
5. Mchedlov-Petrosyan O.P., L.P. Papkova, V.G. Dyachenko et al. Reasons for Durability of certain Construction Materials / Construction Materials and Structures, Kiev, # 1, 1993, p. 12 *{Prichiny dolgovechnosti nekotorykx stroitel;nykh materialov).*
6. INTERNATIONAL ATOMIC ENERGY AGENCY, «Improved cement solidification of LOW and INTERMEDIATE level radioactive wastes», Technical Reports Series № 350, p.12, IAEA, Vienna (1993).
7. Naumov G.B., Ryzhenko B.N., Khodakovsky I.L. Reference book on Thermodynamic Values/M.: Atomizdat, 1971 *(Spravocnik termodinamicheskikh velichin).*
8. Babushkin V.I., Matveev G.M., Mchedlov-Petrosyan O.P. Thermodynamics of Silicates. 3d edition./ M.: Strojizdat, 1972, Vol. 2, p. 350 (Termodinamika silikatov).
9. Mendeleev D.I. Foundations of Chemistry *{Osnovy khimii).* M.-L., 1928, Vol. 2, pp. 94-116.
10. Tejmurov G.S. Physico-Chemical and Chemico-Spectral Studies of Masonry Mortars of Some Medieval Structures in Gyanji/ Chemistry of Azerbaijan, # 2,1988, p. 149-153. *{Physico-chimicheskie i khimiko-spectral'nye issledovaniya kladochnyx rastvorov nekotorykh srednevekovykh sooruzhenij Gyanji).*
11. Znachko-Yavorskij I.L. Essays on History of Binding Materials from Ancient Times to Mid 1800's. *{Ocherkipo istorii vyazhutscikh materialov s drevnejshikh vremen do serediny devyatnadtsatogo veka).* Publishing House of USSR Academy of Science, M.-L., 1963, pp. 300-306.
12. Lagerblad B., TrSgardh J. Conceptual Model for Concrete Long Degradation in a Deep Nuclear Waste Repository / SKB Techn. Rep. 95-21,1994, Stockholm, Sweden.

Mater. Res. Soc. Symp. Proc. Vol. 1107 © 2008 Materials Research Society

The Immobilisation of Clinoptilolite Within Cementitious Systems

L. E. Gordon[(a)]*, N. B. Milestone[(a)], M. J. Angus[(b)]
(a)* Immobilisation Science Laboratory, Department of Engineering Materials, University of Sheffield. Mappin Street, Sheffield, S1 3JD , UK. Phone: + 44 (0) 1142225973, Fax: + 44 (0) 114 222 5943, E-Mail: l.gordon@sheffield.ac.uk
(b) Nexia Solutions, Sellafield, Cumbria, UK.

ABSTRACT

The zeolitic ion exchanger clinoptilolite was encapsulated within various cementitious systems in order to assess their suitability for the retention of the radioelements, Cs and Sr. The pozzolanic reaction of clinoptilolite is reduced in composites containing BFS and PFA and appears not to continue after 7 days of hydration. $Ca(OH)_2$ persists up to 360 days of hydration in a 9:1BFS:OPC system with 10% clinoptilolite added, despite the presence of unreacted pozzolan. This may be due to low pH of the pore solution, if Na and K act as counter cations in the aluminous C-S-H, a product of pozzolanic hydration or are exchanged onto the clinoptilolite. Saturation of the pore solution with Ca may prevent further dissolution of $Ca(OH)_2$. Cs leaching occurs in all samples during accelerated tests due to breakdown of the clinoptilolite structure.

The alternative cement system calcium sulfoaluminate cement (CSA) has a different hydration chemistry and properties to OPC and OPC composites with a lower pore solution pH. Clinoptilolite appears to react in a hydrating CSA system with significant reaction continuing between 28 and 90 days of hydration. Leaching of Cs from CSA is higher than from an OPC system, in which almost all of the clinoptilolite crystallinity is lost. The major product of CSA hydration is ettringite. Cs may be adsorbed within cation sites of the C-S-H in an OPC system but not by ettringite which does not retain Cs so Cs has high mobility and leachability through the CSA matrix.

INTRODUCTION

^{137}Cs and ^{90}Sr can be successfully exchanged on a naturally occurring zeolite, clinoptilolite in a clean up process at Sellafield, UK. No decision has yet been made on the encapsulation route for this ILW, although cementation is the accepted route for ILWs in the UK. Encapsulation of zeolites within a traditional OPC system results in a pozzolanic reaction[1,2] where the crystalline aluminosilicate framework reacts with $Ca(OH)_2$[3]. This work considers the potential encapsulation of clinoptilolite in two types of composite cement formulations based on fly ash and blast furnace slag where high levels of replacement of OPC with fine pozzolans may reduce the pozzolanic reaction of the encapsulated zeolite, and encapsulation within an alternative cementitious system, calcium sulfoaluminate cement, where the hydration reactions are different from those of the OPC based composites.

The major products of CSA hydration are crystalline ettringite and $Al(OH)_3$.[4] The factors that may enhance the stabilisation of clinoptilolite in CSA cements relative to OPC based systems are: Firstly, the pH of the pore solution in CSA cements is reported to be between 10 and 11-5[3], whereas in OPC based systems the pH is >12. Secondly, the production of ettringite quickly binds free pore water reducing the potential for diffusion of

OH- ions and thus alkaline attack on the clinoptilolite framework. CSA cements hydrate quickly which improves the early mechanical properties.

EXPERIMENTAL

Various OPC composite cement formulations containing clinoptilolite were mixed in late 2003 allowing long term studies.. OPC, PFA and BFS were also supplied by BNFL and meet their specifications for encapsulation use at Sellafield. The clinoptilolite used comes from the Mojave desert, USA and was received from Sellafield as part of their shipment. It has a median particle diameter of 628μm. Ion exchange with Cs was conducted by submerging clinoptilolite in a 1M CsCl solution with a 1:4 solution:clinoptilolite ratio for 48 hours at 60°C. The clinoptilolite was then dried and the process repeated.. Table 1 gives the compositional details for each of the raw materials used. CSA was supplied from Tanshan Polar Bear special cement in China.

The cement formulations investigated for this study were OPC, 3:1PFA:OPC, 9:1 BFS:OPC, 3:1BFS:OPC and 1:1BFS:OPC and CSA 1 kg (solid) mixes were made containing 10 wt% clinoptilolite, mixed with a water/binder (w/b) ratio of 0.35 within a Kenwood chef mixer for 15 minutes. Samples were cured at 20°C at 95% humidity within 50ml tubes. For CSA a w/b ratio of 0.5 was used. At specific curing times samples were removed, crushed and hydration was quenched with acetone before vacuum drying.

XRD analyses of the composite cement samples cured for 3, 7, 28, 90 and 360 day are reported. Leaching analysis was carried using two methods. One was soxhlet extraction where 6g of sample with a particle size of 1-2mm diameter was continually replenished with distilled water at 75±5°C for 24 hours (24 cycles). The other was a method of in pot leaching where 6g of sample with a particle size of 1-2mm diameter was placed in 40ml of distilled water at room temperature and continually rotated for a period of 24 hours, followed by analysis of Cs concentration in the leachate by flame atomic emission spectroscopy.

Component	Clinoptilolite (%)	Cs-clinoptilolite (%)	Component	OPC (%)	PFA (%)	BFS (%)	CSA (%)
SiO_2	67.81	57.68	SiO_2	20.96	54.28	34.5	18.25
Al_2O_3	21.54	18.35	Al_2O_3	5.24	25.89	13.74	22.93
Ca	1.27	0.18	CaO	64.58	1.44	42.1	33.23
Na	3.05	0.82	Fe_2O_3	2.61	6.43	0.97	5.94
Mg	0.39	0.11	MgO	2.09	1.41	7.29	2.00
K	3.75	2.74	SO_3	2.46	1.12	0	~15
Sr	0.25	0.02	K_2O	0.59	2.83	0.49	0.65
Cs	0.02	17.09	Na_2O	0.28	1.91	0.22	0.14

Table 1. Compositional details (weight %) of the cement raw materials and the clinoptilolite before and after Cs exchange.

RESULTS

Pozzolanic reaction occurs in a pure OPC system. The only peak on the XRD traces of OPC containing 20% clinoptilolite was the broad, low intensity peak for clinoptilolite at 22.5°2θ (figure 1). The intensity of this peak did not change between 3 and 360 days of hydration indicating that pozzolanic reaction of the clinoptilolite occurs in the early stage of hydration (<3days).

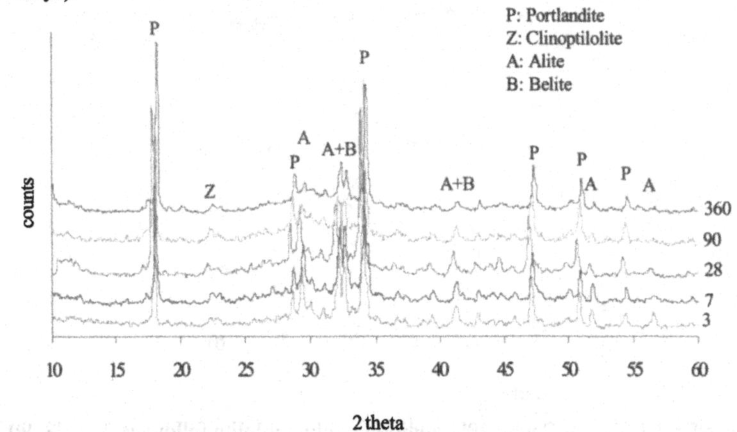

Figure 1. XRD traces of OPC with 10wt% clinoptilolite following 3, 7, 28, 90 and 360 days of hydration at 20°C. (the peak at 44.5°2θ in the 28 days sample is due to Al in the sample holder)

When clinoptilolite was encapsulated within a 3:1 PFA:OPC system (figure 2), the major clinoptilolite peak at 22.5°2θ had an increase in intensity relative to that of the OPC system, (figure 1), indicating that reduced pozzolanic reaction of the clinoptilolite had occurred. Once again, the intensity of the clinoptilolite peaks remained the same between three and 360 days of hydration indicating that further significant pozzolanic reaction of the clinoptilolite did not continue past three days of hydration. $Ca(OH)_2$ was detected by XRD up to 360 days of hydration despite the presence of unreacted pozzolan.

The BFS/OPC composites shown in the XRD traces in figures 3-5 also indicated that very little crystalline clinoptilolite remained in these systems with the only clinoptilolite reflection remaining the peak at 22.5°2θ. Again there is little change in pozzolanic reaction after 3 days of hydration. $Ca(OH)_2$ persists in all systems up to 360 days of hydration.

The XRD traces of the CSA system (figure 6) showed that ettringite had formed but this tends to mask the clinoptilolite peaks present, particularly the major clinoptilolite peak at 22.5°2θ. However, this peak in the samples is of low intensity and appears to reduce between 28 and 90 days of hydration suggesting that reaction of the clinoptilolite continues at later hydration ages than occurs in OPC and OPC based composites.

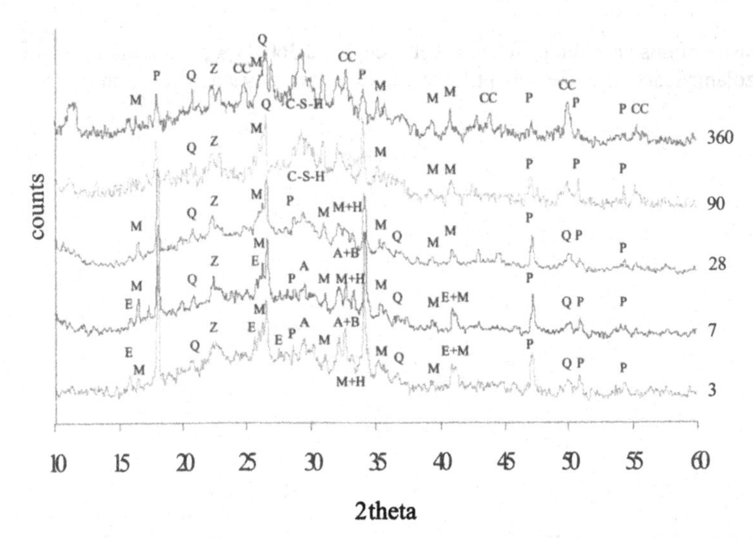

Figure 2. XRD traces for 3:1PFA:OPC with 10% addition of clinoptilolite following 3, 7, 28, 90 and 360 days of hydration at 20°C.
Key: A: Alite, B: Belite, CC: Calcium carbonate (vaterite), C-S-H: Calcium silicate hydrate, E: Ettringite, H: Hematite, M: Mullite, P: Portlandite, Q: Quartz, Z: Clinoptilolite

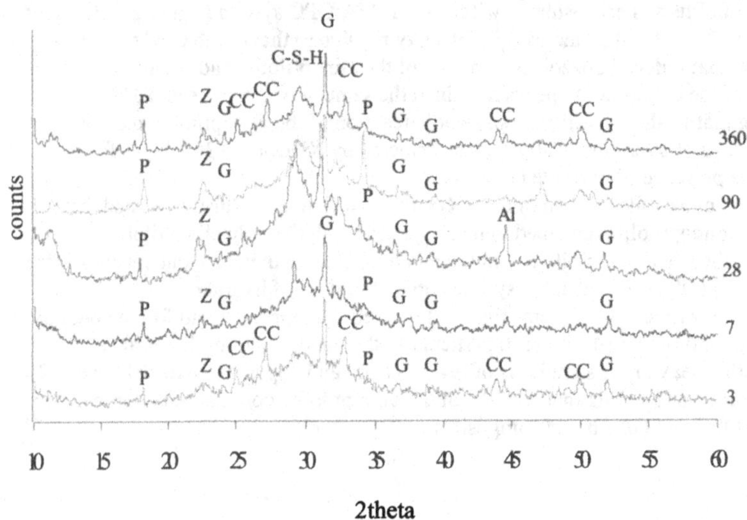

Figure 3. XRD traces of 9:1BFS:OPC with 10wt% clinoptilolite following 3, 7, 28, 90 and 360 days of hydration at 20°C.
Key: CC: Calcium carbonate (vaterite), C-S-H: Calcium silicate hydrate, G: Gehlenite, P: Portlandite, Z: Clinoptilolite

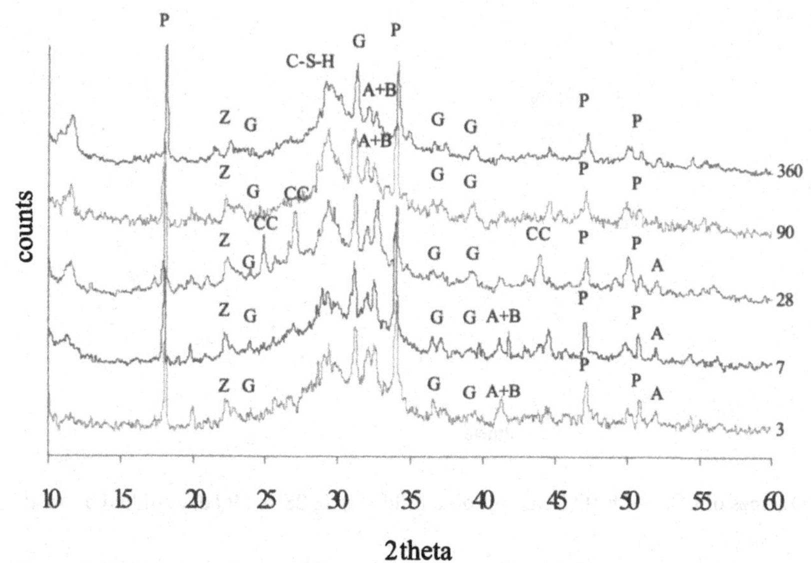

Figure 4. XRD traces of 3:1BFS:OPC with 10wt% clinoptilolite following 3, 7, 28, 90 and 360 days of hydration at 20°C.
Key: A: Alite, B: Belite, CC: Calcium carbonate (vaterite), C-S-H: Calcium silicate hydrate, G: Gehlenite, HY: Hydrotalcite, P: Portlandite, Z: Clinoptilolite

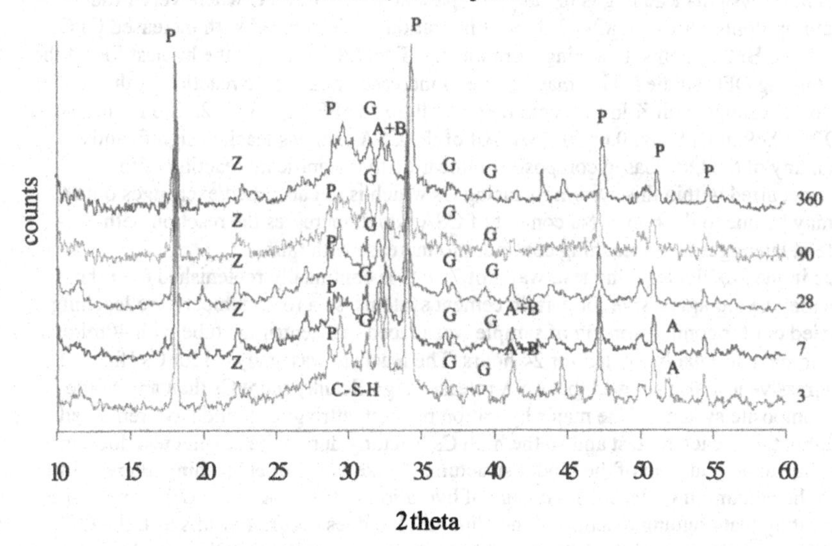

Figure 5. XRD traces of 1:1BFS:OPC with 10wt% clinoptilolite following 3, 7, 28, 90 and 360 days of hydration at 20°C.
Key: A: Alite, B: Belite, C-S-H: Calcium silicate hydrate, G: Gehlenite, P: Portlandite, Z: Clinoptilolite

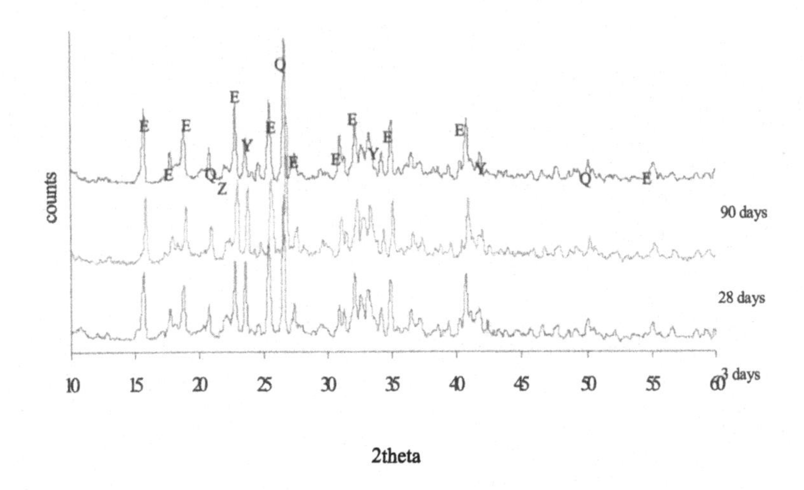

Figure 6. XRD traces of CSA with 10% clinoptilolite following 3, 28 and 90 days of hydration at 20°C.

Cs leaching from all the cement systems studied using soxhlet extraction is given in figure 7 and appeared to increase with age of hydration. This could be due to further reaction of the clinoptilolite at later ages with the increased temperature as $Ca(OH)_2$ is present. For all of the composite cement systems leaching is broadly comparable to that of OPC where very little crystalline clinoptilolite remained after 3 days of hydration but is reduced with increased OPC substitution in the BFS systems. Leaching from the 3:1PFA:OPC system is the highest for all the systems containing OPC studied. This may be due to increased pozzolanic reaction of the zeolite or possibly ion exchange with K in the cement pore solution as PFA contains 2.83% K compared to 0.59 in OPC, 0.49 in BFS and 0.65 in CSA. All of the CSA samples leached significantly more Cs than any of the OPC based composites indicating that significant reaction of the clinoptilolite occurred within this system releasing Cs which is not absorbed/exchanges onto C-S-H. This may be due to the higher Na content of CSA which increases the reaction with clinoptilolite at later ages as Cs leaching continues to increase with time.

Since in the Soxhlet leaching test, water of 75±°C is continually replenished over the sample, this may be too aggressive for certain cement systems, so a room temperature leaching test was carried out for comparison. 5g of sample was placed in a centrifuge tube with 40ml of distilled water and continually rotated for 24 hours. The leachate was analysed for Cs (figure 8). This less aggressive test also showed that CSA released significantly more Cs than any of the OPC based composite systems. The major hydration product, ettringite should have remained stable throughout this leaching test and so the high Cs leaching during the soxhlet test does not appear to be due to degradation of the binder structure. Unlike the soxhlet leaching, there is little difference in the amount of Cs leached with age of hydration in the CSA or the OPC composite systems indicating that ongoing reaction of the clinoptilolite does occur. For this test, the OPC system leached most Cs of all of the OPC composites As the amount of $Ca(OH)_2$ available for reaction with the clinoptilolite increased with increasing OPC present, the results followed this trend with the amount of Cs released increasing with the OPC proportion present.

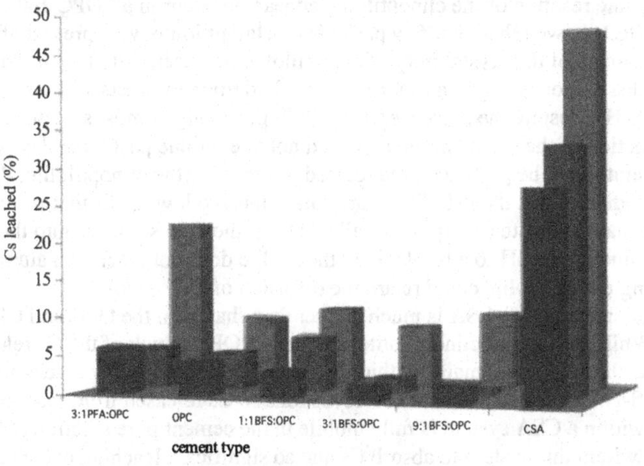

Figure 7. Cs leaching for all cement systems containing 20% Cs exchanged clinoptilolite following 3 (front), 28 (middle) and 90 (back) days of hydration at 20°C (unground clinoptilolite median particle diameter: 627.7μm) by soxhlet extraction.

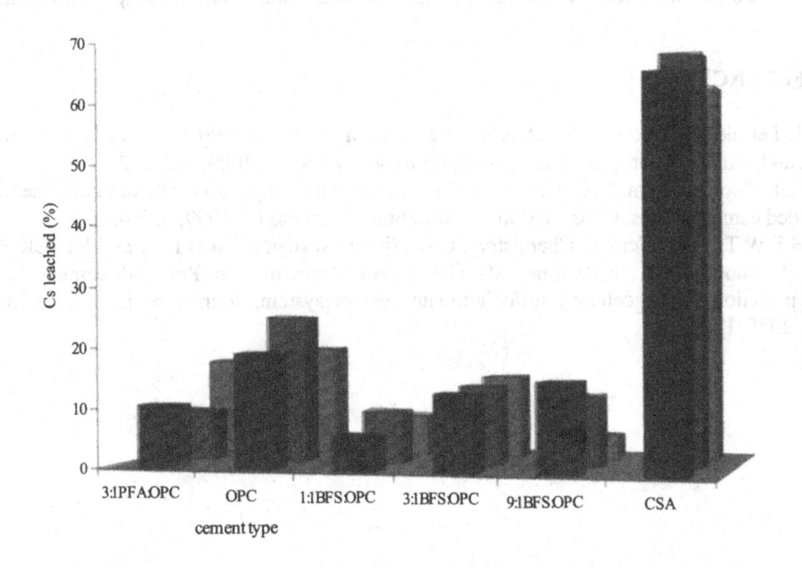

Figure 8. In pot Cs leaching for all composite cement systems containing 20% Cs exchanged clinoptilolite following 3, 28 and 90 days of hydration at 20°C (unground clinoptilolite, median particle diameter: 627.7μm).

DISCUSSION AND CONCLUSION

Pozzolanic reaction of the clinoptilolite appears to occur in all OPC and OPC based composites. XRD showed that very few peaks from clinoptilolite were present after 3 days, indicating that much of the crystallinity of clinoptilolite had been lost. The products from clinoptilolite dissolution will go on to form cement hydration products which will result in an aluminous C-S-H. Despite the presence of $Ca(OH)_2$ persisting in most systems to 360 days, pozzolanic reaction of the clinoptilolite appeared not to continue past three days of hydration at ambient temperatures. The presence of unreacted pozzolana, (the clinoptilolite, PFA and BFS components) indicates that the pH of the pore solution is too low for further pozzolanic reaction to occur. This may be due to adsorption of alkalis from the pore solution onto the aluminous C-S-H formed reducing the pH to a level where the zeolite does not react.. An aluminosilicate coating forming on the zeolite could retard the diffusion of OH^- ions.

Cs leaching from the CSA is much greater than that from the OPC and OPC based composites. While reaction of clinoptilolite is higher in OPC, much of the Cs released from the reaction of the clinoptilolite remains within the cement matrix, possibly associated with an aluminous C-S-H. Ettringite does not absorb Cs ions so Cs released from reaction of clinoptilolite within a CSA system is fully mobile in the cement pore solution as there is no C-S-H component within this system to absorb Cs and so significant leaching occurs..

ACKNOWLEDGEMENTS

The authors would like to thank the EPSRC and Nexia Solutions as sponsors of this project.

REFERENCES

[1] T. Perraki, G. Kakali, F. Kontoleon, The effect of natural zeolites on the early hydration of Portland cement, Microporous and Mesoporous Materials, 61, 2003, 205-212.
[2] C. S. Poon, L. Lam, S. C. Kou, Z. S. Lin, A study on the hydration rate of natural zeolite blended cement pastes, Construction and Building Materials, 13, 1999, 427-43.
[3] H F W Taylor, "Cement Chemistry", first edition, Academic Press Inc., London, UK, (1990).
[4] Q. Zhou, N. B. Milestone, M. Hayes, An alternative to Portland cement for waste encapsulation – The calcium sulfoaluminate cement system, Journal of Hazardous Materials, 136, 2006, 120-129,.

Mater. Res. Soc. Symp. Proc. Vol. 1107 © 2008 Materials Research Society

Investigation on the Immobilisation of Carbon in OPC-BFS and OPC-PFA Systems

Hajime Kinoshita[1], Paulo H. R. Borges[1], Claire A. Utton[1], Neil B. Milestone[1] and Cyril Lynsdale[2]
[1] Department of Engineering Materials, The University of Sheffield, Mappin Street, Sheffield, S1 3JD, U.K.
[2] Department of Civil and Structural Engineering, The University of Sheffield, Mappin Street, Sheffield, S1 3JD, U.K.

ABSTRACT

The reaction of CO_2 gas with OPC, OPC-BFS and OPC-PFA composite cement systems were studied using XRD, SEM and TG to investigate the applicability of these materials to immobilise carbon arising from graphite waste. XRD results suggested that calcite formed in OPC system after the carbonation reaction, whereas calcite and vaterite were observed in OPC-BFS and OPC-PFA systems. In OPC system, nearly half of $Ca(OH)_2$ was consumed to form $CaCO_3$. In OPC-BFS and OPC-PFA systems, the amount of $CaCO_3$ formed, corresponded to the consumption of greater than 100% of $Ca(OH)_2$ initially present, suggesting that other hydration products e.g. C-S-H were also consumed, either directly or indirectly during the carbonation process. The OPC-BFS system became more porous after carbonation. OPC-PFA system indicated a high efficiency on the conversion of Ca in the system into $CaCO_3$.

INTRODUCTION

A large amount of graphite waste exists and will increase in UK from the decommissioning of Magnox stations. It has been reported that such graphite can go through self-sustaining oxidation at temperatures of about 550°C [1], which would eventually produce CO_2 gas. It is known that the reaction of cementitious systems with CO_2 occurs even under ambient condition due to the high thermodynamic stability of the carbonates of alkali-earth elements e.g. $CaCO_3$ and $MgCO_3$. It may be possible to apply this reaction to immobilise the carbon originating from the Magnox graphite waste by deliberately oxidise graphite wastes into CO_2. The reaction between cementitious materials and CO_2 has been of interest, especially in recent years, due to its potential as a raw material for the sequestration of CO_2 [2-4].

Present work is a preliminary study focused on the reaction of CO_2 gas with neat ordinary Portland cement (OPC), OPC-blast furnace slag (BFS) and OPC- pulverised fuel ash (PFA) composite cement systems. These cementitious systems were selected as they are often used to immobilise low and intermediate level nuclear wastes [5].

Experimental

Experimental conditions are detailed in Table I. A neat OPC grout, a mixture of BFS-OPC (9:1 BFS:OPC) and that of PFA-OPC (3:1 PFA:OPC) were prepared, with a water to solids ratio (w/s) of 0.33. The grouts were moulded in prisms of 25 x 25 x 285 mm and cured at 40 °C

Table I. Experimental details.

Composition (wt%)	W/S ratio	Curing condition	Size	Carbonation condition
OPC=100 BFS:OPC=90:10 PFA:OPC=75:25	0.33	Temp: 40°C R.H.: 100% Time: 15months	Particles: 3.36-5.00mm Powder: <63μm	Temp: 25-30°C R.H.: 50-60% CO_2 conc.: 15% Time: 21days

with 100 % relative humidity for 15 months. All of the cured samples were crushed and sieved into either particles of 3.36-5.00 mm or powder of <63 μm prior to the carbonation. The carbonation was performed in a chamber filled with 15 % CO_2 at 25-30 °C with 50-60 % relative humidity for 21 days. The detailed chemical compositions of the OPC, BFS and PFA used in the present study are shown in Table II.

The carbonation products were identified by XRD using Siemens D500 Reflection Diffractometer with Cu $K\alpha 1$ and $K\alpha 2$ radiation (average λ = 1.54178 Å). Effect of the carbonation on the microstructure was studied via SEM in backscattered electron imaging (BSEI) mode, using a Camscan MkII. A Perkin Elmer Pyris1 TGA was utilised to estimate the amount of CO_2 immobilisation in the samples.

RESULTS AND DISCUSSION

XRD results

It is known that the reaction of CO_2 with cements produces $CaCO_3$ in the form of calcite [2-4]. Our XRD results for the neat OPC system (not shown in the present paper) indicated the presence of calcite in the system, before carbonation. After carbonation, a significant increase in calcite content in the OPC particle sample was not clear. In contrast, a significant increase in calcite content was observed in the powder sample, which is reasonable as carbonation should progress more rapidly in a powder due to the larger surface area.

Figure 1 (a) shows XRD patterns of the 9:1 BFS:OPC system. Differing from neat OPC, the 9:1 BFS:OPC system reacts well with CO_2. Even the particle sample formed a significant amount of $CaCO_3$ after carbonation. Two polymorphs of $CaCO_3$, calcite and vaterite (Figure 2 [6]), were found after carbonation both in particle and powder samples: vaterite was dominant in the former, and calcite in the latter. Vaterite is a metastable polymorph of the calcium carbonate. The mechanism of the formation of this hexagonal phase is still unclear, and various factors could contribute to its formation such as temperature and humidity [7]. It is also known that the

Table II. Chemical compositions of raw materials (wt%).

	CaO	SiO_2	Al_2O_3	Fe_2O_3	SO_3	MgO	K_2O	Na_2O	LOI*
OPC	64.58	20.96	5.24	2.61	2.46	2.09	0.59	0.28	0.73
BFS	42.10	34.50	13.74	0.97	**	7.29	0.49	0.22	1.05
PFA	1.62	49.53	26.45	8.70	0.88	1.56	4.58	<0.01	4.10

* Loss of ignition ** sample contains reduced sulphur

growth rate of calcite crystals is proportional to the number of growth sites, i.e. to the surface area of the particle, which is very rough and active if created by crushing [7].

Similar to 9:1 BFS:OPC system, the 3:1 PFA:OPC has both calcite and vaterite, as shown in Figure 2 (b). However, a significant amount of calcite existed even before carbonation, and the effect of carbonation appeared as the increase in vaterite content.

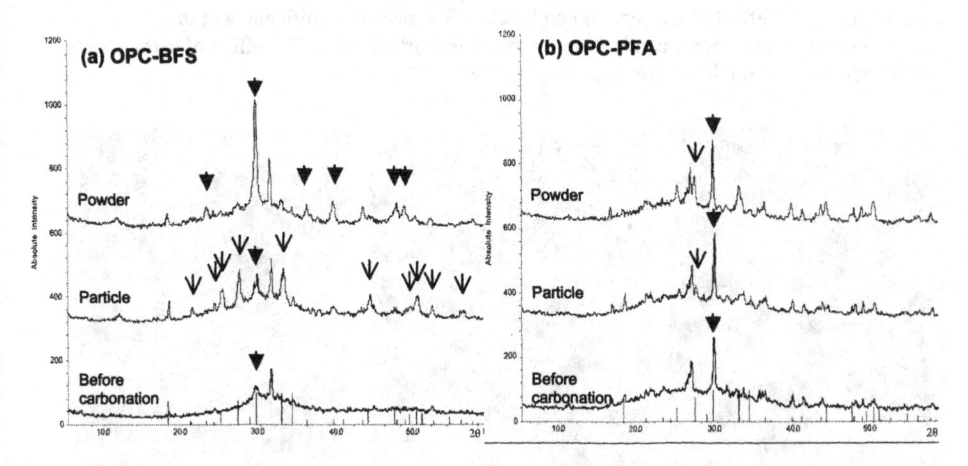

Figure 1. XRD patterns of (a) 9:1 BFS:OPC and (b) 3:1 PFA:OPC systems: ▼ and ⩔ indicate peak positions of two different polymorphs of $CaCO_3$, calcite and vaterite, respectively.

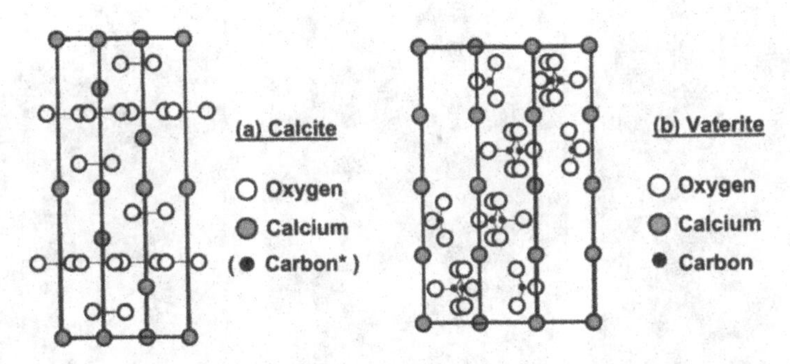

Figure 2. Schematic representation of the structural frameworks of (a) calcite and (b) vaterite after Maciejewski et al. [6]. Carbon atoms are not visible in the Figure (a) as they are located on the same plane as the oxygen atoms.

SEM / BSEI results

Figure 3 shows BSE images of particle samples before and after carbonation, obtained via the SEM observation. For the OPC system, the effect of carbonation was not too obvious although the microstructure seems less fine. $CaCO_3$ is larger than $Ca(OH)_2$ by 11.8 % in solid volume and should therefore be expected to fill the gaps existing in the microstructure [4]. On the other hand, the effect of carbonation on OPC-BFS system is significant, and the microstructure became significantly more porous after carbonation. The effect of carbonation was not too obvious on the microstructure of the OPC-PFA system.

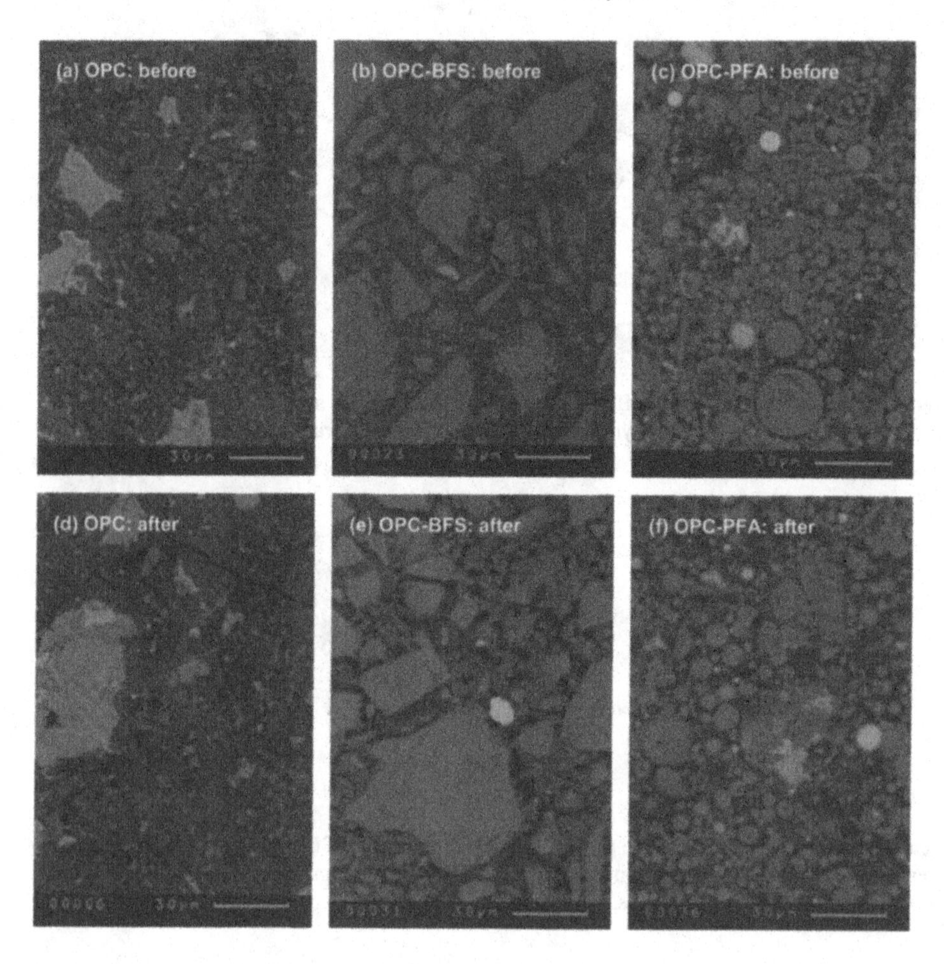

Figure 3. BSE images of samples before carbonation (top) and after carbonation (bottom): (a) and (d) OPC system, (b) and (e) OPC-BFS system, (c) and (f) OPC-PFA system.

TG results

Figures 4, 5 and 6 show TG curves of the powder samples before and after carbonation. In each figure, derivative thermogravimetric (DTG) curves are also shown. The DTG curves reflect changes in the slope of TGA curves as marked peaks and help to identify the different processes difficult to identify on TG curves [8]. DTG curves are not for the quantitative analysis.

TG/DTG curves for OPC system, both for before and after carbonation (Figure 4 (a) and (b)), indicate three main processes occurring upon heating, which corresponds to the previous studies in the literature [8-10]: the weight loss up to around 300~350 °C is attributed to the loss of water and dehydration of such hydrates as C-S-H gel (Calcium Silicate Hydrate) and AFt/AFm phases such as ettringite and monosulfate; the weight loss within a range of 450 - 550 °C is due to the dehydration of $Ca(OH)_2$; and weight loss due to the decomposition of $CaCO_3$ occurs with in a range of 600 - 800 °C. TG curve for the powder sample has a significant weight loss due to the decomposition of $CaCO_3$ as seen Figure 4 (b). The approximate content of $Ca(OH)_2$ and $CaCO_3$ was calculated both for before and after carbonation from the weight loss between 400 - 500 °C (loss of H_2O) and 580 - 780 °C (loss of CO_2), respectively, and indicated in Table III. The calculation results suggest that 4.1 g of CO_2 was absorbed by 100 g of OPC sample in the carbonation condition of the present study.

The effect of carbonation on TG of the OPC-BFS system was very clear, as shown in Figure 5 (a) and (b). Correspondingly, the calculation shows a large increase in amount of $CaCO_3$ (Table III). The calculation was performed in the same manner as for the OPC system: from the weight loss between 400 - 500 °C for $Ca(OH)_2$ and 580 - 780 °C for $CaCO_3$. The results suggest that 4.6 g of CO_2 was absorbed by 100 g of OPC-BFS sample. As the 5.2 g of $Ca(OH)_2$ originally existed per 100 g of OPC-BFS sample can absorb 3.1 g of CO_2, this calculation result suggests that other hydration products e.g. C-S-H were also consumed either directly or via conversion to $Ca(OH)_2$. The latter is more likely as the $Ca(OH)_2$ content increased after the carbonation process, thought to be the result of further hydration of the samples during the process due to the fresh surface area produced by the grinding process. The consumption of C-S-H may be the reason for the increase of porosity after carbonation as observed in BSEI micrographs.

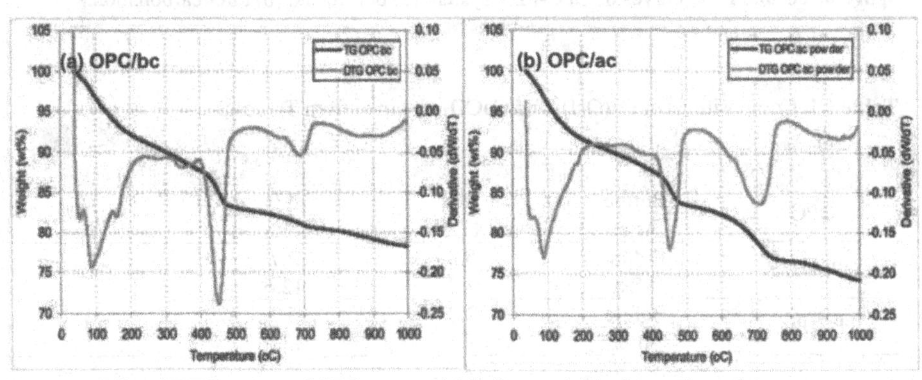

Figure 4. TG and DTG curves of OPC system (a) before and (b) after carbonation.

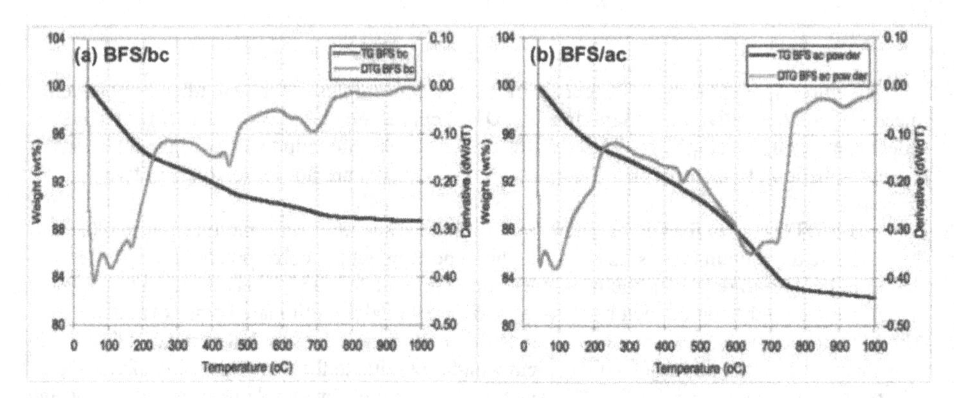

Figure 5. TG and DTG curves of OPC-BFS system (a) before and (b) after carbonation.

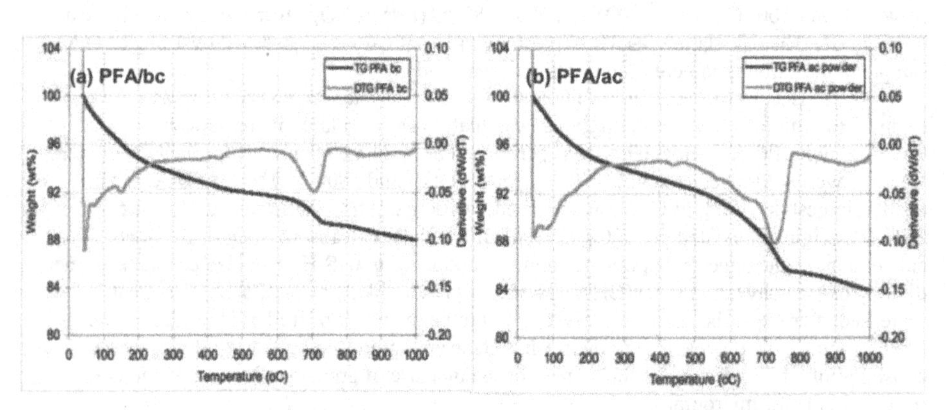

Figure 6. TG and DTG curves of OPC-PFA system (a) before and (b) after carbonation.

Table III. Compositions of $Ca(OH)_2$ and $CaCO_3$ obtained from TG results.

		$Ca(OH)_2$ (wt%)	$CaCO_3$ (wt%)	CO_2 absorbed by 100 of sample (g)
OPC	Before	18.8	5.0	4.1
	After	17.5	13.8	
OPC-BFS	Before	5.2	2.7	4.6
	After	7.3	12.6	
OPC-PFA	Before	2.8	5.7	3.4
	After	3.7	13.0	

OPC-PFA systems also indicated an increase in $CaCO_3$ content on TG/DTG curves, as shown in Figure 6 (a) and (b). It is noticeable that the quantity of $Ca(OH)_2$ is very small in this system. Calculations from the weight loss suggest that 3.4 g of CO_2 was absorbed by 100 g of OPC-PFA sample. Similar to OPC-BFS system, this value exceeds the value that 2.8 g of $Ca(OH)_2$ originally existed per 100 g of OPC-PFA sample can absorb (1.7 g). Although the final quantity of $CaCO_3$ is similar to other systems, the conversion of overall Ca into $CaCO_3$ is very effective in this system because the Ca concentration in this system is very low (Table I and II).

The amount of $Ca(OH)_2$ in all samples did not decrease significantly, and in the case of the composite cement systems, actually increased, despite its consumption to form $CaCO_3$. As already mentioned, the further hydration during carbonation process is thought to be due to the fresh surface area produced by the grinding process. Probably the hydration process occurs faster than carbonation process.

SUMMARY

The reaction of CO_2 gas with OPC, OPC-BFS and OPC-PFA systems were studied to investigate the applicability of these materials to immobilise carbon arising from graphite waste. XRD results suggested that calcite formed in OPC system after the carbonation reaction whereas calcite and vaterite were observed in OPC-BFS and OPC-PFA systems.

The final quantity of $CaCO_3$, the product of carbonation was similar in all systems. The calculation results suggested that 4.1 , 4.6 and 3.4 g of CO_2 was absorbed by 100 g of OPC, OPC-BFS and OPC-PFA systems, respectively, in the carbonation condition of the present study. The results of OPC-BFS and OPC-PFA systems suggested that other hydration products e.g. C-S-H was also consumed either directly or indirectly together with $Ca(OH)_2$ during carbonation process.

OPC-BFS system became much more porous after carbonation, and this may be unfavourable to use the final product as a waste form. OPC-PFA system, on the other hand, did not show a clear change in microstructure after carbonation. OPC-PFA system also indicated a very high efficiency on the conversion of Ca in the system into $CaCO_3$.

REFERENCES

1. M. Bowry and B. J. Handy, CoRWM Document **1453**, (2005).
2. J. K. Stolaroff, G. V. Lowry and D. W. Keith, Energy Conversion and Management, **46**, 687-699 (2005).
3. A. Iizuka, M. Fujii, A. Yamasaki, and Y. Yanagisawa, Ind. Eng. Chem. Res. **43**, 7880-7887 (2004).
4. M. Fernández Bertos, S. J. R. Simons, C. D. Hills and P. J. Carey, J. Hazardous Mater. **B112**, 193-205 (2004).
5. N. B. Milestone, Y. Bai, P. R. Borges, N. C. Collier, J. P. Gorce, L. E. Gordon, A. Setiadi, C. A. Utton and Q. Zhou, Mater. Res. Soc. Proc. **932**, 673-680 (2006).
6. M. Maciejewski, H. R. Oswald and A. Reller, Thermochimica Acta **234**, 315-328 (1994).

7. E. T. Stepkowska, J. L. Pérez-Rodríguez, M. J. Sayagués and J. M. Martínez-Blanes, J. Thermal Anal. Calorimetry **73**, 247-269 (2003).
8. L. Alarcon-Ruiza, G. Platretb, E. Massieub and A. Ehrlacher, Cement and Concrete Res. **35**, 609-613 (2005).
9. R. Vedalakshmi, A. Sundara Raj, S. Srinivasan, and K. Ganesh Babu, Thermochimica Acta **407**, 49-60 (2003).
10. I. Pane and W. Hansen, Cement and Concrete Res. **35**, 1155-1164 (2005).

Mater. Res. Soc. Symp. Proc. Vol. 1107 © 2008 Materials Research Society

Processes Related to the Water Uptake by EUROBITUM Bituminised Radioactive Waste: Theoretical Considerations and First Experimental Results

An Mariën[1], Steven Smets[1], Xiangling Li[2] and Elie Valcke[1]
[1] W&D Expert Group, SCK•CEN, Boeretang 200, 2400 Mol, Belgium
[2] EURIDICE EIG, SCK•CEN, Boeretang 200, 2400 Mol, Belgium

ABSTRACT

According to the present Belgian radioactive waste management program, EUROBITUM bituminised radioactive waste will be disposed of in a geologically stable underground clay formation. The Boom Clay is studied as a potential host formation because of its low diffusion and high retention properties towards radionuclides. The presence of the radioactive waste should not disturb these properties. Due to the presence of hygroscopic salts (25 to 30 weight% $NaNO_3$), EUROBITUM will take up pore water which will result in a swelling and possibly in a very high swelling pressure. First scoping calculations suggest that the swelling pressure exerted to Boom Clay should remain below 7 to 8 MPa to avoid the formation of fractures. If the bitumen in EUROBITUM behaved like a perfect semi-permeable membrane and if no swelling were allowed after the dissolution of $NaNO_3$ into a saturated solution of 10.8 M, osmotic pressures of ~50 MPa could be attained. To better understand the interaction between the swelling EUROBITUM and the host formation, coupled hydro-chemical-mechanical constitutive laws for EUROBITUM have to be developed. To this purpose, water uptake tests under constant volume ('confined') and constant stress ('semi-confined') conditions are being performed. After ~2 years of hydration of small inactive EUROBITUM samples in constant volume conditions, the swelling pressure has raised to ~12 MPa. The volume of samples that can swell against counter pressures of 2.2, 3.3, or 4.4 MPa (constant stress tests) increased with ~5 to 11 volume%, independently of the applied counter pressure. Approximately 10 weight% of the initial $NaNO_3$ content has been leached.

INTRODUCTION

EUROBITUM is a bituminised radioactive waste form, of which ~3000 m^3 has been produced by the EUROCHEMIC/BELGOPROCES reprocessing facility (Mol-Dessel, Belgium) to immobilize precipitation sludges from the reprocessing of spent fuel. This medium-level long-lived waste form consists of ~60 weight% hard bitumen Mexphalt R85/40[1] and ~40 weight% waste, of which the latter is mainly composed of $NaNO_3$ (25 to 30 weight% of EUROBITUM) [2]. According to the present radioactive waste management program of NIRAS/ONDRAF – the Belgian agency for the management of radioactive waste and fissile materials – EUROBITUM will be disposed of in a geologically stable clay formation. The Boom Clay is studied as a potential host formation because of its low diffusion and high retention properties towards radionuclides [3].

[1] The nomenclature for bitumen can be found in [1]

The presence of the radioactive waste in the Boom Clay should not disturb its performance. For that reason the compatibility of EUROBITUM with the Boom Clay formation has to be studied. These compatibility studies start from preliminary designs of the final repository. Several 220 litre drums, generally filled to 80% of their volume with EUROBITUM, will be placed together in a concrete 'overpack' container, which in turn will be placed in a concrete-based disposal gallery. For such a repository design in the Boom Clay at a depth of –220 m, preliminary scoping calculations suggest that within 100 years after the emplacement of the waste, pore water will have filled all the voids in the disposal gallery [4]. The penetration through the concrete based barriers between the Boom Clay and the drums increases the alkalinity of the pore water. The hygroscopic salts inside EUROBITUM will take up the alkaline pore water, which will result in a dissolution and subsequent leaching of the salts and radionuclides. The relative magnitude of the water and salt fluxes, which are influenced by the mechanical behaviour of the EUROBITUM, determines to what extent the water uptake will cause swelling (if swelling is possible) and/or a swelling pressure (if swelling is (partially) inhibited). Preliminary scoping calculations have shown that fractures can be created in the Boom Clay when the pressure applied to the Boom Clay exceeds 8 MPa [5]. Therefore the repository has to be designed in such a way that it will allow the swelling of the EUROBITUM to prevent the generation of pressures above 8 MPa. In the mean time the free volume in the disposal gallery should be maintained at a minimum. To better understand the interaction between the swelling EUROBITUM and the host formation, coupled hydro-chemical-mechanical constitutive laws for EUROBITUM (and for the Boom Clay) have to be developed [5].

The constitutive model COLONBO has been developed by the French Atomic Energy Commission (CEA) to predict the leaching and swelling behaviour of bituminised waste in both free swelling and totally confined conditions. The model assumes the bituminised waste to be a two-phase porous medium that cannot be compressed, and considers only diffusion and non-osmotic flow [6, 7]. The International Centre for Numerical Methods and Engineering (UPC-Polytechnical University of Cataluña, Barcelona, Spain) is developing a fully coupled hydro-mechanical-chemical constitutive model for EUROBITUM that considers the coupled osmotic flow and ultrafiltration [8]. To support the development of constitutive laws for EUROBITUM, laboratory water uptake tests are being performed to study the swelling, swelling pressure build-up, and $NaNO_3$ leaching of small inactive EUROBITUM samples, hydrated in constant volume or constant stress conditions. The first results of these tests are presented and discussed in this paper. To be able to extrapolate the results of the leaching behaviour of small inactive EUROBITUM samples (diameter 38 mm, height 10 mm) to the behaviour of real EUROBITUM drums, two additional water uptake tests will be started up at SCK•CEN with large inactive EUROBITUM samples (diameter 33 cm, height 8 cm).

THEORETICAL CONSIDERATIONS

To have an idea of the extent of swelling and swelling pressure that can be attained, it is useful to reflect on the process that causes the swelling and the swelling pressure build-up, namely osmosis. Some basic calculations can be performed for a bitumen matrix that is considered as a perfect semi-permeable membrane in which the $NaNO_3$ solution is assumed to be homogeneously distributed in the waste matrix.

A full scale 220 litre drum, filled with 216 kg EUROBITUM with ~26 weight% $NaNO_3$, contains ~56.5 kg $NaNO_3$ and ~54 litre free volume. The volume of water needed to dissolve the whole $NaNO_3$ content to a saturated solution (10.8 M) is ~40 litre. In the hypothetical case that no free volume would be available for the EUROBITUM to swell further after the dissolution of the complete $NaNO_3$ content, an osmotic pressure of ~50 MPa would be generated. In contrast, the osmotic pressure will stay below 8 MPa if the bitumen matrix is able to swell until the concentration of the $NaNO_3$ solution has decreased to a ~1.7 M solution. To achieve this goal, ~370 litre of free volume has to be foreseen per drum of EUROBITUM.

In reality the water uptake is accompanied by a release of salt into the 'overpack' container, the disposal gallery, and the surrounding Boom Clay. If the leaching of $NaNO_3$ out of EUROBITUM and the removal of $NaNO_3$ by diffusion through the Boom Clay is fast enough, the free volume that results in a 1.7 M $NaNO_3$ solution inside the 'overpack' container is overestimated by considering bitumen as a perfect semi-permeable membrane.

EXPERIMENTAL DETAILS

The laboratory tests to study the processes related to the water uptake by EUROBITUM are performed in so-called water uptake cells (see Figure 1).

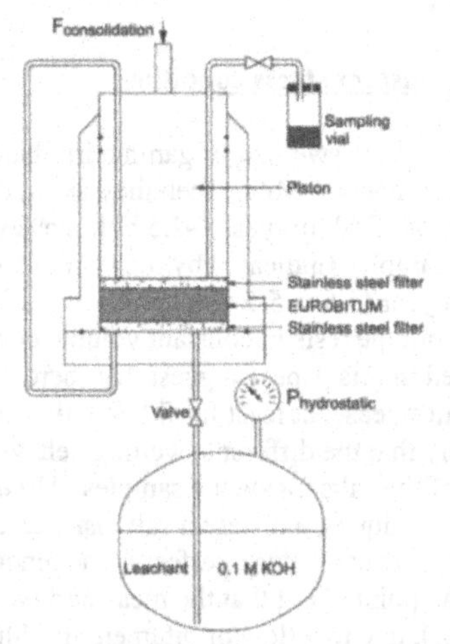

Figure 1. Experimental set-up for water uptake tests with EUROBITUM samples (diameter 38 mm, height 10 mm), contacted at both sides with 0.1 M KOH (SA/V[2] = 0.32 or 0.45 mm^{-1}) that is regularly sampled and renewed to maintain the water activity above 0.95.

Cylindrical, inactive EUROBITUM samples (diameter 38 mm, height 10 mm) are hydrated with a 0.1 M KOH leachant, which is chosen to represent young cement water. The solution is supplied through two filters at both sides of the samples (SA/V = 0.32 or 0.45 mm^{-1}). On top of the upper filter, a movable piston is present, which can be either fixed (constant volume

[2] SA/V = ratio of the external geometrical surface area in contact with the leachant to the leachant volume.

conditions) or movable under a constant stress of 2.2, 3.3, or 4.4 MPa (constant stress conditions). Pressures of 2.2 and 4.4 MPa are representative for respectively the hydrostatic pressure and the total pressure in the Boom Clay at a depth of –220 m. In the constant volume tests, the EUROBITUM samples will exert an increasing swelling pressure, which is measured by means of a load cell. In the constant stress tests, the swelling of the samples is followed by measuring the displacement of the piston with a dilatometer. In all these tests the leaching of $NaNO_3$ is followed by replacing the leachant with a fresh 0.1 M KOH solution at times $t = (2.x)^2$, with x = 1, 2, 3,... and t in days (e.g. the first sampling days are 4, 16, 36,... days after the start of the hydration). The solution is regularly renewed to maintain the water activity[3] higher than 0.95, which is important to study the leaching behaviour of EUROBITUM in conditions where the water uptake is maximal (as observed by CEA [6, 7]). The EUROBITUM samples in the constant stress tests and a few samples in the constant volume tests were gamma irradiated before the water uptake experiments started, to study the influence of the physico-chemical ageing of the bitumen matrix on the extent of swelling, swelling pressure build-up, and $NaNO_3$ leaching. The gamma irradiated samples were irradiated at ~140 Gy/h in the absence of oxygen to a total absorbed dose of 0.8 MGy. This dose corresponds to the calculated total dose absorbed by the bituminised waste after about 1,000 years of auto-irradiation [9, 10].

RESULTS AND DISCUSSION

Swelling of EUROBITUM in constant stress conditions

Figure 2 shows the evolution of the swelling of gamma irradiated EUROBITUM samples under constant stresses of 2.2, 3.3, and 4.4 MPa, when they are hydrated with 0.1 M KOH.

Most samples swell at a rate of ~250 µm/year (~2.5 volume%/year), independently of the applied counter pressure. For 3 samples (indicated by the symbols ▲, + and ■ in Figure 2), swelling rates of 350 to 550 µm/year (3.5 to 5.5 volume%/year) are measured. The SA/V ratio is lower for these 3 samples, but both the tests in constant volume conditions as tests that are started up recently (not discussed in this paper) suggest that there is no influence of the SA/V ratio on the water uptake and salt release, at least for SA/V ratio's in the range of 0.32 to 0.45 mm^{-1}. Therefore, it is more likely that the different swelling behaviour is related to a different spatial and/or size distribution of the salts inside the samples. Although this has not yet been confirmed by characterization techniques, a different salt distribution is a plausible explanation since the bituminization process cannot produce perfect homogenous drums (from which the samples are cut out). It should be pointed out that the measured swelling degrees are somewhat underestimated (systematic error), due to a flow of bitumen into filter pores and into voids between the piston and the cell wall. The volume of bitumen that consequently does not contribute anymore to the displacement of the piston, can be calculated only at the end of the test.

Sneyers *et al.* reported a volume increase of ~15% for EUROBITUM samples after 240 days (~22.5 volume%/year) of leaching in clay water solutions in free swelling conditions (no counter pressure). Although the different geometry of their experimental set-up (~2 times larger external geometrical surface area in contact with a ~25 times larger leachant volume, which was not renewed regularly) allows a higher $NaNO_3$ leaching rate, it is unlikely that the higher leaching

---

[3] The chemical acitivity of water (a_w) in a solution is defined as [6]: $a_w = p/p_0$ (with p = water vapour pressure in the vapour in equilibrium with the solution; p_0 = water vapour pressure in the vapour in equilibrium with pure water.)

rate can only be ascribed to their different experimental geometry [11]. It is most probable that the rate of the water uptake by the hygroscopic salts inside the bitumen is lower when swelling is partially inhibited. It is also conceivable that a part of the water in an almost leached layer will be pressed either outside the sample or more inwards (comparable with reversed osmosis), caused by the fact that the water pressure in an almost completely leached layer is smaller than the applied counter pressures.

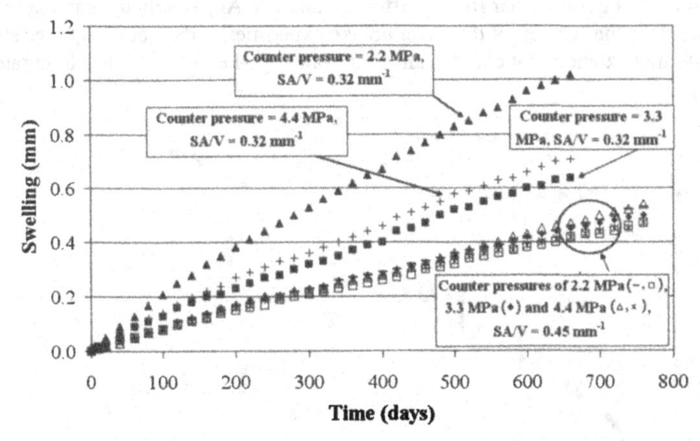

Figure 2. Evolution of the swelling of gamma irradiated EUROBITUM samples (diameter 38 mm, height 10 mm), in contact at both sides with 0.1 M KOH that is regularly renewed to maintain the water activity above 0.95. The samples are allowed to swell against a constant stress of 2.2, 3.3, or 4.4 MPa. (Uncertainty on swelling = 0.012 mm; 95% confidence[4])

Swelling pressure build-up by EUROBITUM in constant volume conditions

In Figure 3 the evolution of the swelling pressure is shown for both gamma irradiated and non-irradiated inactive EUROBITUM samples, hydrated in constant volume conditions. For most samples, swelling pressures of ~12 MPa are already attained within less than 800 days of hydration. As calculated above, a maximum swelling pressure of ~50 MPa is possible if $NaNO_3$ cannot leach out of the waste. As with the swelling measurements, the measured swelling pressures are also underestimated (systematic error), due to a flow of bitumen into the filter pores and other voids, and due to a small expansion (~150 µm) of the frame around the water uptake cell because of the very high pressures.

For one sample (indicated by the symbols ◆ in Figure 3) the initial increase of the swelling pressure was much faster than for the other samples, but after ~150 days the rise of the swelling pressure proceeded at a lower rate. It is likely that the difference in pressure evolution can be explained by a different spatial and/or size distribution of the salts inside the samples, as pointed out before. In two experiments, the leachant was not replaced. The corresponding EUROBITUM samples show a faster increase of the swelling pressure. As for these experiments the water activity has not been maintained above 0.95, the amount of salts that is still inside the

[4] There is also a systematic error on the magnitude of the swelling, as described in the text above figure 2.

EUROBITUM samples after a certain leaching period, will be higher and therefore also the driving force to take up water. This pressure increasing effect is partially counteracted by a slower uptake of water due to the lower water activity gradient between the hydrating EUROBITUM sample and the leachant. At this moment it is impossible to conclude whether these leaching conditions are more or less relevant for the real disposal scenario than the conditions where the leached $NaNO_3$ is regularly removed. For the latter conditions, the swelling pressure evolution is quite similar for the different samples. Apparently the gamma irradiation of the samples, before the start-up of the water uptake experiments, did not change the structure of the bitumen to that extent that it could change the water uptake or $NaNO_3$ leaching rate [12].

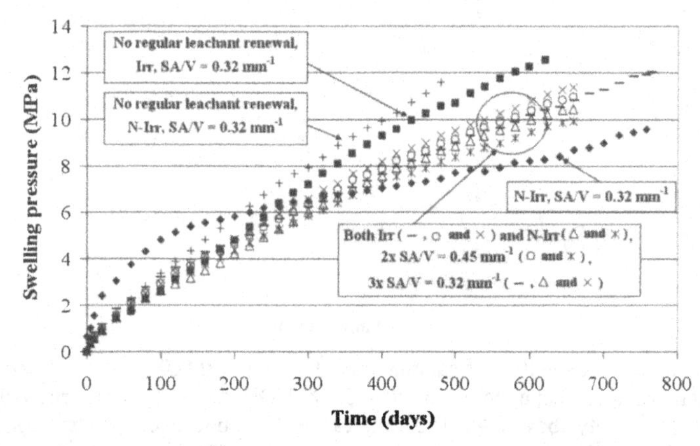

Figure 3. Evolution of the swelling pressure exerted by both gamma irradiated and non-irradiated EUROBITUM samples (diameter 38 mm, height 10 mm), present in a constant volume and contacted at both sides with 0.1 M KOH that is regularly renewed (with 2 exceptions) to maintain the water activity above 0.95. Legend: Irr = gamma irradiated sample, N-Irr = non-irradiated sample. (Uncertainty on swelling pressure = 0.5%; 95% confidence[5])

Leaching of $NaNO_3$

Figure 4 shows the evolution of the cumulative amount of $NaNO_3$ that has leached out of the EUROBITUM samples since the beginning of the experiments. The cumulative amount (at time t) was calculated by counting up the mass of $NaNO_3$ in all the samples with leachant solution, removed from the cells (before time t). For the sake of clarity some replicates (same leaching conditions, comparable leaching behaviour) are not shown in this figure.

Since the initial $NaNO_3$ content in the EUROBITUM samples is ~4.2 g, ~6 to 12 weight% of the initial $NaNO_3$ content has leached in ~1.5 to 2 years. Out of one sample even 800 mg $NaNO_3$ (19 weight% of the initial $NaNO_3$ content) has leached in ~1.5 years. It is very difficult to sample completely the leachant solution that has been in contact with the bituminised waste sample during the time period between two sampling days. Therefore the calculated cumulative amount

[5] There is also a systematic error on the magnitude of the swelling pressure, as described in the text above figure 3.

of leached $NaNO_3$, as shown in Figure 4, is somewhat underestimated (systematic error). The underestimation of the cumulative leached amount of $NaNO_3$ decreases with the number of samplings and can be corrected at the end of the test. The higher leaching rate for the sample from which 800 mg $NaNO_3$ has already leached in ~1.5 years, as well as the variability in leaching rates between the other samples is probably caused by a different size and/or spatial distribution of the salt crystals in the samples. Taken into account the influence of the salt distribution on the leaching of $NaNO_3$, the leaching rate in constant stress conditions is comparable with the leaching rate in constant volume conditions. Therefore the magnitude of the applied counter pressures does not seem to have an effect on the leaching rate, at least for counter pressures higher than 2.2 MPa. Nevertheless there seems to be an influence of the test conditions on the time dependency of the leaching process. The leaching in constant volume conditions proceeds rather linearly with the square root of the time, pointing to a diffusion controlled process [6, 7]. In contrast, the leaching in constant stress conditions evolves more linearly with the time, suggesting a contribution of dissolution controlled leaching of $NaNO_3$. From the results of the leaching it is not possible to conclude on the effect of gamma irradiation. Only for two of the three gamma irradiated samples in the constant volume tests, a lower leaching rate is observed than for the non-irradiated samples.

To be able to relate the $NaNO_3$ leaching data to the swelling and/or swelling pressure-build up, as discussed above, it is important to know the water uptake rate. Therefore the decrease of the leachant volume as a consequence of the water uptake by the EUROBITUM samples is followed in some cells. To ensure a constant leachant volume, very small volumes of new leachant are added at regular times. However at this moment it is too difficult to interpret the results of the water uptake curves, which are consequently not shown in this paper.

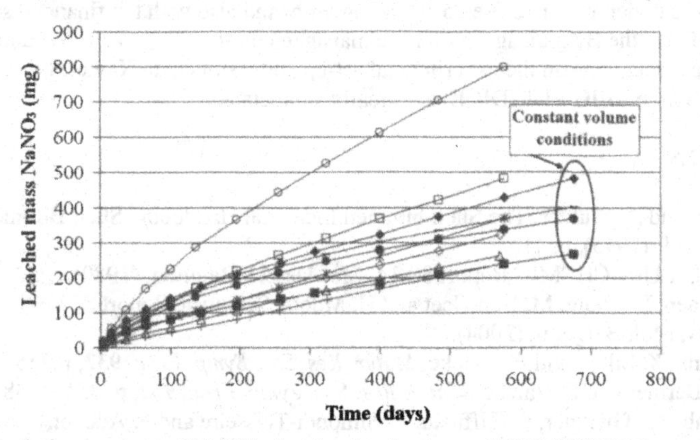

Figure 4. Cumulative amount of $NaNO_3$ that has leached out of EUROBITUM samples (diameter 38 mm, height 10 mm) contacted at both sides with 0.1 M KOH that is regularly renewed to maintain the water activity in the leachant solution higher than 0.95. (Uncertainty on leached $NaNO_3$ = 10%; 95% confidence[6])
Legend: (1) EUROBITUM samples in constant volume conditions: —◆—, —●—: non-irradiated samples; —■—, ——: gamma irradiated samples; (2) gamma irradiated EUROBITUM samples in constant stress conditions: —◇—, —□—: counter pressure = 4.4 MPa; —△—, —✳—: counter

[6] There is also a systematic error on the leached masses of $NaNO_3$ in figure 4, as described in the text above figure 4.

pressure = 3.3 MPa, –+–, –⊖–: counter pressure = 2.2 MPa; SA/V = 0.45 mm^{-1} for ——, –◇–, –△–, –+–, and SA/V = 0.32 mm^{-1} for –◆–, –●–, –■–, –⊟–, –✳–, –⊖–.

CONCLUSIONS

Compatibility studies are required to assess if EUROBITUM can be disposed of in the Boom Clay without disturbing the performance of the latter. The uptake of pore water by the hygroscopic salts in EUROBITUM will lead to a swelling and, if swelling is (partially) inhibited, a swelling pressure. Both preliminary calculations and experimental results show that high swelling pressures (more than 10 MPa) will be created if swelling is not possible. This can lead to the formation of fractures in the Boom Clay. The experiments in constant pressure conditions might help to calculate the free volume in the repository for EUROBITUM that is necessary to allow the latter to swell and in addition to prevent the build-up of swelling pressures above ~8 MPa. After hydrating small inactive EUROBITUM samples during two years at constant counter pressures of 2.2, 3.3, and 4.4 MPa, swelling degrees of ~5 to 11 volume% are measured (swelling rates of ~2.5 to 5.5 volume%/year). Since only ~6 to 19 weight% of the initial $NaNO_3$ content has leached out of the EUROBITUM samples, the swelling and swelling pressure build-up will still proceed for several years. Meanwhile the preliminary results contribute to the development of coupled hydro-chemical-mechanical constitutive laws for EUROBITUM.

ACKNOWLEDGMENTS

This work is undertaken in close co-operation with, and also with the financial support of NIRAS/ONDRAF, the Belgian agency for the management of radioactive waste and fissile materials. The discussions on the experimental set-up and results with Xavier Sillen (SCK•CEN) and Frédéric Bernier (EIG EURIDICE) are greatly appreciated.

REFERENCES

1. P. Morgan and A. Mulder, The Shell bitumen Industrial Handbook, Shell Bitumen, ISBN-0-95 16625-1-1, (1995), p. 41.
2. H. Eshrich, EUROCHEMIC Report 80-14, Mol-Dessel, Belgium, (1980).
3. M. De Craen, L. Wang, M. Van Geet and H. Moors, Scientific Report SCK•CEN-BLG-990, SCK•CEN, Mol, Belgium, (2004).
4. E. Weetjens, X. Sillen and E. Valcke, *Mater. Res. Soc. Symp. Proc.* **932**, p.735 – 742, (2006).
5. X. Li, F. Bernier and E. Valcke, *Mater. Res. Soc. Symp. Proc.* **932**, p. 751 – 758, (2006).
6. J. Sercombe, B. Gwinner, C. Tiffreau, B. Simondi-Teisseire and F. Adenot, *J. Nucl. Mat.* **349**, p. 96 – 106, (2006).
7. X. Lefebvre, J. Sercombe, A. Ledieu, B. Gwinner and F. Adenot, *Mater. Res. Soc. Symp. Proc.* **932**, p. 681 – 688, (2006).
8. N. Mokni, S. Olivella, X. Li, S. Smets and E. Valcke, presented at the International Conference 'Clays in Natural and Engineered Barriers for Radioactive Waste Confinement', Lille 2007 (Organisation: ANDRA, Paris, France).
9. S. Kowa, N. Kerner, D. Hentschel and W. Kluger, KfK Report 3241, Karlsruhe, Germany, (1983).

10. F. Rorif, E. Valcke, P. Boven, H. Ooms, J. Peeters and S. Smets, *Mater. Res. Soc. Symp. Proc.* **932**, p. 689 – 696, (2006).
11. A. Sneyers and P. Van Iseghem, *Mater. Res. Soc. Symp. Proc.* **506**, p. 565 – 572, (1998).
12. E. Valcke and R. Gens, presented at ICEM'07 Conference, Bruges, Belgium, (2007).

Mater. Res. Soc. Symp. Proc. Vol. 1107 © 2008 Materials Research Society

What Will We Do With the Low Level Waste From Reactor Decommissioning?

Dr Adam R Meehan, Dr Stephen Wilmott, Miss Glenda Crockett and Dr Nick R Watt
Magnox Electric Ltd, Berkeley Centre, Berkeley, Gloucestershire, GL13 9PB, UK.

ABSTRACT

The decommissioning of the UK's Magnox reactor sites will produce large volumes of low level waste (LLW) arisings. The vast majority of this waste takes the form of concrete, building rubble and redundant plant containing relatively low levels of radioactivity. Magnox Electric Ltd (Magnox) is leading a strategic initiative funded by the Nuclear Decommissioning Authority (NDA) to explore opportunities for the disposal of such waste to suitably engineered facilities that might be located on or adjacent to the site of waste arising, if appropriate and subject to regulatory acceptance and stakeholder views. The strategic issues surrounding this initiative are described along with an update of progress with stakeholder consultations in relation to the proposed licensing of the first such facility at Hinkley Point A, which could be viewed as a test case for the development of similar disposal facilities at other nuclear sites in England and Wales.

INTRODUCTION

In March 2007 a new Government policy on the long-term management of the UK's solid LLW was published [1]. The policy calls for waste producers to carry out comprehensive and consultative option studies into the disposal of LLW arising at their sites, including consideration of a wide range of options. In anticipation of such a non-prescriptive, high level policy being adopted, Magnox undertook a series of studies into the options for the disposal of the LLW that will arise during the first phase of clean-up at Hinkley Point A, Bradwell, Sizewell A and Dungeness A. This options assessment process included the following site-specific components:

- a consultation-based 'multi-attribute decision analysis' ('MADA');
- a review of all relevant local and national Government planning policies;
- the collation of other issues relevant to any decision raised at a stakeholder workshop;
- consideration of compatibility with preferred site end-uses;
- a review of conventional (non-radiological) worker safety;
- a review of international practice for LLW disposal; and
- consideration of matters of proportionality i.e. the balance between risks and costs.

Government policy aims, for example to reserve remaining capacity at the UK's national LLW Repository (LLWR), as well as the NDA's aspirations were also taken into account.

This paper describes this options assessment process, along with interim outcomes, for Hinkley Point A, which is the lead site in Magnox's programme of LLW disposal option studies.

OPTIONS ASSESSMENT

Multi-attribute decision analysis

Given the wide range of potentially complex issues affecting the selection of a disposal route for LLW, multi-attribute decision analysis (MADA) with wide stakeholder engagement was selected as part of the overall options assessment process. Environment Agency guidance [2] was followed to help to ensure a systematic and transparent approach. MADA is a well-established formal process for putting a list of possible courses of action, or '*options*', in an order of preference, known as the '*ranking*'. It is often used as part of a decision-making process where there are a number of issues or factors in any decision, not all of which point to the same choice, i.e. it is a decision aid tool. MADA involves the identification of '*attributes*' (criteria or factors) against which the performance of each option is assessed. This assessment of options against attributes is by means of a numerical score allocated using a scoring scheme predefined for each attribute. The scores for each option are summed over all the attributes, taking account of the relative importance of each attribute by means of numerical weighting factors. This provides an overall score for the performance of each option, which provides the basis for option ranking. The robustness of this ranking is explored through 'sensitivity analysis', which involves varying data and assumptions to see how the rankings are affected.

A long-list of possible options was initially drawn up on the basis of: the extant Government policy [3] as well as the proposed Government policy [4] at the time that the option study was carried out; international guidance and practice *e.g.* [5]; and the output of a generic workshop [6]. Only practicable and licensable options were taken forward. Some options were considered as highly unlikely to be practicable or licensable and were therefore 'screened out' at an early stage. The options not taken forward for further consideration included: export for disposal overseas (considered to be contrary to Government policy [1] and not economic for large volumes of low level waste); disposal at sea (contrary to the London Convention); and alternative forms of below-ground disposal, such as deep geological disposal (ruled out on the basis of disproportionate costs) or borehole disposal (not feasible for large volumes of waste). A summary of options taken forward for consideration in the option study is given in table I.

Table I. Options for detailed consideration in the MADA.

Option	Description
1	On-site disposal: a) Fit-for-purpose near-surface disposal on site using the turbine hall basement void. For Hinkley Point A this option is taken to require coastal defences. b) Fit-for-purpose near-surface disposal on site not using the turbine hall void, at a location not requiring coastal defences. c) Below-surface (tens of metres) disposal on site, not requiring coastal defences
2	Fit-for-purpose near-surface disposal at a site-specific facility near to site.
3	Fit-for-purpose near-surface disposal in a national or regional facility.
4	Disposal in a near-surface national facility designed to accept the full range of wastes covered by the term LLW: a) The existing Low Level Waste Repository (LLWR) near Drigg in Cumbria. b) A new national facility engineered to the same standard and with the same waste acceptance criteria.

Stakeholder involvement in the MADA was achieved by means of a number of workshops, independently facilitated by SERCO Assurance. At the first generic workshop held on 27th February 2006, which included stakeholders from Hinkley Point A, Bradwell, Sizewell A and Dungeness A, the '*options*', MADA process and '*attributes*' were reviewed by stakeholders

[6]. This workshop included representatives from the regulators, local planning authorities, the NDA, the Hinkley Point A Site Stakeholder Group (SSG) and the nuclear industry. A Hinkley Point A site-specific workshop was held on 15th May 2006 to obtain attribute '*weightings*' for the MADA [7]. At a second site-specific Hinkley Point A workshop on 5th September 2006, the results of the MADA were discussed with stakeholders and their views on the outcome captured. Other issues that stakeholders felt should be taken into account in selecting a preferred option were collated [8]. The options study process has been discussed regularly at meetings of the Hinkley Point A SSG and periodic updates provided to the NDA's National Stakeholder Waste Issues Group. Options assessment documents have also been posted on the Company's website.

A set of attributes was developed to allow the options under consideration to be distinguished. These were selected and defined to ensure that they: reflect regulatory guidance [2] and proposed Government policy [4]; covered the important issues; distinguished usefully between options; were not so numerous that they unnecessarily complicated the assessment process; and were, in so far as is possible, independent of each other. In compiling the list of attributes information from a wide range of sources was used [9, 10, 11, 12]. As a result of discussions at the generic workshop, several modifications were made to the attribute list and definitions. The final attribute list is given in table II, where the attributes are shown allocated to groups (*i.e.* safety, transport, *etc.*) on the basis of the nature of the issues they cover.

Table II. Attributes selected for the MADA.

Group	Attribute
Safety	Public safety
	Workforce safety
Secondary Waste (spoil, etc.)	Spoil / secondary waste
Transport	Long-distance transport
	Local transport
Environment Impacts	Local disturbance
	Long-term coastal erosion/flooding
	Very long-term environmental impacts
Facility Performance	Relative radiological performance
	Relative chemical performance
	Relative long-term intervention risk
Cost & Schedule	Timing of disposal route availability
	Lifecycle cost

Option Scores

The options were scored against the attributes by suitably qualified and experienced largely 'in-house' personnel. This approach was taken since attribute scoring is a relatively objective process. Nevertheless, some sensitivity analysis on the scoring was carried out, as discussed later. An overview of the scores allocated to the options is provided below:

- **Safety** (public and workforce safety): This category is principally a measure of the risks posed by construction activities and the transport of materials and wastes. The best performing options are 1b and 1c: these do not involve the risks to safety arising from the construction of coastal defences (as option 1a does) or the associated materials transport, nor do they involve the long-distance transport of LLW (as options 2-4b do), and they also involve less transport of excess spoil etc. (see below).

- **Secondary Waste** (spoil, etc.): This category is a measure of the quantity of bulk non-radioactive spoil and rubble, including that generated on the reactor site (after re-use on site of spoil arising there has been taken into account). Options 1a-1c perform the best as regards this issue.
- **Transport** (long-distance transport and local transport): This category is a broad measure of the extent to which LLW is moved from the site to a disposal facility (the product of quantity and distance), together with the volume of HGV traffic (in movements per day) required at the reactor site and, if relevant, at an off-site location. The on-site options 1b and 1c perform well against these attributes as they do not require transport of LLW or the use of any off-site location. However, option 1a necessitates the movement of additional construction material for required coastal defences and options 2-4b involve transport of LLW significant distances for final disposal as well as traffic movements at an off-site location.
- **Environmental Impacts** (local disturbance, long-term coastal erosion/flooding and very long-term environmental impacts): This category is a measure of the various negative environmental consequences that could arise from the disposal operations including any associated coastal defences. On-site options perform well in terms of disturbance as no separate disposal facility is required to be operated. In the case of option 1c the facility will be virtually immune from disruption by natural forces after closure.
- **Facility Performance** (relative radiological performance, relative chemical performance, and relative long-term intervention risk): This category is a measure of how well the disposal facility is expected to perform after closure in terms of isolating the radionuclides and potentially harmful chemical substances from the environment, and the risk that unplanned for mitigation may need to be carried out in the future to ensure adequate performance. The best performing options are 1c and 4b which both offer a relatively stable environment and high level of containment. Option 1a scores relatively poorly in part because of the assumed relatively high risk associated with a future need for intervention due to the coastal defences associated with this option.
- **Cost and Schedule** (timing of availability of disposal route and lifecycle cost). This category is a measure of programme (speed of implementation) and cost issues as they impact upon the disposal process. Apart from option 4a (disposal to the existing facility near Drigg) all off-site options score relatively poorly in terms of timing owing to the need to wait a significant period for a facility to become available, whereas on-site disposal options fare better. Option 1c (below-surface disposal) and options 4a and 4b (disposal at a facility able to take the full range of LLW) are disadvantaged in terms of cost, whereas fit-for-purpose near surface facilities perform well on this issue.

Attribute Weightings

In this study, the approach taken to determining weightings was to hold a site-specific workshop with representation from a wide range of stakeholders. For the purpose of the weighting exercise, participants were provided with a specified level of 'impact' (*e.g.* the risk of a fatality, the amount of long-distance waste transport, the cost, *etc.*) for each attribute. It was the relative importance of these which was used to elicit attribute weightings. Three sets of weightings were obtained, these representing the general consensus views of an 'industry' group (including representatives from Magnox Electric Ltd, British Energy and the NDA), a 'statutory organisations' group (including representatives from Somerset County Council, West Somerset District Council, the Environment Agency and the then English Nature) and a 'Site Stakeholder Group' (SSG) group (including local councillors, former employees and other members of the local community who are members of the SSG). Some key points arising from the weightings exercise are summarised for each group below.

Industry Group

Public safety was accorded a 'high' weighting by the industry group, although not the highest possible on the grounds that the risk to the public was, in absolute terms, very low. Workforce safety was accorded the same weighting as public safety on the basis of equal value of human life (public and workers).

The timing of disposal route availability attribute was also allocated a 'high' weighting: the group viewed that the early availability of a disposal route was very important. The group considered that the lifecycle cost was very important because of the need to make prudent use of available resources if the decommissioning process is to be carried forward effectively and because of the amount of money involved; accordingly a 'very high' weighting was allocated.

Statutory Organisations

Safety was considered to be relatively important overall, with that of the public weighted more strongly ('high') than that of the workforce ('medium') on the basis that the risks to the latter are voluntary.

Local transport and disturbance at the reactor site were agreed as being of 'high' importance, reflecting the concern felt, i.e. that significant efforts should be seriously considered to minimise such environmental effects. Local transport and disturbance at a new location were given a higher weight (i.e. 'very high') than at the reactor site as the nuisance would be much greater at what was assumed to be a location where there was no significant transport already.

There was a strongly held view that it is very important to protect the local environment over the long term (to take account of future generations who did not benefit from the generation of electricity from the power station), and therefore the long-term coastal erosion/flooding and very long-term environmental impacts attributes were both allocated 'high' weightings.

SSG

The significance of radiological performance generated much debate. Although all participants rated it as significant, no strong consensus was achieved. The median view was for 'high' weighting, with one participant assigning 'very high' because of perceived uncertainties in the prediction of harm from radiation, and one assigning medium because these are very low activity wastes. The majority of the syndicate rated cost as an issue of significance, although there was no consensus between 'medium' and 'high' weightings.

Results

Figure 1 summarises the relative overall favourability of the scored options using the weighting sets from each of the three stakeholder groups.

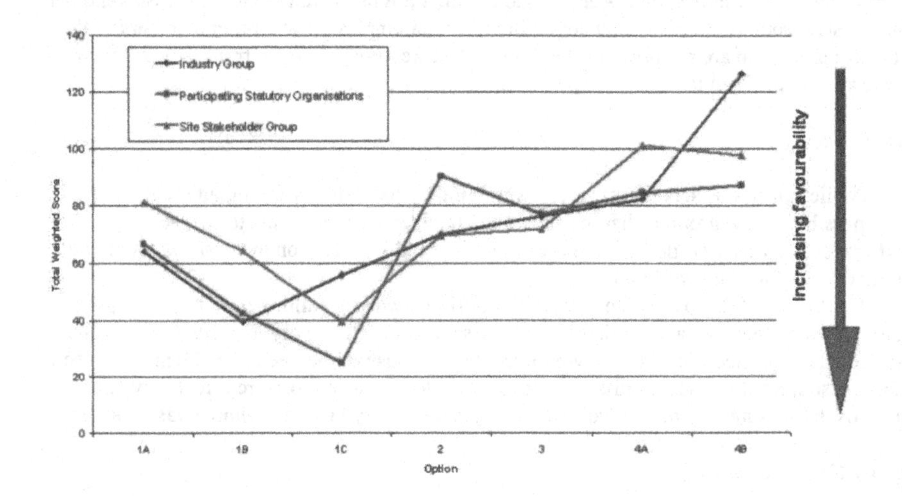

Figure 1. Weighted scores of options considered in the Hinkley Point A MADA.

It is clear that whilst there are small differences between the option rankings using the three different weighting sets, the relative favourability of the options is similar. Of the on-site options, 1b (near-surface disposal not in the turbine hall and not requiring coastal defences) and 1c (below-surface disposal) were found to be the most favoured with all three sets of weightings. Option 1a (disposal in the turbine hall with coastal defences) was the least favoured of the on-site options. This arises because option 1a involves disposal on the seaward side of the site and has been assumed to require the construction of substantial coastal defences with consequent safety, transport, environmental and cost impacts. Off-site options are less favoured principally for reasons relating to safety, transport, cost and schedule (timing of disposal route availability).

Sensitivity Analysis

A sensitivity analysis on the results of the MADA was carried out. The objective of the sensitivity analysis was to determine the robustness of the overall conclusions of the MADA as summarised above, i.e. to determine how easily the relative ranking of the options could be changed by using different assumptions, data or weightings.

The main results described above are based on the following assumptions, the effect of each of which was tested in turn:

- that coastal defences would be needed for option 1a (turbine hall), these being defences which would otherwise not be required;

- that any off-site transport of LLW would be by road;
- that any new off-site disposal facility would be on a green-field site;
- that in the absence of a disposal route being available, decommissioning would not be delayed but rather all waste arisings would be stored on site pending disposal route availability; and
- that the volume of waste was the maximum volume estimated at the time of the study.

The only significant change in the option rankings related to the first of the assumptions listed above, i.e. if the assumption was made that option 1a (turbine hall) did not require coastal defences (or at least not significant defences in terms of material quantities or construction effort), then that option became about as favourable as the other variants of on-site disposal.

As regards parameter values and scores, changes studied included the following: the degree of change to future coastal erosion/flood development due to the presence of coastal defences in option 1a (turbine hall); the distances to off-site disposal locations; and the time period until disposal facilities that do not yet exist become available. However, none of the conclusions were found to be sensitive to these parameters.

The sensitivity of the conclusions to considerations of disposal facility performance was also studied by removing the three facility performance attributes from the analysis, i.e. relative radiological performance, relative chemical performance and long-term intervention risk. The removal of the facility performance attributes from the analysis did not have a marked impact upon the option scores overall. However, the preferred option using the weightings for the SSG did change from option 1c (below-surface disposal) to option 1b (near-surface disposal)[1]. This is because the initial preference for option 1c was, in part, a result of the high weighting given to the radiological performance attribute by the SSG.

In addition, it was considered useful to examine the rankings of the options without considering the effects of cost. Removing cost as a relevant factor had the effect of improving the relative favourability of option 1c markedly (and to a lesser degree options 4a and 4b) such that option 1c becomes the most favoured option with all three sets of weightings.

Conclusions of the MADA

The key determining factor in the overall outcome of the MADA was found to be the need for coastal defences; specifically, on-site disposal was favoured only as long as there is no associated need for significant coastal defences (that would not otherwise be present). This conclusion was found to be robust against changes to assumptions, data and weightings. Therefore, the MADA is supportive of option 1b (near-surface disposal not requiring coastal defences) and of option 1c (below-surface disposal). The relative preference of option 1b and option 1c within the MADA was not found to be robust, *i.e.* it could be relatively easily changed by changes to assumptions, data or weightings. This suggests that further considerations would be required to select the preferred option, as discussed in the remainder of this paper. A fuller description of the MADA including all of the attribute definitions, the reasoning behind all of the attribute scores and weightings, and a description of all results and sensitivity analyses, is available [7].

---

[1] The preferred option did not change using the other two weighting sets.

Planning policy review

A review of national and local planning policies relevant to the study was carried out [13]. The general direction of planning policies, with increasing emphasis on the proximity principle, is in line with and supports the conclusions of the MADA. In addition, policies showing a preference for management of waste at the point of arising or at the nearest appropriate site can be considered unsupportive of regional or national disposal facilities. For each variant of on-site disposal considered, both favourable and unfavourable policy areas were identified. Policies discouraging coastal development were considered to apply to all variants of on-site disposal, as was the Local Waste Plan policy against radioactive waste disposal in Somerset. There are also many policies discouraging development that would require the use of coastal defences, such as have been assumed to be required in the case of option 1a. There are a number of policies concerning uncertainties in predictions of coastal evolution and which encourage application of a precautionary approach when determining any planning applications for such development; these again were assumed only to apply to option 1a. The policies that are generally supportive of an on-site disposal option were found to be those concerned with sustainability and a reduction in long distance transport of waste. Use of the nearest appropriate location, management of waste at the point of arising, co-location of disposal facilities with complementary activities and the re-use of industrial sites are all policies that can be regarded as favourable to the on-site disposal of waste. Except where on-site disposal would require the construction and maintenance of coastal defences, Magnox came to the view that those local and national policies that are supportive of on-site disposal outweighed those that do not support it. In other words, on balance, the planning policies are likely to be seen as supportive of on-site disposal in the form of option 1b and in the form of option 1c.

Wider issues raised by local stakeholders

At the second stakeholder workshop, to which all those participants in the first Hinkley Point A workshop were invited, stakeholders were presented with the output of the MADA for discussion and given the opportunity to raise issues which they believed should be taken into account before any final decision is made. Participants were placed in syndicates for the main discussions. All groups were concerned that the chosen option should be consistent with the preferred site end-use, with it being noted that variants of on-site disposal with smaller footprints would allow greater flexibility. Other key points raised, which were considered relevant to any decision, were that: reliance on long-term monitoring, to ensure on-going safety, should be minimised; the waste volume should be minimised [as far as reasonably practicable]; and there may be benefits of having fewer but larger LLW disposal facilities e.g. in terms of ease of monitoring, post-disposal security and lifecycle costs. All participants discussed the possible effects of on-site disposal on local employment levels and on the general social and economic well-being of the area, with mainly positive points being made. However, the possibility of the 'blight' effect, for example on property prices, was also raised. Participants were also given the opportunity to comment on the outcome of the MADA. Their responses are described in [8] but, in brief, disposal on site in some form as indicated by the MADA was in broad accord with the preferences of stakeholders present at the workshop provided that it is compatible with the final end-use for the site.

Compatibility with preferred site end-uses

The 'end-use' for a nuclear licensed site is the use (or set of uses) to which the land may be put once decommissioning is complete. The 'end-state' of a site is the physical condition of the site at that point. Magnox has considered the compatibility of an end-state with an on-site disposal facility with various potential site end-uses. An on-site disposal facility would have a footprint that would occupy a relatively small portion of a site, this being in the form of a grassed over mound once the facility's vaults had been filled and capped. Environmental monitoring for re-assurance purposes would take place for a period at least up until final site clearance (currently assumed to be in about 100 years time). It is unlikely that there would be any need on radiological protection grounds for access restrictions to the landscaped capping mound. Magnox therefore believes that, with the possible exceptions of residential use and crop-growing, end-states involving an on-site disposal facility are potentially compatible with a wide range of site end-uses including those preferred options identified by stakeholders in the Hinkley Point A site end-use consultation exercise [14].

Conventional worker safety review

One issue that requires consideration before a final option is selected is worker safety, principally during facility construction, maintenance and operation. Government policy [1] advocates "*use of a risk-informed approach to ensure safety, with consideration given to radiological and conventional risks, striking an appropriate balance between the two*". The principal safety issue discriminating between options is the risk to workers associated with conventional *i.e.* non-radiological hazards. For this reason, a risk assessment in relation to conventional worker safety was carried out. This concluded that option 1b (near-surface disposal) is safer for workers than option 1c (below-surface disposal). Preliminary radiological performance assessments have shown that, as may be reasonably expected, option 1c affords a higher standard of radiological performance than option 1b. That said, the preliminary radiological performance assessment modelling also shows that for both variants of on-site disposal the predicted doses and risks to the most exposed members of the public are significantly better than regulatory requirements for the most likely future development of the disposal system (*i.e.* eventual leakage of contaminants to groundwater). Therefore, the better long-term radiological performance of option 1c does not provide overriding justification for preferring it over option 1b.

International practice for LLW disposal

It is noted that the safe disposal of LLW to near surface facilities is widely practised. Magnox considers that near surface disposal represents best practice for the disposal of high volume, low hazard wastes of the type that will arise at Hinkley Point A. Examples include the landfills for the disposal of high volume, low hazard LLW at Morvilliers in France and El Cabril in Spain, the LLW landfill at Clive in the US and the on site land-raise for LLW at Ringhalls in Sweden. Even facilities designed to accept higher hazard low level wastes, typically of concrete vault design, are generally near surface facilities *e.g.* Centre de L'Aube in France, LLWR near Drigg in UK, *etc.*

Consideration of matters of proportionality i.e. the balance between risks and costs

The principle of optimisation is applied to ensure that risks to the public are as low as reasonably achievable. Once this is demonstrated, additional improvements to reduce risks further that would result in significant time, cost or effort being incurred need not be sought. Given the radiological performance of the two lead variants of on-site disposal and their relative costs, it is therefore considered by Magnox Electric that there is no radiological performance reason for preferring option 1c (below-surface disposal). There is, however, a significant financial reason for preferring option 1b to option 1c.

Government Policy and NDA Strategy

With the LLWR near Drigg nearing its current Authorised disposal capacity, and with uncertainty regarding its future availability, it is likely that the use of this national resource will need to be prioritised in the future to reserve its capacity and maximise its lifetime. NDA Strategy [15] highlights that a situation could occur after 2008 where there is insufficient capacity to take LLW from all nuclear sites. In order to help to mitigate this risk, the NDA state that waste producers will be encouraged to place more emphasis on application of the waste management hierarchy and waste minimisation. In addition, in relation to waste disposal the NDA state that it will address "*what opportunities exist to dispose of LLW on sites where it arises, subject to considerations on coastal erosion and climate change and in consultation with stakeholders*" and that "*In general, our preferred approach would be to build upon the principle established at Dounreay that, where possible, sites should host their own LLW facilities. In many cases, this will be specifically for higher volume low activity LLW. Local stakeholders will be consulted on any proposals and, so far, we have found that local communities are generally open-minded.*" Government policy [1] stipulates that new disposal facilities should be fit-for-purpose i.e. that they should be engineered to a degree commensurate with the types of waste they will receive. In relation to application of the proximity principle, it states that "*The need to consider alternatives to long distance transport where possible applies in particular to large quantities of lower activity soil and rubble that will arise that will arise from large nuclear site decommissioning activities.*" Government's stated overall objective is to provide a clear policy framework for the management of LLW, which will address the foreseen shortage of disposal capacity and ensure the availability of better and more cost-effective ways of dealing with LLW while still meeting the appropriate environmental and safety standards. On this basis, option 1b (near surface disposal) is preferred over option 1c (below surface disposal).

CONCLUSION

A comprehensive options assessment for the disposal of LLW arising at Hinkley Point A, including wide stakeholder participation, has been conducted. On-site disposal in a near-surface facility has been selected as the preferred option, for the following reasons: it is a preferred option within the MADA; on balance it is consistent with local and national planning policies; it is largely consistent with stakeholder views; it is consistent with the majority and most likely of the suggested site end-uses; it is 'fit-for-purpose' but nevertheless has a radiological performance far better than current standards require; it is safer for workers than other variants of on-site disposal; it reserves remaining capacity at the LLWR near Drigg in Cumbria for those wastes which require such a facility; it is in accordance with international LLW disposal practice; it is in accordance with NDA Strategy; and it is consistent with Government policy. Subject to ongoing stakeholder engagement, Magnox Electric Ltd intends to submit a planning application for an on-site disposal facility at Hinkley Point A in the near future.

REFERENCES

1. Defra, DTI and the Devolved Administrations, Policy for the Long Term Management of Solid Low Level Radioactive Waste in the United Kingdom, 2007.
2. EA/SEPA, Guidance for the Environment Agencies' Assessment of Best Practicable Environmental Option Studies at Nuclear Sites', 2004.
3. DTI, Review of Radioactive Waste Management: Final Conclusions, Cm2919, 1995.
4. Defra, DTI and the Devolved Administrations, A Public Consultation on Policy for the Long Term Management of Solid Low Level Radioactive Waste in the United Kingdom, 2006.
5. International Atomic Energy Agency, The Principles of Radioactive Waste Management, IAEA Safety Series No 111-F, 1995.
6. Magnox Electric Ltd, Minutes of a Generic LLW Optioneering Workshop, London, 27th February 2006.
7. Magnox Electric Ltd, Analysis of Options for the Disposal of Decommissioning Low-Level Radioactive Waste Arising from Hinkley Point A Care and Maintenance Preparations, 2006.
8. Magnox Electric Ltd, Summary of Stakeholder Views on the Disposal of Low-Level Radioactive Waste from Hinkley Point A Care and Maintenance Preparations: Multi-attribute Decision Analysis Feedback and Collation of Other Issues, 2006.
9. H M Government, The UK Government Sustainable Development Strategy, Presented to Parliament by the Secretary of State for Environment, Food and Rural Affairs by Command of Her Majesty, Cm 6467, 2005.
10. Defra, Achieving a Better Quality of Life, Review of Progress Towards Sustainable Development, 2004.
11. Enviros Consulting Ltd, Site Decommissioning: Sustainable Practices in the Use of Construction Resources, Guidance on the Application of Sustainable Practices to the Management of Decommissioning Wastes from Nuclear Licensed Sites, CIRIA W009, 2005.
12. CoRWM, Managing our Radioactive Waste Safely, CoRWM's Recommendations to Government, 2006.
13. Magnox Electric Ltd, Planning Policy Issues Relating to On-Site Disposal of Low Level Radioactive Waste at Hinkley Point A, 2006.
14. Hinkley Point A Site Stakeholder Group, Hinkley Point A Site End State, Final Report, HINA/R/E&PE/023, 2007.
15. Nuclear Decommissioning Authority, NDA Strategy, 2006.

Mater. Res. Soc. Symp. Proc. Vol. 1107 © 2008 Materials Research Society

Cement Based Encapsulation Trials for Low-Level Radioactive Effluent Containing Nitrate Salts

Atsushi Sugaya, Kenichi Horiguchi, Kenji Tanaka and Kentaro Kobayashi
Tokai Reprocessing Research and Development Centre, Japan Atomic Energy Agency (JAEA)
4-33 Muramatsu, Tokai-Mura, Naka-Gun, Ibaraki, 319-1194 Japan

ABSTRACT

The operation of nuclear fuel reprocessing plants generates a radioactive effluent containing nitrate salts as the major constituent. This waste must be disposed of safely and economically, and to achieve this aim the Japan Atomic Energy Agency (JAEA) is developing a cement based encapsulation method to immobilise this waste. Non-radioactive development work has been performed at both small and large-scale (up to 200 litres) to investigate the optimum cement formulation. The results from these studies demonstrate that nitrate waste that has been concentrated by evaporation of the water to a predetermined level can be successfully encapsulated up to a waste loading corresponding to 50 wt% sodium nitrate. It has been identified that high concentrations of bicarbonate ions in the effluent can have a detrimental effect on the strength of the cement encapsulated waste; however, provided this concentration is controlled, successful encapsulation of the effluent is still achieved.

1. INTRODUCTION

Nuclear fuel reprocessing plants generate a large volume of low-level radioactive effluent waste. It is necessary that this waste effluent is disposed of both safely and economically. In Japan this waste must be disposed of on land, and not at sea, by appropriate pre-treatment and burial underground. JAEA treats this low-level radioactive effluent waste by passing it through a nuclide separation process that produces two types of effluent waste:

- A nitrate effluent (~90% original effluent volume) with a comparatively low radioactivity level that can be disposed of by shallow burial at a lower cost;
- A slurry effluent (~10% original effluent volume) including several kinds of salts and with a relatively high radioactivity level that will require disposal deep underground.

The typical composition of the nitrate and slurry effluent is shown in Table I. Separation into two effluents reduces the cost of disposal because it allows the larger volume of waste to be disposed of by shallow burial at a much lower cost than deep underground burial. The latter is required for the smaller volume of slurry effluent or would have been needed for the untreated effluent.

Table I: Typical salt composition of a nitrate and a slurry effluent

Waste	$NaNO_3$	$NaNO_2$	$NaHCO_3$	Na_2SO_4	$Fe(OH)_3$
Nitrate effluent (wt%)	98	-	-	2	-
Slurry effluent (wt%)	75.5	10	12	1	0.5

Following separation into two types of waste effluent, water is evaporated from both types of waste to minimise their volume (the total salt content of the two effluents after evaporation is about 70 wt%). So that the effluents can be disposed of safely it has been proposed that they are encapsulated in a cement matrix. This approach produces a solid material that is easier to handle and transport than the liquid effluent, and can be stored safely prior to eventual disposal.

To investigate methods for encapsulating these two effluent wastes, non-radioactive trials were performed using 'waste simulants'; for the nitrate effluent a solution of sodium nitrate was used, while for the slurry effluent the appropriate salt ratios (Table I) in solution were used.

The work presented here gives some of the results from our study of the ability of three cement formulations to encapsulate these two waste simulants.

The cements used were ordinary Portland cement (OPC) (conforming to JIS 5210[1)]), a blast furnace slag (BFS) and OPC mixture (conforming to JIS 5211), and 'Super Cement' (SC) that is produced by the JGC Corporation (Japan). The importance of the water-to-cement ratio used to immobilise the waste effluent has also been analysed, and the effect of the bicarbonate ion concentration in the slurry effluent has been investigated. In these trials the optimum conditions for encapsulation have been determined by small-scale trials and then demonstrated at full 200-litre scale.

2. EXPERIMENTAL

2.1 Small-scale trials

Small-scale trials were used to investigate the optimum cement type and the effect of the water-to-cement ratio on encapsulating nitrate effluent; in addition, the effect of the bicarbonate concentration in the slurry effluent was analysed at small-scale. Small-scale trials (~2 litres) were performed using a low-shear mixer (described in the Japanese Industrial Standard R 5201). Waste simulant was mixed with a hardener (aqueous sodium hydroxide), disperser ('WORK500' that is produced by the Zeon Corporation) and antifoamer ('NS-DEFOAMER 171' that is produced by the San Nopco Limited) at low speed, followed by the addition of cement powders. Once all powders had been fed into the mixer, high speed mixing was performed for a further five minutes. The mixing paddle was removed and the cement was sealed in a container to prevent loss of water and cured at ambient temperatures.

Each sample was tested to assess:

- Fluidity of the cement – the cone flow table test was used: 344 cm^3 grout is poured into a flow cone on a table; the cone is lifted and the cement allowed to flow from its base across the table. The diameter of the resulting circle of grout in two perpendicular directions is averaged to give a measurement of the fluidity. The acceptance criterion for fluidity is a minimum diameter of 250 mm.
- Time for the cement to set – the acceptance criterion is within one day, which is required because of the restricted curing space available.
- Presence of bleed – the acceptance criterion is no bleed present, because the presence of bleed can indicate insufficient cement hydration. In addition, it is undesirable to have free liquid present in a waste package.
- Compressive strength – the acceptance criterion is over 10 MPa after 28 days curing. This gives a significant margin over the waste transportation requirement of approximately 8 MPa.

2.2 Full-scale trials

Full-scale trials were performed using 200-litre drums. The waste simulants were prepared by mixing the salt solutions in 200-litre drums using a mixing paddle, while heating the solution to 50°C to simulate the elevated waste temperature produced from the evaporation and concentration of the effluent. Sodium hydroxide, disperser and antifoam were added before feeding the cement powders into the drum while mixing. Mixing was continued for a further five minutes once the addition of cement powders had been completed.

Fluidity measurements, setting time and bleed observations were performed as described for the small-scale trials, and after 28 days of curing cores were bored from the 200-litre drums and tested for compressive strength tests and density. Homogeneity of the sample was assessed by the difference in strength and density along the length of the cores (at 10 different points) to give an indication of the performance of the mixing paddle and operating conditions.

3. RESULTS AND DISCUSSION

3.1 Small-scale trials – Treating nitrate effluent: the effect cement type and nitrate loading

Three cement types were used in these trials: 100% OPC, BFS/OPC mixture in a ratio 70:30, and 100% SC; in each case a water-to-cement ratio was chosen so that a fluidity of about 280 mm was achieved. For each cement type a series of different nitrate loadings were prepared: 30, 40, 50, 60 and 70 wt% $NaNO_3$ (given as the percentage weight of sodium nitrate in the cemented product). Thus, in total 15 trials were performed.

In the cases of OPC and 70:30 BFS/OPC, bleed was present in all trials, with this bleed remaining after 4 or 5 days of curing. This compares with no bleed present for the SC trials. Pictures of samples are shown in Figure 1, while Figure 2 shows the compressive strength of the samples after about 28 days curing. For all three cements a decrease in compressive strength was observed with increasing nitrate salt loadings. In the case of OPC and BFS/OPC 70:30 cements, with a nitrate loading of 70 wt% the compressive strength fell below the acceptance criterion of 10 MPa. In contrast, the SC trials still showed acceptable compressive strength even for the highest nitrate salt loading, for which its strength was 20 MPa.[2)]

Owing to the superior performance of SC in these trials, this cement powder has been chosen to be investigated further for the encapsulation of nitrate and slurry effluent at JAEA.

(a) 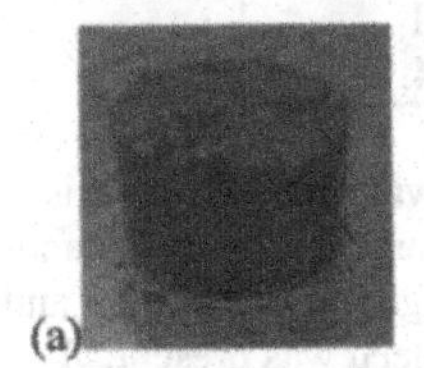
(b) 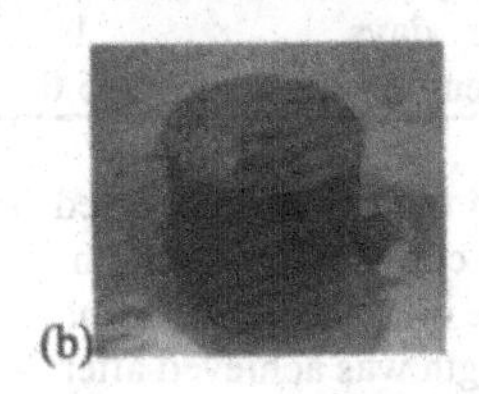
(c) 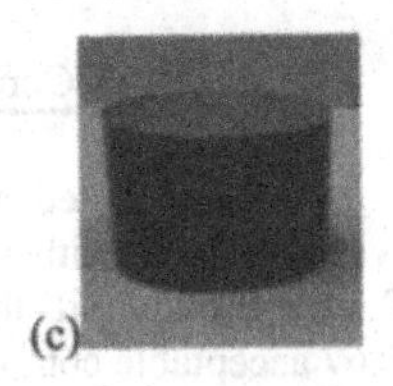

Figure 1: Cement samples (a) OPC, $NaNO_3$ loading 30 wt%, 4 days curing; (b) 70:30 BFS/OPC, $NaNO_3$ loading 30 wt%, 5 days curing; (c) SC, $NaNO_3$ loading 70 wt%, 1 day curing.

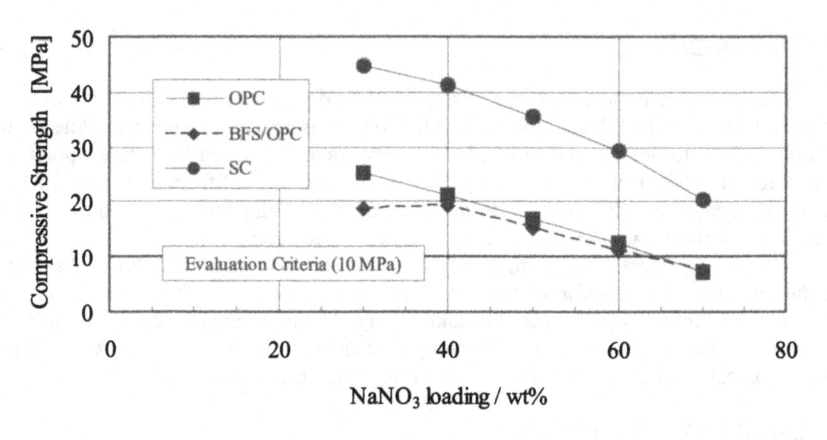

Figure2: Compressive strength measurements after 28 days curing.

3.2 Small-scale trials – Treating nitrate effluent: the effect of water-to-cement ratio

Water-to-cement (w/c) ratios in the range 0.50 to 0.83 were investigated using a sodium nitrate loading of 50 wt% and effluent temperature 80°C to reflect the typical nitrate concentration that can be achieved by the evaporator. The details of the trials and results are summarised in Table II.

Table II: The effect of water-to-cement ratio for SC and nitrate effluent simulant; nitrate loading 50wt%, effluent temperature 80°C.

	Water-to-cement ratio			
	0.82	0.67	0.59	0.50
$NaNO_3$ / wt%	50	50	50	50
Water / wt%	22.5	20	18.5	16.7
SC / wt%	27.5	30	31.5	33.
Fluidity test / mm	293	266	208	154
Bleed present (yes/no)	No	No	No	No
Setting time / days	1	1	1	1
Compressive strength / MPa	35.0	40.4	44.1	42.3

As expected, a reduction in fluidity was observed as the w/c ratio was reduced, but this was associated with an increase in compressive strength. For a water-to-cement ratio of 0.6 and 0.5 the fluidity fails the acceptance criterion of 250 mm, however at the higher ratios of 0.83 and 0.67 acceptable compressive strength was achieved after 28 days and no bleed was present. The acceptable range of the w/c ratio for SC is high compared to the usual value for OPC of 0.35.

This may be important because there is a limit to the salt concentration that can be achieved by the evaporator, and so the higher w/c ratio used with SC will minimize the amount of cement required and hence increase the sodium nitrate loading of the encapsulated waste.

3.3 Small-scale trials – Treating slurry effluent: the effect of bicarbonate concentration

A further set of trials investigated the effect of the bicarbonate concentration in the slurry effluent waste on the properties of the cemented waste. The nitrate waste loading, w/c ratio, and the $NaHCO_3$ concentration prior to evaporation (to a total salt concentration of 70 wt%) used in these trials are given in Table III together with the results. As the sodium bicarbonate concentration is increased there is a significant reduction in the fluidity and compressive strength and an undesirable increase in the setting time for the cement. These results suggest that the $NaHCO_3$ concentration must be below 10 g/l (prior to evaporation) in order to achieve the required strength, fluidity and setting time.

Table III: Effect of $NaHCO_3$ on the slurry effluent encapsulation; effluent temperature 50°C.

	$NaHCO_3$ Concentration before evaporation (g/l)		
	10	49	85.7
$NaNO_3$ / wt%	41	38	37
Other salts / wt%	9	12	13
Water / wt%	20	25	23
SC / wt%	30	25	27
w/c	0.67	1.00	0.85
Fluidity test / mm	301	276	209
Bleed present (yes/no)	No	No	No
Setting time / days	1	2	4
Compressive strength / MPa	18.6	12.6	10.9

3.4 Full-scale 200 litre trials

Full-scale 200-litre trials were performed using SC to immobilise nitrate effluent simulant and slurry effluent simulant. The formulations used for these trials are summarised in Table IV.

Table IV: Details of full-scale encapsulation trials.

Composition	Nitrate effluent	$NaHCO_3$ Concentration before evaporation(g/l)		
		10	45.6	45.6
$NaNO_3$ / wt%	50	41	38	38
Other salts / wt%	0	9	12	12
Water / wt%	15	20	25	20
SC / wt%	35	30	25	30
w/c	0.43	0.67	1.00	0.67

A summary of the results are given in Table V, while Figure 3 shows an example of a 200-litre drum trial after core sampling. These results demonstrate that homogeneous samples were achieved using the chosen mixing method to encapsulate the nitrate effluent, because only a small range of compressive strengths and specific gravities were measured along the length of the cores extracted from the trial drums. As the concentration of $NaHCO_3$ was increased it was found that the compressive strength decreased, which is consistent with the small scale trials undertaken, and it was confirmed that adequate compressive strength is achieved provided the $NaHCO_3$ concentration does not exceed 10 g/l (prior to evaporation) with a total salt loading of 50 wt% at full 200-litre scale.[3)]

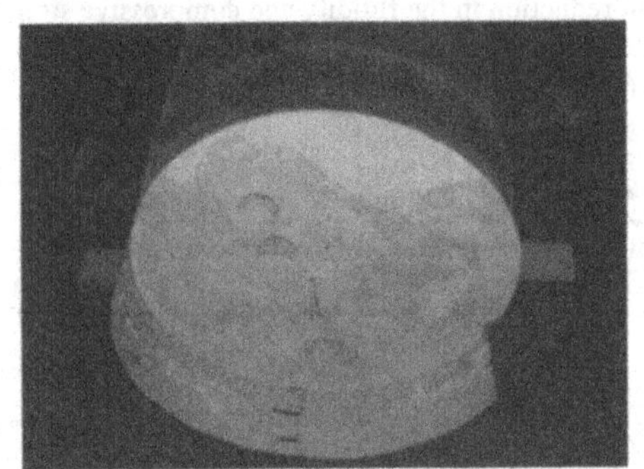

Figure 3: Example of 200-litre drum trial after core sampling

Table V: Results of full-scale trials. The range of compressive strength and specific gravity values recorded along the length of cores are shown.

Measurement	Nitrate effluent	$NaHCO_3$ Concentration before evaporation(g/l)		
		10	45.6	45.6
Fluidity test / mm	450	282	348	252
Bleed present (yes/no)	No	No	No	No
Setting time / days	1	1	6	5
Compressive strength / MPa	18.5 - 19.9	11.9 - 14.0	3.9 - 8.3	4.7 - 15.6
Setting temperature (maximum) / °C	60.8	55.5	69.4	73.1
Specific gravity	1.87 - 1.89	1.76 - 1.81	1.70 - 1.78	1.71 - 1.79

4. CONCLUSIONS

SC gives better encapsulation performance than either OPC or BFS/OPC formulations for the nitrate effluent, with the latter two cement formulations leading to bleed and a lower compressive strength than SC.

SC has been demonstrated to be capable of encapsulating nitrate effluent up to a sodium nitrate loading of 70 wt% at small scale and 50 wt% at full 200-litre scale. In these cases, adequate compressive strength, setting time and an absence of bleed is achieved.

Owing to the limit on the salt concentration that can be achieved by the evaporator, the capacity of SC to set with high w/c ratios is beneficial in minimizing the amount of cement required, and hence raising the overall sodium nitrate loading in the encapsulated waste.

In the case of the slurry effluent, where $NaHCO_3$ is present, encapsulation of the waste can be achieved provided the concentration of $NaHCO_3$ does not exceed 10 g/l prior to evaporation of the effluent.

REFERENCES

1. Japanese Industrial Standards : developed and published by the Japanese Industrial Standards Committee.
2. Y.SUZUKI *et al.* : *2003 Autumn Meeting of Atomic Energy Society of Japan, Shizuoka, Japan,* I65, (2003).
3. K.HORIGUCHI *et al.* : *2007 Autumn Meeting of Atomic Energy Society of Japan, Kitakyushu, Japan*, in print, (2007).

Glasses and Glass Composite Materials

Mater. Res. Soc. Symp. Proc. Vol. 1107 © 2008 Materials Research Society

High Level Waste (HLW) Vitrification Experience in the US: Application of Glass Product/Process Control to Other HLW and Hazardous Wastes

Carol M. Jantzen and James C. Marra
Savannah River National Laboratory
Aiken, SC 29808

ABSTRACT

Vitrification is currently the most widely used technology for the treatment of high level radioactive wastes (HLW) throughout the world. At the Savannah River Site (SRS) actual HLW tank waste has successfully been processed to stringent product and process constraints without any rework into a stable borosilicate glass waste since 1996. A unique "feed forward" statistical *process* control (SPC) has been used rather than statistical *quality* control (SQC). In SPC, the feed composition to the melter is controlled *prior* to vitrification. In SQC, the glass product is sampled *after* it is vitrified. Individual glass property models form the basis for the "feed forward" SPC. The property models transform constraints on the melt and glass properties into constraints on the feed composition. The property models are mechanistic and depend on glass bonding/structure, thermodynamics, quasicrystalline melt species, and/or electron transfers. The mechanistic models have been validated over composition regions well outside of the regions for which they were developed because they are mechanistic. Mechanistic models allow accurate extension to radioactive and hazardous waste melts well outside the composition boundaries for which they were developed.

INTRODUCTION

Borosilicate glasses have been used in the US and in Europe to immobilize radioactive HLW for ultimate geologic disposal. Vitrification has also been developed as a technology to immobilize low activity waste, low-level wastes, mixed (radioactive and hazardous) wastes, and TRU wastes in durable glass formulations for permanent disposal and/or long-term storage. Waste glass formulations must maximize the amount of waste to be vitrified so that waste glass volumes and the associated storage and disposal costs are reduced. Moreover, glass formulation optimization for HLW [1,2,3] or other wastes must simultaneously balance multiple product/ process (P/P) constraints (Table I).

Table I. Waste Glass Product and Process Constraints

Product Constraints	Process Constraints
chemical durability	melt viscosity
glass homogeneity	liquidus
thermal stability	waste solubility
regulatory compliance	melt temperature/corrosivity
mechanical stability	radionuclide volatility
	REDOX*

* controls foaming and melt rate

Most P/P properties, other than melt temperature, cannot be measured directly. The waste streams are often highly variable and difficult to characterize. In addition, in the US, the P/P constraints must be satisfied to a very high degree of certainty (>95%) as the canister geometry makes rework (remelting) of the product impossible. This requires a "systems approach" so that the P/P constraints can be optimized simultaneously [1]. The "systems approach" ensures that the final product safeguards the public, and that the production process used is safe to operate.

The successful "systems approach" used at the Savannah River Sites HLW Defense Waste Processing Facility (DWPF) for the past 11 years is based on "feed forward statistical process control." The feed composition to the melter is controlled *prior* to vitrification and a confirmatory glass sample is taken only once every 2-3 years. The feed composition is used to calculate the P/P properties of a melter feed from mechanistic P/P models that relate the melt composition to the P/P properties [2,3]. These models are the foundation of the SPC system used to monitor and control glass composition for HLW (Product Composition Control System) [4]. Over the last 11 years of radioactive operation ~8.8 million liters of HLW sludge have been vitrified at the DWPF into 4.2 millions kilograms of borosilicate glass.

The mechanistic models can be extrapolated well outside the glass composition range for which they were developed as will be shown in this study. These models can, therefore, be directly applied to other types of HLW, MW, and TRU wastes.

PRODUCT CONSTRAINT: DURABILITY

The most important glass product property is the glass durability. The durability of a waste glass is the single most important variable controlling release of radionuclides and/or hazardous constituents. The intrusion of groundwater into, and passage through, a waste form burial site in which the waste forms are emplaced is the most likely mechanism by which constituents of concern may be removed from the waste glass and carried to the biosphere. Thus it is important that waste glasses be stable in the presence of groundwater.

The DWPF durability model is known as the Thermodynamic Hydration Energy Reaction MOdel (THERMO™). [5,6] THERMO™ estimates the relative durability of silicate and borosilicate glasses based on their compositions. THERMO™ calculates the thermodynamic driving force of each glass component to hydrate based on the mechanistic role of that component during dissolution, e.g. ion exchange, matrix dissolution, accelerated matrix dissolution, surface layer formation, and/or oxidative dissolution. The overall tendency of a given glass to hydrate is expressed as a preliminary glass dissolution estimator, e.g. the change in the free energy of hydration of a glass (ΔG_p) based solely on its composition. For glasses that undergo accelerated matrix dissolution, an accelerated hydration free energy, ΔG_a, is calculated from known strong base [SB] weak acid [WA] equilibrium. The ΔG_a term is additive to ΔG_p such that the overall durability of the glass, expressed as the final hydration free energy (ΔG_f), can be predicted, e.g. $\Delta G_f = \Delta G_p + \Delta G_a$. The more negative the ΔG_f the more readily the hydration reaction will occur and the less durable the glass.

Recently, Jantzen and Pareizs [7] have proposed an Activated Complex Theory (ACT) model based on the early work of Helgeson [8] and the more recent work of Oelkers [9] on basalt glass dissolution. This approach attempts to define the activated complexes that participate in the irreversible formation of the glass gel layer. The formation of the hydrated gel layer is the irreversible step. The leached layer exhibits acid/base properties which are manifested as the pH dependence of the thickness and nature of the gel layer. The gel layer has been found to age into either clay mineral assemblages or zeolite mineral assemblages. The formation of one phase preferentially over the other has been experimentally related to changes in the pH of the leachant and related to the relative amounts of Al^{+3} and Fe^{+3} in a glass. The formation of clay mineral assemblages on the leached glass surface layers (lower pH and Fe^{+3} rich glasses) causes the dissolution rate to slow to a long-term "steady state" rate. The formation of zeolite mineral assemblages (higher pH and Al^{+3} rich glasses) on leached glass surface layers causes the dissolution rate to increase and return to the initial high forward rate. The return to the forward dissolution rate is undesirable for long-term performance of glass in a disposal environment.

The ACT approach [7] models the role of glass stoichiometry, in terms of the quasi-crystalline mineral species (mineral moieties) in a glass. The stoichiometry of the mineral moieties in the parent glass appears to control the activated surface complexes that form in the leached layers, and these "mineral" quasi-crystals (some Fe^{+3} rich and some Al^{+3} rich) play a role in whether or not clays or zeolites are the dominant species formed on the leached glass surface. The chemistry and structure, in terms of Q distributions of the parent glass, are well represented by the atomic ratios of the glass forming components. Thus, glass dissolution modeling using simple atomic ratios is shown to represent the structural effects of the glass on the dissolution and the formation of activated complexes in the glass leached layer. This provides two different methods by which a linear glass durability model can be formulated. One based on the quasi-crystalline mineral species in a glass and one based on cation ratios in the glass: both are related to the activated complexes on the surface by the law of mass action.

The ACT model was based on some of the same data used in the development of the THERMO™ model including glasses from a round robin on the Waste Compliance Plan (WCP) glasses that span the entire range of the glasses anticipated for processing at the SRS [10], and glasses from a round robin conducted on the Environmental Assessment glass [11,12]. The glasses modeled included glasses made in crucibles and glasses made in large scale pilot scale melters. In addition, data was included in ACT from full scale canisters poured during the non-radioactive startup of the DWPF during Qualification Runs (sections and grab samples), and radioactive glasses from the SRS M-Area facility [13]. While the DWPF glasses are enriched in Fe_2O_3 compared to Al_2O_3, the M-Area glasses are enriched in Al_2O_3 and deficient in Fe_2O_3. The ranges of glass compositions modeled in THERMO™ and ACT are given in Table II.

The geochemical code EQ3/EQ6 was used to model the leachate compositions from short and long term ASTM C1285 (PCT) tests to determine what phases could precipitate, e.g. what phase was each leachate supersaturated with respect to. The EQ3/EQ6 predictions were coupled with the glass composition data in ACT and this provided a link between the atomic ratios of the glasses and the leachate super saturation with respect to either analcime or ferrite phases [7]. Thus glass composition, in terms of quasi-crystalline structural ratios could be used to determine if a glass would form analcime and return to the forward rate or not. Since the pH of a static leachate is also driven by the glass composition and is a parameter entered into the EQ3/EQ6 model, it was not considered as a separate parameter during modeling. The use of the glass atomic ratios determined in this manner correctly predicted the well studied [14] PAMELA glasses SM58 and SAN60 glasses (Table II). The former did not return to the forward rate and the latter glass did.

The data in Table II demonstrates that the ACT durability model covers a wider composition range than both the THERMO™ durability model data and the THERMO™ validation data. This allows either the ACT model or the THERMO™ model to be applied to broader composition ranges of LLW, TRU, and mixed waste glasses than either was developed for since both models are based on known dissolution mechanisms.

Table II. Oxide Ranges of Durability Model and Validation Glasses Compared to those of Van Isenghem and Grambow [14]

Oxide (wt%)	THERMO™ Model Range	THERMO™ Validation	ACT Model Range	SM58	SAN60
Al_2O_3	1.36‡-13.90	0.56‡-22.80	2.99‡-25.04	1.20‡	18.10
B_2O_3	6.10-13.30	4.31-21.19	3.48-13.65	12.30	17.00
BaO	0.00-0.66	0.00-0.19	0.00-0.25	0.00	0.00
CaO	0.38-2.23	0.00-8.68	0.00-8.68	3.80	3.50
Ce_2O_3	0.00-1.44	0.00-0.02	0.00-0.14	0.00	0.00
Cr_2O_3	0.00-0.55	0.00-0.86	0.00-0.86	0.00	0.00
Cs_2O	0.00-1.16	0.00-0.26	0.00-0.12	0.00	0.00
FeO	0.00-8.81	0.00-2.57	0.00-3.99	0.00	0.00
Fe_2O_3	0.00-14.30	0.00-20.77	0.00-20.77	1.20	0.30
K_2O	0.00-5.73	0.00-7.21	0.00-4.81	0.00	0.00
La_2O_3	0.00-0.42	0.00-0.03	0.00-0.42	0.00	0.00
Li_2O	2.59-5.16	0.00-5.41	0.00-11.15	3.70	5.00
MgO	0.00-3.24	0.00-3.79	0.00-1.86	2.00	0.00
MnO	0.00-3.36	0.00-5.09	0.00-5.09	0.00	0.00
MoO_3	0.00-1.67	0.00-0.02	0.00-0.22	0.00	0.00
Na_2O	6.42-16.80	4.26-24.43	2.84-24.43	8.30	10.70
Nd_2O_3	0.00-5.96	0.00-0.36	0.00-0.67	0.00	0.00
NiO	0.00-2.97	0.00-2.57	0.00-1.77	0.00	0.00
P_2O_5	0.00-0.65	0.00-0.59	0.00-3.08*	0.00	0.00
PbO	0.00-0.25	0.00-0.28	0.00-0.25	0.00	0.00
SiO_2	39.80-59.80	38.72-63.75	37.79-68.50	56.90	43.40
SrO	0.00-0.45	0.00-0.05	0.00-0.16	0.00	0.00
TiO_2	0.00-3.21	0.00-1.05	0.00-1.71	4.40	0.00
U_3O_8	0.00	0.00	0.00-5.66	0.00	0.00
ZnO	0.00-1.46	0.00-0.33	0.00-0.44	0.00	0.00
ZrO_2	0.00-1.80	0.00-1.46	0.00-1.25	0.00	0.00
Fission Products and Actinides			N/A	6.20	2.0

‡ During development of THERMO™ it was determined that a minimum of 3 wt% Al_2O_3 was necessary in high Fe_2O_3 containing and high Na_2O containing glasses to avoid phase separation [15]. This is consistent with the known miscibility gap in the Al_2O_3-Fe_2O_3-Na_2O-SiO_2 quaternary system that defines the crystallization of basalt [15].

* During development of glass durability models, glasses with P_2O_5 values in excess of 2.6 wt% were shown to exhibit crystalline phase separation (CPS) [13,16,17].

PROCESS CONSTRAINTS: VISCOSITY AND RESISTIVITY

The viscosity of a waste glass melt as a function of temperature is the single most important variable affecting the melt rate and pourability of the glass. The viscosity determines the rate of melting of the raw feed, the rate of glass bubble release (foaming and fining), the rate of homogenization, and thus, the quality of the final glass product. If the viscosity is too low, excessive convection currents can occur, increasing corrosion/erosion of the melter materials (refractories and electrodes) and making control of the waste glass melter more difficult. Waste glasses are usually poured continuously into steel canisters or cans for ultimate storage. Glasses with viscosities >500 poise do not readily pour. Moreover, too high a viscosity can reduce product quality by causing voids in the final glass. Therefore, a range of viscosities between 20 and 110 poise at T_{melt}, are currently being used for Joule heated waste glass melters.

The approach taken in the development of the viscosity and resistivity process models [2,18,19] was based on glass structural considerations, expressed as a calculated non-bridging oxygen (NBO) term. This NBO parameter represents the amount of structural depolymerization in the glass. Calculation of the NBO term from the glass composition was combined with quantitative statistical analyses of response surfaces to express glass viscosity and glass resistivity as a function of both melt temperature and glass composition. The model was developed on as made compositions and recently revised [19] based on analyses of the same non-radioactive glasses and frits (220 viscosity-temperature measurements). During revision the model was validated [19] on an additional 200 glasses (radioactive and non-radioactive and 1004 viscosity-temperature pairs) (Table III). Uranium was shown to have no impact on glass viscosity and ThO_2 at <1 wt% had no impact on glass viscosity.

Table III. Oxide Ranges of Viscosity Model and Validation Glasses

Parameters and Oxide (wt%)	Viscosity Model	Viscosity Validation	Parameters and Oxide (wt%)	Viscosity Model	Viscosity Validation
Temperature (°C)	873-1491	803.35-1491	La_2O_3	0.00-0.36	0.00-8.62
Viscosity (poise)	10.23-1122.02	10.23-11,000	Li_2O	2.59-6.96	0.00-17.74
Al_2O_3	0.00-13.90	0.00-29.02	MgO	0.49-2.92	0.00-4.80
B_2O_3	6.41-12.20	4.33-13.25	MnO	0.00-3.26	0.00-4.02
BaO	0.00-0.20	0.00-0.50	Na_2O	5.80-15.80	5.80-16.8
CaO	0.00-1.47	0.00-1.79	NiO	0.00-2.97	0.00-3.01
Cr_2O_3	0.00-0.09	0.00-1.18	SiO_2	45.60-77.04	34.15-77.04
Cs_2O	0.00-0.15	0.00-0.67	SrO	0.00-0.07	0.00-0.18
CuO	0.00-0.33	0.00-0.51	ThO_2	0.00	0.00-0.06
Cu_2O	0.00-0.30	0.00-0.82	TiO_2	0.00-1.78	0.00-3.10
FeO	0.00-7.14	0.00-7.14	U_3O_8	0.00	0.00-5.76
Fe_2O_3	0.00-14.20	0.00-16.86	ZnO	0.00	0.00-0.21
K_2O	0.00-5.73	0.00-5.84	ZrO_2	0.00-0.99	0.00-2.00

The viscosity model has been validated over composition and temperature regions (800-1500°C) well outside of the regions for which it was developed (Table III) because it is based on known glass structural mechanisms. This affords the ability to use the viscosity model for the broader composition ranges of LLW, TRU, and mixed wastes.

PROCESS CONSTRAINTS: LIQUIDUS

A liquidus temperature model prevents melt pool or volume crystallization during operation. Volume crystallization needs to be avoided because it can involve almost simultaneous nucleation of the entire melt pool as volume crystallization can occur very rapidly. Furthermore, once spinel crystals are formed (the most ubiquitous liquidus phase occurring in US defense HLW), these crystals are refractory and cannot be redissolved into the melt pool. The presence of either the spinel or nepheline liquidus phases may cause the melt viscosity and resistivity to increase which may cause difficulty in discharging glass from the melter as well as difficulty in melting via Joule heating. Once a significant amount of volume crystallization has occurred and the resulting crystalline material has settled to the melter floor, melting may be inhibited and the pour spout may become partially or completely blocked making pouring difficult.

The crystal-melt equilibria were modeled based on quasicrystalline concepts [20,21]. A pseudobinary phase diagram between a ferrite spinel (an incongruent melt product of transition metal iron rich acmite) and nepheline was defined. The pseudobinary lies within the Al_2O_3-Fe_2O_3-Na_2O-SiO_2 quaternary system that defines the crystallization of basalt glass melts (note that the basalt glass system is used as an analogue for waste glass durability, liquidus, and the prevention of phase separation). The liquidus model developed based on these concepts has been used to prevent unwanted crystallization in the DWPF HLW melter for the past six years while allowing >10 wt% higher waste loadings to be processed. The liquidus model and the pseudobinary are shown [20,21] to be consistent with all of the thermal stability data generated on DWPF HLW glasses. The model ranges developed on 105 different glass compositions and validation ranges (161 glasses) are given in Table IV.

The liquidus model has been validated over composition regions well outside of the regions for which it was developed (Table IV) because it is based on quasicrystalline melt theory. This affords the ability to use the liquidus model for the broader composition ranges waste glasses.

Table IV. Oxide Ranges of Liquidus Model and Validation Glasses

Oxide Species (wt%)	Liquidus Model Ranges	Validation Ranges	Oxide Species (wt%)	Liquidus Model Ranges	Validation Ranges
Al_2O_3	0.99-14.16	0.00-16.734	Li_2O	2.49-6.16	0.00-7.499
B_2O_3	4.89-12.65	0.00-19.996	MgO	0.47-2.65	0.00-7.31
CaO	0.31-2.01	0.00-10.3	MnO	0.74-3.25	0.00-4.00
Cr_2O_3	0.00-0.30	0.00-1.2	Na_2O	5.99-14.90	4.996-22.737
FeO	0.02-6.90	0.02-6.90	NiO	0.04-3.05	0.00-3.05
Fe_2O_3	3.43-16.98	3.43-16.98	SiO_2	41.80-58.23	29.979-58.23
K_2O	0.00-3.89	0.00-4.002	TiO_2	0.00-1.85	0.00-5.003

PROCESS CONSTRAINTS: SULFATE SOLUBILITY

Sulfate and sulfate salts are not very soluble in borosilicate waste glass. The alkali and alkaline earth sulfate salts, in conjunction with alkali chlorides, collect on the melt surface as a low melting (600-800°C), low density, low viscosity melt phase known as gall. At moderate concentrations, the salts have a beneficial effect on melting rates. At excessively high feed concentrations, molten alkali sulfates float on the surface of the melt pool or become trapped as inclusions in the glass. Soluble sulfate salts are often enriched in cesium and strontium, which can impact radionuclide release from the cooled glass if the salts are present as inclusions or a frozen gall layer.

The results of sulfate solubility measurements from both dynamic melter tests and static crucible tests performed with HLW waste simulants were compared. This data was also compared to Slurry-Fed Melt Rate (SMRF) data generated on HLW melts. In addition, a survey was made of both dynamic and crucible tests for Low Activity Wastes (LAW) and crucible tests performed with commercial soda-lime-silica glasses. Phenomenological observations in the various studies, e.g. completeness or lack of gall and secondary sulfate phases, were categorized into melt conditions representing "at saturation, over saturation, and super saturation." This enabled modeling of the most desirable "at saturation" conditions, e.g. no appearance of a sulfate layer on the melt pool, in relation to undesirable conditions of over saturation (partial melt pool coverage) and super saturation (almost complete melt pool coverage). Sulfate solubility was determined to be related to melt polymerization and so the HLW viscosity model given in this study was used to define the sulfate solubility [22,23].

PROCESS CONSTRAINT: REDuction/OXidation (REDOX) EQUILIBRIUM

Control of the REDuction/OXidation (REDOX) equilibrium in the DWPF melter is critical for processing high level liquid wastes. Foaming, cold cap roll-overs, and off-gas surges all have an impact on pouring and melt rate during processing of waste glass. All of these phenomena can impact waste throughput and attainment. These phenomena are caused by gas-glass disequilibrium when components in the melter feeds convert to glass and liberate gases such as steam, CO_2, O_2, H_2, NO_x, and/or N_2. In order to minimize gas-glass disequilibrium a REDOX strategy is used to balance feed reductants and feed oxidants while controlling the REDOX between $0.09 \leq Fe^{2+}/\Sigma Fe \leq 0.33$. A $Fe^{+2}/\Sigma Fe$ ratio ≤ 0.33 prevents metallic and sulfide rich species from forming nodules that can accumulate on the floor of the melter. Control of foaming, due to deoxygenation of manganic species, is achieved by converting oxidized MnO_2 or Mn_2O_3 species to MnO during melter preprocessing. At the lower REDOX ratio of $Fe^{+2}/\Sigma Fe \sim 0.09$ about 99% of the Mn^{+4}/Mn^{+3} is converted to Mn^{+2} and foaming does not occur.

The REDOX model relates the $Fe^{+2}/\Sigma Fe$ ratio of the final glass to the molar concentrations of the oxidants and reductants in the melter feed. The REDOX model is based on Electron Equivalents (EE) that are exchanged during chemical reduction (making an atom or molecule less positive by electron transfer) and oxidation (making an atom or molecule more positive by electron transfer). Therefore, the number of electrons transferred for each REDOX reaction can be summed and an Electron Equivalents term for each organic and oxidant species defined [24,25,26]. The model accounts for reoxidation of the manganese by nitrate salts in the cold cap.

CONCLUSIONS

At the Savannah River Site (SRS) actual HLW tank waste has been processed into a stable borosilicate glass waste form since 1996 using a unique "feed forward" statistical *process* control (SPC) without the necessity of rework or melter outages due to incorrect processing parameters. The property models that form the basis of the SPC are mechanistic. The mechanistic models have been validated over composition regions well outside of the regions for which they were developed because they are based on known mechanisms. This affords the ability to use these models for the broader composition ranges of LLW, TRU, and mixed wastes. In particular, THERMO™ and VISCOMP™ were applied to the stabilization of the mixed high alumina M-Area wastes at the SRS in borosilicate glass.

ACKNOWLEDGMENTS

This paper was prepared in connection with work done under Contract No. DE-AC09-96SR18500 with the U.S. Department of Energy (DOE).

REFERENCES

1 C.M. Jantzen, J. Non-Crystalline Solids, 84, 215-225 (1986).
2 C.M. Jantzen, Ceram. Trans., V. 23, 37-51 (1991).
3 C.M. Jantzen, and K.G. Brown, Am. Ceramic Society Bulletin, 72, 55-59 (May, 1993).
4 K.G. Brown, R.L. Postles, and R.E. Edwards, Ceram. Trans., V. 23, 559-568 (1991).
5 C.M. Jantzen, K.G. Brown, T.B. Edwards, and J.B. Pickett, U.S. Patent #5,846,278, (December 1998).
6 C.M. Jantzen, J.B. Pickett, K.G. Brown, T.B. Edwards, and D.C. Beam, U.S. DOE Report WSRC-TR-93-0672, Westinghouse Savannah River Co., Savannah River Technology Center, Aiken, SC, 464p. (Sept. 1995).
7 C.M. Jantzen, and J.M. Pareizs, (accepted J. Nucl. Mat.).
8 H.C. Helgeson, W.M. Murphy, and P. Aagaard, Geochimica et Cosmochimica Acta, 48, 2405-2432 (1984).
9 E.H. Oelkers and S.R. Sislason, Geochim. Cosmochim. Acta, 65 [21], 3671-3681 (2001).
10 S.L. Marra, and C.M. Jantzen, U.S. DOE Report WSRC-TR-92-142 (May, 1992).
11 C.M. Jantzen, N.E. Bibler, D.C. Beam, and M.A. Pickett, U.S. DOE Report WSRC-TR-92-346, Rev.1, (February, 1993).
12 C.M. Jantzen, N.E. Bibler, D.C. Beam, D.C. and M.A. Pickett, Ceram. Trans. V. 39, 313-322 (1994).
13 C.M. Jantzen, K.G. Brown, J.B. Pickett, and G.L. Ritzhaupt, U.S. DOE Report WSRC-TR-2000-00339, (September 30, 2000).
14 P. Van Iseghem and B. Grambow, Sci. Basis for Nuclear Waste Mgt. XI, Mat. Res. Soc., Pittsburgh, PA, 631-639 (1987).
15 C.M. Jantzen, C.M. and Brown, K.G., Ceram. Trans., V. 107, 289-300 (2000).
16 C.M. Jantzen, U.S. DOE Report 86-389 (1986).
17 C.M. Jantzen, K.G. Brown, and J.B. Pickett, Ceram. Trans., V. 119, 271-280 (2001).
18 C.M. Jantzen, U.S. Patent #5,102,439, (April, 1992).
19 C.M. Jantzen, U.S. DOE Report WSRC-TR-2004-00311 (February 2005).
20 C.M. Jantzen, and K.G. Brown, J. Am. Ceramic Soc., 90 [6], 1866-1879 (2007).
21 C.M. Jantzen, and K.G. Brown, J. Am. Ceramic Soc., 90 [6], 1880-1891 (2007).
22 C.M. Jantzen, and M.E. Smith, U.S. DOE Report WSRC-TR-2003-00518 (January 2004).
23 C.M. Jantzen, D.K. Peeler, and M.E. Smith, M.E. Ceram. Trans. 168, 141-151 (2005).
24 C.M. Jantzen, J.R. Zamecnik, D.C. Koopman, C.C. Herman, and J.B. Pickett, U.S. DOE Report WSRC-TR-2003-00126, Rev.0 (May 2003).
25 C.M. Jantzen, D.C. Koopman, C.C. Herman, J.B. Pickett and J.R. Zamecnik, Ceram. Trans. V. 155, 79-91 (2004).
26 C.M. Jantzen, and M.E. Stone, US DOE Report WSRC-STI-2006-00066 (2007).

Mater. Res. Soc. Symp. Proc. Vol. 1107 © 2008 Materials Research Society

Nuclear Waste Immobilization by Vitrification in a Cold Crucible Melter: 3D Magnetic Model of a Melter

A. Bonnetier
Commissariat à l'Énergie Atomique (CEA) Marcoule, BP 17171, 30200 Bagnols-sur-Cèze Cedex, France
E-Mail: armand.bonnetier@cea.fr

Abstract – *The design and development of prototype cold crucible melters for waste vitrification are based on models of the basic physical phenomena, including electromagnetic induction and the thermal and hydraulic properties in natural or forced convection. The complexity of new nuclearized facilities results in significant errors on the results of predictive models based on 2D axisymmetric geometry that can only be resolved by modeling the device in 3D geometry. This document discusses the specification and electromagnetic design of a melter carried out using electromagnetic computation software, FLU3D, developed in 3D geometry by Cedrat. The principles and results of this study are directly applicable to nuclear facilities with allowance for the particular requirements of a nuclearized environment.*

INTRODUCTION

Direct induction cold crucible melters have been developed for more than 20 years to vitrify high- and medium-level nuclear waste. The design and dimensions of the prototype melters depend on the operational requirements and on the desired melting capacity. The specific features of each melter depend on whether it will be used to vitrify solid or liquid feed, or as part of a sequential incineration-vitrification process.

INDUCTION MELTING OF GLASS

Nuclear Waste Vitrification

The nuclear waste vitrification process must meet multiple constraints both on the melting facility itself and on the end product. A mixture of fission products and glass frit produces a molten glass matrix that is poured into a stainless steel canister. Ideally it must be as stable and homogeneous as possible, since the waste containment properties depend on the matrix quality.

Although vitrification techniques have changed considerably over the years, they still require large, complex industrial facilities and must comply with stringent safety criteria concerning reliability and wear because they are designed for continuous operation. With the existing hot crucibles the physical and chemical reactions occurring at the glass/crucible interface and in the molten glass limit the melter lifetime to about 5000 hours before corrosion becomes critical.

In industrial operation the atmosphere around the melter is highly radioactive and must be isolated from the exterior by very thick concrete containment enclosures. Any operations, including maintenance, must therefore be performed using telemanipulator arms, complicating the tasks and requiring prior "nuclearization".

Principle of Direct Induction Melting

Direct induction melting implies supplying energy directly to the molten glass by induction heating. Direct induction avoids any contact between the energy source (the inductor) and the product being heated. Induction melting is based on magnetic induction, a phenomenon discovered by Michael Faraday in 1831. Currents from a frequency converter are fed to a copper inductor, generating a magnetic flux. When a conducting material is placed in such a magnetic field, induced currents flow inside it. These eddy currents cause the temperature of the material to rise by Joule effect.

When a solenoid is supplied with an AC voltage a magnetic flux is produced in and around the solenoid. When a conducting plate is exposed to a portion of this magnetic flux, currents are induced in the plate. These eddy currents in turn produce a secondary magnetic flux that tends to oppose the initial flux.

Glass is a good electrical insulator at room temperature, and its electrical conductivity increases with the temperature. The chemical elements constituting the basic molten glass matrix are therefore selected to obtain physical properties sufficient to be considered electrically conducting.(1)

THE CONVENTIONAL COLD CRUCIBLE MELTER

Description

The molten glass is contained in a sectorized circular cold crucible; the metal sectors are internally cooled by flowing water maintained at a temperature between 20°C and 100°C depending on the application. The crucible sectors are separated by an electrical insulator. The sectors are held together by strapping that also ensures gas-tightness of the crucible. The complete melter is mounted on a refractory concrete hearth in which water-cooled metal sectors are embedded. It also contains the pouring spout.

The inductor is supplied by a generator via a conductor passing through the wall of the radioactive containment cell. High-frequency power is supplied by a frequency converter oscillating at the $LCw^2 = 1$ circuit resonance frequency.

Operating Principle

The crucible is designed for maximum transparency to the magnetic field created by the inductor. When the glass is in the molten state a thin shell or "skull" of solidified, electrically insulating glass 5 to 10 mm thick forms in contact with the cold crucible wall. This is the key innovative aspect of the cold crucible melting process. The glass skull protects the melter components that would otherwise be subjected to severe corrosion by the molten glass. The temperature gradients resulting from the magnetic field create natural convection cells that cause a stirring action in the melt.

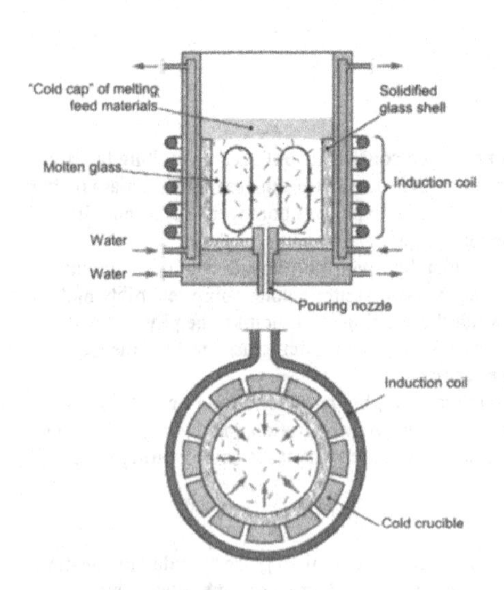

***Figure 1**. Schematic diagram of a cold crucible induction melter*

The advantage of the cold crucible melter lies in the fact that molten glass is never in contact with any structural component. The crucible service life is increased accordingly to between 3 and 5 years of operation. The temperature of the glass melt is no longer limited (up to 2000°C at the center of the crucible) and allows the use of refractory glass if it is a sufficiently good electrical conductor. The glass skull guarantees enhanced final glass quality and allows industrial production of non-nuclear products without impurities because the process prevents material transfers at the crucible/molten glass interface that would otherwise occur in a hot metal or refractory crucible. Pollution of the glass is thus virtually nil.(2)

PRELIMINARY STUDIES

Before undertaking the electromagnetic analysis discussed below, several prior studies must be carried out:

- The waste stream must be characterized very precisely, as it affects the specified composition of the glass frit that is mixed with the waste to obtain the final glass. The glass must exhibit satisfactory resistance to leaching and internal irradiation over time. Its processing temperature must be acceptable for a cold crucible melter between 800 and 1350°C (the upper limit is set by the thermomechanical properties of the receiving canister, as the cold crucible melter is capable of melting glass at more than 2000°C). The glass physical characteristics must be compatible with the cold crucible melter.
- The glass physical parameters (electrical resistivity, thermal conductivity, viscosity) must be measured as a function of the temperature. All these parameters are measured by the CEA at Marcoule.

The electrical resistivity of the glass from the cold crucible wall temperature to the operating temperature is the first physical parameter to be analyzed prior to the electromagnetic analysis. In order to allow induction heating, the electrical resistivity must be compatible with the induction frequency and the melter diameter. Given the frequency limitations of existing RF converters, glass with an electrical resistivity exceeding 0.1 ohm-meter at the nominal melting temperature will be very difficult to heat by direct induction.

ELECTROMAGNETIC PHENOMENA

The cold crucible is represented in three-dimensional geometry with a periodic θ angle in the azimuthal direction. The glass domain is axisymmetric. The single-turn induction coil comprises several parallel or braided strands. The scope of the electromagnetic study is three-dimensional.

The induction coil carries a sinusoidal alternating current. Currents are thus induced in the system components. The phenomenon can be characterized by one of Maxwell's equations:

$\Delta\vec{B} = \mu \cdot \sigma \cdot \dfrac{\partial \vec{B}}{\partial t}$ where $\vec{B}$ represents the magnetic field in the medium, μ is the magnetic permeability of the material, σ is the electrical conductivity, and t the time. Magnetic induction varies as a sine-wave function of time; this equation demonstrates the characteristic magnetic field diffusion length in the material, known as the electromagnetic skin depth δ,

$$\delta = \sqrt{\frac{2}{\mu \cdot \sigma \cdot \omega}}$$ where ω is the induction current pulsation.

This quantity represents the magnetic induction penetration depth in the material. Frequencies of about 100 to 600 kHz are used for cold crucible induction melting of glass.

Glass being a poor electrical conductor, the skin depth is roughly equivalent to the glass melt radius. For example, for an induction frequency of 150 kHz, the skin depth in glass with a resistivity of 5×10^{-2} Ω·m is about 290 mm. As a result, induced currents are formed throughout the glass mass, which is thus heated by Joule effect.

The crucible sectors, hearth and inductor consist of materials with high electrical conductivity (stainless steel, copper, etc.); their skin depth is minimal, and induced currents develop virtually on the surface of these components) (for 150 kHz the skin depth is 0.18 mm in copper). For these reasons the magnetic analysis is based on a volume calculation in the glass and a surface calculation for the melter structure metals.

ELECTROMAGNETIC ANALYSIS OF A NON-NUCLEAR MELTER

The following description is limited to the cold crucible sectors and the impact of a metal flange at the top of the melter on power dissipation in the cold crucible and on the melting efficiency.

A few essential aspects of the electromagnetic analysis will be discussed for the cold crucible sectors of a non-nuclear industrial melter. The analysis for a nuclear application follows the same principles, but with additional nuclearization constraints. In our example, the cold crucible melter can be supplied with cooling water by electrically insulating hoses; we therefore first analyze the electrically insulated sectors.

Electrically Insulated Cold Crucible Sectors

After specifying the melter dimensions and the induction frequency depending on the glass characteristics and process parameters, the first step is to design the cold crucible sectors.

The cold crucible sectors must meet the following criteria for an industrial melter: containment of the glass and off-gases in the melter, satisfactory electromagnetic efficiency, easy maintenance, and acceptable manufacturing cost. Based on these criteria, the number of sectors, their shape and the gap width are determined to obtain the highest performance. The initial calculations are performed in 3D geometry on two half-sectors with symmetry for the remainder of the melter to accelerate the computations.

Figure 2. *View in 3D geometry of two cold crucible half-sectors, the induction coil, and the glass*

The five turns of the induction coil are shown in turquoise, the half-sectors in green, the sector gap in yellow, the glass in turquoise, and the atmosphere above the glass in gray. Electromagnetic calculations are performed throughout the modeled system as well as in air and in an "infinite cell" to reach the zero magnetic field boundary condition.

The computational variables are the number, size and shape of the sectors; the surface power density, the currents induced on the sector faces sectors, and the volume power density in the glass are determined to optimize the crucible. All the plots shown below were calculated with an induction current of 1000 amperes.

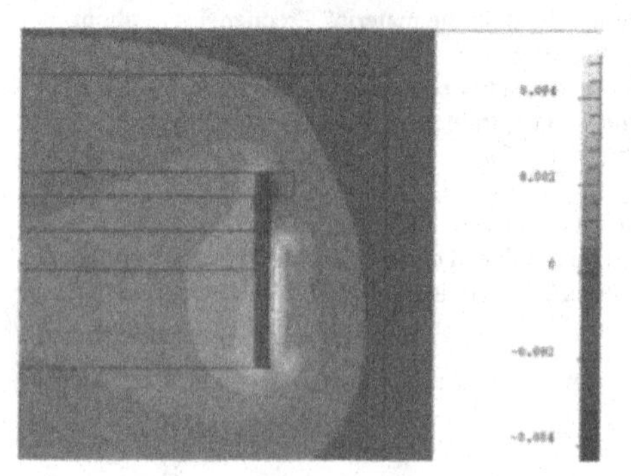

Figure 3. *Cross sectional view of complex magnetic induction (in Teslas)*

The cross sectional view of complex magnetic induction ensures that the value at the outer limits of the calculation domain in air is low enough for the infinite cell to reach zero field strength, and identifies the location of the strong magnetic fields. In this case the magnetic field is smoothly distributed with maxima at the upper and lower edges of the cold crucible.

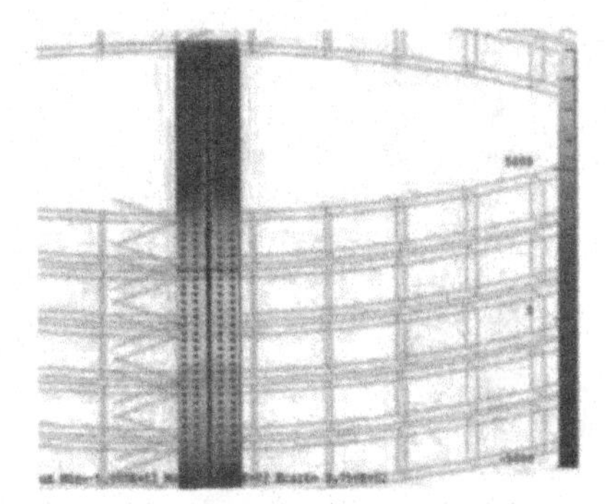

Figure 4. Power densities (W/m²) and currents on outer faces of the two half-sectors

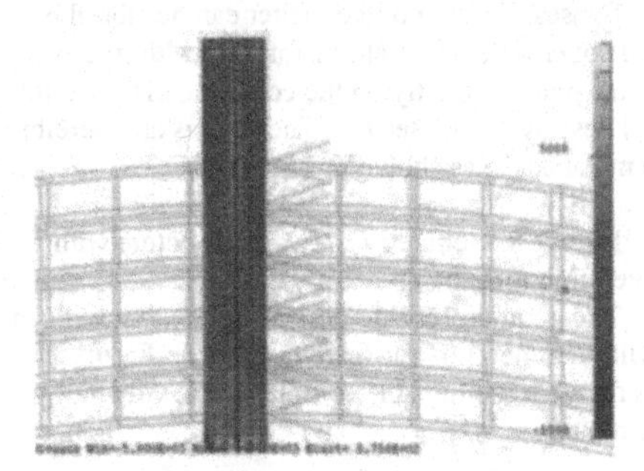

Figure 5. Power densities (W/m²) and currents on inner faces of the two half-sectors

Figure 4 shows the power densities on the sector faces. The surface power densities must be minimized to limit electrical losses in the cold crucible. The arrows represent the surface current vectors.

The direction of current flow on the outer faces is opposite that in the induction coil, and the maximum surface power density is indeed situated at mid-height on the induction coil. Vertical currents are observed near the edge of the sector gap. The maximum surface power density on the outer faces is about 10 000 W/m^2 with a 1000 A inductor current.

The direction of current flow on the inner faces is opposite that on the outer faces. Vertical currents are observed near the edge of the sector gap and more substantial currents at the lower edge of the sectors. The power densities are lower on the inner faces than on the outer faces.

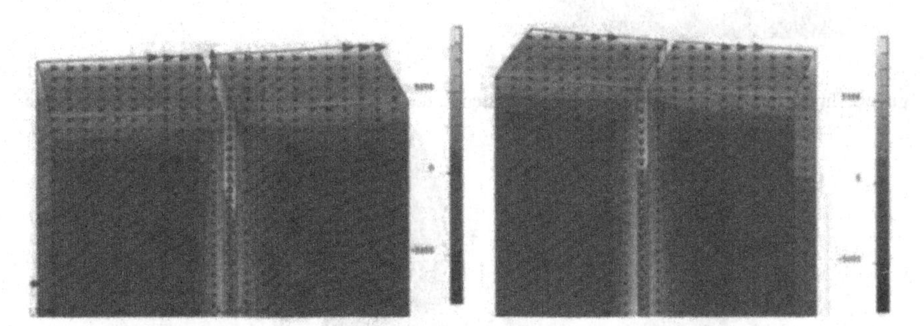

Figure 6. Power densities (W/m²) and currents on upper faces of the gap between the two half-sectors

Figure 6 clearly shows that the induced currents do not simply flow around each electrically insulated sector, but flow upward on the right-hand gap face and downward on the left-hand face. The induced currents are very strong on the sector gap faces, and increase as the gap width diminishes.

Figure 7 shows that the power density is highest on the gap faces—nearly four times greater than on the outer face. The high power density is localized a few centimeters above the last turn of the induction coil. The gap width must be optimized at this point to diminish unnecessary losses in the cold crucible.

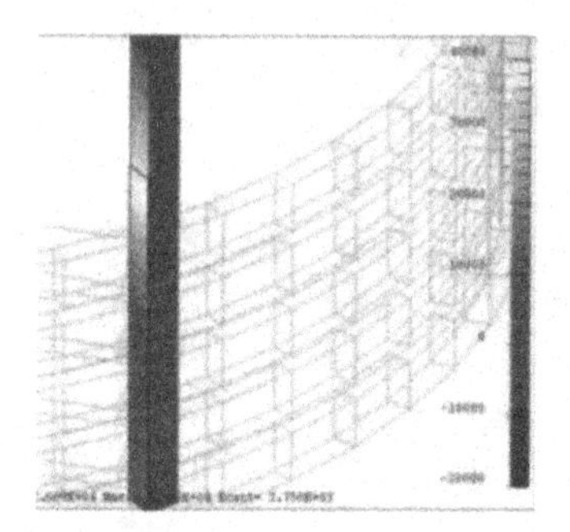

Figure 7. Power densities (W/m^2) on a gap face and an outer face of the two half-sectors

Cold Crucible Sectors Electrically Bonded by a Cover Mounting Flange

The melter must be covered to confine the melting off-gases. The top of the melter can be closed by an electrically insulating refractory or water-cooled metal cover with penetrations for the melting feed materials and for the off-gas stream. If the metal cover were placed directly on the cold crucible it would short-circuit the sectors, causing electrical discharges and destroying the sectors. The sectors are therefore often connected by an upper flange to accommodate the metal cover as shown in **Figure 8**.

Figure 8. Geometry of two half-sectors with cover mounting flange

The geometry is the same as before, but with an additional metal flange (shown in red in **Figure 8**); the flange surrounds the complete circumference of the melter

Figure 9 shows the surface power density calculated in the cold crucible sectors with the metal flange.

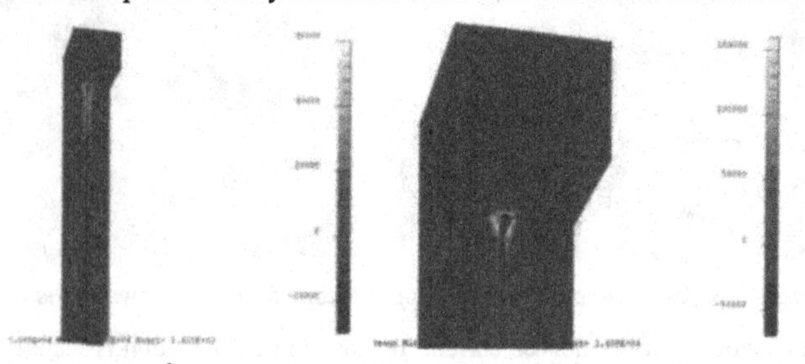

Figure 9. Power densities (W/m^2) in outer faces of the two half-sectors and flange

The flange causes high power consumption at the top of the sectors.

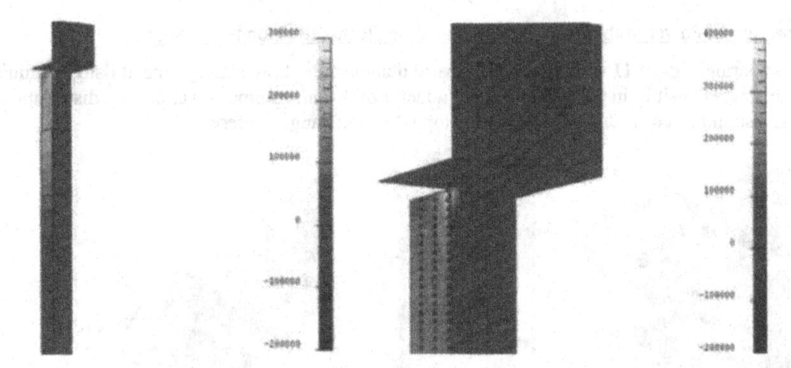

Figure 10. *Power densities (W/m²) and currents on a sector gap face and flange*

The local power is clearly higher when the cover mounting flange is added. The efficiency of the two versions described can be compared directly by integrating the local power density over the glass volume and all the crucible surfaces. The power dissipated on the sector faces with the flange is significantly greater than with insulated sectors. The power integral must be computed for the melter components to determine the melter efficiency and basic electrical data (inductance and resistance).

Table I and **Table II** show the calculated results; the data initially observed were the power in the glass and in the surfaces with an inductor voltage of 2000 V. In the following tables, from left to right: I represents the current supplied to the induction coil, P the resulting active power (kW), Q_{VAR} the reactive power (kVAR), R the melter resistance seen by the induction coil (ohms), L the melter inductance seen from the induction coil (henrys), P_{glass} the power in the glass (kW), $P_{surfaces}$ the power in the crucible and in the flange (kW). Then, for 2000 V: the total power P_{total} (kW), P_{glass} and $P_{surfaces}$ for 2000 V.

Table I. *Electrical parameters for a crucible with sectors electrically insulated from one another*

I (A)	P (kW)	Q_{VAR}	R (Ω)	L (H)	P_{glass} (kW)	$P_{surfaces}$ (kW)
1000	192.01	**1259.75**	0.1920	1.337×10^{-6}	**169.28**	**22.73**
(for 2000 V)	**472.97**				**416.98**	**55.59**

Table II. *Electrical parameters for a crucible with sectors electrically bonded via the upper flange*

I (A)	P (kW)	Q_{VAR}	R (Ω)	L (H)	P_{glass} (kW)	$P_{surfaces}$ (kW)
1000	128.76	**817.74**	0.1288	8.676×10^{-7}	**46.38**	**82.38**
(for 2000 V)	**751.58**				**270.72**	**480.86**

The calculations were performed by supplying a current to the induction coil in order to obtain the melter inductance and resistance based on the calculated active and reactive power. However, the best controlled parameter allowing power comparisons is the voltage across the inductor; this is why the values have also been indicated with respect to an inductor voltage of 2000 volts.

Table I and **Table II** provide the following comparative data:

- Adding a metal flange considerably increases the losses in the cold crucible surfaces, which rise from 55.99 to 480.86 kW for an inductor voltage of 2000 volts.
- Adding a metal flange diminishes the effective power supplied to the glass from 416.98 to 270.72 kW for an inductor voltage of 2000 volts.
- Adding a metal flange diminishes the melter resistance and inductance, which changes the generator settings for the impedance and operating frequency.

Comparison of Power in the Glass with Insulated or Electrically Bonded Sectors

Comparing **Figure 11** with **Figure 12** clearly demonstrates how adding a metal flange diminishes the volume power densities in the glass by about a factor of 3. The volume power density distribution in the glass is also modified, with less power at the top when the flange is present.

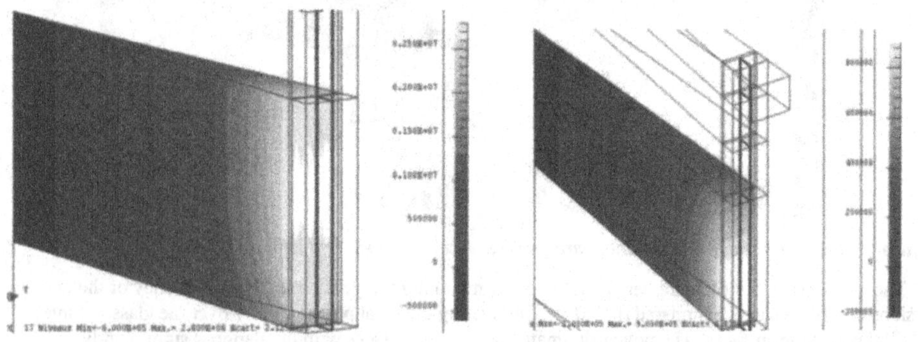

***Figure 11**. Volume power densities in the glass (W/m^3) with electrically insulated sectors*

***Figure 12**. Volume power densities in the glass (W/m^3) with electrically bonded sectors*

When sectors have been optimized, the calculations are repeated for the complete melter with the cold hearth. When the melter has been specified, we then calculate the shielding necessary to ensure electromagnetic compatibility.

CONCLUSION

The electromagnetic analysis of the direct induction cold crucible melter described here allowed us to better identify the power distribution and the induced currents that directly impact the melter efficiency and can result in local overheating of the cold crucible and the glass. Each additional metal component in the melter can considerably modify its electromagnetic characteristics.

A comprehensive analysis of a direct induction cold crucible melter requires wide-ranging physical expertise. The work performed by the CEA at Marcoule covers the chemistry of the glass and waste, and measurements of physical and electromagnetic properties.

REFERENCES

1. A. Gagnoud, *Modélisation des installations de fusion en creuset froid*, PhD thesis (1986).
2. L. Jacoutot, *Modélisation numérique de phénomènes couplés dans des bains de verre brassés mécaniquement et élaborés en creuset froid inductif*, PhD thesis (2006).

Mater. Res. Soc. Symp. Proc. Vol. 1107 © 2008 Materials Research Society

NMR Investigation of Cation Distribution in HLW Wasteform Glass

Diane Holland[1], Ben G Parkinson[1], Moinul M Islam[1], Adam Duddridge[1], Jonathan M Roderick[1], Andy P Howes[1] and Charlie R Scales[2]
[1]Department of Physics, University of Warwick,
Coventry CV4 7AL, UK
[2]Nexia Solutions, Sellafield,
Seascale, Cumbria, CA20 1PG, UK

ABSTRACT

Magic-angle-spinning NMR has been used to establish the structural roles of various cations added to the borosilicate glass which is used for the vitrification of high-level nuclear waste (HLW). Representative surrogate oxides with nominal valencies of +1, +2 and +3 have been studied which span the range of oxides from modifier to intermediate and conditional glass-former. NMR has been carried out on those nuclei which are accessible and the species observed have been correlated with the physical and chemical behaviour. The controlling factor is the manner in which the alkali cations partition between the various network groups, changing the distribution of silicon Q^n species and the boron N_4 ratio. Identifiable superstructural units are also present in these glasses. The aqueous corrosion rate increases with Q^3 content, as does the weight loss due to evaporation from the melt. The activation energy for DC conduction scales with N_4. Values of N_4 obtained for these glasses deviate significantly from those predicted by the currently accepted model (Dell and Bray) and are strongly affected by the modifier or intermediate nature of the surrogate oxide and also by its effect on the distribution of non-bridging oxygens between the silicate and borate polyhedra.

INTRODUCTION

The BNFL mixed alkali borosilicate glass wasteform (MW), in its fully lithiated form, has composition 10.29 mol% Li_2O, 10.53 mol% Na_2O, 18.57 mol% B_2O_3, 60.61 mol% SiO_2. It has a fairly high glass transition temperature (Tg), so that reasonable levels of waste loading can be achieved before radiogenic heating becomes a problem. It also has the ability to dissolve a wide range of oxides without phase separation and good chemical stability to prevent the leaching of radioactive elements. These properties depend on glass composition and structure, both of which are changed by addition of waste components. We have used nuclear magnetic resonance (NMR) to relate the addition of waste components to changes in structure and thus to changes in properties. For such complex, amorphous materials, structure determination requires the use of techniques such as NMR which are nucleus specific and sensitive to the short-range order (SRO) structural units which could occur in these glasses, including $[SiO_4]$ tetrahedra, Q^n, where n is the number of bridging oxygens $[BO_4]$ tetrahedra, and $[BO_3]$ planar triangles.

The structures of alkali borosilicate glasses, of general formula $RMe_2O.KSiO_2.B_2O_3$, where $K = [SiO_2]/[B_2O_3]$ and $R = [Me_2O]/[B_2O_3]$ have been successfully modelled by Dell et al. [1] who produced formulae for calculating the fractions of the individual boron species as a function of R for given values of K. The most important of these fractions is N_4, the fraction of boron atoms which are 4-coordinated $[BO_4]$. Some deviations from the Dell model are observed, including sharing of modifiers between the two networks; better mixing of the $[BO_3]$ and Q^n

networks; and strong association of $[BO_4]$ units with Si. This latter gives rise to medium-range order (MRO) units based on reedmergnerite and danburite, as observed by Utegulov *et al.* [2] and Bunker *et al.* [3]. MRO units have also been linked with the observation of multiple $[BO_4]$ sites [4]. Reedmergnerite units are rings of three Q^4 and one $[BO_4]$ polyhedra $[BSi_3O_8]^-$ and the boron environment is $[B(OSi)_4]$. Danburite units are rings of two Q^4 and two $[BO_4]$ polyhedra $[B_2Si_2O_8]^{2-}$ and the boron environment is $[B(OSi)_3(OB)]$.

EXPERIMENTAL DETAILS

Sample preparation

The nominal sample compositions were of the general form (A) $xMe_2O_n.(100-x)MW$ and (B) $xCs_2O(100-x)[yMe_2O_n(100-y)MW]$. Me has valence n and y = 0.95 (Al_2O_3), 1.68 (La_2O_3), 2.55 (MgO). System B was designed to enable measurement of caesium volatility from the melt. K remains constant at 3.26 but R (= 1.1 in the base glass MW) depends on the structural role of Me. High purity Wacomsil® quartz (SiO_2), and >99.9% pure sodium tetraborate ($Na_2B_4O_7$), boric acid (H_3BO_3), sodium carbonate (Na_2CO_3), lithium carbonate (Li_2CO_3) and Me (oxide or carbonate) were combined stoichiometrically in 100 g batches. An addition of 0.1 mol% Fe_2O_3 was made to all batches to reduce ^{29}Si T_1 and improve S/N. The batches were melted between 1050°C and 1600°C for ~ 20 minutes and either splat-quenched or cast into moulds followed by annealing. Chemical analysis of selected samples showed little deviation from nominal compositions - typically ± 0.05 in K and ± 0.01 in R. All samples were X-ray amorphous.

Chemical durability and volatilisation measurements

A Soxhlet method [5] was used to determine chemical durability. Samples of system A, with Me = Ca, Sr, Ba, Zn, Pb, Mn, were cut as 25mm diameter and 1 mm thick discs, with surfaces polished to a 1 micron finish. Each sample was placed in a Soxhlet extractor fitted with a condenser and a round-bottomed flask containing 300ml of deionised water maintained at 100°C. The sample in the Soxhlet was at ~75°C. After 14 days, the sample weight loss was measured.

Volatilisation from the melt was determined by measuring the weight lost by 0.2 g samples from system B after 4 hours at 1000°C.

Ionic (DC) conductivity studies

DC ionic conductivity was determined using the ac impedance technique for temperatures between 200–350°C and a frequency range of 5Hz-1MHz. Sputtered platinum electrodes were applied to the disc sample (system A, with Me = Ca, Sr, Ba, Zn, Pb, Mn) which was then pressed firmly between two ceramic discs, with a piece of gold foil inserted between the sample and each disc. Each gold electrode was connected to a Hewlett Packard 4921 Impedance Analyser using platinum wire. Impedance plots of the real part, Z′, versus the imaginary part, Z″, of the complex impedance plane, yielded a semi-circle, which was fitted to give the real axis intercepts at 0 and R, the resistance of the glass sample. The dc ionic conductivity was then calculated using the relation

$$\sigma = \frac{1}{R} \cdot \frac{t}{A} \qquad (1)$$

where t is the sample thickness and A is the electrode surface area of the glass sample

NMR

^{11}B MAS NMR spectra were acquired using a Varian/Chemagnetics Infinity 600 NMR spectrometer operating at 192.04 MHz with a Chemagnetics 4 mm probe with rotor spinning at 15 kHz. A pulse delay of 1s and a pulse width of 0.7μs (B_1~ 60 kHz) were used and spectra were referenced against solid BPO_4 (−3.3 ppm wrt primary reference $Et_2O{:}BF_3$). ^{29}Si MAS NMR spectra were acquired using a CMX Infinity 360 NMR spectrometer operating at 71.54 MHz with a Varian 6 mm probe with rotor spinning at 5-6 kHz. A 5s pulse delay and 2μs pulse width (30° tip angle) were used and spectra were referenced to tetramethylsilane.

RESULTS

Chemical durability

The weight loss after 14 days of Soxhlet test is shown in Figure 1 as a function of MeO content. For the alkaline earths, weight loss increases with the electropositive character of the cation, reaching 100 % at 15 mol% of BaO. Amorphous corrosion layers (confirmed by glancing angle XRD) are formed on the samples. Adding ZnO or PbO increases chemical durability of the glass and no corrosion layer is observed. In the case of MnO a surface deposit of Mn_7SiO_{12} is formed at high concentration of oxide.

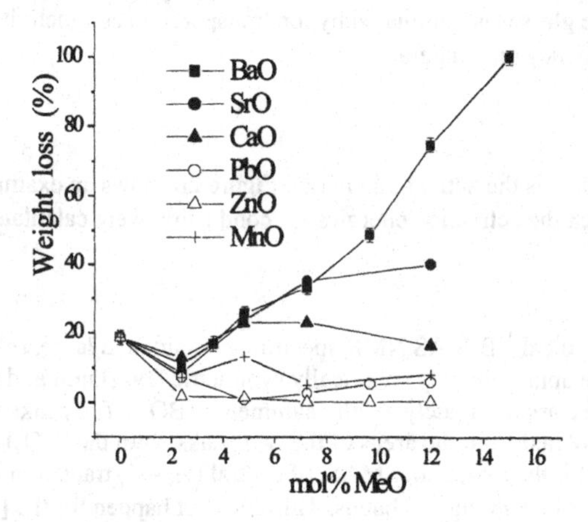

Figure 1. Weight loss after 14 days of Soxhlet test as a function of mol% MeO additive for system A, $xMe_2O_n.(100-x)MW$, where Me = Ca, Sr, Ba, Zn, Pb, Mn.

Ionic (DC) conductivity studies

Figure 2a shows values of conductivity measured at 300°C for selected MeO. The values decrease by more than an order of magnitude with increase in MeO content; from 0.93 ± 0.01 μScm^{-1} for the base glass MW to, for example, 0.053 ± 0.01 μScm^{-1} for the glass containing

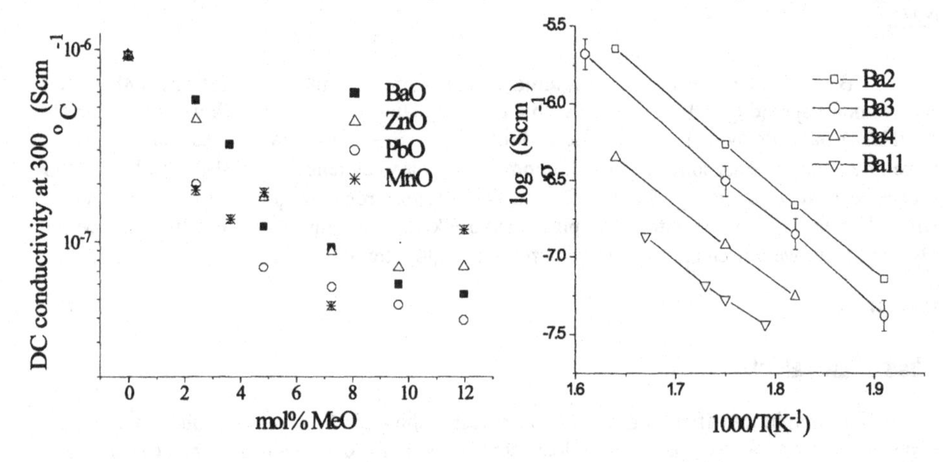

Figure 2. (a) DC conductivity at 300°C in system A, $xMe_2O_n.(100\text{-}x)MW$, where Me = Ba, Zn, Pb, Mn; (b) Arrhenius plots for Me = Ba and x = 2.4, 3.62, 4.83 and 11.98 mol%.

11.98 mol% of BaO. All systems show a decrease in conductivity with increased addition of MeO. Conduction in these glasses is dominated by ion transport and conductivity σ follows a simple exponential dependency on temperature,

$$\sigma = \sigma_0 \exp(-E_a/RT) \qquad (2)$$

where σ_0 is a constant and E_a is the activation energy. Figure 2b shows an example of the Arrhenius plots from which the activation energies for conduction were calculated.

^{11}B MAS NMR

Figure 3 shows a typical ^{11}B MAS NMR spectrum containing a large, relatively narrow peak at ~ 0 ppm due to the approximately spherically symmetric $[BO_4]$ unit and small broader peak at ~ 13 ppm due to the approximately axially symmetric $[BO_3]$. The peaks were fitted [6] and the N_4 value was calculated from the areas of the two peaks, with the $[BO_3]$ area being increased by 4% to allow for the loss, under MAS, of central (½, –½) transition intensity from the $[BO_3]$ centreband into the spinning sidebands. This does not happen for the $[BO_4]^-$ site with its much smaller quadrupole interaction so that all of the central (½, –½) transition intensity appears in the centreband [7]. The $[BO_4]$ peak was also fitted with two Gaussian-Lorentzian lineshapes, after the procedure of Du *et al.* [4], a good approximation where the quadrupole coupling constant is very small. These represent the separate contributions from $[B(OSi)_4]$ and

$[B(OSi)_3(OB)]$ and allow calculation of the relative fractions of reedmergnerite and danburite units.

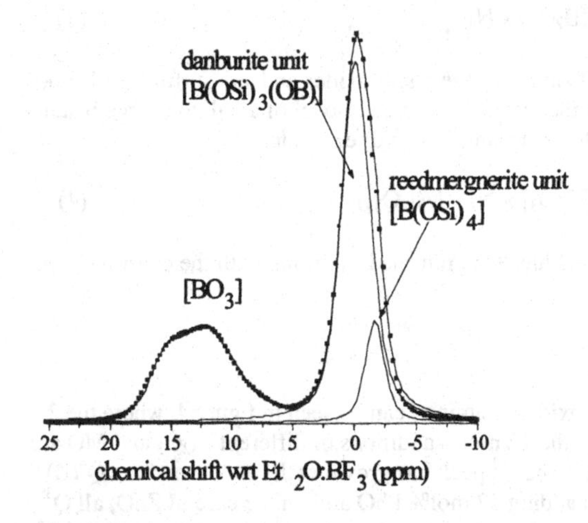

Figure 3. Typical ^{11}B MAS NMR spectrum showing fitting of $[BO_4]$ resonance to two peaks from reedmergnerite and danburite units.

^{29}Si MAS NMR

^{29}Si MAS NMR spectra for system A, with 12 mol% additive, are shown in figure 4. The broad, asymmetric lines contain contributions from several species which are simplified to $Q^4(B)$ – a Q^4 species having predominantly one $[BO_4]$ next-nearest neighbour and Q^3 with any next nearest neighbour. The initial fit was constrained by estimating Q^3 from the number of non-bridging oxygens (NBO) associated with the silicate polyhedra using the value of N_4 and the total modifier content

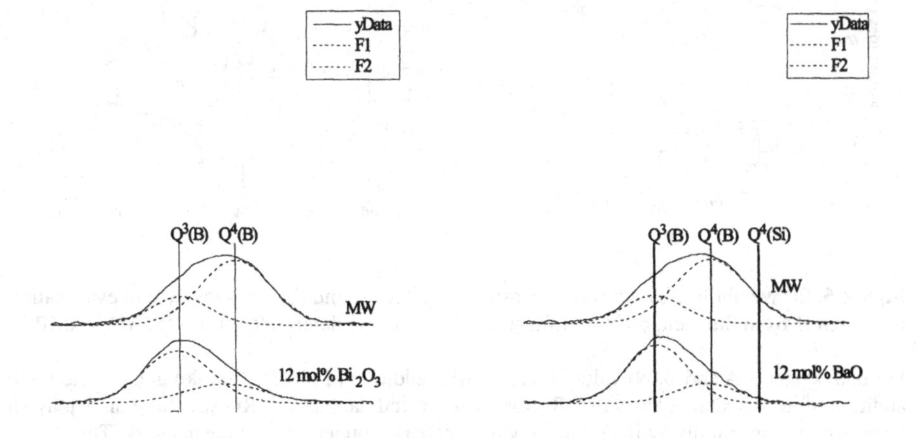

Figure 4. Example ^{29}Si MAS NMR spectra and fits for 12 mol% Me_2O_3 and MeO additions to system A, $xMe_2O_n.(100-x)MW$.

$$NBO(silicate) = 2 \times ([modifier\ oxide] - [B_2O_3] \times N_4) \qquad (3)$$

The [modifier oxide] referred to in eqn (1) includes the alkali oxides and any surrogates deemed to be modifiers. NBO(silicate) will be further reduced by the removal of alkali to charge balance any network units formed by intermediate oxide additions. For example

$$NBO(silicate) = 2 \times ([modifier\ oxide] - [B_2O_3] \times N_4 - [Al_2O_3]) \qquad (4)$$

Once acceptable peak positions and half-widths were returned by fitting with the constraint, the constraint was removed for the final fit.

DISCUSSION

The different roles of the various oxides examined can be seen in figure 4, where the 2 peak fits to the ^{29}Si spectra from MW and the 12 mol% additions of different M_2O_3 and MO are compared. In the case of the modifier BaO, the Q^3 peak has grown at the expense of the $Q^4(B)$ peak. In contrast, there is little change on adding 12 mol% PbO and, in the case of ZnO, all Q^3 has disappeared and some Q^4 has been formed. In the case of Al_2O_3, Q^3 has reduced and the Q^4 peak has moved to a chemical shift which reflects the presence of Al next nearest neighbours (^{27}Al NMR also gives a 4-coordinated Al peak at ~ 53 ppm, typical of $[Al(OSi)_4]$ rather than $[Al(OB)_4]$). The amount of Q^3 present appears to be an important factor in controlling physical properties, as shown by Figure 5 for corrosion loss from system A and volatilisation loss from system B.

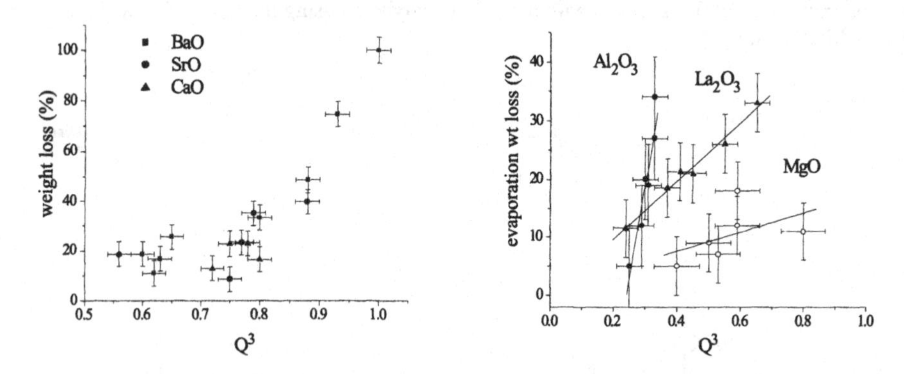

Figure 5. (a) weight lost by aqueous corrosion of system A and (b) mass loss due to evaporation of system B from the melt, both as a function of Q^3 content determined from ^{29}Si MAS NMR.

For both systems A and B, N_4 values increase with addition of modifier oxides and decrease with addition of intermediate oxides [8]. Reedmergnerite and danburite MRO structural units play an important part in stabilising $[BO_4]^-$ units which increase connectivity of the network. The fraction of danburite MRO units increases as a function of caesium oxide content when intermediate oxides such as aluminium or lanthanum are present, and decreases when either MgO or group I alkali oxides alone are present [9]. Figure 6 shows the relation between fraction

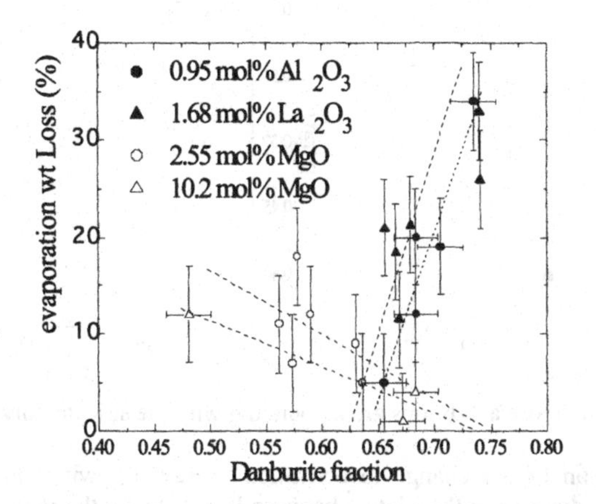

of danburite and volatilisation loss. This may reflect the energetically unfavourable $[BO_4]^-[BO_4]^-$ link in the danburite unit.

Figure 6. Trends in weight loss due to evaporation from the melt with danburite content determined from ^{11}B MAS NMR.

Values for E_a are typical of alkali ion transport and conductivity will be determined by their number and mobility. Introduction of MeO into the MW glass system reduces the conductivity because less mobile Me^{2+} ions replace alkali ions and also block transport pathways through the glass. The similar values for E_a suggests that the reduction in conductivity with increasing addition of MeO is largely controlled by the reduction in the number of mobile ions rather than significant change in inter-site barrier height. In the case of additions of intermediate oxides such as PbO and ZnO, a further alkali ion environment is created in the charge neutralisation of network species such as $[ZnO_4]^{2-}$ or $[PbO_4]^{2-}$. These alkali ions may require a significantly different activation energy to change site. Figure 7(a) compares the activation energies measured for the BaO-MW and Zn-MW systems and shows that there are subtle changes in activation energy with composition. The divergence of values beyond 4 mol% addition of MeO might indicate that the association of some alkali ions with $[BO_4]^-$ and $[ZnO_4]^{2-}$ results in higher average inter-site barriers for these ions. Figure 7(b) shows that there is a relationship between E_a and N_4 for these samples.

CONCLUSIONS

The glasses formed by adding oxides to the basic wasteform composition (MW) show extensive changes in structure which depend on how the added oxide affects the association of modifier cations with the borate and silicate polyhedra. Alkali metal oxides and alkaline earth oxides increase both Q^3 and N_4, with $[BO_4]$ being stabilised by formation of danburite as well as reedmergnerite MRO units. In contrast, additions of PbO or ZnO reduce N_4 and Q^3 is either unchanged or even reduced to zero. In the case of Al_2O_3, N_4 and Q^3 are reduced as expected by its intermediate character. However, in the case of La_2O_3, although there is a reduction in N_4 which matches the prediction for intermediate behaviour, there is a strong increase in Q^3 which

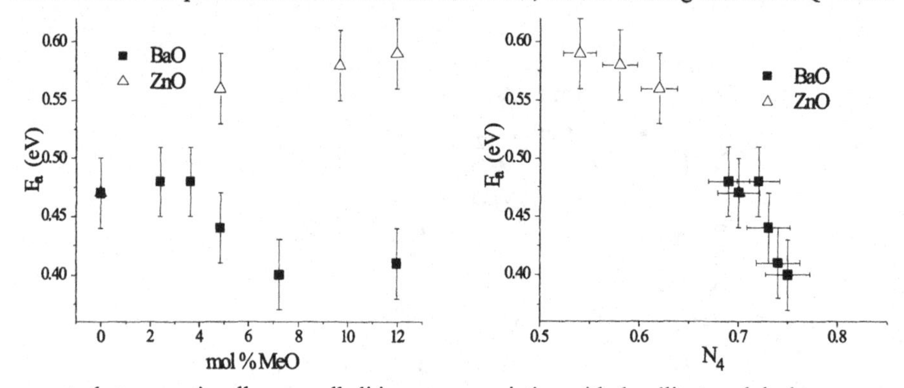

suggests that proportionally more alkali ions are associating with the silicate polyhedra.

Figure 7. (a) comparison of the change in activation energy, E_a, with mol% additive for modifier BaO and intermediate ZnO; (b) relation between E_a and N_4 for the same samples.

ACKNOWLEDGEMENTS

The authors thank Nexia Solutions for support of this work and EPSRC for partial funding of the NMR facility at Warwick.

REFERENCES

1. W. J. Dell, P. J. Bray and S. Z. Xiao, *J. Non-Cryst. Solids* **58,** 1-16 (1983).
2. Z. N. Utegulov, J. P. Wickstead and G.-Q. She, *Phys. Chem. Glasses* **45** (3), 166 (2004).
3. B. C. Bunker, D. R. Tallant, R. J. Kirkpatrick and G. L. Turner, *Phys. Chem. Glasses* **31** (1), 30 (1990).
4. L.-S. Du and J. F. Stebbins, *J. Non-Cryst. Solids* **315**, (2003) 239.
5. High-Temperature Processes – Leach Tests, BNFL Research Technology, Operating Instructions HTP 07, Issue 2 (1999)
6. D. Massiot, F. Fayon, M. Capron, I. King, S. Le Calve, B. Alonso, O. J. Durand, B. Bujoli, Z. Gan and G. Hoatson, *Magn. Reson. Chem.* **40,** 70 (2002).
7. K. J. D. Mackenzie and M. E. Smith, *Multinuclear Solid-state NMR of Inorganic Materials*, Pergamon Materials Series, Vol. 6, 2002.
8. D. Holland, B. G. Parkinson, M. M. Islam, A. Duddridge, J. M. Roderick, A. P. Howes and C. R. Scales, *App. Mag. Reson.* (in press) (2007)
9. B. G. Parkinson, D. Holland, M. E. Smith, A. P. Howes and C. R. Scales *J. Phys.: Condens. Matter* **19** 415114 (2007)

Mater. Res. Soc. Symp. Proc. Vol. 1107 © 2008 Materials Research Society

Changes to Alkali Ion Content Adjacent to Crystal-Glass Interfaces

Michael J. D. Rushton[1], Robin W. Grimes[1] and Scott L. Owens[2]
[1]Department of Materials, Imperial College London, London SW7 2AZ, UK
[2]Nexia Solutions Ltd., Warrington, Cheshire WA3 6AS, UK

ABSTRACT

Atomic scale molecular dynamics simulations have been used to predict the location of glass modifying Na, Li and Mg species in a borosilicate Magnox type waste glass adjacent to interfaces with the (100) and (110) surfaces of MgO, CaO and SrO crystals. These simulations show a considerable increase in alkali and alkali earth concentration adjacent to specific interfaces. In particular, there are significant, systematic changes in Na, Li and Mg position and concentration as a function of both the crystal's terminating surface and composition.

INTRODUCTION

Despite our best efforts, glass-crystal interfaces occur within "vitrified" nuclear waste [1]. These arise as a result of a lack of reactivity of refractory material in the process feed, or due to crystallisation of complex phases from the vitrification melt. Vitrification of high activity waste from nuclear processes is a desirable treatment route, as glass is a suitable host for high volume fractions of waste of widely varying composition. However these widely varying compositions lead to a greater likelihood of introducing insoluble phases into the glass melt. A better understanding of the structures and processes that occur at the interfaces between these insoluble fractions and the bulk glass during high-temperature processing would greatly aid our ability to optimize feed compositions to the vitrification process and mitigate against any detrimental effects that having a glass-crystal interface might have on, for example, transport of radionuclides. However, due to the multi-element compositions involved and the complexity of their disorder, the atomic structures of glass-crystal interfaces have not been well characterised experimentally nor have they been well modelled. In particular, variations in the composition of a mixed alkali glass, towards the interface, have only been reported in one modelling study [2, 3] and the influence of other additions (such as Mg) have never been described. This study also forms a precursor to examining the interactions between other, more complex crystalline products and the bulk glass melts.

The aim of this work is therefore to understand glass ceramic interfaces in the context of vitrified nuclear wasteforms. For this reason, the glass compositions examined are related to the composition used by British Nuclear Fuels Ltd. (BNFL) for vitrification of wastes resulting from reprocessing Magnox fuel. The basic glass composition used during the vitrification of Magnox waste (MW) [4] is given in table 1. It is a borosilicate composition which contains essentially equal amounts (in atomic percent) of sodium and lithium as glass structure modifiers.

Magnox fuel assemblies are clad in an alloy of aluminium and magnesium; although efforts are made to remove this cladding during reprocessing, Al and Mg are present in relatively large amounts in the resulting vitrified waste. In an effort to characterise the effect of these additions on interfaces, a Magnox waste glass containing Al and Mg has also been considered. The composition of this MW+Al+Mg glass, reported in table 1, was derived from the compositional analysis given in [4] with fission products and iron contributions removed.

Table I. Glass compositions: upper Magnox waste glass (MW), lower Magnox glass + Al + Mg.

	Oxide Components (wt %)					
	SiO_2	B_2O_3	Al_2O_3	Na_2O	Li_2O	MgO
MW	64.7	18.8	-	11.2	5.3	-
MW+Al+Mg	54.6	19.6	5.6	9.7	4.4	6.1

Interfaces were constructed between the glass blocks and the (100) and (110) surfaces of alkali oxides with the rock-salt structure: MgO, CaO and SrO. Simulation cell dimensions were chosen to give cell volumes consistent with the experimentally determined glass densities [3].

METHODOLOGY

Crystal-glass interfaces were generated using a melt-quench approach based on molecular-dynamics (MD) simulations. In all cases, simulations were performed using Accelrys' Discover code [5]. Initially each system was heated to 7000K for 15ps, in an NVT ensemble using a time-step of 1fs. This high temperature (i.e. much higher than the melting points of any of the glasses) was used to ensure that a random, liquid like structure for the glass was obtained in a relatively short time. Subsequently, the randomised structure was quenched to 1000K by decreasing temperature in 10K increments and performing 100fs of NVT MD at each step. Temperature was controlled during dynamics runs using the Nose- Hoover thermostat [6]. The quench rate for the procedure outlined above, $1\times10^{14}Ks^{-1}$, is many times quicker than experimental quench rates, but compares favourably with other contemporary simulation studies (e.g. 1×10^{14} - $1\times10^{15}Ks^{-1}$ [7] and $1\times10^{13}Ks^{-1}$ [8]).

Following the quench to 1000K, another 15ps of NVT dynamics was performed. This equilibration stage was provided to allow the network and modifier structures to develop. From 1000K the glass was quenched to 300K in the NPT ensemble. This was followed by 5ps of equilibration at 300K. An additional 5ps of NPT dynamics were performed, during which atomic coordinates and velocities were collected at 50fs intervals.

The crystal-glass interface was generated by simply splitting the glass and inserting the ceramic block. Consequently, the full melt-quench was performed with the ceramic block *in-situ*. However, due to the presence of the ceramic, the glass structure will form in a modified way to accommodate the ceramic. As the room temperature structure of the interface was of interest, and to prevent the ceramic block melting or high temperature mixing of the glass and ceramic components, the atoms of the ceramic were held fixed as the system was randomised and quenched to 1000K. Although immobile, interactions between glass and ceramic atoms were still calculated such that the glass could react to the ceramic's presence. The constraint on the ceramic block was released at 1000K during the equilibration and data collection stages of the simulation. Strictly speaking, this means these simulations reflect the influence that the crystal has on the glass structure but not to the same extent the effect the glass has on the crystal surface structure.

RESULTS

Figure 1 shows results for the interface between MgO (100) and the MW glass (A, the upper portion) and MgO (100) and the MW+Al+Mg glass (B the mid and lower portions). On the far left of the top portion of the figure a snapshot of the whole glass is given. On the right of this are snapshots of the Li and then the Na ions in the glass but with all other ions removed. In the lower two portions of the figure the whole glass and then Li, Mg and then Na only are shown.

To quantify the magnitude of the changes to alkali content adjacent to the interface, the density of alkali atoms was calculated as a function of distance from each interface. These density distributions were obtained in the following manner. The simulation cell was decomposed into a series 0.19 Å histogram bins along the z-axis, perpendicular to the interface. The z-coordinate of an atom was thus a measure of the perpendicular distance to the glass-ceramic interface. The total number of each species in each slice was calculated and divided by the slice volume to give the localised number density for that species. This procedure was repeated for all the data collection frames of the simulation to produce a time averaged concentration profile. The resulting density values for Li and Na (in the case of MW) or for Li, Mg and Na species (in the case of MW+Al+Mg glass) are reported on the right hand side of Figure 1. It is important to note that these are averages from four separate simulations. In all cases, the average alkali or alkali earth value for the corresponding bulk glass is represented by a line though the density profile at 100%.

The upper portion of figure 1 shows that there is an increase in the local concentration of Li and Na ions in the interface region. Both alkali species exhibit a similar increase, a factor of six. For the MW+Al+Mg glass there is also an increase in alkali concentration, again for both species to a very similar extent, although it is slightly less, a factor of five. This may reflect the finding that for the MW+Al+Mg glass there is also an increase in Mg concentration at the interface which may tend to block potential sites for alkali species. Indeed, the increase in Mg concentration is a factor of nine so that the total increase in glass modifier content at the MW+Al+Mg interface is greater than that for the MW glass alone. Whether this reflects a change in the proportion of modifier ions in the glass or resulting changes in the physical properties is not yet been determined.

Figure 2 shows equivalent results to those presented in figure 1 but now for the CaO (100) interface. Considering first the interface with the MW glass (upper portion, A, of figure 2), again increases in alkali content are predicted but this time, the increase in Li concentration, a factor of more than seven is significantly greater than that for Na, a factor of just over four. This illustrates that the change in alkali content adjacent to an interface depends on the composition of the crystal and furthermore that the change in alkali concentration may not be the same for different alkali species.

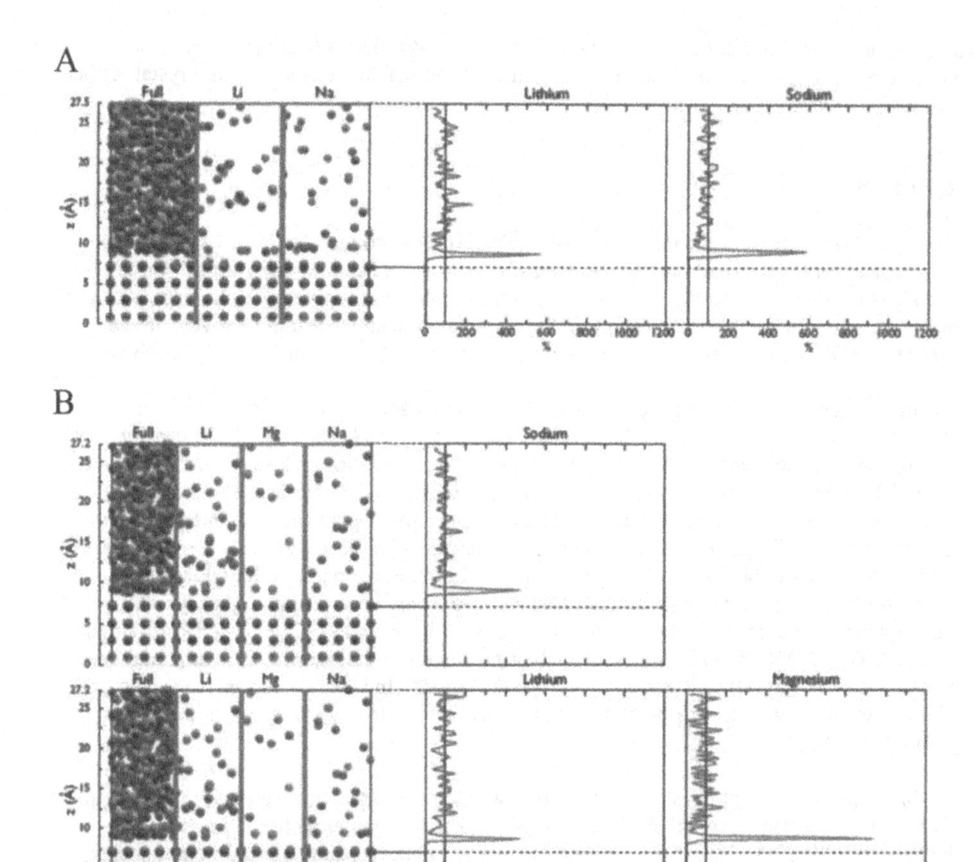

Figure 1. Density of alkali and Mg species in glass as function of distance from interfaces between MgO-(100) and waste glasses: A. Magnox waste glass; B. Magnox glass + Al + Mg (density expressed relative to each species' homogeneous density in bulk glass).

Considering next the lower portions (B) of figure 2 that relate to the MW+Al+Mg glass, again there is a still marked increase in alkali content compared to the glass average but it has been reduced compared to that observed with the MW glass interface. This time the increase in Mg concentration at the interface though a substantial factor of six, is not as great as it was for the MgO interface. Clearly changes to the alkali earth content in the glass adjacent to the crystal surface also depend on the composition of the crystal.

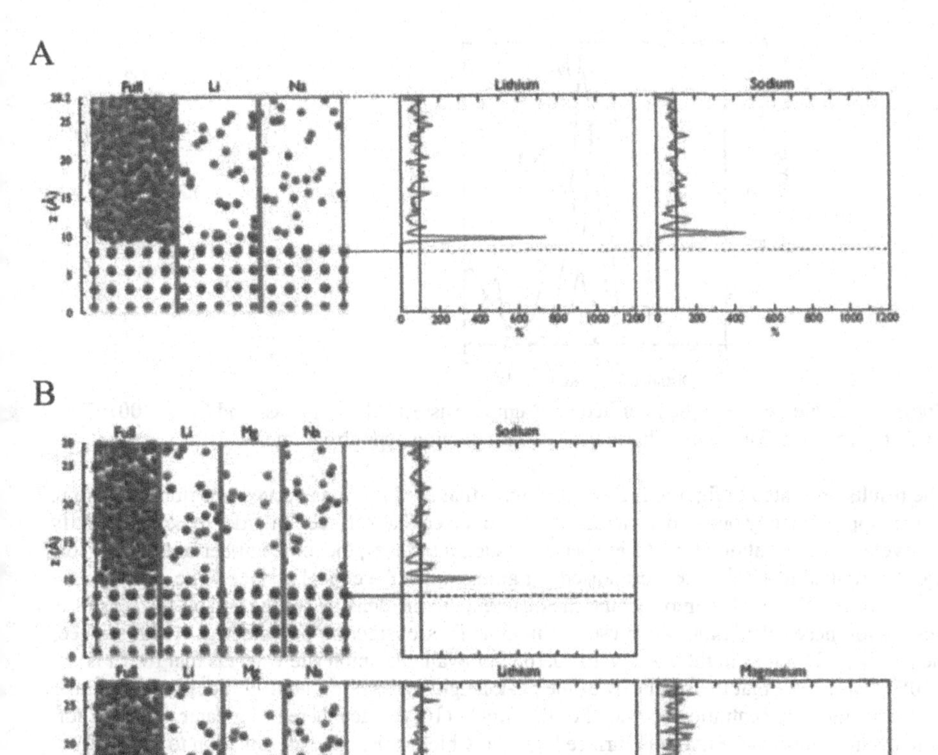

Figure 2. Density of alkali and Mg species in glass as function of distance from interfaces between CaO-(100) and waste glasses: A. Magnox waste glass; B. Magnox glass + Al + Mg (density expressed relative to each species' homogeneous density in bulk glass).

The order in which alkali concentration peaks occurred at the interface was found to be consistent for all the interfaces reported here and for other related systems [3]. In systems with glasses containing Li and Na, the Li peak was always closer to the interface than the Na peak. In MW+Al+Mg glasses the order was Li, Mg then Na (as shown for the interface of MW+Al+Mg and CaO-(100) in figure 3). This ordering seems to scale with the size of the different alkali species. The Shannon [9] VI coordinate radii for Li, Mg and Na are 0.72, 0.76 and 1.02 Å respectively. These show that, when moving away from the interface, the peaks occurred in order of increasing alkali radius. This would be consistent with these modifier ions being more or less confined to layers that correspond to a continuation of the crystal into the glass.

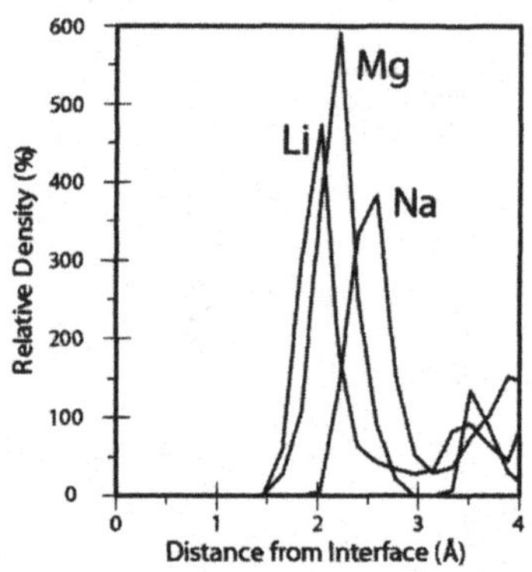

Figure 3. Region close to the interface of Magnox waste + Al + Mg glass and CaO-(100). The peaks reflect the effective modifier ion size when moving away from interface.

The results presented in figures 1, 2 and 3 suggest that modifier ions occupy positions that are in a layer approximately one lattice parameter above the crystal surface. In order to better quantify the overall concentration of modifier species at such interfaces, the total number of modifier ions was determined in a 3 Å slice, normalised per anion site in the crystal surface. The normalization factor takes into account the different atomic densities of the crystal surfaces. Results for these calculations are presented in table II. Considering first the MgO (100) surface, the number of Na ions in the 3 Å layer is 0.165 per available anion site whereas that for Li is 0.105. This shows that the similarity of the peak heights shown in figure 1A does not necessarily reflect the total concentration profile. For the MgO (110) surface there is a greater tendency for Li to occupy sites adjacent to this surface than Na, which is the opposite situation to the (100) surface. Furthermore, the total alkali values for MgO surfaces shows that the (100) surface attracts a 30% lower concentration of alkali species per anion site than the (110) surface.

Table II. Number of alkali atoms in the MW glass adjacent to interfaces as function of number of anion sites in top layer of ceramic block.

		Atoms/anion site in surface	
Crystal	**Alkali**	**(100) Interface**	**(110) Interface**
MgO	Na	0.165	0.158
	Li	0.105	0.235
	total	0.270	0.393
CaO	Na	0.135	0.167
	Li	0.159	0.215
	total	0.294	0.382
SrO	Na	0.093	0.189
	Li	0.160	0.325
	total	0.253	0.514

Values for interfaces that form on surfaces of CaO are also presented in table II and suggest that again the total alkali species in the layer adjacent to the (110) surface is greater than that for the

(100) surface. However, adjacent to the CaO (100) surface the relative concentrations of Li and Na are opposite to that determined for the MgO (100) surface i.e. for CaO (100) there is more Li than Na whereas for MgO (100) there is more Na than Li. Finally, above SrO (100), the amount of Li is 1.7 times that of Na, and essentially the same ratio holds for the (110) surface. Nevertheless, the total concentration of alkali above for all three (100) surfaces did not vary to a significant extent. The same was not true for the (110) surfaces where values for MgO and CaO were almost the same but the value for SrO was a third larger.

CONCLUSIONS

The results presented in this paper suggest that there is a change in the composition of MW glass adjacent to interfaces with crystalline ceramics. This may have important implications for the durability of the glass vicinity of the interface since it no longer has the composition of the bulk glass, which was itself formulated to be a chemically durable host material. In particular, it was found that:

- Na, Li and Mg (if present) do not necessarily segregate to glass/crystal interfaces to the same extent.
- The presence of an ion such as Mg will change the extent to which alkali species may segregate to a crystal surface.
- A specific modifier ion may segregate to different crystal surfaces to differing extents.
- The total concentration of modifier ions in layers above specific crystal surfaces depends upon the crystal orientation and composition.

ACKNOWLEDGEMENTS

Computing resources were provided by the MOTT2 facility (EPSRC Grant GR/S84415/01) run by the CCLRC eScience Centre. MJDR is grateful to Nexia Solutions for providing additional support for his EPSRC studentship.

REFERENCES

1. M. I. Ojovan and W. E. Lee, An Introduction to Nuclear Waste Immobilisation. Amsterdam: Elsevier, 2005.
2. M. J. D. Rushton, R. W. Grimes and S. L. Owens, submitted to J. Am. Ceram. Soc.
3. M. J. D. Rushton PhD. Thesis, University of London, 2006.
4. P. K. Abraitis, F. R. Livens, J. E. Monteith, J. S. Small, D. P. Trivedi, D. J. Vaughan, and R. A.Wogelius, *Appl. Geochem.* **15**, 1399 (2000).
5. Accelrys DISCOVER Molecular Simulation Program v2002.1. Accelrys, 2002.
6. W. G. Hoover, *Phys. Rev. A*, **31**, 1695 (1985).
7. A. Abbas, J. M. Delaye, D. Ghaleb, and G. Calas, *J. Non-Cryst. Sol.* **315**, 187 (2003).
8. A. N. Cormack, J. Du, and T. R. Zeitler, *Phys. Chem. Chem. Phys.* **4**, 3193 (2002).
9. R. D. Shannon, *Acta Crystallogr.* **A32**, 751 (1976).

Mater. Res. Soc. Symp. Proc. Vol. 1107 © 2008 Materials Research Society

A Mössbauer Study of Iron in Vitrified Wastes

Oliver M. Hannant, Paul A. Bingham, Russell J. Hand and Sue D. Forder[1]
Immobilisation Science Laboratory, Department of Engineering Materials, University of Sheffield, Mappin Street, Sheffield, S1 3JD, UK
[1] Materials and Engineering Research Institute, Sheffield Hallam University, Howard Street, Sheffield, S1 1WB, UK

ABSTRACT

^{57}Fe Mössbauer spectroscopy has been used to study the environment and oxidation state of iron in a series of vitrified sewage sludge ash (SSA) wastes, which are broadly similar in composition and variability to some intermediate-level legacy wastes (ILLW). The SSA wastes studied here are incapable of forming an homogeneous glass when melted at 1450 °C although increasing additions of CaO reduce the crystalline content, which consists of $Ca_3(PO_4)_2$ and $Ca_3Mg_3(PO_4)_4$. Bulk glass transition temperatures of 670 - 850 °C have been measured, the value decreasing with increasing CaO content. Tetrahedrally coordinated Fe^{3+} appears to exist in vitrified SSA only at high CaO contents (> ~ 29 mol. %) and whilst broadly similar to our previous results for simulated vitrified SSA, behavioural differences have been noted between fitted Mössbauer parameters for real and simulated vitrified SSA. We suggest that these could be attributable to the buffering action of carbon and other reducing agents in the real SSA wastes.

INTRODUCTION

The application of vitrification in the nuclear waste industry is well documented [1] but it is less well known for its role in the safe and hygienic disposal of toxic wastes such as sewage sludge incinerator ashes (SSA) [2]. Despite some obvious differences between nuclear and other hazardous wastes, many similarities can be found between SSA wastes and some intermediate level legacy wastes (ILLW) in terms of compositional type and variation. The safe disposal of waste is a challenge faced by all sectors of society. Vitrification is seen as a safe and well understood technology for the disposal of dangerous waste streams and has long been used for the immobilisation of nuclear wastes. Recently this technology has become increasingly viable for the immobilisation of other non-nuclear wastes such as SSA due to increasing landfill taxes and gate fees [3] and stricter legislation on the classification of waste types [4]. Although different processes may be involved, whether considering nuclear or hazardous waste immobilisation the challenges remain the same: to create a stable product with well-understood and desirable physical properties. Furthermore, the broad compositional similarities between SSA and some ILLW - namely wastes for which SiO_2, Al_2O_3, Fe_2O_3, MgO and CaO are major constituents - mean that studies of the vitrification of SSA are applicable to some nuclear decommissioning and remediation activities. A key problem with the vitrification of many wastes including SSA and comparable nuclear legacy wastes is the large variability of the composition of the waste [1,2]. Thus, to generate a viable and commercial-scale method for tackling these wastes a process must be capable of creating a desirable product from a non-uniform input. Of fundamental concern to the creation of such a method is an understanding of the structure and properties of the products formed. It is through a thorough understanding of the structure, properties and the processes that form them, that the vitrification process may be optimised.

Previous research has shown that the presence of iron can strongly affect the physical properties of a glass, including its melting behaviour, mechanical properties, viscosity, chemical durability and melting behaviour [5, 6]. Of particular interest are the structural environments and oxidation states of iron in glassy waste forms, their interrelationships and their effects on physical properties. Iron can exist in most oxide glasses in tetrahedrally coordinated network forming sites and / or octahedrally coordinated network modifying sites [5-8]. In addition to studying the melting behaviour and crystallinity of a series of SSA wastes mixed with CaO as a melting flux and subsequently vitrified, the research presented here has determined the coordination environments and oxidation states of iron and their compositional dependency using ^{57}Fe Mössbauer spectroscopy. The characterization of such structural features and, in particular, the factors which affect them, is important to the vitrification of any non-uniform wastestream such as SSA and some ILLW's.

EXPERIMENTAL DETAIL

SSA samples were collected from Blackburn Meadows Waste Water Treatment Plant, Sheffield, UK, and batched with increasing additions of calcium carbonate ($CaCO_3$) to give the analysed sample compositions shown in Table I. The base sample, consisting of 100 % SSA, is sample SSA 1. Additions of increasing levels of CaO have been made between 6 and 21 mol % in approximately 2.5 % increments in an effort to produce an amorphous sample: CaO was chosen due to its low cost and its effectiveness as a melting flux. Samples were melted in an electric furnace at 1450 °C for 5 hours in Al_2O_3 crucibles. Samples were cast on a warmed steel plate and were then annealed at 700 °C for 1 hour before cooling to room temperature.

Table I. Sample compositions by mol. % as analysed by EDS.

Sample	SiO_2	Al_2O_3	Fe_2O_3	CaO	P_2O_5	MgO	Na_2O	K_2O	TiO_2
SSA 1	48.73	12.25	5.44	14.31	8.12	5.93	1.81	1.82	1.29
SSA 2	45.17	12.44	5.15	20.36	7.56	5.09	1.36	1.44	1.18
SSA 3	40.79	12.13	4.52	27.80	6.35	4.26	1.53	1.43	0.94
SSA 4	41.47	11.15	4.42	28.20	6.47	4.36	1.61	1.40	0.84
SSA 5	39.32	14.18	4.05	28.53	6.07	4.14	1.25	1.31	0.89
SSA 6	40.62	10.73	4.26	29.95	6.42	4.31	1.55	1.34	0.79
SSA 7	38.16	11.54	3.97	33.05	5.90	4.18	0.98	1.21	0.74
SSA 8	34.15	15.51	3.43	34.24	5.21	3.93	1.39	1.04	0.81
SSA 9	36.47	10.33	3.78	35.90	5.65	4.19	1.58	1.14	0.82

Elemental analysis of the vitrified samples was conducted using energy-dispersive X-ray spectroscopy (EDS) on a Phillips P500 SEM. Trace quantities (not shown) of SO_3, CoO and CuO were measured at levels of 0 – 0.27 mol %. Errors associated with EDS measurements are estimated to be ± 2 mol % for SiO_2, ± 1 mol % for Al_2O_3, Fe_2O_3, P_2O_5, MgO and CaO, and ± 0.2 mol % for Na_2O, K_2O and TiO_2.

X-ray diffraction (XRD) was performed using a Phillips PW1710 spectrometer at 0.5 °/min between 5 ° and 60 ° 2θ with Co-Kα radiation. Differential thermal analysis (DTA) was performed using a Perkin Elmer DTA 7 to study the melting behaviour of the samples. Approximately 25 mg of crushed samples (< 75μm diameter) were heated from room temperature to 1400 °C at 10 °C/min in Pt crucibles.

^{57}Fe Mössbauer spectroscopy was carried out at room temperature using an in-house spectrometer. Experiments were performed in transmission mode using a ^{57}Co source embedded in a Rh matrix calibrated relative to α-Fe. Spectra were measured using a constant acceleration waveform at a velocity range of ± 6 mm/s or ± 10 mm/s where magnetic hyperfine field splitting was observed. Spectral deconvolution was performed using the commercially available software RECOIL [9] by the application of a separate extended Voigt (xVBF) lineshape per iron environment *i.e.* one component per coordination environment per redox state. This method was used to obtain a quantitative analysis of the individual environments. Extended Voigt lineshapes were chosen as the convolution of Lorentzian and Gaussian components can represent the natural lineshape of the emission/absorption process and the normal distribution of sites, characteristic of an amorphous material, respectively. The isomer shift and quadrupole splitting of each lineshape was initially constrained and given starting seeds suggested by Dyar [7] and Tomandl [8]. Where appropriate these constraints were removed after an initial fit, although in some cases this could not be done due to difficulties in the fitting of many lineshapes and the presence of small amounts of hyperfine field splitting (hfs) with an unknown profile.

RESULTS

X-Ray Diffraction

Samples with CaO contents of <~34 mol % exhibit crystallinity. XRD of sample SSA 1 (Figure 1) shows strong evidence for a $Ca_3Mg_3(PO_4)_4$ phase. With increasing CaO content this phase is replaced by $Ca_3(PO_4)_2$. The Bragg peaks (Al peaks arise from the sample holder) are at the detection limit for sample SSA 8 (~34 mol % CaO), which is largely amorphous with only traces of $Ca_3(PO_4)_2$. Only when CaO > ~ 34 mol % is a largely amorphous product obtained.

Differential Thermal Analysis

Glass transition temperatures for all measured samples fall between 657 °C and 773 °C depending upon the level of CaO addition. Selected sample spectra are shown in Figure 2 and the corresponding measured T_g values are shown in Table II.

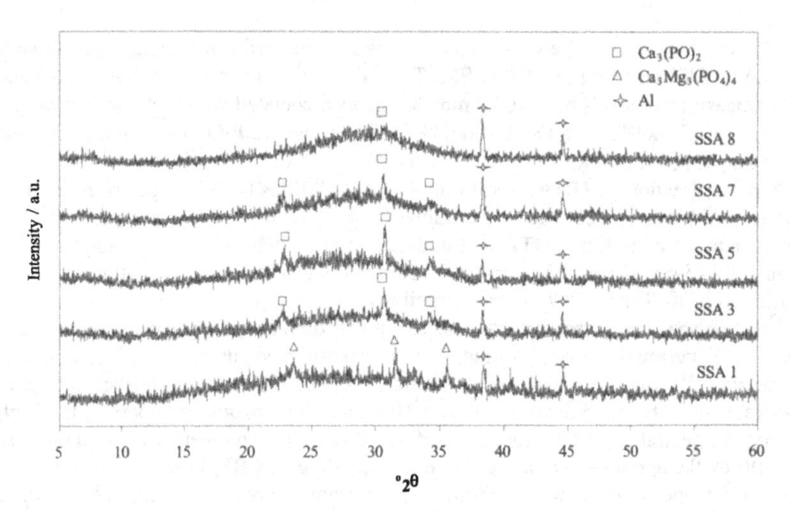

Figure 1. XRD traces for samples SSA 1, 3, 5, 7 and 8. Al peaks are attributed to the sample holder.

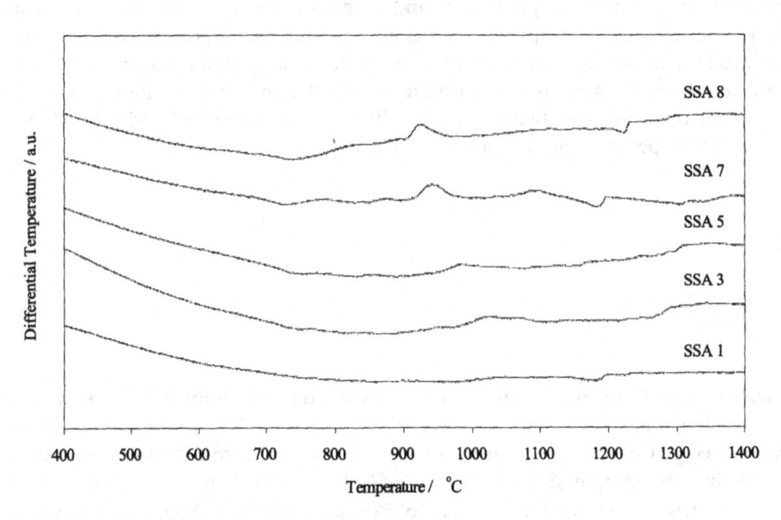

Figure 2. DTA analysis of samples SSA 1, 3, 5, 7 and 8.

^{57}Fe Mössbauer Spectroscopy

^{57}Fe Mössbauer spectroscopy was used to identify changes of iron redox and coordination as a function of CaO content. Sample spectra are shown in Figure 3. Spectra were deconvoluted with the application of up to 4 Voigt-based lineshapes representing Fe^{2+} and Fe^{3+} cations occupying octahedral and tetrahedral sites. Recoil-free fractions for Fe^{3+} and Fe^{2+} have been measured previously by variable temperature Mössbauer spectroscopy and a ratio of $f(Fe^{3+})/f(Fe^{2+}) = 1.18$ has been used when assessing redox ratios and site occupancies. Mössbauer spectra for samples SSA 1 and SSA 2 display hyperfine splitting (hfs), a phenomenon usually indicative, in room-temperature spectra, of the presence of a crystalline paramagnetic phase, although fitting of these spectra has yielded limited useful data thus far. When CaO is added to the SSA at levels > ~ 27 mol % and the mixture is subsequently vitrified all Fe in the sample can be represented by a series of doublets, suggesting that Fe is only present in the amorphous matrix. At CaO contents of < ~ 29 mol % no tetrahedral Fe^{3+} is observed and all of the iron occurs as octahedrally coordinated Fe^{3+} and as Fe^{2+}. With increasing CaO loading, Fe^{2+} abundance decreases from ~32 % to ~27 % of total Fe. At CaO contents between ~29 mol% and ~34 mol%, an exchange is observed between octahedrally coordinated Fe^{3+} and tetrahedrally coordinated Fe^{3+}. Upon further increases in CaO content this interaction becomes more pronounced until tetrahedrally coordinated Fe^{3+} becomes dominant.

Table II. Fitted ^{57}Fe Mössbauer parameters and T_g onset temperature for vitrified SSA samples

Sample	SSA 1	SSA2	SSA 3	SSA 4	SSA 5	SSA 6	SSA 7	SSA 8	SSA 9
% $Fe^{3+}/\Sigma Fe$	-	-	67.2	68.7	65.6	65.6	68.3	73.0	72.9
% Fe^{3+}_{oct}	-	-	67.2	67.5	65.4	58.9	13.6	5.5	4.0
% Fe^{3+}_{tet}	-	-	0	1.1	0.1	6.9	54.8	67.5	68.8
% Fe^{2+}_{oct}	-	-	16.1	12.0	14.7	12.2	14.8	11.0	10.2
% Fe^{2+}_{tet}	-	-	16.7	19.3	19.7	22.1	16.8	16.0	16.5
T_g onset / °C	773	-	717	-	704	-	695	657	-

DISCUSSION

A completely amorphous product has not been obtained with any CaO addition studied here, although levels above ~ 34 mol% result in a largely amorphous material. Sample SSA 8 appears to have trace levels of $Ca_3(PO_4)_2$ which suggests that the abundance of any crystalline phase is close to the detection limit of the defractometer; approximately 5 % by volume. XRD and DTA results show that samples with lower CaO additions produce multiphase systems with crystals of $Ca_3Mg_3(PO_4)_4$ forming in the vitrified raw SSA waste (sample SSA 1) and $Ca_3(PO_4)_2$ in vitrified wastes with CaO abundances of 20 to 34 mol. %. This suggests that the $Ca_3(PO_4)_2$ phase is preferentially formed when the CaO content exceeds a minimum value, and when it does not, Mg^{2+} substitutes onto a Ca^{2+} site to form $Ca_3Mg_3(PO_4)_4$. In either scenario, the presence of a crystalline phase may be undesirable due to negative effects on melting and forming behaviour and the partitioning of toxic or radioactive components, and potentially upon chemical durability. As a fixed annealing temperature was used this meant that in some samples the annealing temperature used was above T_g. Further experiments with lower annealing temperatures have confirmed that the annealing temperature originally used had no effect upon crystallinity or upon Fe^{3+} coordination. The hfs occurring in the Mössbauer spectrum for vitrified raw waste (sample SSA 1 containing no CaO addition, Figure 3) is indicative of the presence of an iron-bearing phase with some magnetic ordering. The broad nature of the fitted hyperfine components suggests that this phase may not be fully crystalline, a theory supported by XRD analysis and the lack of iron-bearing crystalline phases. Hfs, although often attributable to iron in crystalline phases, can also exist with a similar profile in amorphous alloy systems [10].

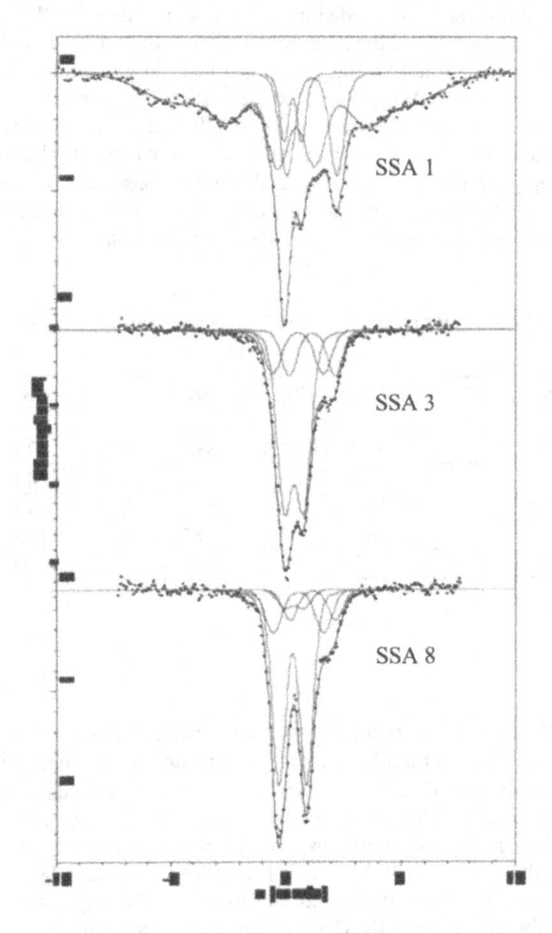

Figure 3. Mössbauer spectra for samples SSA 1 showing the presence of hfs, SSA 3 showing only octahedrally coordinated Fe^{3+} and SSA 8 showing mostly tetrahedrally coordinated Fe^{3+}.

Fitted Mössbauer parameters, shown in Table II, indicate that a clear relationship exists between the $Fe^{3+}/\Sigma Fe$ ratio and the CaO content, suggesting that glass basicity [11] is the source of the redox change. Partitioning of the Fe^{3+} shows interesting behaviour. Fitting indicates that essentially all Fe^{3+} is found in octahedral coordination when CaO content is below ~ 29 mol %

as shown in Figure 4. This result may arise from the necessity for charge balance of Fe^{3+} by the alkaline-earth cation. Such behaviour has been previously noted by Mysen *et al.* [5] for aluminosilicate glasses and melts: their measured Fe^{3+} isomer shifts indicate compositional regions in which only one type of Fe^{3+} coordination site is present. These occurred in their studies at high and low alkali / alkaline earth contents. Rüssel *et al.* [12] observed that Mg^{2+} does not stabilise Fe^{3+} in tetrahedral coordination in aluminosilicate glasses, hence we may infer that Ca^{2+} content directly controls Fe^{3+} partitioning.

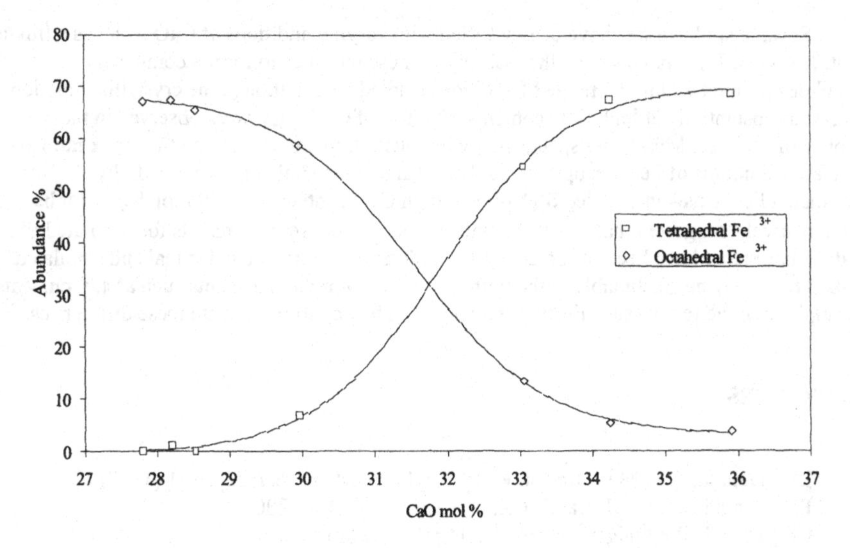

Figure 4. Fe^{3+} coordination as a function of the analysed CaO content of vitrified SSA. Fe^{3+} abundances given as a fraction of total Fe.

At CaO contents greater than ~ 29 mol % a direct interaction is observed between the octahedral and tetrahedral coordinations of Fe^{3+}. Antoni *et al.* [13] observed an increase in the abundance of octahedral Fe^{3+} from 16 % to 34 % following partial molar replacement of Na_2O by CaO in a series of aluminosilicate glasses. However, their glasses were considerably simpler than ours, and in particular they contained no P_2O_5 or MgO. Despite these broad agreements with previous studies, the change in Fe^{3+} partitioning that we have observed over a relatively small compositional region for our vitrified SSA samples does not completely support our previous conclusions based on studies of the vitrification of simulated SSA [14,15]. For the simulated SSA samples the content of octahedrally coordinated Fe^{3+} remains roughly constant at 20-30 % for all $Fe^{3+}/\Sigma Fe$ ratios studied, and interaction is mostly observed between tetrahedrally coordinated Fe^{3+} and octahedrally coordinated Fe^{2+} [14, 15]. Merino *et al.* [16] have measured SSA compositions and found the presence of carbon, water, and sulphide species in significant quantities. It is possible that these components, which are probably present at similar levels in our SSA samples, could act as a redox buffer during melting. This would maintain the redox ratio in the product within tighter limits than would otherwise be expected, based on our previous results obtained from simulated and simplified SSA compositions [14, 15]. Possible reasons for these apparent differences between Mössbauer spectra and fitted parameters from simulated and real vitrified SSA are now being investigated.

CONCLUSIONS

Sewage sludge ashes have been vitrified with varying additions of CaO to act as a fluxing agent. XRD and DTA have shown the inability of these mixtures to form a completely amorphous product within the range of additions studied here, although the crystalline fraction decreased substantially at high CaO contents. Crystals of $Ca_3(PO_4)_2$ were observed in the majority of samples. Mössbauer spectroscopy has quantitatively determined the redox ratio and the relative amounts of Fe^{3+} occupying tetrahedral and octahedral sites. Tetrahedrally coordinated Fe^{3+} exists in vitrified SSA only at high CaO contents (> ~ 29 mol. %) and it has been tentatively suggested that, whilst broadly similar to our previous results for simulated SSA, the differences that have been observed between Mössbauer parameters for real and simulated vitrified SSA may be attributable to the buffering action of reducing agents such as carbon, water and sulphide in the real wastes. Further studies are underway to investigate these differences.

REFERENCES

1. I.W. Donald, B.L. Metcalfe and R.N.J. Taylor, *J. Mater. Sci.* **32**, 5851 (1997).
2. P.A. Bingham and R.J. Hand, *J. Appl. Ceram.* **100**, 120 (2001).
3. Chancellor's Pre budget report 2002. http://www.hm-treasury.gov.uk/pre_budget_report/prebud_pbr02/report/prebud_pbr02_repchap7.cfm
4. Landfill directive briefing paper, Defra, UK. http://www.defra.gov.uk/environment/waste/topics/landfill-dir/pdf/landfilldir.pdf
5. B.O. Mysen, D. Virgo and F.A. Seifert, *Amer. Mineral.* **69**, 834 (1984).
6. M. B. Volf, Chemical Approach to Glass, Elsevier, Amsterdam, 1984.
7. M. D. Dyar, *Amer. Mineral.* **70**, 304 (1985).
8. G. Tomandl, in Glass: Science and Technology, 4B Ed. D.R. Uhlmann, N.J. Kreidl, Academic Press, New York, 1990, ch. 5.
9. RECOIL v.1.05, Intelligent Scientific Applications Inc., Ottowa, Canada.
10. S.H. Kilcoyne, D. Greig and M.N. Gona, *Hyperfine Interact.* **165**, 167 (2005).
11. J. A. Duffy, *J. Non-Cryst. Solids* **297** (2002) 275.
12. C. Rüssel and A. Wiedenroth, *Chem. Geol.* **213**, 125 (2004).
13. E. Antoni, L. Montagne, S. Daviero, G. Palavit, J. L. Bernard, A. Wattiaux and H. Vezin, *J. Non-Cryst. Solids* 345&346, 66 (2004).
14. O.M. Hannant, P.A. Bingham, S.D. Forder and R.J. Hand, *Phys. Chem. Glasses: Eur. J. Glass Sci. Technol. B* **48** (2007) (in press).
15. O.M. Hannant, P.A. Bingham, S.D. Forder and R.J. Hand, *Proc. XXIst Int. Congr. Glass*, Strasbourg, France (2007), CD-ROM paper U30.
16. I. Merino, L.F. Arevalo and F. Romero, *Waste Man.* **25**, 1046 (2005)

Mater. Res. Soc. Symp. Proc. Vol. 1107 © 2008 Materials Research Society

In Situ Characterisation of Model UK Nuclear Waste Glasses by X-ray Absorption Spectroscopy Under Process Conditions

Neil C. Hyatt,[1*] Andrew J. Connelly,[1] Martin C. Stennett,[1] Francis R. Livens,[2] and Robert L. Bilsborrow.[3]

[1] Immobilisation Science Laboratory, Department of Engineering Materials, The University of Sheffield, Mappin Street, Sheffield, S1 3JD. UK.

[2] Centre for Radiochemistry Research, Department of Chemistry, The University of Manchester, Oxford Road, Manchester, M13 9PL. UK.

[3] STFC Daresbury Laboratory, Warrington, Cheshire, WA4 4AD. UK.

ABSTRACT

The local co-ordination environment of Zr in a model alkali borosilicate glass, of relevance to nuclear waste immobilisation, was studied by *in situ* X-ray absorption spectroscopy between 25 – 1060 °C. Analysis of Zr K-edge XANES spectra, in comparison with those of well characterised standards, demonstrated, for the first time, the reversible transformation of ZrO_6 to ZrO_7 co-ordination polyhedra at high temperature. This observation was rationalised on the basis of the combined effects of network modifier cation diffusion and thermal expansion.

INTRODUCTION

PUREX reprocessing of nuclear fuels, in the UK and France, affords a High Level Waste stream comprising fission products, corrosion species, minor actinides, and traces of U and Pu, that is vitrified in an alkali borosilicate matrix. The choice of this vitreous matrix is due to the considerable flexibility of the glass structure with respect to chemical constitution, permitting vitrification of waste streams of variable composition [1, 2]. These alkali borosilicate glasses may be fabricated at reasonable temperatures, *ca* 1060°C, and show superior chemical durability with respect to other vitreous wasteforms (*e.g.* phosphate based glasses) [1, 2]. In Europe, vitrification of high level nuclear waste is currently undertaken at the Sellafield Waste Vitrification Plant (WVP), operated by Sellafield Ltd., and the La Hague vitrification plants, R7 and T7, operated by COGEMA.

An understanding of the structure of nuclear waste glasses is essential in order to fully rationalise the relationship between glass composition, structure, and properties such as durability, viscosity and crystallisation potential. The complex chemical nature of such nuclear waste glasses demands an element selective spectroscopic technique, such as X-ray Absorption Spectroscopy (XAS), in order to determine both oxidation state and local chemical environment. Indeed, XAS investigations have been crucial in establishing the structural role of key elements, such as Mo, Zr, Zn, Sr, Tc, Ce, Pu, Pd and Te, in model nuclear waste glasses [3-12]. XAS studies of quenched glasses have provided an insight into important structure-property relationships; for example, the rapid leaching of Mo from simulant waste glasses is rationalised on the basis of the presence of isolated MoO_4^{2-} species encapsulated by the polymeric glass network [7, 9, 10]. However, the oxidation state and co-ordination environment determined in quenched materials are not necessarily representative of those in high temperature melts. Disproportionation and charge transfer on cooling may prohibit quenching of unusual oxidation states, as demonstrated by XAS for the case of Cr^{2+} in basaltic silicate melts [13]. Furthermore, transition metal cations, such as Ni, have been demonstrated to show rather different environments in quenched silicate glasses, where NiO_5 is dominant, compared to melts studied *in situ*, where NiO_4 is dominant [14].

In contrast to a considerable number of *in situ* XAS studies of silicate glass melts of geological interest, as reviewed by Brown *et al.* [15], there are no similar studies of nuclear waste related glasses to date. Clearly, however, such studies are essential in order to fully understand the behaviour of key waste elements under processing conditions. In this context, the structural role of zirconium is of particular interest since the viscosity of alkali borosilicate glasses is known to be a sensitive function of ZrO_2 content. Previous room temperature XAS studies established that ZrO_6 is the dominant species in quenched alkali borosilicate glasses but with next nearest neighbour silicon atoms indicating the presence of corner-sharing SiO_4 species [5, 8]. Our study of quenched alkali borosilcate glasses with compositions relevant to the immobilisation of UK nuclear wastes has confirmed these findings [16]. However, the structural chemistry of zirconium in (boro)silicate glass melts is essentially an open question. The only reported *in situ* XAS study showed that ZrO_6 species remained dominant in $Na_2Si_3O_7$ glasses doped with 3 wt% ZrO_2, up to 1277 °C [15]. However, given that Zr is co-ordinatively unsaturated as ZrO_6 species in alkali borosilicate glasses at room temperature, the existence of ZrO_7 and ZrO_8 species at high temperature could be anticipated. To investigate this hypothesis, we performed an *in situ* Zr K-edge XAS study of a model nuclear waste glass with composition (mol%): Na_2O – 10.79%, Li_2O – 5.38%, B_2O_3 – 19.08%, SiO_2 – 62.23%, ZrO_2 – 2.52%. This composition corresponds to the alkali borosilicate base glass used for UK HLW vitrification, doped with 2.52 mol% ZrO_2. Our preliminary findings, reported here, provide strong evidence for the co-existence of ZrO_6 and ZrO_7 species at a typical processing temperature of 1060 °C.

FURNACE DESIGN FOR *IN SITU* XAS STUDIES OF GLASS MELTS

The low viscosity of the alkali borosilicate compositions of interest at typical processing conditions (1050 - 1100 °C) precluded the use of the generic "heating wire" type furnace, for this *in situ* XAS study, in which the molten glass is retained by high surface tension, as reviewed by Brown *et al.* [15]. Furthermore, the design of SRS beamline 16.5, selected for this study, does not permit the acquisition of data in energy dispersive mode which is advantageous in using the micro-furnace design reviewed by Brown *et al.* [15]. Consequently, it was necessary to use an alternative furnace design for this study to permit the acquisition of XAS data using a focussed X-ray beam in transmission mode, on beamline 16.5. The furnace arrangement, shown in Figure 1, comprises a water-cooled, vertically mounted, tube furnace with $MoSi_2$ furnace elements, capable of maintaining a sample temperature of up to 1100 °C. The furnace is fixed, on a height adjustable stand, to the beamline optics rail by means of a clamp. The sample containment consists of a "T"-shaped quartz tube, containing a small quantity of glass powder. The tube is mounted, using a clamp, to form an inverted "T", such that the sample is contained in the lower part of the tube stem and the horizontal bar of the upturned "T". The open end of the vertical furnace tube is plugged with insulating wool to minimise radiative heat loss and maintain thermal stability at high temperature. Perpendicular bore tubes permit optical access to the sample, allowing data to be acquired in transmission or fluoresence mode, as desired, by rotating the sample tube. Preliminary alignment of the sample and furnace height is made by adjusting the securing clamps, guided by a laser aligned to the path of the X-ray beam. After alignment is achieved, beryllium windows are fixed to open ends of the optical access tubes to minimise radiative heat loss. Fine adjustment of the sample position is made by remote adjustment of the optical table height (to 100 μm precision) whilst simultaneously monitoring the absorption of the X-ray beam to optimise the edge step. The furnace control system comprises a Eurotherm 2216E controller, power supply and Type K thermocouple. An additional thermocouple is fastened to the tube stem, by means of nichrome wire, with the tip ~3 mm from the sample position for accurate measurement of sample temperature. We estimate, based on off-line tests, that the difference between the actual and recorded sample temperature is <1 °C, due primarily to the long (~20 mm) furnace hot zone. The sample temperature is continuously recorded during the experiment, using a PC with a Pico TC-08 interface, and varies by less than ± 0.5 °C during a programmed isotherm.

Figure 1: High temperature XAS set up on SRS beamline 16.5: A) sample thermocouple monitor; B) transmitted beam ion chamber; C) vertical tube furnace; D) fluoresence detector; E) sample contained in quartz tube viewed perpendicular to transmitted beam; F) crucible to capture sample in event of loss of containment.

EXPERIMENTAL

Zr K-edge X-ray absorption spectra were collected on beamline 16.5 at the Synchrotron Radiation Source, Daresbury, UK. The storage ring operates at 2 GeV with a typical beam current of 150 mA. A double crystal Si (220) monochromator was used, detuned to 50% of maximum intensity, for harmonic rejection. The Zr K absorption edge was calibrated by measuring the K-edge from a Zr foil at E_0 = 17998.0 eV. Spectra were recorded in transmission mode using ion chambers filled with a mixture of Ar (698 mbar) / balance He (incident beam, I_0) and Kr (368 mbar) / balance He (transmitted beam, I_t). Data were acquired from $NaLiZrSi_6O_{15}$, $ZrSiO_4$ and ZrO_2 standards at room temperature, in the form of fine powders, diluted with BN (in a *ca* 1:10 volume ratio), packed into aluminium sample holders. Two spectra were acquired and summed for each standard. Data from a NLBSZ glass sample of composition (mol%), Na_2O – 10.79%, Li_2O – 5.38%, B_2O_3 – 19.08%, SiO_2 – 62.23%, and ZrO_2 – 2.52%, were obtained at 25 – 1060 °C, using the furnace described above. Between two and eight spectra were acquired at each temperature and summed. Acquisition time for all XAS spectra was ~20 minutes, at high temperature the glass sample was allowed to equilibrate for 30 minutes before acquisition of the first spectrum. Comparison between successive spectra acquired at each temperature revealed no discernable changes. All glass spectra were acquired with the sample in equilibrium with air, with the open end of the quartz tube loosely plugged with insulating wool. The sample of glass with composition specified above was prepared by dissolving ZrO_2 in a pre-prepared frit at 1060 °C for 6 h, the melt was quenched and the sample was annealed at 500°C for 1 h before cooling to room temperature.

RESULTS AND DISCUSSION

X-ray absorption spectra were acquired at 25 °C for three crystalline standards, $NaLiZrSi_6O_{12}$ (zektzerite), ZrO_2 (baddelyite) and zircon ($ZrSiO_4$) and from the NLBSZ glass sample, as shown in Figure 2a. Zr is co-ordinated as near regular octahedral ZrO_6 species in $NaLiZrSi_6O_{12}$ (Zr-O distances: 2.06 – 2.09 Å); distorted ZrO_7 polyhedra in ZrO_2 (Zr-O distances: 2.05 – 2.27 Å; and as ZrO_8 polyhedra in $ZrSiO_4$ (Zr-O distances: 2.13 – 2.27 Å) [17-19]. The XANES spectra of the standard materials are in excellent agreement with those previously reported [4, 5]. As shown in Figure 2a, two common edge features are observed in these spectra: feature C, assigned as simple and multiple scattering within the Zr-O co-ordination polyhedron; and feature B, assigned to multiple scattering involving first (O) and next nearest (Na, Si, Zr) neighbours [20]. The intensity of feature A has been shown to be dependent on the distortion of the Zr co-ordination polyhedron, the weak intensity of this feature is consistent with essentially regular polyhedra [20]. A further feature, D, of unknown origin is apparent only the XANES spectrum of $ZrSiO_4$.

The relative intensity of features B and C varies with Zr co-ordination number, as shown in Figure 2a: for ZrO_6 polyhedra C > B; whereas for ZrO_8 polyhedra, B > C; in the case of ZrO_7 polyhedra features B and C cannot be clearly distinguished. Comparison of the XANES spectra of the standards and the NLBSZ glass acquired at 25 °C, reveals a remarkable similarity between the spectra of the NLBSZ glass and $NaLiZrSi_6O_{12}$ standard. Based on this XANES fingerprint, the presence of ZrO_6 as the dominant co-ordination environment in the NLBSZ glass is inferred. This conclusion is supported by complete analysis of the full EXAFS data from the glass, as discussed by Connelly *et al.* [16].

A clear difference is apparent between the XANES spectra acquired from the NLBSZ glass at 25 °C and 1060 °C, as shown in Figure 2a: whereas feature C is more intense than feature B at 25 °C, the intensities of these features are approximately equal at 1060 °C. The difference between these spectra is indicative of a change in the Zr co-ordination environment at high temperature. The increase in intensity of feature B with respect to feature C suggests the presence of more highly co-ordinated ZrO_7 and / or ZrO_8 species at 1060 °C. Meneghini *et al.* demonstrated that quantification of zirconium co-ordination polyhedra in calcium silicate glasses may be achieved by fitting linear combinations of the XANES spectra of standard materials to that of the XANES spectrum of a glass [21]. Following this methodology, we attempted to fit binary and ternary combinations of $NaLiZrSi_6O_{12}$, ZrO_2 and $ZrSiO_4$ XANES spectra to that of NLBSZ glass at 25 °C and 1060 °C, using a non-linear least squares algorithm to minimise the difference between observed and summed spectra. Linear combinations of $NaLiZrSi_6O_{12}$ and ZrO_2 spectra always provided a better fit to the glass spectra than combinations of $NaLiZrSi_6O_{12}$ and $ZrSiO_4$; this was due, in part, to the presence of the additional feature D in the $ZrSiO_4$ spectrum which could not be accounted for in the spectra of the glass. Convergence of the least squares algorithm could not be achieved using combinations of $NaLiZrSi_6O_{12}$, ZrO_2 and $ZrSiO_4$ spectra. Based on this analysis, the proportions of ZrO_6 and ZrO_7 polyhedra in NLBSZ glass at 25 °C are, respectively, 83 and 17%; increasing to 53% and 47%, respectively, at 1060 °C.

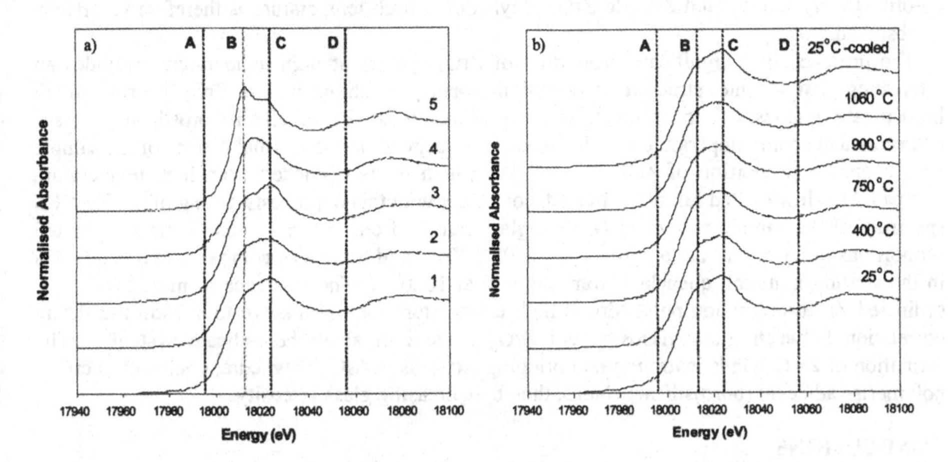

Figure 2: a) Zr K-edge XANES, from 1) NLBSZ glass at 25 °C, 2) NLBSZ glass at 1060 °C, 3) $NaLiZrSi_6O_{12}$, 4) ZrO_2, 5) $ZrSiO_4$; b) variable temperature Zr K-edge XANES spectra from NLBSZ glass. See discussion for origin of features A-D.

Figure 2b shows a series of XANES spectra acquired initially at 25 °C, and then at temperatures between 1060 °C and 25 °C on cooling. Comparison of these spectra reveals that the intensity ratio of features C and B varies smoothly on cooling, with the XANES spectrum of the recovered glass being similar to that acquired before heating. The reversibility of the observed change in the XANES spectra indicates that the temperature induced change in Zr co-ordination is reversible.

Using bond valence calculations, Brown and Farges *et al.* showed that co-ordination of Zr to non-bridging oxygens of network forming MO_4 tetrahedra (M = Al, Si), was favoured for ZrO_6 polyhedra in (alumino)silicate glasses provided that additional charge compensation was made by network modifier cations (*e.g.* Na^+, Li^+) [15, 22]. This Zr co-ordination cluster, similar to that present in $NaLiZrSi_6O_{12}$, is dominant in NLBSZ glass at 25 °C as confirmed by comparison of XANES spectra in Figure 2 and full analysis of EXAFS data presented elsewhere [16]. For ZrO_8 polyhedra, Brown and Farges *et al.* showed that co-ordination of Zr to bridging oxygens of MO_4 tetrahedra was favoured, without charge compensation by network modifier cations – due, essentially, to the longer Zr-O bond length associated with larger polyhedra [15, 22]. The lower bond valence of these longer Zr-O bonds, involving bridging oxygens, requires an increase in co-ordination number to charge compensate the Zr cation. Based on this insightful analysis, the temperature induced change in Zr co-ordination polyhedra may be rationalised. At high temperature, rapid diffusion of network modifier cations would be expected to de-stabilise Zr-O bonds involving non-bridging oxygens of network forming MO_4 polyhedra. In addition, thermal expansion of Zr-O bond lengths would be expected to favour formation of Zr-O bonds involving non-bridging oxygens. These two factors operate in tandem, leading to conversion of ZrO_6 to ZrO_7 species at high temperature. On cooling, ZrO_7 species capture more slowly diffusing network modifier cations resulting in the conversion of ZrO_7 to ZrO_6 species. The conversion of co-ordinatively unsaturated ZrO_6 to ZrO_7 polyhedra at high temperature is therefore reversible, as observed.

The presence of a significant proportion of ZrO_7 species at high temperature, provides an interesting insight into structure-composition-property relationships in ZrO_2 bearing alkali borosilicate glasses. For example, the co-ordination of Zr in alkali borosilicate glasses determined at room temperature may be expected to depend on the thermal history of the sample with a greater proportion of ZrO_7 species retained in melts quenched from high temperature compared to slow cooled samples. Indeed, consistent with this hypothesis, fitting of the XANES spectrum of the sample recovered NLBSZ glass sample from our *in situ* experiment estimated proportions of Zr co-ordination polyhedra as 92% ZrO_6 and 8% ZrO_7, compared to 83 and 17% in the starting material, quenched from the melt at 1060°C. The existence of more highly co-ordinated Zr co-ordination polyhedra at high temperature may also serve to explain the strong correlation between glass viscosity and ZrO_2 content in alkali borosilicate systems. The formation of Zr-O bonds involving non bridging oxygens of MO_4 polyhedra, could act to cross-polymerise adjacent (boro)silicate chains, thereby increasing glass viscosity.

CONCLUSIONS

The temperature dependence of the Zr co-ordination environment in an alkali borosilicate glass was investigated by *in situ* X-ray Absorption Spectroscopy. Comparison of XANES spectra from the glass and standard materials demonstrated the presence of ZrO_6 as the dominant species at 25 °C. At high temperature a significant proportion of ZrO_6 polyhedra were converted to ZrO_7 species, however, on cooling this conversion proved reversible with ZrO_6 polyhedra forming the dominant species at 25 °C. At high temperature the formation of Zr-O bonds involving non-bridging oxygens may be favoured, due to the combined effects of thermal expansion and high mobility of network modifier cations destabilising Zr-O bonds involving non-bridging oxygens. This study highlights the need for further *in situ* investigation of co-ordinatively unsaturated cations in molten vitreous systems of technological and geological importance.

ACKNOWLEDGEMENTS

We thank Nexia Solutions Ltd. and EPSRC for the provision of a CASE studentship to AJC. This work was carried out as part of the TSEC programme KNOO and as such we are grateful to the EPSRC for funding under grant EP/C549465/1.

REFERENCES

1. W. Lutze and R.C. Ewing, "Nuclear wasteforms for the future", Elsevier (1998).
2. I.W. Donald, B.L. Metcalfe and R.N.J. Taylor, *J. Mater. Sci.*, **32**, 5851-5887 (1997).
3. L. Galoisy, G. Calas, G. Morin, S. Pugnet and C. Fillet, *J. Mat. Res.* **13**, 1124-1127 (1998).
4. L. Galoisy, E. Pelegrin, M.A. Arrio, P. Idldefonse and G. Calas, *J. Am. Ceram. Soc.* **82**, 2219-2224 (1999).
5. D.A. McKeown, I.S. Muller, A.C. Buechele and I.L. Pegg, *J. Non-Cryst. Solids* **258**, 98-109 (1998).
6. M. Le Grand, A.Y. Ramos, G. Calas, L. Galoisy, D. Ghaleb and F. Pacaud, *J. Mater. Res.* **15**, 2015-2019 (2000).
7. C. Calas, M. Le Grand, L. Galoisy and D. Ghaleb, *J. Nucl. Mater.* **322**, 15-20 (2003).
8. D.A. McKeown, W.K. Kot and I.L. Pegg, *J. Non-Cryst. Solids* **317**, 290-300 (2003).
9. R.J. Short, R.J. Hand, N.C. Hyatt and G. Mobus, *J. Nucl. Mater.* **340**, 179-186 (2005).
10. N.C. Hyatt, R.J. Short, R.J. Hand, W.E. Lee, F.R. Livens and J.M. Charnock, *Ceram. Trans.* **168**, 179-178, (2005).
11. J.N. Cachia, X. Deschanels, C. Den Auwer, O. Pinet, J. Phalippou, C. Hennig and A. Scheinost, *J. Nucl. Mater.* **352**, 182-189 (2006).
12. W.W. Lukens, D.A. McKeown, A.C. Buechele, I.S. Muller, D.K. Shuh and I.L. Pegg, *Chem. Mater.* **19**, 559-566 (2007).
13. A.J. Berry, M.G. Shelley, G.J. Foran, H.St.C. O'Neill and D.R. Scott, *J. Synchr. Rad.* **10**, 332-336 (2003).
14. F. Farges, G.E. Brown, G. Calas, L. Galoisy, G.A. Waychunas, *Geophys. Res. Lett.*, **21** 1931-1934 (1994).
15. G.E. Brown, F. Farges and G. Calas, *Rev. in Miner. and Geochem.* **32**, 317-409 (1995).
16. A.J. Connelly, PhD Thesis (University of Sheffield).
17. D.K. Smith and H.W. Newkirk, *Acta Cryst.* **18**, 983-991 (1965).
18. K. Robinson, G.V. Gibbs and P.H. Ribbe, *Am. Miner.* **56**, 782-790 (1971).
19. S. Ghose and C. Wan, *Am. Miner.* **63**, 304-310 (1978).
20. F. Farges, *Chem. Geol.* **127**, 253-268 (1996).
21. C. Meneghini, S. Mobilio, L. Lusvarghi, F. Bondioli, A.M. Ferrari, T. Manfredinic and C. Siligardic, *J. Appl. Cryst.* **37**, 890-900 (2004).
22. F. Farges, C.W. Ponander and G.E. Brown, *Geochim. et Cosmochim. Acta* **55**, 1563-1574, (1991).

Mater. Res. Soc. Symp. Proc. Vol. 1107 © 2008 Materials Research Society

Glass Formulation Development in Support of Melter Testing to Demonstrate Enhanced High Level Waste Throughput

James C. Marra, Kevin M. Fox, David K. Peeler, Thomas B. Edwards, Amanda L. Youchak, James H. Gillam, Jr., John D. Vienna[1], Sergey V. Stefanovsky[2], and Albert S. Aloy[3]
Savannah River National Laboratory, Aiken, SC, U.S.A.
[1] Pacific Northwest National Laboratory, Richland, WA, U.S.A.; [2] SIA Radon Institute, Moscow, Russia; [3] V. G. Khlopin Radium Institute, St. Petersburg, Russia

ABSTRACT

The U.S. Department of Energy (DOE) is currently processing high-level waste (HLW) through a Joule-heated melter (JHM) at the Savannah River Site (SRS) and plans to vitrify HLW and Low activity waste (LAW) at the Hanford Site. Over the past few years at the Defense Waste Processing Facility (DWPF), work has concentrated on increasing waste throughput. These efforts are continuing with an emphasis on high alumina concentration feeds. High alumina feeds have presented specific challenges for the JHM technology regarding the ability to increase waste loading yet still maintain product quality and adequate throughput. Alternatively, vitrification technology innovations are also being investigated as a means to increase waste throughput. The Cold Crucible Induction Melter (CCIM) technology affords the opportunity for higher vitrification process temperatures as compared to the current reference JHM technology. Higher process temperatures may allow for higher waste loading and higher melt rate.

Glass formulation testing to support melter demonstration testing was recently completed. This testing was specifically aimed at high alumina concentration wastes. Glass composition/property models developed for DWPF were utilized as a guide for formulation development. Both CCIM and JHM testing will be conducted so glass formulation testing was targeted at both technologies with a goal to significantly increase waste loading and maintain melt rate without compromising product quality.

INTRODUCTION

Vitrification of high level defense wastes has been underway in the United States since 1996 with operations at the DWPF at the Savannah River Site. A recent focus of these operations is on increasing waste throughput. To achieve higher waste throughputs, both improvements in waste loading and increases in melt rate have been targeted [1]. Glass composition development efforts have resulted in increases in waste loading from nominally 28 wt % to 38 wt %. Glass formulation efforts have also resulted in the development of frits that permit higher melt rates. Melter system enhancements such as the incorporation of a melter glass pump have also increased melter production rates [2]. Construction of the Waste Treatment Plant (WTP) has begun at the Hanford Site for facilities to vitrify high-level and low activity radioactive wastes, both of which have throughput or glass production rate goals [3].

Future waste compositions will present challenges to continued process improvements. Specifically, DWPF waste compositions with high alumina concentrations have the potential to increase nepheline ($NaAlSiO_4$) crystal formation in the glass [4]. The formation of nepheline can have a detrimental impact on glass durability because it decreases the amount of the glass forming oxides Al_2O_3 and SiO_2 in the residual glass matrix. In addition to the durability concerns associated within this compositional range, the refractory nature of alumina may have a negative impact on melt rate for high alumina concentration feeds. Future processing at the WTP will also be impacted by high alumina concentration feeds [3].

The reference JHM technology has a maximum process temperature of 1150° C. This temperature limitation, a result of materials of construction, combined with the refractory nature of the high alumina feeds hampers continued waste throughput improvements in JHMs. The CCIM technology offers the potential for higher vitrification process temperatures which could permit increased waste loading and/or melt rates for high alumina concentration feeds.

A focused test program is in progress with two primary goals: i) demonstrate a maximum attainable alumina concentration in a glass without detrimentally impacting glass properties or melter processing; and ii) gain insight into effects of alumina concentration on waste throughput. The test program involves two phases: a glass composition testing phase and a melter testing phase. Two approaches were utilized in the glass formulation testing phase. The first involved utilization of the positive attributes of the CCIM and formulating glasses specifically for processing at the higher process temperatures afforded by the CCIM. The second approach involved more detailed formulation and testing of glasses to evaluate compositional effects on glass properties. The results of this testing will be used to direct JHM testing on high alumina concentration glass compositions.

EXPERIMENTAL DETAILS

Glass models

Glass composition models have been utilized extensively as prediction tools for waste glass formulation and for control of vitrification processes [5-7]. A Product Composition Control System (PCCS) was developed for DWPF based on work by Jantzen, et al. [7]. A Measurement Acceptability Region (MAR) approach was developed by Peeler and Edwards to facilitate formulation of waste glasses for DWPF [8]. The MAR approach allows for efficient evaluation of glass compositions against the PCCS constraints for various glass quality and processing properties. A "nepheline discriminator" is included as one of the MAR terms. The nepheline discriminator is based on work by Li, et al. [9] and utilizes waste glass composition to predict the potential for nepheline formation. Specifically, glasses with $SiO_2/(SiO_2+Na_2O+Al_2O_3) > 0.62$, where the chemical formula represents mass fractions in the glass, do not precipitate nepheline. The MAR approach was utilized in the formulation efforts in this study with specific emphasis on the nepheline discriminator function.

Glass formulation development for CCIM testing

As previously mentioned, the CCIM technology allows for processing at higher temperatures than those available in the JHM. Therefore, candidate glass compositions for CCIM testing at the SIA Radon Institute in Russia were targeted for melting temperatures of 1250° C and the MAR constraint for liquidus temperature (associated with the lower processing temperatures in the JHM) was ignored. A DWPF Sludge Batch 4 (SB4) surrogate waste composition was identified for testing (Table I). This sludge had a high alumina concentration and was previously found to present waste loading and melt rate challenges for the JHM [10].

A previously developed frit composition (Frit 503) was used as a basis for identification of candidate glass formulations. Frit 503 was modified by lowering the sodium content and/or increasing the boron content, or both. It was hypothesized that removing sodium from the glass would mitigate nepheline formation based on the nepheline discriminator equation. A decrease in alkali content would also result in an increase in SiO_2 concentration and would be expected to be favorable per the nepheline discriminator equation. Increasing boron content was also postulated to reduce nepheline formation based on previous observations [9]. The candidate frit compositions are shown in Table II. The MAR assessment tool was used to evaluate the compositions for waste loadings from 30 to 65 wt % waste loading (on a calcined oxide basis). Based on favorable MAR assessment results, it was decided to fabricate glasses using all five candidate frit compositions at 40, 45, 50, and 55 wt % waste loading. The compositions were batched using oxide chemicals and melted at 1250° C in Pt/Rh crucibles. After nominally two hours at temperature, the glasses were quenched by pouring on a steel plate. A portion of the glass was heat treated to simulate cooling along the centerline of the DWPF canister (i.e. Canister Centerline Cooling (CCC) profile) [11]. The CCC cooling profile was adjusted to a higher starting temperature consistent with the 1250° C melt temperature. Both quenched and CCC glass samples were then evaluated for crystallization using X-ray diffraction (XRD). Durability testing was conducted on quenched and CCC glass samples using the Product Consistency Test (PCT) [12].

Glass formulation development for JHM testing

A two-phased approach was used to develop glass compositions to support future JHM testing at the V. G. Khlopin Radium Institute (KRI) in Russia. The initial phase involved development of a test matrix of glasses to evaluate the solubility of aluminum, chromium and sulfur over a range of compositions that were considered bounding for future sludge batches to be processed through DWPF and WTP. A total of 75 glass compositions were developed. The 75 glasses were divided into a "U.S. matrix" (45 glasses) for glass fabrication and analysis at the Savannah River National Laboratory (SRNL) and the Pacific Northwest National Laboratory (PNNL) and a "KRI matrix" (30 glasses) for glass fabrication and testing at KRI. Details on these glass compositions are described elsewhere [13]. The results of this testing were used to direct glass composition development for future melter testing at KRI. The future melter testing will include small-scale melter testing to evaluate melt rate followed by testing in a larger-scaled JHM. To support DWPF needs, a sludge batch 5 (SB5) simulant composition was identified for testing. Two possible flowsheets are under consideration for SB5 processing at DWPF. One involves typical tank blending and washing strategies to produce a suitable feed for processing in the DWPF. The second involves an additional process where aluminum is dissolved and removed from the sludge in order to reduce the mass of material that must be processed through the DWPF. Table I shows compositions for both scenarios. To support Hanford WTP interests, a high alumina concentration sludge surrogate was selected for future melter testing (Table I).

Development of the specific glasses for subsequent melter testing was based in part on the results of the first portion of this study (i.e., the "U.S. Matrix" and "KRI Matrix" glasses) and on leveraging existing data that may support the development process. In fact, the frit composition identified for melter testing to support WTP (Table II) was based on these previous studies and, for brevity, will not be discussed further. Candidate glasses for JHM testing to support DWPF were assessed using the MAR assessment tool for an array of frit formulations and the two SB5 compositions identified in Table I for varying waste loadings. This screening assessment led to 4 primary frit formulations for more detailed assessment. Test glasses based on these 4 frit formulations were fabricated at 35 and 40 wt % waste loading (calcined oxide basis), melted at 1150° C for nominally 2 hours, and poured on a steel plate. Both as-fabricated and CCC glasses were characterized as described previously.

RESULTS AND DISCUSSION

MAR assessment for CCIM glasses

A MAR assessment for SB4 candidate glasses for testing in the CCIM was conducted for waste loadings from 30 to 65 wt %. The candidate compositions failed the current liquidus temperature constraint for JHM melter processing (1050° C) at all but the lowest waste loadings. As mentioned, this constraint was ignored due to the ability of the CCIM to achieve high process temperatures. When ignoring the liquidus constraint, the constraints that limited waste loading were "high viscosity" at low waste loadings and the "nepheline discriminator" at high waste loadings. It should be noted that the high viscosity limitation would likely be overcome by the higher processing temperatures in the CCIM. The MAR acceptability regions for the 5 candidate frits are summarized in Table III.

MAR assessment for DWPF JHM glasses

MAR assessments were performed using the two sludge compositions described earlier, along with the candidate frit compositions. In general, the ranges of waste loading (WL) over which an acceptable glass was predicted were larger for the SB5 "without Al-dissolution" sludge composition. All of the selected frits provided WLs of 40% or better for this sludge. The WLs are all limited by predictions of nepheline crystallization. Glass prepared with Frit 503 was also limited by a predicted liquidus temperature of more than 1050° C at a WL of 42%. The range of WLs over which an acceptable glass was predicted was smaller for the frits with the SB5 "with Al-dissolution" sludge composition. The WL ranges were limited by either predictions of "high liquidus temperature", "low viscosity", or both. The MAR acceptability region for these candidate frits for the two sludge processing conditions is presented in Table IV.

Experimental results for CCIM candidate glasses

Samples evaluated by XRD were analyzed under conditions providing a detection limit for crystalline content of ~0.5 vol %. The XRD results for the quenched CCIM candidate glasses showed only the presence of spinel crystals (magnetite and trevorite) at the highest waste loading (55 wt %). Spinel crystals were evident in the CCC glasses for all waste loadings. Nepheline was detected in CCC glasses at 55 wt % waste loading for compositions 503-R1, 503-R2, 503-R3, and 503-R5. Nepheline was not found in any glasses for the 503-R4 composition. This was contradictory to the MAR assessment of predicted nepheline formation at 52 wt % waste loading for the 503-R4 composition. This anomaly was due to either uncertainty in the nepheline discriminator near the discriminator value or being at the XRD detection limit for this glass.

The PCT results for the 50 and 55 wt % waste loading glasses are shown in Table V. The normalized release values were all significantly below that of the Environmental Assessment (EA) glass. The EA glass is the reference standard used to assess repository acceptability of a waste glass. An increase in relative release rate was evident for the 55 wt % CCC glasses as compared to the 50 wt % CCC glasses likely due to the presence of nepheline in the 55 wt % CCC glasses.

Based on the results of the experimental testing, composition 503-R4 was selected for follow-on CCIM testing with the SB4 composition. The durability of this composition exceeded that of the EA glass and no evidence of nepheline formation was found in 55 wt % waste loading glasses, including those subjected to the CCC heat treatment.

Table I. Surrogate sludge compositions identified for testing.

Oxide	SB4	SB5 w/o Al-dissolution	SB5 w/ Al-dissolution	Hanford High Alumina
Al_2O_3	28.03	33.25	16.62	53.27
CaO	3.04	2.09	2.92	2.39
Cr_2O_3	0.22	0.20	0.28	1.16
Fe_2O_3	31.89	26.42	36.85	13.11
K_2O	0.07	0.16	0.22	0.32
MnO	6.36	5.20	7.25	--
Na_2O	20.57	24.62	24.62	7.96
NiO	1.82	2.31	3.22	0.89
SiO_2	2.98	1.82	2.54	10.88
TiO_2	0.04	0.52	0.72	0.02
ZnO	0.05	0.07	0.10	0.18
ZrO_2	0.10	0.23	0.32	0.88
BaO	0.08	0.11	0.15	0.12
Ce_2O_3	0.24	0.23	0.32	--
CuO	0.06	0.07	0.10	--
La_2O_3	0.03	0.03	0.04	--
MgO	3.05	1.41	1.97	0.26
PbO	0.42	0.10	0.14	0.91
SO_4^{2-}	0.95	1.16	1.62	0.44
B_2O_3	--	--	--	0.42
Bi_2O_3	--	--	--	2.54
CdO	--	--	--	0.05
Li_2O	--	--	--	0.38
P_2O_5	--	--	--	2.34
F^-	--	--	--	1.48
Total	100.00	100.00	100.00	100.00

Table II. Candidate frit formulations for melter testing (oxide concentrations in wt %).

Melter	Frit ID	B_2O_3	CaO	Li_2O	Na_2O	K_2O	SiO_2	Total
CCIM	503-R1	14.0	--	8.0	2.0	--	76.0	100.0
	503-R2	14.0	--	8.0	--	--	78.0	100.0
	503-R3	16.0	--	8.0	2.0	--	74.0	100.0
	503-R4	16.0	--	8.0	--	--	74.0	100.0
	503-R5	18.0	--	8.0	--	--	74.0	100.0
JHM (DWPF)	503	14.0	--	8.0	4.0	--	74.0	100.0
	517	17.0	--	10.0	3.0	--	70.0	100.0
	520	8.0	1.0	10.0	4.0	--	77.0	100.0
	521	10.0	1.0	8.0	6.0	--	75.0	100.0
JHM (WTP)	HAL-17	31.0	12.0	7.4	4.3	5.0	40.3	100.0

Table III. MAR assessment for candidate glasses for CCIM testing.

Frit ID	SB4	
	WL Range (wt %)	Limiting Model
503-R1	34-49	High η, Neph
503-R2	42-51	High η, Neph
503-R3	30-49	Neph
503-R4	37-51	High η, Neph
503-R5	33-50	High η, Neph

Table IV. MAR assessment for candidate glasses for JHM testing to support DWPF.

Frit ID	SB5 w/o Al-dissolution		SB5 w/ Al-dissolution	
	WL Range (wt %)	Limiting Model	WL Range (wt %)	Limiting Model
503	26-41	T_L, Neph	26-36	T_L
517	26-40	Neph	26-28	Low η
520	25-42	Neph	25-39	T_L, Low η
521	25-40	Neph	25-39	T_L, Low η

Experimental results for JHM candidate glasses

In general, all of the quenched glasses (both 35 and 40 wt % waste loading) were either X-ray amorphous (no crystallization at the XRD detection limit) or contained small amounts of magnetite, trevorite, or both. XRD results for the CCC glasses were similar to those for the quenched glasses, although all of the CCC glasses at 40% waste loading were found to contain trevorite. Nepheline formation was not identified in any of the study glasses up to 40 wt % waste loading, consistent with the nepheline discriminator constraint that was included in the MAR assessments.

The PCT results for the 40 wt % waste loading (Table VI) show that each glass has a durability that is considered acceptable, with normalized releases for boron (NL [B] in g/L) that are better than an order of magnitude below that of the EA glass standard, regardless of heat treatment. The PCT results for the 35 wt % waste loading glasses were comparable to the 40 wt % waste loading glasses.

Because the crystallization and PCT data were comparable for all frit compositions, the selection of a frit for testing at KRI was made with the intent of better determining the effect of frit composition on melt rate. Recent frit development efforts for DWPF have identified frits with a higher concentration of B_2O_3 as being beneficial for improving melt rate [10]. Frits 520, 503 and 517 were, therefore, recommended for the initial melter testing since they cover a relatively wide range of B_2O_3 concentrations (8, 14 and 17 wt%, respectively). This initial testing is, thus, intended to identify differences in melt rates among the frits and to see if there are significant differences in melt rate or throughput between the two flowsheets. These results will guide selection of conditions (frit type and flowsheet conditions) for subsequent larger-scale melter testing. The results will ultimately guide future DWPF frit development efforts.

Table V. PCT Results for CCIM candidate glasses at50 and 55 wt % waste loading.

Glass ID	Waste Loading	Normalized Release (g/L) B	Li	Na	Si
503-R1 Quench	50	0.62	0.70	0.60	0.37
	55	0.48	0.60	0.60	0.37
503-R1 CCC	50	0.35	0.47	0.38	0.28
	55	0.84	0.92	0.64	0.34
503-R2 Quench	50	0.56	0.68	0.52	0.35
	55	0.45	0.59	0.53	0.36
503-R2-CCC	50	0.32	0.48	0.34	0.28
	55	0.53	0.68	0.45	0.27
503-R3 Quench	50	0.65	0.72	0.63	0.36
	55	0.47	0.58	0.59	0.37
503-R3 CCC	50	0.40	0.52	0.44	0.31
	55	0.87	0.93	0.65	0.34
503-R4 Quench	50	0.55	0.68	0.53	0.34
	55	0.42	0.56	0.53	0.34
503-R4 CCC	50	0.33	0.48	0.34	0.28
	55	0.59	0.75	0.49	0.32
503-R5 Quench	50	0.61	0.71	0.55	0.35
	55	0.47	0.58	0.54	0.35
503-R5 CCC	50	0.35	0.48	0.35	0.28
	55	0.49	0.67	0.44	0.28
EA		18.11	9.99	13.78	4.04

Table VI. PCT Results for JHM candidate glasses at 40 wt % waste loading.

Frit ID	Sludge Type	WL	Heat Treatment	NL [Li] (g/L)	NL [B] (g/L)	NL [Na] (g/L)	NL [Si] (g/L)
520	SB5 w/o Al-diss.	40%	Quench	0.67	0.60	0.69	0.42
			CCC	1.14	1.02	0.85	0.55
503			Quench	0.65	0.55	0.57	0.40
			CCC	0.62	0.52	0.56	0.39
517			Quench	0.66	0.69	0.62	0.43
			CCC	0.66	0.66	0.60	0.43
521			Quench	0.58	0.56	0.64	0.38
			CCC	1.19	1.32	0.92	0.51
520	SB5 w/ Al-diss.		Quench	1.02	1.13	1.21	0.62
			CCC	1.16	1.15	1.20	0.67
503			Quench	0.80	0.85	0.88	0.48
			CCC	0.82	0.86	0.87	0.49
517			Quench	1.01	1.13	1.11	0.56
			CCC	1.10	1.18	1.15	0.61
521			Quench	0.92	1.03	1.15	0.56
			CCC	0.97	1.05	1.14	0.58

CONCLUSIONS

The MAR assessment approach combined with experimental testing of select glasses proved to be successful in efficiently identifying glass composition for subsequent melter testing. A candidate frit (503-R4) for testing at 50 wt % waste loading was identified for testing in the

CCIM. Three candidate frits (frits 503, 520 and 517) were identified for initial small-scale melter testing at KRI to evaluate melt rate. This initial testing will lead to selection of a preferred frit for follow-on larger-scale melter testing, Glass produced in the melter testing will be used to evaluate the validity of the models, especially with respect to nepheline formation.

REFERENCES

1. M.E. Smith, A.B. Barnes, J.R. Coleman, R.C. Hopkins, D.C. Iverson, R.J. O'Driscoll and D.K. Peeler, "Recent Process and Equipment Improvements to Increase High Level Waste Throughput at the Defense Waste Processing Facility," *MS&T'06 Conference Proceedings*, American Ceramic Society, Westerville, OH (2006).
2. M.E. Smith, A.B. Barnes, D.F. Bickford, K.J. Imrich, D.C. Iverson and H.N. Guerrero, "DWPF Glass Air-lift Pump Life Cycle Testing and Plant Implementation,", *Ceramic Transactions*, Vol. 168, American Ceramic Society, Westerville, OH (2005).
3. J.M. Perez, S.M. Barnes, S. Kelly, L. Petkus and E.V. Morrey, "Vitrification Testing and Demonstration for the Hanford Waste Treatment Plant," *Ceramic Transactions*, Vol. 168, American Ceramic Society, Westerville, OH (2005).
4. D.K. Peeler, T.B. Edwards, D.R. Best, I.A. Reamer and R.J. Workman, "Nepheline Formation Study for Sludge Batch 4 (SB4): Phase 2 Experimental Results," WSRC-TR-2006-00006, Washington Savannah River Company, Aiken, SC (2006).
5. P. Hrma, G.F. Piepel, J.D. Vienna, P.E. Redgate, M.J. Schweiger and D.E. Smith, "Prediction of Nuclear Waste Glass Dissolution as a Function of Composition," *Ceramic Transactions*, Vol. 61, American Ceramic Society, Westerville, OH (1995).
6. P.Hrma, G.F. Piepel, P.E. Redgate, D.E. Smith, M.J. Schweiger, J.D. Vienna and D.S. Kim, "Prediction of Processing Properties for Nuclear Waste Glasses," *Ceramic Transactions*, Vol. 61, American Ceramic Society, Westerville, OH (1995).
7. C.M. Jantzen, J.B. Pickett, K.G. Brown, T.B. Edwards and D.C. Beam, "Process/Product Models for the Defense Waste Processing Facility (DWPF): Part I Predicting Glass Durability from Composition Using a Thermodynamic Hydration Energy Reaction Model (THERMO)," WSRC-TR-93-672, Westinghouse Savannah River Company, Aiken, SC (1995).
8. D.K. Peeler and T.B. Edwards, "Frit Development for Sludge Batch 3," WSRC-TR-2002-00491, Westinghouse Savannah River Company, Aiken, SC (2002).
9. H. Li, P. Hrma, J.D. Vienna, M. Qian, Y. Su and D.E. Smith, "Effects of Al_2O_3, B_2O_3, Na_2O, and SiO_2 on Nepheline Formation in Borosilicate Glasses: Chemical and Physical Correlations, *J. Non-Crystalline Solids*, 331 (2003).
10. D.K. Peeler and T. B. Edwards, "High B_2O_3/Fe_2O_3-based Frits: MAR Assessments for Sludge Batch 4 (SB4)," WSRC-TR-2006-00181, Washington Savannah River Company, Aiken, SC (2006).
11. S. L. Marra and C. M. Jantzen, "Characterization of Projected DWPF Glass Heat Treated to Simulate Canister Centerline Cooling," WSRC-TR-92-142, Revision 1, Westinghouse Savannah River Company, Aiken, SC (1993).
12. "Standard Test Methods for Determining Chemical Durability of Nuclear Waste Glasses: The Product Consistency Test (PCT)," *ASTM C-1285,* (2002).
13 A. Aloy, J. D. Vienna, K. M. Fox, T. B. Edwards and D. K. Peeler, "Glass Selection Strategy: Development of US and KRI Test Matrices," WSRC-STI-2006-00205, Washington Savannah River Company, Aiken, SC (2006).

Mater. Res. Soc. Symp. Proc. Vol. 1107 © 2008 Materials Research Society

Electron Irradiation and Electron Tomography Studies of Glasses and Glass Nanocomposites

G. Möbus, G. Yang,
Z. Saghi, X. Xu, R.J. Hand, A. Pankov, M.I. Ojovan
Immobilisation Science Laboratory, Department of Engineering Materials, University of Sheffield, Sheffield, S1 3JD, UK

ABSTRACT

Characterization of glasses and glass nanocomposites using modern transmission electron microscopy techniques is demonstrated. Techniques used include: (i) high-angle-annular dark field STEM for imaging of nanocomposites, (ii) electron tomography for 3D reconstruction and quantification of nanoparticle volume fractions, and (iii) fine structure electron energy loss spectroscopy for evaluation of boron coordination. Precipitation of CeO_2 nanoparticles in borosilicate glasses is examined as a function of glass composition and redox partner elements. A large increase in the solubility of Ce is found for compositions where Ce retains +IV valence in the glass. Irradiation experiments with electrons and γ-rays are summarized and the degree of damage is compared by using changes in the boron K-edge fine structure, which allows the gradual transition from BO_4 to BO_3 coordination to be followed.

INTRODUCTION

Some of the most important benefits of transmission electron microscopy for the study of nuclear waste materials lie in the high-spatial resolution imaging of inhomogeneities, such as nanoscale precipitates in glass composite materials. Another important application is in high spatial resolution analysis, *e.g.* to track glass coordination parameters as a function of temperature and/or irradiation. In this paper we present examples of each of these fields, namely 2D and 3D mapping of nanocomposites by STEM in part 1, and irradiation induced transformation of boron coordination in glass in part 2.

Borosilicate glass chemistry is diverse with respect to phase separation phenomena and crystallization behavior and has therefore attracted major interest in both the glass community as well as in nuclear and hazardous waste immobilization community. Understanding the chemistry of precipitation aids design of glass compositions that avoid unwanted precipitates, but also enables the design of glass composite materials, that can achieve high waste loadings, by deliberately developing selected precipitates [1-3]. The glasses considered in this paper are summarized in tables 1 and 2.

PART I: GLASS COMPOSITE MATERIALS: IMAGING & VALENCE ESTIMATION

The solubility and redox chemistry of Ce (a common Pu surrogate) was studied by adding a constant fraction of 4mol% CeO_2 to 5 borosilicate glasses, varying in glass composition or other waste loading elements, so as to provide a variety of redox active partner elements for Ce, as detailed in Table 1. The microstructure of glass fragments was examined by dark-field STEM

(Fig. 1 a-d), where heavy particles appear bright (Z-contrast). Glasses (i) and (ii) differ in alkali choice, where Na/Li is more industrially relevant, while K has been chosen for its higher electron beam resistance. In previous work we have identified 3 dissimilar Ce-rich precipitates of different Ce valence within glass (i). The valence has been determined by EELS as described in [4], by quantifying the M-edge line ratio for the M5 and M4 sub-levels.

Table 1: Glass compositions for precipitation studies (part 1) (mol%)

Glass-Acronym	Network formers & modifier elements		Waste simulant elements
(i) KBS-CFC:	K_2O: B_2O_3: SiO_2	= 12.9: 25.7: 51.4	CeO_2: Fe_2O_3: Cr_2O_3 = 4:4:2
(ii) NLBS-CZC:	Li_2O: Na_2O: B_2O_3: SiO_2	= 4.3: 8.6: 25.7: 51.4	CeO_2: ZrO_2: Cr_2O_3 = 4:4:2
(iii) KBS-CZC:	K_2O: B_2O_3: SiO_2	= 12.9: 25.7: 51.4	CeO_2: ZrO_2: Cr_2O_3 = 4:4:2
(iv) KBS-CH:	K_2O: B_2O_3: SiO_2	= 12.9: 25.7: 51.4	CeO_2: HfO_2 = 4:6
(v) KBS-ACFN:	K_2O: B_2O_3: SiO_2	= 15: 15: 60	CeO_2:Ag_2O:Fe_2O_3:Nd_2O_3= 4:1:3:2

Table 2: Glass compositions for irradiation studies (part 2) (mol%)

Glass-Acronym	Glass matrix & Alkali	
(vi) KBS:	K_2O: B_2O_3: SiO_2	= 12.9: 25.7: 51.4
(vii) Nuclear glass:	Li_2O: Na_2O: K_2O: B_2O_3: SiO_2	= 7.2: 8.3: 0.4: 12.6: 55.2:
	Other elements Al_2O_3 3.0; MgO 9.8; Fe_2O_3 1.6; P_2O_5 0.05, BaO 0.24, La_2O_3 0.7; MoO_3 0.8	

Here, we extended the Ce-chemistry analysis to the other glasses with some unexpected findings. Ce was supplied as Ce(+IV) O_2 powder, but was expected to switch to +III due to the high melting temperature of 1400°C (see e.g. [5]) and the relative ease by which it may be reduced [6]. We found:

1) The valence of Ce in glasses (i) - (iii) was +III, while it was +IV in glasses (iv) and (v).
2) The solubility of Ce was significantly increased (no Ce-precipitates formed) in glasses (iv) + (v) and we believe that the Ce(+IV) valence is the main reason for this, although the change in waste simulants (at constant waste loading) and glass composition will have contributed to this change in preferred valence.
3) The Ce-Hf (iv) glass was the only fully amorphous/homogeneous glass in our series.
4) Precipitation of round nanoparticles of metallic Ag in glass (v) was assumed to be triggered by Ce(+IV), an effect which is exploited in photonic applications [7].

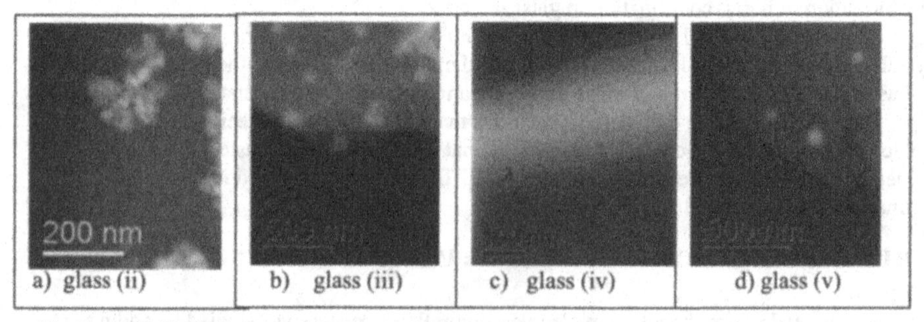

Fig. 1: ADF-STEM images of 4 glasses with numbers as detailed in table 1.

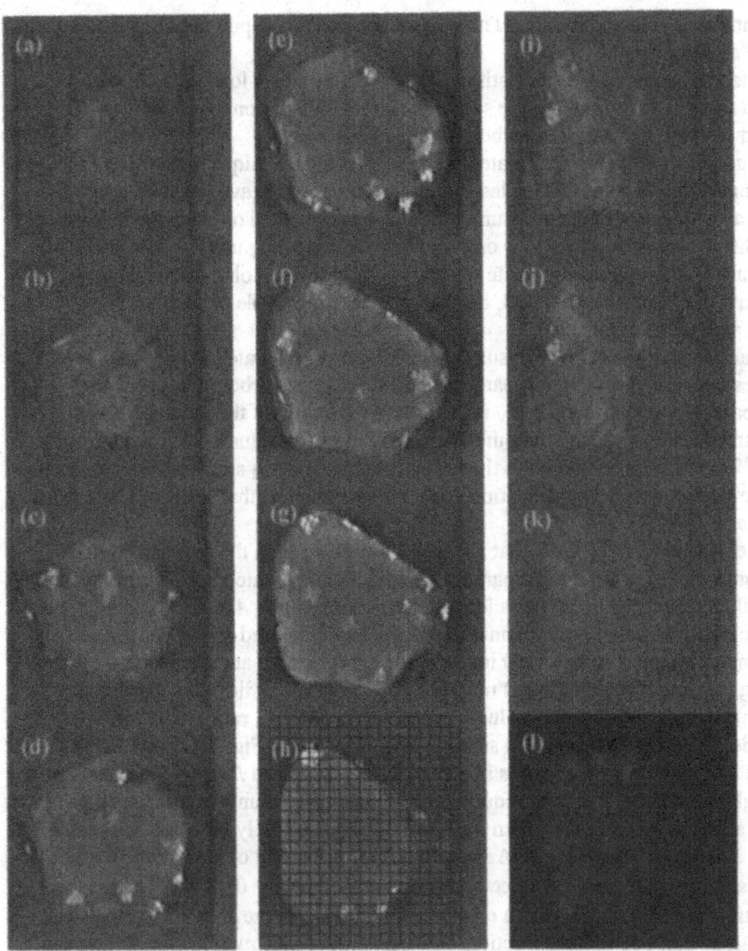

Fig. 2: A series (a-k) of parallel plane sections through a tomographically reconstructed glass fragment (glass ii), acquired from a tilt series of dark field STEM images, as shown in [4,11], using Z-contrast to highlight CeO_2 nanoparticles. Image (h) has a line raster superimposed for illustrating stereological principles (see text) and (l) shows a re-projected image by summation of the stack. This corresponds to what a TEM image would show and contains the exaggerated particle density appearing upon projection.

Electron tomography has been shown to be a potentially powerful tool to generate 3D microstructure mapping of glass volumes on the nanoscale [4]. We extended our previous work by applying stereological principles to quantitatively evaluate 3D volume reconstructions for a

glass fragment containing ~ 20 precipitates. The importance and future potential of such an analysis lies in the ability to:
(i) calculate volume fraction and therefore estimate the increased waste loading achieved in the GCM as compared to a glass restricted to the solubility limit. An additional benefit is that the percentage of residual dissolved cations can be estimated.
(ii) estimate the homogeneity of the precipitate distribution, if the technique is applied to various glass fragments from different parts of the glass block; in the case of heavy particles in a light glass matrix, such as with CeO_2, particle accumulation by gravity could occur in the melter if cooling is slow. This in turn raises questions on criticality if Ce is being used as a Pu surrogate.
(iii) estimate how near/far the glass composite material is from the percolation threshold. At this volume fraction limit, the precipitates would, on average, start to provide a direct diffusion/leaching path to the surface [2].
The 3D volume stack can be binarised with suitable thresholds to separate the high-atomic number precipitates from the glass, or to separate the glass from the carbon support film background. The particle volume fraction V_V is then simply the ratio of the white pixels in the two binarised volumes. Due to thresholding uncertainties a range of volume fractions of 3.0 – 3.5% is estimated. This is to be compared to the molar fraction of CeO_2 supplied, which was 4%, and would indicate (assuming for simplification equal molar volumes) that ~ 1% of Ce is still dissolved in the glass.

Stereological principles [8] show that a single 2D slice through the material can be evaluated in the same way to achieve a 2D particle area fraction A_A, which upon averaging over sufficient slices will be equal to the 3D particle volume fraction: $V_V = \langle A_A \rangle$. This can be seen by looking at the 1D case of particle line fractions (dark lines superimposed on Fig. 2h) which play the role of tomographic sections across a 2D image. This grid provides another (rough) estimate for V_V by counting the point (line crossings) ratio P_{slice} of points which lie inside particles relative to the point count in the entire sample. For example in Fig 2g a ratio of 10/266 = 0.037 is found. To emphasise the difference between sections and projections, Fig 2(l) shows the 2D particle density on a projection which results in an artificially enhanced P_{Proj} ~ 33%. A model of both particle and glass fragment shapes is required to estimate true volume fractions from the projection images as the relationship between P_{slice} and P_{Proj} (respectively between $A_{A,slice}$ and $A_{A,proj}$) is given by a factor $f = \delta z(\text{particle}) / \Delta z(\text{specimen})$ for the extent of an average particle δz and the entire glass fragment Δz in the z-direction. Using scale bars, $f = 0.08$ is estimated for isotropic particle and fragment shape, which results in a reduction of the above 33% projected point ratio into roughly 2.6% of volume fraction. Tomographic reconstruction of this kind can also be applied to single free standing [9] or embedded [10] nanoparticles. Reconstruction of one of the dendrites seen in Fig. 2 as a 3D object has been demonstrated in [11].

PART II: GLASS COORDINATION AND IRRADIATION PHENOMENA

We used changes in boron coordination characterized by N4 = [BO4]/[BO3+BO4], as a finger-print for irradiation induced damage in a borosilicate glass matrix. The changes were obtained from the fine structure of the electron energy loss spectrum (EELS) of the B-K-edge. In previous work we studied N4 as function of $\{t, T, Z_{alkali}\}$, where t is the electron irradition time (dose), T is the in-situ temperature of TEM specimen T, and Z_{alkali} is the atomic mass of the alkali network modifier in the glass [12-14]. In the present work we extended these studies by undertaking gamma-pre-irradiation of glass vii (see table 2) before TEM-analysis. All glasses undergo some BO_4 to BO_3 conversion during TEM irradiation. Glass samples were irradiated under a γ-source (Isotron ^{60}Co-source) up to a dose of 1 MGy. It was found that the boron conversion rate was different after irradiation suggesting that the glass structure was partially converted by γ-irradiation (Fig. 3b *vs* c). This could provide evidence of a mobilisation of cations via the reaction: $[BO_4]^{-}$ $Cation^{+}$- + γ → e^{-} + $[BO_3]$ + $Cation^{+}$, which releases alkalis from the negatively charged $[BO_4]^{-}$ tetrahedra in a similar fashion to radiation-induced reactions leading to liberation of cations from negatively charged species in silicate glasses [2]. The effect is compared to electron irradiation in a pure ternary borosilicate glass (vi of table 2) in Fig. 3a.

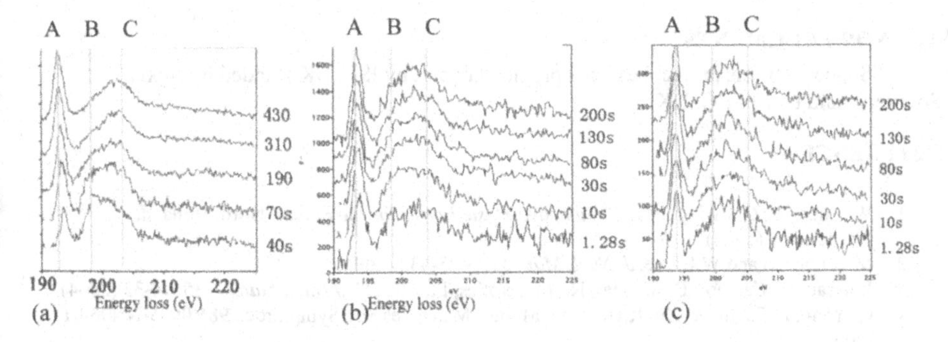

Fig. 3: electron irradiation series of B-K-edge fine structure. (a) K-BS glass, (b) Simulated waste glass un-irradiated, (c) after pre-irradiation with gamma. Peak A indicates BO_3, Peak B BO_4 while peak C is of mixed nature.[14]

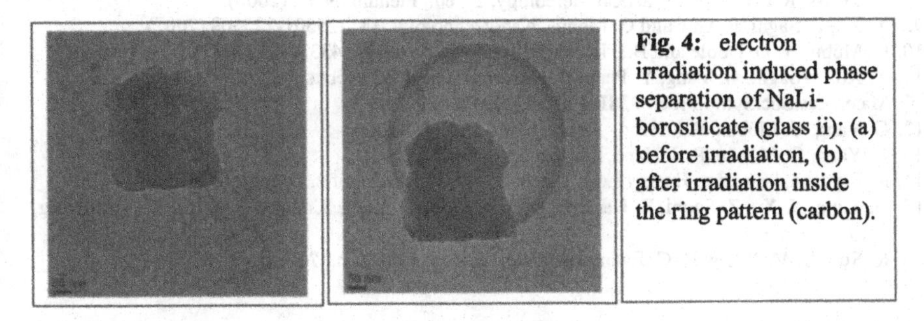

Fig. 4: electron irradiation induced phase separation of NaLi-borosilicate (glass ii): (a) before irradiation, (b) after irradiation inside the ring pattern (carbon).

As complement to the EELS studies of electron irradiation we also examined glass (ii) of table 1 after extensive focused electron irradiation in bright field TEM; partial phase separation is visible inside the ring of carbon-contamination indicating the exposed area (Fig 4). This glass is nominally outside the vycor region where phase separation is expected, and the effect is believed to be electron irradiation specific (see also [16]).

CONCLUSIONS

We have analyzed precipitation in glass nanocomposite materials, where the waste elements are partially immobilized in the crystalline particle phase and partially in the glass matrix, using a combination of EELS spectroscopy and STEM imaging. We found significant variation in Ce solubility in borosilicate glasses depending on Ce valence. We also used tomography on the nanoscale to estimate particle volume fractions of CeO_2 and residual Ce in solution. We have also examined boron coordination changes in borosilicate due to γ irradiation and electron irradiation primarily by EELS spectroscopy. Both γ and electron irradiation led to similar structural changes

ACKNOWLEDGMENTS

Support for this project has been provided in part by ISL, UK (funded by Nexia Solutions) and by EPSRC, UK.

REFERENCES

1. W. Lutze, R. C. Ewing, eds., *Radioactive Waste Forms for the Future*, North Holland, Amsterdam, (1988)
2. M. I. Ojovan and W.E.Lee, *J. Nucl. Mat.,* **335**, 425-432 (2004).
3. Loiseau, P., Caurant, D., Baffier, N., Mazerolles, L.,Fillet, C. *J Nucl. Mater.* 335, 14-32 (2004).
4. G. Yang, Z. Saghi, X. Xu, R.Hand, G. Möbus Mater.Res.Soc.Symp.Proc., **985** 0985-NN06-01, (2007)
5. C. Lopez, X. Deschanels, J. M. Bart, J. M. Boubals, C. Den Auwer, E. Simoni, *J. Nucl. Mater.* **312** 76-80 (2003)
6. H.D. Schreiber and A.L. Hockman, J. Am. Ceram. Soc., 70, 591-594 (1987)
7. Y. Dai, J. Qiu, X. Hu, L. Yang, X. Jiang, C. Zhu and B. Yu *Appl. Phys. B - Lasers O.* **84**, 501 505 (2006).
8. J.C. Russ, R.T. Delhoff, Practical Stereology, 2nd ed., Plenum, N.Y., (2000).
9. X. Xu, Z. Saghi, R. Gay and G. Möbus, Nanotechnology, **18**, 225501-225508 (2007)
10. G. Möbus, R.D. Doole and B.J. Inkson. Ultramicroscopy, **96** , 433 - 451 (2003).
11. X. Xu, Z. Saghi, G. Yang, Y. Peng, B. Inkson, R. Gay, G. Möbus Mater.Res.Soc.Symp.Proc., **928E** 0982-KK02-04, (2007)
12. G. Yang, G. Möbus, R.J. Hand, *Phys Chem Glass*, **47**, (2006)
13. G. Yang, G. Möbus, R. J. Hand, *Micron*, **37**, 433-441(2006)
14. R. Brydson, "Electron Energy Loss Spectrocopy", Bios Scientific, Oxford, (2001).
15. G. Yang, X. Xu, Z. Saghi, R. Hand, G. Möbus; Proceed. Intern. Congr. Glass, ICG17, Strasbourg, 2007
16. K. Sun, L. M. Wang, R. C. Ewing, *Mat. Res. Soc. Symp. Proc.*, **757**, II5.3 (2003).

Mater. Res. Soc. Symp. Proc. Vol. 1107 © 2008 Materials Research Society

Glass Composite Materials for Nuclear and Hazardous Waste Immobilisation

Michael I. Ojovan and Jariah M. Juoi
Immobilisation Science Laboratory, Department of Engineering Materials,
University of Sheffield, Mappin Street, Sheffield S1 3JD, UK

Aldo R. Boccaccini and William E. Lee
Department of Materials, Imperial College London, South Kensington Campus,
London SW7 2AZ, UK

ABSTRACT

Glass composite materials (GCM) are versatile wasteforms for immobilising various types of both radioactive and hazardous wastes. We review current research on the utilisation of GCMs for hazardous and radioactive waste immobilisation. Compared to homogeneous glassy materials GCMs can incorporate larger amounts of waste elements and, in the case of glass matrix composites, they can be produced using lower processing temperatures (by viscous flow sintering) than those of conventional melting.

INTRODUCTION

Hazardous and nuclear waste vitrification is attractive because of its high versatility, i.e. the large number of elements that can be incorporated in the glass, and the potential for a highly durable and low volume wasteform. Recently, there has been a trend to develop waste host systems which are intermediate between completely glassy or crystalline materials [1-6]. Both hazardous and nuclear waste constituents can be immobilised not only by direct chemical incorporation into the glass structure in a classical vitrification approach but also by the physical encapsulation of the waste in a glass matrix, forming a glass composite material (GCM) consisting of both vitreous and crystalline phases. The major component may be the crystalline phase with a vitreous phase acting as a binding agent or alternatively the vitreous phase may be the major component, with particles of a crystalline phase dispersed in the glass matrix. Standard glass-ceramics, which are obtained by the controlled crystallisation of a parent glass, can be considered a special case of GCMs. There have been significant recent research efforts on the utilisation of GCMs for immobilising both radioactive and non-radioactive wastes, which will be reviewed in this paper.

Radioactive waste arises from three main sources, the Nuclear Fuel Cycle (NFC) used for power generation as well as military purposes, non-NFC institutes (including non-nuclear industries, medical and research institutions) and accidents [7]. Non-radioactive wastes include incinerator ashes and air pollution control residues, sewage and dredging sludge, iron and steel industry slag and ashes, Zn hydrometallurgical process effluents and coal ash from power stations [3, 4]. Some of these wastes are highly hazardous containing significant concentrations of heavy metals, and/or organic pollutants. Vitrification to form glassy materials including glass-ceramics for treatment and recycling of non-radioactive wastes has been recently discussed [3, 4]. Moreover, there have been several recent studies on radioactive waste immobilisation in GCMs [1, 8, 9]. Previous reports have considered manufacturing processes, microstructure and properties of different wasteforms. However no review has dealt specifically with the different aspects of processing and characterisation of GCMs for both nuclear and hazardous (and mixed) waste immobilisation.

Several processing routes have been examined to produce GCMs, e.g. controlled crystallisation of a glass, powder methods and sintering, petrurgical method and sol-gel precursor glass, which give flexibility in the choice of the optimal immobilisation technology for the target waste. We give here a brief overview of previous work on using GCMs to immobilise radioactive and hazardous wastes emphasizing the potential of these materials as final wasteforms for immobilisation of radionuclides and mixed hazardous and radioactive species.

GCM AS RADIOACTIVE WASTEFORMS

General considerations

GCMs are of interest for radioactive waste immobilisation due to their versatility as a final wasteform enabling them to accommodate legacy, currently generated and expected future radioactive wastes. GCMs can incorporate a range of radioactive nuclides present in Low or Intermediate Level Waste (LILW) and High Level Waste (HLW). GCMs include: glass-ceramics where a glassy waste form is crystallised in a separate heat treatment [10]; materials in which a refractory crystalline phase is encapsulated in glass, such as hot pressed lead silicate or sodium borosilicate glass matrix encapsulating up to 30 vol% of $La_2Zr_2O_7$ or $Ga_2Zr_2O_7$ pyrochlore crystals to immobilize actinides [11-14]; borosilicate glasses for encapsulation of TRISO-UO_2 particles from pebble bed modular reactor [15]; materials in which spent clinoptilolite from aqueous waste reprocessing is immobilized by pressureless sintering [16]; U/Mo-containing materials immobilized in a GCM termed U-Mo, glass formed by cold crucible melting (which partially crystallise on cooling) [17]; yellow phase containing up to 15 vol.% of sulphates, chlorides and molybdates [18]; and materials to immobilize ashes from incineration of solid radioactive wastes [19]. Powder glass sintering is thus an alternative technology to the traditional glass melting method for immobilisation of high-level nuclear waste [20]. Figure 1 outlines GCMs that have been used to immobilise various radioactive wastes [8, 9].

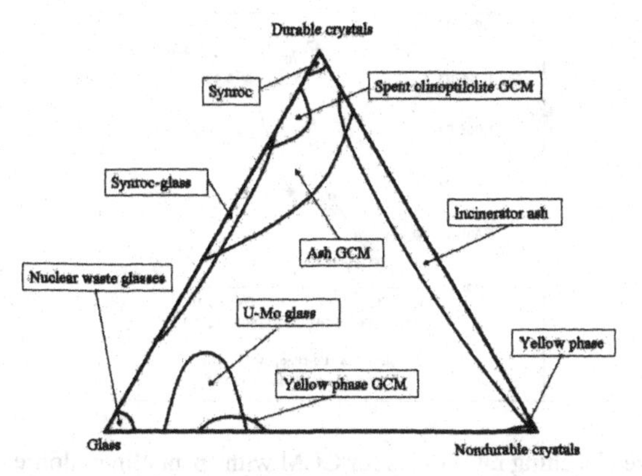

Figure 1. Schematic of radioactive wastes and glassy nuclear wasteforms for their immobilisation.

Generally, host glasses used to develop GCMs should have the ability to incorporate waste constituents, be processed at relatively low processing temperatures (below 1000 °C), as well as be chemically durable and radiation resistant. High silica-content glasses including borosilicates, alumina silicates, alumina borosilicates, soda-lime silicates, lead silicates, lead borosilicates, phosphates, copper phosphates, silver phosphates, lead-iron phosphates and soda aluminaphosphate glasses have been examined as host matrices for GCM production. Several processing routes have been utilised to produce GCMs for radioactive waste immobilisation [see ref. 7 for details]. GCMs may be used to immobilise long-lived radionuclides (such as actinide species) by incorporating them into the more durable crystalline phases, whereas the short-lived radionuclides may be accommodated in the less durable vitreous phase. Acceptable durability will result if the active species are locked into the crystal phases that are encapsulated in a durable, low activity glass matrix. The GCM option is currently being investigated in many countries including Australia, France, UK, USA and Russia [7-9]. The processing, compositions, phase assemblages and microstructures of GCMs may be tailored to achieve the necessary properties for improved performance of the wasteform. As an example of a recently developed GCM system, we present below details of the GCM designed for spent clinoptililite waste arising from low-level aqueous waste treatment facilities.

Case study: GCM for spent clinoptilolite waste

Simulated spent clinoptilolite was immobilised in a GCM produced by pressureless sintering for 2 h at 750 °C [7]. Waste loading ranging from 1:1 up to 1:10 of glass to clinoptilolite volume ratios corresponding to 37- 88 mass % were studied. Water durability of the GCMs assessed by 7 day leaching tests in deionised water at 40 °C based on ASTM C1220-98 standard is shown in Figure 2. The Cs normalised leaching rates remain below 6.35 10^{-6}g/(cm^{2}day) in a GCM with 73 mass % waste loading. When the waste loading exceeds 80 mass %, the normalized Cs leaching rate became as high as 9.06 10^{-4}g/(cm^{2}day) which defines this waste loading as the critical threshold loading (Figure 3). This order of magnitude increase can be explained by the formation of clusters made of interconnected clinoptilolite particles.

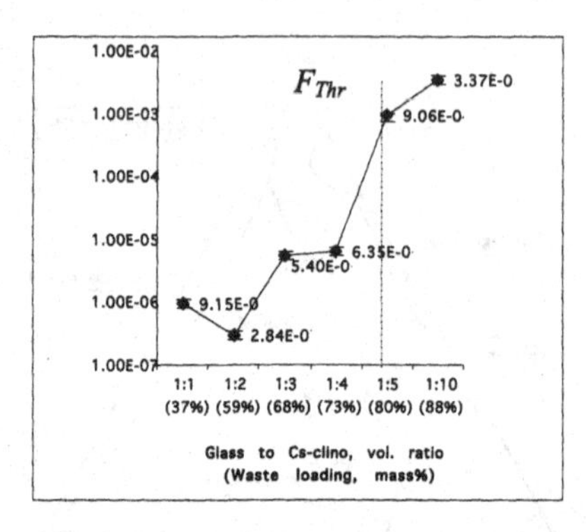

Figure 2: Normalized leaching rates of Cs for GCM with spent clinoptilolite quantifying the critical threshold F_{Thr}.

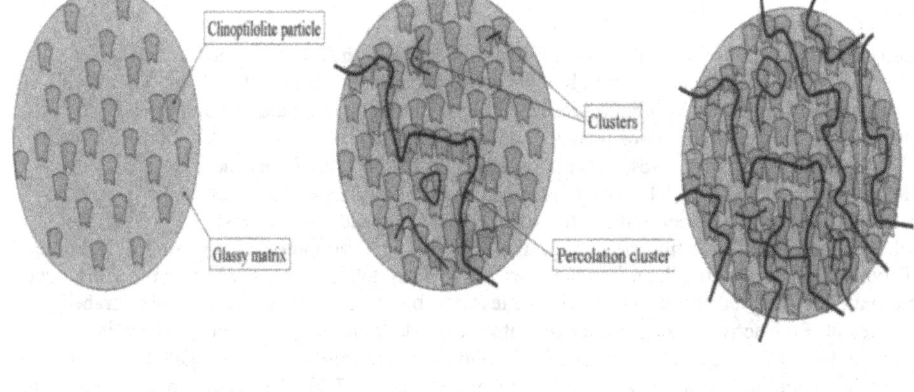

Figure 3: Schematic of GCMs microstructure at different clinoptilolite waste loading. Clusters made of mutually-contacting clinoptilolite particles provide pathways for radionuclide leaching.

At low waste loading below the critical threshold, ($F < F_{Thr}$) the GCM leaching property is governed by the water durable glass host matrix. Full encapsulation of clinoptilolite particles by the glass matrix (Figure 3a) ensures that there are no readily available escape routes for the mobile radionuclides into contacting water. At a certain (termed percolation) threshold waste loading $F = F_{Thr}$ there is not enough glass matrix to fully encapsulate the clinoptilolite particles which contact each other and form a large, percolation cluster providing an easy pathway for radionuclides to leach from the GCM (Figure 3b). At waste loading equal or above critical loading ($F \geq F_{thr}$), the GCM leaching behaviour is governed by universal percolation theory laws [21]. The normalised leaching rate near the threshold of formation of percolating clusters is given by

$$NR_{Cs} = \frac{NR_{Cs}(0)}{(f_{Cs} - f_{Thr})^{\beta}},$$

where $NR_{cs}(0)$ is the normalised leaching rate far from the percolation threshold, which we assessed as $f_{Thr} \sim 0.8C_{cs}/100$ wt.% and β=0.41 is the universal critical exponent [21]. Here, f_{Cs} is the mass fraction of Cs in the unleached specimen and it is directly proportional to the waste as calculated from $f_{Cs} = F_{waste} \times C_{Cs}/100$ wt.%, where, F_{waste} is the mass fraction of waste and C_{Cs} is the concentration of Cs in the clinoptilolite. At waste loading exceeding f_{Thr} there is limited glass matrix volume and hence the GCMs leach behaviour is fully governed by the clinoptilolite particles (Figure 3c).

GCMs FOR HAZARDOUS WASTE ENCAPSULATION

Disposal of solid industrial wastes (e.g., fly ash from coal power stations, slag and ashes from waste incinerator plants, slag from iron and steel production, waste water treatment sludge) in landfill is a costly and environmentally-unfriendly method. There is increasing interest in optimization of the composition and processing routes to develop cost-effective new materials from vitrified waste [3, 4]. With hazardous wastes the requirement is to immobilise the harmful elements in an inert, chemically-durable matrix. In the category of new products developed from hazardous and non-hazardous waste GCMs take a prominent role. In this context GCMs include both traditional glass matrix composites and glass-ceramics fabricated from waste, both with and without hazardous elements (e.g. heavy metals) [22, 23].

The versatility of the glass-ceramic production process is manifested by the many wastes that have been used as raw materials, which include coal fly ash, mud from zinc hydrometallurgy, slag from steel production, ash and slag from waste incinerators, red mud from alumina production, waste glass from lamp and other glass products as well as electric-arc furnace dust and foundry sands, as reviewed elsewhere [3]. To produce an appropriate parent glass for crystallisation, additions to the (silicate) wastes are often required considering that there is always a trade-off between the amount of waste used and the optimisation of properties of the new products. In general, since the main objective is to reuse the maximum possible amount of waste material in new products, the quantity of raw materials or non-waste additions introduced for improving performance must be kept as low as possible.

GCMs constitute multibarrier systems which have been specifically proposed for immobilisation of residues containing hazardous elements. The proposed approach [24] considers formation of a biostable glass matrix using hazardous element–free waste and to incorporate the hazardous residue (e.g. heavy-metals containing material) into the glass matrix in particulate form. Thus the stable matrix provides a true physical barrier to leaching of the heavy metals and improves the chemical durability of the products. In general, the dispersed phase should also contribute to improving the mechanical properties and reliability of the material, acting as a reinforcement phase by exploiting the composite approach, e.g. forming a glass matrix composite [25], thus broadening the application potential of the waste-containing products. It has been shown, for example, that the incorporation of alumina platelet reinforcement enhances the mechanical properties and wear resistance of silicate matrices obtained from appropriate mixtures of wastes (e.g. coal fly ashes and recycled soda-lime glass cullet) [26]. The products have been shown to be suitable for applications such as building or decoration materials, e.g. as wall partition blocks, pavements, floor and wall tiles, thermal insulation, fire protection elements, acoustic tiles or roofing granules [3]. Other possible uses of these composites, which usually exhibit high hardness, include abrasive media for blasting and polishing applications. An interesting technological advantage of using the sintering process to obtain these products is that conventional ceramic technology which is readily available in ceramic tile factories (semi-dry pressing, firing) can be adapted to the new raw materials from silicate waste without major or costly alteration of the processes. Finally, the design of layered composite materials or functionally graded materials by smart combination of hazardous waste (as inclusion) and non-hazardous waste (as the matrix) has been proposed [24], as schematically shown in Figure 4.

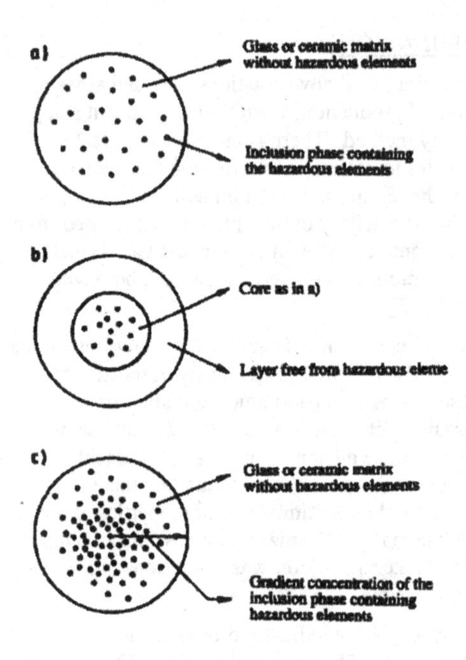

Figure 4: Schematic diagram of GCM systems exhibiting layered and functionally graded microstructure [24]. These GCMs are matrix type composite materials with a well-dispersed discontinuous inclusion phase containing the hazardous elements, e.g. heavy metals: (a) microstructure after conventional processing with uniform distribution of inclusions, (b) microstructure comprising a core highly loaded with hazardous elements surrounded by a shell free from dangerous components, (c) optimised graded microstructure in which the content of hazardous elements progressively decreases from the highly loaded center of the component to its surface, allowing the maximum efficiency in terms of hazardous residue content and safe leaching behaviour. Figure published with permission of Selper Limited.

The use of graded and layered microstructures in multi-phase waste containing composite materials has two major objectives: i) to increase the leaching resistance and chemical durability of GCMs made from hazardous waste by maintaining the highest possible waste content, and ii) to improve the effective properties of the GCM products in relation to their potential industrial applications. The microstructure of a typical GCM in which the hazardous waste in particulate form is uniformly encapsulated in an inert glassy matrix is shown in Figure 4(a). As an alternative to the use of protective coatings to prevent chemical reactions of the external surface with the environment, GCMs can be produced exhibiting a progressive variation in composition in continous (graded) or discrete (layered) form. The simplest approach is shown in Figure 4(b) in which a core highly loaded with hazardous elements is surrounded by a stable layer free from dangerous componets. An optimisation of this structure is shown in Figure 4(c), in which the content of hazardous elements progressively decreases from the highly loaded centre to its surface. This kind of graded and layered microstructures have been earlier proposed for the safe immobilisation of radioactive waste in glass [27]. Powder technology and sintering are the processing techniques of choice for fabrication of these GCMs.

CONCLUSIONS

Waste immobilisation in GCMs has emerged as a versatile technology enabling reliable immobilisation of complex and varying composition waste streams, including both radioactive and hazardous residues, which are otherwise difficult to immobilise using traditional vitrification technology. The optimisation of GCM phase assemblages and microstructures is important for achieving high waste loadings and highly-durable wasteforms. In the case of GCMs from hazardous waste the ultimate goal is to produce engineered materials which are safe to use in several applications including building and decoration materials as well as in special machinery and wear resistant parts.

REFERENCES

[1] M.I. Ojovan and W.E. Lee, *An Introduction to Nuclear Waste Immobilisation* (Elsevier, Amsterdam, 2005).
[2] I.W. Donald, *Glass Technology*, **48**, 155-163 (2007)
[3] R.D. Rawling, W. P. Wu, A. R. Boccaccini, *J. Mater. Sci.* **41**, 733-761 (2006).
[4] P. Colombo, G. Brusatin, E. Bernardo, G. Scarinci, *Current Opinion in Solid State and Material Science* **7**, 225-239 (2003).
[5] W.E. Lee, M.I. Ojovan, M.C. Stennett and N.C. Hyatt, *Adv. Applied Ceramics* **105** [1] 3-12 (2006).
[6] M.A. Lewis. and D.F. Fischer, "Properties of glass-bonded zeolite monoliths," *Ceram. Trans.*, 45, 277-286, 1994.
[7] J.M. Juoi, M.I. Ojovan, W.E. Lee. *Proc. ICEM'05*, Glasgow, Scotland, ICEM05-1069, ASME (2005).
[8] M.I. Ojovan and O.G. Batyukhnova. *Proc. WM'07,* Tucson, AZ, WM7061 (2007).
[9] M.I. Ojovan, W.E. Lee, *New Developments in Glassy Nuclear Wasteforms* (Nova Science Publishers, New York, 2007).

[10] P. Loiseau, D. Caurant, O. Majerus and N. Baffier. *J. Mater. Sci.* **38**, 843-864 (2003).
[11] A. R. Boccaccini, E. Bernardo, L. Blain and D.N. Boccaccini., *J. Nucl. Mater.* **327**, 148-158 (2004).

[12] A.A. Digeos, J.A. Valdez, K.E. Sikafus, S.Atiq, R.W. Grimes and A.R. Boccaccini, *J. Mater. Sci.* **38**, 1597-1604 (2003).

[13] A.R. Boccaccini, T. Berthier, S. Seglem, *Ceram. Int.* **33** 1231-1235 (2007).

[14] S. Pace, V. Cannillo, J. Wu, D. N. Boccaccini, S. Seglem, A. R. Boccaccini, A. R., *J. Nuclear Mater.* **341** 12-18 (2005).

[15] A. Abdeluas, S. Noirault, B. Grambow. *J. Nucl. Mat.*, **358**, 1-9 (2006).

[16] N. Henry, P. Deniard, S. Jobic, R. Brec, C. Fillet, F. Bart, A. Grandjean and O. Pinet, *J. Non. Cryst. Solids* **333**, 199-205 (2004).

[17] I.A. Sobolev, M.I. Ojovan, T.D. Scherbatova, O.G Batyukhnova. *Glasses for radioactive waste* (Energoatomizdat, Moscow, 1999).

[18] I.A. Sobolev, S.A. Dmitriev, F.A. Lifanov, A.P. Kobelev, S.V. Stefanovsky and M.I. Ojovan, *Glass Technology* **46**, 28-35 (2005).

[19] J.M. Juoi, M.I. Ojovan. *Glass Technology*, **48**, 124-129 (2007).

[20] S. Gahlert, G. Ondracek, in; *Radioactive waste forms for the future*, Lutze, W., Ewing, R. C., eds. (North Holland Publ., pp. 161-166, 1988)

[21] M. Sahimi, *Application of Percolation Theory* (Taylor & Francis Publisher, London, 1994).

[22] A.R. Boccaccini, J. Janczak, H. Kern, G. Ondracek, in *Proc. 4th. International Symposium on the Reclamation, Treatment and Utilization of Coal Mining Wastes*. Vol. II, Skarzynska, K. M. ed. (1993), pp. 719-726.

[23] A.R. Boccaccini, M. Kopf, W. Stumpfe, *Ceram. Int.* **21**, 231-235 (1995).
[24] A.R. Boccaccini, J. Janczak, D. M. R. Taplin, M. Köpf, *Environmental Technol.* **17**, 1193-1203 (1996).
[25] J.A. Roether, A. R. Boccaccini, Chapter 20 in *Handbook of Ceramic Composites*, ed. by N. P. Bansal, (Kluwer Academic Publ., Boston, Dordrecht, London, 2005).
[26] A.R. Boccaccini, M. Bücker, J. Bossert, K. Marszalek, *Waste Management* **17**, 39-45 (1997).
[27] C. Bauer, S. Gahlert, G. Ondracek, in *The geological disposal of high level radioactive waste*, Brookings D. G., ed. (Theoprastus Publications, pp. 187-195, 1987).

Mater. Res. Soc. Symp. Proc. Vol. 1107 © 2008 Materials Research Society

Waste Loading of Actinide Chloride Surrogates in an Iron Phosphate Glass

James M. Schofield, Paul A. Bingham and Russell J. Hand
Immobilization Science Laboratory, Department of Engineering Materials, University of Sheffield, Sir Robert Hadfield Building, Mappin Street, Sheffield, S1 3JD, UK.

ABSTRACT

Pyrochemical and aqueous processing of impure Pu metal and Pu oxide results in a waste stream that contains calcium, plutonium and americium chlorides ($CaCl_2$, $PuCl_3$ and $AmCl_3$) that tend to form insoluble salt layers on top of melts during vitrification. By adding chlorides to ammonium dihydrogen phosphate, ammonium chloride is evolved during vitrification, leaving behind the Ca, Pu and Am cations as oxides. HfO_2 and Sm_2O_3 were used as surrogates for PuO_2 and Am_2O_3 respectively. The effects of waste loading and melt duration on some basic physical properties for different melt durations were investigated. Synthetic batches containing CaO, HfO_2 and Sm_2O_3 in molar ratios 93.34 : 5.91 : 0.74 respectively, reflecting the waste composition, were mixed with a fixed glass composition of P_2O_5 (75 mol%) and Fe_2O_3 (25 mol%). $NH_4H_2PO_4$ was used as a precursor material. The solubility of calcium, hafnium and samarium oxides into phosphate glasses is reported and elemental analysis of the glasses is compared with expected glass compositions.

INTRODUCTION

Pyrochemical and aqueous reprocessing of impure plutonium metal and plutonium oxide results in calcium chloride, plutonium chloride and americium chloride waste [1]. Ordinarily the heavy metals would be treated further and precipitated as hydroxides [2]. When handling radioactive waste, increasing the number of processing steps, will also increase the amount of contaminated waste that must be dealt with during the decommissioning of radioactive waste treatment plants. The current project is aimed at identifying potential vitrification routes for the chloride waste stream in order to decrease processing steps.

Conventionally, borosilicate glasses are used to vitrify nuclear wastes. However, these glasses have low chloride solubilities (typically <1.5 wt% Cl) before liquid-liquid immiscibility occurs [3,4]. Immiscibility in nuclear waste disposal glasses is undesired as one of the phases may possess unacceptable chemical durability. Whilst the chloride waste could be reprocessed to an oxide (CaO, PuO_2 and Am_2O_3) and then vitrified in a borosilicate glass, the additional reprocessing step would be costly and produce more waste for disposal due to additional processing steps. Hence we have examined iron phosphate glasses as an alternative matrix.

Donze *et al.* [5] have reported that lead, tin, and cadmium chlorides ($PbCl_2$, $SnCl_2$ and $CdCl_2$) react with ammonium dihydrogen phosphate ($NH_4H_2PO_4$) to produce $MO\text{-}P_2O_5$ glasses (where M = Pb, Sn and Cd), and off-gases of H_2O and NH_4Cl where the reactions are stoichiometrically balanced [5]. Ammonium chloride volatilization could provide a way to remove all chlorine from the actinide chloride containing waste. For a given metal chloride

(MCl_x), if sufficient ammonium dihydrogen phosphate is present to volatilise all of the chloride then

$$MCl_x + xNH_4H_2PO_4 \rightarrow MO_{x/2} \cdot 0.5xP_2O_{5\,(glass)} + xNH_4Cl + 3xH_2O \quad (1)$$

If there is an excess of $NH_4H_2PO_4$ then ($y > x$) and

$$MCl_x + yNH_4H_2PO_4 \rightarrow MO_{x/2} \cdot 0.5yP_2O_{5\,(glass)} + xNH_4Cl + (y-x)NH_3 + 0.5(3y-x)H_2O \quad (2)$$

Finally if there is insufficient $NH_4H_2PO_4$ to volatilise all of the chloride, then ($x > y$) and

$$MCl_x + yNH_4H_2PO_4 \rightarrow y/xMO_{x/2} \cdot 0.5yP_2O_{5\,(glass)} + yNH_4Cl + (1-y/x)MCl_x + yH_2O \quad (3)$$

The durability of phosphate glasses can be increased with the addition of iron oxide [3,6-16]. Thus the addition of heavy metal chlorides to an iron phosphate melt, producing a heavy metal iron phosphate glass, resulting in the evolution of ammonium chloride via reactions (1) or (2), could produce a potentially viable wasteform. It should be noted that reaction (3) is undesirable and must be avoided.

Plutonium and americium salts are highly radioactive, therefore the use of surrogates is required. Based on equations (1) and (2) we can assume that all the chlorides will be volatilised and thus the initial compositional work has been undertaken using oxide simulants. Further it can be assumed that Pu will be in the 4+ oxidation state in oxide glasses. Thus in this work hafnium has been used as the plutonium surrogate and samarium as the americium surrogate: Hf^{4+} provides a good model for Pu^{4+} [17] and Sm^{3+} has a similar ionic radius to Am^{3+}. The initial work is intended to identify possible waste loadings of an 80 wt% $CaCl_2$, 16 wt% $PuCl_3$ and 4 wt% $AmCl_3$ waste stream in a durable iron phosphate glass. Melt duration has been varied to study the effect of time on the final glass composition and waste solubility.

EXPERIMENTAL DETAILS

Synthetic mixtures were prepared based on a 75 mol% P_2O_5 ($NH_4H_2PO_4$ used as a precursor) 25 mol% Fe_2O_3 glass, to which we have added CaO ($CaCO_3$ used as a precursor material), HfO_2 and Sm_2O_3 as the waste components. Dry powders were used in the batching process. Each batch weighed 180g and was mixed within a polythene bag for 10 minutes prior to being transferred to a crucible. The molar ratios of CaO, HfO_2 and Sm_2O_3 were 93.34 : 5.91 : 0.74; these were chosen to correspond to the molar content of the cations in the processed wastes. I shows the nominal compositions of the prepared batches. Batching errors occurred from batch water content (~1 wt%), weighing scale accuracy (0.01g for P, Fe and Ca, 0.001g for Hf and Sm oxides) and from the batch unavoidably adhering to weighing equipment and mixing bags (estimated to be ~1 wt%).

The prepared batches were heated in re-crystallized alumina crucibles at a rate of 1.4 °C min^{-1} followed by a 3 or 9 hour dwell period at 1325 °C. Glasses were obtained by casting these melts into blocks on a stainless-steel plate. Glasses were annealed at 510 °C for one hour then cooled to room temperature at a rate of 1 °C min^{-1}. Batches were considered to have been at the melt temperature for the required duration ± 6min, due to the time taken to pour each melt and slight delays in ramp temperature due to the burden put on the furnace elements by the extraction fans.

Table I. Batch compositions. Phosphorous pentoxide was supplied using ammonium dihydrogen phosphate. Calcium oxide was supplied using calcium carbonate.

Nominal (Batch) Composition (Mol%)						
Sample	**A**	**B**	**C**	**D**	**E**	**F**
P_2O_5	75	71.25	67.5	63.75	60	56.25
Fe_2O_3	25	23.75	22.5	21.25	20	18.75
CaO	0	4.667	9.335	14.002	18.669	23.336
HfO_2	0	0.296	0.591	0.887	1.183	1.479
Sm_2O_3	0	0.037	0.074	0.111	0.148	0.185

Density measurements were repeated 10 times on ~40g glass samples using the Archimedes method. Measurements were taken using a four-figure Oertling model R20 balance. Density measurements were assessed as being accurate to $\pm 1\times10^{-4}$ g/cm^{-3}. A Netzsch DIL 402C dilatometer was used to measure thermal expansion coefficient over the range 50 to 300 °C (α_{50-300}) and glass transition mid-point temperature, T_g, were measured. Sample dimensions were 15 × 4 × 4 m. Samples were cut and polished to 45 μm on all surfaces, and heated at 5 °C min^{-1} to 20 °C above the dilatometric softening point, T_d. Thermal expansion coefficient is accurate to ± $2x10^{-7}$ taking into account the ±1 °C variation in the parallel end surfaces of the sample, furnace temperature accuracy of ~0.5 °C, and sample thermocouple accuracy of ~0.5 °C. Glass transition temperature, T_g, is reported to be accurate to ± 2°C.

Elemental analysis of the glass samples was carried out using Energy Dispersive X-ray Spectroscopy (EDS) (Link Analytical) on a Philips 500 Scanning Electron Microscope (SEM). The sample microstructures were also examined using SEM and backscattered electron imaging (BEI). The EDS was calibrated using metallic or ceramic standards of stoichiometric phases of the appropriate elements. Glass samples for EDS were cut to 15 x 15 x 4mm, mounted in epoxy resin and polished to ~1μm using diamond pastes and a colloidal silica water based polish. Polished samples were prepared for SEM with silver paint and carbon coated to provide a conductive pathway. Three semi-quantitative EDS measurements were made on regions near the top, in the middle and towards the bottom of the cross-sectioned, as-poured glass samples. Estimated errors in EDS measurements are dependent upon elemental abundances: ± 0.5 wt % for major constituents (P_2O_5, Fe_2O_3 and CaO) and ± 0.15 wt % for minor constituents (HfO_2, Sm_2O_3, Al_2O_3). Errors were then converted into mol %. A variation of ± 1° in sample tilt angle and variations in the carbon coating thickness of the prepared sample were also taken into consideration. Deviation in waste loading in mol% was calculated from the deviations in mol% for the individual waste components (Ca, Hf and Sm oxides) using:

$$\text{Deviation in waste loading (mol\%)} = \sqrt{(\Delta CaO)^2 + (\Delta HfO_2)^2 + (\Delta Sm_2O_3)^2} \quad (4)$$

RESULTS AND DISCUSSION

Figure 1 shows that glass density is lowest at ~ 5 mol % waste loading, with density increasing at lower and higher waste loadings from this point onwards. Brow *et al.* [18] have observed similar density "anomalies" on making alkali network modifier additions to binary phosphate glasses finding the density minima at ~20 mol % additions. It is believed that density decreases as the network expands to accommodate the isolated alkaline earth polyhedra. However, when the number of terminal oxygens available per modifying ion is exceeded by the coordination number of Ca-polyhedra, neighbouring alkaline earth polyhedra must share the terminal oxygens leading to a density increase. This mechanism is suggested but it cannot be confirmed without further structural analysis on these glasses.

The thermal expansion coefficient of the glasses increases gradually with increasing waste loading (Figure 2). The glass transition temperature (T_g) obtained with increasing melt duration tend towards each other as waste loading increases. The reduction in the difference in T_g between the 3hr and 9hr melts could be due to the lower phosphate contents in the more highly waste loaded glasses (see also the following discussion of Figure 3a).

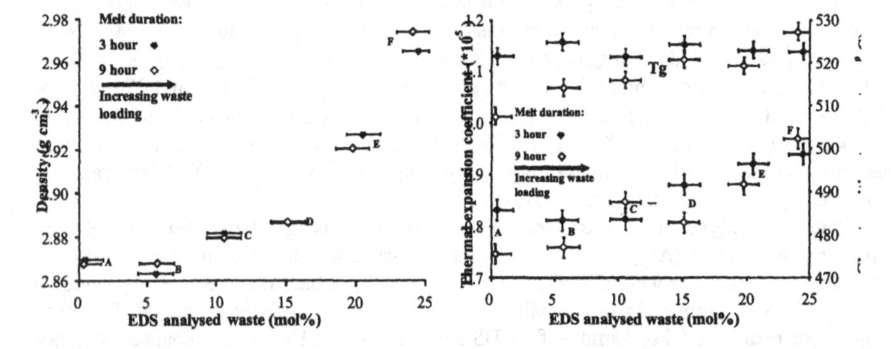

Figure 1. Glass density plotted against EDS mol% waste (CaO, HfO_2, Sm_2O_3).

Figure 2. Glass transition temperature (T_g) (right axis), and thermal expansion coefficient (α) (left axis), plotted against EDS analysed mol % waste (CaO, Hf_2O_3, Sm_2O_3). Data obtained by dilatometry.

Figure 3a shows that the difference between the batched and EDS measured phosphate contents increases with melt duration. With all glasses P_2O_5 loss during a 9hr melt time is greater than that for a 3hr melt. Figures 3b and 3c show that there is little significant variation in Fe_2O_3 or CaO on increasing the melt duration from 3 hours to 9 hours. Figure 3d shows that the measured HfO_2 contents are very similar to batched contents up to a batched 1.48 mol% HfO_2 (glass F) when the EDS analysis shows the glass matrix to contain ~1.2 mol% HfO_2. This result combined with the SEM BEI image of the glass F (Figure 4) indicates that the solubility limit of HfO_2 in this glass has been exceeded.

At 0 mol% waste the EDS analyzed Fe_2O_3 content is higher than that calculated from the batch (Figure 3b) and there is also an increase in Sm_2O_3 mol% glass content during melting

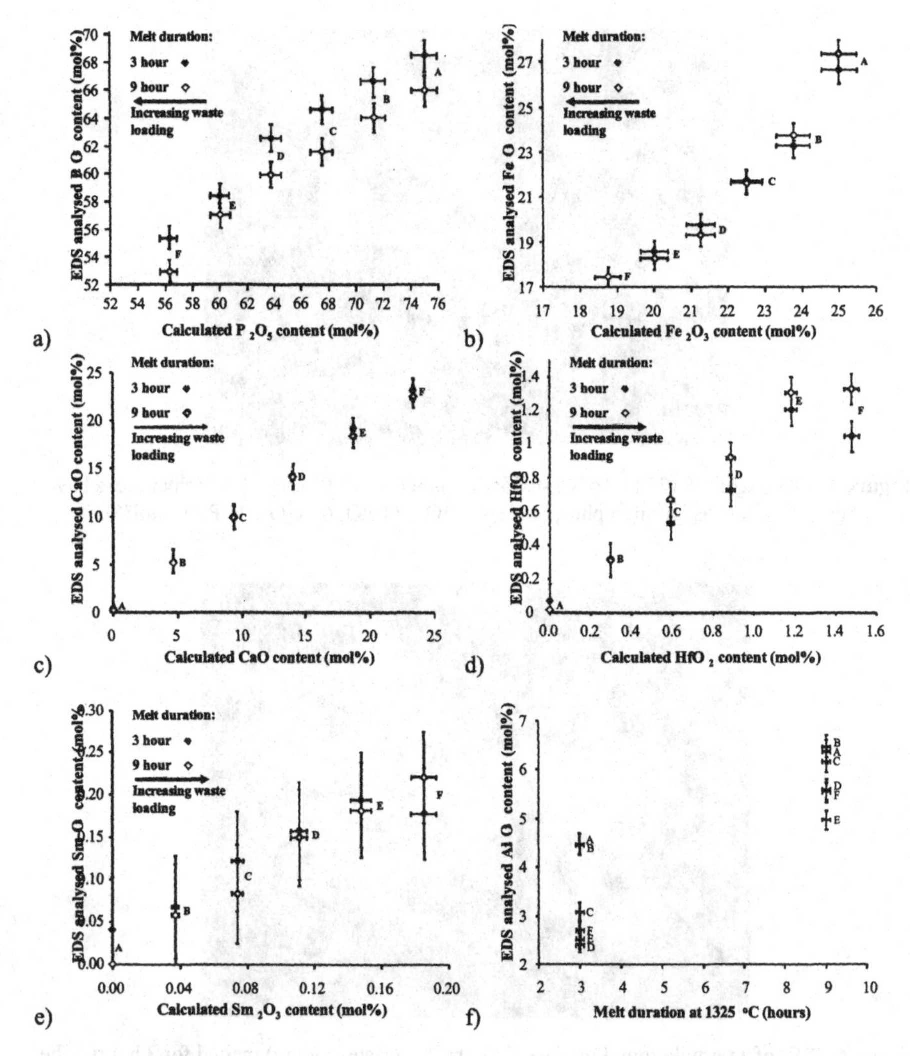

Figure 3. Analysed (EDS) oxide contents: a) P_2O_5; b) Fe_2O_3; c) CaO; d) HfO_2; and e) Sm_2O_3 analysed contents versus batch contents and f) Al_2O_3 content versus time.

(Figure 3e). These increases may be due to the evolution of phosphorous from the melt increasing the molar percentage of the other elements in the glass although, as shown in Figures 3c and 3d, the measured CaO content and HfO_2 content (for lower waste loadings) are very similar to the batched contents. Furthermore, as shown in Figure 3f, increasing the melt duration also increased the amount of Al_2O_3 dissolved from the crucibles; the majority of dissolved Al_2O_3 will result from flux line corrosion. There is no clearly discernible relationship between glass

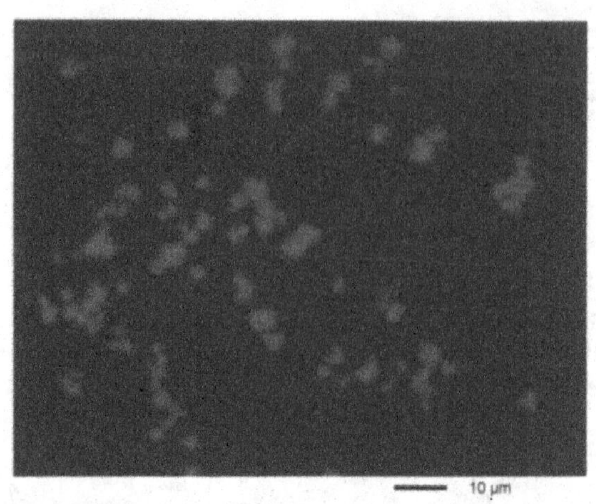

Figure 4. BEI of glass F (25 mol% waste loading) melted for 9 hours. The lighter areas have been identified as hafnium phosphate (49 HfO_2, 2 CaO, 6 Fe_2O_3, 43 P_2O_5 mol%).

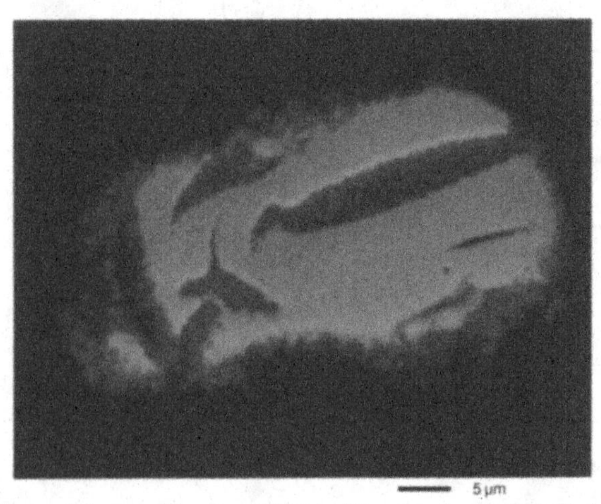

Figure 5. BEI of a particle found in glass E (20 mol% waste loading) melted for 3 hours. The lightest areas are identified as hafnium oxide, the darker grey as hafnium phosphate (49 HfO_2, 2 CaO, 6 Fe_2O_3, 43 P_2O_5 mol%) and the black is the bulk glass.

composition (waste loading) and the level of Al_2O_3 contamination (Figure 3f), although Figure 3f may suggest that those glasses with lower waste loading attack the alumina crucibles more aggressively.

A hafnium phosphate phase was found in the 25 mol% (glass F) but not in the 20 mol% (glass E) waste loaded glasses. Semi-quantitative EDS analysis indicates that the composition of this phase is 49 HfO_2, 2 CaO, 6 Fe_2O_3, 43 P_2O_5 mol%; the iron and calcium contents may be due to the interaction volume associated with the electron beam. The angular geometry suggests that this phase has crystallized from the melt further indicating the solubility limit of hafnium in this glass been exceeded. The same phase was found whether glass F was melted for 3 hours or 9 hours. Differences in the EDS analysed HfO_2 content of these glasses over the two different melt durations are within the calculated errors (see Figure 3d).

Figure 5 shows a hafnium rich deposit found at the bottom of a crucible section from the glass E (20 mol% waste loading) melted for 3 hours. The deposit was analysed as HfO_2 in the brightest areas, with a hafnium phosphate reaction rim (49 HfO_2, 2 CaO, 6 Fe_2O_3, 43 P_2O_5 mol%). This particle is relatively large and reflects potential problems arising from settling effects of dense batch components.

As no differences were found in the HfO_2 solubilities in either glasses E or F melted for 3 hours or 9 hours, we conclude that increasing the melt duration from 3 to 9 hours has little detectable effect on HfO_2 solubility in these iron phosphate glasses at 1325 °C. No hafnium containing deposits or crystalline phases were detected in glasses A to D (0 to 15 mol% waste loading). Glasses E and F clearly contain undissolved HfO_2. Glasses F clearly contains crystallized hafnium phases. Therefore it is not correct to assume that the reported EDS values for glasses E and F are as accurate as those reported for glasses A to D as the glasses E and F are not homogenous, thus the reported results are not representative of the entire glass sample.

Our results demonstrate that the limiting factor constraining waste loading for this waste is the solubility of HfO_2 in these glasses. HfO_2 (simulating PuO_2) is present in larger quantities than Sm_2O_3 (simulating Am_2O_3) in the waste stream, thus it is perhaps not surprising that it should be the limiting factor. HfO_2 is a highly refractory oxide with a melting point of 2774 °C which is, in fact, higher than that of PuO_2 (2400 °C); and its chemically similar relative, ZrO_2, is known to have limited solubility in many glasses. As the glass melts were not stirred, settling of the high density HfO_2 (9.68 Mg m^{-3}) is also more likely to occur (as indicated by the relict shown in Figure 5); this may also be an issue with PuO_2 which has a density of 11.50 Mg m^{-3}. It is, of course, possible that it will be easier to incorporate Hf or Pu into the glass starting from the chloride waste, rather than an oxide surrogate precursor, as it will not be necessary to actually dissolve an oxide particle directly into the melt.

Based on our results to date it appears that the waste loading in these glasses is limited to less than 20 mol% in oxide terms (~10 wt%), (equivalent to ~12 wt% in oxide terms or ~18 wt% in chloride terms of the original chloride waste stream) in order to achieve a homogeneous wasteform.

CONCLUSIONS

Additions of CaO, HfO_2 and Sm_2O_3 in the molar ratios 93.34 : 5.91 : 0.74, simulating the expected oxide products from vitrifying a $CaCl_2$, $PuCl_3$, $AmCl_3$ waste stream, have been made to a 75 P_2O_5 – 25 Fe_2O_3 (mol %) glass at loading levels up to 25 mol% waste. Increasing melt time led to both increased phosphate loss and increased attack of the alumina crucibles used in this research. Hafnium oxide is the limiting factor for the production of a homogeneous wasteform, leading to a waste loading limit of less than 20 mol% waste in oxide terms. This is equivalent to ~18 wt% waste in terms of the original chloride waste stream.

ACKNOWLEDGMENTS

JMS thanks AWE, UK and EPSRC, UK for a studentship and PAB thanks EPSRC for financial support.

REFERENCES

1. I.W. Donald, B.L. Metcalfe, S.K. Fong, L. A. Gerrard, D. M. Strachan and R.D. Scheel, *J. Nucl. Mater.* **361**, 76-93 (2007).
2. L. Xiaomei, R. C. Burns and G. A. Lawrance, *Water, Air, & Soil Pollut.* **165**, 131-152 (2005).
3. H. Li, J. G. Darab, P. A. Smith, M. J. Schweiger, D. E. Smith and P. R. Hrma, *Proc. 36th Ann. Meeting of the Inst. of Nucl. Mater. Manage.* 466-471 (1995).
4. P. A. Bingham, R.J. Hand, M.C. Stennett, N.C. Hyatt and M.T. Harrison, This volume.
5. S. Donze, L. Montagne and G. Palavit, *Chem. Mater.* **12**, 1921-1925 (2000).
6. M. G. Mesko and D. E. Day, *J. Nucl. Mater.* **273,** 27-36 (1999).
7. B. L. Metcalfe, S. K. Fong and I. W. Donald, *Glass Technol.* **46**, 130-133 (2005).
8. X. Yu, D. E. Day, G. J. Long and R. K. Brow, *J. Non-Cryst. Solids* **215**, 21-31 (1997).
9. C. S. Ray, X. Fang, M. Karabulut, G. K. Marasinghe and D. E. Day, *J. Non-Cryst. Solids* **249,** 1-16 (1999).
10. X. Fang, C. S. Ray, A. Mogus-Milankovic and D. E. Day, *J. Non-Cryst. Solids* **283,** 162-172 (2001).
11. D. L. Griscom, C. I. Merzbacher, N. E. Bibler, H. Imagawa, S. Uchiyama, A. Namiki, G. K. Marasinghe, M. Mesko and M. Karabulut, *Nucl. Instrum. Methods Phys. Res., Sect. B* **141,** 600–615 (1998).
12. M. Karabulut, G. K. Marasinghe, C. S. Ray, D.E. Day, G. D. Wadhill, C. H. Booth, P. G. Allen, J. J. Bucher, D. L. Caulder and D. K. Shuh, *J. Non-Cryst. Solids* **306,** 182–192 (2002).
13. M. Karabulut, E. Metwalli, D. E. Day and R. K. Brow, *J. Non-Cryst. Solids* **328,** 199–206 (2003).
14. G. K. Marasinghe, M. Karabulut, C. S. Ray, D. E. Day, M. G. Shumsky, W. B. Yelon, C. H. Booth, P. G. Allen and D. K. Shuh, *J. Non-Cryst. Solids* **222,** 144–152 (1997).
15. M. G. Mesko, D.E. Day and B.C. Bunker. *Waste Manage.* **20**, 271-278 (2000).
16. I. W. Donald, B. L. Metcalfe, S. K. Fong and L. A. Gerrard, *J. Non-Cryst. Solids* **352,** 2993-3001 (2006).
17. C. Lopez, X. Deschanels, J. M. Bart, J. M. Boubals, C. Den Auwer and E. Simoni, *J. Nucl. Mater.* **312**, 76-80 (2003).
18. R. K. Brow, C. A. Click and T. M. Alam, *J. Non-Cryst. Solids* **274**, 9-16 (2000).

Mater. Res. Soc. Symp. Proc. Vol. 1107 © 2008 Materials Research Society

Devitrified and Phase Separated Material Found in Simulated High Level Nuclear Waste Glasses Containing Ca and Zn Additions

R. Short[1], E. Turner[1], B. Dunnett[1], A. Riley[2]

1. Nexia Solutions, Sellafield, Seascale, Cumbria, CA20 1PG, Cumbria
2. British Nuclear Group, Sellafield, Seascale, Cumbria, CA20 1PG, Cumbria

ABSTRACT

In the UK, blended high level nuclear waste (HLW) streams from the Magnox and THORP reprocessing plants are currently vitrified using a lithium sodium borosilicate base glass frit. Laboratory and full size non-radioactive simulations (produced on the Vitrification Test Rig at Sellafield [1]) of these compositions have shown that these glasses need to be melted at circa 1050°C to obtain a reasonable viscosity for pouring. Also, at high waste loadings an alkali molybdate phase (termed "yellow phase") can form in these glasses [e.g. 2, 3]. Vitrification flowsheets are set to avoid yellow phase formation as this phase is highly corrosive to the inconel melter in the molten state and is partially water soluble at ambient temperature and so may challenge product quality.

Ca and Zn additions to the base glass frit have been found to reduce viscosity and allow melt homogeneity and pouring at lower temperatures. It was also theorised that Ca additions could increase the solubility of Mo and thus reduce the likelihood of yellow phase formation. The composition of the phase separated material in as-cast and heat treated specimens of Ca and Zn HLW glasses produced at both laboratory and full scale is examined in this work

INTRODUCTION

Vitrification of HLW produced by the reprocessing of used fuel in UK has long been performed using a lithium sodium borosilicate base glass composition as the vitrification matrix. This composition provides a stable and chemically durable host for radionuclides at ambient temperatures, and can be processed at relatively low temperatures (circa 1050°C) in the Waste Vitrification Plant (WVP). Laboratory testing and trials performed on the Vitrification Test Rig (VTR), a full scale inactive vitrification plant, have shown that at waste incorporation rates of up to 28wt%, a homogeneous single phase product glass can be produced.

At higher waste loadings in feeds with high Mo contents, a very small amount of phase separation can occur due to the low solubility of Mo in borosilicate compositions. The resulting "yellow phase", a mixture of alkali molybdates and chromates, accounts for less than 0.0015wt% of the product. However, its formation is undesirable as it is highly corrosive to the inconel melters used on WVP on the molten state, and is partially water soluble in the solid state. Thus, feed specifications for high Mo waste streams on WVP are set to avoid the formation of yellow phase.

Ca and Zn containing alkali borosilicate glasses offer the potential for multiple benefits in the vitrification process. These formulations generally have lower viscosities than the standard alkali borosilicate base glass used on WVP allowing processing of the waste at lower temperatures. Several Ca and Zn containing glass formulations were trialed at laboratory scale before selection of a composition that gave the best overall performance in terms viscosity, durability, waste incorporation and corrosion rate (of inconel). This formulation was then trialed at full scale on the VTR. This work reviews the crystallisation behavior of the chosen formulation.

EXPERIMENT

Glasses with the compositions shown in table I were prepared. Blend 25 and Blend 31 were both prepared in the laboratory and represent waste glasses containing 25wt% and 31wt% waste oxides of a Highly Active Liquor (HAL) waste stream comprised of a mixture of 75% oxide and 25% Magnox HALs. Magnox 38 was produced on the VTR and represents a waste glass containing 38wt% waste oxides of a Magnox only waste stream. The elements shown are those that comprise the base glass and the major components of yellow phase. Blend and Magnox waste streams do differ considerably in terms of compositions of major elements and at waste loadings previously investigated with Magnox glasses (<28wt%) yellow phase has not proved problematic. However, the levels of Mo, Li, Cs and Sr (key elements in yellow phase formation [4]) in Magnox 38 are very similar to those found in the Blend compositions studied here and thus this material was suitable for use as a full-scale comparison.

Table I. Compositions of Ca and Zn containing glasses.

	Weight %		
	Blend 25	**Blend 31**	**Magnox 38**
B_2O_3	16.89	15.09	12.71
CaO	1.33	1.23	1.25
Cs_2O	1.43	1.75	1.35
Li_2O	3.99	4.21	3.95
MoO_3	2.35	2.77	2.6
Na_2O	8.29	7.39	6.55
SiO_2	41.76	37.47	32.16
SrO	0.39	0.49	0.47
ZnO	3.99	3.59	3.76
Other waste oxides	19.58	26.01	35.20
Total	100	100	100

All glasses were prepared from a base glass frit containing all of the Si, B, Na, Ca and Zn present in the final glass product, and approximately ½ of the Li content. The remaining elements were added as a mixture of solid metal nitrates and oxides (approximately 75% oxide, 25% nitrates) as produced by the calcination process [2] on the VTR. In the case of the Blend glasses, the calcine was mixed with the frit in appropriate quantities to give 100g product of the composition shown in table I, then melted in a silica crucible in a muffle furnace for 4 hours at 1050°C and stirred with a silica rod after 3 hours. The glass was then cast and annealed at 500°C for 3 hours, then cooled to room temperature at 0.5°C min^{-1}. A separate 100g batch of each of the Blend glasses was prepared and also melted at 1050°C for 4 hours in silica crucibles. These batches were annealed (as above) in their crucibles then sectioned to allow for a visual inspection for the presence of yellow phase. The Magnox 38 glass was made on the VTR using the normal continuous feeding procedure. The total feed cycle time was 6.75 hours to give a total pour weight of 199.2kg. The average internal melter temperature at the completion of the feed cycle was 1017°C and the glass was cast into a preheated steel container.

Specimens from each of the Blend compositions were heat treated at 650°C and 750°C for 14 days. They were then cooled to room temperature at 1°Cmin^{-1}. A specimen of Magnox 38 was heat treated at 650°C for 14 days then cooled to room temperature at 1°C min^{-1}.

Samples for Scanning Electron Microscopy (SEM) were sectioned using a diamond saw, and mounted in resin. The samples were polished to a 1μm surface finish using successively finer grades of SiC and diamond paste, then gold coated in preparation for analysis. SEM and Energy Dispersive Spectroscopy (EDS) analyses were carried out on the samples using a Jeol JSM-5600 SEM and PGT Prism Digital spectrometer (model OPJ095-1029) to determine the composition of phase separated/devitrified material in the samples.

Further phase identification was also carried out using powder X-ray Diffraction (XRD). Samples were scanned using a Philips automatic powder X-ray diffractometer using Copper K alpha radiation generated at 40kV and 55mA from 2 to 70°2θ with a step size of 0.02°.

DISCUSSION

As-cast products

Visual inspection of Blend 25 and Blend 31 glasses revealed that after 4 hours at 1050°C, no yellow phase was present on a macroscopic scale in either composition (compare figures 1a, a homogeneous waste product and 1b, a glass containing yellow phase).

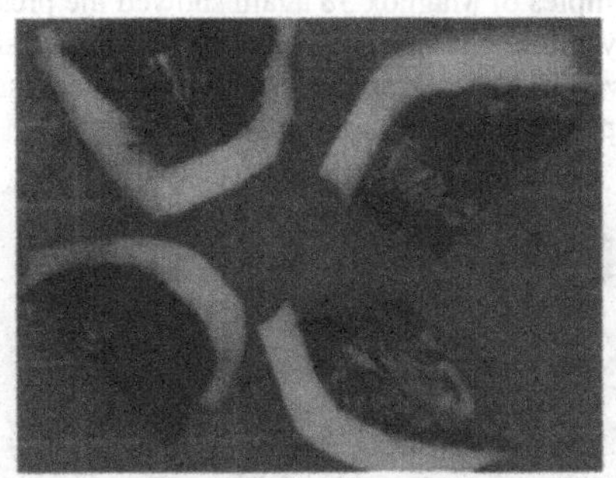

Figure 1a. Blend 31 (scale markings = 1cm).

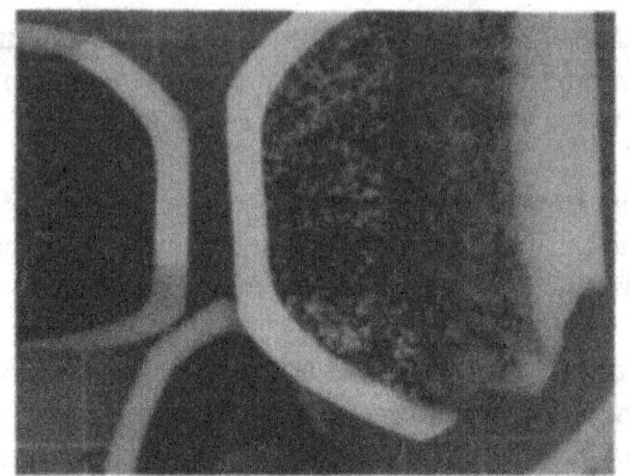

Figure 1b. Example of HLW glass at high waste loadings showing presence of yellow phase (scale markings = 1cm).

SEM analysis of the as-cast Blend 25 and 31 samples revealed the presence of RuO_2 crystals and small amounts of rare earth phases. Platinum group metals have negligible solubility in HLW glass compositions but are reasonably inert and have very low leach rates from the product. Lanthanide elements are included in simulated HLW streams to represent the actinides found in active streams. Some of these elements are present in concentrations of >5wt% in the simulated waste glass products and have been observed to phase separate to a small degree on the microscopic scale, although again these phases do not adversely effect the physical or chemical properties of the products. However, no crystalline phases containing Ca or Mo were observed in the as-cast products.

Visual analysis of the as-cast Magnox 38 glass did not reveal any yellow phase (see figure 2). SEM analysis of as-cast samples of Magnox 38 again showed the presence of RuO_2 and also of Mg and Al based refractory spinel phases. These phases incorporate small amounts of Fe, Cr and Ni and are regularly observed in Magnox products with high waste incorporation rates, but have no negative impact on the properties of the solid product.

Figure 2. The as-cast Magnox 38 product glass (length of product ≈1m).

Heat treated products

SEM analysis of the Blend 25 and Blend 31 product glasses at temperatures ≥650°C did generate some additional crystalline phases to those observed in the as-cast product. EDS analysis revealed that these phases contained Ca and Mo (figure 3). SEM EDS analysis of the heat treated sample of Magnox 38 also revealed devitrification of a Ca and Mo containing crystalline phase.

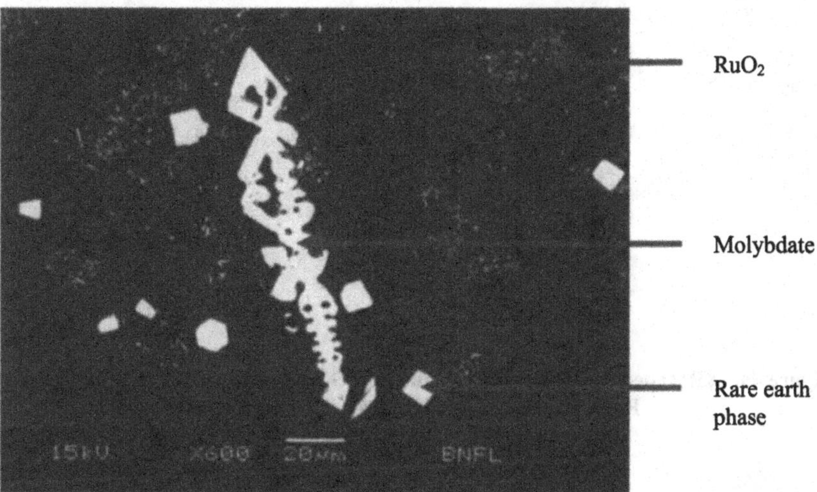

Figure 3. Backscattered SEM image of a heat treated Blend 31 sample (typical of those found in all of the samples in this study that were heat treated at ≥650°C).

XRD analysis (figures 4a and 4b) of the as cast and heat treated samples revealed that the Ca and Mo containing phase was also present in the as cast sample of Magnox 38, indicating that the Ca was associating with the Mo during the melting process in this material. No Mo containing crystalline phase was observed in the as cast Blend glasses. The Mo containing crystalline phase could not be identified using the International Centre for Diffraction Data (ICDD) database, but had close matches to $SrCa(MoO_4)_2$, $NaCe(MoO_4)_2$, $LiCe(MoO_4)_2$, $NaLa(MoO_4)_2$, $Nd_2Mo_3O_9$ (ICDD cards: [30-1287], [79-2242], [84-539], [24-1103] and [33-936] respectively). All of these phases share a scheelite type structure (space group I41/a) which is able to accommodate a wide range of alkali, alkali earth and rare earth elements in various stoichiometries to charge balance the $[MoO_4]^{2-}$ group (e.g. [5]). Thus it is likely that the actual unit cell is a complicated structure containing several elements in small concentrations along with Ca, Mo and O. RuO_2 (ICDD card [40-1290]) was also identified in all of the as cast and heat treated products, and in the Magnox 38 samples a third phase was identified that gave close matches for $MgAlFeO_4$, $MgAl_{0.8}Fe_{1.2}O_4$ and Mg $Al_{0.6}Fe_{1.4}O_4$ (ICDD cards [11-9], [71-1235] and [71-1234] respectively). All of these phases are cubic spinel type structures, phases which are regularly found in Magnox waste products and can accommodate a range of Mg, Al, Fe, Cr and Ni stoichiometries.

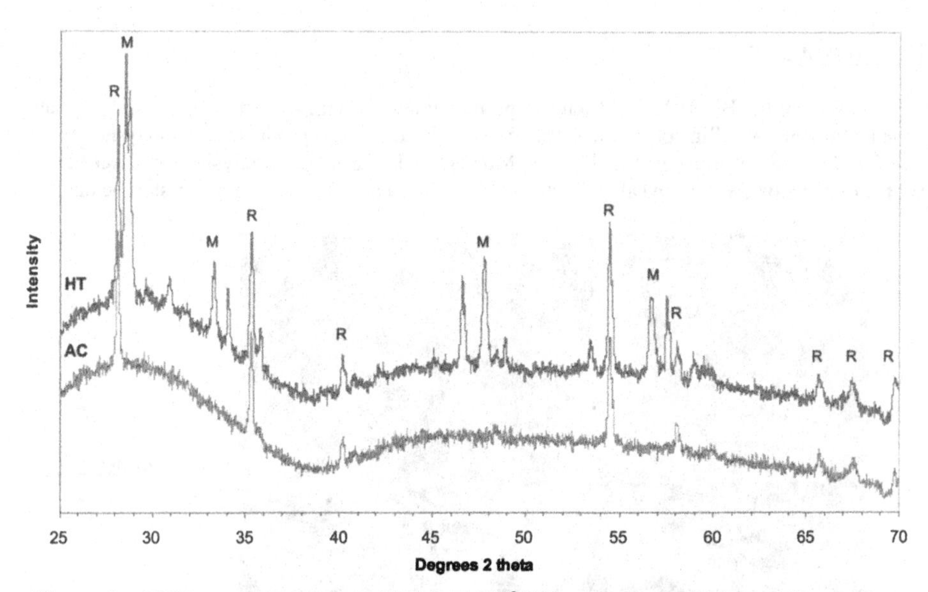

Figure 4a. XRD spectra of as-cast (AC) and 650°C heat treated (HT) Blend 31 samples. RuO_2 (R) and molybdate (M) phases are labeled

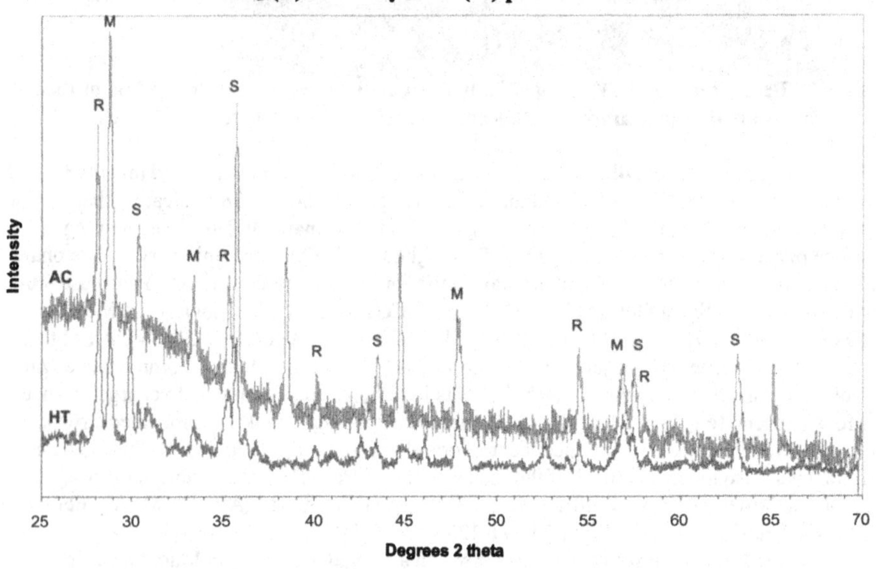

Figure 4b. XRD spectra of as-cast (AC) and 650°C heat treated (HT) Magnox 38 samples. RuO_2 (R), molybdate (M) and spinel (S) phases are labeled

XRD results also indicated the presence of other crystalline phases in both the lab scale and full scale samples that did not provide any close matches from the ICDD. Examination of the SEM EDS results indicated that the unidentified phases were likely to be comprised chiefly of combinations of lanthanide elements.

CONCLUSIONS

The lack of yellow phase presence in the as cast Blend product at laboratory scale, and the association of Mo with Ca in the as cast full scale Magnox glass indicated that Ca played a role in increasing the solubility of Mo (and thus decreasing the formation of yellow phase) in alkali borosilicate HLW glasses. This is reinforced by the formation of a Ca and Mo containing phase under heat treatment of these glasses, rather than alternative Mo containing phases. However, full scale inactive trials of the Ca and Zn containing base glass with a high loading (>31wt%) of a Blended waste would be required to confirm the yellow phase reducing properties of Ca.

ACKNOWLEDGMENTS

The authors would like to thank British Nuclear Group plc for funding this work.

REFERENCES

1. K. Bradshaw, N. Gribble, P. Mayhew, M. Talford & A. Riley, Wa. Man. Sym Proc, Touson, AZ, 2006

2. *Radioactive Wasteforms for the Future*, edited by E. Lutze & R. Ewing (North Holland, Amsterdam, 1988)

3. E. Schiewer, H. Rabe, Weisenburger, in *Scientific Basis for Nuclear Waste Management V*, edited by W. Lutze, (Mater. Res. Soc. Symp. Proc. **11**, Berlin, 1982)

4. S. Morgan, P. Rose, R. Hand, N. Hyatt, W. Lee, C. Scales, in *Environmental Issues and Waste Management Technologies in the Ceramic and Nuclear Industries IX*, edited by J. Vienna and D. Spearing. (Ceram. Trans., **155**, IN, 2004)

5. R. Short, R. Hand, N. Hyatt, G. Möbus, J. Nuc. Mat. 340, (2005)

Mater. Res. Soc. Symp. Proc. Vol. 1107 © 2008 Materials Research Society

Cold Crucible Vitrification of Uranium-Bearing High Level Waste Surrogate

S.V. Stefanovsky, A.G. Ptashkin, O.A. Knyazev, M.A. Zen'kovskaya, J.C. Marra[1]
Center of Advanced Technologies, SIA Radon,
7th Rostovskii lane 2/14, Moscow 119121 RUSSIA
[1] Materials Science and Technology Directorate, Savannah River National Laboratory, Building 773-42A, Aiken, SC 29808, U.S.A.

ABSTRACT

Three tests on vitrification of U-bearing DWPF SB2 and SB4 waste surrogates (40-50 wt% waste loading) in a lab-scale (56 mm inner diameter) copper cold crucible with melt surface area of 24.6 cm^2 energized from a 5.28 MHz/10 kW generator were conducted. At the feeding of the air-dry batch (3-5 wt% moisture) glass productivity and specific melt productivity achieved under steady-state conditions were 0.64 kg/h and 256 kg/(m^2×h) [6150 kg/(m^2×day)], respectively. Similar to previously prepared U-free products, the U-bearing vitrified SB2 (high-ferrous) waste product was composed of U-enriched glassy matrix and high-ferrous spinel structure phase. As expected, the vitreous phase in the vitrified SB2 waste surrogate was enriched with Na, Al, Si, P, S, Ca, Sr, Pb, U oxides and Cl, whereas spinel preferentially accumulated transition metals (iron group elements: Mn, Fe, Ni) as well as Cu and Zn oxides. Iron oxide content in the spinel phase was ~67-71 wt %. This result was consistent with XRD data and implies that this spinel was close to magnetite with respect to chemical composition and structure. This was consistent with our previous data on characterization of glass crystalline materials containing both U-bearing and U-free SB2 waste surrogate produced in resistive furnace and cold crucibles. The vitrified SB4 (high-alumina) waste products were predominantly amorphous at waste loading of up to 50 wt %. Traces of quartz occurred in the "skull" zone. In all the products, uranium was found to be uniformly distributed within the vitreous phase. Leach rates of Na, Si, Fe, and U from the vitrified SB2 waste surrogate (50 wt % waste loading) determined by a SPFT procedure were found to be 1.38, 1.42, 0.122, and 0.841 (g×m^{-2}×d^{-1}), respectively.

INTRODUCTION

Currently a Joule heated ceramic melter (JHCM) is the baseline technology for vitrification of Savannah River Site high-level waste (HLW) at the Defense Waste Processing Facility (DWPF) [1]. A cold crucible induction melter (CCIM) is considered an alternative to the JHCM. SIA Radon performed testing to determine if the CCIM could achieve high waste loading (target 45-60 wt %) and maintain high throughput for Savannah River Site high-level waste (HLW) feeds. The first step of the work was to perform CCIM Testing to demonstrate maximized waste loading with Sludge Batch 2 (SB2) [2-4]. A maximum target SB2 waste loading of 60 wt % with the Frit 320 formulation (in wt %: 8 Li_2O, 8 B_2O_3, 12 Na_2O, 72 SiO_2) was tested in the SIA Radon large-scale CCIM melter (400 mm diameter unit) [4]. Specific attention was given to glass quality and processing issues associated with the high sulfate concentrations at this waste loading.

The next step was to determine the maximum waste loading and throughput achievable for the Sludge Batch 4 (SB4) composition (SB4 is expected to be processed in the DWPF in FY07). As followed from SRNL research work on selection of frit and glass composition and Radon lab-scale melting tests in alumina crucibles in a resistive furnace [5], a frit 503-R4 with chemical composition (wt %): B_2O_3 – 16, Li_2O – 8, SiO_2 – 76, was determined to be an appropriate glass forming additive to vitrify the SB4 waste. Tests in small (50 mL) alumina crucibles showed that the waste/frit mixtures at waste loading of 40 to 60 wt % yielded primarily homogeneous glass with minor crystal formation at a temperature of 1250 °C. Traces of spinel type crystalline phase were observed only at the highest waste loading of 60 wt % in glass.

In this paper we describe results of the tests on vitrification of the DWPF SB2 and SB4 wastes in a lab-scale cold crucible to evaluate and compare process variables and vitrified product phase composition, structure and properties.

FEED PREPARATION AND MELTING RUN DESCRIPTION

The U-bearing SB2 and SB4 waste surrogates were vitrified in a lab-scale unit with a 56 mm inner diameter copper cold crucible with melt surface area of 24.6 cm^2 energized from a 5.28 MHz/10 kW generator (Figure 1). The external surface of the crucible was coated with ground fireclay based putty.

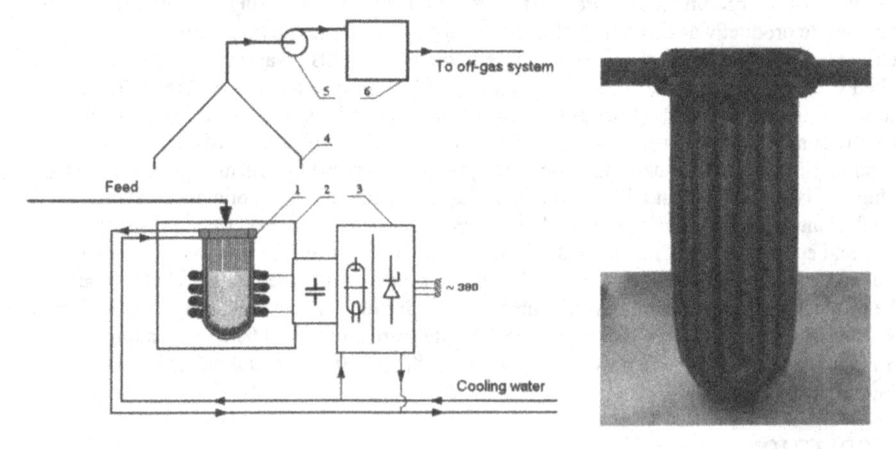

Figure 1. Flow Sheet of the Radon ICCM Unit (left) and Lab-Scale Cold Crucible (right).

1 – cold crucible with inductor, 2 – protective screen, 3 – high-frequency generator, 4 – off-gas unit, 5 – fan, 6 – off-gas filter.

Chemical compositions of the SB2 and SB4 waste surrogates and target glasses are given in Table I. The feed was prepared by a simplified procedure consisting of preparation of waste surrogate by intermixing of salts and oxides of either SB2 or SB4 waste constituents (uranium was introduced as U_3O_8). The waste surrogate (40 or 50 wt % on oxide basis) and glass forming chemicals to simulate Frit 320 or Frit 503-R4 were intermixed in a ball mill. A 50 wt % waste loading was targeted for SB2 tests while 40 and 50 wt % waste loadings were targeted for SB4 tests. In total, oxide mixtures in the amount of approximately 500-800 g for each composition were prepared. Each batch had moisture content of 3-5 wt %.

The melting runs consisted of five steps: 1) melting initiation and starting melt production, 2) operation volume formation and achieving steady-state conditions, 3) operation under steady-state conditions, 4) melt soaking at temperature, and 5) controlled cooling by gradual power reduction followed by turning-off the generator and cooling down to room temperature. When cooling was completed the crucible was dismantled and the solidified product was removed to be analyzed. The major process variables are given in Table II.

The vitrified products were examined by X-ray diffraction using a DRON-4 diffractometer (Fe Kα – radiation, scan range of 10 to 90 ° 2-θ), scanning electron microscopy with energy

Table I. Chemical compositions (wt %) of the waste surrogates and target glasses.

Oxides	SB2 waste	SB2 Glass	SB4 waste	SB4 Glass	
		50 wt % WL		40 wt % WL	50 wt % WL
Li_2O	-	4.00	-	4.80	4.00
B_2O_3	-	4.00	-	9.60	8.00
F	0.01	0.01	-	-	-
Na_2O	12.08	12.04	18.71	7.48	9.35
MgO	0.24	0.12	2.77	1.11	1.39
Al_2O_3	16.83	8.41	25.49	10.20	12.75
SiO_2	1.98	37.01	2.71	46.68	39.36
P_2O_5	0.14	0.07	-	-	-
SO_3	0.83	0.41	0.87	0.35	0.43
Cl	1.51	0.75	-	-	-
K_2O	0.09	0.04	0.07	0.03	0.03
CaO	3.76	1.88	2.77	1.11	1.38
TiO_2	-	-	0.04	0.01	0.02
Cr_2O_3	0.37	0.19	0.20	0.08	0.10
MnO	3.89	1.94	5.78	2.31	2.89
Fe_2O_3	42.26	21.14	28.99	11.60	14.49
NiO	2.17	1.08	1.66	0.66	0.83
CuO	0.20	0.10	0.05	0.02	0.03
ZnO	0.39	0.19	0.05	0.02	0.02
SrO	0.10	0.05	-	-	-
ZrO_2	0.79	0.39	0.09	0.04	0.05
I	0.04	0.02	-	-	-
Cs_2O	-	-	-	-	0.22*
BaO	0.27	0.13	0.07	0.03	0.03
La_2O_3	-	-	0.03	0.01	0.02
Ce_2O_3	-	-	0.21	0.09	0.11
PbO	0.32	0.16	0.38	0.15	0.19
ThO_2	-	-	0.03	0.01	0.02
U_3O_8	11.75	5.87	9.03	3.61	4.52
Total	100.00	100.00	100.00	100.00	100.00

* added to batch over 100 wt %.

dispersive spectroscopy (SEM/EDS) using a JSM-5300 + Link ISIS unit (voltage is 25 KeV, beam current is 1 nA, probe diameter is 1 to 3 μm, dwell time is 100 s; metals, oxides and fluorides were used as standards).

Table II. Major process variables in the lab-scale tests on SB2 and SB4 wastes vitrification.

Major process variables	SB2	SB4	
Waste loading, wt %	50	40	50
Average vibration power, kW	6.4	6.8	4.5
Weight of batch (oxide mixture) fed into the cold crcible, kg	0.800	0.490	0.651
Weight of produced melt, kg	0.680	0.414	0.540
Melt surface temperature, ^0C	1320-1440	1200 ÷ 1400	1250 ÷ 1320
Glass productivity during steady-state operation, kg/hr	0.64	0.40	0.24
Specific glass productivity during steady-state operation, kg/(m^2×hr) [kg/(m^2×d)]	260 [6240]	~163 [3912]	~98 [2352]
Melting ratio during steady-state operation, kW×hr/kg	10.2	17.0	19.2

VITRIFIED PRODUCT CHARACTERIZATION

The vitrified product had black color in bulk and ranged from green to dark-green color in thin layers in transmitted light. The product solidified in the cold crucible was not uniform over the bulk (Fig. 2, *left*). It consisted of dense core and intermediate near-"skull", and "skull" areas.

The XRD pattern of the vitrified U-bearing SB2 waste produced in the cold crucible was similar to those of the U-bearing sample produced in an alumina crucible in a resistive furnace and the U-free samples produced in the large-scale cold crucible at the same waste loading (50 wt %) [2,3]. The vitrified waste product was glass-crystalline and composed of crystalline spinel magnetite type phase distributed within the glassy matrix (Fig. 2, *right*). Reflection d = 3.318 Å may be attributed to residual partly unreacted quartz (probably captured from the "skull" zone). XRD patterns of all the vitrified SB4 waste samples exhibited a "hump" typical of a vitreous state. The glass with 40 wt % waste loading sampled in the zone close to rim contained minor quartz. Traces of quartz were likely present in this sample due to unreacted quartz in the feed material. The glass with 50 wt % waste loading contained both minor quartz and traces of another crystalline phase, probably magnetite-type spinel. Content of the latter does not exceed a few volume percent.

The products with 50 wt % waste loading were investigated in more detail by SEM/EDS (Fig. 3). Both the vitrified products were composed of matrix glass and individual crystals and their aggregates distributed within. As seen from Fig. 3*c*, the amount of the crystalline phase is negligible in the core and increases towards the "skull" area. In the skull region, both large individual crystals and aggregates of fine crystals were evident. Chemical compositions of co-existing phases are given in Table III.

The chemical composition of the core glasses (1 on Fig. 3*b*, s1 on Fig. 3*d*) was rather close to the target values taking into account that the total was ~88-90 wt % and Li_2O + B_2O_3 contributed about 10-12 wt %.

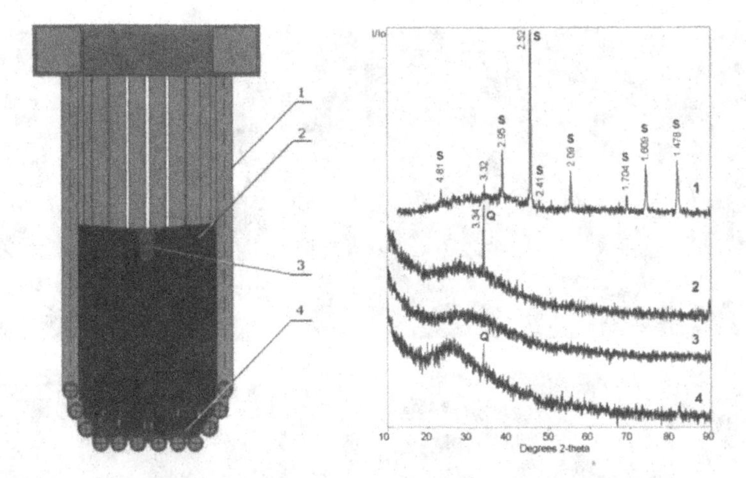

Figure 2. Cross-section of the cold crucible after cooling to room temperature (*a*, 1 - pipes forming cold crucible walls, 2 – solidified product, 3 – shrink hole, 4 – "skull" and unmelted batch) and XRD patterns (*b*) of the vitrified SB2 (1) and SB4 (2-4) wastes at 50 (1, 4) and 40 wt % (2, 3) waste loadings (2 – near-"skull" area, 3 – core).

Homogeneous glass regions (s1 on Fig. 3*d* and t2 on Fig. 3*e*) were somewhat depleted in Fe_2O_3 and other transition elements as compared with the target due to accumulation of Fe and transition metals in the spinel phase. The spinel phase in the "skull" area was strongly enriched with transition metals, especially iron (t1 on Fig. 3*e*). Fig. 3*f* also demonstrated a segregation of a second generation spinel phase crystallized from the residual melt in the cold crucible due to oversaturation of the melt with transition metal oxides. The very small size of the crystals of this second generation spinel (≤ 1 μm) made it difficult for precise determination of the chemical composition (t2 on Fig. 3*f*).

As expected, the vitreous phase was enriched with Na, Al, Si, P, S, Ca, Sr, Pb, U oxides and Cl, whereas spinel preferentially accumulated transition metals (iron group elements: Mn, Fe, Ni) as well as Cu and Zn oxides. Iron oxide content was ~67-71 wt %, and consistent with XRD data, indicated that this spinel was close to magnetite in chemical composition and structure. This was consistent with our previous data on characterization of glass and crystalline materials prepared from both U-bearing and U-free SB2 waste surrogates melted in a resistive furnace and in a cold crucible.

Uranium oxide (recalculated to U_3O_8) was present in the vitreous phases for both SB2 (~5.8-6.5 wt %) and SB4 (~4.5-4.8 wt %) vitrified wastes. The trace amounts detected in the spinel phase (~1.4-1.5 wt %) were due to beam penetration into the surrounding glass during electron probing. The same effect occurred for Na, Si, S and probably Ce.

The degree of crystallinity of the vitrified U-bearing SB2 and SB4 waste surrogates was determined from SEM data (four replicate measurements). In both the materials, the average content of the spinel phase did not exceed ~8 vol %.

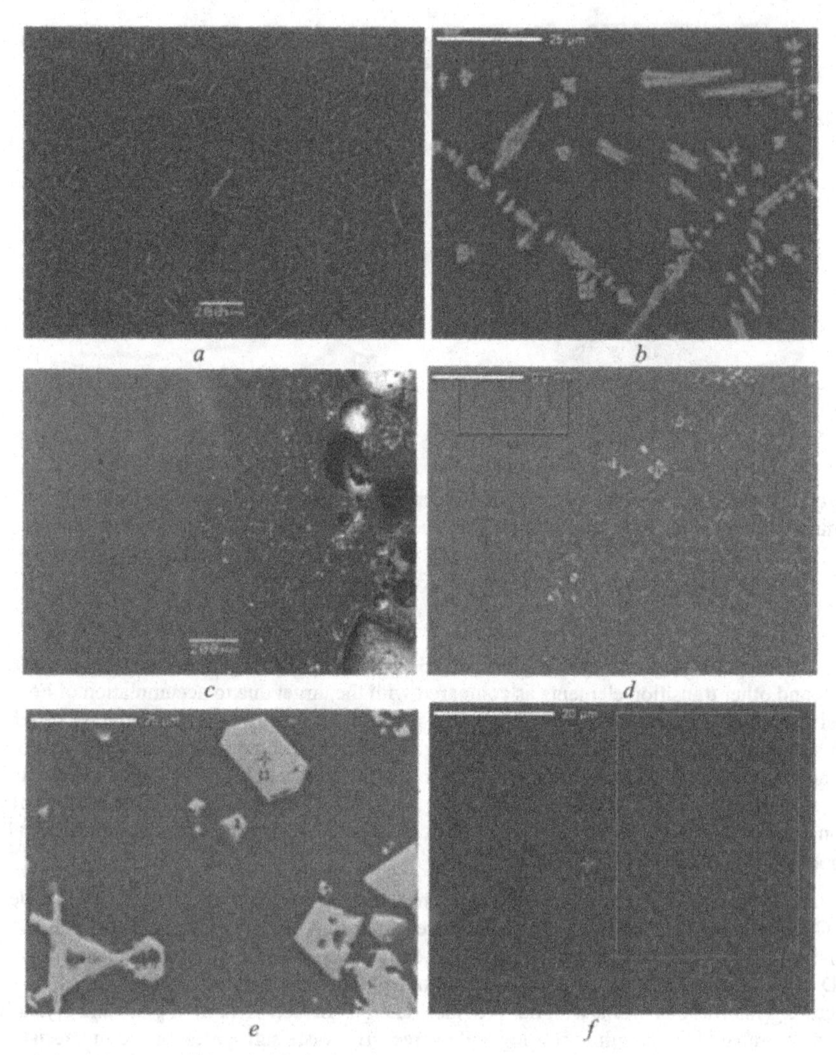

Figure 3. SEM images of the vitrified SB2 (*a*, *b*) and SB4 (*c*-*f*) wastes.

SPFT TESTING OF THE VITRIFIED SB2 WASTE

Leach resistance of the vitrified waste was estimated using a single-pass-flow-through (SPFT) procedure. The SPFT test was performed in an apparatus delivered by Pacific Northwest National Laboratory, USA, under a previous contract between US DOE and SIA Radon. The procedure based on contact between powdered specimen and continuously flowing leachant at various temperatures and pH values is described in detail elsewhere (see, for example, [6]).

Table III. Chemical Compositions of co-existing phases in the vitrified SB2 and SB4 wastes.

Oxides	Point 1 on Fig.3*b*	Point 2 on Fig.3*b*	Pts 3-5 on Fig.3*b*	Area s1 on Fig.3*d*	Point t2 on Fig.3*e*	Point t1 on Fig.3*e*	Area s1 on Fig3*f*	Point t2 on Fig.3*f*
Na_2O	12.16	11.91	4.19	8.96	8.44	0.54	8.07	8.56
MgO	0.08	<0.01	0.32	1.50	1.06	1.09	1.68	1.14
Al_2O_3	9.41	9.74	3.15	12.15	12.42	2.54	11.69	11.41
SiO_2	43.79	42.53	11.23	40.68	42.16	0.51	40.38	39.91
P_2O_5	0.27	0.15	0.01	<0.01	<0.01	<0.01	<0.01	<0.01
SO_3	0.20	0.35	0.05	0.29	0.43	0.11	0.04	0.29
Cl	0.30	0.31	0.10	-	-	-	-	-
CaO	2.33	2.19	0.47	1.22	1.27	0.1	0.86	0.84
Cr_2O_3	<0.01	<0.01	0.15	<0.01	0.08	3.47	<0.01	0.08
MnO	1.74	1.58	2.99	3.12	3.20	5.52	2.78	2.21
Fe_2O_3	11.71	12.20	69.32	14.03	12.58	71.03	14.92	21.33
NiO	0.14	0.03	6.35	0.97	0.40	13.91	0.71	2.8
CuO	0.39	0.41	0.11	<0.01	<0.01	0.02	<0.01	<0.01
ZnO	0.31	0.30	0.44	<0.01	<0.01	0.32	<0.01	<0.01
SrO	1.26	0.70	0.17	-	-	-	-	-
Cs_2O	-	-	-	0.36	0.15	<0.01	0.32	0.23
BaO	0.03	0.21	0.10	<0.01	<0.01	0.01	<0.01	<0.01
Ce_2O_3	-	-	-	0.18	0.08	0.32	<0.01	0.13
PbO	0.15	0.27	0.08	0.41	<0.01	<0.01	0.41	<0.01
U_3O_8	6.46	5.85	1.48	4.53	4.83	<0.01	3.74	3.20
Total	90.74	88.73	100.72	88.40	87.12	99.48	85.60	92.14
Phase	Glass	Glass	Spinel	Glass	Glass	Spinel	Glass with spinel	

The major target parameters of the test were as follows: pH = 2, leachant – 0.01 M HNO_3, leachant flow rate – 3 mL/hr (8.33×10^{-10} m^3/s), T = 90±2 °C, particle size – 150-315 μm, surface area – 0.01458 m^2. Variations in concentrations and leach rates of Na, Si, Fe, and U with time are shown on Fig. 4.

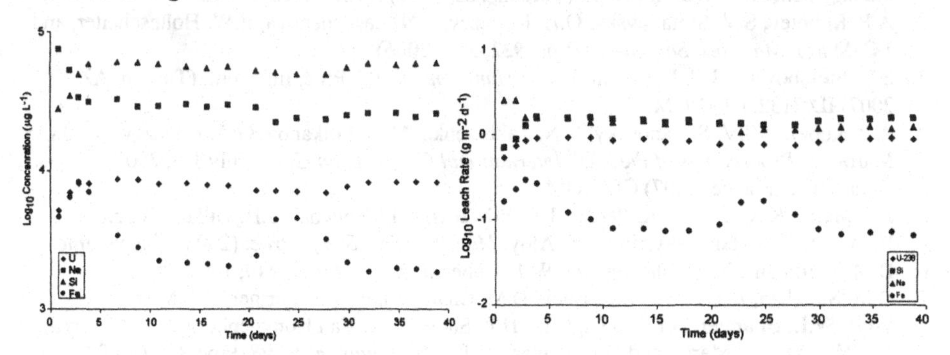

Figure 4. Plot of the elemental concentrations and leach rates versus time for SPFT test with the vitrified SB2 waste.

Average elemental leach rates were found to be 1.38, 1.42, 0.122, and 0.841 $g \times m^{-2} \times d^{-1}$ for Na, Si, Fe and U, respectively. The leach rate value for Si was comparable to the dissolution rate for Si in a lanthanide borosilicate (LaBS) glass under similar test conditions (pH = 2, T = 70 °C) [6] and somewhat higher than the leach rates of similar elements for a Pu-doped alkali borosilicate glass at 90 °C and pH = 9-12 [6]. It should be noted that these borosilicate glasses developed for Pu immobilization are considered to be durable [6,7]. A feature of Fe leaching was the increase of the Fe leach rate between 20th and 25th days of testing following by its reduction to steady-state level (Figure 4).

CONCLUSION

Vitrification of both the SB2 and SB4 HLW surrogates with borosilicate Frits 320 and 503-R4, respectively, in the lab-scale (56 mm inner diameter) copper cold crucible yielded chemically durable glassy products containing traces of unreacted quartz and minor (<10 vol %) magnetite-type spinel structure phase (at ~50 wt % waste loading). The vitreous phase was found to be enriched with Na, Al, Si, P, S, Ca, Sr, Pb, U oxides and Cl, whereas transition metals (iron group elements: Mn, Fe, Ni) as well as Cu and Zn oxides preferentially portioned to the spinel phase.

ACKNOWLEDGEMENTS

The work was performed under financial support from US DOE in the framework of collaborative work between SRNL and SIA Radon "Maximizing Waste Loading for Application to Savannah River High Level Waste". Authors are grateful to Mr. B.S. Nikonov (IGEM RAS) for SEM/EDS examination and Dr. G.A. Varlakova and I.V. Startseva for SPFT testing of the glassy products.

REFERENCES

1. S.L. Marra, R.J. O'Driscoll, T.L. Fellinger, J.W. Ray, P.M. Patel, J.E. Occhipinti, in: *Waste Management '99.* Proc. Int. Conf. (Tucson, AZ, 1999) ID 48-5. CD-ROM.
2. A.P. Kobelev, S.V. Stefanovsky, O.A. Knyazev, T.N. Lashchenova, E.W. Holtzscheiter, and J.C. Marra, *Mat. Res. Soc. Symp. Proc.* **932**, 353 (2006).
3. S.V. Stefanovsky, J.C. Marra, in: *Waste Management '07.* Proc. Int. Conf. (Tucson, AZ, 2007) ID 7132. CD-ROM.
4. A.P. Kobelev, S.V. Stefanovsky, V.N. Zakharenko, M.A. Polkanov, O.A. Knyazev, and J.C. Marra, in: *Proceedings of the XXIst International Congress on Glass*, July 1-6, 2007 (Strasbourg, France, 2007) CD-ROM.
5. J.C. Marra, K.M. Fox, D.K. Peeler, T.B. Edwards, A.L. Youchak, J.H. Gillam, Jr., J.D. Vienna, S.V. Stefanovsky, and A.S. Aloy, *Mat. Res. Soc. Symp. Proc.* (2008), this volume.
6. D.M. Wellman, J.P. Icenhower, and W.J. Weber, *J. Nucl. Mater.*, **340**, 149 (2005).
7. D.M. Strachan, A.J. Bakel, E.C. Buck, D.M. Chamberlain, J.A. Fortner, C.J. Mertz, S.F. Wolf, W.L. Bourcier, B.B. Ebbinghaus, H.F. Show, R.A. Van Konynenburg, B.P. McGrail, J.D. Vienna, J.C. Marra, and D.K. Peeler, in: *Waste Management '98.* Proc. Int. Conf. (Tucson, AZ, 1998) ID 65-08. CD-ROM.

Leaching

Mater. Res. Soc. Symp. Proc. Vol. 1107 © 2008 Materials Research Society

Immobilization of Radioactive Iodine Using AgI Vitrification Technique for the TRU Wastes Disposal: Evaluation of Leaching and Surface Properties

Tomofumi Sakuragi, Tsutomu Nishimura, Yuji Nasu, Hidekazu Asano, Kuniyoshi Hoshino[1] and Kenji Iino[1]
EBS Technology Research Project, Radioactive Waste Management Funding and Research Center, Pacific Square Tsukishima, 1-15-7, Tsukishima, Chuo City, Tokyo 104-0052, Japan
[1] Hitachi-GE Nuclear Energy, 3-1-1 Saiwai, Hitachi, Ibaraki 317-8511, Japan

ABSTRACT

Iodine-129 collected from a reprocessing plant is regarded as the dominant nuclide in terms of safety for TRU wastes disposal in Japan. AgI vitrification ($AgI-Ag_2O-P_2O_5$) is a potential iodine immobilization technique, which has the advantages of less iodine volatilization (low-temperature vitrification) and high volume reduction efficiency (approx. 1/25 the original waste volume). The iodine immobilization property can be evaluated by examining the surface condition of the AgI glass immersed in water. In this study, immersion tests have been performed on AgI glass in pure water in a 3% H_2-N_2 atmosphere at room temperature, and the surface characteristics have been examined. The thin layer (<4.3 μm) that is formed has been found to consist of AgI, which may act as a barrier, preventing leaching of glass components. The concepts behind the iodine release model have been proposed based on diffusion and the solubility of the components at the glass surfaces.

INTRODUCTION

Radioactive iodine (I-129, half-life of 1.6×10^7 years) generated during the operation of reprocessing plant in Japan is collected using an alumina-base iodine filter onto which silver is deposited and iodine adsorbs as a form of silver iodide (AgI). This filter is referred to as spent silver-sorbent and is expected to be disposed of deep underground as TRU waste [1]. For TRU disposal safety, I-129 from the spent silver-sorbent, which is treated as being released instantaneously, has the largest impact on dose, due to its long half-life and small distribution coefficient onto barrier materials. These factors imply moreover that iodine migration is easily affected by geological conditions such as groundwater flow. In case of average transmissibility is 10^{-10} m^2/s (reference case) the maximum dose is evaluated to be 2 μSv/y, whereas over 10 μSv/y is calculated when one orders larger transmissibility, 10^{-9} m^2/s (alternative case) [1]. Therefore, it has been suggested that an iodine immobilization technique over 10^5 years is feasible to reduce the maximum dose in the alternative case [2].

Several techniques have been applied to iodine immobilization [1, 2] and AgI vitrification using an $AgI-Ag_4-P_2O_7$ system, developed as a new solid electrolyte [3], is one potential technique. This candidate material was investigated for its basic properties [4, 5] and was found to have the advantages of less iodine volatilization (low-temperature vitrification) than borosilicate, more chemical stability than lead oxide or vanadate glass [6], and a high volume reduction efficiency (approximately 1/25 the original waste volume). However, long-term performance of this candidate material, which is required to have a leaching period over 10^5 years, has not yet been investigated.

This paper is intended to evaluate the leaching period for AgI glass. Leaching tests have been conducted in pure water, followed by tests to characterize the glass surface after leaching. Based on the results, the leaching mechanism is modeled.

EXPERIMENTAL

The $AgI-Ag_2O-P_2O_5$ glasses (AgI glass) used for the leaching tests were prepared from silver iodide (AgI, Kojima Chemicals Co., Ltd., purity > 99 %) and silver pyrophosphate ($Ag_4P_2O_7$, Rare Metallic Co., Ltd., purity > 99 %). After being rinsed with deionized water and dried, the AgI and $Ag_4P_2O_7$ were mixed homogeneously in a ratio of 538:462. The mixture was then melted in an alumina crucible in an electric furnace for 3 hours at 773 K. A stepwise decrease in temperature around 473 K allowed molding using a columnar molder (10 mm diameter, 10 mm height). Finally, the AgI glass was cooled at room temperature. Basic properties of the glass [4, 5] are shown in Table I.

Aqueous leaching tests were performed on the AgI glass in pure water, following the MCC-1 static low temperature test [7]. Samples were prepared and kept in a glove box purged with a 3% H_2-N_2 gas mixture at room temperature. No reducing agents were used, no pH control was performed and the solid-to-liquid mass ratio was 1:10. (S/V ratio was 0.1 cm^{-1}) Leached elements were measured using ICP-MS (Agilent 7500a), ICP-AES (Optima 3300XL) and the molybdenum blue colorimetric method for I, Ag and P, respectively. The normalized leach amount Ln (g/cm^2) for each element was obtained with the following equation:

$$Ln = \frac{C \times V}{f \times S} \qquad (1)$$

where C is the concentration in the leachant, f is the initial fraction in the glass, V is the volume of leachant, and S is the surface area of the glass. The final pH and Eh were also measured (Toko Chemical Laboratories Co., Ltd., TPX-90i).

After the leaching tests, the glass surface was characterized using XRD (Rigaku RINT2500) and SEM (Hitachi S-4200). The surface layer thickness was determined through SEM inspection at several random points on the glass section.

Table I. Basic properties of AgI glass [4, 5]

Composition	$3AgI-2Ag_2O-P_2O_5$
Melting temperature	< 773 K
Iodine content	30wt%
Density	5.92 g/cm^3
Compressive strength	90 MPa
Thermal expansion coefficient	2×10^{-5} deg^{-1}

RESULTS AND DISCUSSION

Leaching test

Table II shows the results of the leaching tests. As the immersion period is longer, the concentration of silver and phosphorus in solution increases slightly and there is little leaching of iodine. In congruent dissolution of the glass matrix, the molar ratio is as follows; Ag:P:I = 7:2:3, but it is observed that iodine dissolves incongruently with phosphorus and silver, suggesting that the dissolution mechanism is not simple.

Table II. Results of MCC- static leaching test for AgI glass

immersion period, days	pH	Eh, mV	concentration, mol/L Ag	I	P	surface layer thickness, μm
28	6.4	400	7.6×10^{-5}	N.D. ($<3.9 \times 10^{-8}$)	6.3×10^{-5}	0.68
84	6.4	407	7.7×10^{-5}	1.3×10^{-7}	6.5×10^{-5}	1 *

* Estimated value, maximum thickness 4.3 μm. See text for details.

Characterization of surface layer

Figure 1 shows SEM observation of the glass surface after 28 days of immersion in pure water. A thin layer is observed at the interface between the glass and the solution. Thicknesses are summarized in Table II. Note that due to discrepancies, the thickness after 84 days is considered to be 1 μm. The thickest point was 4.3 μm. This layer consists of AgI, confirmed by XRD patterns as shown in Fig. 2. The AgI deposition may be due to the incongruent dissolution mentioned above, and the AgI layer could be playing a barrier role in the dissolution of the inner glass matrix.

Although the tests were performed under reducing conditions (3% H_2-N_2 gas mixture), the solution values of Eh indicates oxidation conditions (400 and 407 mV), leading to the possibility that dissolution of AgI is limited even in reducing condition such as those found deep underground.

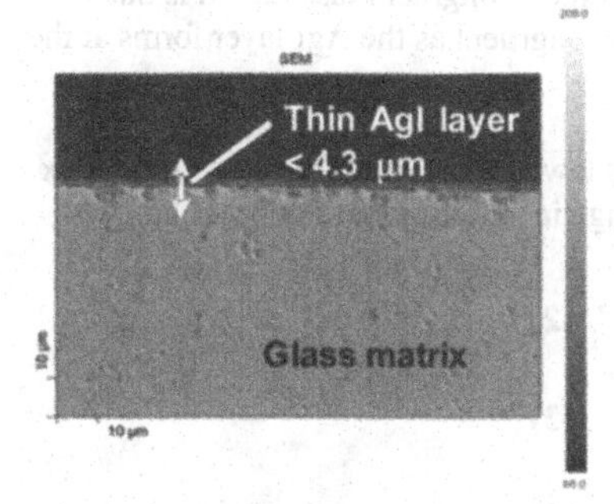

Figure 1. SEM observation at the solid-water interface of AgI glass after 28 days of immersion in pure water.

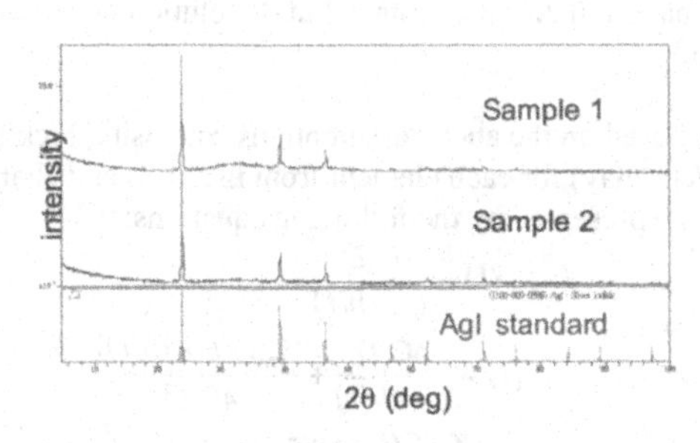

Figure 2. XRD patterns for the surface layer of AgI glass after immersions of 28 days of immersion in pure water. Comparison with an AgI standard reagent.

Leaching model

From the experimental observations, that is, the incongruent dissolution of iodine in the beginning and the formation of the surface AgI layer, possible leaching mechanisms can be proposed. Initially, the incongruent dissolution is due to silver phosphate in the glass matrix dissolving selectively, with iodine dissolution suppressed by the excess silver and forming an AgI precipitate. The generation of the AgI layer is promoted through the phosphate dissolution, but diffusion of ions through the AgI layer should be retarded. After a certain period, AgI generation is comparable to dissolution and they then dissolve congruently. The dissolution of AgI gives I^- and additional Ag^+ leaching. The model is shown schematically in Fig. 3.

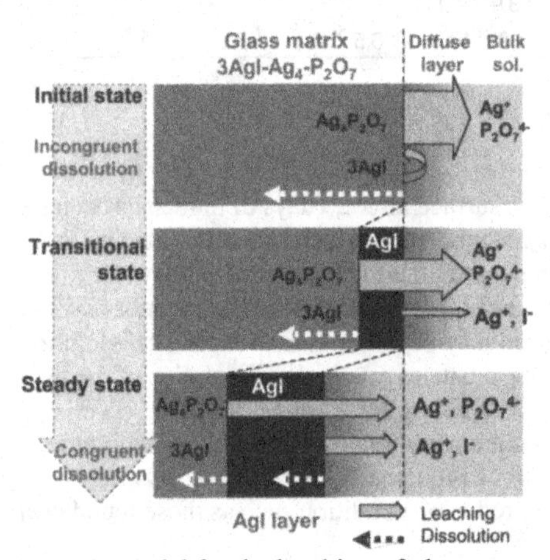

Figure 3. Model for the leaching of glass components. Initial incongruent dissolution is due to the formation of AgI precipitate, but dissolution becomes congruent as the AgI layer forms at the surface.

Based on the above assumptions, and using Fick's law and the solubility of AgI, the flux J ($mol/cm^2/day$) for each element from the glass and change in surface layer thickness h(t) with time t is represented by the following equations:

$$J_{P_2O_7^{4-}}(t) = \frac{C_0 D}{L + \tau h(t)} \quad (2)$$

$$J_{Ag^+}(t) = \frac{4C_0 D}{L + \tau h(t)} + \frac{K_{sp} D(L + \tau h(t))}{4C_0 L^2} \quad (3)$$

$$J_{I^-}(t) = \frac{K_{sp} D(L + \tau h(t))}{4C_0 L^2} \quad (4)$$

$$\frac{\partial h(t)}{\partial t} = \left\{ \frac{C_0 D}{L + \tau h(t)} - \frac{K_{sp} D(L + \tau h(t))}{12 C_0 L^2} \right\} \frac{M}{\rho} \quad (5)$$

where C_0 is the saturated concentration of phosphate at the glass surface, D is the diffusion coefficient of the ion, K_{sp} is the solubility product of AgI, L is thickness of the diffuse layer at the solid-water interface, τ is the delay factor for the AgI layer, while M and ρ represent the molecular weight and density of the glass matrix, respectively. Here, the Ag^+ concentration at the AgI layer surface is set by linear interpolation between the glass surface and the bulk solution.

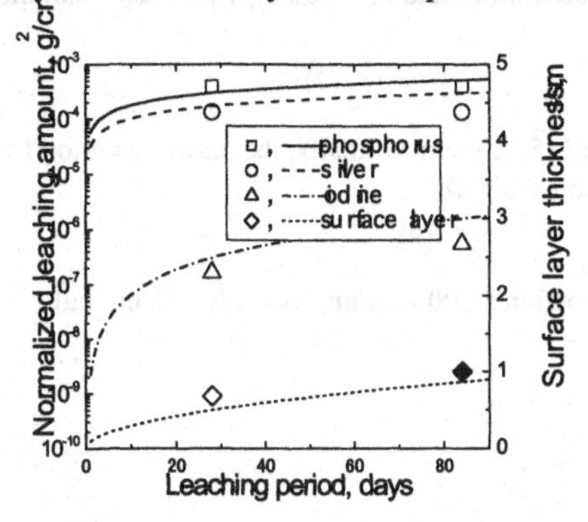

Figure 4. Fitting models (lines) to experimental results (symbols) obtained from equations (2) through (5) in the text. The triangle in 28 days represents the detection limit. The filled symbol is an estimated value, due to the distribution of the AgI layer. Used parameters are summarized in Table III.

Figure 4 shows the results of fitting the models to the experimental results for leaching (normalized leaching amount Ln) and surface layer thickness. The used and fitted parameters are presented in Table III. The values of C_0 and L are given by extrapolation from preliminary experiments. The model is in fair agreement with the experimental results. The value of K_{sp} obtained by fitting, 8.3×10^{-17} mol^2/L^2, is consistent with data in the literature [8, 9]. The delay factor τ of 50 000 indicates that the surface AgI layer acts as a strong barrier.

Table III. Input data and fitted parameters for model using equations (2) to (5) in the text

Fixed (input) data		Fitted parameters	
C_0: Saturated P conc.	6.0×10^{-6} mol/L		
D: Diffusion coeff.	2.0×10^{-5} cm^2/s	K_{sp}: solubility product	8.3×10^{-17} mol^2/L^2
L: Diffuse layer	2.0×10^{-2} cm	τ: delay factor	50 000
M: Molecular weight	1310		
ρ: Density	5.92 g/cm^3		

Estimation of leaching period

In order to evaluate the long-term performance of AgI glass, the leaching period, or the lifetime of the waste, is estimated. Normalized leaching rates, Rf (g/cm^2/day), in a transition state can be obtained by differentiation of Ln and are presented in Fig. 5, which shows that the Rf approach congruent dissolution with time. In a steady state, the congruent dissolution rate is represented as the equilibrated leaching rate Re, which can be obtained using Rf as:

$$Re = \sqrt{Rf(P) \times Rf(I)} \tag{6}$$

The calculated Re is also shown in Fig. 5. The dissolution rate of the glass, v (cm/day), can then be obtained using Re:

$$v = \frac{Re}{\rho} \tag{7}$$

where ρ is the AgI glass density shown in Table I (5.92 g/cm^3). Finally, the leaching period Lt for a waste form with initial radius r is calculated as follows:

$$Lt = \frac{r}{v} \tag{8}$$

This work assumes that the glass waste is contained in a 200 L drum, which is a 66 cm tall cylinder with a diameter of 48 cm.

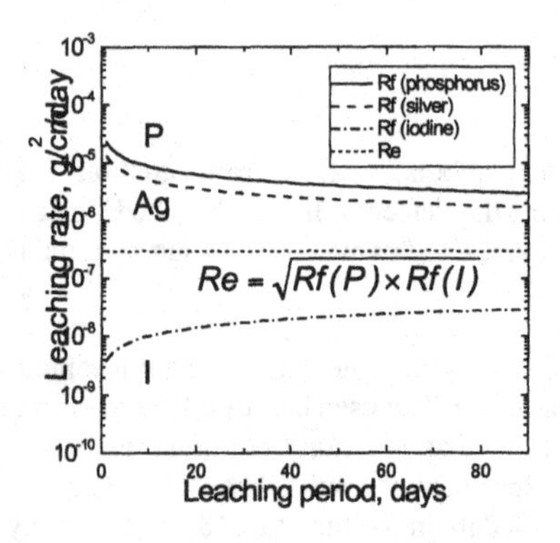

Figure 5. Change in leaching rate (Rf) with time obtained from Fig. 4. Equilibrated leaching rate (Re) is calculated with equation (6) in the text.

Table IV summarizes the leaching period estimates. The evaluated leaching period of 1.3 × 10^6 years is sufficient for our goal of over 10^5 years. AgI glass has already been shown to have properties favorable for a deep underground repository such as strong mechanical durability [4, 5] Overall, AgI vitrification is found to be a candidate technique for iodine immobilization in TRU waste disposal.

Table IV. Estimation of iodine leaching period from the assumed waste form in pure water

Assumed waste form size	Leaching rate, Re (g/cm^2/day)	Dissolution rate, v (cm/day)	Leaching period, Lt (year)
48 cm diameter cylinder	3.0×10^{-7}	5.1×10^{-8}	1.3×10^{6}

CONCLUSION

Immobilization using an AgI vitrification technique has been evaluated through leaching tests and examination of the surface properties of the glass immersed in pure water. A thin layer (<10 μm) of AgI has been found to be generated. Based on the diffusion and solubility at the glass-water interface, leaching has been modeled and the leaching period has been estimated to be 1.3×10^6 years.

For practical use of this technique, evaluation of the glass material in a wide range of ground water conditions is needed. Surface properties are expected to be more complicated in seawater than in pure water because the low solubility of AgCl ($K_{sp} \cong 10^{-10}$) causes another precipitate at the solid-water interface. Modeling the leaching mechanism in seawater is a topic for future research.

AgI vitrification, by means of immobilization with a leaching period over 10^5 years, is judged to be one technique feasible for reducing the maximum dose from TRU wastes.

ACKNOWLEDGMENT

This research is a part of the result "Development of Iodine Immobilization Technique" under a grant from the Ministry of Economy, Trade and Industry, Japan.

REFERENCES

1. FEPC and JAEA, Second Progress Report on Research and Development for TRU Waste Disposal in Japan (2007).
2. T. Nishimura, T. Sakuragi Y. Nasu, H. Asano, H. Tanabe, Proc. International Workshop on Mobile Fission and Activation Products in Nuclear Waste Disposal, La Baule, France (2007).
3. T. Minami, Y. Takuma, M. Tanaka, Electrochem. Sci. Technol. 124 (1977) .
4. H. Fujihara, T. Murase, T. Nishi, K. Noshita, T. Yoshida, M. Matsuda, Mat. Res. Soc. Symp. Proc. 556, 375 (1999).
5. K. Noshita, T. Nishi, T. Yoshida, H. Fujihara, T. Murase, Proc. The 7th International Conference on Radioactive Waste Management and Environmental Remediation (1999) .
6. T. Nishi, K. Noshita, T. Naitoh, T. Namekawa, K. Takahashi, M. Matsuda, Mat. Res. Soc. Symp. Proc. 465, 221 (1999).
7. Pacific Northwest Laboratory, MCC Workshop on Leaching of Radioactive Waste Form, NL Rep.-3318 (1980).
8. W. M. Latimer, The oxidation states of elements and their potentials in aqueous solutions, 2nd ed., Prentice Hall, New York (1952).
9. W. Stumm and J. J. Morgan, Aquatic chemistry: an introduction emphasizing chemical equilibria in natural waters, 2nd ed. Wiley & Sons, New York (1981).

Mater. Res. Soc. Symp. Proc. Vol. 1107 © 2008 Materials Research Society

Determination of the Forward Rate of Dissolution for SON68 and PAMELA Glasses in Contact With Alkaline Solutions

K.Ferrand, K.Lemmens
SCK•CEN, Boeretang 200, B-2400 Mol, Belgium

ABSTRACT

In the new Belgian disposal design, the nuclear waste glass will be surrounded by a 3 cm thick carbon steel overpack and a 70 cm thick concrete buffer. An initially high pH is expected after water intrusion in the concrete buffer and this may have an effect on the radionuclide release from the waste glass. This study was performed in order to determine the forward rate of dissolution for SON68 and PAMELA glasses (SM513 LW11 and SM539 HE 540-12), conducting dynamic tests at 30°C in contact with alkaline solutions. In these experiments, the silicon concentration in solution was determined by UV/Visible spectrophotometry according to the blue β-silicomolybdenum method. The forward rates of dissolution were quite similar for the three glasses except at the highest pH for which a slightly higher value was found for SM539 glass. For SON68 glass, a good agreement with the previously established interpolation law was observed until pH 11.5, but at higher pH, the interpolation law slightly overestimates the dissolution rate [1].

INTRODUCTION

In Belgium, the study of a preliminary reference design for the disposal of vitrified HLW and spent fuel in an underground repository described in detail in the SAFIR 2 report has revealed some uncertainties regarding the engineered barrier system performance. The main uncertainties were due to the possibility of localised corrosion or stress corrosion that might threaten the integrity of the overpack during the thermal phase [2]. Consequently, NIRAS/ONDRAF has recently selected a new reference design based on the so-called Supercontainer design. It comprises three main components: the stainless steel liner, a Portland Cement concrete and the carbon steel overpack. The canisters of HLW or spent fuel are enclosed in a carbon steel overpack with a thickness of 3 cm, preventing the contact with (near field) ground water during the thermal phase. A Portland Cement concrete of about 70 cm of thickness has been chosen for the buffer because it will provide a highly alkaline chemical environment allowing to passivate the surface of this overpack and to inhibit its corrosion. On the other hand, once the overpack is locally perforated, the high pH of the incoming water may have an impact on the lifetime of the vitrified waste [3]. Most published data and national programs are related to clayey backfill materials and few studies are reported in alkaline media [4, 5]. Hence, to evaluate the durability of the glass matrix in a such environment, a series of experiments has to be conducted. This paper presents the results of the experiments performed to determine the forward rate of dissolution for SM539 and SM513 PAMELA glasses and for the SON68 glass at 30°C in alkaline solutions. The forward dissolution rate is defined as the dissolution rate in diluted conditions, and can be considered as the maximum dissolution rate possible at the given temperature and pH.

EXPERIMENTAL PROCEDURE

Dynamic glass experiments were performed at a constant temperature of 30°C and at imposed pH values (at 25°C) of 9 ± 0.2 ; 11.5 ± 0.2 ; 13 ± 0.2 and 14 ± 0.2. The leachants were composed of ultrapure water and KOH and were placed in polypropylene bottles. The pH was adjusted at room temperature and was checked and adjusted manually for experiments conducted at pH 9 and 11.5. A 53-125 μm size fraction for SM539 HE 540-12 and SON68 glasses and a 125-250 μm size fraction for SM513 LW11 glass were obtained by milling and sieving glass fragments after crushing sample rods with a hammer. Powders were cleaned 10 times ultrasonically in deionized water and dried at 105°C. The specific surface areas were determined by BET using Kr, and were equal to 0.330 ± 0.008 $m^2.g^{-1}$ for SM539 HE 540-12, 0.247 ± 0.002 $m^2.g^{-1}$ for SON68, and 0.063 ± 0.001 $m^2.g^{-1}$, for SM513 LW11. These specific surface areas are higher than the expected values probably due to the presence of fine particles. However, as the experiments were conducted at only 30°C and had a short duration, the small particles should have no influence on the glass dissolution rate. The glass compositions are given in table I. Glass powder was placed in a PTFE cell with a volume equal to 360 mm^3 to reach rapidly the steady-state silicon concentration. The experimental set-up is given in figure 1. Samples were taken at intervals of 40 minutes during 9 hours and diluted with ultrapure water. The silicon concentration was measured by UV/Visible spectrophotometry according to the blue β-silicomolybdenum method, with an uncertainty around 7%, and then plotted as a function of time. The steady-state concentration was used to calculate the corresponding dissolution rate r (equation 1).

$$r = \frac{[Si]_{steady-state} \times \left(\frac{F}{S}\right)}{f_i} \qquad (1)$$

Where $[Si]_{steady-state}$ is the steady-state silicon concentration in the outlet solution in $g.m^{-3}$; F is the solution flow rate in $m^3.d^{-1}$; S is the surface area of the glass sample exposed to the solution in m^2 and f_i is the mass fraction of silicon in the glass.

Experiments were performed at several flow rates ranging from 0.05 to 0.8 $mL.min^{-1}$ and the resulting rates were plotted as a function of the measured silicon concentrations. The forward dissolution rate was obtained by extrapolation of the regression line to zero silicon concentration.

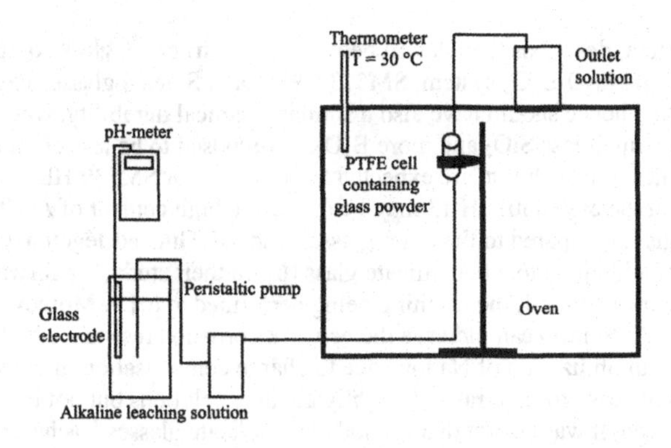

Figure 1. Experimental set-up.

Table I. Composition of SM513 and SM539 PAMELA glasses and SON68 glass.

Oxide	Glass			Oxide	Glass		
	SM513 LW11	SM539 HE 540-12	SON68		SM513 LW11	SM539 HE 540-12	SON68
SiO_2	52.15	35.273	45.48	TiO_2	4.54		
B_2O_3	13.08	25.575	14.02	MgO	2.05		
Na_2O	9.12	8.773	9.86	ZrO_2			2.65
Al_2O_3	3.61	20.237	4.91	Nd_2O_3			1.59
CaO	4.54	5.045	4.04	MoO_3			1.70
Li_2O	4.19	3.491	1.98	ZnO			2.50
Fe_2O_3	1.70		2.91	Cs_2O			1.42

RESULTS AND DISCUSSION

In the experiments conducted at pH 13 and pH 14, the pH of the outlet solutions were similar to the pH of the inlet solutions. However, for those conducted at pH 9 and pH 11.5, the pH was decreased by 2 units.

Figures 2 and 3 indicate the dissolution rates in $g.m^{-2}.d^{-1}$ as a function of the silicon concentrations in $mg.L^{-1}$ for the experiments conducted with PAMELA glasses. In the case of the experiments performed at pH 9 and 11.5 with SM513 LW11 glass, only 3 flow rates were used because a lower flow, necessary to obtain experimental data at higher silicon concentrations, could not be fixed. For SON68, a similar graph was plotted (not shown). Table II summarizes the forward rates of dissolution determined for the three glasses with the associated uncertainty (95% confidence interval), the uncertainties due to the possible evolution of the glass surface area with time are not taken into account. For each glass, the forward rate increases at higher pH , which is in agreement with the solubility of the main oxides forming the glass. Until pH 13, quite similar values are obtained for the three glasses. However, at the highest pH, the forward rate of dissolution determined for SM539 HE 540-12 glass is about 4 times higher than for the other glasses and is equal to 1.6 $g.m^{-2}.d^{-1}$.

The understanding of the corrosion resistance of borosilicate glass is based on the glass structure. The model of Dell, including results for the ternary system as well as for the binary alkali borate and alkali silicate system, describes the relation between glass structure, glass composition and glass stability. In a SiO_2-Na_2O-B_2O_3 system, SM513 LW11 and SON68 glasses have a very similar composition and hence should have also a similar chemical durability, whereas SM539 HE 540-12 glass, containing less SiO_2 and more B_2O_3, is supposed to be less durable. According to this, the forward rates of dissolution are expected to be higher for SM539 HE 540-12 glass. The fact that this is not the case until pH 13 may be due to the high content of Al_2O_3 (20.237 wt%) in the SM539 glass compared to the other glasses. Indeed, Gin and Jégou have also shown the favorable effect of adding Al to a borosilicate glass [6]. In their study, the forward rate of dissolution decreased by a factor 4, the leaching being performed at pH 9. Moreover it is known that an increase of the Al/Si ratio can increase the network corrosion resistance by formation of Si-O-Al bridges and immobilization of Na ions due to charge compensation in the vicinity of these bridges. This is the case for neutral and slightly alkaline solutions but not in strongly alkaline solutions. Indeed, it was shown that for soda lime silicate glasses leached at 100°C in a 2.5 M NaOH solution, corrosion is observed to increase with increasing Al contents due to the ease of hydrolysis of Al-O bonds and the gradual change from tetrahedral to octahedral coordination of the Al ion under these alkaline conditions [7]. We also observed this lower resistance to corrosion at pH 14, which explains the higher forward rate of dissolution for SM539 HE 540-12.

For SON68 glass, a previous study was conducted by Gin and Chouchan to determine the forward rate of dissolution at 30 °C in pure water [8]. A Soxhlet device was used and a forward rate of dissolution equal to 2.8×10^{-3} $g.m^{-2}.d^{-1}$ was determined. It has been shown that the interpolation law (equation 2) proposed by Minet [1] and validated in a range of temperature between 50-100 °C and in a range of pH between 6 and 10 can also be applied for experiments performed at 30°C.

$$r_0 (T, pH) = r_{0T0} \exp [(-A(1/T-1/T_0)] \, 10^{n0 * \max(pH-7;0)} \quad (2)$$

With $T_0 = 373$ K; $r_{0T0} = 0.89$ $g.m^{-2}.d^{-1}$; $A = 10800$; $n_0 = 0.4$

Figure 4 indicates the forward rates of dissolution for SON68 glass as a function of pH obtained in our study (noted as 'experimental data') and those calculated according to the equation 2 (noted as 'calculated data'). Until pH 11.5, an excellent agreement is observed. At pH 13, the difference is larger. For experiments conducted at the highest pH, the experimentally determined rate is 2.6 times lower than the calculated value. At this very alkaline pH, the presence of fine particles could explain this discrepancy.

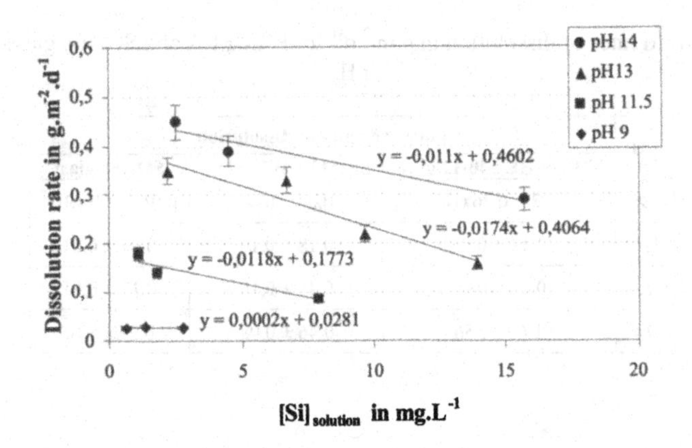

Figure 2. Dissolution rates for SM513 LW11 glass as a function of silicon concentrations.

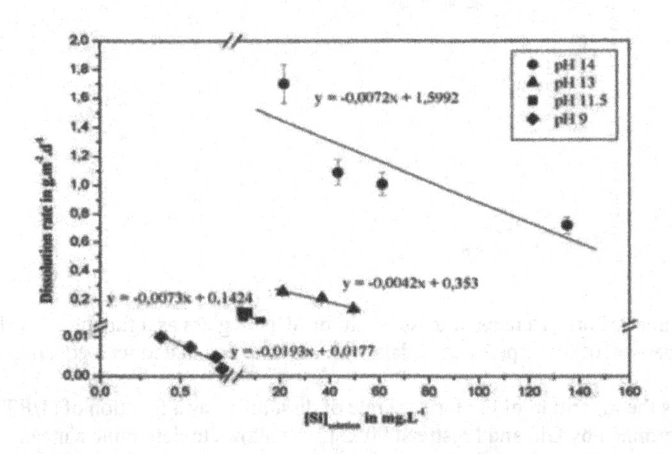

Figure 3. Dissolution rates for SM539 HE 540-12 glass as a function of silicon concentrations.

Table II. Forward rates of dissolution in $g.m^{-2}.d^{-1}$ for PAMELA and SON68 glasses at alkaline pH.

pH	Forward rates of dissolution		
	SM539 HE 540-12 glass	**SM513 LW11 glass**	**SON68 glass**
9	0.02 ± 0.26x10^{-2}	0.03 ± 0.01	0.01 ± 0.14x10^{-2}
11.5	0.14 ± 0.05	0.18 ± 0.16	0.07 ± 0.01
13	0.35 ± 0.01	0.41 ± 0.10	0.17 ± 0.01
14	1.60± 0.56	0.46 ± 0.08	0.35± 0.10

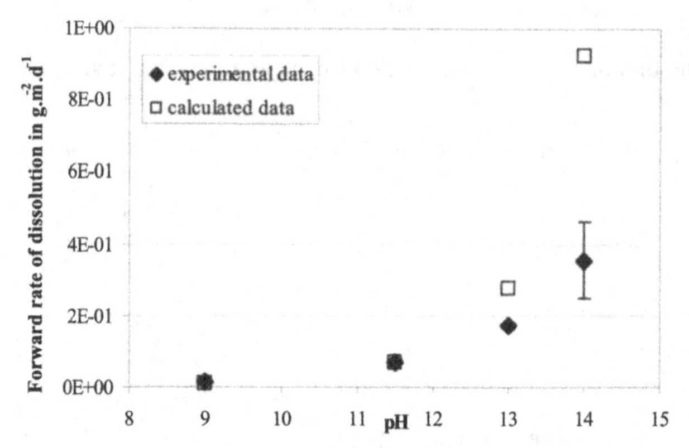

Figure 4. Forward rates of dissolution for SON68 glass as a function of pH. Comparison of the experimental data with the rates calculated with equation (2)

Figure 5 shows the logarithm of the forward rate of dissolution as a function of (1/RT) including the values determined by Gin and Mestre at 90°C [5]. It allows to determine a mean activation energy of 86 $kJ.mol^{-1}$ which is typically characteristic of a surface reaction. This result is in agreement with the activation energy of 75 ± 15 $kJ.mol^{-1}$ for SON68 at temperatures ranging from 30 to 250°C [9].

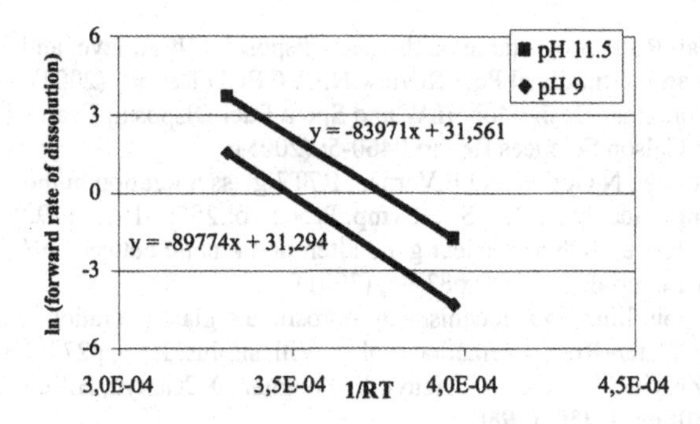

Figure 5. Determination of the activation energy for SON68 glass.

CONCLUSION

Dynamic glass leaching and quantification of silicon concentration by UV/Visible spectrophotometry using the blue β-silicomolybdenum method has allowed to determine the forward rates of dissolution of PAMELA and SON68 glasses at 30°C in contact with alkaline solutions. Regardless of the glass composition, quite similar forward rates of dissolution were determined, which increased at higher pH. The largest impact of glass composition was observed for the SM539 HE 540-12 glass at the highest pH (14). In this case, the rate was about 4 times higher than for the other glasses, probably due to the easier hydrolysis of the Al-O bonds. Rather unexpectedly , the ratio between the forward rate of dissolution obtained at pH 14 and that at pH 9 is not very high. The largest difference is observed for SON68, with a factor of 35 . For SON68 glass, it was shown that, if the previous interpolation law determined by CEA is applied for experiments conducted at high pH and used for modelling, the forward rate will be slightly overestimated for pH higher or equal than 13.

The forward rates of dissolution determined in this study have been confirmed conducting experiments with a synthetic cementitious water at pH 13.5. Indeed, similar forward rates of dissolution as those determined at pH 13 have been obtained for SM539 glass and SON68 glass.

ACKNOWLEDGMENTS

This work was supported by the Belgian Agency for Radioactive Waste and Enriched Fissile Materials (NIRAS/ONDRAF) and was performed with SON68 glass provided by CEA.

REFERENCES

1. Y. Minet; Synthèse des connaissances sur la vitesse initiale d'altération en eau pure (V_0) des verres des domaines R7T7, AVM, CEA. Conversion en fraction annuelle de verre altéré en tenant compte de la surface utile; CEA internal report; (1999).

2. SAFIR2: Belgian R&D programme on the deep disposal of high-level and long-lived radioactive waste- an International Peer Review.NEA OECD Report; (2003).
3. Belgian Supercontainer Design for HLW and Spent Fuel Disposal: Evaluation of the Reference Design; Galson Sciences Report 0460-5; (2005).
4. Z. Andriambololona, N.Godon and E.Vernaz; R7T7 glass alteration in the presence of mortar: effect of the cement grade; Mat. Res. Soc. Symp. Proc.; vol.257; p151; (1992).
5. S. Gin and J.P. Mestre; SON68 nuclear glass alteration kinetic between pH 7 and pH 11.5; Journal of Nuclear Material; Vol.295; p83-96; (2001).
6. S. Gin and C.Jégou; limiting mechanisms of borosilicate glass alteration kinetics: effect of glass composition; Water–Rock Interaction, vol. 1, Villasimius, Italy, p279; (2001).
7. G. Dunshi, H. Zhiying, Y.Yuxia, C. Hanyi, L. Zonghan, X.Xianyu, H.Ren, Y. Liqun; J.Non-Cryst.Solids; vol.80; p341-350; (1986).
8. S. Gin, J-L Chouchan; vitesse de dissolution initiale du verre inactif R7T7 de référence à la température de 30 °C; CEA internal report; (2001).
9. F.Delage and J.L. Dussossoy; R7T7 glass initial dissolution rate measurements using a high-temperature Soxhlet device; Mat.Res.Soc.Symp.Proc.; vol.212; p41-47; (1991).

Mater. Res. Soc. Symp. Proc. Vol. 1107 © 2008 Materials Research Society

Single Idealized Cracks: A Tool for Understanding Fractured Glass Block Leaching

Laure Chomat[1], Frédéric Bouyer[1], Stéphane Gin[1] and Stéphane Roux[2]
[1]CEA Marcoule, DEN/VRH/DTCD/SECM/LCLT, BP 17171, F-30207 Bagnols-sur-Cèze Cedex
[2]LMT Cachan, 61 Avenue du Président Wilson F-94 235 Cachan Cedex

ABSTRACT

Within the scope of the long term behaviour of the R7T7 glass, which is the French nuclear glass, leaching and its coupling with transport mechanisms is studied. Experiments carried out on a SON 68 glass (inactive R7T7 type glass) model cracks in static basic conditions show a strong coupling between solution transport and glass leaching, depending on crack aperture. Moreover, gravity driven convective transport was evidenced for vertical model cracks, whereas only molecular diffusion was detected for horizontal model cracks under the same alteration conditions. In addition, an original device was developed to study the influence of temperature gradients on alteration kinetics as a convective driving force. These experiments show conclusively that thermally- or gravity-induced convective flow must be taken into account, even if such convective effects have not been established experimentally in neutral condition, which is more realistic condition for geological storage. A modeling, based on a porous geochemical software (HYTEC) accounting for both chemistry and transport, has been successfully applied to describe alteration within simple silicate glass cracks. It will be extended to study SON 68 glass model cracks, and more complex fracture networks.

INTRODUCTION

In France, vitrification is the process used for the immobilization of HL radwastes. A large amount of glass with radioelement is cast into an iron canister. During the cooling process, thermal stresses (unavoidable for such meter scale canisters) lead to a multiple fragmentation and a complex crack network appears in the glass block, thereby increasing its specific surface. Considering the aim of geological disposal, this aspect may have a major influence on the long term behaviour, which must be ensured during several tens of thousand years. Indeed, when exposed to water, the glass leaching rate is proportional to the reactive surface. Moreover, glass dissolution involves complex couplings between incongruent leaching, secondary phase precipitation [1] and transport. Thus the study of the glass block alteration requires to consider separately different scales to single out the relevant elementary phenomena. Therefore, experiments were conducted on single ideal cracks in static and dynamic alteration conditions. The dynamic conditions were produced by a temperature gradient, which induced well defined flow rate. This paper presents the experimental approach designed for the study of an ideal crack and aims to highlight the different transport mechanisms and their couplings with the chemical reactions taking place during water alteration. As a perspective, the alteration modeling within cracks will be briefly presented.

EXPERIMENTS

Static alteration condition on single model crack

Most of these experiments were conducted on SON 68 glass and consisted in leaching simplified model cracks (two well polished glass pieces of 25×25 mm^2 or 25×50 mm^2 area, separated by a calibrated Teflon spacer or polyamide yarn of variable thickness) maintained in vertical or horizontal position in basic condition (NaOH 0.27 ± 0.03 mol/l) at a constant prescribed temperature of 90 ± 1 °C, inducing a pH value above 11. The use of a basic pH and high temperature enhances chemical reaction rates [2], so that alteration can be conveniently and accurately followed thanks to SEM imaging of the altered surface.

Table 1 presents the range of experiments carried out, indicating crack apertures, lengths and positions.

Table 1. Studied crack apertures and lengths according to the position imposed to the model crack (uncertainty of measurement is 10 µm (resp. 20 µm) for distances below (resp. above) 160 µm). When the upper aperture is indicated as free, only one clamping device was applied on the bottom of the vertical model crack so that the upper aperture was not ensured to be at the same bottom value.

Length (mm)		25	25	25	25	25	25	25	25	25	50
Vertical : *aperture (µm)*	***Top***	free	free	free	200	free	free	40	60	60	60
	Bottom	60	80	160	200	220	550	40	60	120	60
Horizontal : *aperture (µm)*	***Right***	40	60	80	220	550	x	110	200	x	60
	Left	40	60	80	220	550	x	80	100	x	60

In addition, control experiments were conducted on a 60 µm aperture model crack in pure water at 90 ± 1 °C to check whether conclusions derived in basic condition may be extended to neutral conditions.

Dynamic alteration conditions in model cracks : Thermoconvection experiments

The assessment of the influence of a temperature gradient, as a convective transport drive, on alteration within the model crack is studied thanks to a specific leaching device. The latter is designed to generate a convective transport between two cells filled with pure water and maintained at different controlled and regulated temperatures. The bridging between these two cells consists in two model cracks in parallel with a fixed aperture separated by a 94 mm vertical distance; consequently a convection loop is induced into these two model cracks (see figure 1). In contrast to the previous static experiment, neutral conditions were used.

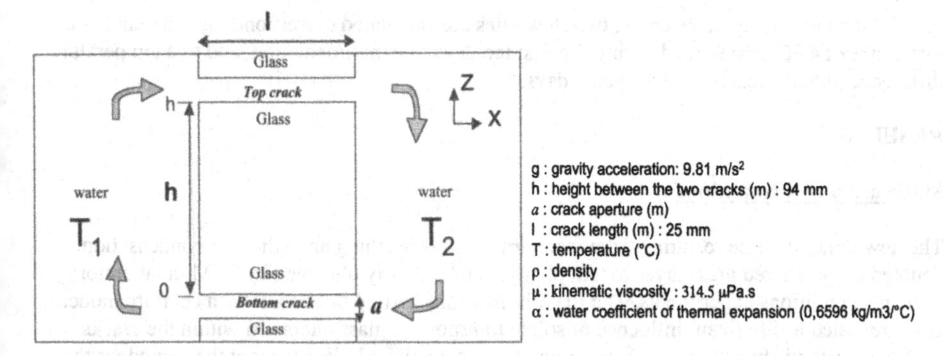

Figure 1. Schematic description of the thermo-convective device (left) with arrows representing the flow loop for $T_1 > T_2$ and table of relevant parameters.

Because of the small size of the cell and the slow flow rates induced by the thermoconvection, inertial effects can be safely neglected (small Reynolds number) so that a Stokes regime holds. Moreover, the aspect ratio of the model crack is large enough to allow for a further simplication to the so-called Reynolds or lubrication approximation. Henceforth, the flow in the crack is of Poiseuille type (parabolic velocity profile through the thickness). Finally the model crack flows can be described as obeying a simple Darcy law, whereas the pressure in the two lateral isothermal cells is essentially hydrostatics. Yet, because of the temperature difference, both cracks are subjected to a pressure difference, and hence a circulation loop sets in. The solution of this problem is elementary and the mean velocity in the cracks reads [3] (with notations explicited in Fig. 1).

$$\bar{v} = -\frac{\alpha g h a^2}{24 \mu \ell}(T_1 - T_2) \quad (1)$$

Only few degree temperature differences will be studied for the 25 millimeter length model crack to ensure that it won't change alteration mechanisms. Table 2 summarizes the different conditions studied with the thermo-convective set-up.

Table 2. Considered experimental conditions in the thermo-convective set-up.

Aperture (μm)	Time (days)	ΔT (°C)	Calculated flow rate (m.s^{-1})
60	44	4 – 5 *+/- 1*	4.6 10^{-5} / 5.8 10^{-5}
82	36	5 *+/- 1*	1.03 10^{-4}
200	28	4 *+/- 1*	5.2 10^{-4}

For a 60 μm aperture crack, two flow rates are calculated corresponding to a temperature difference of 4 °C, maintained during the first ten days of experimentation, and to a temperature difference of 5 °C, reached after twenty days.

RESULTS

Static alteration conditions

The new mineral phase resulting from incongruent glass leaching and silicon recondensation, denoted as the altered glass layer for simplicity, can be clearly observed by SEM in laboratory basic pH conditions experiments. The first observations carried out on a 60 μm aperture model crack revealed a significant influence of solute transport on glass alteration within the cracks. This conclusion is based on the observation of a thicker altered glass layer at the exit edges than in the middle of the crack. Figure 2 shows the altered layer thickness for a horizontal and vertical model crack.

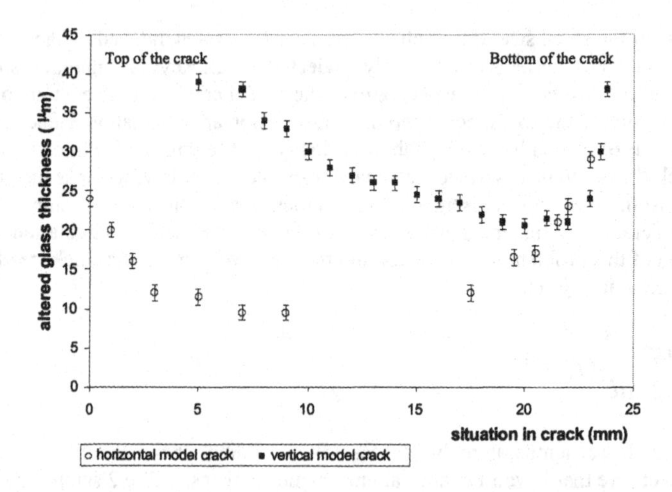

Figure 2. Altered glass layer thickness as a function of the coordinate along the crack length for a horizontal (empty symbol) and vertical (filled symbol), measured by SEM (model crack of 60 μm aperture).

The dissymmetric altered glass layer thickness profile observed for the vertical crack indicates that a convective (gravity driven flow) transport mechanism prevails, whereas for the horizontal crack a simple diffusion mechanism may hold. Indeed, the minimum alteration thickness, where local saturation is reached at an earlier time, is close to the bottom of vertical crack, indicating that convection takes place downward. All experimental observations comfort this conclusion. Figure 3 proposes a synoptic presentation of all alteration thickness profiles.

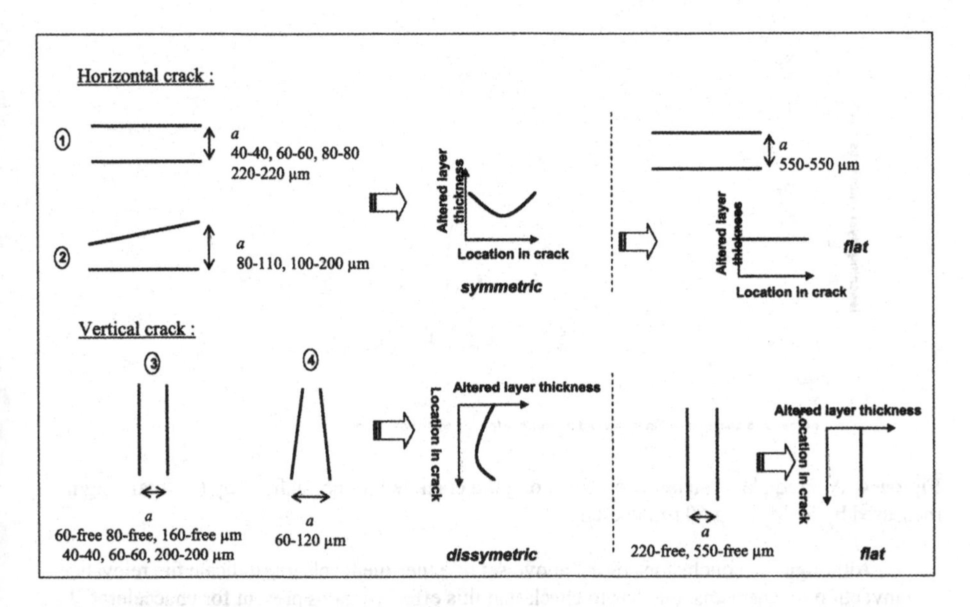

Figure 3. Synoptic figures presenting different altered glass layer profiles as a function of the aperture and orientation of the model crack.

As previously mentioned, experiments carried out in a constant aperture vertical model crack present a minimum alteration position located in the last one-tenth of the crack length.

Moreover, those different experiments show that, in basic condition, model crack aperture increase lead to flatter altered glass layer thickness profile. Consequently, the significance of the coupling between transport and chemistry depends on a form factor (*a*, see figure 1). However, model crack types labelled as 2 and 4 in figure 3 present similar shape profiles to those observed respectively for model crack types labelled as 1 and 3. Therefore, an aperture variation smaller than 100μm, inducing different local leaching conditions, cannot account for a dissymmetric profile. A convective transport appears to be the only way to understand that phenomenon. The whole range of experiments described previously demonstrates in an original way that, in vertical configuration, convective transport induced by water density gradient must be considered. As compared to gravity, temperature gradient seems to be an unlikely cause of convective transport, as experiments were performed under regulated temperature.

25 mm and 50 mm long cracks showed a similar behaviour. For the vertical model crack, the minimum alteration position lies approximately at the same relative position as shown in Figure 4.

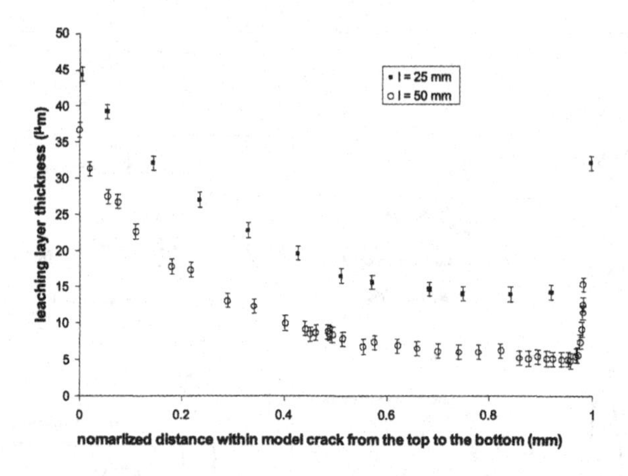

Figure 4. Altered glass thickness profile along the crack length normalized by the total length, measured by SEM, for a 60 µm aperture.

Although the conclusions of the above set of experiments clearly indicate the relevance of convection mechanisms, one has to check that this effect remains present for unaccelerated (i.e. non-basic) leaching conditions. The slow kinetics of neutral leaching did not allow us to conduct similar experiments, however a discussion of this point is proposed in Section 4.

Dynamic alteration conditions: Thermoconvection experiments

Maintaining temperature gradient with a one degree accuracy is difficult, because of room temperature variations. So, the imposed temperature gradient was chosen to be about 4 or 5 °C. Both temperature and the evolution of a KCl tracer introduced in the hotter cell were monitored. The latter was used to measure the flow rate and is compared in Table 2 to the theoretical expectation as given in Eq. 1 for the three experiments.

Table 2. Experimental data of the three different experiments, carried out in thermo-convective device, compared to the theoretical flow rate given from Eq. 1.

Aperture (µm)	Measured flow rate ($m.s^{-1}$) +/- 15 %	Calculated flow rate ($m.s^{-1}$)
60	$1.8\ 10^{-5}$	$4.6\ 10^{-5}$ / $5.8\ 10^{-5}$
82	$3.5\ 10^{-5}$	$1.03\ 10^{-4}$
200	In analysis	$5.2\ 10^{-4}$

Considering the large uncertainty of the temperature monitoring, the agreement between measured and calculated flow rates can be considered as fair, thereby validating our set-up and procedure.

After leaching, the altered glass layer thickness was intended to be quantified through SEM imaging. The low alteration thickness and the damaging of the layer during coating and polishing lead to measurement and interpretation difficulties. Yet, the analysis of the altered glass layer into the two 60 μm aperture cracks whose profile is shown in Fig. 5 suggests that convective transport plays a significant role because of the absence of typical symmetric shape of the alteration layer thickness profile characteristic of diffusion transport. However, the peculiar w-shape of these profiles is not yet fully understood. This type of experiments constitutes a very discriminating test case for modeling.

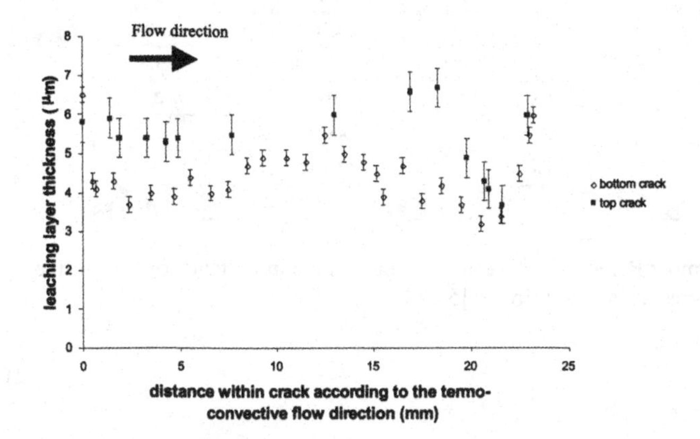

Figure 5. Measurement of the leaching layer thickness within the two cracks altered in the thermo-convective device (60 μm aperture).

DISCUSSION

Static (carried out in basic conditions) and thermo-convective (neutral) experiments show a significant influence of convective transport on the alteration kinetics within cracks, whatever the convective driving force, gravity or temperature gradient.

One may wonder why such a coupling has not, so far, been observed in the case of neutral condition and vertical crack in static conditions? The difficulty is that the altered glass layer thickness obtained after 290 alteration days is close to the SEM detection limit. Moreover, theoretical estimates of the convection flow rates are a subtle issue due to the intricate coupling between chemistry and transport. However, a model, based on simplified assumptions, was developed to get a crude order of magnitude of the involved flow rates. The free surfaces within the crack and out of the crack are considered to be altered at constant (but different) rates, deduced from leached layer thickness within the crack (respectively the average and the maximum deduced rate). This alteration releases different amounts of chemical species in the crack and in the surrounding solution (out of the crack), according to the incongruent leaching [2][4]. They induce different fluid densities within the crack and the surrounding solute, and so convection flow rate. This latter is determined considering Poiseuille law in model crack, diffusion transport negligible and stationary state. This model gives consistent values for flow rates induced by density variations in experiments carried out in basic condition (see table 3).

Table 3. Average flow rates determined by a simple model evaluating average flow rate induced by gravity in vertical model crack in basic and neutral conditions.

	aperture (μm)	average flow rate ($m.s^{-1}$)	Pe
NaOH 0.27 mol/L	40	$3.6\ 10^{-6}$	0.14
	60	$7.1\ 10^{-6}$	0.3
	80	$8.33\ 10^{-6}$	0.44
	160	$1.25\ 10^{-5}$	1.3
	200	$1.70\ 10^{-5}$	2.3
	220	$1.48\ 10^{-5}$	2.2
	550	$2.46\ 10^{-5}$	9
Pure water	60	$6.32\ 10^{-7}$	0.03

The Péclet number Pe, which characterizes the relative importance of convective versus diffusive flux, is classically defined through [5] [6]

$$\mathrm{Pe} = \frac{a\bar{v}}{D_m} \tag{2}$$

This dimensionless number is constructed from a characteristic length, which is the aperture (a) for cracks, the average flow rate ($\bar{v}$) and the diffusion coefficient ($D_m = 10^{-9}\ m^2.s^{-1}$). A Péclet number larger (resp. smaller) than 1 indicates the predominance of convection (resp. molecular diffusion).

The application of this crude model to neutral conditions for a 60 μm aperture vertical crack gives an approximate average flow rate value of one order of magnitude lower than in basic conditions. This model also shows up that, for the pure water experiment on 60 μm aperture vertical crack, the Péclet number is very low compare to 1 which indicates than the dominant transport mechanism, in this case, could be diffusion. Consequently, the flow rate in neutral conditions may be too low to be observable from the altered glass layer thickness dissymmetry. However, convection certainly takes place and should be taken into account in a more detailed modeling of the kinetics of alteration.

PERSPECTIVES ON MODELING

Resorting to numerical modeling is essential in order to address more complex and realistic geometries, and to investigate long term behavior in a faithful way, and hence understanding the interplay between chemical reaction and transport mechanism. However, glass alteration is very complex as it involves incongruent leaching of different species, secondary phase precipitation and transport. At present, no commercial software addressing both flow and chemistry is available. Since we have seen that in confined geometries such as crack, a simple lubrication approximation is sufficient, the use of a porous geochemical software (HYTEC) (whose ability to take account of the whole chemistry has been validated) appears to be the best compromise. HYTEC [7] is the association of a chemical module and a transport module. Chemical reactions are based on thermodynamic equilibriums and additional kinetic reactions, which is necessary to represent the glass alteration [8]. Transport involves diffusion and advection, but only through Darcy's law. The extension to a non porous medium (such as a realistic glass and its cracks) can be resolved within the same framework. Firstly, only surface interaction is allowed by imposing molecular diffusion values corresponding to bulk glass and choosing a high contrast of permeabilities between the glass and the solution. Secondly, the dissolution rate depends on a Monod term, which limits chemical interaction to the surface, and on a saturation term. The glass dissolution results in the precipitation of a secondary phase (so-called gel) thermodynamically more stable than the glass itself.

The flow rate, determined from the lubrication approximation, is introduced as a constant value within the geochemical model. Kinetics describing the glass alteration as well as the gel formation for a simple pure silicate glass has been investigated, in order to reproduce the alteration profile within a fracture. The introduction of a flow rate in simulation predicts dissymmetric profiles for pure silicate glass. These first results are encouraging for modeling SON 68 glass alteration, for which chemistry has to be refined.

CONCLUSION

Original experiments performed on SON 68 model cracks in static and basic alteration conditions proved that transport has a strong impact of alteration in model crack, for small aperture cracks. For vertical cracks, the dissymmetric altered glass layer thickness profile shows evidence of a gravity driven convection. The effect of a thermal gradient on leaching in model cracks was also investigated thanks to a specifically designed device. Results indicate the relevance of thermo-convection, in quantitative agreement with the theoretical expectation, which influences notably the glass alteration.

Although, vertical crack convection has only been observed in basic conditions, and not for neutral ones, its occurrence has been argued for through a crude modeling. This study shows, in an original way, that convective transport, induced either by concentration variation or by thermal gradient, can have a great influence on the alteration and hence *must* be taken into account.

Modeling with a modified geochemical model is an encouraging direction which is currently investigated for SON 68 glass, after a first validation on pure silicate glass. Another considered perspectives is the extension of the modeling and experiments to more complex geometries, i.e. crack networks involving several hundreds cracks.

REFERENCES

1. E. Vernaz, S. Gin, C. Jégou, I. Ribet, J. Nucl. Mat. 298, 27-36 (2001).
2. S. Ribet, S. Gin, J. Nucl. Mat. 324, 152-164 (2004).
3. P. M. Adler, J. F. Thovert, Fractures and Fracture Networks, Kluwer Academic Publishers (mars 1999), p. 75.
4. N. Godon, Technical Report No DIEC/2003/02 (2003).
5. V. V. Mouzenko, S Békri, J-F. Thovert and P. M. Adler, Chem. Eng. Comm. Vols 148-150, 431-464 (1996).
6. R. L. Detwiler and H. Rajaram, R. J. Glass, Water Resour. Res. 36, 1611-1625 (2000).
7. J. Van der Lee, L. De Windt, V. Lagneau and P. Goblet, Comp. & Geosc. 29, 265-275 (2003).
8. G. De combarieu, PhD thesis, Université Paris XI- ORSAY, France (2007).

Mater. Res. Soc. Symp. Proc. Vol. 1107 © 2008 Materials Research Society

Accelerated Weathering of Composite Cements Used for Immobilisation

Paulo H. R. Borges[1], Neil B. Milestone[1] and Roger E. Streatfield[2]
[1]Immobilisation Science Laboratory, Engineering Materials, University of Sheffield, Mappin St, Sheffield S13JD, UK
[2]Magnox Electric Ltd., BNG, Berkeley Centre, Berkeley, Gloucestershire GL13 9PB

ABSTRACT

Trying to estimate the long-term durability of cemented wasteforms is a difficult task as the cement matrix is a reactive medium and interactions can occur with the encapsulated waste as well as with the environment. There are few studies of samples that have been stored under controlled conditions for more than 10-15 years. Wasteforms are now being expected to last hundreds of years, much of that likely to be in some form of storage where sample integrity is important. There is also the concern that results from any long-term samples may only be indicative as both formulations and materials change with time.

This paper discusses changes in physical properties that occur in composite cements when some of the short-term accelerated procedures employed in construction testing are applied to encapsulating matrices. Changes after increased temperature of curing, wetting/drying and accelerated carbonation are discussed.

Many of the encapsulating formulations currently used are composite cements where large replacement levels of OPC with supplementary cementing materials (SCMs) such as PFA or BFS are made, primarily to reduce heat output. Accelerating the exposure conditions, either by increasing temperature or through wetting/drying has the effect of changing the hydration pattern of the composite cement by generating more hydration in the SCMs than would normally occur. The large amount of porosity that occurs because of limited hydration allows intrusion of gases and ready movement of water, so the samples subjected to accelerated testing do not appear as durable as expected if stored at ambient.

INTRODUCTION

Blended cements containing blastfurnace slag (BFS) and pulverized fly-ash (PFA) are specified for encapsulation of low level waste (LLW) and intermediate level radioactive waste (ILW) in the UK for technical and economic advantages. High levels of replacements of up to 90 wt% of BFS and 75 wt % of PFA are specified in the formulation of grouts used for waste immobilisation. This helps preventing thermal gradients and cracking, which could lead to loss of durability inside steel drums used to confine the ILW and LLW. Previous research [1-3] has shown that the early hydration in these systems containing high levels of OPC replacement is limited. The nuclear industry has studied samples that have been stored under controlled conditions for more than 10-15 years. However, the behaviour of such grouts after longer term is still unknown and the changes in the microstructure can be only assessed if some form of accelerated tests or modelling were carried out. Artificially ageing these materials by submitting them to accelerated tests could help investigations of the changes in microstructure and properties of those composites on a considerably shorter term. The reactions that are likely to occur during accelerated testing cannot be anticipated because the various processes are coupled, creating an unpredictable scenario [4]. Nevertheless, the use of accelerated testing is still recommended to evaluate the post-closure behaviour of cemented waste.

Accelerated tests used in this study compare the current formulations with other possible systems, indicating which ones are more durable to different types of attack. Characterisation techniques, such as TGA, SEM and water porosity together with physical measurements were used to determine the changes in the microstructure and physical properties of current grouts when different curing temperatures and/or chemical attacks such as carbonation and wet-dry cycling take place.

EXPERIMENTAL DETAILS

Encapsulation grouts studied

A 100% OPC sample hydrated with water to solid ratio (w/s) of 0.33 was used as a reference sample (PC). Grouts 3:1 PFA:OPC (3F1C) and 5:4 PFA:OPC (5F4C) contained 75 wt% and 55.5 wt% PFA, respectively, the upper and lower limits used in the formulation envelope of PFA/OPC grouts. Grouts 9:1 BFS:OPC (9S1C) and 3:1 BFS:OPC (3S1C) contained 90% and 75% of BFS, respectively, the upper and lower limits of the envelope for BFS/OPC grouts in the UK. The w/s used, 0.42 and 0.33, are the same as currently employed in PFA/OPC and BFS/OPC grouts, respectively.

In addition to these grouts made within the OPC replacement envelope limits, 3:2 BFS:OPC (3S2C) and 2:3 PFA:OPC (2F3C) were also studied. They contained 60% BFS and 40% PFA respectively. The last grout was 9S1CA, a modification of grout 9S1C activated with a sodium silicate solution (Na_2SiO_3). The activator solution contained 6.1% Na_2O, 7.5% SiO_2 and 86.4% H_2O (Ms = 1.2). The objective of the alkaline activation was to overcome the low degree of hydration of the 9:1 BFS:OPC grouts. In summary, the grouts studied contained 60%, 75% and 90% OPC replacement by BFS and 40%, 55.5% and 75% OPC replacement by PFA.

Pure systems were studied, containing no active wastes or simulants. The grouts were mixed using standard procedures and moulded in 30 ml screw capped plastic cylinders (22 mm diameter by 70 mm height) for preparation of samples for TGA and SEM. Cylinders of the same dimensions were moulded for apparent porosity and accelerated carbonation measurements. They were cured at 20°C and 60°C for up to 90 days. The effect of temperature on the physical properties was assessed on samples cured for 7 to 90 days. Accelerated carbonation and wet-dry cycling were performed on samples after 90 days of curing.

Characterisation of sample after increased temperature of curing

At the time of characterisation (7, 14, 28 and 90 days), one cylinder of each grout was crushed using a hammer and quenched in acetone to stop further reactions. The samples were then transferred to a vacuum desiccator to dry before preparation for SEM and TGA. SEM was performed using a JEOL JSM 6400 scanning electron microscope. Small pieces were moulded into an epoxy resin and vacuum impregnated. After setting, the moulded samples were ground and polished down to 0.25 μm with diamond paste. Before analysis, the samples were stored in vacuum desiccators to avoid carbonation. One day prior to SEM analysis, the edges of the samples were painted with silver paint to provide a conductive path and the samples were carbon coated.

For TGA, all the samples were crushed using a pestle and mortar and sieved to pass 63 µm. TGA was performed in a Perkin Elmer Pyris 1 TGA, heating from 30°C to 1000°C at 10°C per minute under nitrogen atmosphere. The weight loss from TGA and the first derivative (DTG) were used to estimate the amount of bound water as well as the calcium hydroxide and calcium carbonate content. It was assumed that all the calcium carbonates found by TGA originated from calcium hydroxide and, therefore, the CO_2 loss in the decarbonation region (dc_{loss}) was converted into a calcium hydroxide loss ($CH_{dc\text{-}loss}$) and added to the calcium hydroxide calculated from the dehydroxylation region (dh_{loss}) to give a total calcium hydroxide (CH_{total}). CH_{total} and the bound water content, Bw, were calculated, respectively, as in equations (1) and (2).

$$CH_{total} = 1.68 \cdot dc_{loss} + 4.11 \cdot dh_{loss} \quad (1)$$

$$Bw = loss\,(105^{\circ}C - 1000^{\circ}C) - dc_{loss} + 0.41 \cdot dc_{loss} \quad (2)$$

Three cylinders were used to determine the apparent porosity using the water density method. According to this test, the samples should be dried at 105°C until constant weight. The dry weight (W_1) is then recorded using a scale. After this, the samples are evacuated and allowed to saturate with water inside a desiccator. After complete saturation, the saturated weight under water (W_2) and the saturated surface dry weight (W_3) are recorded. The apparent porosity (p, in percentage) is calculated by the equation (3).

$$p = \frac{W_3 - W_1}{W_3 - W_2} \cdot 100 \quad (3)$$

Accelerated carbonation

All the grouts were subjected to accelerated carbonation using 5% CO_2, 60% relative humidity and 25±5°C. The samples used in this section were small cylinders of 22 x 45 mm. The carbonation rate at 5% CO_2 was calculated using the phenolphthalein method. The depth of carbonation was measured spraying the phenolphthalein solution onto the freshly broken surfaces. Five depths of carbonation were measured with a ruler with precision of 0.5 millimetres. The mean of these five readings is reported as the depth of carbonation at the specific age, in conjunction with the standard deviation. The carbonation rate was calculated after linear regression of the curve of depth of carbonation plotted against square root of time.

Accelerated wet-dry cycling

Prisms of 25 x 25 x 285 mm were moulded to permit measurements of length changes after accelerated wet-dry cycling. An accelerated wetting and drying tank was constructed and the complete method of use is described elsewhere [5]. The cycle established was 15 hours wetting, 30 minutes for draining the tank and 8:30 hours drying at 45°C. Length changes were measured using a comparator, from 7 days of accelerated wetting and drying onwards. The length change at any age was calculated as in equation (4), where L is the length change %; L_x is the reading of the specimen minus reading of the reference bar at age x; L_i is the initial reading of the specimen minus reading of the reference bar. Positives figures show that the sample has expanded, whereas negative figures correspond to shrinkage of the sample.

$$L = \frac{L_x - L_i}{L_i} \cdot 100 \quad (4)$$

RESULTS AND DISCUSSION

Effect of the temperature of curing on BFS/OPC grouts

Figure 1 shows the bound water content for the BFS/OPC grouts and the reference grout, PC. The numbers 20 and 60 after the grouts correspond to the temperature of curing. In general, high temperature curing promoted the hydration of BFS if the grout contained sufficient $Ca(OH)_2$ from OPC hydration, i.e 3S2C (60% BFS) and 3S1C (75% BFS). Both the non-activated and activated 9:1 BFS:OPC systems contained relatively low bound water content, as these grouts did not contain sufficient $Ca(OH)_2$ to promote the hydration of the slag after 90 days (Table I). However, high temperature promoted the hydration in 9S1CA-60.

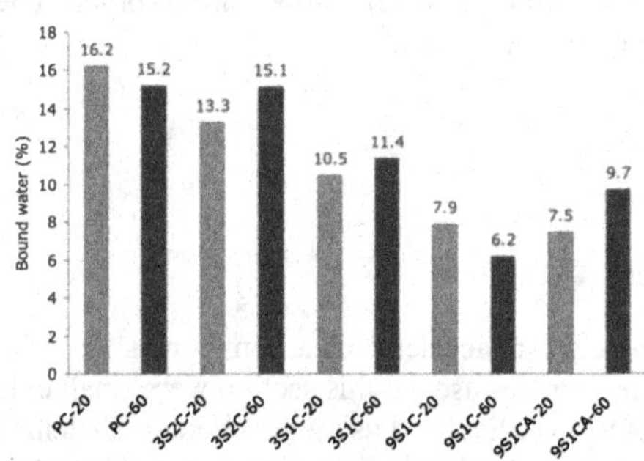

Figure 1. Bound water content for BFS/OPC grouts cured at 20°C and 60°C at 90 days

Table I. Calcium hydroxide content (%) in PC and BFS/OPC grouts at 90 days

PC-20	PC-60	3S2C-20	3S2C-60	3S1C-20	3S1C-60	9S1C-20	9S1C-60	9S1CA-20	9S1CA-60
23.6	22.8	10.1	11.2	6.6	7.2	3.9	3.9	0.9	0.6

The porosity results (Figure 2) agreed with TGA results, as grouts with higher bound water content had lower porosity. The porosity did not change for 9S1C but significantly decreased for grouts 3S2C and 3S1C when samples were cured at higher temperature (60°C).

Figure 3 shows the SEM micrographs of grouts 9S1C and 9S1CA cured at 20°C and 60°C for 90 days. The effect of high temperature curing is particularly noticeable with the 9:1 BFS:OPC grouts. The slag particles in grouts cured at 60°C showed distinct hydration rims (Figs 3b and d) compared to those at 20°C where the edges remain sharp (Figs 3a and d). The activated grout, 9S1CA, had reduced porosity only when 60°C curing was performed.

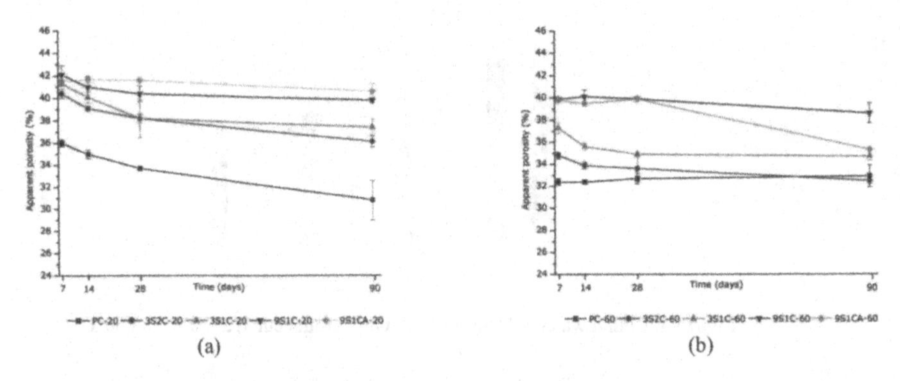

Figure 2. Apparent porosity of PC and BFS/OPC grouts cured at (a) 20°C and (b) cured at 60°C

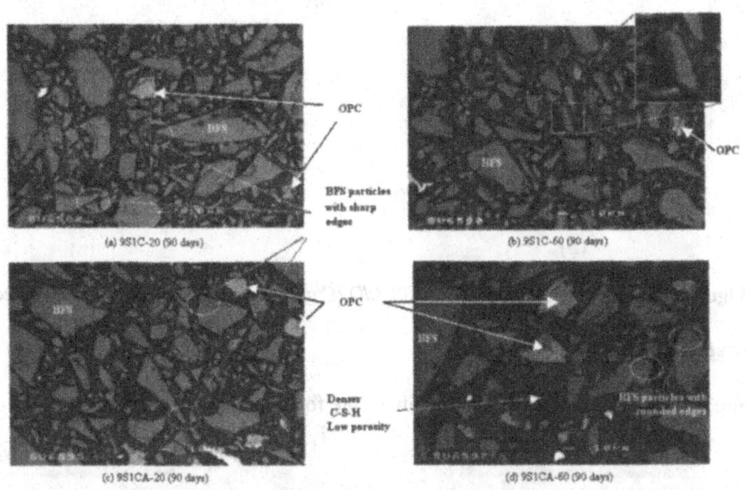

Figure 3. SEM of grouts 9S1C and 9S1CA grouts cured at 20°C and 60°C

Effect of the temperature of curing on PFA/OPC grouts

Figure 4 shows the bound water content for PFA/OPC grouts. There wss a marked reduction in bound water content as the amount of PFA replacement increased. Curing at 60°C slightly decreased the amount of hydrates formed in the PFA/OPC grouts. The apparent porosity results (Figure 5) show that generally, the effect of curing temperature caused less change than for the BFS/OPC grouts. The high water to solids ratio (0.42) in PFA/OPC grouts is responsible for the high porosity at all ages, no matter the temperature of curing.

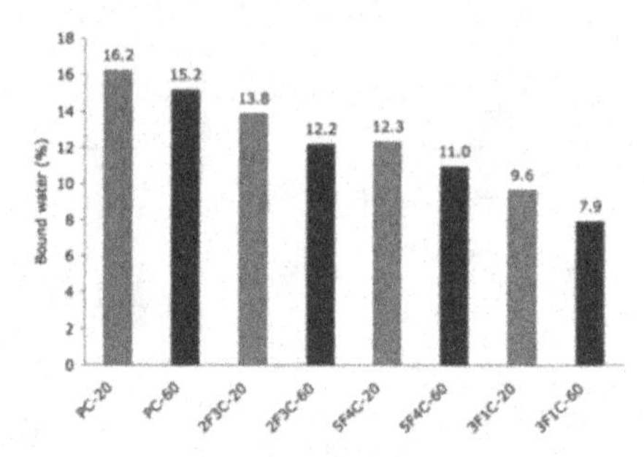

Figure 4. Bound water content for PFA/OPC grouts cured at 20°C and 60°C

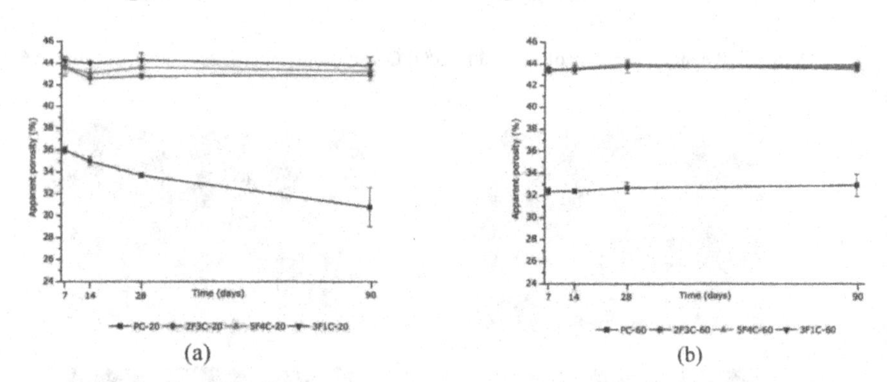

(a) (b)

Figure 5. Apparent porosity of PC and PFA/OPC grouts cured at (a) 20°C and (b) cured at 60°C

Accelerated carbonation

Figure 6 shows the accelerated carbonation rate for the grouts studied.

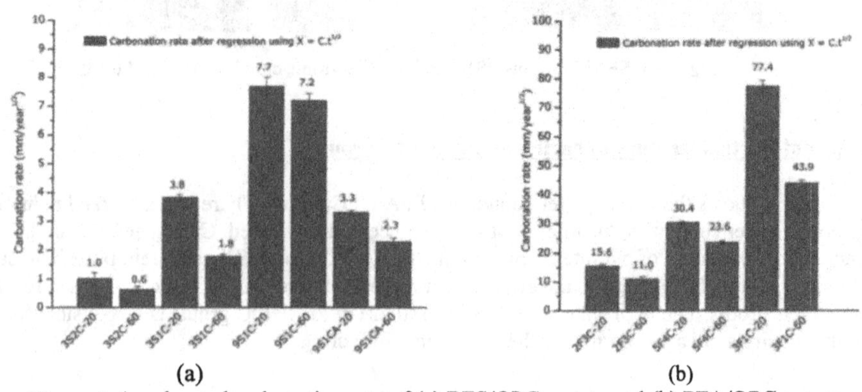

(a) (b)

Figure 6. Accelerated carbonation rate of (a) BFS/OPC grouts and (b) PFA/OPC grouts

Curing at high temperature reduced the carbonation rate for all grouts. This is in agreement with the previous results for BFS/OPC grouts, where high temperature curing reduced the overall porosity of the grouts. In the PFA/OPC matrices, high temperature curing also decreased the accelerated carbonation rates despite comparable porosities for both 20°C and 60°C cured samples. It is supposed that at 60°C a more refined or tortuous microstructure, which reduced the diffusion rate of CO_2 has been formed improving the durability of these grouts to carbonation. An alternative is that the C-S-H formed is denser and more resistant to carbonation.

In general, BFS/OPC grouts are much more resistant to carbonation than PFA/OPC grouts. The carbonation rate showed a good correlation with the amount of calcium hydroxide (CH) present in the grouts before the test (Figure 7). In the grouts with low CH amount, such as 3F1C and 9S1C, the carbonation rate is much higher than in grouts with high CH content.

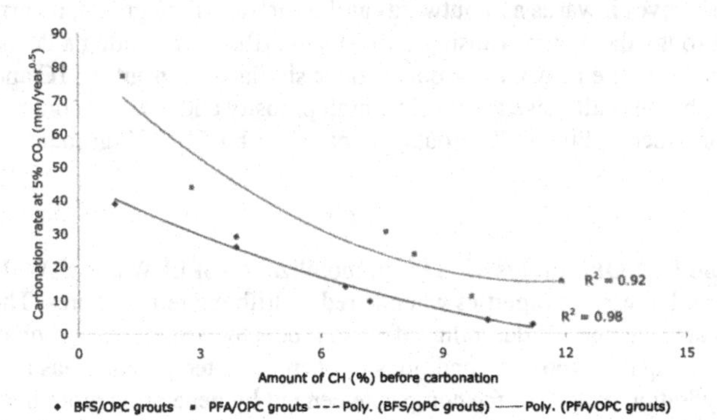

Figure 7. Correlation between accelerated carbonation rate and amount of CH before carbonation in BFS/OPC and PFA/OPC grouts

Accelerated wetting and drying

Figure 8 shows the expansion of PC, BFS/OPC and PFA/OPC grouts after wet-dry cycling.

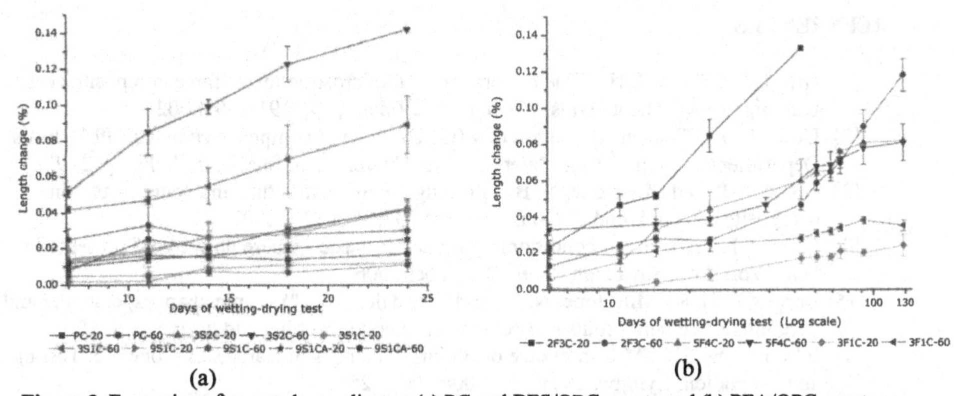

Figure 8. Expansion after wet-dry cycling on (a) PC and BFS/OPC grouts and (b) PFA/OPC grouts

The BFS/OPC samples cured at 60°C proved less resistant to wet-dry cycling than those cured at 20°C. In addition, the grouts containing 90% BFS (activated or not) were the most resistant to wet-dry cycling test and their expansions were relatively low. As the amount of BFS in the matrix decreased (75% and 60%), the expansion and cracking increased. grouts 3S2C-60 and 3S1C-60, which are relatively resistant to carbonation, expanded the most during 24 days of wet-dry cycling.

The mechanism of change in wet-dry cycling is different from that in carbonation. Expansion, cracking and loss of durability is a combination of low porosity and delayed hydration of SCM containing grouts. Low porosity *per se* does not cause any loss of durability because the PC grouts were the least porous but withstood the cycles very well. However, grouts 3S2C and 3S1C had a compact microstructure but sufficient CH present to cause further hydration of the slag. When the water moves inwards and outwards at the surface of the grouts, it carries soluble Ca^{2+} and OH^- ions through the matrix, causing $Ca(OH)_2$ to diffuse, promoting a delayed hydration of BFS near the surface. The PFA/OPC grouts behave similarly to grouts 9S1C and 9S1CA under wet-dry cycling because all these grouts have high porosity and low CH content. This explains the better performance of PFA/OPC grouts compared with BFS/OPC grouts.

CONCLUSIONS

BFS/OPC and PFA/OPC grouts used for immobilization of LLW and ILW develop different microstructure and physical properties when cured at different temperatures. These different curing regimes also change the durability of these grouts to accelerated carbonation and wet-dry cycling. Accelerating the exposure conditions by increasing temperature has the effect of changing the hydration pattern of the composite cement by generating more hydration in the SCMs than would normally occur. This could be detrimental for the durability to wet-dry cycling but advantageous in terms of durability to carbonation.

ACKNOWLEDGEMENTS

The authors would like to thank Magnox Electric for funding this PhD project.

REFERENCES

[1] Hill, J. and Sharp, J. H., "The mineralogy and microstructure of three composite cements with high replacement levels," *Cem. Conrc. Comp.*, 24,. 191-199, 2002.

[2] Utton, C. A., "The encapsulation of a $BaCO_3$ waste in composite cements," PhD thesis Department of Engineering Materials., The University of Sheffield, 2006, pp. 249.

[3] Gorce, J.-P. and Milestone, N. B., "Probing the microstructure and water phases in composite cement blends," *Cem. Con. Res.*, 37,. 310-317, 2007.

[4] Glasser, F. P., "Cement conditioning of nuclear waste - where do we go?" presented at DoE Workshop, Savannah River, December 2006.

[5] Borges, P. H. R., Milestone, N. B., and Lynsdale, C. J., "Volume changes, cracking and durability of cement grouts subjected to accelerated wetting and drying cycles,", International RILEM Conference on Volume Changes of Hardening Concrete: Testing and Mitigation, Lyngby, Denmark, 2006, 241 – 250

Mater. Res. Soc. Symp. Proc. Vol. 1107 © 2008 Materials Research Society

Mechanistic Investigation of Internal Corrosion in Nuclear Waste Containers Over Extended Time Periods

Elsie Onumonu[1] and Dr Nicholas P.C. Stevens[1]
[1] The University of Manchester, Materials Performance Centre, Sackville Street, P.O. Box 88, Manchester M60 1QD, UK.

ABSTRACT

Storage of the UK's Intermediate Level Wastes (ILW), which comprises Magnox fuel cladding, uranium and small items of equipment exposed to radiation, is currently achieved via encapsulation within cementitious grout housed in 500 litre 316L stainless steel drums. The cements used display a high pH; in such an environment many metals form surface hydroxides or oxides. Magnox reacts with free water at high pH with the liberation of hydrogen whilst undergoing corrosion to form hydroxide species.

Corrosion of Magnox cladding has previously been monitored by measuring the rate of hydrogen evolution and/or weight loss. Recent work by our group has shown impedance techniques may also be useful in monitoring early corrosion behaviour. In this project electrochemical polarisation techniques will be employed to examine the corrosion behaviour of Magnox fuel in situations where it is in electrical contact with other metals, including uranium, and hence determine how galvanic effects influence corrosion behaviour. In this paper we describe the background to such experiments along with some preliminary results.

INTRODUCTION

A Magnox reactor is a nuclear reactor fuelled by uranium metal. UF_4 (uranium tetrafluoride) is reduced to uranium metal, which is then cast and machined into rods. The uranium rod is then sealed into a can of Magnox Al80 (a magnesium alloy with small amounts of aluminium), which is often referred to as the fuel cladding. Magnox is used as fuel cladding because of its low neutron absorption cross-section.

After discharge from the nuclear reactor, fuel elements are initially stored in cooling ponds of high alkalinity to allow the decay of short lived radioactive isotopes for a minimum of 90 days. The ponds are dosed with 200 g m^{-3} of sodium hydroxide to give pH > 11.5 [1]. Such conditions are employed to prevent soluble fission products being released as a result of corrosion of the cladding [2].

Magnesium and its alloys display highly anodic corrosion potentials and corrode easily in water below pH 9. However they quickly passivate in more alkaline conditions due to the formation of a magnesium oxide/hydroxide layer [3].

$$Mg^{2+} + H_2O <=> Mg(OH)_2 \quad (1)$$

This film has a high stability due to its low solubility and rate of dissolution but can be disrupted by the action of aggressive anions, most notably chlorides, which promote localised corrosion [4].

Aluminium quickly develops a thin, dense and strongly adherent alumina layer on exposure to air which protects the metal from rapid reaction with moisture. However, in alkaline conditions, such as those encountered in cement pore waters, OH^- ions attack and dissolve this passive layer. As a result the protection of the air formed oxide layer that forms on this highly electropositive metal is diminished. In pore water the oxides thins according to the reaction:

$$Al_2O_3 + 2OH^- + 3H_2O \rightarrow 2Al(OH)_4^- \quad (2)$$

Aluminium corrodes across a thin/porous oxide layer via attack by OH^- ions and water to produce soluble hydroxyl aluminate ions and hydrogen gas.

$$Al + 2OH- + 2H_2O \rightarrow Al(OH)^-_4 + H^2 \quad (3)$$

When these metals are immersed in cooling ponds more complex electrochemical behaviour may occur should they come into galvanic contact. Stainless steel (316L) and uranium may also be involved in these galvanic interactions.

Electrochemical investigations of the metal/cement interface only provides information regarding corrosion characteristics. They cannot provide information regarding local topography. Surface features were, therefore, investigated via X-ray tomography; a technique which allows the production of 3D images.

EXPERIMENTAL DETAILS

Galvanic corrosion

Three types of materials were used in our experiments: magnesium 99.99% Mg000350 Goodfellow, the composition of the magnesium is given in table 1. Aluminium 99.999% Al1020 Advent, the composition of the aluminium is given in table 2. Stainless steel (316L) Advent, the composition of the stainless steel is given in table 3.

Element	Fe	Cr	Ni	Mo	Mn	Si	C	S	Mg
ppm	69	16-18	10-14	2-3	<2	<1	<0.03	<0.03	bal

Table 1 Composition of Mg000350 magnesium

Element	Mg	C	N	O	P	Si	others	Al
ppm	1.325	<1.0	<1.0	<1	1287	1228	<3.508	bal

Table 2 Composition of Al1020 aluminium

Element	C	Si	Mn	Ni	Cr	Mo	S	P	Fe
wt%	0.03	1	2	10-14	16-18	2-3	0.03	0.045	bal

Table 3 Composition of 316L stainless steel

Each test sample consists of 2 or 3 metal plates of cross-sectional area 5 mm x 10 mm. Each metal plate is spot welded to copper wire (providing electrical contact to the potentiostat) and the plates are encapsulated in araldite resin, as shown in Figure 1. A space of 2-3 mm was maintained between each plate so as to ensure there was no direct electrical contact between the plates. These multi-metal electrodes were polished with 800 grit SiC paper prior to use in order to largely remove any surface film and provide a reproducible starting point. In all experiments only the cross-sectional face of the metal plates were exposed to the alkaline conditions. A Standard Reference Electrode (SCE) was employed in all tests.

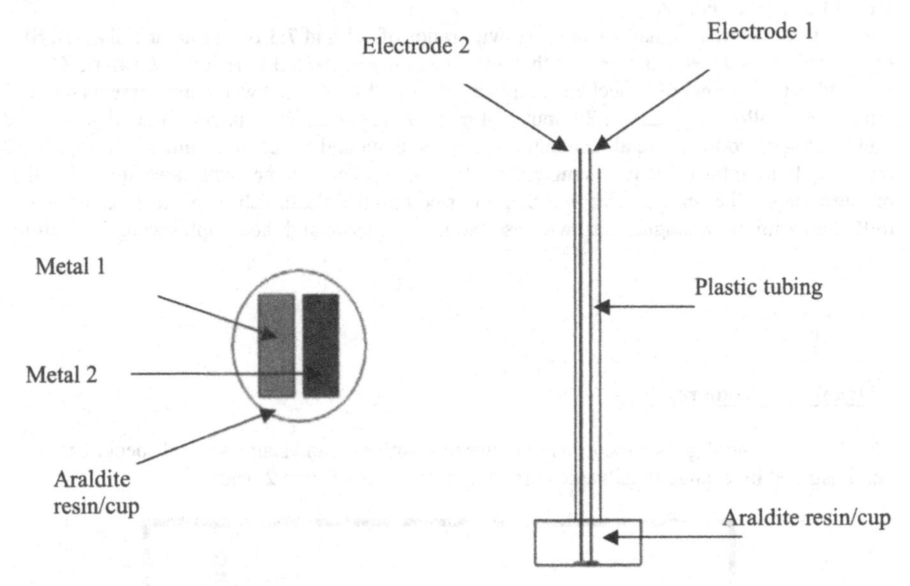

Figure 1: Schematic of 2 electrode sample

All electrochemical tests were performed in a 1000 ml electrochemical cell using 0.01M $Ca(OH)_2$ as the test electrolyte it produces alkaline conditions similar to those encountered in concrete. All solutions were prepared using de-ionised water and AR grade chemicals.

Electrochemical measurements were made using an eight channel dynamic Zero Resistance Ammeter (ZRA) manufactured by ACM instruments coupled to a Pico Logger ADC-20 digital acquisition card which allowed logging of the galvanic current as a function of time. The ZRA enables tests to be performed on multiple metal systems as it allows the maintenance of the galvanic current flowing between coupled electrodes whilst measuring the individual coupling current at each test electrode.

X-ray tomography

A set of experiments were designed to gain images of Al and Mg in cement using X-ray tomography. The metal specimens employed in this work were 5 cm lengths of 5 mm diameter rods; these specimens were not pre-treated. The tubes in which the cement was cast were see-through plastic cylinders 1 cm in diameter and 10 cm in length. The specimen was positioned in the middle of the cement.

Two different cements were used with ratios of 9:1 and 7:3 Blast Furnace Slag (BFS): Ordinary Portland Cement (OPC) with a water cement ratio of 0.33 (de-ionised water). OPC was added to the bowl of a mechanical mixer followed by BFS and water and contents mixed for 5 minutes. Following a further 2 minutes of mixing the paddle was taken off the mixer and the bowl was scraped to ensure all contents were in the bowl and the contents mixed for a further 3 minutes. In total the cement was mixed for 10 minutes in accordance with times quoted in the mixture trials. The cement paste was then inserted into the plastic tubes. Next the electrode (either aluminium or magnesium) was inserted into the paste and the samples were then left to set.

DISCUSSION

Galvanic corrosion results

The development of galvanic corrosion between aluminium and stainless steel specimens is demonstrated by change in galvanic current over time, see Figure 2 below.

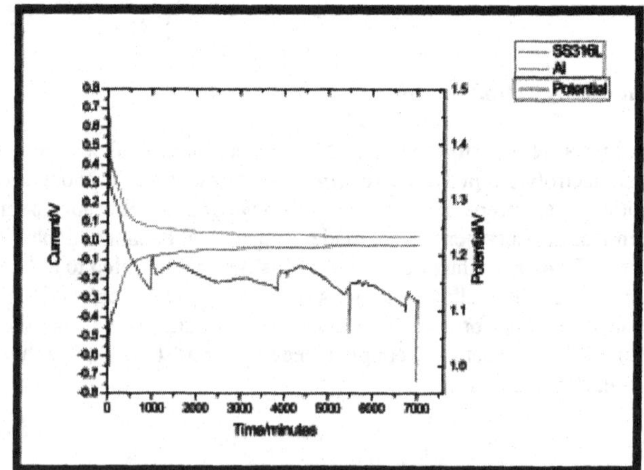

Figure 2. Change in galvanic current over time for coupled Al and SS316L in 0.01M $Ca(OH)_2$

We see an initial increase in current produced at the stainless steel 316L (SS316L) electrode which suggests a decrease in the rate of oxygen reduction at this surface. The measured current is negative and as such this material is the cathode of the couple. Conversely, the current produced at the aluminium (Al) electrode is, as expected, the anode of the couple with a current equal in magnitude but of opposite sign to that produced at the stainless steel electrode. The current displays an initial decrease suggesting a decrease in its rate of corrosion with time. These features may be explained by proposing that either the aluminium electrode develops a more protective layer with time and cannot supply electrons to the stainless steel at the rate displayed at short times, or that the stainless steel develops a layer that cannot support oxygen reduction at the initial rate which could be due to the consumption of oxygen by cathodic reaction.

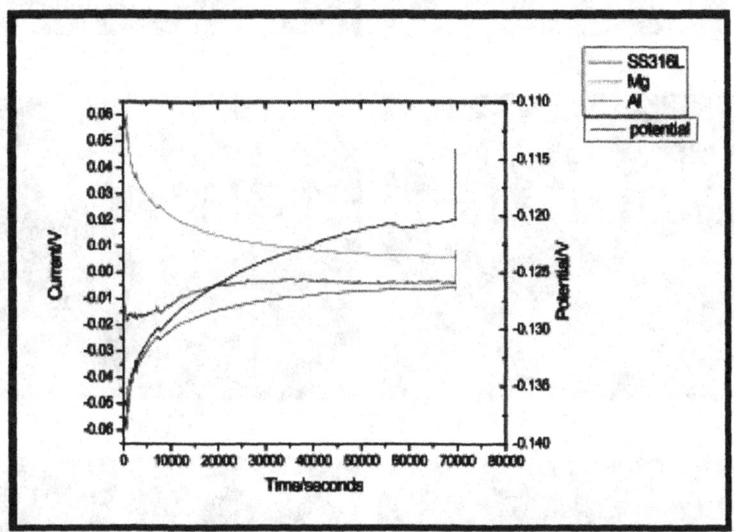

Figure 3: Changes in galvanic current over time for coupled Al, Mg and SS316L in 0.01M $Ca(OH)_2$

From Figure 3, we may conclude that Mg provides electrons to both the stainless steel and aluminium electrode during galvanic corrosion in 0.01 M $Ca(OH)_2$ and is therefore acting as the anode with electron flow from Mg to Al and SS316L, i.e. electrons are being produced by the anodic reaction and consumed, albeit at different rates at the Al and SS316L electrodes which function as cathodes. The effect of coupling these metals at this pH has been to increase the rate of metal loss (corrosion) at the Mg electrode and rate of oxygen reduction at the Al and SS electrodes; both processes are seen to decrease with time however as evidenced in Figure 3. The result is confirmed by the optical images displayed in Table I, these show a thickened white hydroxide layer on Mg while Al and SS316L are shiny and, therefore possess either none or much thinner surface layers.

Table I. Optical images representing Al, Mg and SS316L

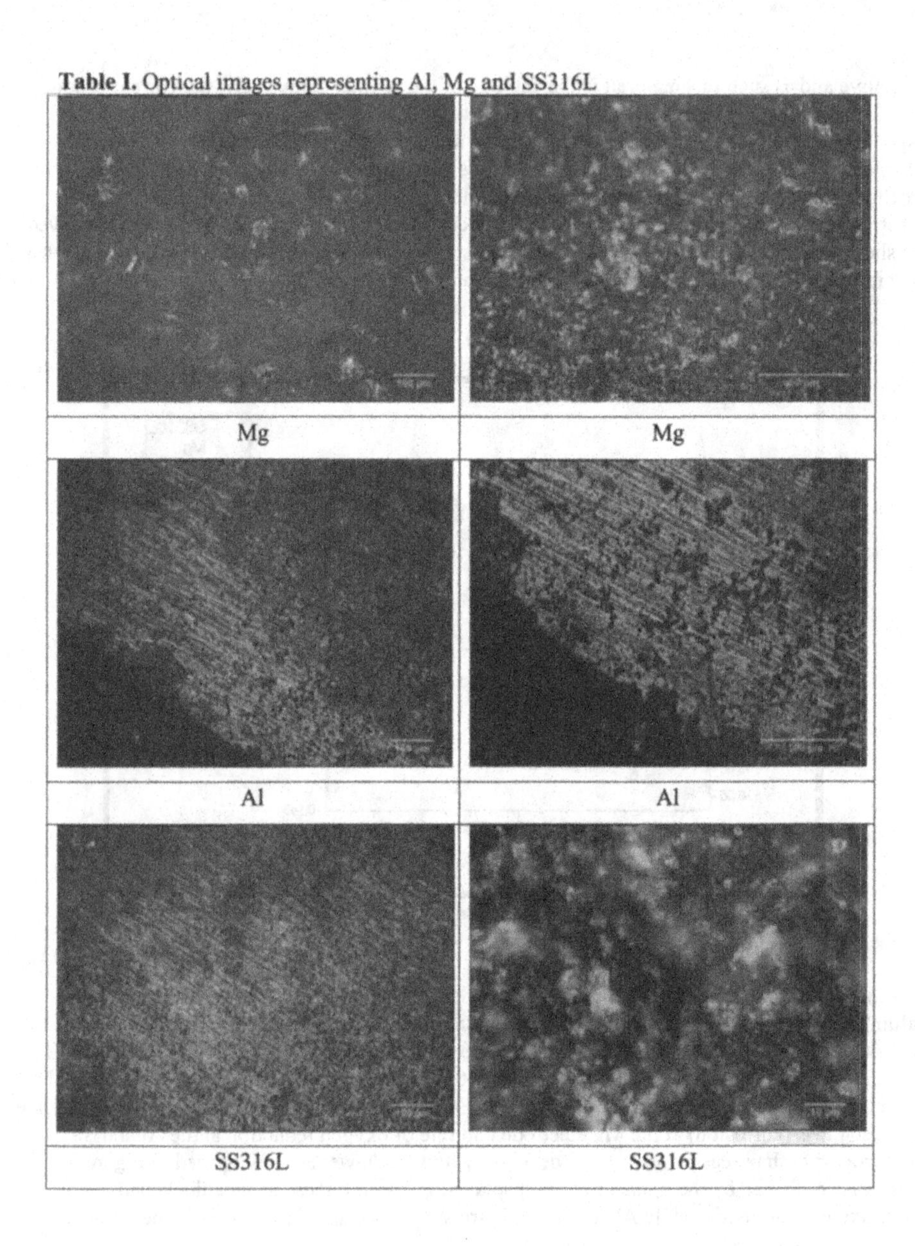

Tomography results

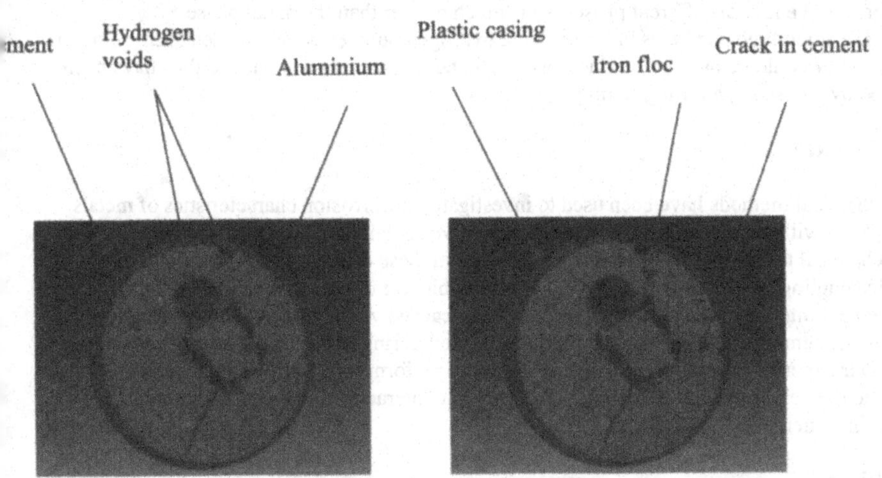

Figure 4: Amira images of aluminium in cement

In the images displayed in Figure 4 the dark circular area is the plastic in which the cement was cast. The lighter area is the cement and the bright flecks are iron flocs from BFS. The long arms are cracks in the cement and the rectangular area surrounded by the dark grey voids (produced hydrogen bubbles) is the aluminium.

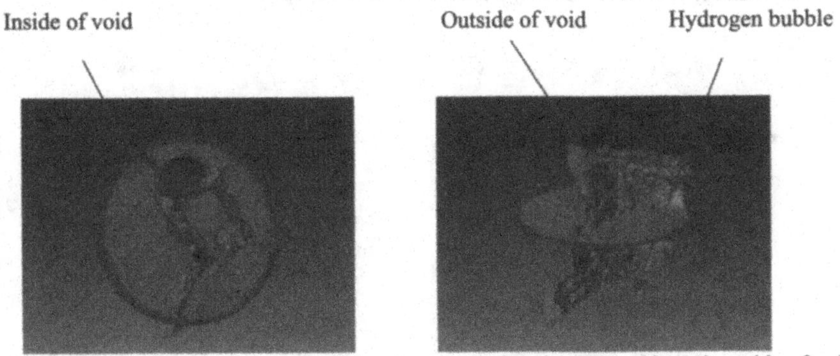

Figure 6: Amira images of aluminium in cement showing the inside and outside of voids

Using the program Amira we can segment the image according to the density of the different components (Figure 5). The inside of the void is coloured purple while the outside is gold.

There is a problem with cement and light metals, as each have densities between approximately 2 to 2.5 g cm^{-3}. Therefore the outlining must be done manually, based on noise images; the cement is comprised of a mix of different phases, so is much noisier than the metal phase.

The smooth gold spheres in Figure 5 are hydrogen bubbles which are detached from the surface. They could be bubbles produced as we did not dearate the cement and they have been produced by air escaping during setting.

CONCLUSIONS

Electrochemical methods have been used to investigate the corrosion characteristics of metals encapsulated with cement and enclosed within SS ILW containers. Data produced via electrochemical techniques reveal the likely changes in these characteristics as a result of galvanic coupling. With X-ray tomography it is possible to examine the features of the metal/cement interface. These features include those caused via hydrogen bubble formation, cracks in the cement and the corrosion products formed during exposure for known time periods

Our experimental set up provides an ability to perform long term galvanic tests at controlled oxygen levels and hence long-term galvanic interactions Magnox and uranium samples in cement may be studied.

REFERENCES

1. C. Kirby, *Corr.Sci.* **27**, 6 (1987).
2. R. Burrows, S. Harris and N.P.C. Stevens, *IChemE*, **83**, A7 (2005).
3. P.M. Bradford, B. Case, G. Dearnaley, J.F. Turner and I.S. Woolsey, *Corr. Sci*, **16**, 10 (1976).
4. M. Thomas and J. Weber, J, *Werkstoffe und Korrosion*, **19**, 10 (1968).

Ceramic Wasteforms

Mater. Res. Soc. Symp. Proc. Vol. 1107 © 2008 Materials Research Society

HIPed Tailored Ceramic Waste Forms for the Immobilisation of Cs and Sr

Melody L. Carter and E. R. Vance
Australian Nuclear Science and Technology Organisation,
Institute of Materials Engineering,
Menai, NSW 2234, Australia

ABSTRACT

This paper illustrates the benefits of hot isostatically pressed (HIPed) tailored ceramic waste forms for the immobilisation of Cs and Sr separated from spent nuclear fuel. Experimental data on microstructure and aqueous durability are presented for Cs- and Sr-bearing hollandite-rich tailored ceramics prepared with 12-18 wt% waste (on an oxide basis). MCC-1 type leach testing, on the sample containing 12 wt% waste at 90°C for 28 days revealed extremely low normalised 7-28-day Cs and Sr release rates of 0.003 and 0.004 g/m^2day respectively.

INTRODUCTION

The U. S. Advanced Fuel Cycle Initiative (AFCI) is conducting research on aqueous separations processes for the nuclear fuel cycle. This research includes development of solvent extraction processes for the separation of cesium (Cs) and strontium (Sr) from spent nuclear fuel to reduce the short-term decay heat load. The thermal load on the repository is greatly reduced if the emplacement of Cs/Sr waste is delayed until these relatively short-lived (principal half-lives of ~30 years) materials decay. The baseline process for separation of Cs and Sr from dissolved spent light water reactor fuel, as part of the Uranium Extraction Plus (UREX+) process [1], is a solvent extraction method utilising chlorinated cobalt dicarbollide and polyethylene glycol (CCD/PEG) in a phenyltrifluormethyl sulfone (FS-13) diluent. Cs and Sr are stripped from the CCD/PEG organic stream with an aqueous solution of guanidine carbonate and diethylenetriaminepentaacetic acid (DTPA) yielding a metal carbonate product solution (Table 1).

The Cs/Sr strip solution from the UREX+ process will require treatment and solidification for managed storage. Storage will require a waste form that is stable, does not produce radiolysis products (e.g., buildup of potentially explosive gases like hydrogen), has high product density (to minimise storage volume and disposal costs), and has properties that enhance heat management.

The selection and integration of appropriate waste form processing technology is an essential component of developing a waste form solution for this separated waste stream. Generically, synroc is an advanced crystalline ceramic comprised of geochemically stable natural titanate mineral analogues, some of which have immobilised uranium and thorium in the natural environment for many millions of years [2]. However synroc can take various forms depending on its specific use and can be tailored to immobilise particular components in the high level waste (HLW) by incorporating them into the crystal structure of the mineral analogue phases.

The principal advantage of the synroc-C ceramic, targeted to PUREX HLW, is that the waste ions are incorporated in durable titanate mineral phases (HLW loading can be varied between 0 and 35 wt.%), which are considerably more insoluble in water (at least two orders of magnitude) than the silicates and phosphates in supercalcine ceramics [3], as well as borosilicate glass [2]. The main titanate minerals in synroc-C are hollandite ($BaAl_2Ti_6O_{16}$), zirconolite

($CaZrTi_2O_7$) and perovskite ($CaTiO_3$). Zirconolite and perovskite are the major immobilisation hosts for long-lived actinides such as plutonium (Pu) and the rare earths, whereas perovskite mainly immobilises Sr. Hollandite principally immobilises Cs, along with rubidium (Rb) and barium (Ba).

Hart *et al.* [4] designed a hollandite-rich waste form composition consolidated by hot isostatic pressing at 1200 °C at 200 MPa. The waste form consisted of 70 wt% hollandite ($Ba_{0.8}Cs_{0.4}(Ti_{1.5}Al_{0.5})Ti_6O_{16}$) + 20 wt% perovskite $Ca_{0.79}Sr_{0.21}TiO_3$ + 10 wt% rutile, with 5 wt% Cs nominally substituted in the hollandite and 2.5 wt% of Sr substituted in the perovskite. Hart *et al.* [4] also demonstrated that up to 10 wt% Cs could be added to synroc-B, the standard synroc-C precursor [2], and that this could be hot-pressed to form a durable product.

Carter *et al.* [5] studied the aqueous durability of Rb-doped Synroc and hollandite. Rb hollandite and Synroc containing a modified Purex-type high level nuclear waste were produced by the alkoxide/nitrate route. The samples were hot-pressed at 21 MPa, with and without 2 wt% Ti metal added, for 2-10 hours at temperatures between 1200-1250°C in a graphite die. The samples were leached at 90°C and showed the Ba, Cs and Rb leach rates to be broadly similar to those of reference grade synroc-C [2].

In the current work we investigated immobilisation of the separated Sr and Cs waste streams resulting from the waste streams generated during separations in advanced aqueous reprocessing. Two hollandite-rich formulations were developed to immobilise the feed stream (see Table I). The first formulation, utilising higher Cs/Sr waste loadings than workers had investigated previously [4, 6, 7] was designed as an academic exercise with the Cs, Ba and Rb to be immobilised by hollandite and the Sr to be immobilised by perovskite. Following the examination of the sample produced by this formulation, a second formulation was developed where the Cs, Ba, Rb and Sr were targeted to all be immobilised by the hollandite.

EXPERIMENTAL DETAILS

The initial formulation was targeted as 84.5 wt% $Ba_{0.5}Cs_{0.55}Rb_{0.14}Al_{1.69}Ti_{6.31}O_{16}$, 5.5 wt% $SrTiO_3$ and 10 wt% TiO_2 , corresponding to a 21 wt% waste loading on an oxide basis. The second sample was designed to contain 85 wt% $Ba_{0.65}Sr_{0.16}Cs_{0.31}Rb_{0.08}Ti^{3+}{}_{0.5}Al_{1.51}Ti_{5.99}O_{16}$ and 15 wt% TiO_2 (waste loading of 12 wt%). In this second formulation Ba was added in excess of its availability in the waste feed. Table II lists the compositions on an oxide basis.

CCD/PEG Feed (Table I) was used in the following experiments. The organic materials were not added, as they would all be removed during the calcination step (see below). The sample was produced from nitrates (Ba, Sr, Cs and Rb) and alkoxides of Al and Ti. After drying, the powder was calcined in air at 750°C for 1 hour. The powder was ball milled prior to HIPing. Two wt% of Ti metal to reduce some of the Ti^{4+} to Ti^{3+}, was added to the powder before HIPing. The samples were sealed in stainless steel cans and HIPing was carried out at 1275°C/30MPa/1h under argon. Ti^{3+} assists the hollandite to accommodate the Cs and Rb at the required levels.

Table I. Projected composition of Cs/Sr solidification feed streams from the separation process.

Component	CCD/PEG Feed
Guanidine carbonate (g/L)	100
DTPA (g/L)	20
Cesium (g/L)	0.30
Strontium (g/L)	0.10
Barium (g/L)	0.28
Rubidium (g/L)	0.05

A JEOL JSM6400 scanning electron microscope (SEM) equipped with a Noran Voyager energy-dispersive spectroscopy system (EDS) was operated at 15 keV for microstructural work.

Table II. Hollandite-rich composition

21 wt% waste loading		12 wt % waste loading	
Oxide	Wt%	Oxide	Wt%
BaO	8.57	BaO	11.4
Cs_2O	8.66	Cs_2O	4.95
Rb_2O	1.46	Rb_2O	0.85
SrO	3.13	SrO	1.84
Al_2O_3	6.63	Al_2O_3	8.75
TiO_2	71.55	TiO_2	72.18

The X-ray diffraction (XRD) was carried out using a Philips PW 1050 instrument using Cu Kαradiation. Leach testing was carried out using the both the Product Consistency Test (PCT-B) [8] and a modified MCC-1 [9] leach test. The PCT-B protocol involved crushing the samples and sieving them to obtain particles 75-150mm in diameter (100-200 mesh). The particles were washed in non-polar cyclohexane to remove the fines instead of water, to prevent pre-leaching of Cs that would influence the leach results, and 1g samples were leached in 10 ml of water at 90°C for 1 and 7 days. Specimens for the MCC-1 test were polished (~ 0.25 μm finish) disks (8x8x2 mm) and were leached using a modified version of the ASTM C – 1998 standard [9], in which the leachates were completely replaced with fresh water at the end of each time interval (1, 7, and 28 days) to avoid steady-state conditions being reached. An aliquot of each leachate was analysed for elemental releases.

RESULTS AND DISCUSSION

Examination by XRD and SEM of the first sample containing a waste loading of 21 wt% revealed the sample had not formed the phases predicted in the design, though hollandite was the major phase. Figure 1 shows the back scattered electron image. The sample consisted of hollandite, partly oxidised Ti metal relics, a small amount of Al_2O_3, a small amount of $SrAl_{12}O_{19}$ and a Cs rich phase which occurred as submicron grains which were too small to be analysed in the SEM. The EDS analysis revealed that the Sr had not formed perovskite as would be predicted from the results of previous workers [2-4] but had entered the hollandite as well as forming the $SrAl_{12}O_{19}$. With Sr entering the hollandite structure the entire system became deficient in Ti due to the fact that in the original design, every mole of Sr required 1 mole of Ti to form perovskite but when the Sr entered the hollandite the ratio was ~1:6 for Sr:Ti. The overdemand by hollandite for Ti destabilised the system allowing the formation of the Cs-rich phase. Following this work the second formulation was developed to incorporate the Sr in the hollandite, reducing the waste loading to a more practical (from the aspect of radiogenic heat production [4]) level of 12 wt%.

The XRD pattern of the hollandite-rich ceramic containing 12 wt% waste after HIPing showed peaks corresponding to hollandite and rutile, in agreement with expectations from the overall design of the starting composition.

The scanning electron microscope (SEM) analysis showed that Cs, Ba, Rb and Sr all entered the hollandite phase as predicted in the design. The backscattered SEM image (Figure 2) of a polished surface showed the sample to consist of hollandite, rutile and a small amount of Al_2O_3. The presence of alumina indicates that the redox environment in the HIP can was slightly more reducing than anticipated, insofar as some of the Al targeted to the B site of the hollandite was displaced by Ti^{3+}. This has no detrimental effect on the hollandite rich ceramic waste form (see below).

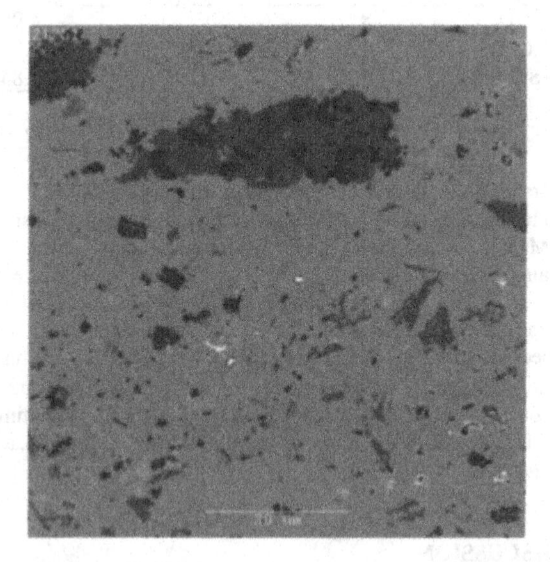

Figure 1. Backscattered electron image of sample containing 21 wt% waste. The sample consists of a hollandite matrix (light gray), Ti relics (dark gray), Al_2O_3 (black), $SrAl_{12}O_{19}$ (mid-grey) and a Cs rich phase (white).

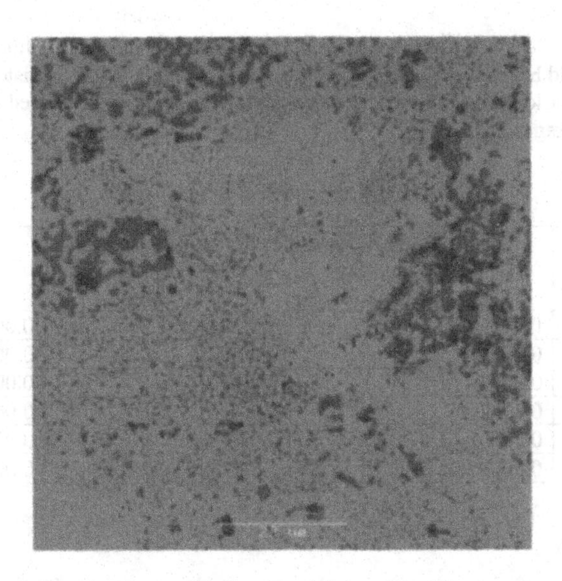

Figure 2. Backscattered electron image of hollandite-rich ceramic. The sample consists of a hollandite matrix (light gray), TiO_2 (mid gray) and Al_2O_3 (black).

The 1 and 7 –day PCT leach results showed (Table III) the normalised extractions of all waste elements to be below 0.7g/L, corresponding to 0.08 g/m^2/d, assuming a density of ~4.2 g/cm^3 for the hollandite-rich monolith. This result is in line with the Cs leach rate for hollandite-rich ceramics [6]. Note that the normalised PCT-B leachate concentrations for the reference EA glass for Na, Li and B are 13-16 g/L and the extractions of all elements in the hollandite-rich ceramic are well below these limits.

Table III. PCT-B results for 12 wt% waste loading composition.

Element	Normalised Concentration (g/L) 1 day	Normalised Concentration (g/L) 7 days
Al	0.004	0.05
Ba	0.05	0.07
Cs	0.6	0.7
Rb	0.02	0.3
Sr	0.1	0.2
Ti	0.000002	0.00002

Errors on all elements were < 7%.

The MCC-1 leach results are presented in Table IV. We note that the 7-day PCT result agrees quite well with the MCC-1 datum even though the SA/V ratios in the two forms of leach test differ by a factor of ~100. For ultimate disposal of this waste form, it may be thought that storage for ~ 300 years might allow the form to qualify as low-level waste, but the ongoing activity from the decay of ^{135}Cs would need to be taken into account: additional shielding of the waste form could be necessary to reduce the contact dose to a value consistent with the specifications for low-level waste. Such considerations are also influenced by the size of the waste form packages.

Table IV. MCC-1 results for 12 wt% waste loading composition.

Element	Normalised Release Rate ($g/m^2/day$) 0-1 days	Normalised Release Rate ($g/m^2/day$) 1-7 days	Normalised Release Rate ($g/m^2/day$) 7-28 days
Al	0.021(1)	0.004(1)	0.001(1)
Ba	0.058(3)	0.0075(1)	0.0016(3)
Cs	0.28(4)	0.016(1)	0.0032(6)
Rb	0.11(1)	0.011(1)	0.0023(4)
Sr	0.093(6)	0.010(1)	0.004(3)
Ti	0.00023(1)	0.000040(1)	0.00001(1)

CONCLUSION

These experiments confirm that immobilisation of the separated cesium and strontium waste stream from advanced fuel reprocessing in a flexible hollandite-rich ceramic is technically a very feasible process, resulting in a dense monolith with excellent aqueous durability. MCC-1 type leach testing, on the sample containing 12 wt% waste at 90°C for 28 days revealed extremely low normalised 7-28-day Cs and Sr release rates of 0.003 and 0.004 g/m^2day respectively. The use of hot isostatic press technology also avoided any potential cesium loss during the hot consolidation step. In addition, the ability to tailor the waste form design chemistry in combination with the flexible hot isostatic press process enables a common synroc processing line to immobilise a range of different waste streams from the proposed advanced recycling initiatives.

ACKNOWLEDGEMENTS

We wish to thank K. Olufson and P. Yee for carrying out the leach testing, and T. Eddowes for carrying out the HIPing of the samples.

REFERENCES

1. G. F. Vandegrift, M. C. Regalbuto, S. B. Aase, H. A. Arafat, A. J. Bakel, D. L. Bowers, J. P. Byrnes, M. A. Clark, J. W. Emergy, J. R. Falkenberg, A. V. Gelis, L. D. Hafenrichter, R. A. Leonard, C. Pereira, K. J. Quigley, Y. Tsai, M. H. Vander Pol, and J. J. Laidler, WM'04 Conference, *'Lab-Scale Demonstration of the UREX+ Process'* February 29, – March 4, 2004, Tucson, AZ.
2. A. E. Ringwood, S. E. Kesson, N. G. Ware, W. Hibberson and A. Major, *Nature*, **278**, 219 (1979).

3. E. R. Vance, *J. Austr. Ceram.* Soc. **38,** [1], 48 (2002).
4. K. P. Hart, E. R. Vance, R. A. Day, B. D. Begg, and P. J. Angel, in *Scientific Basis for Nuclear Waste Management XIX*, ed. W. M. Murphy and D. A. Knecht (Mater. Res. Soc. Symp. Proc. **412**, Pittsburgh, PA, 1996) pp. 281-287.
5. M. L. Carter, E. R. Vance, G. R. Lumpkin, and G. R. Loi, M. L. Carter, E. R. Vance, G. R. Lumpkin, and G. R. Loi, in *Scientific Basis for Nuclear Waste Management XXIV*, ed. K. P. Hart and G. R. Lumpkin (Mater. Res. Soc. Symp. Proc. **663**, Pittsburgh, PA, 2001) pp. 381-388.
6. M. L. Carter, E. R. Vance, and H. Li, in *Scientific Basis for Nuclear Waste Management XXVII*, ed. V. M. Oversby and L. O. Werme, (Mater. Res. Soc. Symp. Proc. **807**, Pittsburgh, PA, 2004) pp. 249-254.
7. M. L. Carter, E. R. Vance, and H. Li, in *Environmental Issues and Waste Management Technologies in the Ceramic and Nuclear Industries IX,* ed. J. Vienna and D.Spearing. (Ceram. Trans. **155,** Indianapolis, IN, 2004) pp. 21-30.
8. PCT is based on the ASTM Designation: C 1285-02 Standard Test Methods for Determining Chemical Durability of Nuclear, Hazardous, and Mixed Waste Glasses and Multiphase Glass Ceramics: The Product Consistency Test (PCT). (2002)
9. ASTM C 1220 - 98. "Standard Test Method for Static Leaching of Monolithic Waste Forms for Disposal of Radioactive Waste". ASTM International. (1998).

Mater. Res. Soc. Symp. Proc. Vol. 1107 © 2008 Materials Research Society

Neutron and Resonant X-ray Diffraction Studies of Zirconolite-2M

Karl R. Whittle [1,2], Katherine L. Smith[1], Neil C. Hyatt[2], and Gregory R. Lumpkin[1]

[1] Institute of Materials Engineering, ANSTO, Private Mail Bag 1, Menai 2234, NSW, Australia
[2] Department of Engineering Materials, University of Sheffield, Sheffield, S1 3JD, UK

ABSTRACT

Zirconolite (nominally $CaZrTi_2O_7$) is a constituent phase of potential waste forms for the safe immobilisation of actinide wastes. Structural studies of such materials provide important information about cation ordering, lattice parameters, and strain effects, and provide input into the modelling of alpha decay damage and the development of future wasteform designs. A suite of zirconolites based on the replacement of Ti with Nb and Fe has been studied using high resolution neutron diffraction and resonant X-ray diffraction to determine the degree of disorder across the available cation sites. Resonant X-ray diffraction is a unique method which allows the location of certain cations to be determined accurately by taking advantage of the change in scattering power close to an absorption edge (e.g., Nb-K and Zr-K). Using standard X-ray diffraction alone this is not possible and there is little scattering difference between Nb and Zr.

Raman spectroscopy and measured lattice parameters have shown that the exchange of Ti with Nb and Fe has a non-linear effect on the unit cell dimensions and Raman peak positions while retaining the 2M polytype. Mössbauer spectroscopy has shown that the Fe preferentially fills the Ti split (C2) site. The results from this study provide a more complete picture of the cation order-disorder problem and are generally consistent with the behaviour of lattice parameters across the series.

INTRODUCTION

Zirconolites [1-5], based on $CaZrTi_2O_7$ are an important class of accessory minerals which are able to accommodate a wide range cations within the structure. The chemical composition of natural zirconolite can vary extensively, with the main substitutions involving lanthanides (Ln), actinides (Act), Nb, and Fe. Zirconolite is one of the systems proposed to safely immobilise highly radioactive actinide waste. In natural samples, lanthanides and actinides often partially replace the Ca^{2+}cations, with charge balance maintained by partial replacement of Ti^{4+} by lower charged cations, such as Al^{3+}, Mg^{2+} and Fe^{3+}.

The zirconolite structure can be considered as a pyrochlore derivative formed by the contraction of the structure perpendicular to one of the pyrochlore (111) planes, resulting in a layered structure consisting of alternating HTB and Ca/Zr layers parallel to (001) in the new monoclinic space group C2/c, example images are shown in Figure 1. There is also more ordering of the cations in zirconolite compared to pyrochlore, giving a general formula of ABC_2X_7, where A = Ca, Ln, Act; B = Zr, Hf, Ln, Act; and C = Ti, Zr, Nb, Fe, Mg, Al, W, and other minor elements. The C site consists of three different sites, two octahedral sites (C1, and C3) which make up the HTB motif and an unusual, five coordinated, split site lying on an off centred position within the six membered ring (C2). This latter site replaces one of the A site cation positions in the parent pyrochlore structure. Several polytypes are possible in zirconolite, depending on how the HTB and Ca/Zr layers are stacked upon one another. In addition to monoclinic 2M (C2/c) zirconolite studied here, hexagonal 3T, and monoclinic 4M (C2/c) polytypes have been identified and structurally refined.

We have previously studied the cation distribution[3,5] within a zirconolite-2M by continuous replacement of Ti^{4+} by Fe^{3+} and Nb^{5+}, in the manner $CaZrTi_{2-2x}Fe_xNb_xO_7$ where x = 0 to 1 in steps of 0.1 using synchrotron X-ray diffraction. In this work, the cation locations are investigated using a coupled Rietveld refinement of neutron and resonant X-ray diffraction data.

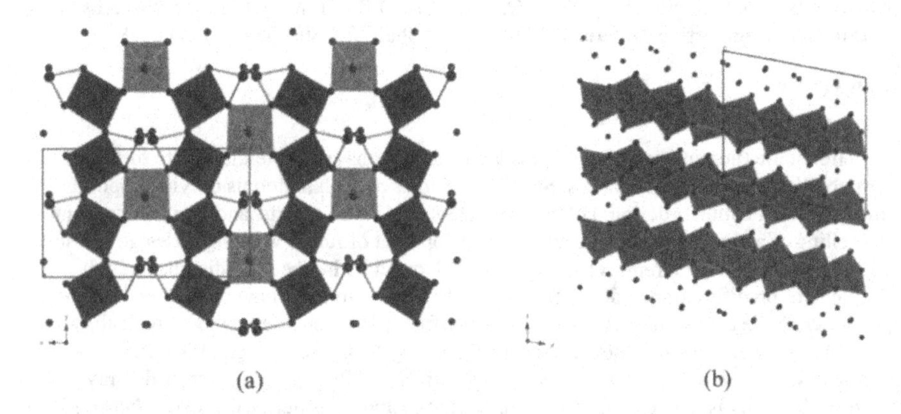

(a) (b)

Figure 1. Crystal structure images showing (a) the HTB layers containing the Ti-O polyhedra, and (b) the CaO_8 and ZrO_7 polyhedral layers in $CaZrTi_2O_7$.

RESONANT X-RAY DIFFRACTION BACKGROUND

The accepted formula for the intensity of X-ray diffraction peaks is shown below:

$$F_{HKL} = \sum_{i=1}^{n}\sum_{j=1}^{m} k_i f_{ij} e^{-2\pi i(Hxi+Kyi+Lzi)} . T_{ij}^{iso} \quad (1)$$

where (HKL) is the reflection, (x,y,z) the atomic co-ordinates, k_i the occupancy, f_{ij} the scattering power, and T_{ij} thermal vibration parameters. The scattering power for an element is different for neutrons and X-rays. Essentially in neutron diffraction the scattering power is determined by the isotope of the element under investigation and is relatively insensitive to energy and angle. The X-ray scattering power is determined predominantly by the electron number of the element under investigation (due to X-rays being scattered by orbital electrons). The X-ray scattering power for any element is shown below:

$$f(Q,\omega) = f^{\circ}(Q) + f'(\omega) + if''(\omega) \quad (2)$$

where Q is a unit related to angle ($4\pi\sin\theta/\lambda$) and ω is the energy of the X-rays. In a laboratory diffractometer the wavelength is usually fixed, thus there is no energy variance and the scattering power is related entirely to the angle of measurement. If the energy of the incident X-rays is changed then the X-ray scattering of a specific element changes. The largest changes are seen when the energy of the incident X-rays is close to an absorption edge. Examples of the change in f' and f'' for the Zr K-edge are shown in Figure 2a. The change in scattering power can be seen in the recorded data, shown in Figure 2b. Such changes once quantified, can be used in Rietveld refinements to model a unit-cell that can be reconciled with the observed data.

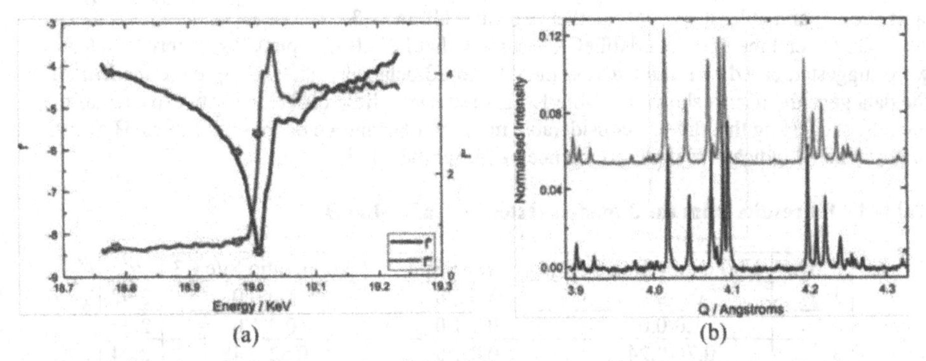

(a) (b)

Figure 2. (a) f' and f'' measured for the Zr K-edge in these samples. The asterisks indicate those energies/wavelengths of X-rays used in this analysis. (b) Plots of normalised data collected at 17.9974 keV (bottom) and 18.0192 keV (top), a region showing an observed difference is indicated.

EXPERIMENTAL

The samples used in this study were those prepared and examined in previous studies[3,5]. Time-of-flight powder neutron diffraction data were collected using the HRPD diffractometer at the UK pulsed spallation neutron source ISIS, Rutherford Appleton Laboratory [6,7]. For each composition studied, between 1 g and 10 g of powdered sample were loaded into a thin-walled, cylindrical vanadium sample can and diffraction data collected for between 185 and 1500 μAh of integrated beam current, depending on sample quantity. Summed and normalised data collected in the two available detectors located at 90° and 168° incident to the incoming flux over the time-of-flight range ~50 to 130 ms (corresponding to a *d*-spacing range of ~1 to 3.2 Å, and 0.8 to 3.2 Å respectively) were analysed

High resolution X-ray diffraction data were collected using the ID31 diffractometer at the European Synchrotron Radiation Facility. Data were collected at the energies, 17.912, 17.990, and 18.015 KeV (near the Zr K-edge), and 18.988, and 19.002 KeV (near the Nb K-edge). A third measurement at the Nb K-edge was undertaken but due to absorption from the Zr component the resultant data was poor and unreliable. The changes in both f' and f'' were determined using a Kramers-Kronig transformation of measured sample absorption across the Zr and Nb K-edges. The exact wavelengths, and thus the correct f' and f'', were determined using a silicon standard (NBS640c) to determine one wavelength at each edge, and then use this wavelength to determine the unit cell size. Since the unit cell size does not change as a function of wavelength, this can be used to correctly determine the wavelength of the measurement. Coupled refinement using multiple data sets were carried out using GSAS with EXPGUI toolkit [8,9].

RESULTS AND DISCUSSION

The data once collected were analysed using models previously suggested using Mössbauer spectroscopy [3,5]. Mössbauer spectroscopy indicated the Fe preferentially occupied the split site C2, and at least one of the other two sites, C1 and C3.

In the end member $CaZrNbFeO_7$, assuming that Fe fills the split C2 site, three models were suggested based on mixing across the C1 and C3 octahedral sites. Coupled refinement of the data gave the results shown in Table 1. There was very little difference between each of the models, suggesting that there is considerable mixing of Nb and Fe on these two sites. However, the best model indicates a slight preference of Nb for the C1 site.

Table 1 - Fit results from the 3 models tested on $CaZrNbFeO_7$

Model	Nb/Fe ratio Site C1	Nb/Fe ratio Site C2	Nb/Fe ratio Site C3	χ^2
1	0.5/0.5	0.0/1.0	1.0/0.0	2.92
2	1.0/0.0	0.0/1.0	0.0/1.0	2.95
3	0.76/0.24	0.0/1.0	0.52/0.48	2.74

When the results for $CaZrTiNb_{1/2}Fe_{1/2}O_7$ were analysed it was determined there were at least 10 models for cation ordering available! Four examples are listed below in Table II.

Table II - Fit results for 4 test models for data collected on $CaZrTiNb_{1/2}Fe_{1/2}O_7$

Model	Nb/Fe/Ti C1	Nb/Fe/Ti C2	Nb/Fe/Ti C3	χ^2
1	0.0/0.17/0.83	0/0.66/0.34	1.0/0.0/0.0	5.21
2	0.17/0.07/0.76	0/0.66/0.34	0.66/0.2/0.14	4.30
3	0.1/0.17/0.73	0.14/0.66/0.2	0.66/0.0/0.34	3.85
4	0.17/0.17/0.66	0.0/0.66/0.34	0.66/0.0/0.34	3.74

At this point in the refinement it became obvious that even though there was extra information from the resonant X-ray diffraction coupled with the neutron diffraction, further information would greatly aid the deduction of the cation ordering. Such information came from the previously collected Mössbauer spectroscopy. As outlined previously the Mössbauer spectroscopy 3,5 showed evidence of at least 1 extra cation site where the Fe could reside. Surveying the literature it was found that a mineral, Sapphirine-2M, had a closely related structure with similar values for recorded Mössbauer spectra, values shown in Table III.

The data from the Mössbauer spectra showed that the recorded data for these samples agreed with the data recorded from sapphirine-2M. However, there was still uncertainty over the exact site (C1 or C3) that was occupied. In order to determine the exact site the asymmetry of the octahedra were calculated and compared with the values for sapphirine-2M, shown in Table IV.

As can be seen in Table IV, the octahedral site that agrees best with the results previously reported for sapphirine-2M is C1, i.e. it is the most asymmetric. Using this extra information it was possible to reduce the number of potential models for the location of the cations within the structure. To this end the occupancy of the Fe was fixed on the C1 and C2 sites using fractions determined from the Mossbauer results, e.g. in the system $CaZrTiNb_{1/2}Fe_{1/2}O_7$, C1 was set to be 17% while C2 was set to 33%, remembering that this site can only be half-filled due to the spatial proximity of the split sites.

Table III. Results from the analysis of recorded Mössbauer spectra. The ratio of the 5 co-ordinate site to the 6 co-ordinate site was ~ 2:1 in the zirconolite. This value was used in setting the occupancy of the Fe on the C2 site. The upper/lower lines for $CaZrTi_{1.2}Fe_{0.4}Nb_{0.4}O_7$ within the table refer to the different crystal sites within the structure.

System	Isomer shift/mm s^{-1}	Quadrupolar Split/mm s^{-1}	Co-ordination
$CaZrTi_{1.2}Fe_{0.4}Nb_{0.4}O_7$	0.26	2.35	5
	0.34	0.74	6
Sapphirine-2M	0.30	0.76	6

Table IV. Results from the asymmetry calculations for the octahedral in zircnolite-2M and sapphirine-2M. The upper/lower lines for each sample within the table refer to the different crystal sites within the structure.

System	Site	Angular Variance	Bond Length Variance
$CaZrTi_{1.2}Fe_{0.4}Nb_{0.4}O_7$	C1	63.95	1.61
	C3	60.71	0.45
Sapphirine-2M	1	69.75	1.53
	2	61.65	2.55

The first model to be tested was one whereby the Nb^{5+} and Ti^{4+} were fixed in their fractional occupancies. This model immediately refined to a lower value than those previously tested and provided a basis for a further refinement, results shown in Table 5. The second refinement based on this methodology involved the refinement of the occupancies of the Nb^{5+} and Ti^{4+} across these two sites. Such a refinement reduced the observed χ^2 significantly, values shown in Table V, a graphical example of the refinement is shown in Figure 3.

Table V. The occupancies for the three cation sites using the new model based on Mossbauer data.

Model	Nb/Fe/Ti C1	Nb/Fe/Ti C2	Nb/Fe/Ti C3	χ^2
Nb/Ti Fixed	0.25/0.17/0.58	0/0.66/0.34	0.5/0.0/0.5	3.35
Nb/Ti Refined	0.36/0.17/0.47	0/0.66/0.34	0.27/0.0/0.73	3.12

With the extra data from the Mossbauer spectroscopy and allowing the site refinement of Nb and Ti over the C1 and C3 sites, it can be seen that there is also a slight preference of Nb for the C1 site in the sample with x = 0.5. This is in general agreement with the data for the sample with x = 1.0.

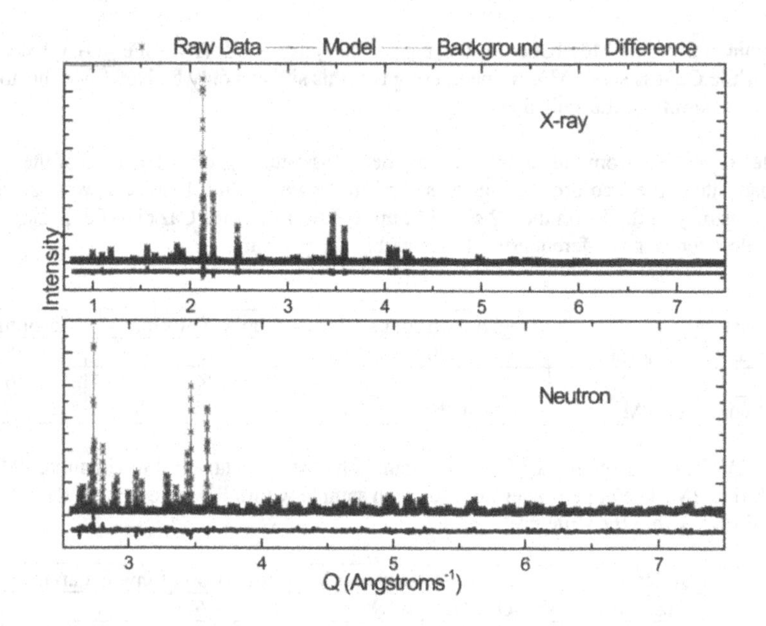

Figure 3. Graphical representation of the refinements on the system $CaZrTiNb_{1/2}Fe_{1/2}O_7$. The results show that model is consistent with the recorded data. The data is collected over slightly different ranges.

CONCLUSIONS

The distribution of Nb, Fe, and Ti within the HTB layer of the zirconolite-2M structure has been determined. The results have identified where Fe and Nb partition in zirconolite-2M, i.e. C1, C2 for Fe and C1, C3 for Nb. The techniques employed herein, namely resonant X-ray diffraction combined with neutron diffraction, can be used to determine cations locations within the zirconolite structure. These techniques can be applied to other 'real' systems proposed as nuclear wasteforms (e.g., pyrochlores).

ACKNOWLEDGEMENTS

The authors wish to acknowledge the expertise and assistance of Dr Irena Margiolaki at ID31, ESRF, Grenoble who provided expertise in the collection and analysis of the resonant X-ray data. The authors wish to acknowledge the expert help from Dr Kevin Knight at HRPD, ISIS Laboratory, UK for help in collection of the neutron diffraction data.

REFERENCES

1. S. V. Yudintsev. *Geology of Ore Deposits* **2003**, *45*, 151.
2. E. R. Vance; G. R. Lumpkin; M. L. Carter; D. J. Cassidy; C. J. Ball; R. A. Day; B. D. Begg. *Journal of the American Ceramic Society* **2002**, *85*, 1853.
3. F. J. Berry; G. R. Lumpkin; G. Oates; K. R. Whittle. *Hyperfine Interactions* **2005**, *166*, 363.
4. G. R. Lumpkin. *Journal of Nuclear Materials* **2001**, *289*, 136.
5. G. R. Lumpkin; K. R. Whittle; C. J. Howard; Z. Zhang; F. J. Berry; G. Oates; C. T. Williams; A. N. Zaitsev. Scientific Basis For Nuclear Waste Management XXIX, 2005, Ghent, Belgium.
6. R. I. Smith; S. Hull; A. R. Armstrong. The Polaris Powder Diffractometer at Isis. In *Epdic 3, Pts 1 and 2 - Proceedings of the 3rd European Powder Diffraction Conference*, 1994; Vol. 166; pp 251.
7. R. I. Smith; S. Hull "User Guide for the Polaris Powder Diffractometer at ISIS," Rutherford Appleton Laboratory, 1997.
8. B. H. Toby. *Journal of Applied Crystallography* **2001**, *34*, 210.
9. A. C. Larson; R. B. Von Dreele "General Structure Analysis System (GSAS)," Los Alamos National Laboratory Report LAUR, 2000.

Mater. Res. Soc. Symp. Proc. Vol. 1107 © 2008 Materials Research Society

Effect of Gallium Oxide on Phase Assemblage in Apatite and Whitlockite Hosts for Waste Immobilization

Lee. A. Gerrard, Shirley. K. Fong, Brian. L. Metcalfe and Ian. W. Donald
Materials Science Research Division, AWE, Aldermaston, Berkshire, UK

ABSTRACT

To immobilize halide and actinide ions present in specific ILW waste a process has been developed that uses mineral phases as the host material. The mechanism of substitution of gallium into these phases will have a large effect on the phase assemblage. This will inevitably affect the total amount of halide that can be immobilized in to total phase mixture.

The full simulated waste stream composition containing varying concentrations (1-40 wt.%) of gallium oxide was studied. Also nominal compositions for gallium doped fluorapatites $(Ca_{10-1.5x}Ga_x)F_2(PO_4)_6$ (x = 0, 0.25, 0.5, 0.75, 1.0) and gallium doped whitlockites $Ca_9Ga_y(PO_4)_{6+y}$ (x = 0.2, 0.4, 0.6, 0.8, 1.0) were prepared at 750-1050 °C.

These were studied by powder x-ray diffraction (XRD) to determine the phase assemblage and solid solution limits of gallium in the apatite and whitlockite phases. It was found that a complete solid solution was formed between whitlockite, $Ca_3(PO_4)_2$, and $Ca_9Ga_y(PO_4)_{6+y}$. In the nominal apatite compositions it was found that gallium did not substitute into the apatite structure but was instead partitioned over $Ca_9Ga_y(PO_4)_{6+y}$, gallium phosphate, and unreacted gallium oxide. At higher temperatures gallium suppressed the formation of the apatite phase and was largely partitioned into the $Ca_9Ga_y(PO_4)_{6+y}$ phase whereas at lower temperature the majority was present as unreacted Ga_2O_3. In the full DCHP compositions it was found that gallium is likely to be partitioned over a number of phases including apatite, cation-doped whitlockite and gallium phosphate.

INTRODUCTION

A number of fluoride, chloride, plutonium and americium containing wastes arising from pyrochemical reprocessing of plutonium require to be safely immobilized at AWE [1-3]. The relatively high proportion of chlorides and fluorides make conventional routes such as vitrification and cementation unsuitable. Consequently, a two stage process has been developed where the actinides and halides are initially immobilized in phosphate mineral phases by a solid state reaction with calcium hydrogen phosphate. The resulting immobilized waste is a fine powder, so is not considered passively safe. The second stage is then to sinter this powder with a sodium aluminophosphate glass to convert it into a solid wasteform.

Apatite and whitlockite were chosen as the main host phases for immobilization due to a number of reasons. Firstly, studies of natural apatites have shown that this mineral phase is highly resistant to radiation damage. This property has also been confirmed in accelerated ageing studies, using the short lived ^{238}Pu isotope, where no radiation damage was detectable by power x-ray diffraction (XRD) after the equivalent of 400 years ageing [4]. Secondly, both apatites and whitlockites are known to accommodate a range of actinides. Finally, apatites and whitlockites are known to accommodate non-stoichiometric compositions, which allow the structures to cope with variations in the waste stream composition.

Apatite [5] can be described as a tunnel structure composed of corner connected CaO_6 and PO_4 polyhedra where the tunnels, extending through the structure in the *c*-direction, are filled by calcium ions and anions (F^-, Cl^- or OH^-). Studies have shown that the calcium ions can be substituted by a range of cations.

Whitlockite is a calcium phosphate mineral of the form $Ca_3(PO_4)_2$ and has two crystalline modifications ($\alpha + \beta$). The structure undergoes a phase transition at 1120 °C from $\beta \leftrightarrow \alpha$. Studies of whitlockite have shown that this mineral phase is also able to accommodate a wide range of cations again allowing this phase to cope with the wide variation of cations present in the full waste stream. The β-whitlockite (low temperature modification) structure [5,6] is composed of columns that run parallel to the *c*-axis made up of repeating·[PO_4-Ca-Ca-Ca-PO_4]·units. The waste streams are expected to vary very widely in composition. Therefore it is necessary to determine an upper limit at which it can be ensured that the actinides and major constituents (*i.e.* gallium) will be adequately immobilized in the apatite/whitlockite phases *via* the solid state synthesis route. Although the halides are expected to be immobilized within the apatite phase, it is also necessary to identify which phases the gallium will be present in. In these non-active studies Hf^{4+} has been used as a surrogate for Pu^{4+}.

EXPERIMENTAL

Single cation studies

Samples were prepared by solid state sintering. Appropriate quantities of $CaHPO_4$, Ga_2O_3, $CaCO_3$ and additionally CaF_2, were ground under acetone and calcined for four hours in air. The mixtures were then reground and calcined a further two times to ensure the reaction had reached completion. Gallium-doped whitlockite compositions were prepared at 1050 °C assuming substitution of gallium according to the formulae below:-

$Ca_9Ga_y(PO_4)_{6+y}$ $\quad y = 0.2, 0.4, 0.6, 0.8, 1.0$

Gallium-doped fluorapatite compositions were prepared at 750, 950 and 1050 °C again assuming substitution of gallium *via* cation vacancies according to the formulae below:-

$Ca_{10-3x/2}Ga_xF_2(PO_4)_6$ $\quad x = 0.50, 1.00, 1.25, 1.50, 1.75, 2.00$

Full cation studies

The following experiments were aimed at determining the effect of varying the concentrations of the gallium oxide. The full simulant waste was prepared by finely grinding together the components listed in Table I. Weight percentages for the gallium oxide were varied from 1-40 wt.%. With these amendments the other components' weight percentages change, but the same ratio is kept between them. All the components in the full waste stream were weighed into a ball-mill and ground together for 20 hrs. A small batch of 7.5 g was added to 22.5 g of calcium hydrogen phosphate and the mixture then blended by Turbular® mixing for 30 mins.

Table I. Compositions of the gallium loaded full wastes.

Component	Wt. %
	Average (Actual)
HfO_2	20.8
Sm_2O_3	4.5
Ga_2O_3	28.0 (1-40)
Al_2O_3	9.8
MgO	6.3
Fe_2O_3	1.3
Ta_2O_5	1.3
NiO	1.3
CaF_2	10.4
KCl	16.3

The powder was then transferred to an alumina crucible and calcined with a heating regime of:-

- 20-400 °C at 200 °C/hr
- 400-750 °C at 100 °C/hr
- Hold for 4 hrs at 750 °C
- 750-20 °C at 100 °C/hr

When cold, the calcined batch material was removed from the crucible and lightly ground in a pestle and mortar to give a free-flowing powder. Samples from these batches were removed for analysis by PXRD.

Powder X-Ray Diffraction analysis (PXRD) was performed using a Bruker D8 Advance powder diffractometer with Bragg-Brentano flat plane geometry. Diffraction patterns were generated over a two theta range of 10 to 90 °, step width of 0.02 ° and a time per step of 4 s, using monochromatic Cu $K\alpha_1$ radiation ($\lambda = 1.54056$ Å). Powdered samples were dried overnight at 105 °C and 50 wt.% dried Al_2O_3 was added as a standard before analysis. Samples for PXRD were prepared using a simple top pack loading method to acquire a smooth surface. These PXRD patterns were then normalized to the main Al_2O_3 {1,1,1} reflection for both intensity and 2θ.

Patterns analyses were carried out using Diffrac Plus Evaluation [7] and the Rietveld refinements were carried out using the PC-GSAS refinement program and EXPGUI add on [8].

RESULTS AND DISCUSSION

Gallium doped whitlockite

The substitution mechanism assumed that gallium phosphate is added to pure whitlockite $Ca_3(PO_4)_2$ until the creation of the double phosphate $Ca_9Ga(PO_4)_7$ (isostructural to the natural whitlockite mineral). At composition y = 0, phase pure whitlockite was obtained. On substitution of gallium into the system phase pure samples were obtained and could be indexed to either $Ca_3(PO_4)_2$ or $Ca_9Ga(PO_4)_7$. No other peaks were observed and there was a clear progression of the reflections from $Ca_3(PO_4)_2$ to $Ca_9Ga(PO_4)_7$, (Fig.1).

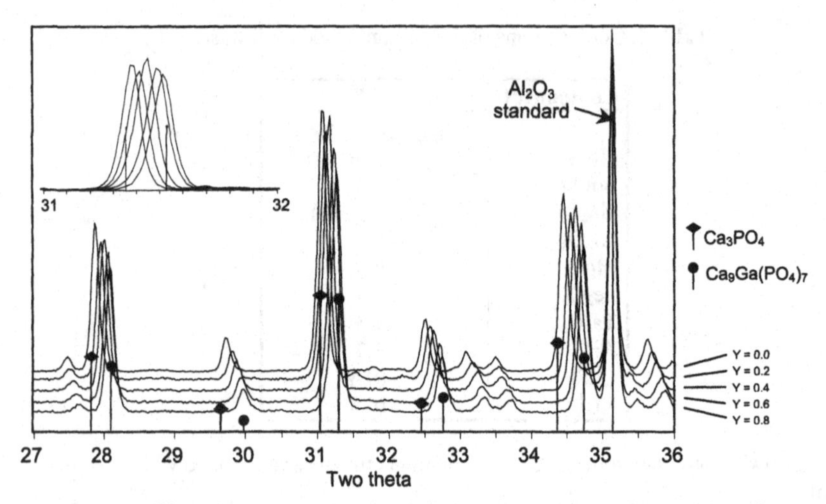

Figure 1. X-ray diffraction pattern showing the phase assemblage of the gallium doped whitlockite.

This result indicated that a complete solid solution forms between these two phases. Refinement of the cell parameters indicated a steady 'linear' contraction of the unit cell of $Ca_9Ga_y(PO_4)_{6+y}$ as y increases until composition y ~ 0.8 after which there is a slight expansion, as shown in (Fig.2).

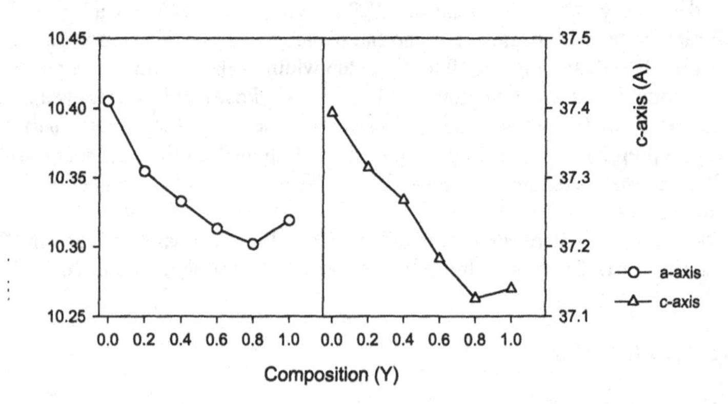

Figure 2. Calculated unit cell parameters of gallium doped whitlockite of composition $Ca_9Ga_y(PO_4)_{6+y}$.

A possible explanation for this trend could be due to the ionic radius of Ga^{3+} (0.62 Å) being smaller than Ca^{2+} (1.00 Å) hence the initial contraction of the unit cell parameters on increased gallium oxide loadings. At y ~ 0.8 the initial site for the gallium placement could now possibly be fully occupied (or as much is favourable) and consequently the gallium starts to fill a second site leading to an expansion of the unit cell.

Gallium doped fluorapatite 1050 °C

The gallium doped fluorapatite compositions assumed substitution of calcium with gallium *via* cation vacancies (*i.e.* $3Ca^{2+} \rightarrow 2Ga^{3+} + v_{\square}$). At 1050 °C phases were indexed to fluorapatite, calcium gallium phosphate (isostructural with natural whitlockite) and gallium oxide, (Fig.3).

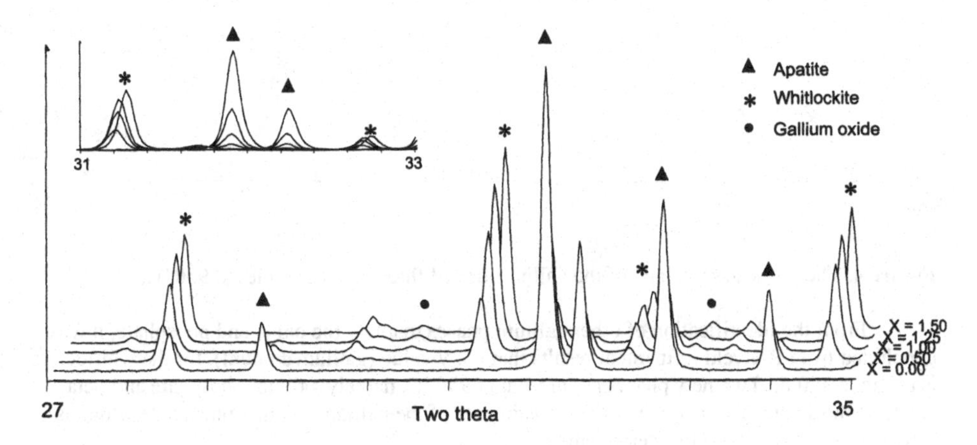

Figure 3. The phase assemblage of the gallium-doped fluorapatite samples at 1050 °C.

On increasing the gallium content, apatite formation was suppressed, while proportions of $Ca_9Ga(PO_4)_7$ and Ga_2O_3 increased, until at composition x = 1.50 only these two phases were left present. Fluorapatite was the only fluoride-containing phase observed in these compositions. However, as the weight percentage of fluorapatite drops to zero, it is probable that fluoride has been volatised during the high temperature synthesis.

The unit cell parameters of the fluorapatite phase show no significant change over the composition range suggesting no gallium is substituted for the calcium ions. On increasing the gallium content the proportion of the whitlockite phase increases in conjunction with a shift in reflection positions. This suggests that the gallium is becoming incorporated into this phase. Small amounts of unreacted gallium oxide are also present.

Gallium doped fluorapatite 950 °C

At the synthesis temperature of 950 °C phases formed were indexed to fluorapatite, calcium gallium phosphate, gallium phosphate, β-calcium phosphate and gallium oxide. As observed for the 1050 °C synthesis gallium suppresses for formation of the apatite phase while proportions of the other phases in general increase, (Fig.4). Again, at this lower temperature, the unit cell parameters of the fluorapatite do not change on increasing gallium content. This suggests that no gallium is incorporated into this structure. The whitlockite reflections do not shift throughout the compositional range but have moved significantly away from those of the phase pure whitlockite indicating gallium inclusion.

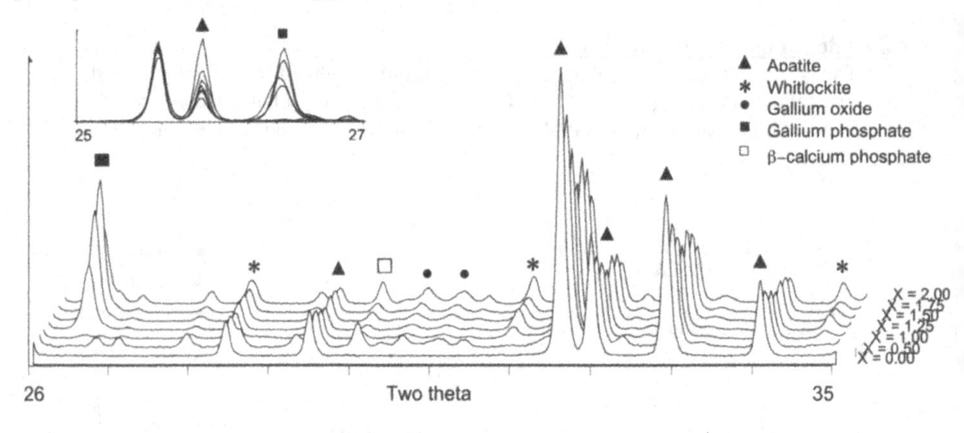

Figure 4. The phase assemblage of the gallium-doped fluorapatite samples at 950 °C.

From the identification of other gallium phases, namely the oxide and phosphate, it was concluded from Rietveld refinement results that the y value (gallium amount), for the whitlockite is constant at 0.6. Two new phases are present at 950 °C, namely gallium phosphate and beta-calcium phosphate. These are the basic constituents of the gallium calcium phosphate which is synthesised at higher reaction temperatures.

Gallium doped fluorapatite 750 °C

At the synthesis temperature of 750 °C phases formed were indexed to fluorapatite and gallium oxide in addition to small quantities of calcium gallium phosphate, β-calcium phosphate and calcium fluoride. However, un-indexed reflections were observed at 2θ = 26.720, 29.220 and 30.653°, indicating that at least one additional unidentified phase was also present (Fig.5).

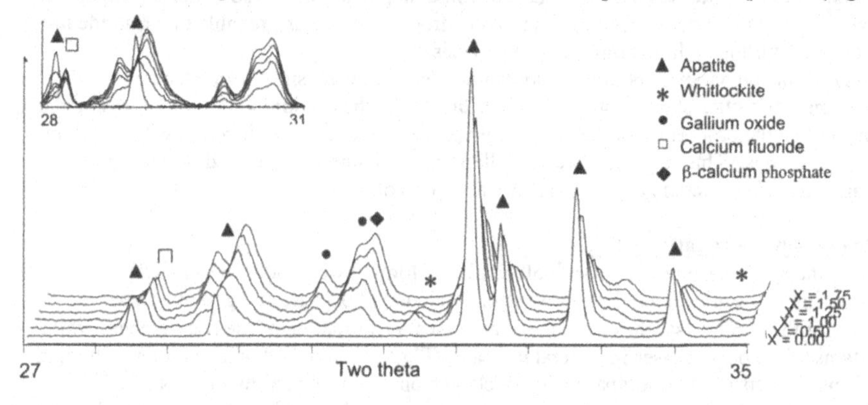

Figure 5. The phase assemblage of the gallium-doped fluorapatite samples at 750 °C.

Gallium doped full waste (750 °C)

In these initial non-active studies, using hafnium oxide as a surrogate for plutonium oxide, the concentration (weight percentage) of gallium was varied from 1-40 wt.%. This allowed an examination of the effect that differing amounts of gallium in the waste would have on the final phase composition and ultimately on the durability of the final wasteform.

The main phases identified were apatite (mixed fluor/chlor), β-$Ca_2P_2O_7$, $GaPO_4$ and Ga_2O_3 in addition to a whitlockite-type phase which was indexed to ($Ca_9Ga_y(PO_4)_{7+y}$). In the earlier section (Gallium doped whitlockite) it was shown that a complete solid solution can form between $Ca_9Ga(PO_4)_7$ and $Ca_3(PO_4)_2$ so the precise composition will fall within the two end members.

On increasing total gallium concentration in these compositions, in general the formation of the apatite phase was suppressed while proportions of other phases β-$Ca_2P_2O_7$, whitlockite type phase $Ca_9Ga_y(PO_4)_{6+y}$ and $GaPO_4$ increased. On closer inspection the strongest reflections of the main chlor-fluorapatite phase shift with increasing additions of gallium oxide. A clear shift to lower two-theta positions, from 1→40 wt.% shows an increase in the unit cell lengths, (Fig.6). This suggests that either more gallium is becoming incorporated in the apatite structure changing the unit cell parameters, or as gallium forms secondary phases other cations, in the full waste composition, become integrated into the apatite.

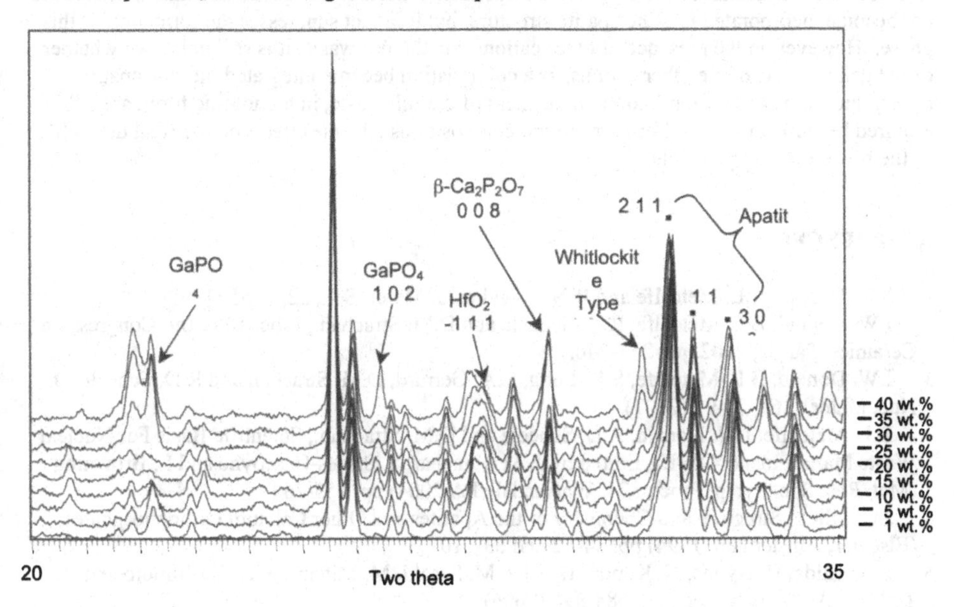

Figure 6. The phase assemblage of the gallium-doped fluorapatite samples at 950 °C.

It can be seen that with increasing additions of gallium oxide the whitlockite phase becomes more prevalent. Although no refinement was carried out to determine the unit cell parameters of this phase, the peaks positions indicate that the composition closely matches

$Ca_9Ga(PO_4)_7$. No significant shift in the peak positions was observed indicating no change in the composition with increasing gallium additions. Although other cations can be incorporated into the whitlockite phase, given that the concentration of gallium is increasing in these compositions, it is likely that main cation substituting into this structure is gallium.

One interesting region of the series is centred about $2\theta = 22$ °. This area shows two reflections that are tentatively assigned to $CaHfO_3$ and $AlPO_4$. On increasing gallium additions these peaks disappear but simultaneous growth of another phase is observed at $2\theta = 20.5$ and 23.2 °. This phase has yet to be assigned following an in depth literature and database search. On further gallium additions this phase is lost and one or two new phases appear almost on top of the original peaks centred on $2\theta = 22$ °. This phase(s) is likely to be due to a gallium containing phase, possibly a monazite, cristobalite type or a gallium-doped whitlockite.

CONCLUSIONS

It has been shown that a complete solid solution is formed from ($Ca_3(PO_4)_2$) to $Ca_9Ga(PO_4)_7$ at 1050 °C in the whitlockite studies. This phase is also formed in the apatite studies at all temperatures. From Rietveld refinement studies it was concluded that gallium does not become incorporated into the apatite structure, but it in fact suppresses the formation of this phase. However, in the presence of other cations, *i.e.* the full waste, it is still unknown whether or not gallium, one of the other cations, or a combination become integrated into the apatite phase. Further investigation into the robustness of the full waste, in monolithic form, are still required because with the differing resultant compositions adverse effects on the final durability of the wasteform are plausible.

REFERENCES

1. I.W. Donald, B.L. Metcalfe and R.N.J. Taylor., J. Mater. Sci., 32, 5851 (1997).
2. I.W. Donald, B.L. Metcalfe, R.D. Scheele and D.M. Strachan., Proc. 10th. Int. Congress on Ceramics, Part D, 2002, pp. 233-240.
3. I.W. Donald, B.L. Metcalfe, S.K. Fong, L.A. Gerrard, D.M. Strachan and R.D. Scheele., J. Nuc. Mater., 361, 78-93 (2007).
4. B.L. Metcalfe, I.W. Donald, R.D. Scheele and D.M. Strachan., Scientific Basis For Nuclear Waste Management XXVIII, Edited by J. M. Hanchar, S. Stroes-Gascoyne and L. Browning (Mat. Res. Soc. Symp. Proc., 824, Warrendale PA 2004) pp. 255-260.
5. J.C. Elliot., Structure and Chemistry of the Apatites and Other Calcium Orthophosphates, (Elsevier, Amsterdam, 1994) pp. 34-52 and 63-110.
6. K.Yoshida, H. Hyuga, N. Kondo, H. Kita, M. Sasaki, M. Mitamura, K. Hashimoto and Y. Toda., J. Am Ceram. Soc., 89, 688-690 (2006).
7. Bruker AXS, "DIFFRAC^{PLUS} Evaluation Package Release 2006 – EVA V.12.
8. A.C. Larson and R.B. Von Dreele., Los Alamos National Laboratory Report LAUR, 2000, 86-748.; B.H. Toby, J. Appl. Cryst., 2001, 34, 210-213.

Mater. Res. Soc. Symp. Proc. Vol. 1107 © 2008 Materials Research Society

High Temperature Behaviour of Polyoxometalates Containing Lanthanides

Hajime Kinoshita[1], Marcus Brewer[1], Caytie E. Talbot-Eeckelaers [2], Nik Reeves[1], Roy Copping[2], Clint A. Sharrad[2] and Iain May[2]
[1] Department of Engineering Materials, The University of Sheffield, Mappin Street, Sheffield, S1 3JD, U.K.
[2] Department of Chemistry, The University of Manchester, Oxford Road, Manchester, M13 9PL, U.K.

ABSTRACT

The possibility of a simple heating process of POM to obtain tungsten bronze was investigated for nuclear waste immobilisation via DTA/TG and high temperature XRD. Heating process up to 900 °C caused the decomposition of structure for both systems. Cooling process seemed to have little effect on the final product for the $K_{11}[Nd(PW_{11}O_{39})_2]\cdot xH_2O$, whereas the cooling profile showed a significant effect on the $K_{13}[Nd(SiW_{11}O_{39})_2]\cdot xH_2O$. Nd formed two types of tungsten bronzes, namely Nd_2WO_6 and $Nd_4W_3O_{15}$ in $K_{11}[Nd(PW_{11}O_{39})_2]\cdot xH_2O$ and $K_{13}[Nd(SiW_{11}O_{39})_2]\cdot xH_2O$, respectively.

INTRODUCTION

Recent studies suggest a possibility of a heating process of polyoxometalates (POM) at a temperature significantly lower than the classical methods to obtain tungsten bronzes, $MxWO_3$ (M: alkali, alkaline earth and rare earth, x: between 0 and 1) containing lanthanide [1], which may be applied for nuclear waste immobilisation [2].

The majority of oxide bronzes are tungsten bronzes of general form M_xWO_3 which exhibit a cubic structure based on WO_6 octahedra as shown in Figure1 (a). It is known that one of the tungsten bronzes, ZrW_2O_8 possesses a negative thermal expansion property [3]. Such a property can be highly advantageous for the nuclear waste forms. Until recently the synthesis of tungsten bronzes containing lanthanides has required extensive heating e.g. around 1050 °C for 100 hours [4]. Such a time and energy consuming process could be simplified by using POM as precursors which would be thermally decomposed at lower temperatures and for shorter times yielding the same results [1, 2].

POM are best described as symmetrical and compact groups of edge and corner shared $M'O_6$ octahedra [5] with a general composition of $[M_xM'_yO_z]^{n-}$ where M (the heteroatom) is a positive metal or non-metal and M' (the addenda) is W or Mo in many cases. With the addition of further heteroatoms, preferably atoms with large radii and high coordination numbers, two or more of the core POM units can be linked as shown in Figure 1 (b). It suits the properties of the lanthanide and actinide elements and hence they are ideal for integrating into POM structures.

Present work investigates the possibility of a simple heating process of POM to obtain tungsten bronze. The high temperature behaviour of various POM containing lanthanide elements was studied to investigate 1) the change of specimen upon heating, 2) phases formed in high temperature, and 3) effect of cooling profile on the final products.

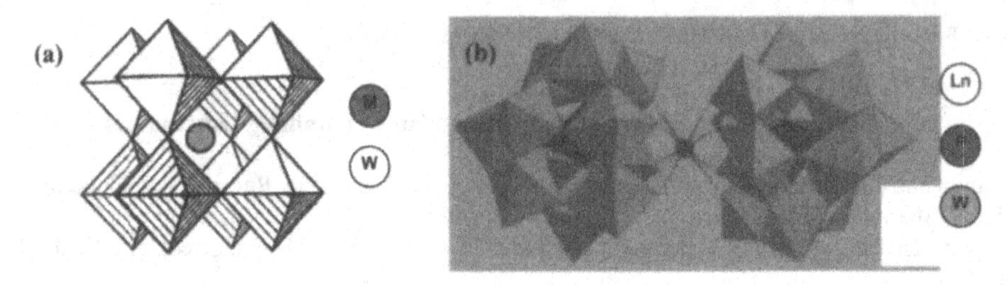

Figure 1. WO_6 octahedra based structures: (a) tungsten bronze M_xWO_3 with a cubic structure [1] and (b) POM units integrated with lanthanide atom $[Ln_1(P_1Mo_{11}O_{39})_2]^{11-}$ [4].

EXPERIMENTAL

We used Nd as a representative of Lanthanide elements, and two types of POM, $K_{11}[Nd(PW_{11}O_{39})_2]\cdot xH_2O$ and $K_{13}[Nd(SiW_{11}O_{39})_2]\cdot xH_2O$ were synthesised based on the procedures of Copping et al. [5] and Peacock et al. [6]. In brief: 1) a stoichiometric amount of $Nd(NO_3)_3\cdot 6H_2O$ solution was added to $H_3PW_{12}O_{40}$ or $H_4SiW_{12}O_{40}$ solution at pH 4.3; 2) K_2CO_3 and NH_4Cl were added to maintain the pH at 4.3; 3) the solution was left to be stirred for 1 hour; 4) CH_3CN was added drop by drop with shaking until it was no longer immediately miscible; 5) a portion of the solution was then stored for several days at 5 °C and more CH_3CN was added to aid crystallisation.

As shown in Figure 2, the high temperature behaviour of the obtained POM powders was investigated via Thermogravimetry (TG), Differential Thermal Analysis (DTA) and High-temperature X-ray Diffraction (XRD). TG and DTA were performed on Perkin Elmer Pyris1 TGA and Perkin Elmer DTA7, respectively. Samples were heated from 50°C to 900°C at a rate of 5°Cmin^{-1} under N_2 flow. High-temperature XRD were conducted on STOE STADI X-ray powder diffractometer equipped with an image plate position sensitive detector (IP-PSD), a Mo X-ray source (λ = 0.70926 Å) and a graphite furnace element. A 15 minute XRD measurement was repeated four times at each temperature, starting at 25 °C, heating to 100 °C and then at every 100 degree interval up to and down from 900 °C.

A portion of the synthesised POM powders were separately heat treated: the powders were ground and pressed into pellets of 7mm diameter; heated in an electric furnace at a rate of 5°Cmin^{-1} to 900 °C under air with a dwell of 120 minutes and cooled back to room temperature at the same rate. Room-temperature XRD measurement was undertaken for these samples using STOE STADI X-ray powder diffractometer used for the high temperature measurements.

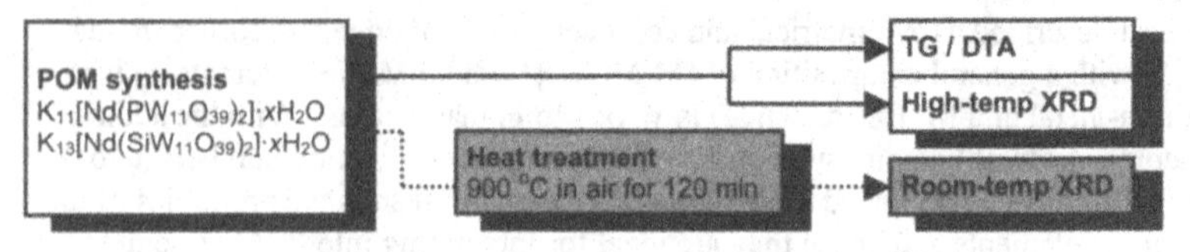

Figure 2. Schematic diagram of basic experimental procedure.

RESULTS AND DISCUSSION

Effect of heating

Figure 3 shows DTA and TG curves for $K_{11}[Nd(PW_{11}O_{39})_2]\cdot xH_2O$ and $K_{13}[Nd(SiW_{11}O_{39})_2]\cdot xH_2O$ on heating from room temperature to 900 °C. Figures on the left are DTA and TG for $K_{11}[Nd(PW_{11}O_{39})_2]\cdot xH_2O$, and those on the right are for $K_{13}[Nd(SiW_{11}O_{39})_2]\cdot xH_2O$. Derivative thermogravimetric (DTG) curves are also shown with TG curves. Both samples lose water of about 6 wt% up to 200 °C.

The major difference between two samples appears around 400 °C: the weight of $K_{11}[Nd(PW_{11}O_{39})_2]\cdot xH_2O$ sample decreases much more than that of $K_{13}[Nd(SiW_{11}O_{39})_2]\cdot xH_2O$. Besserguenev et al. studied the high temperature behaviour of $K_9[(ReO)_3(AsW_9O_{33})_2]\cdot nH_2O$ and obtained a very similar TG profile [7]. They stated that the weight loss at this temperature region was attributed to the decomposition of the anion via loss of As_2O_3. Probably the anion in our $K_{11}[Nd(PW_{11}O_{39})_2]\cdot xH_2O$ sample also decomposed around this temperature in a similar manner via loss of P_2O_5. It should be noted that there are a few more process contributed to the weight loss of $K_{11}[Nd(PW_{11}O_{39})_2]\cdot xH_2O$ sample at this temperature region. Considering that $K_{13}[Nd(SiW_{11}O_{39})_2]\cdot xH_2O$ sample also shows a slight weight loss, it may be due to KOH (melting temperature: 406 °C) which might have existed in both samples as an impurity or formed via decomposition of anions.

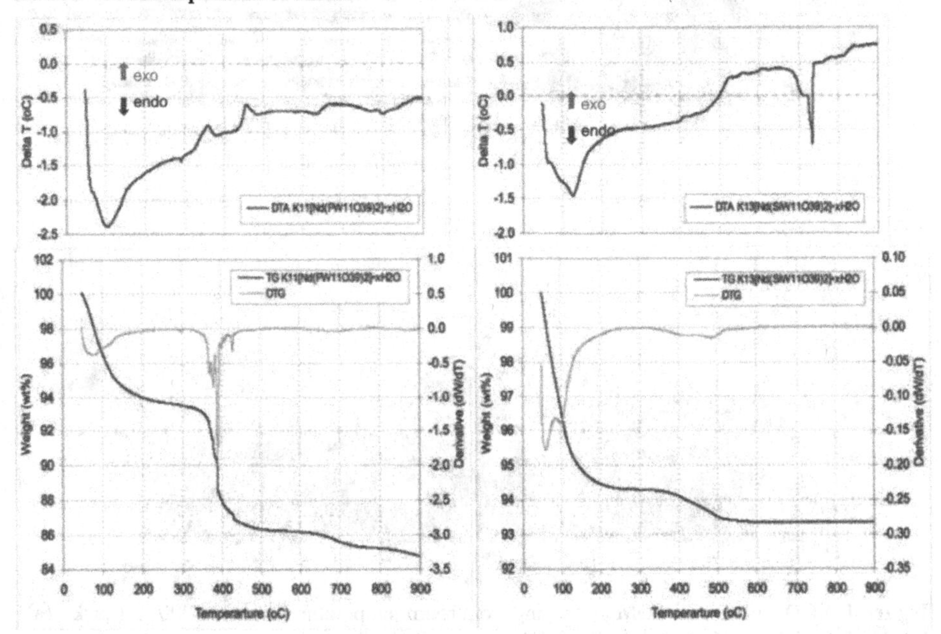

Figure 3. DTA and TG/DTG curves for $K_{11}[Nd(PW_{11}O_{39})_2]\cdot xH_2O$ (left) and $K_{13}[Nd(SiW_{11}O_{39})_2]\cdot xH_2O$ (right) on heating from room temperature to 900 °C.

Another significant difference between our two samples appears around 700 °C. A sharp endothermic reaction with no weight loss suggests a phase transformation in $K_{13}[Nd(SiW_{11}O_{39})_2]\cdot xH_2O$ sample. This sample might have been decomposed at least partially by this temperature region as WO_3 has a phase transformation at 776 °C.

Figures 4 (a) and (b) show the high temperature XRD results for $K_{11}[Nd(PW_{11}O_{39})_2]\cdot xH_2O$ and $K_{13}[Nd(SiW_{11}O_{39})_2]\cdot xH_2O$ samples, respectively. Corresponding to the DTA and TG results, $K_{11}[Nd(PW_{11}O_{39})_2]\cdot xH_2O$ changes the phases around 400 °C and then after 700 °C. Similarly, $K_{13}[Nd(SiW_{11}O_{39})_2]\cdot xH_2O$ shows phase changes corresponding DTA/TG results: a slight change after 500 °C and a significant change after 700 °C. The identification of the high temperature phases is detailed in the following sections.

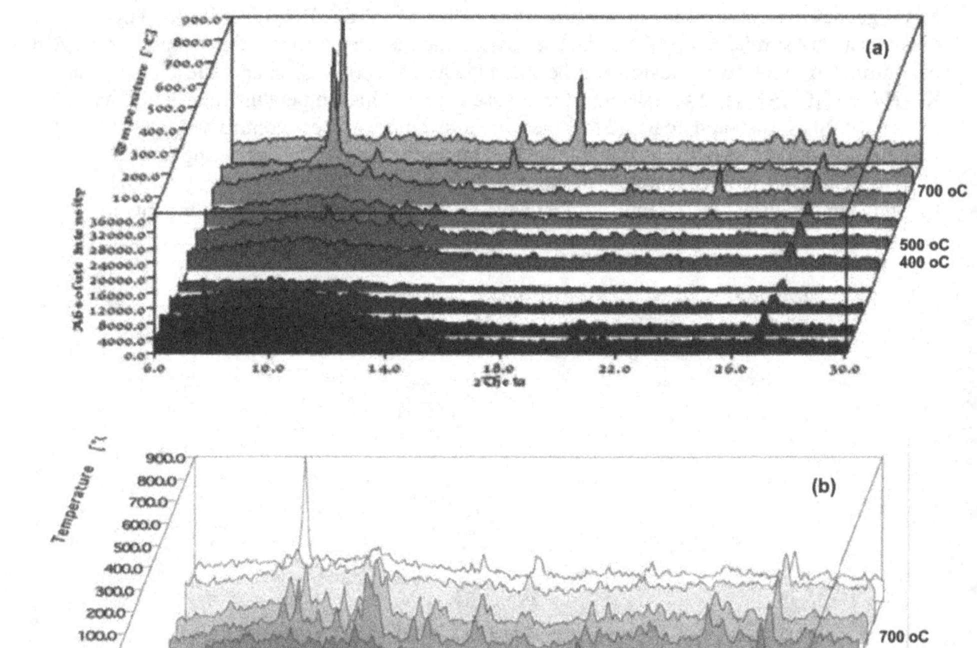

Figure 4. XRD patterns for POM on heating from room temperature (front) to 900 °C (back): (a) $K_{11}[Nd(PW_{11}O_{39})_2]\cdot xH_2O$, and (b) $K_{13}[Nd(SiW_{11}O_{39})_2]\cdot xH_2O$.

Effect of cooling

Figures 5 (a) and (b) show XRD patterns of $K_{11}[Nd(PW_{11}O_{39})_2]\cdot xH_2O$ on cooling from 900 °C to 600 °C and $K_{13}[Nd(SiW_{11}O_{39})_2]\cdot xH_2O$ from 900 °C to 200 °C. The both measurements did not proceed down to room temperature due to technical problem in the heating element on XRD machine. Each figure also indicates, at the bottom, the results from the room-temperature XRD measurements for the samples separately heat treated in an electric furnace up to 900 °C for comparison.

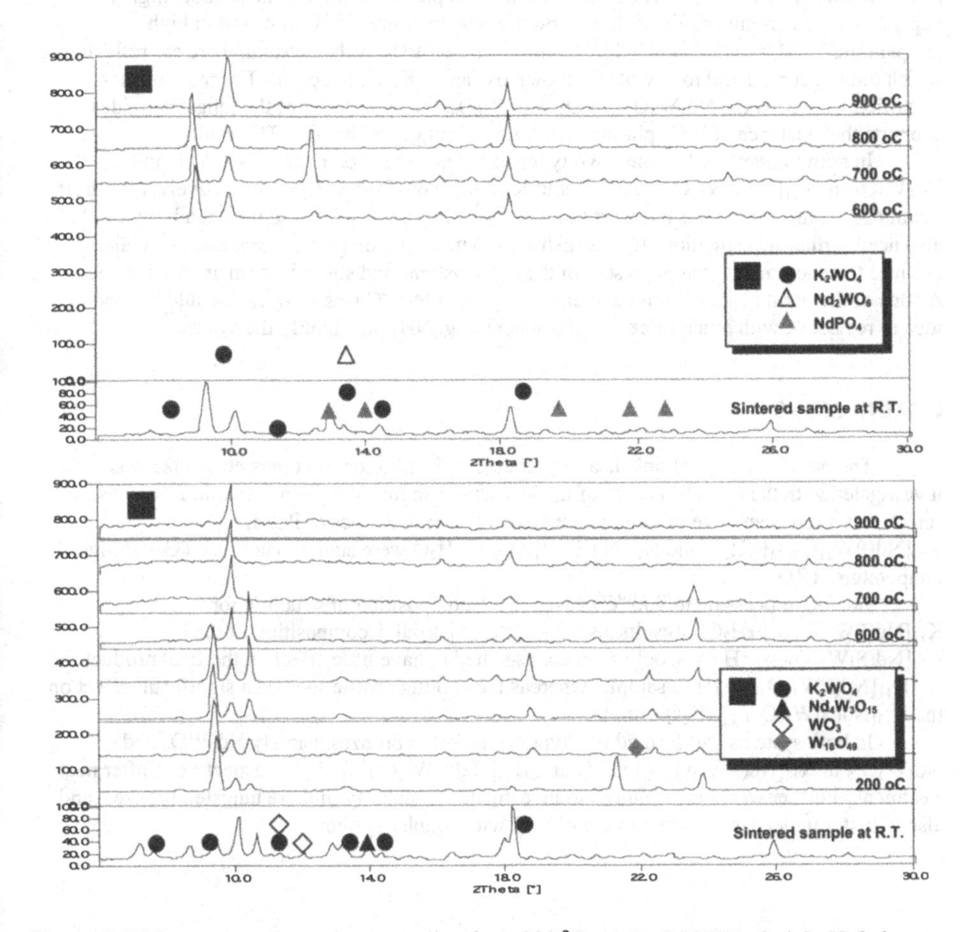

Figure 5. XRD patterns of samples on cooling from 900 °C: (a) $K_{11}[Nd(PW_{11}O_{39})_2]\cdot xH_2O$ down to 600 °C, and (b) $K_{13}[Nd(SiW_{11}O_{39})_2]\cdot xH_2O$ down to 200 °C. Each figure also indicates the results from the room-temperature XRD measurements for the samples separately heat treated up to 900 °C for comparison.

In the results for $K_{11}[Nd(PW_{11}O_{39})_2]\cdot xH_2O$, there is a very good similarity between the XRD pattern for 600 °C and that for heat treated sample. High temperature phases in this system must be quite stable. The main phases identified were K_2WO_4, Nd_2WO_6 and $NdPO_4$. There are reflections could not be indexed. They did not correspond to any of the known oxides of K, W, Nd or P. Besserguenev et al. experienced a similar difficulty in indexing all of the reflections during their study on $K_9[(ReO)_3(AsW_9O_{33})_2]\cdot nH_2O$ [7].

As shown in Figure 5 (b), XRD pattern for the heat treated $K_{13}[Nd(SiW_{11}O_{39})_2]\cdot xH_2O$ is quite different from any of patterns for high temperatures. It suggests that the high temperature phases in this system are not very stable, and that the phases formed in the product highly depends on the cooling profile of the sample (in heat treatment: 5 °C/min, and in high temperature XRD measurement: ~1.7 °C/min). Similar to the other system, there are reflections which did not correspond to any of the known oxides of K, W, Nd or Si. The main phases identified were K_2WO_4, $Nd_4W_3O_{15}$, WO_4 and $W_{18}O_{49}$. The presence of the tungsten oxides supports the existence of WO_3 phase around 700 °C suggested in DTA/TG results.

In both systems, Nd formed two types of tungsten bronze, namely Nd_2WO_6 and $Nd_4W_3O_{15}$ in $K_{11}[Nd(PW_{11}O_{39})_2]\cdot xH_2O$ and $K_{13}[Nd(SiW_{11}O_{39})_2]\cdot xH_2O$, respectively, after heat treatment. A further work is required to investigate the stability of these tungsten bronzes. We also need further investigations to establish a solid understanding of the process, especially because there are many phases present in the both systems and some of them are unidentified. As one of the main phases obtained after heating up to 900 °C was K_2WO_4, it could be a good idea to replace K with much more volatile species e.g. NH_4 to simplify the system.

CONCLUSIONS

The possibility of a simple heating process of POM to obtain tungsten bronze was investigated with the immobilisation of nuclear wastes in tungsten bronze in mind. Nd was selected as a representative of Lanthanide elements and two types of POM, $K_{11}[Nd(PW_{11}O_{39})_2]\cdot xH_2O$ and $K_{13}[Nd(SiW_{11}O_{39})_2]\cdot xH_2O$ were studied via DTA/TG and high temperature XRD.

Heating process up to 900 °C caused the decomposition of structure for the $K_{11}[Nd(PW_{11}O_{39})_2]\cdot xH_2O$. Results also suggested a partial decomposition of the $K_{13}[Nd(SiW_{11}O_{39})_2]\cdot xH_2O$. Cooling process seemed to have little effect on the final product for the $K_{11}[Nd(PW_{11}O_{39})_2]\cdot xH_2O$ sample, whereas the cooling profile showed a significant effect on the $K_{13}[Nd(SiW_{11}O_{39})_2]\cdot xH_2O$ sample.

In both systems, Nd formed two types of tungsten bronzes, namely Nd_2WO_6 and $Nd_4W_3O_{15}$ in $K_{11}[Nd(PW_{11}O_{39})_2]\cdot xH_2O$ and $K_{13}[Nd(SiW_{11}O_{39})_2]\cdot xH_2O$, respectively, after heat treatment. Further works are required to investigate the stability of there tungsten bronzes, and also to better understand the process, probably with simpler systems.

REFERENCES

1. K.Wassermann, M. T. Pope, M. Salmen, J. N. Dann and H. J. Lunk, J. Solid State Chem. **149**, 378-383 (2000).
2. M. T. Pope, U.S. DOE Report **54716** (2000).
3. J. S. O. Evans, T. A. Mary, T. Vogt, M. A. Subramanian and A. W. Sleight, Chem. Mater. **8**, 2809-2823 (1996).
4. W. Ostertag, Inorg. Chem **5** (5), 758-760 (1966).
5. R. Copping, A. J. Gaunt, I. May, M. J. Sarsfield, D. Collison, M. Helliwell, I. S. Denniss and D. C. Apperley, Dalton Trans., 1256-1262 (2005).
6. R. Peacock and T. Weakley, J. Chem. Soc. **A**, 1836-1839 (1971).
7. A. Besserguenev, and M. T. Pope, C. R. Chimie **8**, 933–955 (2005).

REFERENCES

1. [illegible]
2. [illegible]
3. [illegible]
4. [illegible]
5. [illegible]
6. [illegible]

Mater. Res. Soc. Symp. Proc. Vol. 1107 © 2008 Materials Research Society

Immobilization of Liquid Radioactive Waste in Mineral-Like Matrices Derived From Inorganic Cenosphere-Based Sorbents as Precursors

Tatiana A. Vereshchagina[1,2], Nataly N. Anshits[1,2], Elena V. Fomenko[1], Sergei N. Vereshchagin[1,2] and Alexander G. Anshits[1,2]
[1]Institute of Chemistry and Chemical Technology SB RAS, 42 K. Marx Street, Krasnoyarsk, 660036, Russia
[2] Siberian Federal University, 79 Svobodnyi Avenue, Krasnoyarsk, 660041, Russia

ABSTRACT

The progress in study of immobilization of liquid radioactive waste in mineral-like aluminosilicate matrices (feldspar and feldspathoids) derived from cenosphere-based microspherical zeolite sorbents is reported. Based on the analysis of 'composition – morphology – characteristics' correlation obtained for more than 70 cenosphere products with a high content of one morphological type of imperforated cenospheres, optimal ranges of cenospheres composition were identified for the synthesis of zeolites with the predetermined structural type. Given that the cenosphere-based zeolite sorbents are precursors of the final waste form, further activity proceeds toward developing a highly efficient zeolite sorbent of NaP1 type, which would ensure the effective trapping of Cs^+ and/or Sr^{2+} ions and the phase transformation under relatively mild conditions (T≤1000°C). The paper details the main criteria for realization of the 'precursor' approach for fixation of Cs-Sr fraction using cenosphere-based systems; also shown are experimental results indicating practical realization of the proposed approach.

INTRODUCTION

The cenosphere-based sorbents for immobilization of liquid radioactive waste, such as block-type porous materials, microspherical zeolites and encapsulated sorbents (AMP/cenosphere systems, etc.) were reported earlier [1-3]. Specific functions of these materials are displayed at each step of the multi-stage process of handling liquid radioactive waste, whereas using these materials allows for conversion of water-soluble radionuclide compounds into mineral water-insoluble forms under quite mild conditions at a low cost.

As it was demonstrated at a qualitative level [1], due to the similarity of chemical composition of cenospheres to the granite composition, the cenosphere-derived materials (SiO_2–Al_2O_3–Fe_2O_3–CaO–MgO–Na_2O–K_2O–TiO_2 systems) can be transformed into crystalline aluminosilicate compounds of feldspar and feldspathoid types after concentrating Cs^+ or Sr^{2+} ions according to the process 'impregnation/sorption-drying-calcination'. Taking into account that cenospheres is the primary precursor of the intermediate cenosphere-derived sorbent and the final mineral-like waste form, the macrocomponent composition of an initial cenosphere material is a determining criterion for realization of this 'precursor' approach.

The present paper is devoted to the quantitative description of the 'precursor' approach to the conditioning liquid radioactive waste by the example of Cs^+ and Sr^{2+} fixation in the structure of mineral-like aluminosilicates with the use of microspherical zeolite sorbents obtained by hydrothermal zeolitization of coal fly ash cenospheres.

EXPERIMENTAL DETAILS

Close-cut fractions of cenospheres were isolated from cenosphere concentrates obtained from fly ashes at the Novosibirsk Power Plant #5 (N series) and the Moscow Power Plant #22 (Series M) when burning coals of the Kuznetsk coal field. A four-stage scheme including magnetic separation, granulometric and gravimetric separation, and removal of perforated globules and chips was used to separate the cenosphere concentrates [4].

Cenosphere fraction with 0.40 g/cm^3 bulk density and 0.08-0.18 mm diameter was used as an initial material for preparation of microspherical zeolite sorbents (MZS). The zeolite sorbents were produced from cenospheres under hydrothermal conditions in an alkaline medium at 50-250°C, C_{NaOH}=1-4 M [1].

The composition of the cenosphere fractions and MZS was determined by the chemical analysis according to the GOST 5382-91 [5]. The specific surface area was determined by the Brunauer–Emmett–Teller method from the volume of adsorbed argon at 77 K (GKh-1 Gazometr) using the standard BET techniques. The morphology of cenospheres, MZS and MZS-derived materials was studied with a Tesla BS-350 scanning electron microscope and optical microscope Biolam (LOMO).

Cesium and strontium immobilization in the MZS was carried out by subsequent steps of sorption, drying and calcination at 700-1000°C, STA 449C Jupiter DSC was used to study initial MZS and Cs, Sr-loaded samples.

Phase composition of the cenosphere close-cut fractions, MZS and MZS-derived materials was studied by the quantitative full-profile XRD refinement with the use of the Rietveld method [6]. Structural parameters and the relative content of crystal phases were determined by the derivative difference minimization method [7].

The rates of ^{137}Cs leaching from solidified compounds were determined according to the GOST P 52126-2003 [8].

RESULTS AND DISCUSSION

Composition and Morphology of Cenospheres

Microspheres of a wide range of morphological types are formed from the mineral component of coals when burning at power plants. For the purpose of novel materials preparation, hollow aluminosilicate microspheres (cenospheres), which can be easily isolated from fly ashes due to the low (0.3–0.6 g/cm^3) density, are of the most interest. The content of cenospheres in fly ashes derived from various coals varies from 0.03 to 1.2 wt. %, the highest cenosphere yield (1.2 wt. %) being observed when burning black coals of the Kuznetsk coal field. Concentrates of cenospheres of different power stations that utilize Kuznetsk coals have different characteristics, such as chemical composition (wt. %): SiO_2 – 63.1÷65.2, Al_2O_3 – 20÷26.4, Fe_2O_3 – 4.2÷5.1, Ca – 0.9÷2.1, MgO – 1.0÷2.6, K_2O – 2.3÷4.0, Na_2O – 0.5÷1.2, globule distribution by size, shell width and bulk density [9].

The variability of cenosphere concentrate composition and, therefore, unpredicted characteristics do not allow using the concentrates without preliminary stabilization of the composition to obtain materials with preset characteristics. Using the methods of optical and scanning electron microscopy, it was established that the original concentrates contained three types of perforated cenospheres and five types of imperforated cenospheres. As an example,

Figure 1 shows morphological types of imperforated cenospheres identified via the optical microscopy.

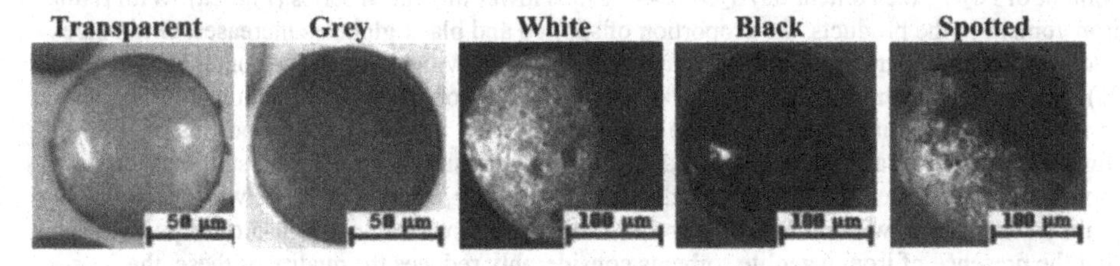

Figure 1. Types of cenospheres by data of the optical microscopy in the reflected light.

74 close-cut fractions of cenospheres with a prevailing content of one type of globules were produced from two concentrates. A systematic study of the 'composition – morphology – characteristics' correlation was conducted for the product series with different magnetic susceptibilities and globule sizes.

The macrocomponent chemical composition of close-cut fractions of imperforated cenospheres for N series (samples H; H1A; MH 80-180 μm) and M series (samples H/MM 1A; MM 80-100 μm) is shown in Figure 2 as paired correlations of Al_2O_3/Fe_2O_3 (Fig. 2 a) and SiO_2/Al_2O_3 (Fig. 2 b) contents.

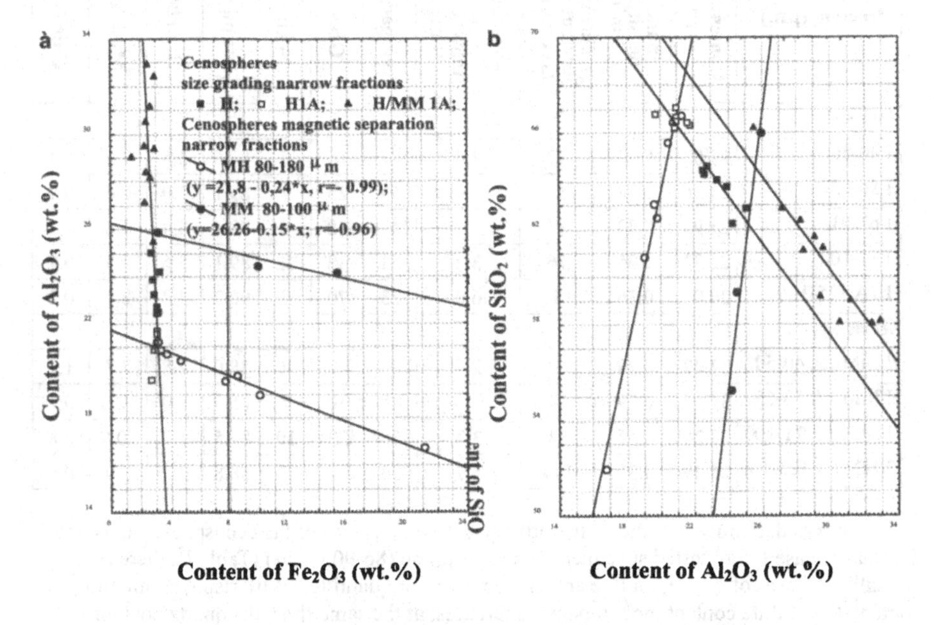

Figure 2. Macrocomponent composition of cenosphere narrow fractions from coal fly ashes of Novosibirsk Power Plant #5 and the Moscow Power Plant #22.

The data indicates that the composition of close-cut fractions of each series of cenospheres can be described with two composition trends. In magnetic products of N series, at the same content of Fe_2O_3, the content of Al_2O_3 is 4-5 wt. % lower than in M series (Fig. 2a). With rising iron content in the products, the proportion of spotted and black globules increases. Grain-size and gravimetric separation of imperforated cenospheres having a low Fe_2O_3 content (2.6÷3.2 wt. %) results in concentrating up to 80-90 % of transparent globules in fine fractions (40-50, 50-63 μm) with a low density (Table I). As the aluminum content increases, the proportion of transparent globules also increases and the width and porosity of the glass crystalline shell monotonously reduce. The N series products with the Al_2O_3 content of 19-22 wt. % contain mainly cenospheres with thick highly porous shells, relief surface and open surface pores. Given that the presence of iron in zeolite sorbents considerably reduces the quality of these, the magnetic products of both series were excluded as a feed for making zeolite sorbents. To produce the cenosphere-derived zeolite sorbents, the cenosphere composition range with an iron content of 2.6-3.2 wt.% Fe_2O_3 is the most preferable.

Table I. Physicochemical characteristics and morphological types of cenosphere close-cut fractions.

Product, fraction (μm)	Content of Al_2O_3 (wt. %)	Physical characteristics			Types by optical microscopy (%)			Phase composition by XRD (wt. %)			
		Bulk density (g/cm³)	Globule size distribution maximum (μm)	Average shell thickness (μm)	Transparent	Grey	White	Glass phase	Quartz	Mullite	Calcite
Series H											
H 40-50	25.03	0.32	47	2.4	77	16	7	87.6	3.5	7.5	1.1
H 50-63	23.26	0.34	58	2.4	68	25	7	89.1	4.0	5.6	0.9
H 63-71	23.91	0.33	67	2.7	47	48	5	88.4	4.4	5.8	1.1
H 71-100	22.73	0.35	76	3.2	51	38	12	89.1	5.6	4.3	0.7
H 1A 125-160	20.91	0.36	136	6.0	23	70	7	90.2	7.2	2.1	0.2
Series M											
H/MM 1A 40-50	32.47	0.29	44	2.4	93	3	4	88.0	2.8	7.7	1.1
H/MM 1A 50-63	32.95	0.29	54		88	10	2	85.9	1.0	12.3	0.7
H/MM 1A 71-100	28.20	0.31	84	3.2	58	32	10	86.5	1.7	10.9	0.8
H/MM1A 125-160	25.49	0.31	144	5.7	33	59	7	88.8	2.3	8.0	0.6

As it was determined by the quantitative phase composition of the cenosphere shells, the dominant phase for all tested specimens is a glass phase (86-90 wt. %) (Table I); there are crystalline phases of quartz, mullite and calcite in minor quantities. With rising aluminum content, the mullite content monotonously increases; at the same time, the quartz content monotonously decreases for N series. It should be noted that the higher mullite content and the lower quartz content are typical of products of M series.

Microspherical zeolite sorbents

A cenosphere fraction of 0.08-0.18 mm with an iron content of 3.0 wt.% in terms of Fe_2O_3 (see Table II) was used for MZS fabrication to provide maximum CEC (cation exchange capacity). The process of hydrothermal treatment leads to the transformation of an amorphous cenospheres' material to low silica zeolites of a different framework topology. Depending on the reaction condition a number of zeolite phases were formed as a monomineral phase or a mixture of zeolites - NaP1, NaX, NaA, analcime, hydroxosodalite, and chabasite. As a rule, a mixture of zeolite phases was generated.

The solid formed from cenospheres under hydrothermal condition consists of particles of different sizes ranging from small (1-10 μm) crystals of zeolites to zeolite crystal aggregates and granules of 70-400 μm in size. Hydrothermal treatment does not change significantly the bulk content of minor impurities but slightly decreases the Si/Al ratio due to some silica dissolution (see Table II). Depending on the degree of conversion, the Na_2O content reaches the value of 13 wt.%, with Na/Al ratio close to unity, which is typical for zeolites.

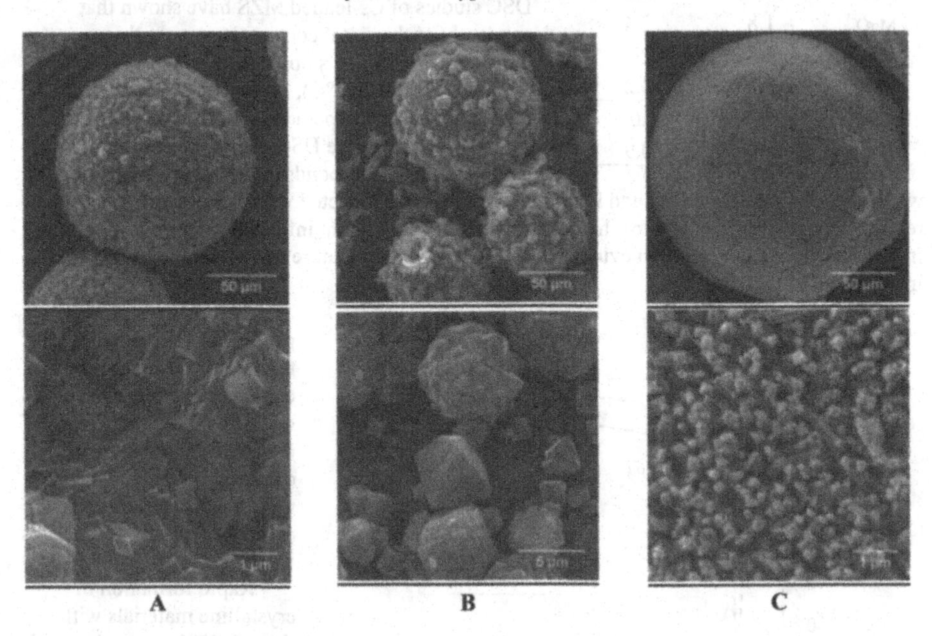

Figure 3. SEM images of hydrothermally treated cenospheres and view from above the particle surface. MZSs containing A – NaA (LTA), B – NaX (FAU) and C – NaP1 (GIS).

As far as low silica zeolites are promising material for sorption of radionuclides we undertook an attempt to fabricate MZS which combine a certain zeolite phase and predefined spherical shape of particles. The method was developed which allows to synthesize the hollow MZS with up to 90% individual zeolite content which completely preserve the spherical shape of the original material [1] (see figure 3).

Table II. Chemical composition (wt.%) of initial cenospheres and MZS formed under hydrothermal treatment (100°C, 1.5M NaOH, 72 h).

	Initial Cenospheres	**NaP1 MZS**
SiO_2	67.6	53.7
Al_2O_3	21	22.2
Na_2O	0.9	12.1
Fe_2O_3	3	3.5
CaO	2.2	2.5
MgO	1.8	2.4
K_2O	2.8	2.1
TiO_2	0.2	0.2
Si/Al	2.25	2.05
Na/Al	0.07	0.9

The composite wall of a particle is 10-20 μm thick and consists of zeolite crystals attached to the remaining not transformed glass material either like the layer (NaA, see figure 3A), aggregates (NaX, see figure 3B) or elongated crystals (NaP1, see figure 3C).

The adsorption characteristics of MZS are identical to those of natural and synthetic zeolites. To illustrate the process of mineral-like matrices formation we choose NaP1 MZS because of known ability of NaP1 to trap Cs^+ and Sr^{2+} ions and relatively low thermal stability of that zeolite.

NaP1 MZS was loaded with caesium or strontium by sorption from nitric solution of 20-250 mg/g. A few samples were prepared with Cs (Sr) content in order to be in the range 0-300 mg/g (0-90 mg/g for Sr) in the final solid.

DSC studies of Cs-loaded MZS have shown that independently of the metal content there were three pronounced thermal effects observed (see figure 4) at the low (endothermic, T<300°C), intermediate (exothermic, 740-750°C) and high temperature (exothermic, 944-1014°C). According to the DSC-MS data the low temperature effect corresponds to the water elimination whereas those at intermediate and high temperatures are connected with a structural rearrangement. The variation of the caesium content has a slight influence on the position of the intermediate peak but cause an evident shift to the high temperature region parallel to the increase of the Cs content.

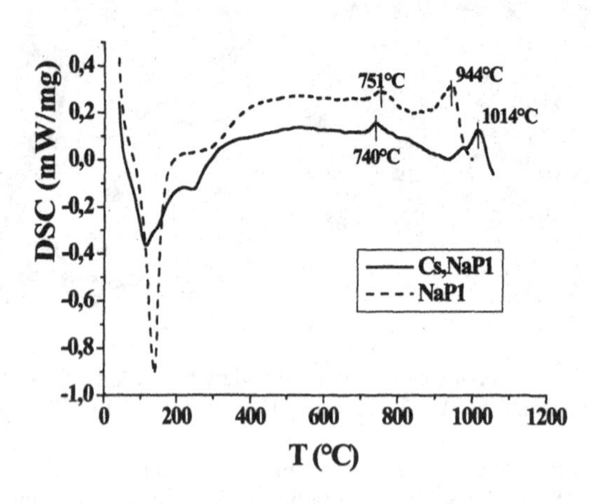

Figure 4. DSC curves for NaP1 MZS and Cs,Na-P1 MZS (300 mg Cs/g).

The changes of XRD profiles of Cs-loaded MZS treated for different time at a certain temperature have displayed that the collapse of the zeolite framework occurred at 550-700°C, but no formation of marked amount of crystal phases was found.

Rapid formation of crystalline materials with sharp XRD lines is observed at 1000°C. Calcinations of the initial NaP1 MZS for 4-5 hours lead to the nepheline phase as a main product with some minor amount of diopside and haematite (see figure 5) while pollucite becomes the dominant mineral phase at high cesium loadings. It should be pointed out that transformation of Cs,Na-P1 to pollucite-nepheline composition can be accomplished also at lower temperatures, but it takes longer time (more than 15 h at 800°C).

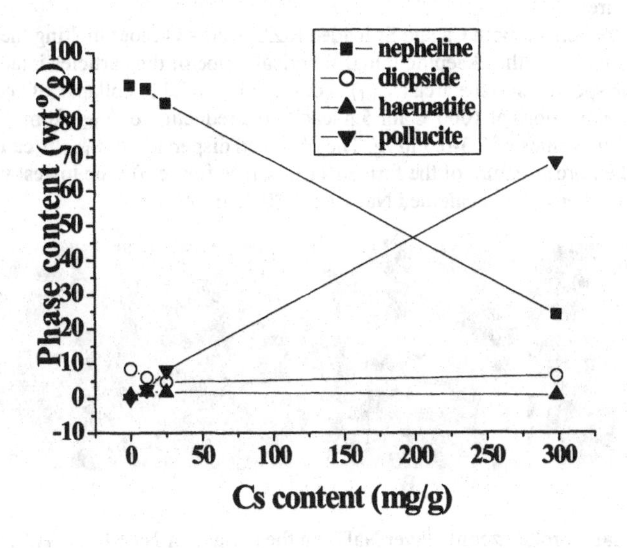

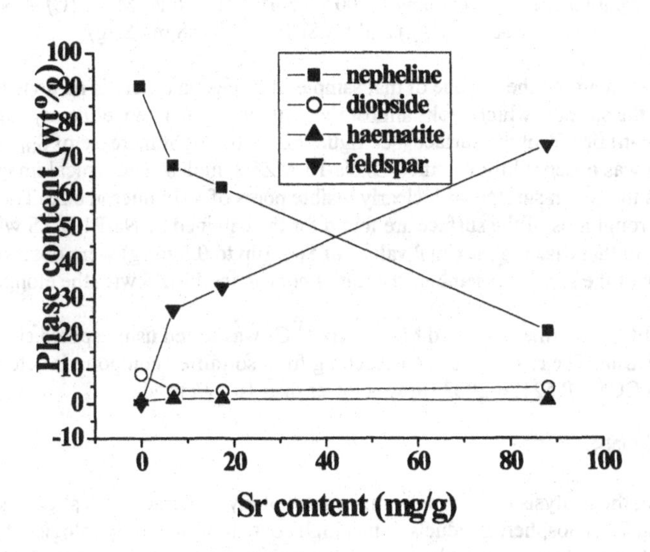

Figure 5. Variation of a relative amount of phases formed for Cs and Sr containing MZS calcined at 1000°C, 5 h as a function of Cs or Sr loading.

The similar dependencies are found for Sr-loaded MZS with the only difference that Sr-feldspar is formed instead of pollucite (see figure 5). Lattice parameters of feldspar formed are sensitive to the Sr content indicating formation of solid solution with incorporation of Sr atom in feldspar structure.

Thermal conversion both Cs and Sr loaded MZS occurs without melting the solids as a solid-phase conversion, thus keeping initial spherical shape of the particles intact. The substantial decrease of the specific surface area (S_{BET}) is observed parallel to collapse of zeolite framework. For instance, calcinations at 1000°C for 5 h lead to the reduction of S_{BET} from 33 m^2/g for initial NaP1 MZS to the values of 0.3-0.5 m^2/g. The observed dispersion of S_{BET} is caused most probably by the pore structure of the formed surface (see figure 6). The lowest value of S_{BET} was found for not exchanged and calcined NaP1 MZS (0.25 m^2/g).

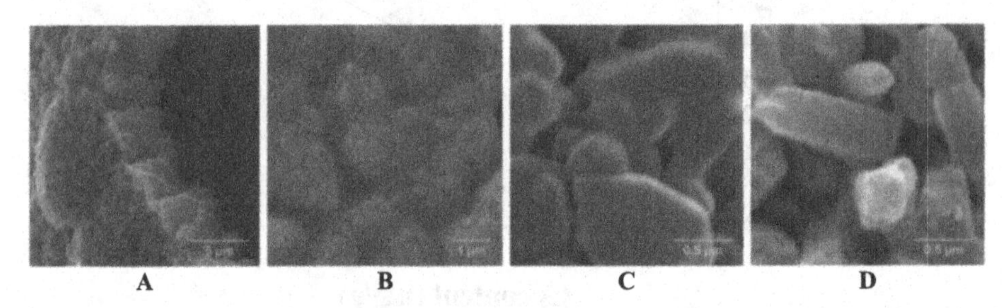

Figure 6. Partially broken zeolite layer NaP1 on the surface of Na-P1 MZS (A) and general view of the surface after thermal treatment (1000°C, 5 h): (B) – NaP1 MZS, (C) – Na,Cs-P1 MZS (300 mg Cs/g) and Na,Sr-P1 MZS (88 mg Sr/g).

The SEM image of the surface of that sample indicates that there is no a visible pore structure on the surface, which looks almost flat with hexagonal nepheline crystals which somewhat protrudes from the surface (see figure 6B). The slight increase of S_{BET} to the values of 0.3-0.4 m^2/g was observed for calcined Cs,Na-P1 MZS samples. The typical image (see figure 6C) displays the rough surface with clearly visible pores of a submicron size. The more pronounced roughness of the surface are found for the calcined Sr,Na-P1 MZS with pores of a micron size. In that case the maximal values of S_{BET} (up to 0.5 m^2/g) were observed and the general view of the surface resembles the initial not calcined MZS with the elongated crystal of NaP1.

The ability of thermally treated MZS to fix ^{137}Cs was tested using spiked simulants of actual radioactive waste. The rate of the ^{137}Cs leaching from solidified compounds determined according to GOST P 52126-2003 [8] was lower than 10^{-6} $g/cm^2 \cdot day$.

CONCLUSIONS

Based on the analysis of 'composition – morphology – characteristics' correlation obtained for more than 70 cenosphere products with a high content of one morphological type of imperforated cenospheres, optimal ranges of cenospheres composition were identified for the synthesis of microspherical zeolites with the predetermined structural type. The zeolite sorbent of NaP1 type having the high ability to trap both Cs^+ and Sr^{2+} ions as well as a low thermal stability providing the phase transformation under relatively mild conditions (T≤1000°C) was developed as a precursor of the final waste form for immobilization of Cs-Sr fraction. The quantitative determination of phase composition of Cs- and Sr-loaded MZSs calcined at different temperatures (700-1000°C) was carried out. It was found that pollucite and Sr-feldspar are the dominant phases (69-74%) in the calcined products at high cesium and strontium loadings.

ACKNOWLEDGMENTS

Authors acknowledge the financial support of the International Science and Technological Center (ISTC Project No. 3535) and the Siberian Branch of the Russian Academy of Sciences (Integration projects Nos. 36, 41, 92) as well as highly appreciate the contribution of Leonid A. Solovyov, Alexei N. Salanov and Nina N. Shishkina to this work.

REFERENCES

1. T.A. Vereshchagina, N.G. Vasilieva, S.N. Vereshchagin, E.V. Paretskov, I.D. Zykova, D.M. Kruchek, L.F. Manakova, A.A. Tretyakov, and A.G. Anshits in *Scientific Basis for Nuclear Waste Management XXIX*, edited by P.V. Iseghem, (Mat. Res. Symp. Proc., **932,** Warrendale, PA, 2006) pp.591-598.
2. T.A. Vereshchagina, E.V. Fomenko, S.N. Vereshchagin, N.N. Shishkina, N.G. Vasilieva, E.N. Paretskov, D.M. Kruchek, T.J. Tranter, A.G. Anshits, *Proc. Intern. Conf. "Coal Science and Technology" (ICCS&T 2005)*, October 12-16, 2005, Okinawa, Japan. - 12 p.
3. N.N. Anshits, A.N. Salanov, T.A. Vereshchagina, D.M. Kruchek, O.A. Bajukov, A.A.Tretyakov, Yu.A. Revenko, and A.G. Anshits, *Int. J. Nucl. Energy Sci.&Tech.*, **2**, Nos.1\2, 8 (2006).
4. T.A. Vereshchagina, N.N. Anshits, I.D. Zykova, A.N. Salanov, A.A.Tretyakov, A.G. Anshits, *Khimiya v interesakh ustoichivogo razvitiya* [*Chemistry for Stable Development*], **9,** 379-391(2001).
5. GOST 5382-91. National Standard of Russian Federation. Cements and materials for cement production. Chemical analysis methods.
6. H. Reitveld, *J. Appl. Cryst.,* **2,** 65 (1969).
7. L.A. Solovyov, *J. Appl. Cryst.,* **37**, 743 (2004).
8. GOST P 52126-2003. National Standard of Russian Federation. Radioactive waste. Long time leach testing of solidified radioactive waste forms.
9. L.I. Kizilshtein, I.V. Dubov, A.L. Shpitsgluz, S.G. Parada, Komponenti zol i shlakov TES (Energoatomizdat, Moscow, 1995) p. 176.

Mater. Res. Soc. Symp. Proc. Vol. 1107 © 2008 Materials Research Society

Synthesis and Characterisation of Ln_2TiO_5 Compounds

Robert D. Aughterson[1], Gregory R. Lumpkin[1], Katherine L. Smith[1], Gordon J. Thorogood[1] and Karl R. Whittle[1,2]

[1] Institute of Materials Engineering, ANSTO, Private Mail Bag 1, Menai 2234, NSW, Australia
[2] Department of Engineering Materials, University of Sheffield, Sheffield, S1 3JD, UK

ABSTRACT

Bulk samples of six Ln_2TiO_5 compounds with Ln = La, Pr, Nd, Eu, Gd and Tb were prepared and characterised. Most of the samples have a phase purity of ~95% (based on BEI and EDS) with the predominant secondary phase primarily being $Ln_2Ti_2O_7$. Using XRD, TEM selected area diffraction and high resolution imaging techniques, we have confirmed the results of previous studies which showed that at room temperature Pr_2TiO_5, Nd_2TiO_5, Eu_2TiO_5 and Tb_2TiO_5 have orthorhombic structures with *Pnma* symmetry. The structure of Tb_2TiO_5 was further monitored as a function of temperature. The relevance of Ln_2TiO_5 compounds to advanced nuclear fuel cycles is discussed.

INTRODUCTION

The Ln_2TiO_5 group of phases are of interest as potential constituents in waste forms due to their potentially high level of actinide incorporation and Gd_2TiO_5 is of specific interest as a burnable neutron absorbers in nuclear fuels. Fuel assemblies loaded at the beginning of a reactor cycle should have a certain excess amount of reactivity to compensate for reactivity loss and fission product build-up during use. This excess reactivity needs to be controlled and this can be achieved by inclusion of neutron absorbers [1]

Previous studies indicated that the structure of these phases varies depending on both the radius of the Ln cation and also temperature. Results indicate that for elements from La to Sm the structure is orthorhombic (*Pnma*), between Er to Lu the structure is cubic (*F-43m*), and from Eu to Ho the structure is temperature dependent ranging from orthorhombic to hexagonal (*P3/mmc*) to cubic [2-5]. In the orthorhombic Ln_2TiO_5 compounds, the Ti4+ cation has a CN of 5, slightly offset from the centre of a pyramidal site (e.g., half of a normal octahedron).

EXPERIMENTAL PROCEDURES

Initial chemical mixing was via the nitrate-alkoxide route. Samples were shear mixed then dried on a hot plate. Dried, agglomerated samples were ground to a fine powder then calcined within an alumina crucible at 750°C for 1 hour in air. The resultant material was removed from the furnace, and milled using yttrium stabilised zirconia milling balls, ~5mm, for 16 hours. The fine powders were then cold uniaxially pressed at 3tonne for 30 seconds and further condensed using a cold isostatic press at 400 MPa for 2 minutes. All samples were sintered at 1400°C for 48 hours in air.

  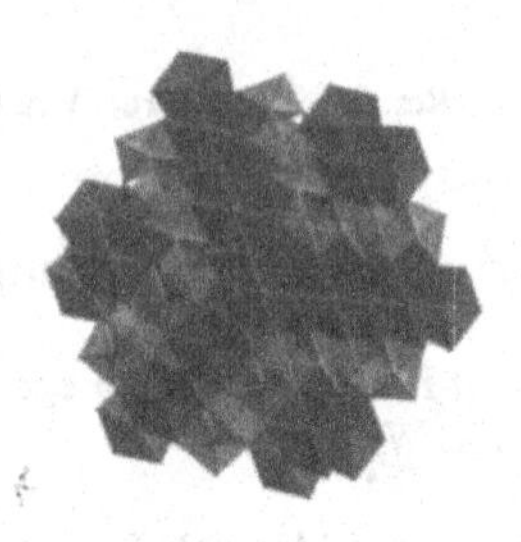

Figure 1. Crystal structures of the 3 phase types for the group Ln_2TiO_5 (Ln=lanthanide). The apices of the polyhedra indicate the positions of oxygen ions, while cations lying close to or at the centres of polyhedra. Ln is represented by the grey and Ti by the blue polyhedra. a) SG Pnma, Ti (CN 5), Tb (CN 7). The Ti is slightly displaced from the base of a square based pyramid. b) SG P3/mmc, Ti (CN 5) is now in a trigonal bi-pyramid form. Gd (CN 6) forms octahedral layers between the trigonal bi-pyramid layers. c) SG; F-43m, Ti (CN 6) is now in the octahedral form whilst Dy (CN 8) is within a slightly deformed cube. This is not too dissimilar to the pyrochlore structure.

Scanning Electron Microscopy (SEM) was carried out on a JEOL JSM-6400 SEM operated at 15 kV and equipped with a Noran Voyager energy dispersive spectrometer (EDX). The instrument was operated in standardless mode; however, the sensitivity factors were calibrated for semi-quantitative analysis using a range of synthetic and natural standard materials. Spectra were usually acquired for 500 seconds, reduced to weight percent oxides using a digital top hat filter to suppress the background, and a library of reference spectra for multiple least squares peak fitting. This was used to determine phases present, examine the microstructure and determine the chemical composition of the phases in each sample.

Transmission Electron Microscopy (TEM) specimens were prepared by crushing small fragments in methanol and collecting the suspension on holey carbon coated copper grids. TEM was carried out using a JEOL 2000 fxII (operated at 100kV), which is fitted with an Oxford energy dispersive X-ray analyser (EDS) and utilises a low-background double-tilt sample holder. Several grains of each sample were examined. Firstly EDS was used to confirm that the grain being examined was of the correct phase (Ln_2TiO_5), then bright field and high resolution (HR) images, and selected area diffraction patterns (SADPs) were taken at suitable crystallographic orientations. Some SADPs were indexed using simulations generated in the software package JEMS. [6]

XRD patterns were collected using a Panalytical X'pert Pro powder diffractometer. Previous to collection of data, the zero point was calibrated by an initial scan of a quartz standard. Each sample was scanned using weighted Cu Ka radiation run from 5 to 95° 2θ with a step size of 0.03°. Powdered Tb_2TiO_5 was heated on a platinum strip with XRD spectra taken at intervals of 100°C over a range between 600 to 1100°C.

Experimental patterns were compared with calculated patterns. Calculated diffraction patterns were generated in the software package CrystalDiffract. [7]

RESULTS AND DISCUSSION

Microstructure and Chemistry

All the fabricated samples are comprised of 90-95 vol.% of the design phases and are porous, with pore sizes of 5-10 microns. Figure 2 shows backscattered electron (BE) and secondary electron (SE) SEM images of Pr_2TiO_5. These are representative of similar images taken of all the fabricated samples. The major (lighter) phase in the BE image in Figure 2 (and equivalent images of the other samples) is the design phase Ln_2TiO_5 (Ln = La, Pr, Nd, Gd, Tb, and Eu). Darker patches indicate the presence of a secondary phase. For samples containing Pr, Nd, Gd and Eu the secondary phase is $Ln_2Ti_2O_7$, in the case of La_2TiO_5 the secondary phase is $La_4Ti_3O_{12}$, while the composition of the secondary phase in the Tb sample is yet to be identified.

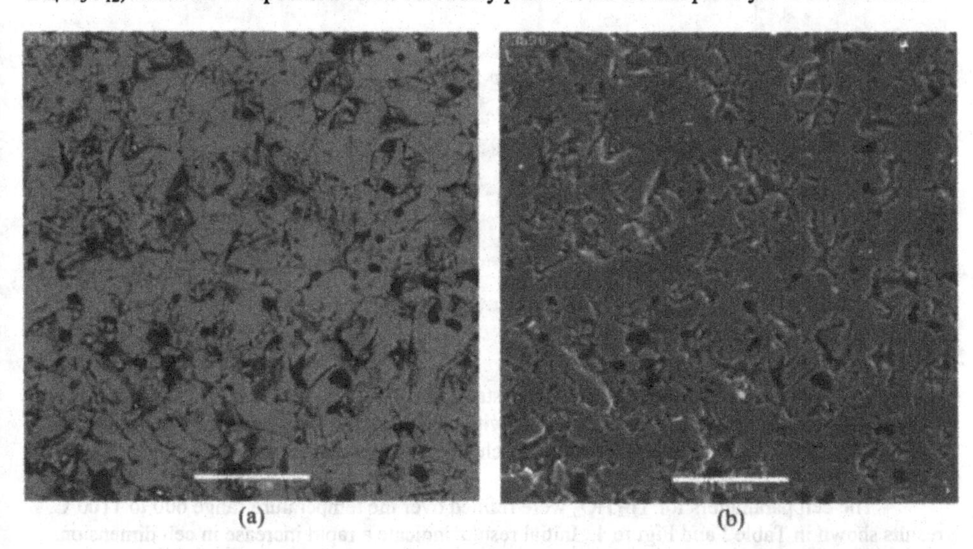

Figure 2. (a) Compositional (BE) and (b) topographical (SE) images of Pr_2TiO_5. Note in the BE image the presence of darker patches showing a second phase. The darkness is due to a lower amount of the heavy lanthanide. The secondary phase was identified using EDS to be $Pr_2Ti_2O_7$. The scale bars are 19 μm.

Unit-Cell Refinements

Refinement of cell parameters from X-ray powder diffraction patterns, shown in Figure 3, were conducted using the software package Jade 7.0. [8]. The initial lattice parameters were taken from Petrova [4] and Zaslavskii [3]. Peaks were fitted using a Pseudo-Voigt peak profile function. These analyses confirm that all the Ln_2TiO_5 compounds have orthorhombic *Pnma* symmetry.

When the change in unit cell size is compared with the change in ionic radius of the 7 co-ordinate lanthanide it is found to follow a linear relationship. This is logical as the lanthanide does not mix appreciably with the Ti 5 co-ordinate sites.

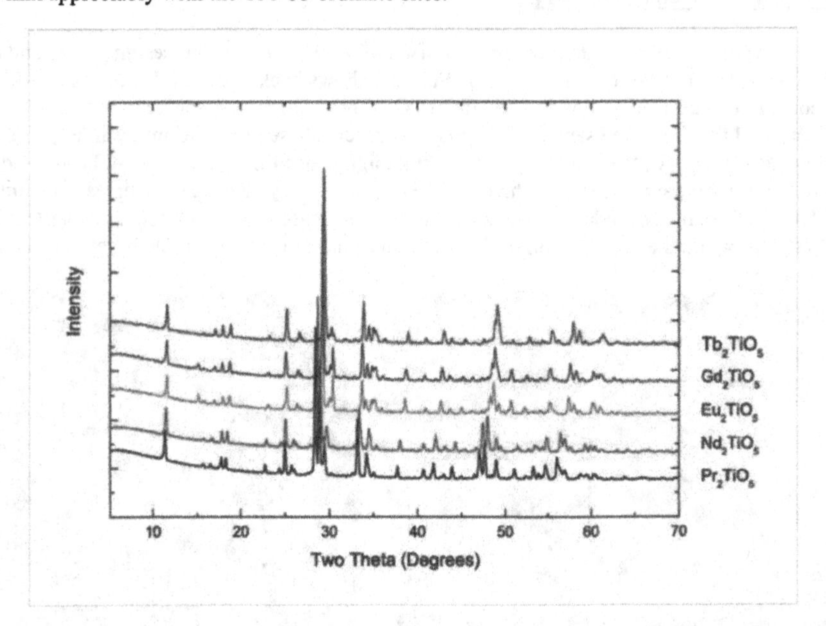

Figure 3. XRD patterns for various Ln_2TiO_5 systems (Ln = Pr, Nd, Gd, Tb, Eu). All patterns were matched with the orthorhombic structure with peak position shifts between patterns caused by variation of the cation size and the resultant change in cell parameters/volume.

The cell parameters for Tb_2TiO_5 were refined over the temperature range 600 to 1100°C, results shown in Table I and Figure 4. Initial results indicate a rapid increase in cell dimensions and hence volume, with this rate tapering off as it approaches 1200°C. Zaslavskii et al reported Tb_2TiO_5 undergoes a phase change from orthorhombic to hexagonal symmetry at 1200°C [2], whereas Petrova et al reported the phase change occurs at 1520°C [3]. Our data are consistent with a phase change in the range 1200-1520°C as one might expect the cell volume expansion to taper off to a maximum just prior to a structural phase change.

Table I – Results obtained from the least squares refinement of unit cell parameters for Tb_2TiO_5 in the range 600-1100°C

Temperature (°C)	a (Å)	error (Å)	b (Å)	error (Å)	c (Å)	error (Å)	vol (Å)	error (Å)
600	10.44905	0.00488	11.27795	0.00472	3.75055	0.00141	441.9793	0.01101
700	10.46579	0.00464	11.28237	0.00541	3.75718	0.00286	443.64374	0.01291
800	10.47309	0.00373	11.29611	0.00435	3.75997	0.0011	444.82392	0.00918
1000	10.47336	0.00506	11.31456	0.00294	3.76793	0.00301	446.50521	0.01101
1100	10.47838	0.00109	11.30756	0.00358	3.77437	0.0018	447.20589	0.00647

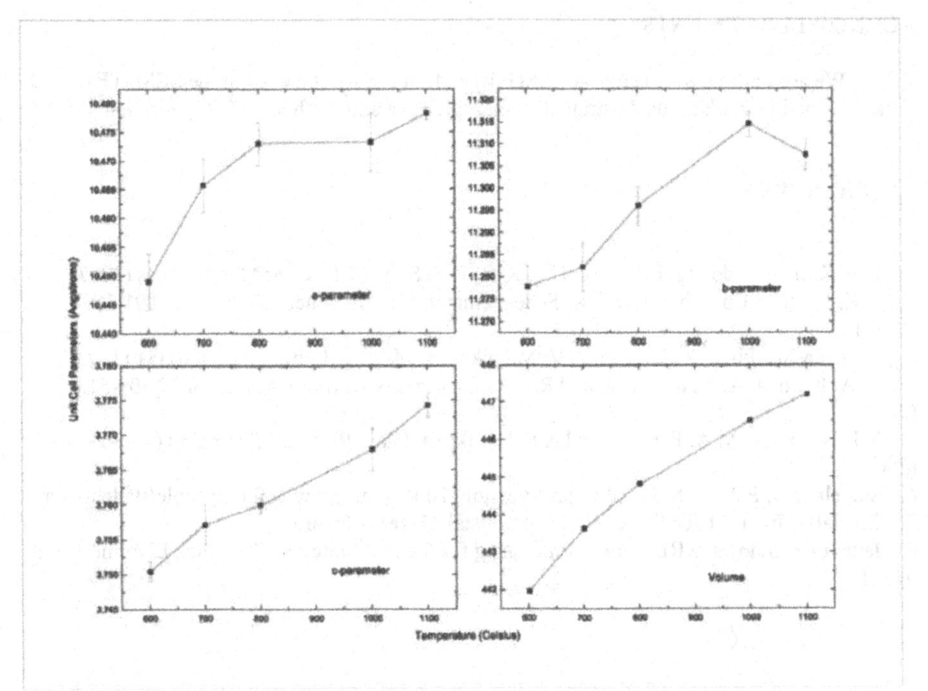

Figure 4. Plots showing cell parameters *a*, *c* and volume increase with temperature for Tb_2TiO_5.

CONCLUSIONS

- Samples of Ln_2TiO_5 compounds with high phase purity are easily obtainable.
- In agreement with previous studies, the results of this study show that at room temperature Pr_2TiO_5, Nd_2TiO_5, Eu_2TiO_5 and Tb_2TiO_5 have orthorhombic structures (*Pnma*).
- Up to 1100°C, Tb_2TiO_5 adopts *Pnma* symmetry. However, the change in cell parameters with temperature is non-linear.
- High resolution neutron and X-ray diffraction will be used together to further refine/establish the structures of the Ln_2TiO_5 compounds fabricated in this study as a function of temperature.
- The radiation tolerance of Gd_2TiO_5 will need to be studied, to further ascertain its suitability as a burnable neutron absorber in nuclear fuels, while the radiation response of the
- The radiation tolerance and aqueous durability of Ln_2TiO_5 compounds needs to be investigated to further ascertain their suitability as potential waste form phases, while the radiation response of Gd_2TiO_5 in particular will affect its projected use as a burnable neutron absorber.

ACKNOWLEDGEMENTS

We are grateful to J. Davis and M.G. Blackford for assistance with the SEM, TEM, and X-ray diffraction work at the Australian Nuclear Science and Technology Organisation.

REFERENCES

1. H.S. Kim, C.Y. Joung, B.H. Lee, H.S. Kim, D.S. Sohn, J. Nuclear Mat. In Press (2007)
2. H.K. Mueller-Buschbaum and K. Scheunemann, J. Inorg. Nucl. Chem., 35, 1091-1098, (1973)
3. A.M. Zaslavskii, A.V. Zverlin, A.V. Melnikov, J. Mat. Sc. Letters, 12 , 350-351 (1993)
4. M.A. Petrova, A.S. Novikova, and R.G. Grebenshchikov, Inorg. Mat., Vol 39 509-513 (2003)
5. Y.F. Shepelev, M.A. Petrova, and A.S. Novikova, Glass Phy. and Chem., 30 (4) 342-344 (2004)
6. Stadelmann, P.A., JEMS – EMS java version, 2004, cimewww.epfl.ch/people/stadelmann
7. CrystalDiffract 1.1 for Windows, 2007 Crystal Maker software
8. Jade for Windows, XRD Pattern Processing for the PC, Materials Data Inc., Livermore, CA, 94551

Radiation Effects in Ceramics

Mater. Res. Soc. Symp. Proc. Vol. 1107 © 2008 Materials Research Society

Characterisation of 20 Year Old Pu^{238}-Doped Synroc C

M.J. Hambley, S. Dumbill, E.R. Maddrell and C.R. Scales.
Nexia Solutions Ltd, Sellafield, Seascale, Cumbria, CA201PG, UK

ABSTRACT

In 1987 samples of Pu^{238} and Cm^{244} doped Synroc C were prepared in the UKAEA Laboratories at Harwell. They were studied for five years before being archived. During decommissioning of the Harwell laboratories the samples were transferred to Sellafield and the opportunity was taken to conduct further studies. Given the age of the samples, they offer a unique insight into the long term radiation stability of Synroc. To date, the three Pu^{238} samples have been examined. The alpha decay dose experienced by the samples is estimated to be 3 x 10^{19} alphas per gram.

The sample allowed to accumulate alpha decay damage (10587) was slightly heterogeneous, with an apparent grain size in the range 3-15μm. EDX analysis confirmed the phases present to be those expected in Synroc C and that the Pu had partitioned predominantly into the perovskite and zirconolite. Microcracking was observed in the hollandite and rutile but cracks arrested once they reached a zirconolite or perovskite grain.

Sample 10588 was annealed after five years, at temperatures up to 1200°C, this sample was microstructurally similar to 10587 at the 20-50μm scale but differs at small scales. A major difference is the presence of both intergranular and intragranular porosity.

Sample 10589 was annealed at the same time as 10588 but differed significantly probably due to actual fabrication temperatures being higher than those recorded.

Information on these samples will be presented, along with a discussion of the implications for the expected long term stability of titanate ceramic wasteforms.

INTRODUCTION

About 85% of the radiation damage to solidified high level radioactive waste is caused by the recoiling nuclei following alpha decay of the incorporated actinide elements (Pu, Cm, Am)[1,2].

The dose to the solid occurs over a long period of time, for PWR waste it reaches about 10^{18} alpha decays in around 1000 years and about 10^{19} alpha decays in around 1 million years.

One of the methods of testing wasteforms for their stability to alpha decays is to dope them with high concentrations of a short half life, high specific activity actinide, so that they receive as high a dose in a few years as the real wasteform will receive in hundreds of millennia.

Samples of aged Pu^{238}-bearing Synroc, which were made at Harwell in the late 1980s and studied for five years before being archived, have recently been examined using Scanning Electron Microscopy (SEM) and Transmission Electron Microscopy (TEM) techniques. The aim of the work was to investigate and document the state of these samples which had been stored for a considerable time.

EXPERIMENT

The samples (identified in earlier work as 10587, 10588 and 10589[3]) were from the second series of Pu^{238}-bearing, Synroc C formulated specimens, which were made in about 1987 at Harwell. They were made using Sandia precursor with all actinides and rare earths in the simulant waste replaced by Pu^{238}, leading to a target concentration of 10.4 wt % Pu in the wasteform. Each of the three samples was individually pressed from 0.8g of powder, taken from the same batch of material, in a graphite die held under a pressure of 23.2 MNm^{-2} at 1170 °C for 2 hours. The activity of the samples was estimated to be 55 GBq alpha g^{-1} or 1.74×10^{18} alpha disintegrations y^{-1} g^{-1}[3].

After initial characterisation in the late 1980's (mainly using XRD and density mensuration), the samples were periodically leach tested and their densities monitored until, in the early 1990's, after about 9.7×10^{18} alpha disintegrations g^{-1}, two of the three Pu^{238} Synroc samples (10588 and 10589) were isochronally annealed[4].

To date, the alpha decay dose to the samples is estimated to be 3×10^{19} alphas per gram. In Synroc C, the Pu^{238} host phases are zirconolite and perovskite which together constitute around 50% of the sample. Thus, the alpha decay dose experienced by Pu^{238} host phases is estimated to be of the order of 6×10^{19} alphas per gram. In practice, as the Pu partitions preferentially into the perovskite, the alpha decay dose to the perovskite will be larger than that to the zirconolite.

In the current work, fragments of material taken directly from the storage containers were examined in the as received state by SEM. Small fragments (1 mm square or less) of the same material were mounted and polished before also being examined in the SEM. The SEM used was a Leica S440 fitted with a GEM detector and Oxford Instruments ISIS energy dispersive x-ray analysis system. Qualitative energy dispersive X-ray analysis (EDX) was used to evaluate the phases present in the samples.

Grid samples for TEM were prepared in a glovebox by crushing a very small piece of the Synroc in alcohol and grinding the powder until a very fine suspension was produced. The samples were examined using a Philips EM400T TEM operated at 120kV, interfaced to an Oxford Instruments energy dispersive X-ray (EDX) detector and INCA analysis system. Convergent beam diffraction patterns and EDX spectra were used to identify the phases found in these samples and assess their self-irradiation damage.

DISCUSSION

Sample 10587

Sample 10587 was the only specimen that was not annealed during the 1990's and as such, was considered to be most indicative of a wasteform which had aged naturally.

Backscattered SEM imaging (Figure 1) and qualitative EDX analysis on the polished fragment of sample 10587, showed that the microstructure was fairly heterogeneous, consisting primarily of rutile (TiO_2) or Magnéli phase interspersed with areas of mixed perovskite ($CaTiO_3$), zirconolite ($CaZr_xTi_{3-x}O_7$) and hollandite ($BaAl_2Ti_8O_{16}$), on a scale of around 20 to 80 microns. Alloy particles containing Ru, Fe, Ni, Mo and sometimes Pu were also present. As expected, the Pu was primarily located in the zirconolite and perovskite phases, with the perovskite having the highest Pu content. The alloy particles often appeared spherical, were generally less than 1 micron across and were mainly seen to be present at triple points. Small islands (generally of the order of 1 to 3 microns) of a higher atomic number phase, probably zirconolite, were present within the rutile areas.

The actual grain sizes of each of the ceramic phases were difficult to ascertain with any certainty. Although the hollandite grains initially seemed to be of the order of between 3 and 15 microns from SEM examination, the TEM work seems to suggest that many grains were actually smaller than this. This also applied to the rutile grains which seemed to vary from about 2 to around 20 microns from SEM evaluation but TEM investigation showed that grains existed down to 30 nm.

Secondary electron SEM imaging showed that there was very little porosity in this sample. There was no evidence of bubbles or interlinked voids on any of the grain or phase boundaries.

Cracking was present in the hollandite and rutile / Magnéli phase regions but stopped abruptly at the interfaces between these and the zirconolite / perovskite regions. The cracking appeared at first to be located along grain boundaries in both these areas, although given the sheer range of grain sizes apparently present in the sample this is not certain. There was no cracking in the zirconolite or perovskite phases. Boundaries between the zirconolite and perovskite appeared diffuse and not particularly well defined. The perovskite was generally located within the zirconolite areas. There was no evidence of Pu or other actinides within the hollandite or rutile. Figure 1(a) shows the typical microstructure of the sample and 1(b) shows detail of the cracking.

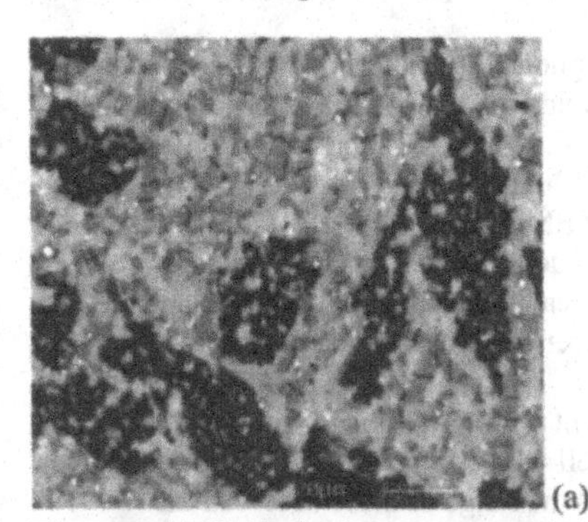
(a)

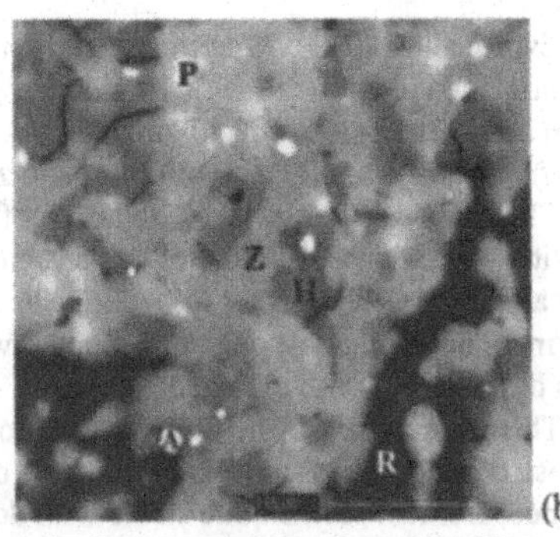

(b)

Figure 1 (a) Typical microstructure of 10587 (b) Detail of microcracking. R= rutile, H= hollandite, Z=zirconolite, P= perovskite, A=alloy.

The cracking pattern seems consistent with swelling in the Pu bearing phases due to radiation damage. Perovskite has been shown to swell by around 15% at saturation and zirconolite by around 6% at saturation[5]. The hollandite and rutile phases which do not accommodate the Pu, remain crystalline and do not swell appreciably. The resulting differential strains appear to have caused cracking in the hollandite and rutile.

In the TEM, six major matrix phases were identified: rutile, a Ti-Al rich phase (possibly a Magnéli phase ($(Ti,Al)_xO_{2-x}$), zirconolite, perovskite, hollandite and a silicon-rich phase believed to be titanite ($CaTiSiO_5$). A number of minor phases were also found including Fe_2O_3, CaO and

$BaSO_4$. Many of the EDX spectra were found to include a contribution from either lead or from metallic nanoparticles containing most of the elements Fe, Ni, Mo, Ru, U and Pu. The major phases were identified by the characteristic appearance of their X-ray spectra and their structure was confirmed by the analysis of convergent beam diffraction patterns acquired from the same location.

The diffraction patterns obtained from typical rutile grains were seen to be consistent with the tetragonal structure of rutile but with lattice parameters of a=0.38 and c=0.296nm, compared to published rutile lattice parameters of a=0.459 and c=0.296[6]. Rutile always showed near-zero concentrations of U and Pu and was present as grains of 30nm up to ~1 µm diameter.

The phase tentatively identified here as a Magnéli phase has a consistent composition of $(Ti_{0.9}Al_{0.1})_xO_{2-x}$ and has diffraction patterns similar to rutile.

The zirconolite phase is one of the major actinide host phases in Synroc[7]. This phase typically contained U and several atomic percent Pu. The actinide content of the zirconolite phase varied in both the ratio of U to Pu and in the absolute concentration of both species (Table I).

The zirconolite diffraction pattern showed the phase to be amorphous. Close inspection of the diffraction patterns shows the presence of two concentric rings corresponding to interatomic distances of 0.246nm and 0.165nm in convergent beam diffraction patterns, or 0.28 and 0.15nm in selected area patterns. The reason for the different position in the two types of pattern is not understood at present. In the literature, the reported studies all used (nearly) single phase samples and consequently they were able to analyse the samples using selected area patterns.

Ewing and Headley[8] report a strong inner amorphous ring alongside two weak outer amorphous rings in zirconolite. The inner ring in Ewing and Headley's diffraction patterns is at 0.302nm and they contrast this to other Ti-Nb-Ta oxides (0.302 to 0.315nm), silicates (0.306 to 0.323nm), a borosilicate glass (0.342nm) and zircon (0.352nm).

Clinard et al[9] observed high levels of strain in TEM samples of aged ^{238}Pu-bearing zirconolite where dark-field images and selected area diffraction pattern measurements were made at alpha doses up to 6.1 $x10^{25}$ α/m^3. The material became metamict between 4.2 and 6.1 $x10^{25}$ α/m^3. The diffraction rings in Clinard's work occurred at a distance of 0.28nm and the position did not vary with alpha dose.

The measurements on 10587 are in good agreement with the published literature. The alpha dose of the present samples (~1.6 $x10^{26}$ α/m^3) is well in excess of the doses reported above so this measurement seems quite stable with dose beyond the range of those observations.

Hollandite is readily distinguished from the other phases by the Ba content. Its diffraction patterns show that it is crystalline and the structure is consistent with a tetragonal lattice with parameters a,b =0.802nm, c =0.2865nm. Comparing this with the data of Cheary[10] who studied a range of barium hollandite compositions, the lattice parameters are low, even allowing for the fact that the lattice parameters of the present hollandites would be expected to be smaller given that they are not as Ba rich as those studied by Cheary.

In the TEM, perovskite was the least commonly found of the major phases, despite being one of the Pu bearing phases. Only a few grains were located in each of the TEM grids examined and no satisfactory diffraction patterns were obtained. A reliable method for quantitation of the Pu is still being developed. However, Table I summarises the current assessment of the composition of the zirconolite and perovskite phases

Table I : Current assessment of the composition of the Pu bearing phases

	10587	10588	10589
Perovskite $CaTiO_3$	Balance Ti 20-30at% Ca 2-4at% Pu 0-1.0at% U	Balance Ti 23-27at% Ca 3-5at% Pu ~0.5at% U	Balance Ti 20-40at% Ca 1-8at% Pu 0.4-1.2at% U
Zirconolite $CaZr_xTi_{3-x}O_7$	Balance Ti 18-30at% Zr 12-20at% Ca 2.5-6.0at% Al 1-3at% Pu 0-3at% U	Balance Ti 17-22at% Zr 18-20at% Ca 2.5-6.0at% Al 2-8at% Pu 1-4at% U	Balance Ti 24-30at% Zr 15-20at% Ca 2-5at% Al 2-3.5at% Pu 0.5-1.5at% U

The fifth major phase found was very silicon-rich compared to the other phases and is believed to be titanite ($CaTiSiO_5$). Typically, very low concentrations of lead were associated with this phase.

Annealed sample 10588

SEM imaging (Figure 2) and EDX analysis, showed that microstructurally, 10588 broadly resembled 10587 in terms of inhomogeneity, phases present, location of the actinides and cracking. Although less cracking was observed in the rutile phase of 10588 than in that from the un-annealed sample 10587

In terms of porosity however, 10588 was noticeably different. A considerable amount of porosity was present in the annealed sample. The porosity appeared to be present as voids (sometimes interlinked) at triple points and along the phase boundaries. The porosity did not generally appear as small round bubbles or interlinked chains of bubbles on the grain boundaries of each phase as might have been expected had the porosity originated from purely radiogenic generation of helium. In one or two areas, tiny bubbles had started to form in the perovskite. Figure 2(a) shows the typical microstructure of sample 10588 and 2(b) shows detail of cracking and porosity.

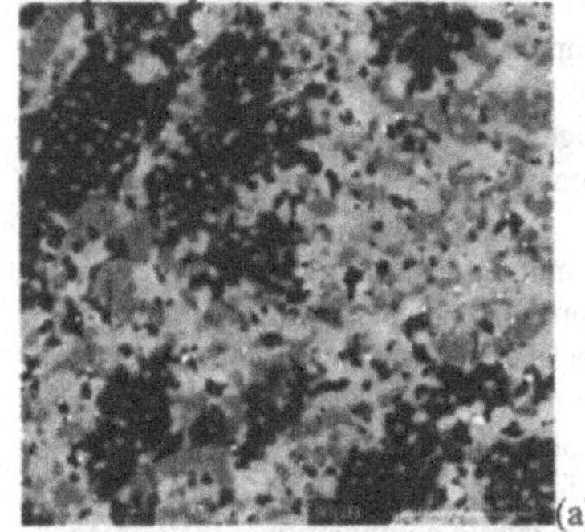
(a) (b)

Figure 2 (a) Typical microstructure of 10588 (b) Detail of microcracking and porosity. R= rutile, H= hollandite, Z=zirconolite, P= perovskite, A=alloy.

Given that the main difference between 10587 and 10588 lies in the annealing treatment, it is most probable that the porosity in 10588 largely occurred on annealing of the sample. This is likely to be due to gases, previously adsorbed onto the surfaces of the calcined material and trapped during hot pressing, being subsequently released during annealing, as well as to the potential release of any radiogenic He formed. Significant porosity formation by the subsequent release on annealing of gases trapped during hot pressing has been seen at ANSTO in non-radioactive samples[11]

Although the grains in this sample varied from about 8 to 25 microns, areas that had grains in the 20 to 25 micron range occurred with more frequency, giving the overall impression of a larger grain size in 10588 than 10587.

Examination of this sample in the TEM showed essentially the same features as in sample 10587. Any differences in grain size, between 10588 and 10587, were not apparent in the TEM samples, probably because the variability referred to above is in a size range above that of the TEM observations. Images of the zirconolite phase show the presence of small (3-10nm diameter) bubbles which may be radiogenic helium (Figure 3).

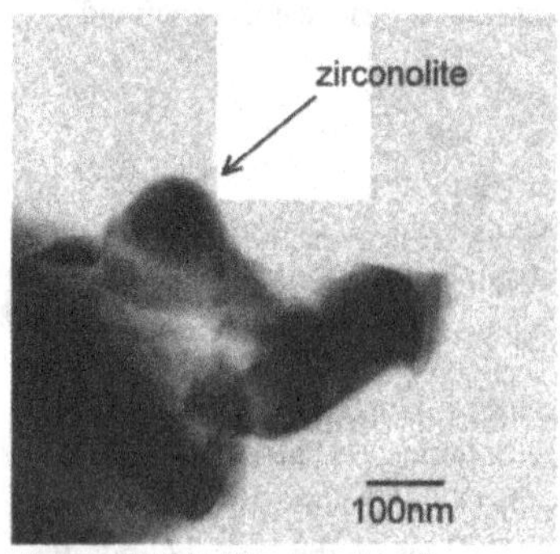

Figure 3 TEM image showing small bubbles in the zirconolite phase.

Annealed Sample 10589

The microstructure of this third sample (Figure 4) was seen to differ strikingly from the first two samples examined, despite the fact that all three samples were pressed from the same batch of powder. In the SEM, low magnification backscattered imaging showed that microstructure was very heterogeneous. The scale of inhomogeneity and the apparent grain sizes of all phases in this sample seemed larger than those in the previous two samples. Larger grain sizes were also observed in the TEM examinations.

The phases and phase compositions found, though, remained the same as in the previous two samples. The alloy phases in sample 10589, whilst still primarily located at triple points, were no longer spherical in shape and many were of the order of 5 microns or more.

The bulk of the sample consisted of areas of rutile (or related Magnéli phases) surrounded by areas of mixed zirconolite and perovskite. Although tiny islands of a higher atomic number phase were still present within the rutile areas, they were far fewer in number and mostly sub micron in size. Backscattered SEM imaging also showed some subtle compositional differences within the rutile. In the fragment examined, the hollandite was concentrated in and confined to just a few specific areas and was not intimately mixed with the zirconolite and perovskite as in the 10587 and 10588 fragments. Consequently there appeared to be less hollandite in this sample. However this may well just be a reflection of the extremely inhomogeneous distribution of hollandite in the 10589 sample as a whole. There was a more distinct delineation of the zirconolite and perovskite phases in this sample than in the previous samples. Diffraction patterns from a few zirconolite grains appeared to show some crystallinity.

In the SEM, Secondary electron imaging showed clear evidence of cracking in all the ceramic phases except perovskite. Many of the cracks in the rutile and hollandite areas were well developed and quite wide. The cracks in the zirconolite appeared to be the result of interlinkage of small spherical bubbles on the grain boundaries. Interlinked bubbles were also seen on some of the hollandite grain boundaries. Considerable amounts of small spherical bubbles were present in the perovskite areas. The bubbles in the perovskite regions seemed to be situated on the boundaries of grains which were around 2-3 microns in size. Figure 4(a) illustrates the typical microstructure of sample 10589 and 4(b) shows detail of the cracking and bubbles.

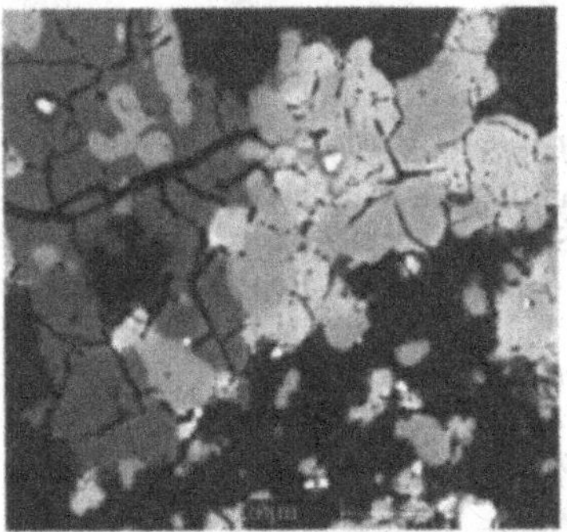
(a)

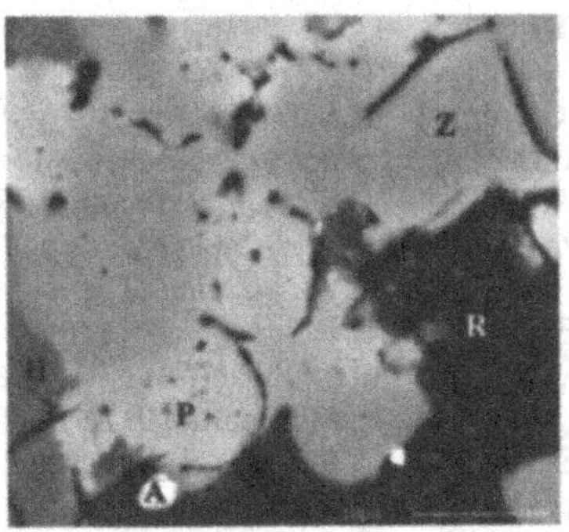

(b)

Figure 4 (a) Typical microstructure of 10588 (b) Detail of microcracking and porosity. R= rutile, H= hollandite, Z=zirconolite, P= perovskite, A=alloy.

Although the general microstructure of 10589 appears to be largely consistent with a structure that has undergone extensive re-crystallisation during the annealing procedure, the significant differences between 10588 and 10589, both of which underwent the same annealing process, suggests that it is more likely that the sample was actually fabricated at a higher temperature than originally intended. This resulted in a wasteform sample which was both too heterogeneous and had too large a grain size.

10589 was the only sample of the three currently examined which had not been previously crushed for XRD analysis. No fragments had spalled away from the pellet during storage, so the sample had to be freshly fractured prior to this examination being undertaken. Despite the extent of cracking seen, the pellet was very resistant to fracture and required significant force to be applied. Thus even in such a large grained sample, the extensive microcracking does not appear to have affected the integrity of the wasteform.

CONCLUSIONS

Extensive damage has been shown on all samples as expected due to the high alpha doses accumulated. The zirconolite and perovskite phases were observed to be amorphous and the other (non Pu-bearing) phases retained their crystalline structure.

The un-annealed sample has indicated that Synroc C is very tolerant of radiation damage effects. No significant porosity was observed despite significant storage times and although the sample exhibited some microcracking, it is not extensive and it is not believed that this would affect the wasteform integrity.

ACKNOWLEDGMENTS

This work was funded by the Nuclear Decommissioning Authority (NDA) as part of the Plutonium Disposition Programme.

REFERENCES

1. W. G. Burns et al. *J. Nucl. Mater.* **107** (1982) 245.
2. J. A. C. Marples. *Nuclear Instruments and Methods in Physics research.* **B32** (1988) 480.
3. K. A. Boult, J. T. Dalton, J. P. Evans, A. R. Hall, A. J. Inn, J. A. C. Marples and E. L. Paige. "The Preparation of Fully-Active Synroc and its Radiation Stability" Oct 87.AERE-R-13318.
4. A. Hough and J. A. C. Marples "The Radiation Stability of Synroc: Final Report". Nov 93. AEA-ES-0201(H).
5. W. J. Weber, J. W. Wald and H. J. Matzke *J. Nucl. Mater.* **138** (1986) 196
6. I E. Grey, C. Li, C.M. MacRae, L.A. Burshill *J. Solid State Chemistry.* **127** (1996) 240-247.
7. D. S. Perera, B. D. Begg, E. R. Vance and M. W.A. Stewart: in 'Advances in Technology of Materials and Materials Processing' **6(4)** (2004) pp214-217.
8. R. C. Ewing and T. J. Headley: *J. Nucl. Mater.* **119** (1983) 102-109
9. F. W. Clinard Jr, D. E. Peterson, D. L. Rohr and L. W. Hobbs: *J. Nucl. Mater.* **126** (1984) 245-254
10. R. W. Cheary, *Acta Cryst.* **B42** (1986) 229-236
11. M. Carter. (Private Communication)

Mater. Res. Soc. Symp. Proc. Vol. 1107 © 2008 Materials Research Society

Self-Irradiation of Ceramics and Single Crystals Doped With Pu-238: Summary of 5 Years of Research of the V. G. Khlopin Radium Institute

Boris E. Burakov, Maria A. Yagovkina, Maria V. Zamoryanskaya, Vladimir M. Garbuzov, Vladimir A. Zirlin and Alexander A. Kitsay
Laboratory of Applied Mineralogy and Radiogeochemistry, The V.G. Khlopin Radium Institute, 28, 2-nd Murinskiy ave., St. Petersburg, 194021, Russia

ABSTRACT

To investigate the resistance of actinide host phases to accelerated radiation damage, which simulates radiation induced effects of long term storage, the following samples doped with plutonium-238 (from 2 to 10 wt. %) have been repeatedly studied using XRD and other methods: cubic zirconia, $Zr_{0.79}Gd_{0.14}Pu_{0.07}O_{1.99}$; monazite, $(La,Pu)PO_4$; ceramic based on Pu-phosphate of monazite structure, $PuPO_4$; ceramic based on zircon, $(Zr,Pu)SiO_4$, and minor phase tetragonal zirconia, $(Zr,Pu)O_2$; single crystal zircon, $(Zr,Pu)SiO_4$; single crystal monazite, $(Eu,Pu)PO_4$; ceramic based on Ti-pyrochlore, $(Ca,Gd,Hf,Pu,U)_2Ti_2O_7$. No change of phase composition, matrix swelling, or cracking in cubic zirconia were observed after cumulative dose $2.77x10^{25}$ alpha decay/m^3. The La-monazite remained crystalline at cumulative dose $1.19x10^{25}$ alpha decay/m^3, although Pu-phosphate of monazite structure became nearly amorphous at relatively low dose $4.2x10^{24}$ alpha decay/m^3. Zircon has lost crystalline structures under self-irradiation at dose $(1.3\text{-}1.5)x10^{25}$ alpha decay/m^3, however, amorphous zircon characterized with high chemical durability. The Ti-pyrochlore after cumulative dose $(1.1\text{-}1.3)x10^{25}$ alpha decay/m^3 became amorphous and lost chemical durability. Radiation damage caused crack formation in zircon single crystals but not in the matrix of polycrystalline zircon. Essential swelling and crack formation as a result of radiation damage were observed in ceramics based on Ti-pyrochlore and Pu-phosphate of monazite structure, but not so far in La-monazite doped with ^{238}Pu.

INTRODUCTION

Crystalline ceramics are the most prospective materials suggested for the immobilization of long-lived radionulcides, in particular, weapons grade Pu and other actinides. The immobilization might be followed by: 1) transmutation (burning) followed by geological disposal of irradiated materials or 2) direct geological disposal of actinide matrices. Different durable host phases have been suggested for actinide (An = U, Pu, Np, Am, Cm) incorporation in the form of solid solutions. These are: different polymorphs of zirconia, $(Zr,An,\ldots)O_2$, in particular, one of cubic fluorite-type structure [1,2]; zircon, $(Zr,Hf,An,\ldots)SiO_4$ [3,4]; monazite, $(La,An,\ldots)PO_4$ [5,6]; Ti-pyrochlore, $(Ca,Gd,Hf,Pu,U)_2Ti_2O_7$ [7] etc. To investigate the resistance of actinide host phases to accelerated radiation damage, which simulates effects of long term storage the ^{238}Pu-doped samples of cubic zirconia and plutonia, zircon, La-monazite, Pu-monazite and Ti-pyrochlore have been repeatedly studied using X-ray diffraction analysis (XRD) and other methods. Main goal of this paper was to summarize principal features of different actinide host materials under self-irradiation from ^{238}Pu during several years.

EXPERIMENTAL DETAILS

Polycrystalline ^{238}Pu-doped samples of Ti-pyrochlore, $(Ca,Gd,Hf,Pu,U)_2Ti_2O_7$ [9,12,16]; gadolinia-stabilized cubic zirconia, $Zr_{0.79}Gd_{0.14}Pu_{0.07}O_{1.99}$ [8,9,14]; zircon/zirconia based ceramic, $(Zr,Pu)SiO_4/(Zr,Pu)O_2$ [10,17]; single crystal zircon, $(Zr,Pu)SiO_4$ [11]; La-monazite, $(La,Pu)PO_4$, and Pu-phosphate, $PuPO_4$, with monazite structure [15] and single crystal Eu-monazite, $(Eu,Pu)PO_4$ were obtained from previous research. The repeated XRD measurements were carried out at ambient temperature after different cumulative dose using special technique developed at the V.G. Khlopin Radium Institute [13]. In this method, the highly radioactive sample is placed into conventional XRD holder, which is then hermetically covered by thin (50-100 μm) Be-window. This avoids contamination of the X-ray diffractometer during analysis. All XRD analyses were carried out under the same conditions: Co K_α irradiation; current–40 mA; tube voltage–30 kV, scan speed–2 degrees/min, step analysis–0.01 degree.

Static leach tests of some samples were carried out after different cumulative doses using modified MCC-1 method. Single ceramic pellets were placed on a thin Pt-support at the bottom of Teflon™ lined stainless steel test vessel with deionized water and then set at 90°C in the oven for 28 days. Normalized Pu mass loss (NL) was calculated as follows:
$NL = A\ W/A_0\ S$, where A – total activity of Pu in the water solution and absorbed on the walls of test vessels after leaching, Bq; A_0 – the initial activity of Pu in the specimen, Bq; W – the initial mass of the specimen, gram; S – specimen surface area without correction for ceramic porosity, m^2.

RESULTS AND DISCUSSION

Ti-pyrochlore

The Ti-pyrochlore became amorphous (from XRD measurements) at cumulative dose of $(1.1\text{-}1.3)\text{x}10^{25}$ alpha decays/m^3. This was accompanied by decrease of ceramic density of approximately 10 % in comparison with initial fresh sample. Many cracks have been observed by optical microscope in ceramic matrix after dose $5.7\text{x}10^{24}$ alpha decays/m^3. Unfortunately, an initial ceramic sample has not been studied by optical microscopy immediately after synthesis. The same situation took place with electron microprobe analysis (EMPA) of this sample. At cumulative dose $5.7\text{x}10^{24}$ alpha decays/m^3 an essential chemical inhomogeneity of pyrochlore phase has been observed. The simplified variation of the estimated pyrochlore formula was $Ca_{0.9}(U_{0.3\text{-}0.5}Pu_{0.2\text{-}0.5}Gd_{0.1\text{-}0.2}Hf_{0.1\text{-}0.2})Ti_2O_7$, however, there are no EMPA data for the initial sample. Some as assumed new-formed inclusions of $(U,Pu)O_x$ and $(Hf,Ti,Ca)O_x$ were identified in ceramic matrix by scanning electron microscopy (SEM). In according to our visual observation from SEM the amount of these inclusions increased with the increase of cumulative dose [16]. This correlates with a shift of main pyrochlore peaks (222) and (400) depending on cumulative dose to the low angle direction that corresponds to position of pure Ti-pyrochlore phase, $Gd_2Ti_2O_7$. We assumed that self-irradiation might cause partial destruction of solid solution, $(Ca,Gd,Hf,Pu,U)_2Ti_2O_7$, into different phases such as: Ti-pyrochlore with less amount of impurities, in particular, U and Pu; $(U,Pu)O_x$ and $(Hf,Ti,Ca)O_x$. However, further investigation is required in order to prove such rapid formation of new formed phases and to explain the mechanism of this process.

Zircon and tetragonal zirconia

Self-irradiation of zircon based ceramic caused amorphization of zircon phase at cumulative dose (1.3-1.5)x10^{25} alpha decays/m^3. In general, such a behavior of zircon under radiation damage has been expectable and it correlated well with increase of unit cell parameters depending on cumulative dose (see table I).

Table I. Unit cell parameters of ^{238}Pu-doped zircon, $(Zr,Pu)SiO_4$, depending on cumulative dose.

Cumulative dose, alpha decay/m^3 x10^{23}	Unit cell parameters, angstroms	
	a	***c***
3.2	6.639(5)	6.014(10)
13	6.642(5)	6.019(10)
30	6.654(5)	6.035(10)
51	6.668(5)	6.04(1)
65	6.679(6)	6.05(1)
91	6.689(5)	6.06(1)
113	6.689(2)	6.07(1)
134, 151, 166	-	-

Tetragonal zirconia, $(Zr,Pu)O_2$, which was a minor phase in zircon based ceramic, has demonstrated significantly higher resistance to radiation damage in comparison with zircon. It remained crystalline when zircon became completely amorphous. No evidence of ceramic matrix swelling or cracking was found by optical microscopy after zircon amorphization. Self-irradiation changed color of ^{238}Pu-doped single crystal zircon from initial pink-brown to yellow-grey and then to grey-green, and caused essential crack formation in crystal matrices (Fig.1). We assumed that main reason of crack formation in zircon single crystals was not just swelling but inhomegeneous (zoned) Pu distribution in crystal matrices. Self-irradiation of zircon crystal caused formation of dispersed particles around crystals (Fig.1, fourth image). We consider these particles as a result of mechanical destruction of zircon crystal matrix under alpha-irradiation. This effect has not been noticed for ceramic samples so far. However, it should be noticed that ceramic samples have not kept in a special glass cassette and repeatedly controlled by optical microscope as single crystals.

Cubic zirconia

Gadolinia-stabilized cubic zirconia, $Zr_{0.79}Gd_{0.14}Pu_{0.07}O_{1.99}$, retained crystalline structure after extremely high dose of self-irradiation, 2.77x10^{25} alpha decay/m^3. No changes in the measured ceramic density, matrix swelling, or cracking were observed. No inclusions of new-formed separate Pu phases have been so far found in ceramic matrix by SEM method. Detailed XRD analysis has demonstrated unusual behavior of main (111) reflection of cubic zirconia depending on cumulative dose. At dose (in alpha decay/m^3x10^{23}): 3; 27; 62 and 134 this peak looked nearly the same, but at 110; 188; 234 and 277 it became more narrow and its intensity increased [14]. We assumed that self-irradiation of cubic zirconia by is accompanied with two processes: accumulations of defects in fluorite-type crystalline structure and repeated self-annealing of those defects at ambient temperature.

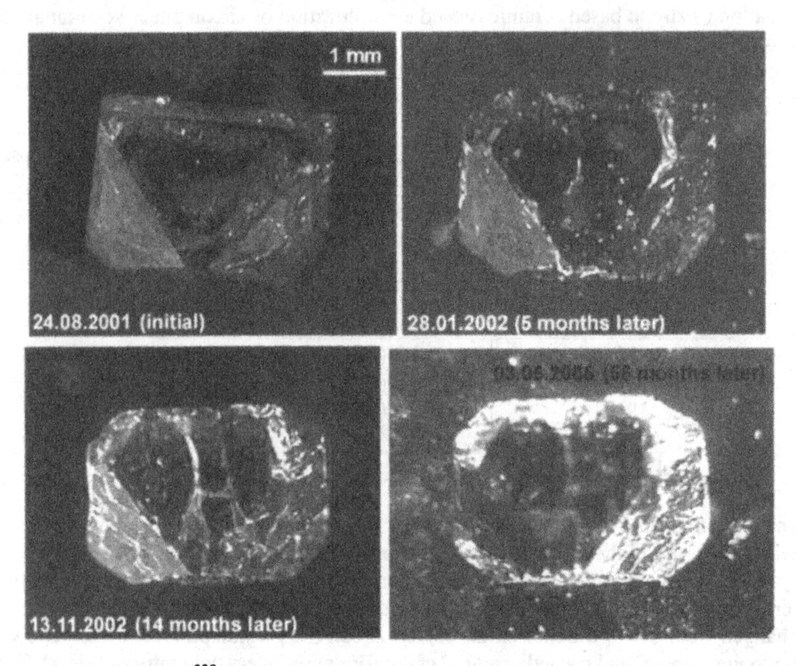

Figure 1. Cracking of ^{238}Pu-doped zircon single crystal under self-irradiation. Average bulk content of ^{238}Pu analyzed by gamma-spectrometry is 2.4 wt.% el. The distribution of all Pu isotopes in the zircon crystals analyzed by EMPA is not homogeneous and ranges from 1.9 to 4.7 wt. % el. The lowest Pu concentrations occur in outer edge regions of the zircon crystals.

Monazite

The La-monazite, (La,Pu)PO_4, remained crystalline at cumulative dose 1.19×10^{25} alpha decay/m^3. Self-irradiation caused repeated change of intensity and width of monazite XRD peaks [15]. No essential changes of unit cell parameters depending on cumulative dose have been observed. No swelling or crack formation in ceramic matrix has so far been observed. Under radiation damage La-monazite changed the color from initial light blue to grey. The Pu-phosphate, $PuPO_4$, with monazite structure became nearly completely amorphous at a relatively low dose 4.2×10^{24} alpha decay/m^3 [15]. Minor phase in the in Pu-phosphate ceramic, e.g. PuP_2O_7, became amorphous earlier than $PuPO_4$. Swelling and crack formation as a result of self-irradiation damage was observed in this ceramic. Also, under self-irradiation this sample completely changed color from initial deep blue to black. Single crystal Eu-monazite,

$(Eu,Pu)PO_4$, doped with 4.9 wt.% was not studied by XRD. No cracks were observed in matrices of these crystals after 36 months; however, the formation of dispersed particles around crystals (Fig.2, second picture) started earlier than for ^{238}Pu-doped zircon crystals and it was clearly observed already after 14 months since crystal synthesis.

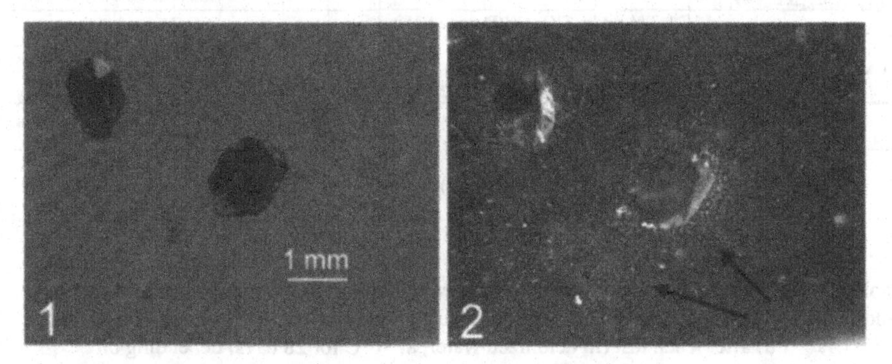

Figure 2. Single crystals of Eu-monazite, $(Eu,Pu)PO_4$, doped with 4.9 wt.% ^{238}Pu: 1) immediately after synthesis and 2) 14 months later. Formation of dispersed particles around crystals is marked by arrows.

This observation correlates with less mechanical durability of monazite matrix in comparison with zircon if we assume that dispersed particles is a result of mechanical destruction under alpha-self-irradiation from ^{238}Pu.

Summary on ceramic properties and leach tests

Principal features of ^{238}Pu-doped ceramics studied are summarized in Table II. The results of static leach tests of ^{238}Pu-doped ceramic based on Ti-pyrochlore, cubic zirconia and zircon /zirconia (see table III) are characterized with significant uncertainties because it is impossible to provide correction for ceramic porosity. There is no doubt; however, that under self-irradiation Ti-pyrochlore has the worst chemical durability in comparison with zircon or cubic zirconia. Zircon based ceramic doped with ^{238}Pu remains chemically durable after significant damage of zircon crystalline structure caused by self-irradiation. Plutonium release from zirconia matrix in deionized water increased approximately 10 times at cumulative dose $11x10^{23}$ alpha decay/m^3 and then stabilized at cumulative doses (in alpha decay/m^3x10^{23}): 56; 81 and 127.

Table II. Principal features of ^{238}Pu-doped ceramics under self-irradiation.

Ceramic, formula of main phase	Minor phases (method of analysis)	^{238}Pu content, wt. % el.	Initial geometric density, g/cm^3	Dose of amorphization, alpha decays/m^3 x 10^{23}

Zircon, $(Zr,Pu)SiO_4$	15 % tetrag. zirconia, $(Zr,Pu)O_2$ (XRD and SEM)	4.6	4.4	130-150 for zircon, but not for tetrag. zirconia
Cubic zirconia, $Zr_{0.79}Gd_{0.14}Pu_{0.07}O_{1.99}$	No (XRD and SEM)	9.9	5.8	No at 277
Ti-pyrochlore, $(Ca,Gd,Hf,Pu,U)_2Ti_2O_7$	1-3 % rutile, TiO_2 (XRD and SEM); new-formed (?) under self-irradiation inclusions of $(U,Pu)O_x$ and $(Hf,Ti,Ca)O_x$ (SEM)	8.7	4.9	110-130
Ln-monazite, $(La,Pu)PO_4$	No (XRD)	8.1	4.7	No at 119
Pu-monazite, $PuPO_4$	30-40 % PuP_2O_7 (XRD)	7.2	4.9	42-45 for Pu-monazite, but less for PuP_2O_7

Table III. Normalized Pu mass losses, NL(Pu), from matrices of ^{238}Pu-doped ceramics based on: gadolinia-stabilized cubic zirconia (9.9 wt.% ^{238}Pu); zircon (4.6 wt.% ^{238}Pu) and Ti-pyrochlore (8.7 wt.% ^{238}Pu) after leach test (in deionized water, at 90°C for 28 days) depending on cumulative dose. Corrections for ceramic porosity were not estimated.

Cumulative dose in ceramic doped with ^{238}Pu, alpha decays/ m^3 $x10^{23}$	NL(Pu), g/m^2	Equal years of storage calculated for the same ceramic but doped with ^{239}Pu
Cubic zirconia ceramic with density 5.6 g/m^3		
11	**0.04**	30
56	**0.35**	140
81	**0.37**	200
127	**0.24**	320
Zircon based ceramic with density 4.4 g/m^3		
7	**0.01**	30
31	**0.04**	150
43	**0.05**	210
66	**0.04**	330
Ti-pyrochlore based ceramic with density 4.9 g/m^3		
29	**0.22**	80
49	**0.28**	140
100	**0.84**	280
133	**1.93**	380

CONCLUSIONS

1. The Ti-pyrochlore, $(Ca,Gd,Hf,Pu,U)_2Ti_2O_7$, doped with ^{238}Pu became amorphous at the cumulative dose $(1.1\text{-}1.3)x10^{25}$ alpha decays/m^3. This was accompanied by decrease of ceramic density of approximately 10 % in comparison to the initial sample. Self-irradiation

caused essential chemical inhomogeneity of ceramic matrix and increase of Pu normalized mass loss under static leach tests.

2. The zircon, $(Zr,Pu)SiO_4$, doped with ^{238}Pu became amorphous at the cumulative dose (1.3-1.5)x10^{25} alpha decays/m^3. At the same dose tetragonal zirconia, $(Zr,Pu)O_2$, which is a minor phase in zircon-based ceramic, has demonstrated significantly higher resistance to radiation damage and remained crystalline.
3. Self-irradiation changed color of ^{238}Pu-doped single crystal zircon from initial pink-brown to yellow-grey and then to grey-green. Radiation damage caused essential crack formation in matrices of ^{238}Pu doped zircon single crystals, although no matrix swelling or cracking was visually observed for completely amorphous polycrystalline zircon sample doped with two times higher amount of ^{238}Pu.
4. Zircon based ceramic doped with ^{238}Pu remains chemically durable after significant damage of zircon crystalline structure caused by self-irradiation.
5. The cubic zirconia of fluorite-type structure has confirmed extremely high resistance to self-irradiation. No change of phase composition, matrix swelling, or cracking in the gadolinia-stabilized cubic zirconia, $Zr_{0.79}Gd_{0.14}Pu_{0.07}O_{1.99}$, were observed after cumulative dose 2.77x10^{25} alpha decay/m^3.
6. The ^{238}Pu-doped La-monazite, $(La,Pu)PO_4$, remained crystalline at cumulative dose 1.19x10^{25} alpha decay/m^3. Under self-irradiation this sample changed the color from initial light blue to gray. No swelling or crack formations have so far been observed.
7. The Pu-phosphate of monazite structure, $PuPO_4$, became nearly completely amorphous at a relatively low dose 4.2x10^{24} alpha decay/m^3. Minor phase in the Pu-phosphate ceramic, e.g. PuP_2O_7, became amorphous earlier than $PuPO_4$. Under self-irradiation Pu-phosphate sample completely changed color from initial deep blue to black. Essential swelling and crack formations as a result of accelerated radiation damage were observed in this ceramic.
8. It is assumed on the basis of XRD data that self-irradiation of cubic zirconia, $Zr_{0.79}Gd_{0.14}Pu_{0.07}O_{1.99}$ and La-monazite, $(La,Pu)PO_4$, is accompanied with two processes: accumulation of defects in crystalline structures and repeated self-annealing of those defects at ambient temperature.
9. It is assumed that increase of Pu content in monazite structured solid solutions, $(REE,Pu)PO_4$, decreases resistance of monazite to self-irradiation.
10. Formation of dispersed particles around crystals of ^{238}Pu-doped zircon and Eu-monazite has been observed. We assume that it is a result of mechanical destruction of crystal matrices under alpha-irradiation.

ACKNOWLEDGEMENTS

Investigation of ^{238}Pu-doped pyrochlore performed under the auspices of the U.S. DOE by the Lawrence Livermore National Laboratory under Contract W-7405-Eng-48. Main research was supported in part by the V.G. Khlopin Radium Institute. Presentation of this paper at MRS'07 was supported by Symposium Organizing Committee.

REFERENCES

1. D. Carroll, *J. Am. Ceram. Soc.*, **46**, [4], 194 (1963).
2. R. Heimann, T. Vandergraaf, *J. Mater. Sci. Lett.*, **7**, 583 (1988).
3. B. Burakov, Proc. SAFE WASTE'93, 13-18/06/1993, Avignon, France, **2**, 19-28 (1993).
4. R. Ewing, W. Lutze and W. Weber, *J. Mat. Res.*, **10**, 243-246 (1995).
5. L. A. Boatner, G. W. Beall, M. M. Abraham, et al., *Scientific Basis for Nuclear Waste Management*, ed. C. J. M. Northrup Jr, Plenum Press, New York, **2**, 289-296 (1980).
6. L. A. Boatner and B. C. Sales, "Monazite", *Radioactive Waste Forms for the Future*, eds. W. Lutze and R. C. Ewing, Elsevier Science Publishers, 495-564 (1988).
7. B. Ebbinghaus, R. VanKonynenburg, F. Ryerson, *et al.*, CD-ROM Proc. Int. Symp. WASTE MANAGEMENT-98, Tucson, AZ, USA, 1998, Rep. 65-04 (1998).
8. B. E. Burakov, E. B. Anderson, M. V. Zamoryanskaya, M. A. Yagovkina, E. V. Nikolaeva, Mat. Res. Soc. Symp. Proc. *Scientific Basis for Nuclear Waste Management XXV*, **713**, 333-336 (2002).
9. B. Burakov, E. Anderson, M. Yagovkina, M. Zamoryanskaya, E. Nikolaeva, *J. Nucl. Sci. and Tech., Suppl.*, **3**, 733-736 (2002).
10. B. E. Burakov, M. A. Yagovkina and A. S. Pankov, "Behavior of Zircon-Based Ceramic Doped with ^{238}Pu under Self-Irradiation", CD-ROM Proc. Int. Conf. Plutonium Future – The Science, Albuquerque, New Mexico, USA, July 6-10, 2003, CP673, 274-275 (2003).
11. J. M. Hanchar, B. E. Burakov, E. B. Anderson and M. V. Zamoryanskaya, Mat. Res. Soc. Symp. Proc. *Scientific Basis for Nuclear Waste Management XXVI*, **757**, 215-225 (2003).
12. B. Burakov, E. Anderson, "Summary of Pu Ceramics Developed for Pu Immobilization (B506216, B512161)", *Review of Excess Weapons Disposition: LLNL Contract Work in Russia,* eds. L.J. Jardine, G.B. Borisov, *Proc.3-rd Annual Meet. for Coordination and Review of LLNL Work,* St. Petersburg, Russia, Jan. 14-18, 2002, UCRL-ID-149341, 265-270 (2002).
13. B. Burakov, "KRI studies of the U.S. Pu ceramics (B506203)", *Excess Weapons Plutonium Immobilization in Russia*, eds. L.J. Jardine, G.B. Borisov, *Proc. Meet. for Coordination and Review of Work*, St. Petersburg, Russia, Nov. 1-4, 1999, UCRL-ID-138361, 251 (2000).
14. B. E. Burakov, M. A. Yagovkina, M. V. Zamoryanskaya, A. A. Kitsay, V. M. Garbuzov, E. B. Anderson and A. S. Pankov, Mat. Res. Soc. Symp. Proc. *Scientific Basis for Nuclear Waste Management XXVII*, **807**, 213-217 (2004).
15. B. E. Burakov, M. A. Yagovkina, V. M. Garbuzov, A. A. Kitsay and V. A. Zirlin, Mat. Res. Soc. Symp. Proc. *Scientific Basis for Nuclear Waste Management XXVIII*, **824**, 219-224 (2004).
16. M. V. Zamoryanskaya and B. E. Burakov, Mat. Res. Soc. Symp. Proc. *Scientific Basis for Nuclear Waste Management XXVIII*, **824**, 231-236 (2004).
17. T. Geisler, B. Burakov, M. Yagovkina, V. Garbuzov, M. Zamoryanskaya, V. Zirlin and L. Nikolaeva, *J. Nucl. Mater.*, **336**, 22-30 (2005).

Mater. Res. Soc. Symp. Proc. Vol. 1107 © 2008 Materials Research Society

Alpha-Decay Damage in Murataite-Based Ceramics

S.V. Stefanovsky[1], A.N. Lukinykh[2], S.V. Tomilin[2], A.A. Lizin[2], S.V. Yudintsev[3]
[1] Centre of Advanced Technologies, SIA Radon, 7th Rostovskii lane 2/14,
Moscow 119121 Russia
[2] Research Institute of Atomic Reactors,
Dimitrovgrad-10 433510 Russia
[3] Institute of Geology of Ore Deposits, Staromonetniy lane 35,
Moscow 119017 Russia

ABSTRACT

Samples of murataite ceramics with the composition (wt.%) 3.8 Al_2O_3, 10.5 CaO, 54.0 TiO_2, 10.6 MnO, 6.0 Fe_2O_3, 4.6 ZrO_2, 8.1 ThO_2, 2.4 Cm_2O_3 (1.8 ^{244}Cm) and a specific activity of 5.5×10^{10} Bq/g were prepared by cold pressing and sintering at 1250 °C for 24 hrs or by melting and recrystallisation in a resistive furnace at 1325 °C and 1350 °C for 1 hr. In the sintered ceramics murataite polytypes with five-fold (5C) or three-fold (3C) repeats of the fluorite unit cell and crichtonite were found to be the major phases. Perovskite, pseudobrookite, and pyrochlore were observed as minor phases. The 5C polytype was rendered X-ray amorphous at a cumulative dose of 2.73×10^{18} α-decays/g (0.21 dpa) whilst the 3C polytype, which contained only traces of Cm, remains crystalline at this dose. In the melted ceramics the 5C and 8C murataite polytypes were found to be the major phases (80-90 % of the bulk) and minor amounts of rutile, crichtonite and perovskite were also observed. Complete amorphization of the murataite polytypes in the ceramics melted at 1325 and 1350 °C was achieved at doses of 2.46×10^{18} α-decays/g (0.19 dpa) and 2.53×10^{18} α-decays/g (0.20 dpa), respectively.

INTRODUCTION

Previous studies in the system Ca-Mn-REE-Zr-An-Al-Fe-Ti-O (REE – rare earth elements, An – actinides) revealed a series of phases with fluorite-related structures that may be suitable matrices for actinide immobilisation [1-3]. The structures can be represented as a combination of alternating pyrochlore and murataite modular units [4] with pyrochlore $A_2B_2O_{7-x}$ (*Fd3m*) and murataite $A_3B_6C_2O_{20-x}$ ($F\bar{4}3m$) being the end-members in the series. In additions to pyrochlore with two-fold (2C) and murataite with three-fold (3C) repeats of the basic fluorite unit cell, other members with seven-fold (7C = 2C/3C/2C), five-fold (5C = 2C/3C), and eight-fold (8C = 3C/2C/3C) repeats have been observed [1-4]. In these structures the eight-coordinated sites [*A*] are occupied by large $An^{3+/4+}$, $REE^{3+/4+}$, Zr^{4+}, Ca^{2+}, Mn^{2+}, Na^+ cations, the octahedral sites [*B*] accommodate smaller sized Ti^{4+}, Fe^{3+}, Al^{3+}, Nb^{5+} cations, and the five-coordinated bipyramidal [*C*] sites are filled with Mn^{3+} and Fe^{3+} cations. Cations populating the [*B*] and [*C*] sites can be partly intermixed, for example up to ¼ of total Fe^{3+} present may enter the [*B*] site [5].

The numerous polytypes in the murataite/pyrochlore series have been proposed as potential wasteforms for actinide and REE fission products [3]. The solubility of actinides and REE's in these materials was shown to decrease as follows: 2C > 7C > 5C > 8C > 3C, while the content of iron group elements, Ti and Al increased. Chemical durability and radiation resistance are major requirements of HLW and actinide waste forms and these properties require detailed study. The chemical durability of murataite-based ceramics under neutral and acid media has previously been reported and was found to be similar to or higher than that of zirconolite- and pyrochlore-based ceramics [6]. The critical amorphization dose of the murataite polytypes under irradiation with 1 MeV Kr^+ ions was between 0.15 and 0.18 dpa [7] which is comparable with the value obtained for titanate-based pyrochlores (~0.20 dpa) [8]. However, ion-irradiation experiments do not adequate simulate α-decay damage in HLW forms because thin films are required and the intensity of ion irradiation (dose rate) is much higher than occurs in actual actinide containing waste forms. From this perspective actinide-doping gives a better indication of the radiation stability during storage [8]. Such investigations of murataite-based ceramics have not been reported to date.

EXPERIMENTAL

Ceramics with the following chemical composition were synthesised (wt.%): 3.8 Al_2O_3, 10.5 CaO, 54.0 TiO_2, 10.6 MnO, 6.0 Fe_2O_3, 4.6 ZrO_2, 8.1 ThO_2, 2.4 Cm_2O_3. The alpha-emitting isotope used in this study was ^{244}Cm ($T_{1/2}$ = 18.1 yrs). Cm_2O_3 with isotopic composition (wt.%) 75.5 ^{244}Cm; 15.2 ^{245}Cm; 8.6 ^{246}Cm; 0.5 ^{247}Cm; 0.2 ^{248}Cm was dissolved in 3M HNO_3. The precursor was prepared by milling an oxide mixture in an AGO-2U planetary mill at an acceleration of 50 g for 5 min. The precursor was then impregnated with the curium nitrate solution, air-dried at ~100 °C, and calcined at 800 °C for 1 hr. The product was re-milled in an agate mortar in ethanol, air-dried, and compacted into pellets 5 mm in diameter and 1-2 mm in thickness. The pellets were put in Pt crucibles, placed in a resistive furnace and heated to a temperature of 1250 °C for 24 hours, 1325 °C for one hour or 1350 °C for one hour.

The fired samples were examined by X-ray diffraction using a RKU-114 photometric chamber and a DRON-3M diffractometer (Cu K_α radiation). For the first technique powdered pellet specimens were placed within thin-walled quartz capillaries and in the second case the pellets was placed into the holder with a Be window.

RESULTS AND DISCUSSION

The pellets sintering at 1250 °C had a geometrical density of 4.05 g/cm^3. The pellets heat-treated at 1325 and 1350 °C had melted and re-solidified as thin films.

The XRD pattern of the ceramic sintered at 1250 °C shows a large number of reflections (Fig. 1 and Table I) due to incomplete phase formation. Chemical composition of individual grains was not determined due to high specific activity of the samples. Nevertheless it is varied widely because the XRD reflections were broad in shape and asymmetric. Therefore only the approximate positions of the reflections could be determined. Murataite 5C and 3C polytypes were responsible for the reflections at (Å) 2.840, 7.03, 5.44, 3.331, 2.462, 2.271, 2.244, 2.135, 1.704, 1.493, 1.434 and 2.801, 2.427, 1.716, 1.460, 1.405 were due to the presence of 5C and 3C murataite polytypes respectively. Other phases present were crichtonite (reflections at 4.159, 3.397, 3.046, 2.991, 2.883, 2.626, 2.581, 2.508, 2.419, 2.244, 2.135, 2.101, 1.913, 1.844, 1.790, 1.770, 1.596, 1.570, 1.546, 1.508 and 1.434 Å), perovskite (reflections at 2.703, 1.913, 1.568 and 1.351 Å), and pseudobrookite (reflections at 7.00, 4.86, 3.138 and 2.436 Å). Some of the broad reflections were superpositions of reflections from two phases, for example those at 3.397 and 3.046 Å could be attributed to murataite and crichtonite, and that at 1.913 Å to crichtonite and perovskite.

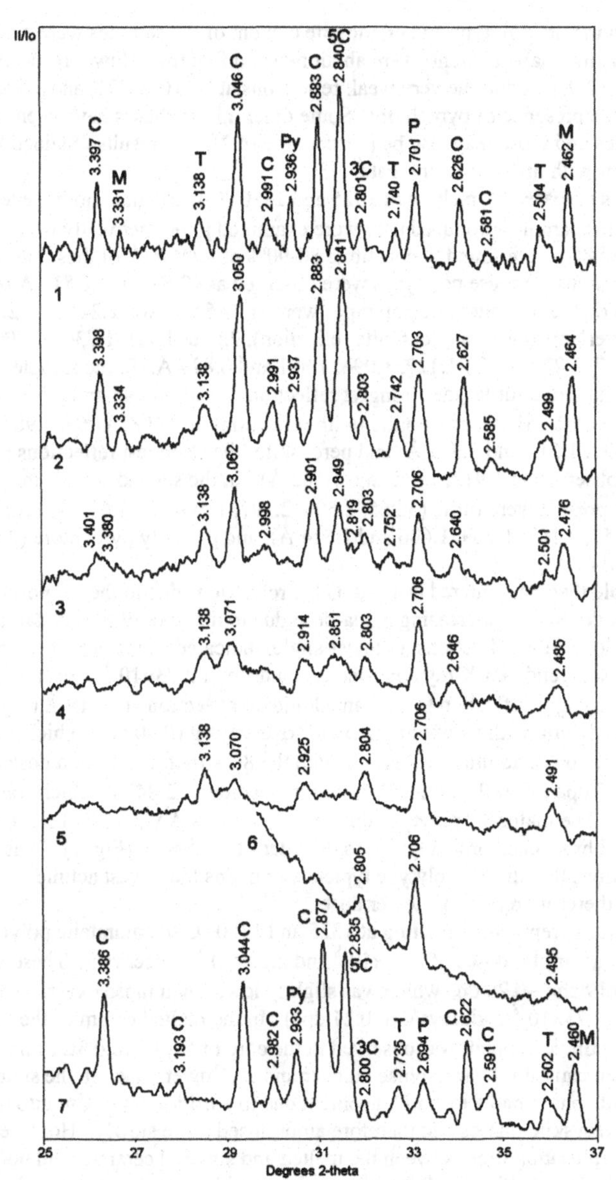

Figure 1. XRD patterns of the ceramic prepared by sintering at 1250 °C: 1 – initial; 2-6 – after α-irradiation, 10^{18} α-decays/g (dpa): 2 - 0.23 (0.02), 3 – 1.23 (0.10), 4 – 2.09 (0.16), 5 – 2.73 (0.21), 6 – 2.80 (0.22); 7 – after annealing at 1250 °C for 5 hrs. 3C, 5C – three- and five-fold murataite polytypes, M – murataite unidentified, C – crichtonite, T – titanate of pyrophanite/ilmenite series, Py – pyrochlore, P – perovskite.

The 5C murataite polytype and crichtonite content of the samples were similar and the perovskite and 3C murataite contents were about two and five times lower respectively. The weak reflection at 2.936 Å and the very weak reflections at 2.316, 1.821, and 1.533 Å were probably due to the presence of pyrochlore. Some other minor phases have been shown to occur in melted ceramics and those may also be present here [6,7]. The results obtained by photometry were in agreement with diffractometric data.

The samples prepared by melting at 1325 and 1350 °C were more homogeneous but due to small amount of the samples produced they were examined photometrically only. Two murataite polytypes (5C and 8C) were found to constitute 80-90% of total bulk of these samples and the major XRD reflections for these polytypes were observed at ~2.840 and 2.817 Å respectively. Other reflections due to the murataite polytypes were at 2.455-2.467, 2.245-2.252, 2.132-2.147 Å (this reflection overlapped with a crichtonite reflection), 2.056, 1.729-1.736, 1.476-1.479, 1.436-1.438, 1.416-1.417, 1.227, 1.122-1.127, 1.094, 1.001 and 0.829 Å. In the sample melted at 1325 °C minor phases included rutile (the strongest reflection was at ~3.244 and others were at ~2.506, 1.704, 1.599, 1.351 Å), crichtonite (with reflections at 3.408, 3.069, 2.989, 2.885, 2.635, 1.843, 1.798, 1.531, 1.511 and 0.901 Å) and perovskite (the strongest reflections was at 2.708 Å and others were observed at 1.912, 1.569 and 1.562 Å). In the second sample melted at 1350 °C the minor phases present were rutile (3.265, 2.506, 2.197, 1.694 and 1.631 Å), crichtonite (3.417, 2.903, 2.642, 2.253, 2.147, 1.769, 1.606 and 1.514 Å), and possibly pyrochlore (2.944, 2.313, 1.820 and 1.533 Å).

In the sample that was sintered at 1250 °C the reflections due to the 5C polytype reduced in intensity and broadened with increasing alpha-dose due to disorder of the crystal lattice and shifts in the position of the reflections, to lower angle, indicated expansion of the crystal lattice. The 5C polytype was rendered X-ray amorphous at a dose of 2.73×10^{18} α-decays/g (0.21 dpa) and splitting of the major reflection to give an additional reflection at 2.819 Å (Fig. 1) was observed after irradiation with a dose of 1.23×10^{18} α-decays/g (0.10 dpa) which potentially indicated the formation of additional polytype with the 8C structure. After a dose of 1.97×10^{18} α-decays/g or 0.15 dpa this reflection had shifted in position to 2.847 Å which was attributed to lattice expansion. The major 3C polytype reflection at ~2.803 Å was still observed after irradiation but had broadened probably due to disorder of the lattice (Fig. 1). This was consistent with the observations that the 3C polytype typically contains the lowest actinide content in these systems and had therefore received a lesser dose.

In the ceramics prepared by melting at 1325 and 1350 °C the murataite polytypes became X-ray amorphous at similar doses of 2.46×10^{18} and 2.53×10^{18} α-decays/g. These values correspond to ~0.19 and ~0.20 dpa which was slightly lower than those experienced by the sintered ceramic (2.73×10^{18} α-decays/g or 0.21 dpa). In the melted ceramics the 5C polytype, with the highest actinide content, was observed in the core of the 8C and 3C grains [3] and the high ^{244}Cm content caused it to amorphise quicker and at a lower dose. In the sintered ceramic the major murataite phase had a lower Cm content due to partitioning of Cm into the minor pyrochlore and perovskite phases and therefore amorphised more slowly. However, the difference in amorphisation time between the melted and sintered ceramics did not exceed ~ 10%. The average dose received, ~ 0.20 dpa, can be considered to be actual critical amorphization dose of the murataite ceramic as a whole. This value is higher than that obtained for ion-irradiated murataites (0.14-0.15 dpa) [7].

With increased accumulated irradiated dose the other phases present in the sintered sample were also eventually amorphized (Fig. 1). The reflection at 2.940 Å possibly due to the presence of a minor pyrochlore phase disappeared at a cumulative dose of 1.23×10^{18} α-decays/g (~0.10 dpa) which is about 2 times lower than that observed for the ^{238}Pu- or ^{244}Cm-doped pyrochlore [8]. The origin of this might be the high Cm concentration in the pyrochlore phase in these samples compared to others reported in the literature. At a cumulative dose of 2.73×10^{18} α-decays/g (0.21 dpa) weak reflections due to crichtonite were still present in the XRD pattern but its lattice parameter had increased by ~ 1.2%. The crichtonite phase became X-ray amorphous at a cumulative dose of 2.80×10^{18} α-decays/g (~0.22 dpa). The perovskite and pseudobrookite structured phases also appeared less susceptible to amorphization. Irradiation of the ceramic by a cumulative dose of 1.70×10^{18} α-decays/g (~0.13 dpa) slightly increased the lattice parameters but no broadening of the reflections were found. These phases, like the crichtonite, probably contained only traces of Cm.

To recover the disordered structure the amorphized sintered ceramic sample was heat-treated at 1250 °C for 5 hrs. The annealed ceramic showed almost the same phase composition as the sample before amorphization (Fig. 1, 7). The only differences observed were the intensity of some peaks and number of weak reflections due to minor phases.

CONCLUSION

Ceramics with a chemical composition (wt.%): 3.8 Al_2O_3, 10.5 CaO, 54.0 TiO_2, 10.6 MnO, 6.0 Fe_2O_3, 4.6 ZrO_2, 8.1 ThO_2, 2.4 Cm_2O_3 (1.8 ^{244}Cm) were prepared by sintering at 1250 °C or by melting at 1325 and 1350 °C. The ceramics were composed primarily of murataite polytypes and crichtonite and minor rutile, pyrophanite/ilmenite, pseudobrookite, and pyrochlore phase. The critical amorphization dose for murataite was found to be $\sim2.5\times10^{18}$ α-decays/g or ~0.20 dpa. This value is some higher than that obtained for ion-irradiated murataites (0.14-0.15 dpa) [7]. Crichtonite which was also present became X-ray amorphous at a cumulative dose of 2.80×10^{18} α-decays/g (~0.22 dpa). Annealing of the samples at 1250 °C for 5 hrs resulted in recovery of the original phase assemblage in the ceramic.

REFERENCES

1. S.V. Stefanovsky, S.V. Yudintsev, B.S. Nikonov, B.I. Omelianenko, A.G. Ptashkin, *Mat. Res. Soc. Symp. Proc.* **556**, 121 (1999).
2. S.V. Yudintsev, S.V. Stefanovsky, B.S. Nikonov, B.I. Omelianenko, *Mat. Res. Soc. Symp. Proc.* **663**, 357 (2001).
3. N.P. Laverov, S.V. Yudintsev, S.V. Stefanovsky, B.I. Omel'yanenko, B.S. Nikonov, *Geology of Ore Deposits* (Transl. from Russian) **48**, 335 (2006).
4. V.S. Urusov, N.I. Organova, O.V. Karimova, S.V. Yudintsev, S.V. Stefanovsky, *Trans. (Doklady) Russ. Acad. Sci./Earth Sci. Sec.*, **401**, 319 (2005).
5. V.S. Urusov, V.S. Rusakov, S.V. Yudintsev, S.V. Stefanovsky, *Mat. Res. Soc. Symp. Proc.* **807**, 243 (2004).
6. S.V. Stefanovsky, S.V. Yudintsev, B.S. Nikonov, A.V. Mokhov, S.A. Perevalov, O.I. Stefanovsky, A.G. Ptashkin, *Mat. Res. Soc. Symp. Proc.* **893**, 429 (2006).
7. J. Lian, S.V. Yudintsev, S.V. Stefanovsky, O.I. Kirjanova, R.C. Ewing, *Mat. Res. Soc. Symp. Proc.* **713**, 455 (2002).

8. W.J. Weber, R.C. Ewing, C.R.A. Catlow, T. Diaz de la Rubia, L.W. Hobbs, C. Kinoshita, Hj. Matzke, A.T. Motta, M. Nastasi, E.K.H. Salje, E.R. Vance, S.J. Zinkle, *J. Mater. Res.* **13**, 1434 (1998).
9. P.E.D. Morgan, F.J. Ryerson, *J. Mat. Sci. Lett.* **1**, 351 (1982).

Plutonium Wasteforms

Mater. Res. Soc. Symp. Proc. Vol. 1107 © 2008 Materials Research Society

Plutonium Feed Impurity Testing in Lanthanide Borosilicate (LaBS) Glass

Kevin M. Fox, James C. Marra, Thomas B. Edwards, Elizabeth N. Hoffman and
Charles L. Crawford
Savannah River National Laboratory, Aiken, SC, U.S.A.

ABSTRACT

A vitrification technology utilizing a lanthanide borosilicate (LaBS) glass is a viable option for dispositioning excess weapons-useable plutonium that is not suitable for processing into mixed oxide (MOX) fuel. A significant effort to develop a glass formulation and vitrification process to immobilize plutonium was completed in the mid-1990s. The LaBS glass formulation was found to be capable of immobilizing in excess of 10 wt % Pu and to be tolerant of a range of impurities. A more detailed study is now needed to quantify the ability of the glass to accommodate the anticipated impurities associated with the Pu feeds now slated for disposition.

The database of Pu feeds was reviewed to identify impurity species and concentration ranges for these impurities. Based on this review, a statistically designed test matrix of glass compositions was developed to evaluate the ability of the LaBS glass to accommodate the impurities. Sixty surrogate LaBS glass compositions were prepared in accordance with the statistically designed test matrix. The heterogeneity (e.g. degree of crystallinity) and durability (as measured by the Product Consistency Test – Method A (PCT–A)) of the glasses were used to assess the effects of impurities on glass quality.

INTRODUCTION

In the aftermath of the Cold War, the United States has identified an excess of up to 50 metric tons (MT) of weapons-useable plutonium. The Department of Energy (DOE) was to construct both a Mixed Oxide Fuel Fabrication Facility (MFFF) and a Plutonium Immobilization Program (PIP) facility to disposition this material. In April 2002, DOE decided not to construct the PIP facility and to solely proceed with the construction of the MFFF facility with a focus only on the disposition of weapons-grade plutonium to meet the non-proliferation agreement between Russia and the United States. This action resulted in up to 13 metric tons of DOE-Office of Environmental Management (DOE-EM) owned, weapons-useable, plutonium-bearing materials having no clear disposition path.

Vitrification utilizing a lanthanide borosilicate (LaBS) glass appears to be a viable option to disposition excess weapons-useable plutonium that is not suitable for processing into mixed oxide (MOX) fuel. A significant effort to develop a glass formulation and vitrification process to immobilize plutonium was completed in the mid-1990s to support the PIP. The LaBS glass formulation was found to be capable of immobilizing in excess of 10 wt% Pu and to be very tolerant of impurities [1-2]. Thus, this waste form could be suitable for the disposition of plutonium owned by the DOE-EM that may not be well characterized and that may contain high levels of impurities. However, the relative tolerance of the glass composition to the various impurities associated with current Pu feeds slated for disposition needs to be studied.

The present study focuses on the development of a composition envelope that describes the solubility of various impurities in the LaBS glass. To define this envelope, a series of glass compositions was selected, fabricated and characterized to evaluate the solubility of various impurity elements and their effects on crystallization and durability. To facilitate laboratory experiments, the glasses were formulated with Hf as a surrogate for Pu. Recent work by French researchers has indicated that HfO_2 is the best surrogate for PuO_2 in borosilicate glasses from a solubility perspective [3]. In their work, HfO_2 matched the solubility of PuO_2 (i.e. Pu in the Pu^{4+} state) much better than CeO_2 in several borosilicate glass compositions. The use of Hf as a surrogate for Pu caused some complication to the testing presented in this paper because HfO_2 is also a component of the LaBS frit. Pu glass testing will be performed on select compositions for comparison with the results of the surrogate testing. Concurrent with the glass formulation studies, melter testing is being conducted to evaluate the effects of impurities on processing [4].

EXPERIMENTAL DETAILS

Development of the Impurity Test Matrix

A detailed analysis of the anticipated Pu feeds to be immobilized in waste glass was provided by Moore and Allender [5]. The projected impurity types and concentrations described in that report were used as the basis for defining the compositions of the glasses to be fabricated for this study.

The report by Moore and Allender projected the concentrations of more than 70 possible elements as impurities in the Pu feed. This list was reduced to seventeen elements based on several criteria. First, all of the elements with a best estimate maximum concentration of 18,000 μg/g and above were included. Impurities that had relatively low concentrations and would be expected to have little impact on glass chemistry were excluded (e.g. the solubility of silicon in the glass should not be an issue and Si was removed from the impurity list). Next, sulfur and lead were included since these elements are known to typically have low solubility in the LaBS glass. Finally, selenium and cesium were included again due to low solubility being expected for these elements in LaBS glass.

Table I lists the impurities that were chosen using these criteria. For each of the elements in this table, an interval of possible concentrations is given. This interval represents the possible concentration of the indicated element as an impurity in the feed. The lower limits were defined by rounding the best estimate concentration for 50% of the projected feeds to either zero or, in the case of chlorine, to 5000 μg/g. The upper limits were defined by the greater of either the best estimate maximum concentration or the best estimate concentration for 98% of the projected feeds. These values were rounded to the nearest thousand μg/g. The concentration values were then converted to mass fractions of the Pu feed, as listed in Table I.

The chemical form of each of these impurity elements in the feed was not necessarily known, but there were some restrictions that were imposed on the approach used in developing the test matrix for this study. The first restriction imposed was a constraint on the overall mass of impurities in the feed. Moore and Allender provide total impurity concentration data for 2200 containers of the anticipated Pu feed based on Prompt Gamma Analysis and chemical estimates from laboratory samples [5]. Using these data, the total mass of the impurities was set to 35% of the overall Pu feed stream. This value was chosen to represent a worst case impurity concentration based on the data provided by Moore and Allender. Thus, on a mass basis, a design point for the study had to satisfy the constraint that the sum of the mass fractions of all of the impurities of that design point had to add to 0.35.

An additional restriction on the composition of the impurities making up a design point was required to address the issue of charge balance for that design point. If each of the impurities of Table I were converted to an oxide as a result of the vitrification process and if the feed were batched in these oxides to introduce the appropriate concentrations of all of the elements of Table I, then there would be no need for a charge balance restriction. However, for Cl, F, and S this is not the case, and the batching of the impurities that involve one or more of these elements imposed a constraint of the amounts of other impurities of Table I that had to be present to provide a charge balance for the impurity concentration. This constraint resulted in the addition of excess cation quantities to allow for the addition of the specified amount of anion per the test matrix. Again, it must be noted that the impurity variability test matrix was designed to support evaluation of impurity combinations and extremes of the impurity concentrations (i.e. "envelope" the range and type of impurities) and not specifically designed to mimic actual feed compositions.

Table I. Impurities and Their Possible Concentrations as Mass Fractions in the Feed

Element	Lower Limit	Upper Limit
Cl	0.05	0.35
Ta	0	0.315
Mg	0	0.35
K	0	0.11
Fe	0	0.08
Na	0	0.096
F	0	0.195
Ca	0	0.048
Ga	0	0.09
Ni	0	0.04
Cr	0	0.038
Cu	0	0.02
S	0	0.005
C	0	0.005
Pb	0	0.006
Se	0	0.005
Cs	0	0.005

Keeping in mind the restrictions identified above, the problem of finding feasible combinations of the impurities was considered as a mixture problem [6]. Statistical software such as JMP Version 6.0.2 is available to assist in working with such problems [7]. Using tools associated with JMP, a matrix of 60 impurity combinations was developed to facilitate the impurity testing to assess individual and interactive effects. The 60 impurity concentrations were combined with HfO_2 as a surrogate for PuO_2 to form the feed material. The feed was then combined with LaBS Frit X (Table II) to represent a 14 wt % waste loading. The 14% waste loading value represented a nominal upper bound for projected for Pu vitrification operations.

Table II. Chemical composition of LaBS Frit X

Component	Al_2O_3	B_2O_3	Gd_2O_3	HfO_2	La_2O_3	Nd_2O_3	SiO_2	SrO
wt%	10.00	13.00	13.50	7.00	19.00	15.00	20.00	2.50

Glass fabrication and characterization

Each glass was batched from the appropriate amounts of reagent metal oxides, carbonates, sulfates, fluorides and chlorides. The batches were thoroughly mixed and melted in Pt/Rh crucibles at 1450 °C for 1 hour. The glass was then quenched by pouring onto a stainless steel plate. The resulting glass patty and the remaining contents of the crucible were ground to a fine powder using a ring pulverizer to further aid in mixing. The glass powder was subsequently re-melted at 1450 °C for one hour and quenched. Testing with actual PuO_2 is conducted using this "double melt" process due to the unavailability of appropriate batch grinding equipment in the SRNL shielded cells facility. Therefore, to facilitate comparison of surrogate and future PuO_2 testing, the double melting procedure was used for the surrogate testing. The double melt process was found in previous testing to simulate fabrication of glasses via a more conventional co-grinding process [1]. Samples of each of the glasses were heat treated to simulate cooling inside a high level waste (HLW) glass canister. This heat treatment allows for the identification of any crystalline phases that may form during slow cooling within the canister.

The chemical composition of each of the study glasses was measured by Inductively Coupled Plasma – Atomic Emission Spectroscopy (ICP-AES). Samples were prepared via two methods: a peroxide fusion dissolution was used for measurement of B concentrations and a lithium metaborate dissolution was used for measurement of the other cation concentrations. Ion Chromatography (IC) was employed to measure the concentration of the anions in each glass. Samples of the quenched and slowly cooled versions of each glass were analyzed for the presence of any crystalline phases by X-ray diffraction (XRD).

The PCT-A [8] was used to measure the durability of samples of each glass, both quenched and slowly cooled. The PCT-A mandates specific leach test conditions and is the current reference test to guarantee the consistency of borosilicate glasses for HLW disposition in the U.S. Samples of each glass were ground to a size fraction of -100 to +200 mesh and placed in stainless steel vessels with de-ionized water following the test procedure. The vessels were sealed and placed in an oven at 90 °C for 7 days. The leachates were then removed from each vessel, filtered, and analyzed via ICP-AES. Normalized elemental release values were calculated using the ICP-AES leach solution data and measured glass composition data.

RESULTS AND DISCUSSION

The results of the chemical composition measurements for each of the study glasses are summarized in Figure 1. A plot is shown for each of the elements (converted to an oxide) that was varied in the glass compositions (i.e. the anticipated impurity elements and the Pu surrogate Hf).

The measured CaO and PbO concentrations were consistently higher than the targeted values. The measured Cr_2O_3 and Fe_2O_3 concentrations are very close to the targets except for the one highest targeted value for each of these components. The measured Cl^-, F^-, SeO_2 and SO_4^{2-} concentrations are well below their target values for all of the study glasses. This is likely due to volatilization of these species during melting of the glass batch. It should be noted that SeO_2 and SO_4^{2-} were tested at two concentration levels and volatility reduced the concentrations in the glass to below the analytical detection limit for all glasses (hence, the appearance of two data points in the plots in Figure 1). The measured HfO_2 concentrations were below their target values for all of the study glasses. It is likely that for HfO_2, the solubility limit in the glass was exceeded because some of the HfO_2 batch material remained in the bottom of the crucibles after pouring the glasses. As stated previously, HfO_2 was also a component of the frit. Therefore, the glasses with impurity compositions involving high PuO_2 (i.e. HfO_2) concentrations had exceedingly high total HfO_2 concentrations when combined with the frit. This apparent solubility limit behavior must be further evaluated through testing with actual PuO_2 to evaluate any potential issues with product quality. It should be noted that undissolved PuO_2 in the production-scale melter should not be a safety issue due to the criticality-safe design (size and shape) of the melter. The measured K_2O concentrations are very close to the target values up to a concentration of about 0.48 wt %, after which the measured concentrations fall below the target values. The measured MgO concentrations are very close to the target values up to a concentration of about 1.1 wt %, after which the measured concentrations fall below the target values. The measured CuO, Ga_2O_3, Na_2O, NiO, and Ta_2O_5 concentrations generally fall very close to their target values across the ranges of concentrations targeted in this study for each of these components.

The XRD results showed that all but two of the quenched glasses were X-ray amorphous. Two of the quenched glasses formulated with 4% impurities in the simulated Pu feed contained some crystalline HfO_2. This is not surprising, as the target HfO_2 concentration in these glasses was ~19.5 wt %. For the slowly cooled glasses, 3 of the 5 glasses with 4% impurities in the simulated Pu feed contained some crystalline HfO_2. This was confirmed by Scanning Electron Microscopy (SEM) as shown in Figure 2. The HfO_2 crystallites appear brighter in this backscattered electron image due to their higher average atom mass, and Energy Dispersive Spectroscopy (EDS) confirmed that the concentration of Hf in these crystallites was much higher than the surrounding glass matrix. Therefore, insolubility of HfO_2 at these concentrations is not unexpected since the solubility limit of PuO_2 in LaBS glass was found to be 13.4 wt % [1]. The basis for the use of HfO_2 as a surrogate for PuO_2 in borosilicate glasses from a solubility perspective was discussed earlier; however, the validity of the surrogate at or near the PuO_2 solubility limit has not been quantified [3]. As mentioned previously, the solubility limit of PuO_2 in the LaBS glass in the presence of impurities must be further evaluated.

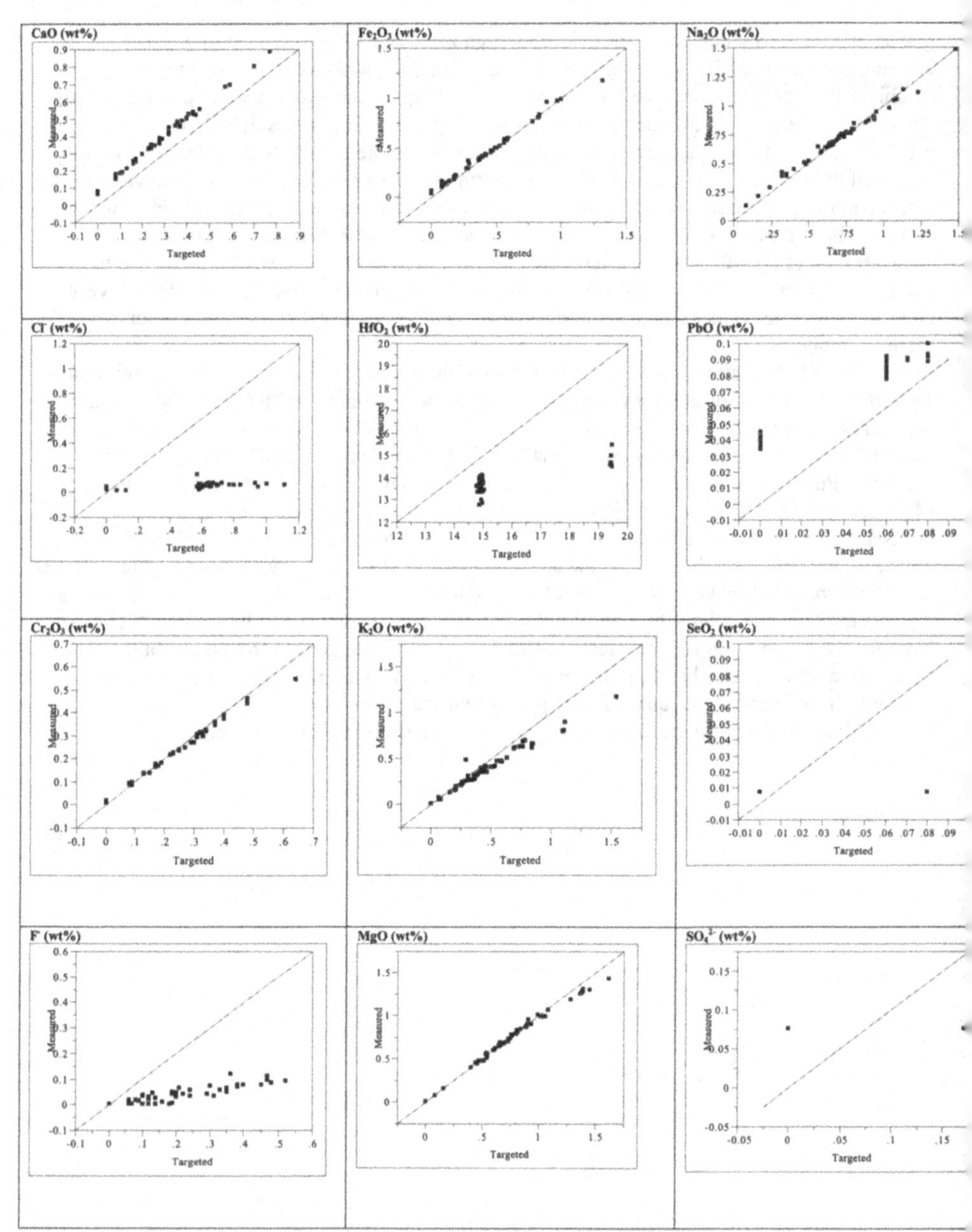

Figure 1. Measured versus Targeted concentrations for select analytes in the study glasses. Plots for CuO, Ga_2O_3, NiO and Ta_2O_3 are omitted – measured concentrations were very close to targeted for these oxides.

The results of the PCTs indicated that each of the study glasses was highly durable as compared to the Environmental Assessment (EA) glass, which is used as a benchmark for defense-related waste glasses destined for the U.S. federal repository [9]. Boron has been found to be released at the same maximum normalized concentration as some high-activity radionuclides, and is therefore typically used to gauge the performance of simulated waste glasses. The concentration of B in the PCT leachates was below the detection limit of the ICP-AES instrument for most of the study glasses. This is consistent with previous testing on a Pu-loaded LaBS glass without impurities where boron leachate values as measured by ICP-AES were also below detection limits [10]. The highest normalized release for boron measured in the study was 0.041 g/L, which is considerably smaller than that of the EA glass (16.695 g/L). The maximum normalized release rates among the 60 glasses for the other elements measured were also quite low: 0.005 g/L for Hf, 0.014 g/L for Gd, 1.229 g/L for Na and 0.090 g/L for Si. The normalized release rates for these elements were too small to attempt to correlate the results with the compositions of the test glasses. Note that the pH of the leachate solutions (typically ~8.0) were generally lower than that of typical high-level waste glasses developed for defense-related radioactive waste sludges (typically ~10.5), which may have had some influence on the PCT results.

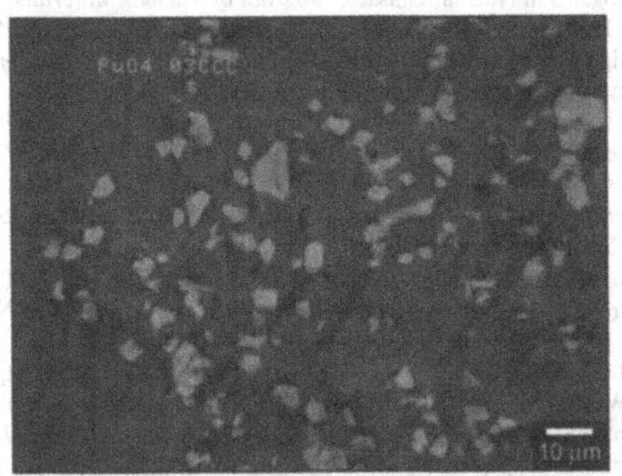

Figure 2. SEM micrograph of a LaBS glass fracture surface in an area containing undissolved HfO_2 (bright crystallites).

CONCLUSIONS

Sixty surrogate glass compositions were developed through a statistical design approach to cover the anticipated ranges of concentrations for the dominant impurity species expected in the Pu feed. These glasses were fabricated and characterized in the laboratory to determine the degree of crystallization that occurred, both upon quenching and slow cooling, and to measure the durability of each glass. XRD and SEM results indicated that some crystalline HfO_2 remained in some of the glasses with the lowest concentration of impurities. No other significant crystalline phases were identified. It is likely that for HfO_2, the solubility limit in the glass was exceeded especially in glasses that had exceedingly high total HfO_2 concentrations. This apparent solubility limit behavior must be further evaluated through testing with actual PuO_2 to evaluate any potential issues with product quality. The PCT results showed that all 60 of the glass compositions tested were very durable (regardless of thermal heat treatment), with the highest normalized release for boron being 0.041 g/L. The normalized release rates for these elements were generally too small to attempt to correlate the results with the compositions of the test glasses. Overall, the LaBS glass system appears to be very tolerant of the impurities projected to be included in the Pu waste stream. There was evidence of volatility especially with the anion species. Corresponding melter process testing is underway to address handling of the volatile species.

REFERENCES

1. J.D. Vienna, D.L. Alexander, H. Li, M.J. Schweiger, D.K. Peeler and T.F. Meaker, "Plutonium Dioxide Dissolution in Glass," *U.S. Department of Energy Report PNNL-11346,* Pacific Northwest National Laboratory, Richland, WA (1996).
2. T.F. Meaker and D.K. Peeler, "Solubility of Independent Plutonium Bearing Feed Streams in a Hf-Based LaBS Frit, contained in: Plutonium Immobilization: The Glass Option - A Compendium of Reports and Presentations," *U.S. Department of Energy Report WSRC-RP-97-00902,* Westinghouse Savannah River Company, Aiken, SC (1997).
3. C. Lopez, X. Deschanels, J.M. Bart, J.M. Boubals, C. Den Auwer and E. Simoni, "Solubility of Actinide Surrogates in Nuclear Glasses," *Journal of Nuclear Materials,* 312 (2003) 76-80.
4. J. R. Zamecnik, T. M. Jones, D. H. Miller, D. T. Herman and J. C. Marra, "Process Testing to Support the Design of a Plutonium Vitrification Facility," *Waste Management '07 Proceedings,* American Nuclear Society, Inc., LaGrange Park, IL, 2007.
5. E.N. Moore and J.S. Allender, "Projected Characteristics of Nominal Feeds to Plutonium Disposition Project," *U.S. Department of Energy Report SRNL-OPD-2007-00008,* Washington Savannah River Company, Aiken, SC (2007).
6. J.A. Cornel, *Experiments with Mixtures: Designs, Models, and the Analysis of Mixture Data*, John Wiley and Sons, New York (2002).
7. *JMP™, Ver. 6.0.2*, [Computer Software] SAS Institute Inc., Cary, NC (2005).
8. ASTM, "Standard Test Methods for Determining Chemical Durability of Nuclear Waste Glasses: The Product Consistency Test (PCT)," *ASTM C-1285,* (2002).
9. C.M. Jantzen, N. E. Bibler, D. C. Beam, C. L. Crawford and M. A. Pickett, "Characterization of the Defense Waste Processing Facility (DWPF) Environmental Assessment (EA) Glass Standard Reference Material," *U.S. Department of Energy Report WSRC-TR-92-346, Revision 1,* Westinghouse Savannah River Company, Aiken, SC (1993).
10. J.C. Marra, C.L. Crawford and N.E. Bibler, "Glass Fabrication and Product Consistency Testing of Lanthanide Borosilicate Frit X Composition for Plutonium Disposition," *U.S. Department of Energy Report WSRC-STI-2006-00318,* Washington Savannah River Company, Aiken, SC (2006).

Mater. Res. Soc. Symp. Proc. Vol. 1107 © 2008 Materials Research Society

Development of Borosilicate Glass Compositions for the Immobilisation of the UK's Separated Plutonium Stocks

Mike T. Harrison and Charlie R. Scales
Nexia Solutions Ltd, The British Technology Centre, Sellafield, Seascale, Cumbria, CA20 1PG, UK.

ABSTRACT

The UK inventory of separated civil plutonium is expected to exceed 100 tonnes by 2010. Whilst the majority of this could be used in the manufacture of MOx (Mixed Oxide) fuel in future power generation scenarios, options for the disposal of surplus plutonium are currently being investigated by Nexia Solutions Ltd on behalf of the UK's Nuclear Decommissioning Authority (NDA). One of the options being considered is immobilisation in a durable glass matrix followed by long term storage and subsequent final repository disposal.

A preliminary experimental survey assessed a selection of potential glass systems on the basis of Pu-surrogate (cerium) loading, durability, and ease of processing. Following this, a number of borosilicate compositions have been taken forward into a more detailed investigation in order to fully qualify their potential for Pu-immobilisation. The selected compositions are lanthanide borosilicate (LaBS), alkali tin silicate (ATS) and high-lanthanide alkali borosilicate (modified-MW). For this second series of experiments, hafnium was selected as the Pu surrogate, and a study of the potential waste loading as a function of temperature for the three selected compositions is described in this paper. Furthermore, several variations of the LaBS composition were fabricated in order to investigate the effect of total lanthanide content on melting temperature. The benchmark of 10 wt% HfO_2 incorporation is achievable for all three glasses with temperatures of 1200, 1300 and 1400 °C required for ATS, modified-MW and LaBS respectively.

INTRODUCTION

Vitrification is one of the alternatives being investigated for the immobilisation of surplus separated PuO_2 stocks in the UK. Most of this material could be used to fabricate Mixed Oxide (MOx) fuel. However, a range of technologies for immobilising separated plutonium is also being explored. One of these technologies is vitrification. Vitrification will be achieved by the assimilation into a durable glass waste form followed by secure above ground storage and subsequent repository disposal. This scenario will be reviewed against a set of specific performance criteria, and compared with a number of other disposition options in order to select the option or options that are capable of being demonstrated as best practice environmentally [1].

A number of glass compositions were assessed for their suitability as a matrix for Pu-immobilisation in the 'First Stage Experimental Assessment' [2], with silicates and phosphates being investigated by Nexia Solutions Ltd and the Immobilisation Science Laboratory (ISL), University of Sheffield, respectively. This initial programme investigated a wide variety of candidates selected from the available literature, and studied a range of glass properties relevant to their application for Pu-immobilisation. On the basis of these results, several candidate compositions have been carried forward to a 'Second Stage Experimental Programme'. This selection was based mainly on the ability of the matrix to incorporate high levels (≥ 10 wt%) of the Pu-surrogate material, and superior chemical durability. Processability was a secondary factor, with some very high melting (≥ 1500 °C) compositions being rejected at this stage despite forming high quality glasses.

The aims of this programme were two-fold: to investigate the temperature dependence of the Pu-surrogate loading, and to extend the durability data to include powder-based PCT-B leach tests. As before, silicate and phosphate glass compositions were divided between Nexia Solutions Ltd and the ISL. The results from the compositional study of Pu-surrogate solubility as a function of temperature completed at Nexia Solutions Ltd are presented below. The durability and leach test results are presented in a separate paper at this conference [3].

Base Glass Compositions

The silicate glass compositions that were the leading candidates from the initial experimental survey, and hence included in the 2nd stage programme were:

- *Modified-MW (MMW)* – One of the surprising results from the 1st stage programme, the addition of a mixture of Al_2O_3, CeO_2, Gd_2O_3 and HfO_2 to MW[1] yielded a highly durable product with surrogate loadings of at least 15 wt%.
- *Lanthanum Borosilicate (LaBS)* – The family of glasses developed in the US for actinide waste immobilisation [4]; Savannah River National Laboratory are currently proposing a LaBS formulation for the immobilisation of excess weapons-usable plutonium [5, 6]. It has been demonstrated as having a high capacity for Pu along with excellent corrosion resistance, although it requires melting temperatures typically in excess of 1400 °C.
- *Alkali Tin Silicate (ATS)* – Another actinide waste glass developed in the US, this formulation is not quite as durable as the other two candidates, but requires a significantly lower melting temperature to achieve the 10 wt% loading benchmark [7].

Choice of Surrogate

One of the key findings from the 1st stage evaluation was the importance of the choice of plutonium simulant in these inactive scoping experimental programmes. Initially, cerium was used, but it was concluded that its redox properties were sufficiently different to Pu to potentially give a misleading indication of solubility and hence waste loading. In glass melts, Pu will exist in either the +3 or +4 oxidation state, with the proportion of the former increasing with temperature [7, 8]. Cerium behaves similarly, but for a given temperature the reduced state is more stable resulting in a greater quantity of the trivalent species [8-10]. In borosilicate glasses, the 3+ ion is generally more soluble than the 4+, hence Ce exhibits higher solubility than Pu [8-11].

Therefore, a metal with a stable +4 oxidation state best simulates the redox chemistry of plutonium, and so hafnium was selected as the surrogate for the 2nd stage evaluation. Hafnium is also the surrogate of choice for the LaBS glass development at Savannah River for excess weapons plutonium [5, 6], which allows a direct comparison to be made between the various compositions under consideration. However, for MMW it appeared that a mixture of oxides gave the glass its beneficial properties, hence CeO_2 was included with the HfO_2 as the Pu-surrogate such that the *total* CeO_2 plus HfO_2 loading was 20 wt%.

---

[1] MW is the standard base glass used for immobilisation of HLW in the UK, and has a typical composition (in wt%) of 62 SiO_2, 22 B_2O_3, 11 Na_2O, 5 Li_2O.

Experimental Strategy

In order to investigate the limits of Pu incorporation for the three selected glass compositions, a matrix of samples was fabricated and assessed using a range of hafnium loading and temperatures. Of particular interest is the minimum temperature required to fully dissolve the benchmark 10 wt% Pu-surrogate, and the maximum loading possible at practical temperatures for industrial scale operation.

The melting strategy adopted for the 2nd stage silicate glass evaluation was the "base glass plus waste" approach. This is consistent with the globally-implemented method for HLW immobilisation, where the active waste material is added to a pre-formed inactive base glass frit. Hence, large quantities of the various base glasses under test were initially fabricated, before being mixed with the relevant amount of Pu-surrogate, and re-melted over the relevant temperature range. Table I details the base glass compositions used for the 2nd stage programme. Several different base LaBS glass compositions were fabricated on the basis of two different approaches to optimising the melting conditions. Marra *et al* [5] predicted that compositions lying along the 1:3 stoichiometric axis in the Ln_2O_3-B_2O_3-(SiO_2+Al_2O_3) ternary phase diagram should have lower melting and liquidus temperatures. Riley *et al* [12] measured a minimum liquidus temperature at a Ln_2O_3 content of ~45 wt% in the Al_2O_3-Ln_2O_3-SiO_2 phase fields. The LaBS_X glass is similar to Marra's improved LaBS composition 'Frit X' [5], LaBS_E and LaBS_F are developments from the 1st stage programme to attempt to minimise the required melting temperature using the approach of Riley [12], and LaBS_G is a hybrid that attempts to fulfil the criteria of both Marra and Riley.

Table I. Base glass compositions used for the 2nd stage experimental programme (weight %).

MMW	ATS	LaBS_E	LaBS_F	LaBS_G	LaBS_X
Al_2O_3 6.3	Al_2O_3 2.7	Al_2O_3 18.2	Al_2O_3 19.2	Al_2O_3 16.10	Al_2O_3 10.0
B_2O_3 17.9	B_2O_3 13.5	B_2O_3 10.4	B_2O_3 10.6	B_2O_3 10.00	B_2O_3 13.0
Gd_2O_3 12.5	Cs_2O 0.6	Gd_2O_3 12.2	Gd_2O_3 11.8	Gd_2O_3 13.33	Gd_2O_3 11.7
Li_2O 4.1	Gd_2O_3 3.8	La_2O_3 20.0	La_2O_3 29.5	La_2O_3 22.22	HfO_2 7.1
Na_2O 8.9	K_2O 5.9	Nd_2O_3 9.0	SiO_2 26.5	Nd_2O_3 13.33	La_2O_3 20.3
SiO_2 50.4	Li_2O 4.7	SiO_2 24.5	SrO 2.4	SiO_2 22.22	Nd_2O_3 15.4
	Na_2O 10.5	Sm_2O_3 3.2		SrO 2.80	SiO_2 20.0
	SiO_2 46.9	SrO 2.5			SrO 2.5
	SnO 3.0				
	TiO_2 2.2				
	ZrO_2 6.2				

EXPERIMENTAL

Batches of the base glasses were prepared by weighing and mixing precursor powders in the correct proportions. The precursors were as-received metal oxides or carbonates, boric acid, and milled MW glass frit[2]. Frits of the base glasses were prepared by loading crucibles with the batched precursors, melting in a furnace at an appropriate temperature for several hours, and then pouring into cold water. Once cooled, the frit was dried at ~60 °C before being ground to a powder (< 300 μm) using a ball mill. A glass monolith was also produced as a reference by

[2] MW glass frit was ground using a ball mill and sieved to obtain particles with size <500 μm.

pouring a portion of the melt into a pre-heated rectangular steel mould. This ingot was removed from the mould whilst still hot and then annealed at a temperature just below the glass transition temperature (T_g). The annealing cycle was typically a 3 hour dwell followed by a cooling to room temperature at 0.5 °C/min. The melting temperatures, times, crucible materials, frit yield, and annealing temperatures used for the base glass melts are shown in Table II.

Table II. Melting temperature, time, crucible, frit yield and annealing temperature for the base glasses.

Base Glass	*Melt Temp. / °C*	*Melt Time / hrs*	*Crucible*	*Frit Yield / %*	*Anneal Temp. / °C*
MMW[3]	n/a	n/a	Silica	n/a	n/a
ATS	1300	3.5	Silica	65	508
LaBS_E	1350	2.0	Alumina	70	750
LaBS_F	1350	3.5	Alumina	75	751
LaBS_G	1395	3.5	Alumina	75	750
LaBS_X	1400	4.0	Alumina	77	751

To make the final surrogate-loaded glass monoliths, ~50 g batches of each base glass mixed with Pu-surrogate was weighed into a suitable crucible, melted in a furnace at for 3-5 hours, poured into a pre-heated rectangular steel mould, and then annealed. The matrix of Pu-surrogate loadings and melting temperatures is given in Table III. The annealing cycle was the same as for the base glasses. Note that for LaBS_X, the HfO_2 loadings in Table III are *in addition* to the 7.1 wt% already in the base glass.

Table III. Melting temperatures and Pu-surrogate loadings used for the base glasses in Table II.

Base Glass	*Pu-Surrogate Loading / wt%*	*Melting Temperatures / °C*	*Crucible*
MMW	5 HfO_2 + 15 CeO_2	1250, 1300, 1350	Silica
	10 HfO_2 + 10 CeO_2	1250, 1300, 1350	Silica
	15 HfO_2 + 5 CeO_2	1250, 1300, 1350	Silica
ATS	10 HfO_2	1200, 1250, 1300	Silica
	15 HfO_2	1200, 1250, 1300	Silica
LaBS_E	10 HfO_2	1300, 1350, 1400	Alumina
LaBS_F	10 HfO_2	1300, 1350, 1400	Alumina
	15 HfO_2	1350, 1400	Alumina
LaBS_G	10 HfO_2	1300, 1350, 1400	Alumina
	15 HfO_2	1300, 1350, 1400, 1450	Alumina
	20 HfO_2	1450	Alumina
LaBS_X	10 HfO_2	1300, 1350, 1400, 1450	Alumina
	15 HfO_2	1350, 1400, 1450	Alumina
	20 HfO_2	1450	Alumina

A number of different techniques were used to characterise the samples produced from the above melting schedule. The bulk density of the glass monoliths was measured using the Archimedes principle in water. Scanning Electron Microscopy (SEM) coupled with Energy Dispersive Spectroscopy (EDS) was used to assess the homogeneity of the glasses, specifically

[3] Note that for MMW, the Al_2O_3 and Gd_2O_3 were added to milled MW glass along with the Pu-surrogates.

looking for undissolved HfO_2. Finally, a sample from each glass ingot was powdered and T_g measured by differential thermal analysis (DTA) using a Perkin Elmer Diamond TG/DTA.

RESULTS AND DISCUSSION

Table IV shows the results from visual and SEM inspection of the Pu-surrogate glasses. Three different types of behaviour were noted:

1. Melting produces a homogeneous[4] glass monolith with no HfO_2 crystals visible in the SEM, and no residual undissolved material in the crucible.
2. Melting produces a largely homogeneous glass with only very small quantities of HfO_2 visible in the SEM, and/or a significant quantity of undissolved material in the crucible.
3. Melting does not produce a homogeneous glass, with a significant quantity of undissolved material visible either optically or in the SEM.

Table IV. Results of Pu-surrogate glass melting: ✗ = undissolved HfO_2 in glass monolith, ? = undissolved HfO_2 remaining in crucible (and/or a few crystals in monolith), and ✓ = homogeneous glass with no HfO_2 visible in monolith or crucible.

Base Glass	*Pu-Surrogate Loading / wt%*	*Melt Time*	*Melting Temperature / °C*					
			1200	*1250*	*1300*	*1350*	*1400*	*1450*
MMW	5 HfO_2 + (15 CeO_2)	3 h		✗	✓	✓		
	10 HfO_2 (+ 10 CeO_2)	3 h		✗	✓	✓		
	15 HfO_2 (+ 5 CeO_2)	3 h		✓	✓	✓		
ATS	10 HfO_2	4 h	✓	✓	✓			
	15 HfO_2	4 h	✓	✓	✓			
LaBS_E	10 HfO_2	4 h			✗	✗	✓	
LaBS_F	10 HfO_2	a) 3 h			✗	✗	?	
	10 HfO_2	b) 5 h			?	✓	✗	
	15 HfO_2	4 h				✗	?	
LaBS_	10 HfO_2	4 h			✗	?	✓	
	15 HfO_2	4 h			✗	✗	?	?
	20 HfO_2	4 h						✗
LaBS_X	10 HfO_2	4 h			✗	✗	?	?
	15 HfO_2	4 h				✗	?	?
	20 HfO_2	4 h						?

As expected, the solubility of HfO_2 increases with temperature for all the base glasses under investigation. In addition, for the LaBS_F glass with 10 wt% HfO_2 added, an improvement in solubility was noted upon increasing the melting time from 3 to 5 hours. This indicates that regardless of the solubility limit of HfO_2 in the glasses, the dissolution can be quite slow and hence kinetic factors will have some influence on the apparent waste loading. The undissolved HfO_2 had a tendency to settle to the bottom of the crucibles due to its relatively high density (9.68 g cm^{-3}), which resulted in distinct bands of HfO_2-rich material at the top of the monolith as the glass was poured into the mould, see Figure 1. This settling is a potential criticality issue both

[4] It should be noted that 'homogeneous' in this context refers to the length scale as limited by the SEM (~ 0.1 μm).

during melting and after solidification as the PuO_2, which is more dense than HfO_2, would not be fully dispersed in the matrix. It is not known whether stirring the melts would improve the dispersion and dissolution of the HfO_2 and hence give an improvement in the incorporation of Pu-surrogate and the criticality concerns.

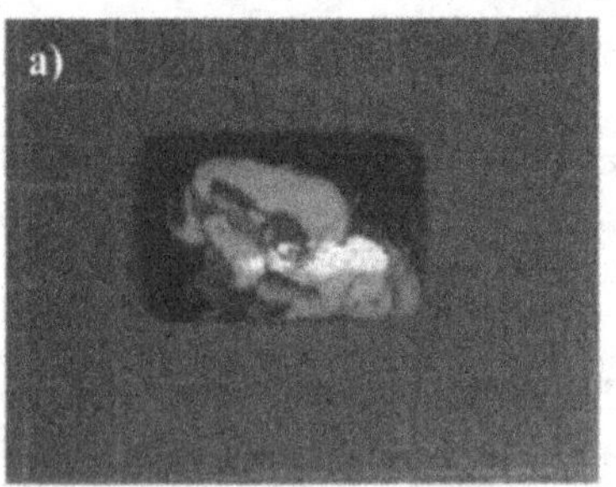

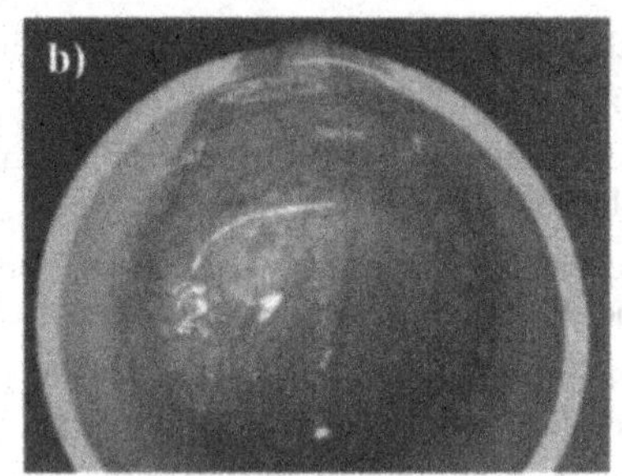

Figure 1. LaBS_X glass melted at 1350 °C with 15 wt% HfO_2 loading showing undissolved HfO_2; a) on top of monolith, and b) remaining in crucible.

MMW Glass

For temperatures of ≥1300 °C, all of the MMW series of melts produced homogeneous amber-coloured glasses with no undissolved material visible in either the monolith or the crucible. The amber colouration is caused by the cerium, and at higher concentrations of CeO_2 (10 and 15 wt%) and lower temperatures (1250 °C) this does not fully dissolve resulting in poor quality inhomogeneous glasses. Figure 2 shows the full monolith and a representative SEM micrograph for the MMW glass with 5 HfO_2 plus 15 CeO_2 wt% melted at 1250 °C. In Figure 2b, the white crystals were identified by EDS as predominantly containing cerium.

Therefore, it appears that for the MMW composition, HfO_2 is more soluble than CeO_2, with up to 15 wt% loading achievable at 1250 °C. By continuing the trend, it should be possible to replace all the Ce with Hf and achieve a homogeneous glass with 20 wt% HfO_2-loading. This is significantly higher than has previously been reported for simple alkali borosilicate glasses [8]. The solubility of the CeO_2 increases with melting temperature as expected from the increase in the proportion of Ce (III) as the Ce (IV) is auto-reduced to a greater extend [8-10]. Furthermore, it should be noted that all of the MMW glasses contain 10 wt% Gd_2O_3, which is added to function primarily as a potential neutron poison, but also as a trivalent surrogate. The presence of this may limit the amount of Ce (III) that can dissolve in the glass.

ATS Glass

The ATS composition was able to full incorporate both 10 and 15 wt% HfO_2 at temperatures as low as 1200 °C. Good quality, homogeneous glasses were obtained using a 4 hour melting time. However, at 1200 °C the melts were reasonably viscous resulting in low pouring yields of ~55%. At 1300 °C the viscosity decreased and the yields increased to ~66%.

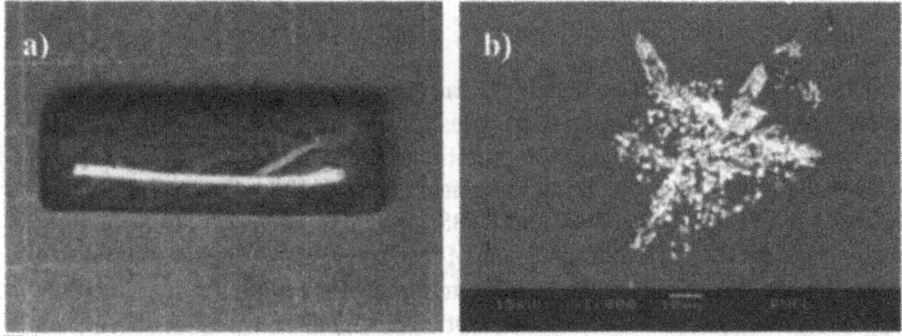

Figure 2. a) MMW glass with 5 HfO_2 and 15 CeO_2 wt% melted at 1250 °C; b) SEM micrograph of a typical region of undissolved CeO_2 in the same glass

LaBS Glasses

For the LaBS_E, LaBS_F and LaBS_G series of glasses, it was found that melting temperatures ≥1400 °C for ≥4 hours were required in order to produce a homogeneous 10 wt%-loaded HfO_2 glass. For the LaBS_X glass, and for ≥15 wt% loading, it was quite common to find undissolved material in the crucible, and small 'clumps' of HfO_2 crystals were visible in the SEM even when melted at 1450 °C in what appeared visually to be homogeneous glasses. It should be noted, however, that the base LaBS_X glass composition already contains 7.1 wt% HfO_2, and hence the actual loadings are ~6 wt% higher than those shown in Table IV. Figure 3 shows an LaBS_X monolith along with a typical SEM micrograph.

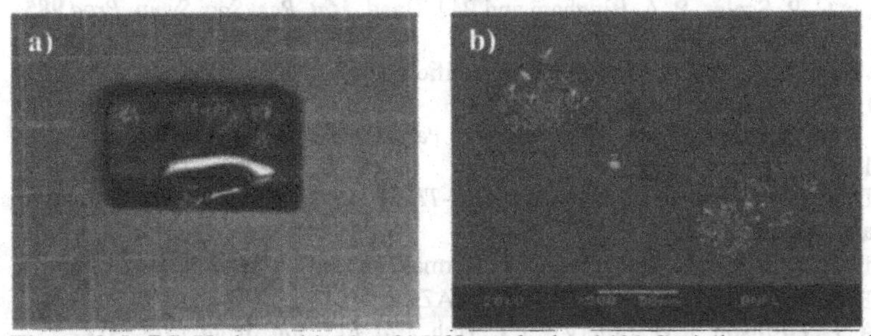

Figure 3. LaBS_X glass with 10 wt% HfO_2 melted at 1400 C; a) 'homogeneous' monolith, and b) SEM micrograph of some typical areas of undissolved HfO_2

Of the four different LaBS compositions fabricated, LaBS_G exhibited the best behaviour in terms of HfO_2 incorporation at 1400 °C, and it appeared to have lower viscosity at this temperature than LaBS_E and LaBS_F as observed from the pour yields. However, in order to fully dissolve ≥15 wt% HfO_2, considerably higher temperatures, or much longer times will be required, e.g. Marra *et al* melted Frit X containing a *total* of ~16 wt% HfO_2 at 1500 °C [5]. This explains the relatively poor performance of the LaBS_X glasses in this limited study. In addition, compositional analysis of these glasses has not been performed, and hence it is possible that the final Al_2O_3 content is higher due to dissolution of the crucible, and this could affect the apparent solubility of HfO_2. Any settled HfO_2 remaining in the crucibles will also effect the actual glass composition when compared to that calculated from the weighed quantities of precursors.

CONCLUSIONS

A more detailed temperature-composition study has been completed for several glass compositions that have been selected as leading candidates for the vitrification of surplus civil plutonium in the UK. Using HfO_2 as a surrogate for plutonium, it has been demonstrated that the benchmark of 10 wt% loading is achievable for all the glasses using melting temperatures of 1200, 1300 and 1400 °C for ATS, modified-MW and LaBS respectively. Higher HfO_2 loadings are certainly achievable for both the ATS and modified-MW glasses without increasing the temperature. However, despite an attempt being made to minimise the liquidus temperature of the LaBS glasses by fabricating several different compositions using different approaches from the literature, loadings higher than 10 wt% will require temperatures significantly greater than 1400 °C to produce a homogeneous product. Finally, in many samples HfO_2 settling was observed, which has criticality implications for Pu-containing glasses. Any potential vitrification process would need to be designed to avoid such problems.

ACKNOWLEDGEMENTS

The author would like to thank the NDA for funding this work.

REFERENCES

1. C. R. Scales, E. R. Maddrell and M. T. Harrison *Options for the Immobilisation of UK Civil Plutonium - 7214*, WM'07 Conference, February 25 - March 1, 2007, Tucson, AZ.
2. M. T. Harrison, C. R. Scales, P. A. Bingham and R. J. Hand, *Mat. Res. Soc. Symp. Proc* **985** (2007).
3. M. T. Harrison and C. R. Scales, MRS 2007: Scientific Basis for Nuclear Waste Management XXXI (presented at this meeting).
4. N. E. Bibler, W. G. Ramsey, T. F. Meaker and J. M. Pareizs, *Mat. Res. Soc. Symp. Proc.* **412**, 65-72 (1996).
5. J. C. Marra, D. K. Peeler, and C. M. Jantzen, *WSRC-TR-2006-00031*, Washington Savannah River Company, Aiken, SC, 2006.
6. J. R. Zamecnik, T. M. Jones, D. H. Miller, D. T. Herman, and J. C. Marra, *WM'07 Conference*, February 25 - March 1, 2007, Tucson, AZ.
7. J. K. Bates, A. J. G. Ellison, J. W. Emery & J. C. Hoh, *Mat. Res. Soc. Symp. Proc.* **412**, 57-64 (1996).
8. C. Lopez, X. Deschanels, J. M. Bart, J. M. Boubals, C. Den Auwer and E. Simoni, *Journal of Nuclear Materials* **312**, 76-80 (2003).
9. J.-N. Cachia, X. Deschanels, C. Den Auwer, O. Pinet, J. Phalippou, C. Hennig and A. Scheinost, *Journal of Nuclear Materials* **352**, 182-189 (2006).
10. J. G. Darab, H. Li, M. J. Schweiger, J. D. Vienna, P. G. Allen, J. J. Butcher, N. M. Edelstein, and D. K. Shuh, *Plutonium Futures: The Science*, Santa Fe, NM, August 25-27, 1997.
11. X. Feng, H. Li, L. L. Davis, L. Li, D. G. Darab, M. J. Schweiger, J. D. Vienna, B. C. Bunker, P. G. Allen, J. J. Bucher, I. M. Craig, N. M. Edelstein, D. K. Shuh, R. C. Ewing, L. M. Wang and E. R. Vance, *Ceramic Trans.* **93**, 409-415 (1999).
12. B. J. Riley, J. D. Vienna and M. J. Schweiger, *Mat. Res. Soc. Symp. Proc.* **608**, 677-682 (2000).

Mater. Res. Soc. Symp. Proc. Vol. 1107 © 2008 Materials Research Society

Towards a Single Host Phase Ceramic Formulation for UK Plutonium Disposition

Martin C. Stennett[1], Neil C. Hyatt[1], Matthew Gilbert[2], Francis R. Livens[2], and Ewan R. Maddrell[3]
[1]Immobilization Science Laboratory, Department of Engineering Materials, University of Sheffield, Sir Robert Hadfield Building, Mappin Street, Sheffield, S1 3JD, UK.
[2]Centre for Radiochemistry Research, School of Chemistry, University of Manchester, Oxford Road, Manchester, M13 9PL.
[3]Nexia Solutions Ltd., Sellafield, Seascale, Cumbria, CA20 1PG, UK

ABSTRACT

The UK has a considerable stockpile of separated plutonium; a legacy of over 50 years of civilian nuclear programmes. This material has been considered both as an asset for future energy generation and a liability due to the proliferation threat. A proportion of the PuO_2 stocks may be consumed by nuclear fission, in mixed oxide (MOx) or inert matrix (IMF) fuels but a quantity of waste PuO_2 will remain which is unsuitable for fuel manufacture and will require immobilisation. A research program is currently underway to investigate the potential of various single phase ceramic formulations for the immobilisation of this waste PuO_2 fraction. In this work a number of synthetic mineral systems have been considered including titanate, zirconate, phosphate and silicate based matrices. Although a wealth of information on plutonium disposition in some of the systems exists in the literature, the data is not always directly comparable which hinders comparison between different ceramic hosts. The crux of this research has been to compile a database of information on the proposed hosts to allow impartial comparison of the relative merits and shortcomings in each system.

INTRODUCTION

The potential for utilising ceramic matrices to immobilise radioactive waste has been recognised for many decades. Ringwood et al. [1] first developed SYNROC (SYNthetic Rock) in the late 1970's as a multiphase ceramic alternative to borosilicate glass for the immobilisation of high level radioactive waste arising from reprocessing of light water reactor fuel. In the mid 1990's a research program was initiated to look at SYNROC based options for disposal of separated plutonium [2]. A multiphase pyrochlore-rich variant of SYNROC was developed to incorporate plutonium and the significant levels of feedstock impurities in the wastestream. Whilst composite ceramics are required for wastestreams with a variety of constituent elements, a number of single host phase ceramic options have been highlighted for pure actinide wastestreams [3-5]. The remit of this research was to evaluate the various ceramic options for the immobilisation of 'pure' PuO_2 containing only actinide impurities arising as a result of the decay of the shorter lived plutonium isotopes. This allowed the focus of the work to be aimed at single host phase ceramic formulations.

WASTEFORM SELECTION

Research into the partitioning behaviour of multiphase wasteforms such as SYNROC [6] and the study of naturally occurring actinide bearing minerals [7] has lead to a range of single phase synthetic ceramic wasteforms being proposed for the immobilisation of actinides. A number of articles reviewing the potential synthetic mineral options have been published [8-13] and the relevant performance criteria which ceramic wasteforms must satisfy have been generally agreed. These criteria are listed below:

- waste loading - high waste loadings result in minimisation of the final wasteform volume and reduces any potential storage costs.
- chemical flexibility - in this context with respect to the addition of neutron absorbing species to mitigate against the possibility of criticality events during the disposal lifetime.
- materials processing - the choice of a system with straightforward processing requirements reduces both the cost associated with the fabrication and the engineering complexity.
- radiation tolerance - with respect to the resistance of the wasteform to amorphisation by alpha decay events and also the effect of any amorphisation on the wasteform properties.
- natural analogues - gives direct information about the long term integrity of potential wasteform materials under geological conditions.

Table I. Summary and comparison of the relevant characteristics of the various host phases under investigation. Information in table adapted from [8]. Ln = lanthanide cation, M = transition metal cation and X = halide anion.

Ceramic phase	Waste loading	Processing temperature	Chemical flexibility	Radiation tolerance	Aqueous durability	Natural analogue(s)
Zirconolite $CaZrTi_2O_7$	Medium	Medium	High	Medium	High	High
Pyrochlore $(Ca,Ln)_2Ti_2O_7$	High	Medium	High	Medium	High	High
Pyrochlore $(Ca,Ln)_2Zr_2O_7$	High	High	Medium	High	High	Low
Fluorite $(Zr,Ln)O_{2-\delta}$	Medium	High	Medium	High	High	Low
Muratite $(Na,Ln)_4M_2Ti_7O_{22}$	Medium	?	High	?	?	Medium
Britholite $Ca_2Ln_3(SiO_4)_6X$	Low	Medium	Medium	Low	Medium	High
Kosnarite $NaZr_2P_3O_{12}$	Medium	Low	Medium	?	Medium	High
Monazite $LnPO_4$	High	Medium	Low	High	High	High

High | Medium | Low

Greg Lumpkin authored a study in 2006 on the characteristics of the various wasteform materials, with respect to the wasteform performance criteria, and ranked the various wasteforms against one another according to these individual criteria [8]. This study provided a useful guide when short listing the most promising materials for current work although it was noted that the data was compiled from samples that were produced via a range of processing techniques. These included wet chemical, solid state, hydrothermal, and melting and recrystallisation processing routes. To directly compare the materials short listed and eliminate any potential bias introduced by variations in processing it was deemed essential to prepare the samples using identical processing methodologies. Table I shows that materials in the titanate-, zirconate-, silicate- and phosphate- based systems were retained after the initial down selection process to allow an objective approach to be maintained and ensured fair evaluation and transparency.

PHASE DEVELOPMENT

The processing methodology chosen had to be compatible with the constraints imposed when working in a high alpha activity environment and also the chemical form of the waste stream. Based on these two conditions, it was decided that samples should be prepared using conventional powder solid state synthesis. After evaluation of the available plutonium oxide surrogates previously used in the literature [14] it was decided to choose one inactive and one active surrogate, CeO_2 and depleted UO_2 respectively. In addition to investigation of the surrogate solid solution limits (waste loading potential) in the systems in question, the flexibility of the systems with respect to the incorporation of neutron absorbing species (Gd^{3+} and Hf^{4+}) was also addressed. Figure 1 shows the processing scheme followed.

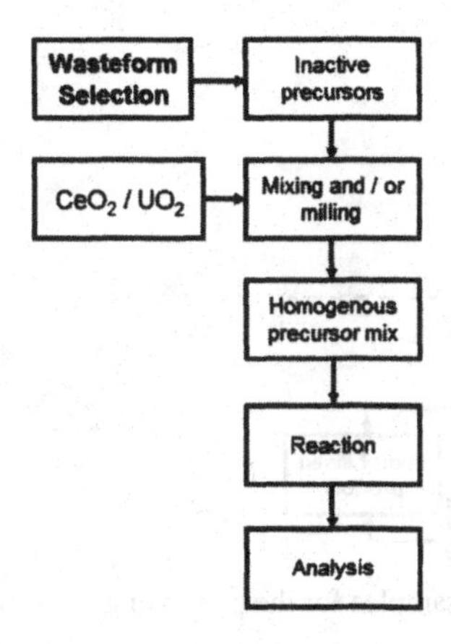

Figure 1. Flow chart outlining the processing stages used to prepare samples for evaluation of surrogate and neutron absorber solid solution limits.

Powder batches were weighing out according to the target stoichiometries and homogenised in a tumbling ball mill for 16 hours, with yttria stabilised zirconia milling media and isopropanol as the liquid medium. The slurries were separated from the media and dried overnight at 100 °C. The resulting dry powder cakes were sieved and reacted in open alumina crucibles in an air atmosphere. The phase assemblages of the powder batches were determined using X-ray diffraction.

PROCESS DEVELOPMENT

For the processing evaluation a single stage reactive sintering route was employed rather than a two stage route consisting of reaction of the powder precursors followed by formation and densification of a pressed powder monolith (see figure 2). Homogenised powder precursors were consolidated by cold uniaxial pressing and sintered at a range of temperatures and times. A hot pressing route, to form immobilised products fit for direct disposal, was also studied but the results presented here relate to the cold pressing scenario only. The inactive precursors were homogenised in a tumbling ball mill for 16 hours, with yttria stabilised zirconia milling media and isopropanol as the liquid medium. The slurries were separated from the media and dried overnight at 100 °C. The resulting dry powder batches were sieved and the PuO_2 surrogate was added. Homogenisation of the surrogate and the precursor batch was done dry, in a rotary tumbling mixer, for 16 hours. Ceramic bodies (10 mm diameter) were formed by cold uniaxially pressing the powders in a hardened stainless steel die at a pressure of ~ 200 MPa. The resulting ceramic bodies were sintered at a range of temperatures and dwell times on zirconia setter plates in an air atmosphere.

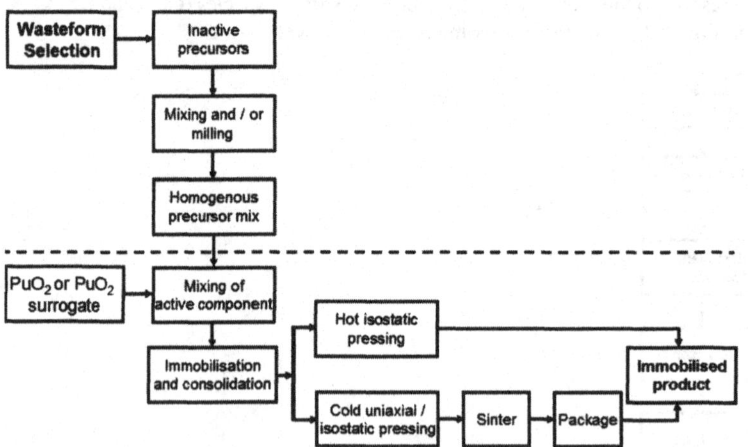

Figure 2. Flow chart outlining processing stages used to prepare samples for the processing evaluation.

SUMMARY OF RESULTS

Table II gives an overview of data obtained from this study on the various host phases [5, 15-16]. Britholite exhibited a high Gd^{3+} solubility and a relatively low processing temperature, albeit with slightly lower maximum sintered densities than the zirconate and titanate based systems. The pyrochlore system demonstrated high Gd^{3+} and Hf^{4+} solubility but required high temperature processing, in excess of 1650 °C. The zirconolite system also readily incorporated Gd^{3+} and Hf^{4+} but required a considerably lower processing temperature. Kosnarite incorporated Hf^{4+} (and Gd^{3+} to a lesser extent) and displayed the most favourable sintering conditions although the maximum density achieved was only just over ~ 90 % theoretical. Another drawback with phosphate based systems is the need to employ organic precursors and incorporating extra steps into the processing procedure to avoid the formation of unwanted intermediate glassy phases.

Table II. Waste loading potential, flexibility for incorporation of neutron poisons and optimum processing temperature of various host phases under investigation. Waste loadings expressed as wt% plutonium equivalent.

Mineral ceramic phase	Structure	Waste loading (wt %) Ce^{4+}	U^{4+}	Flexibility	Optimum sintering temperature (°C)	Density (% theoretical)
$Ca(Zr_{1-x}Ce_x)Ti_2O_7$	2M 4M	< 14 < 33	< 13 n/a	High	1450	~ 98
$(Ca_{1-x}M_x)Zr(Ti_{2-x}Mg_x)O_7$	2M 3O	< 26 < 43	< 24 n/a	High	1450	~ 98
$Gd_2(Zr_{2-x}M_x)O_7$	Pyrochlore Fluorite	< 19 < 54	n/a n/a	High	1650	~ 94
$(Ca_{2+x}Y_{8-2x}M_x)Si_6O_{26}$	Britholite	≤ 67	< 14	Intermediate	1400	~ 93
$Na(Zr_{2-x}M_x)P_3O_{12}$	Kosnarite	< 9	≤ 18	Intermediate	1350	~ 91

For several of the systems the choice of plutonium surrogate gave very different values for the waste loading. The pyrochlore and zirconolite systems both readily accepted cerium and uranium giving similar values for the maximum waste loading in both cases. Cerium was readily incorporated into the Britholite structure but uranium showed very low solubility. Cerium solubility in kosnarite was very low but uranium substituted more readily into the structure. The reasons for the discrepancies can be attributed to the different redox of the surrogates and the effect of the processing conditions employed. The surrogate cation valances have been investigated by analysis of the X-ray absorption near edge structure (XANES). This analysis showed cerium to undergo significant reduction in certain systems ($Ce^{4+} \rightarrow Ce^{3+}$) and uranium to readily oxidise ($U^{4+} \rightarrow U^{5+}/U^{6+}$). In the case of cerium, Ce^{3+} ($CePO_4$) gave a single white line peak whilst Ce^{4+} (CeO_2) gave a doublet. The white line shapes of U^{4+} and U^{6+} were very similar and both consisted of single peaks. A shift was observed in the white line position to higher energy with increasing oxidation state and the U^{6+} white line also displayed a post edge shoulder. Figure 3 showed that cerium remained mainly in the 4+ oxidation state in the zirconolite and pyrochlore systems but underwent a degree of reduction in the britholite system and full reduction in the kosnarite system. Uranium oxidised from U^{4+} to U^{5+} and U^{6+} under air processing in all the systems with a corresponding reduction in ionic radius of ~ 15 to 20 %. Figure 4 shows the XANES for $(Ca_{0.8}U_{0.2})Zr(Ti_{1.8}Mg_{0.2})O_7$ and several standards.

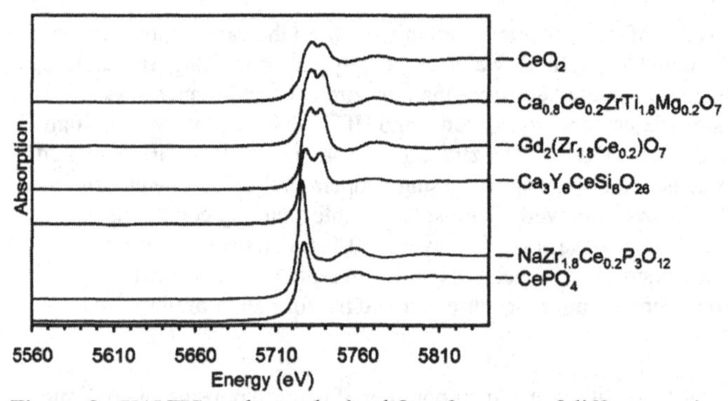

Figure 3. XANES analyses obtained for of a range of different cerium containing samples. CeO_2 and $CePO_4$ included as Ce^{4+} and Ce^{3+} standards respectively.

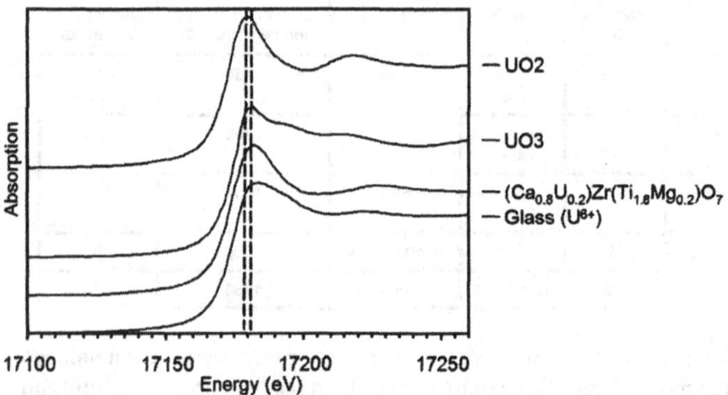

Figure 4. XANES analyses obtained for $(Ca_{0.8}U_{0.2})Zr(Ti_{1.8}Mg_{0.2})O_7$. UO_2, UO_3 and U^{6+} in borosilicate glass included as U^{4+} and U^{6+} standards respectively.

The zirconolite and pyrochlore systems were flexible with respect to changes in surrogate valence due to the availability of crystallographic sites in the structures of the appropriate size to incorporate both smaller U^{6+} cations (Zr or Ti-sites) and also any larger Ce^{3+} cations (Ca or Gd-sites). In addition there existed a number of potential ways to internally charge balancing the change in cerium and uranium oxidation state. In the britholite system the larger Ce^{3+} cations could be easily incorporated on to the large Ca/Y-sites but there were no available site to accept U^{5+} and U^{6+} cations. This resulted in a very low solubility of uranium in the air fired samples. It is predicted that samples fired under neutral (argon) or reducing (H_2/N_2) conditions will show a greater solubility for uranium in the 4+ oxidation state. In the cerium loaded kosnarite system the formation of monazite ($CePO_4$) at very low levels of cerium incorporation indicated that the phosphate systems readily promoted reduction and favoured the formation of simpler orthophosphates. In contrast the oxidised uranium cations (U^{5+} and U^{6+}) could readily be incorporated into the kosnarite structure on the Zr-site since Zr^{4+} and U^{6+} have nearly identical ionic radii in six-fold coordination. It is predicted that in argon or H_2/N_2 fired samples the uranium (U^{4+}) will be too large to fit on the zirconium site and will show very low solubility via that substitution mechanism.

CONCLUSIONS

When all the systems were compared the A-site substituted zirconolite system and the britholite systems offered the best combination of properties: moderate to high waste loading respectively, good to intermediate chemical flexibility and favourable processing conditions. Although discrepancies were recorded in the absolute values of the waste loading for the different surrogate systems it should be pointed out that the observed oxidation of U^{4+}, when processed in air, is not representative of how Pu^{4+} would behave under similar conditions. Under air processing Ce^{4+} was shown to be a much better Pu^{4+} surrogate in the majority of the systems investigated although similar waste loading values would be expected to be recorded for the uranium systems processed in either a neutral or reducing environment. The pyrochlore / fluorite system displayed good waste loading, in the fluorite structured regime but required considerably higher temperature processing. The kosnarite system required the use of organic phosphate precursors which were found not to lend themselves readily to solid state powder processing.

REFERENCES

[1] A. E. Ringwood, S. E. Kesson, N. G. Ware, W. O. Hibberson and A. Major. The SYNROC process: A geochemical approach to nuclear waste immobilisation. Geochemical Journal. 13, 141 (1979).

[2] B. B. Ebbinghaus, G. A. Armantrout, L. Gray, C. C. Herman, H. F. Shaw and R. A. Van-Konynenburg. Plutonium immobilisation project baseline formulation. UCRL-ID-133089 (2000).

[3] R. C. Ewing, W. J. Weber and J. Lian. Nuclear waste disposal – pyrochlore ($A_2B_2O_7$): Nuclear waste form for the immobilisation of plutonium and "minor" actinides. Journal of Applied Physics. 95, 5949 (2004).

[4] W. L. Gong, W. Lutze and R. C. Ewing. Zirconia ceramics for excess weapons plutonium waste. Journal of Nuclear Materials. 277, 239 (2000).

[5] M. C. Stennett, N. C. Hyatt, W. E. Lee, and E. R. Maddrell. Processing and characterisation of fluorite-related ceramic wasteforms for immobilisation of actinides in *Environmental Issues and Waste Management Technologies in the Ceramic and Nuclear Industries XI*, edited by C. C. Herman, S. Marra, D. R. Spearing, L. Vance, and J. D. Vienna. Ceramic Transactions. 176, 81 (2006).

[6] G. R. Lumpkin, K. L. Smith and M. G. Blackford. Partitioning of uranium and rare earth elements in SYNROC: effect of impurities, metal additives, and waste loading. Journal of Nuclear Materials. 224, 31 (1995).

[7] R. C. Ewing. The design and evaluation of nuclear-waste forms: Clues from mineralogy. Canadian Mineralogist. 39, 697 (2001).

[8] G. R. Lumpkin. Ceramic waste forms for actinides. Elements. 2, 47 (2006).

[9] I. W. Donald, B. L. Metcalf and R. N. J. Taylor. The immobilisation of high level nuclear waste using ceramics and glasses. Journal of Materials Science. 32, 5851 (1997).

[10] P. E. Fielding and T. J. White. Crystal chemical incorporation of high level waste species in aluminotitanate-based ceramics: Valence, location, radiation damage, and hydrothermal stability. Journal of Materials Research. 2, 387 (1987).

[11] A. Macfarlane. Immobilisation of excess weapons plutonium: A better alternative to glass. Science & Global Security. 7, 271 (1998).

[12] R. C. Ewing. Plutonium and 'minor' actinides: Safe sequestration. Earth and Planetary Science Letters. 229, 165 (2005).

[13] S. V. Stefanovsky, S. V. Yudintsev, R. Giere and G. R. Lumpkin. Nuclear waste forms in *Energy waste and the environment: A geochemical perspective*, edited by R. Giere and P. Stille. Geological Society of London special publication. 236, (2004).

[14] P. A. Bingham, R. J. Hand, M. C. Stennett, N. C. Hyatt and M. T. Harrison. The use of surrogates in waste immobilisation studies: A case study of plutonium in Scientific Basis for Nuclear Waste Management XXXI (this issue).

[15] M. C. Stennett, N. C. Hyatt, E. R. Maddrell, F. G. F. Gibb, G. Moebus and W. E. Lee. Microchemical and crystallographic characterisation of fluorite-based ceramic wasteforms in Scientific Basis for Nuclear Waste Management XXIX, edited by P. Van Iseghem. Materials Research Society Symposium Proceedings. 932, 623 (2006).

[16] M. C. Stennett, E. R. Maddrell, C. R. Scales, F. R. Livens, M. Gilbert and N. C. Hyatt. An evaluation of single phase ceramic formulations for plutonium disposition in Scientific Basis for Nuclear Waste Management XXX, edited by D. S. Dunn, C. Poinssot, B. Begg. Materials Research Society Symposium Proceedings. 985, (2007).

Mater. Res. Soc. Symp. Proc. Vol. 1107 © 2008 Materials Research Society

The Use of Surrogates in Waste Immobilization Studies: A Case Study of Plutonium

Paul A. Bingham, Russell J. Hand, Martin C. Stennett, Neil C. Hyatt and Mike T. Harrison[1]
Immobilization Science Laboratory, Department of Engineering Materials, University of Sheffield, Sir Robert Hadfield Building, Mappin Street, Sheffield S1 3JD, UK.
[1] Materials and Products, Nexia Solutions Ltd, The Technology Centre, Sellafield, Seascale CA20 1PG, UK.

ABSTRACT

Surrogates are widely used in the research and development of nuclear wasteforms, providing detailed insight into the chemical and physical behaviour of the wasteform whilst avoiding the widespread (restricted and costly) use of radiotoxic elements in the laboratory. However, caution must be exercised when dealing with surrogates since no single element or compound perfectly mimics all aspects of the behaviour of another. In this paper we present a broad discussion of the use of surrogates in waste immobilization, drawing upon and highlighting our research into glass and ceramic wasteforms for the immobilization of bulk PuO_2.

INTRODUCTION

The most widely used methods for the safe immobilization of radioactive nuclear wastes are cement encapsulation (LLW and some ILW) and vitrification (HLW and some ILW) [1, 2]. Other materials, particularly ceramics and glass composites, are also under consideration as hosts for the immobilization of certain specific wastes [1, 2]. The chemical composition of most radioactive and toxic wastes prevents their direct use in early stage laboratory studies. This arises for a variety of reasons: chiefly financial and safety concerns, as illustrated in figure 1. For the majority of radioactive waste immobilization studies, early stage R&D includes desk-based and small-scale laboratory studies; these are followed by larger-scale trials, eventually leading to full-scale inactive and active trials. The use of radioactive components is essential during the latter stages; however, during the earlier stages this may be unnecessary and it can be expeditious to simulate the chemical and physical behaviour of radioactive waste components using inactive surrogates. This has become standard practice in early-stage wasteform development, yet it is a methodology which is by no means perfect, and the research scientist must be aware of its limitations in order to design a sufficiently robust experimental programme.

In many cases, the chemical behaviour of a radioactive element may be simulated using one of its inactive isotopes. For example, the chemical and physical effects of ^{90}Sr, ^{137}Cs and ^{129}I may be safely simulated in laboratory-produced wasteforms using their naturally-occurring isotopes. However, such an approach does not consider the effects of sample irradiation due to radioactive decay. Furthermore, when dealing with wastes containing actinide elements and others such as ^{99}Tc, it is not possible to use inactive isotopes of the same element. In these cases it is necessary to use a surrogate element that mimics aspects of the chemical or physical properties of the element in question. Irradiation effects may be simulated separately using a number of techniques including actinide doping and irradiation by charged particles, by neutrons or by gamma radiation [3-5]. Nevertheless it is the chemical and physical aspects of the behaviour of a waste component which will chiefly determine the range of host materials that are most suitable for its immobilization, and in this paper we have primarily focussed upon these aspects. In the case of actinide elements, the atomic masses of which can be substantially larger than those of the surrogates used, it is also essential that the surrogate loading levels in the wasteform are quoted on a molar basis.

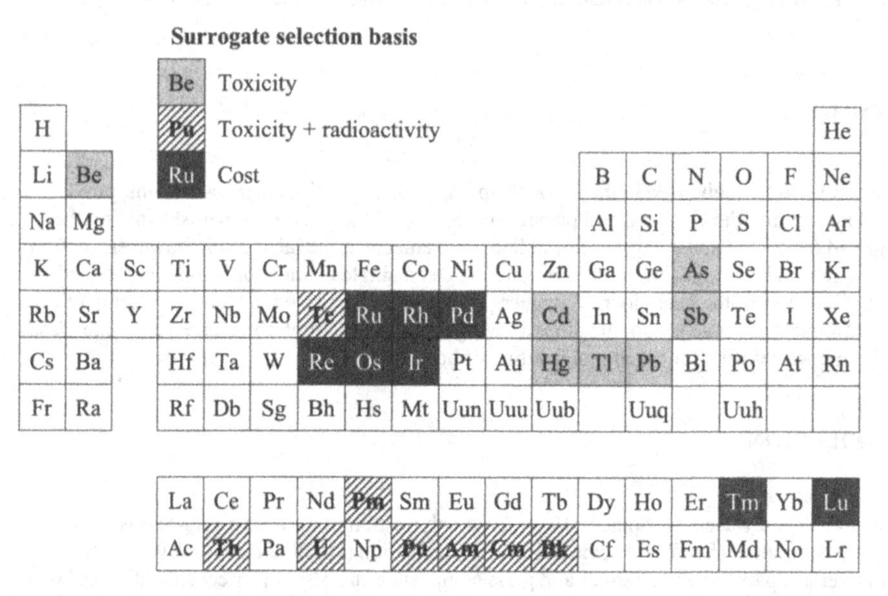

Figure 1. Periodic table adapted to show elements commonly associated with nuclear and toxic wastes that may require the use of surrogates during wasteform development.

Several countries including USA, Russia, UK and France collectively possess hundreds of tonnes of plutonium, usually occurring as PuO_2, which originates from reprocessing and other activities, and which is now surplus to requirements. There is still substantial debate as to what to do with this Pu [6]: some favour its re-use in MOX or IMF fuel, whilst others prefer immobilization and disposal on the basis of greater security. As yet no final decision has been taken, but if the immobilization option is chosen by one or more nations, a range of robust and suitable PuO_2 wasteforms must be available. Our aim in this paper is to impart some of the experience that we have gained through our ongoing research (see, for example, [7, 8]) into potential host materials for the immobilization of bulk PuO_2.

DISCUSSION

In attempting to mimic the behaviour of Pu when developing appropriate wasteforms for its immobilization, the following criteria must be amongst those considered:

- Chemical and physical properties
  - Valence
  - Electronic configuration
  - Ionic radii
  - Coordination
  - Redox potential
  - Density
  - Oxide melting temperature
  - Particle size distribution
- Radioactivity considerations
  - Self-irradiation — Primarily α-decay
  - Radioactive decay products — U and Am
  - Criticality prevention and control — Necessary for immobilization of ^{239}Pu

(Valence, Electronic configuration, Ionic radii, Coordination: These parameters determine the structural, chemical and solubility behaviour of Pu in the final wasteform)

(Redox potential, Density, Oxide melting temperature: These parameters affect wasteform homogeneity and Pu dissolution / reaction kinetics)

Coordination, electronic configuration and ionic radii are all fundamental properties of any waste element that may be under consideration, and they chiefly determine its solubility and local structural environment within any wasteform, for a given set of physical parameters. The oxide melting temperature is of more concern to glass or glass composite routes, in which the oxide melting temperature determines the stages of melting. Plutonium oxides are refractory, therefore it is advisable to use oxides with equivalent refractoriness as surrogates, particularly when studying melting behaviour. Table I summarises the relevant properties of Pu (III) and Pu (IV) and their oxides, and lists a range of elements which may be suitable as surrogates for different aspects of Pu behaviour. All have received some level of prior use as Pu surrogates.

Plutonium valence, redox potentials and their implications

Plutonium occurs in its (III) and / or (IV) valence states in nuclear wasteforms [9] and its higher oxidation states (V, VI, VII) may be neglected as they are only stable under highly oxidising conditions not usually available for wasteforms. Therefore Pu surrogates should be III- or IV- valent unless their purpose is to mimic other behaviour such as density / settling effects. Particularly in the case of glasses or glass-ceramics, the high density of PuO_2 (11.5 g cm^{-3}) can present difficulties in a glass melting vessel, in which its addition to a less dense base glass (typically ~ 2.5 g cm^{-3}) can lead to melt segregation [10]. For ceramic processing this presents less of a problem as the surrogate and other powder precursors are intimately mixed and settling effects during further processing are negligible.

The valence of Pu strongly influences its solubility in glassy wasteforms, with (III) being preferred for higher waste loading [11]. In selecting a Pu surrogate, differences in the (III) / (IV)

Table I. Chemical and physical data for Pu, candidate surrogate elements and their oxides

Element	Valence States	Electronic Configuration	Ionic Radius / pm	Oxide ρ/ g cm^{-3}	Oxide Melt T / °C
Pu	(III)	[Rn] $5f^5$	114 [CN6]	Pu_2O_3 10.50	2240
	(IV)	[Rn] $5f^4$	100 [CN6], 110 [CN8]	PuO_2 11.50	2400
Zr	(IV)	[Kr]	73 [CN4], 86 [CN6], 98 [CN8]	ZrO_2 5.68	2677
Hf	(IV)	[Xe] $4f^{14}$	72 [CN4], 85 [CN6], 97 [CN8]	HfO_2 9.68	2774
Th	(IV)	[Rn]	108 [CN6], 119 [CN8]	ThO_2 10.00	3390
Nd	(III)	[Xe] $4f^3$	112.3 [CN6], 124.9 [CN8]	Nd_2O_3 7.24	2320
Gd	(III)	[Xe] $4f^7$	107.8 [CN6], 119.3 [CN8]	Gd_2O_3 7.10	2420
Sm	(III)	[Xe] $4f^5$	109.8 [CN6], 121.9 [CN8]	Sm_2O_3 7.60	2335
Eu	(II)	[Xe] $4f^7$	131 [CN6], 139 [CN8]	(Eu_3O_4 5.24)	N/A
	(III)	[Xe] $4f^6$	108.7 [CN6], 120.6 [CN8]	Eu_2O_3 7.40	2350
Ce	(III)	[Xe] $4f^1$	115 [CN6], 128.3 [CN8]	Ce_2O_3 6.20	2230
	(IV)	[Xe]	101 [CN6], 111 [CN8]	CeO_2 7.65	2400
U	(IV)	[Rn] $5f^2$	103 [CN6], 114 [CN8]	UO_2 10.97	2827
	(V)	[Rn] $5f^1$	90 [CN6]	(U_3O_8 8.38)	1150-1300 Decomp→UO_2
	(VI)	[Rn]	66 [CN4], 87 [CN6], 100 [CN8]	UO_3 7.30	1150-1300 Decomp→U_3O_8

redox ratio must be considered and it is therefore advisable to study the solubility of surrogates for both Pu (III) and Pu (IV). When a multivalent element is introduced into a glass-forming oxide melt, the ensuing redox equilibrium which arises can be expressed by (1):

$$-\log f(O_2) = (4/m) \log [M^{(n-m)+}] / [M^{n+}] - E^*_M \quad (1)$$

where $f(O_2)$ = oxygen fugacity (i.e. partial pressure), m = number of electrons transferred in reduction, n = charge on the oxidised species, M = the redox element and E^*_M is a monitor of the relative reduction potential for element M under the melt conditions. Therefore, plotting $-\log f(O_2)$ vs. $\log [M^{(n-m)+}] / [M^{n+}]$ yields a straight line, with intercept at $-E^*_M$ and slope $4/m$. This is illustrated using electrochemical data for aqueous solutions [12], in figure 2. Note the anomalous behaviour of the U(VI) – U(V) couple: in aqueous solutions U(V) becomes disproportionated to U(VI) and U(IV), whereas U(V) is stable in silicate melts [13]. Furthermore, redox potentials are affected by host matrix chemistry so redox ratios in aqueous and glass hosts are different. However, redox potentials for Pu in glasses have not yet been measured so here we show aqueous data in order to highlight the differences between Pu and some of its potential surrogates. Similar differences may be expected to arise in glass hosts [13] and therefore no perfect "redox surrogate" exists for Pu. Surrogates such as Ce must therefore be used with caution in this regard.

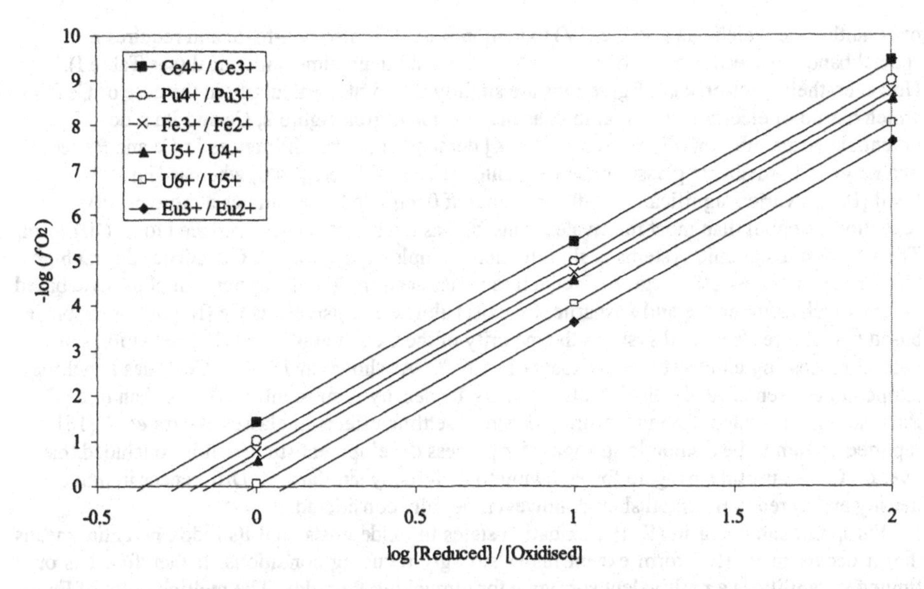

Figure 2. Aqueous electrochemical series for elements relevant to this study, plotted using data from [12], showing log (imposed oxygen partial pressure) vs. log [Reduced] / [Oxidised]

Inactive Pu surrogates

Lanthanides have been widely used as simulants for actinides including Pu. Both groups are Lewis acids, have large ionic radii and possess flexible coordination geometries, preferring high coordination numbers. With the exceptions of Ce, Eu and Pr, suitable lanthanides exhibit only one stable valence state, III, under atmospheric conditions. The redox potentials of Pr and Eu are so strongly positive and negative, respectively, and so dissimilar from that of Pu (see figure 2), that their usefulness as Pu "redox simulants" is limited. The most commonly-used lanthanide Pu surrogates have been Ce, Eu, Nd, Gd, and Sm. Their oxides, CeO_2, Eu_2O_3, Nd_2O_3, Gd_2O_3 and Sm_2O_3 exhibit similar melting temperatures to PuO_2 as shown in Table I (around 2400 °C). Although they have moderately high densities of the order of 7-8 g cm^{-3}, these are substantially lower than that of PuO_2 (11.5 g cm^{-3}) so they may be unsuitable for studying Pu settling effects in glasses. With the exception of Ce (IV), the lanthanides have partially-filled 4f electron shells which allow broad comparability with the partially-filled 5f shells of Pu (III) and Pu (IV). Lanthanide ionic radii also tend towards similarity with those of corresponding Pu oxidation states.

Cerium is a multivalent element which occurs in most oxide hosts in (III) and (IV) valences, analogous to the redox states of Pu in these materials. It has been widely used as a Pu surrogate [7, 8, 11, 14, 15], and trials by Lopez *et al.* [15] have determined that the solubilities of Ce (III) and Pu (III) in borosilicate glasses are similar (2-2.5 mol %) at 1400 °C. Ce also has closely similar solubility to the actinide species Am (III) and Cm (III) at 1200 °C [11]. Cerium is the only lanthanide which has a stable (IV) oxidation state. It is also non-toxic and requires no special handling precautions. The ionic radii of Ce and Pu are almost identical (see Table I). However, their electronic configurations are slightly different, particularly in the case of Ce (IV) which has no 4f electrons. Furthermore, as may be noted from figure 2, Ce and Pu redox potentials differ substantially. Darab *et al.* [14] demonstrated the differences in Ce and Pu redox in a range of borosilicate glasses which exhibited [Ce (IV)]/[Ce_{total}] ≈ 0, whereas [Pu (IV)]/[Pu_{total}] varied significantly with one value of 0.9 quoted. Cerium has a more positive reduction potential than most multivalent ions, and as a result it is easily reduced to its (III) form. The situation in ceramic systems is slightly more complex and although Ce reduces during high temperature synthesis, the degree of reduction varies according to the system. In phosphate based systems such as monazite and kosnarite, Ce exists almost exclusively as Ce (III) but in zirconate based pyrochlore structured systems the majority of the Ce remains in the (IV) oxidation state, even at processing temperatures in excess of 1500 °C. As shown in Table 1, CeO_2 has a melting temperature essentially identical to that of PuO_2. Its density is substantially lower than that of PuO_2 so it is of limited use in measuring density / settling effects in glasses. Marra *et al.* [16] reported cerium to be a suitable surrogate for process development studies. It is concluded that, overall, Ce is a useful surrogate for Pu owing to their many similarities. Differences in oxide density and in redox potential should, however, be fully considered.

Europium can occur in (II, III) oxidation states in oxide hosts, and its redox potential means that it occurs in its (III) form except under strongly reducing conditions. It therefore has only limited suitability as a multivalent surrogate for mimicking Pu redox. The melting point of Eu_2O_3 is similar to that of Pu_2O_3, however, its density is much lower. The electronic configurations of Eu (II, III) are similar to those of Pu but substantial differences in ionic radii occur. Eu has a high thermal neutron cross section [17], allowing it to fill a dual role of surrogate and neutron poison co-dopant, however it has seen only limited investigation in wasteforms, possibly due to its high cost coupled with its shortcomings as a redox simulant.

Gadolinium has only one stable oxidation state, (III), in oxides. Gd_2O_3 has a slightly higher melting temperature and a lower density than Pu_2O_3. As with other lanthanides it has a partially filled f-electron shell, and its ionic radius is similar to that of Pu (III). Gd has been widely used in studies of actinide immobilisation [7, 11, 18] owing to its suitability as a surrogate for Pu (III) and its high thermal neutron absorption cross section [17]. This makes the use of Gd more likely as a co-dopant for criticality control. On this basis a number of studies have utilised Gd (III) in the dual role of Pu (III) surrogate and neutron poison. Deschanels *et al.* [11] have recently studied the solubility of Gd (III) in complex borosilicate glasses. They observed that Gd solubility is only comparable to that of Ce (III) and Pu (III) at high temperatures (1400 °C). At lower temperatures, the solubility of Gd (III) appears to be greater than that of Pu (III).

Neodymium has only one stable ionic form, (III), in inorganic glasses and ceramics. As with other lanthanides it has a partially filled f-electron shell, and its ionic radius is also similar to that of Pu (III). However, Nd has a low neutron cross-section and therefore cannot fill the dual role of Pu (III) surrogate and neutron poison co-dopant. Nevertheless, Nd has received some use as a trivalent actinide surrogate [11, 15]. Its solubility in borosilicate glasses at 1400 °C is similar to, that of Pu (III) [15], but, at lower temperatures it overestimates Pu solubility.

Samarium occurs in only one redox state, (III), in oxides. Its electronic configuration and ionic radius are comparable with those of Pu^{3+}, and its oxide, Sm_2O_3, has a similar melting point. Its density is similar to those of the other lanthanide oxides. Sm is the next strongest thermal neutron absorber after Gd, so its use as a neutron poison enhances its prospects as a Pu (III) surrogate. However, Sm has seen only limited use as a Pu surrogate, possibly on a cost basis.

Zirconium and hafnium have both been employed as surrogates for Pu (IV): both are stable in this redox state and exhibit similar melting temperatures to PuO_2. Both are highly refractory oxides which can mimic the behaviour of PuO_2. HfO_2 has seen particularly widespread use in this capacity owing to an electronic configuration more similar to that of Pu (IV), coupled with its high thermal neutron cross section [17], which presents it as a candidate co-dopant neutron poison, and a higher density (the highest density of any inactive Pu surrogate). It is therefore useful in investigating any Pu settling effects in glasses. Hafnium has been a primary simulant for Pu (IV) [7, 11, 15, 18]. Comparative studies [11, 15] have confirmed that Hf (IV) has similar solubilities in borosilicate glasses to Pu (IV) over a range of temperatures. However, Hf (IV) has a substantially smaller ionic radius than Pu (IV) and therefore has limited suitability in ceramic systems where host sites in the structure accept only cations within a specific size range. For example Hf (IV) has limited solubility in the apatite structure but Pu (IV) is readily accommodated. On the basis of this difference, Hf (IV) may not be as suitable a model for Pu (IV) as is Ce (IV). However, this must be mitigated by the multivalency of Ce and the decision based upon the specific requirements. This adds a note of caution to the use of Hf (IV) as a surrogate for Pu (IV): its use in tandem with other surrogates such as Ce (IV) may be advisable.

Active plutonium surrogates

Uranium and thorium have been used as actinide surrogates, although uranium has received substantially more attention. The high densities of their oxides and similarities in electronic configuration with Pu give them additional benefits over lanthanides. Thorium was used as a Pu (IV) surrogate successfully in the U.S. during the Pu disposition programme in the 1990's. Uranium is multivalent and substantial work has been carried out on uranium redox in glasses [13, 19, 20]. Its combination of electronic configuration, oxide density and melting point and redox behaviour make it a good surrogate for Pu, although the redox chemistry of uranium is more complex. The toxicity and radiotoxicity of Th and U make their use less feasible than inactive surrogates, but the use of Th and depleted U requires far less stringent security and safety measures than Pu. The U. S. DoE research into Pu waste forms during the late 1990's and early 2000's utilised ThO_2 as a surrogate for PuO_2 [20]. Thorium was used as a surrogate based on a comparison of ThO_2 and PuO_2 in terms of density, oxide crystal structure, and valence in LaBS glasses.

CONCLUSIONS

In order to mimic the behaviour of Pu in potential wasteforms, the following properties should be considered: Valence, electronic configuration, ionic radius, coordination, redox potential (for multivalent surrogates), oxide melting temperature, particle size distribution, self-irradiation effects, decay products and criticality prevention. Pu(III) and / or Pu (IV) may occur in oxide wasteforms, with their ratio depending upon host chemistry and oxygen partial pressure

Inactive surrogates for Pu (III) include Ce (III), Gd (III), Sm (III) and Nd (III), and inactive surrogates for Pu (IV) include Ce (IV) and Hf (IV). The most useful redox-active surrogate is Ce (III, IV), although substantial differences in redox potential from Pu (III, IV) must be considered. Active Pu surrogates include U (III, IV, V, VI) and Th (IV). A multi-surrogate approach is advisable for wasteform development programmes.

REFERENCES

1. M. I. Ojovan and W. E. Lee, An Introduction to Nuclear Waste Immobilisation (Elsevier, 2005).
2. I. W. Donald, B. L. Metcalfe and R. N. J. Taylor, *J. Mater. Sci.* **32**, 5851 (1997).
3. W. J. Weber and F. P. Roberts, *Nucl. Technol.* **60**, 178 (1983).
4. W. J. Weber, R. C. Ewing, C. A. Angell, G. W. Arnold, A. N. Cormack, J. M. Delaye, D. L. Griscom, L. W. Hobbs, A. Navrotsky, D. L. Price, A. M. Stoneham and M. C. Weinberg, *J. Mater. Res.* **12**, 1946 (1997).
5. R. C. Ewing, W. J. Weber and J. Lian, *J. Appl. Phys.* **95**, 5949 (2004).
6. R. C. Ewing, *Earth Planet. Sci. Lett.* **229**, 165 (2005).
7. M. T. Harrison, C. R. Scales, P. A. Bingham and R. J. Hand, in *Scientific Basis for Nuclear Waste Management XXX*, edited by D. S. Dunn, C. Poinssot and B. Begg, (Mater. Res. Soc. Proc. **985**, Warrendale, PA, 2007) in press.
8. M. C. Stennett, N. C. Hyatt, W. E. Lee and E. R. Maddrell, *Ceram. Trans.* **176**, 81 (2006).
9. I. Muller, W. J. Weber, E. R. Vance, G. Wicks and D. Karraker, "Glasses, Ceramics and Composites", *Advances in Plutonium Chemistry, 1967-2000*, ed. D. C. Hoffman (American Nuclear Society, 2002) pp. 260-307.
10. V. V. Kushnikov, Y. I. Matyunin and N. V. Krylova, *Soviet Atomic Energy* **70**, 299 (1991).
11. X. Deschanels, S. Peuget, J. N. Cachia and T. Charpentier, *Progress in Nuclear Energy* **49**, in press (2007).
12. http://www.webelements.com
13. H. D. Schreiber, *J. Geophys. Res.* **92**, 9225 (1987).
14. J. G. Darab, H. Li, M. J. Schweiger, J. D. Vienna, P. G. Allen, J. J. Bucher, N. M. Edelstein and D. K. Shuh, *Proceedings of plutonium futures-the science. Topical conference on plutonium and the actinides*, Santa Fe, USA (1997) pp. 143-145.
15. C. Lopez, X. Deschanels, J. M. Bart, J. M. Boubals, C. Den Auwer and E. Simoni, *J. Nucl. Mater.* **312**, 76 (2003).
16. Marra, J. C., Cozzi, A. D., Pierce, R. A., Pareizs, J. M., Jurgensen, A. R. and Missimer, D. M., *Ceramic Transactions* **132**, 381 (2002).
17. R. E. Bastick, *J. Soc. Glass Technol.* **42**, 70T (1958).
18. X. Feng, H. Li, L. L. D. L. Li, J. G. Darab, M. J. Schweiger, J. D. Vienna, B. C. Bunker, P. G. Allen, J. J. Bucher, I. M. Craig, N. M. Edelstein, D. K. Shuh, R. C. Ewing, L. M. Wang and E. R. Vance, *Ceram. Trans.* **93**, 409 (1999).
19. H. D. Schreiber and A. L. Hockman, *J. Amer. Ceram. Soc.* **70**, 591 (1987).
20. T. F. Meaker, U.S. DoE Report WSRC-TR-96-0321. Available electronically at http://www.osti.gov/bridge/servlets/purl/481473-rME0M6/webviewable/481473.pdf

Mater. Res. Soc. Symp. Proc. Vol. 1107 © 2008 Materials Research Society

Durability of Borosilicate Glass Compositions for the Immobilisation of the UK's Separated Plutonium Stocks

Mike T. Harrison and Charlie R. Scales
Nexia Solutions Ltd, The British Technology Centre, Sellafield, Seascale, Cumbria, CA20 1PG, UK.

ABSTRACT

Several glass compositions are currently under investigation for immobilisation of the separated PuO_2 that has been produced as a result of civil nuclear fuel reprocessing in the UK. Whilst a final decision on the fate of what ultimately will be over 100 tonnes of plutonium has yet to be made, all options for the disposition of this material are currently being investigated by Nexia Solutions Ltd on behalf of the Nuclear Decommissioning Authority (NDA).

As one of the immobilisation options, vitrification in borosilicate glass could potentially provide a criticality-safe and stable waste form with durability suitable for long term storage and subsequent repository disposal. From an initial experimental survey of potential candidates, three borosilicate compositions were selected for a more detailed study of the waste loading and chemical durability: lanthanide borosilicate (LaBS), alkali tin silicate (ATS) and high-lanthanide alkali borosilicate (modified-MW). In these inactive tests, hafnium was used as the surrogate for plutonium. This paper describes a range of static leach tests that were undertaken in order to understand the overall durability of the waste forms, as well as the release rates of the Pu-surrogate when compared to any neutrons poisons present in the glass. For the LaBS compositions it was found that the release rate of gadolinium was potentially slightly higher than that of hafnium, although both were as low as 10^{-5} to 10^{-6} g m^{2} day^{-1}. The potential implications for long-term repository behaviour are discussed.

INTRODUCTION

The durability of the waste forms currently under investigation for the immobilisation of surplus civil plutonium in the UK is a key criterion by which their suitability will be assessed. The majority of the separated PuO_2 is suitable for re-use as Mixed Oxide (MOx) fuel, but due to uncertainties in the UK's future nuclear strategy, immobilisation into a durable waste form ultimately followed by deep geological disposal is being investigated in a number of experimental programmes. Ceramics, glasses and 'storage MOx' are all under consideration, and the three scenarios will be reviewed against a set of specific performance criteria in order to identify which waste form is the best environmental option overall [1]. The 'First Stage Vitrification Programme' assessed a number of glass compositions for their suitability as a matrix for Pu-immobilisation. Silicate and phosphate glasses were investigated by Nexia Solutions Ltd and the Immobilisation Science Laboratory (ISL) respectively [2].

Following this initial experimental survey, a number of borosilicate glass compositions were carried forward to a more detailed study of their chemical durability and the temperature dependence of the waste loading. Selection for this '2nd Stage Vitrification Programme' was based on achieving a benchmark of at least10 wt% Pu-surrogate loading coupled with high chemical durability. The latter used simple bulk leach rates (BLRs) as calculated from static leach testing of monolithic glass samples, i.e. a modified MCC-1 test, for comparison. However, for the more durable glass compositions, the concentrations of the individual glass components in the leachate were found to be too small to measure elemental leach rates, especially for the Pu-surrogates. Hence, for the 2nd Stage, a powder-based modified PCT-B test was used in order to increase the sample surface area and allow normalised elemental mass losses $NL_{[i]}$ to be calculated. This would provide further information on the durability of the leading glass candidates and the leach rates of key elements such as the Pu-surrogate and any neutron poisons present. In any disposal scenario, criticality control will be a key element of the safety case.

The results from the 2nd Stage durability study are described below, with the temperature-composition study presented in a separate paper at this conference [3].

Base Glass Compositions

The borosilicate glasses selected for a more detailed investigation of their corrosion properties using a modified PCT-B durability test are described below, with the compositions shown in Table I:

- *Modified-MW (MMW)* – Initially included in the 1st Stage Programme as a baseline for comparison with other glasses, it was found that the addition of Al_2O_3, CeO_2, Gd_2O_3 and HfO_2 to MW[1] yielded a highly durable product with surrogate loadings of at least 15 wt%.
- *Lanthanum Borosilicate (LaBS)* – Developed in the US for actinide waste immobilisation [4]; Savannah River National Laboratory are currently proposing a LaBS formulation for the immobilisation of excess weapons-usable plutonium [5, 6]. It has been shown as having a high capacity for Pu along with excellent corrosion resistance, although melting temperatures in excess of 1400 °C are required.
- *Alkali Tin Silicate (ATS)* – Also developed in the US for actinide waste immobilisation, this formulation is not quite as durable as the LaBS glasses, but requires significantly lower melting temperatures to achieve 10 wt% Pu-loading [7].

Table I. Base glass compositions used for the 2nd Stage Experimental Programme (wt %).

MMW	ATS	LaBS_E	LaBS_F	LaBS_G	LaBS_X
Al_2O_3 6.3	Al_2O_3 2.7	Al_2O_3 18.2	Al_2O_3 19.2	Al_2O_3 16.10	Al_2O_3 10.0
B_2O_3 17.9	B_2O_3 13.5	B_2O_3 10.4	B_2O_3 10.6	B_2O_3 10.00	B_2O_3 13.0
Gd_2O_3 12.5	Cs_2O 0.6	Gd_2O_3 12.2	Gd_2O_3 11.8	Gd_2O_3 13.33	Gd_2O_3 11.7
Li_2O 4.1	Gd_2O_3 3.8	La_2O_3 20.0	La_2O_3 29.5	La_2O_3 22.22	HfO_2 7.1
Na_2O 8.9	K_2O 5.9	Nd_2O_3 9.0	SiO_2 26.5	Nd_2O_3 13.33	La_2O_3 20.3
SiO_2 50.4	Li_2O 4.7	SiO_2 24.5	SrO 2.4	SiO_2 22.22	Nd_2O_3 15.4
	Na_2O 10.5	Sm_2O_3 3.2		SrO 2.80	SiO_2 20.0
	SiO_2 46.9	SrO 2.5			SrO 2.5
	SnO 3.0				
	TiO_2 2.2				
	ZrO_2 6.2				

Four different base LaBS glass compositions were fabricated on the basis of the results from the 1st Stage programme and the approaches of Marra *et al* [5] and Riley *et al* [8] for optimising the melting conditions and minimising the liquidus temperature (T_L). *LaBS_E* and *LaBS_F* have a total Ln_2O_3 content of ~45 wt% (including the added Pu-surrogate) using the

[1] MW is the standard base glass used for immobilisation of HLW in the UK, and has a typical composition (in wt%) of 62 SiO_2, 22 B_2O_3, 11 Na_2O, 5 Li_2O.

study of the Al_2O_3-Ln_2O_3-SiO_2 phase fields by Riley and the observation of a minimum T_L at this composition. *LaBS_X* is similar to Marra's improved LaBS composition '*Frit X*', which lies along the assumed low-melting eutectic trough on the ~1:3 stoichiometric axis in the Ln_2O_3-B_2O_3-(SiO_2+Al_2O_3) ternary phase diagram [5]. *LaBS_G* is a hybrid composition that lies close to this 1:3 axis, but reduces the Ln_2O_3 content to as close to 45 wt% as possible.

Choice of Surrogate

One of the key findings from the 1st Stage programme was the importance of the choice of plutonium surrogate in inactive experimental trials. Initially, cerium was used, but it was concluded that its redox properties were sufficiently different to Pu to potentially give a misleading indication of solubility and hence waste loading [9, 10]. It has been shown that Pu in borosilicate glasses exists predominantly in the tetravalent state [10-13], and therefore a metal with a stable +4 oxidation state best simulates the redox chemistry of plutonium. Hence, for the 2nd Stage evaluation, HfO_2 was selected as the PuO_2 surrogate based on it having a similar solubility in the glass [12], and reasonably high density (9.68 g cm^{-3}). HfO_2 is also the surrogate of choice for the LaBS glass development at Savannah River [5], which allows a direct comparison to be made between the various compositions under consideration. However, for MMW it appeared that a mixture of oxides gave the glass its beneficial properties, hence CeO_2 was included in the Pu-surrogate such that the *total* CeO_2 plus HfO_2 loading was 20 wt%.

EXPERIMENTAL

The durability of all of the glass samples from the 2nd Stage experimental matrix of PuO_2-surrogate loadings and temperatures was measured [3]. The melting strategy adopted for this fabrication process was the 'base glass plus waste' approach, which is consistent with the globally-implemented method for HLW immobilisation, where the active waste material is added to a pre-formed inactive base glass frit. Hence, the six base glasses from Table I were initially produced, before being mixed with the relevant amount of Pu-surrogate, and re-melted at the required temperatures.

The base glasses were prepared by weighing and mixing precursor powders in the correct proportions. The precursors used were as-received metal oxides or carbonates, boric acid, and milled MW glass frit[2] as appropriate. Glass frits were formed by loading crucibles with the batched precursors, melting in a furnace for several hours at the required temperature, and then pouring into cold water. Once cooled, the frit was dried at ~60 °C before being ground to a powder (< 300 μm) using a ball mill. A glass monolith was also produced as a reference for comparison with the loaded glasses by pouring a portion of the melt into a pre-heated rectangular steel mould. This ingot was removed from the mould whilst still hot and transferred to a second furnace for annealing at a temperature just below the glass transition temperature (T_g). The annealing cycle was typically a 3 hour dwell followed by a cooling to room temperature at 0.5 °C/min. The melting temperatures, times, crucible materials, frit yield, and annealing temperatures used for the base glass melts are given in Harrison et al [3].

To make the surrogate-loaded glass monoliths, ~50 g batches of each milled base glass mixed with various quantities of PuO_2-surrogate was weighed into a suitable crucible, melted in a furnace at the required temperatures for 3-5 hours, poured into a pre-heated rectangular steel

[2] MW glass frit was ground using a ball mill and sieved to obtain particles with size <500 μm.

mould, and then annealed. The waste loading, crucible material, melting and annealing temperatures are given in Table II, along with the conditions for the base glass melts.

Table II. Waste loadings, crucible material, melting and annealing temperatures used for the glasses fabricated for durability testing.

Base Glass	*Pu-Surrogate Loading / wt%*	*Crucible*	*Melting Temps. / °C*	*Anneal Temp. / °C*
MMW[3]	5 HfO_2 + 15 CeO_2	Silica	1250, 1300, 1350	n/a
	10 HfO_2 + 10 CeO_2	Silica	1250, 1300, 1350	
	15 HfO_2 + 5 CeO_2	Silica	1250, 1300, 1350	
ATS	Base Glass	Silica	1300	500
	10 HfO_2	Silica	1200, 1250, 1300	
	15 HfO_2	Silica	1200,1250, 1300	
LaBS_E	Base Glass	Alumina	1350	750
	10 HfO_2	Alumina	1300, 1350, 1400	
LaBS_F	Base Glass	Alumina	1350	750
	10 HfO_2	Alumina	1300, 1350, 1400	
	15 HfO_2	Alumina	1350, 1400	
LaBS_G	Base Glass	Alumina	1395	750
	10 HfO_2	Alumina	1300, 1350, 1400	
	15 HfO_2	Alumina	1300, 1350, 1400, 1450	
	20 HfO_2	Alumina	1450	
LaBS_X	Base Glass	Alumina	1400	750
	10 HfO_2	Alumina	1300, 1350, 1400, 1450	
	15 HfO_2	Alumina	1350, 1400, 1450	
	20 HfO_2	Alumina	1450	

Chemical durability measurements were carried out using a modified Product Consistency Test B (PCT-B) leach test. Powdered samples with a size range of 75–150 μm (-100 to +200 mesh) were prepared by ball milling and sieving. Approximately 3.0 g of as-prepared powder was weighed into a 60 mL low density polyethylene (LDPE) bottle, and de-ionised water (DIW, 18.2 MΩ.cm) added equivalent to ten times the mass of glass waste form so that (V_{soln}/m_{solid}) = 10±0.5 mL g^{-1}. Note that using this fixed V_{soln}/m_{solid} ratio meant that the powder surface area to volume ratio (S_{solid}/V_{soln}) in each test varied with glass density. The *S/V* for each of the MMW, ATS and LaBS glasses were calculated, and found to be around 1800, 1850 and 1300 m^{-1} respectively. The bottles were then tightly sealed and placed in an oven at 90 °C for a total of 27 days. After 7 and 27 days, 1.0 mL samples of leachate were removed from the bottles and replaced with 1.0 mL of fresh DIW, acidified with 1.0 mL 1.57 M HNO_3, and diluted to 10.0 mL. The final leachate solution after 27 days was also filtered to isolate the powder. The concentration of the individual glass components in all of the leachate solutions was then analysed via ICP-OES. From the concentration data, normalised elemental mass losses $NL_{[i]}$ and elemental leach rates $ELR_{[i]}$ were calculated.

[3] Note that for MMW, the Al_2O_3 and Gd_2O_3 were added to milled MW glass along with the Pu-surrogates.

RESULTS AND DISCUSSION

MMW Glass

Figure 1 shows the $ELR_{[i]}$ calculated from the 7-day, 27-day, and final filtered leachate samples for the MMW glasses.

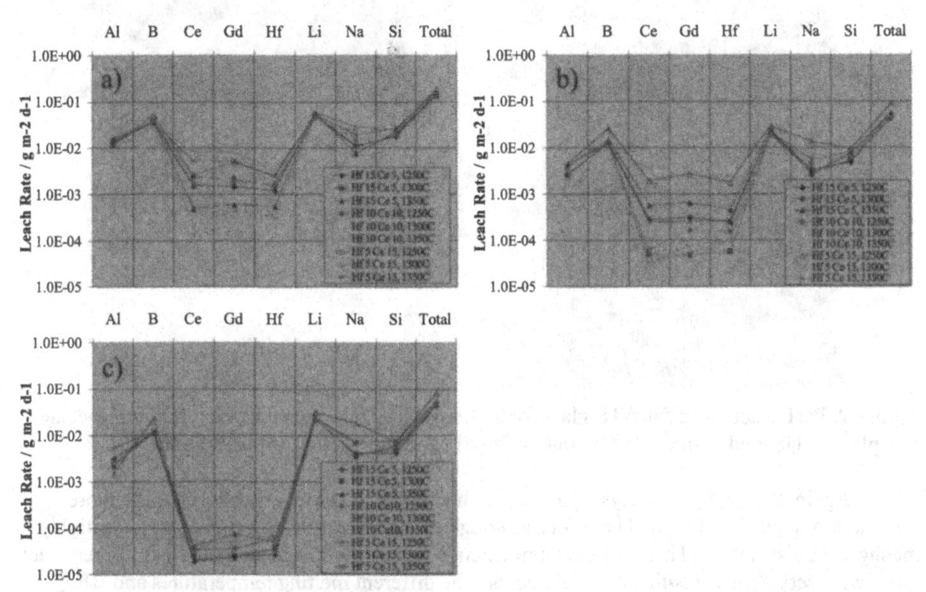

Figure 1. PCT leach rates for MMW glass for a) 7 days, b) 27 days, and c) final filtered leachate. Samples are labelled with the HfO_2 and CeO_2 content in wt%, and the melting temperature in °C.

The general behaviour of the different elements is similar for the 7 day, 27 day, and filtered final leachate, with the $ELR_{[i]}$ for the base glass components (B, Li, Na and Si) significantly higher than for Ce, Gd and Hf. The leach rates at 27 days are lower than at 7 days, but there is an obvious difference between these two and the filtered sample, particularly for the Ce, Gd and Hf. This is possibly due to some insoluble particulates being formed during the glass dissolution, and shows that all the samples should have been filtered. There is no obvious systematic behaviour with melting temperature and Pu-surrogate loading, although the least homogeneous glass (Hf 5 Ce 15, 1250 °C) has, in general, the highest leach rates for all elements. For all of the other MMW glasses, the ELRs are fairly consistent, especially for the final filtered leachate in Figure 1c. This also shows the leach rates of Ce, Gd and Hf are approximately the same within experimental error (~5 x 10^{-5} g m^{-2} d^{-1}).

ATS Glass

Figure 2 shows the $ELR_{[i]}$ for the 7-day, 27-day, and filtered leachate samples taken from the modified PCT-B test for the ATS glasses.

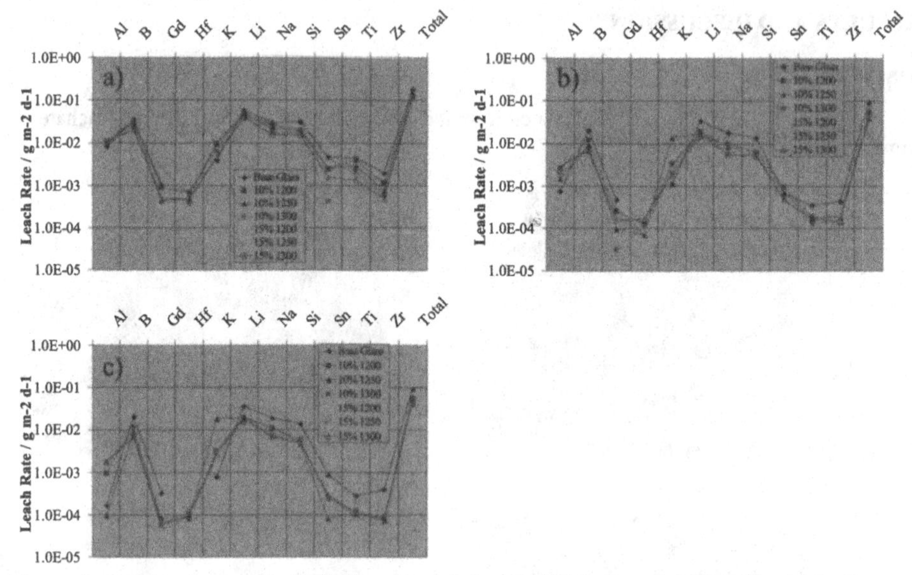

Figure 2. PCT leach rates for ATS glass for a) 7 days, b) 27 days, and c) final filtered leachate. Samples are labelled with the HfO_2 content in wt% and the melting temperature in °C.

Again, the $ELR_{[i]}$ at 7 days all appear higher than at 27 days, especially for the more refractory oxides (Al, Gd, Hf, Ti and Zr), although the overall trends are the same, i.e. a highly incongruent dissolution. There is better agreement between the 27 day and the final filtered leach rates, with very similar results for all elements. The different melting temperatures and HfO_2 incorporation do not produce any significant differences in release rates, apart from at 7 days where the highest loading and lowest temperature glass performs distinctly poorly. This may, however, be caused by an experimental artefact. After 27 days, the leach rates for Hf and Gd are similar at ~10^{-4} g m^{-2} d^{-1}.

LaBS Glasses

Figure 3 shows the $ELR_{[i]}$ for the final filtered leachate samples taken from the modified PCT-B test for all the LaBS glasses. The samples taken after 7 and 27 days were uncharacteristic with very similar, and very high (~5 x 10^{-2} g m^{-2} d^{-1}), $ELR_{[i]}$ for all elements. This is likely to be an artefact of not filtering the leachate when sampling, and the results are not reproduced here.

The release rates for all the elements of the LaBS glasses are in general an order of magnitude better than the MMW and ATS glasses, i.e. ~10^{-3} g m^{-2} d^{-1} for the Si and B, and 10^{-5} to 10^{-6} for the Ln^{3+} and Hf^{4+}. There is also no significant variation in $ELR_{[i]}$ with temperature or HfO_2 loading, even for the glasses containing large quantities of undissolved material. This incongruency suggests that the 'matrix' dissolves, leaving behind highly insoluble Ln_2O_3/HfO_2 phases. There is the possibility of the Ln^{3+} leaching faster than the Hf^{4+}, although the elemental concentrations in the leachates are close to the limits of detection and therefore potentially prone to large errors.

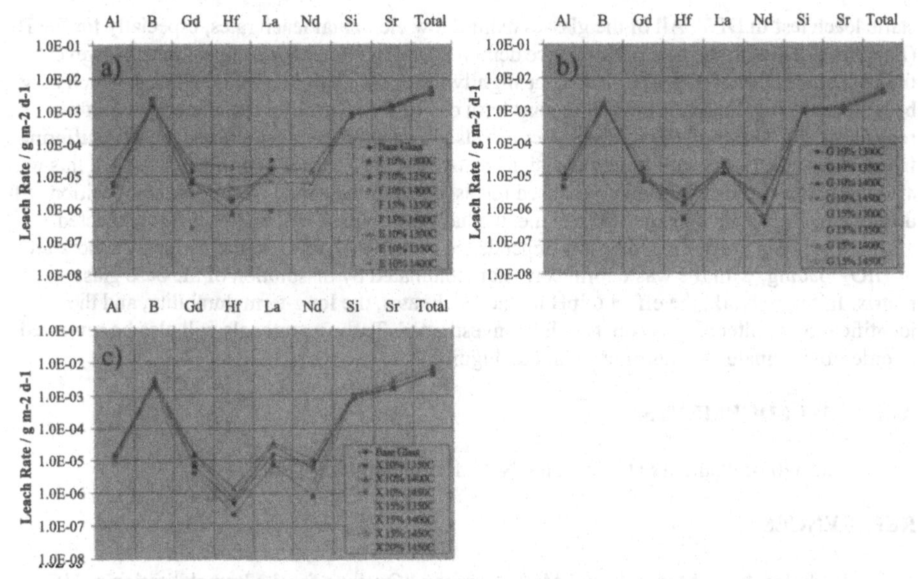

Figure 3. PCT leach rates for the final filtered leachate after 27 days for; a) LaBS_E, F, b) LaBS_G, and c) LaBS_X. Samples are labelled with wt% HfO_2 and melting temperature in °C.

Table III compares an average $NL_{[i]}$ for B and Si for all the glasses under test (taken from the final filtered leachate data after 27 days), compared with the values for Frit 'X' from Marra [5]. Note that the Frit 'X' was leached for 7 days rather than 27, and this will account for the lower $NL_{[i]}$ values along with slightly different test conditions. The superior behaviour of the LaBS glasses in DIW is clear from Table III, with the $NL_{[i]}$ values around an order of magnitude lower for B and Si. However, the behaviour of the candidate glasses under different conditions, e.g. pH, may be modified, and this is currently under investigation.

Table III. Comparison of the averaged $NL_{[B]}$ and $NL_{[Si]}$ for the 2nd stage evaluation glasses.

Glass	*NL [B] / g L^{-1}*	*NL [Si] / g L^{-1}*
MMW	0.646	0.303
ATS	0.428	0.250
LaBS_E	0.058	0.031
LaBS_F	0.077	0.030
LaBS_G	0.061	0.034
LaBS_X	0.100	0.037
FRIT 'X' [5]	0.021	0.014

CONCLUSION

The durability of a number of borosilicate glasses, selected due to their potential for the immobilisation of excess separated PuO_2 stock, has been investigated using a powder based static leach test in DIW. All of the glasses exhibit low elemental leach rates, especially for the Hf (Pu-surrogate), with the best performance demonstrated by the LaBS compositions. However, the lanthanide species appear to leach out slightly faster than hafnium in the glasses, which may be an issue if gadolinium is providing long-term criticality control for the altered glass in the repository. Furthermore, the LaBS glasses are also in general the most inhomogeneous suffering from HfO_2 settling. As well as giving a significant criticality concern during melting [3], this will artificially lower the calculated normalised release rates as the actual wasteform composition will differ from the nominal theoretical one, i.e. the hafnium content will be lower than expected.

There did not appear to be any dependence of the leach rates on the melting temperature or HfO_2 loading, with the waste form corrosion dominated by dissolution of the base glass matrix. In future work, the effect of pH on the leach rates, the long-term durability, and the identification of alteration products will be investigated. Testing protocols will also be reviewed in order to eliminate sources of error and ambiguity.

ACKNOWLEDGEMENTS

The author would like to thank the NDA for funding this work.

REFERENCES

1. C. R. Scales, E. R. Maddrell, and M. T. Harrison "Options for the Immobilisation of UK Civil Plutonium", *WM'07 Conference*, February 25 - March 1, 2007, Tucson, AZ.
2. M. T. Harrison, C. R. Scales, P. A. Bingham, and R. J. Hand, *Mat. Res. Soc. Symp. Proc* **932** (2007).
3. M. T. Harrison and C. R. Scales, MRS 2007: Scientific Basis for Nuclear Waste Management XXXI (presented at this meeting).
4. N. E. Bibler, W. G. Ramsey, T. F. Meaker, and J. M. Pareizs, *Mat. Res. Soc. Symp. Proc.* **412**, 65-72 (1996).
5. J. C. Marra, D. K. Peeler, and C. M. Jantzen, *WSRC-TR-2006-00031*, Washington Savannah River Company, Aiken, SC, 2006.
6. J. R. Zamecnik, T. M. Jones, D. H. Miller, D. T. Herman, and J. C. Marra, *WM'07 Conference*, February 25 - March 1, 2007, Tucson, AZ.
7. J. K. Bates, A. J. G. Ellison, J. W. Emery, and J. C. Hoh, *Mat. Res. Soc. Symp. Proc.* **412**, 57-64 (1996).
8. B. J. Riley, J. D. Vienna, and M. J. Schweiger, *Mat. Res. Soc. Symp. Proc.* **608**, 677-682 (2000).
9. J. G. Darab, H. Li, and J. D. Vienna, *J. Non-Cryst. Solids* **226**, 162-175 (1998).
10. J. G. Darab, H. Li, M. J. Schweiger, J. D. Vienna, P. G. Allen, J. J. Butcher, N. M. Edelstein, and D. K. Shuh, *Plutonium Futures: The Science*, Santa Fe, NM, August 25-27, 1997.
11. J. A. Fortner, E. C. Buck, A. J. G. Ellison, J. K. and Bates, *Ultramicroscopy* **67** , 77-81 (1997).
12. C. Lopez, X. Deschanels, J. M. Bart, J. M. Boubals, C. Den Auwer and E. Simoni, *Journal of Nuclear Materials* **312**, 76-80 (2003).
13. J.-N. Cachia, X. Deschanels, C. Den Auwer, O. Pinet, J. Phalippou, C. Hennig and A. Scheinost, *Journal of Nuclear Materials* **352**, 182-189 (2006).

Spent Nuclear Fuel

Mater. Res. Soc. Symp. Proc. Vol. 1107 © 2008 Materials Research Society

RN Fractional Release of High Burn-Up Fuel: Effect of HBS and Estimation of Accessible Grain Boundary

F. Clarens[1], D. Serrano-Purroy[2], A. Martínez-Esparza[3], D. Wegen[2], E. Gonzalez-Robles[4], J. de Pablo[1,4], I. Casas[4], J. Giménez[4], B. Christiansen[2] and J.P. Glatz[2]
[1]CTM Centre Tecnològic, Av Bases de Manresa 1, 08242, Spain
[2]JRC-ITU, European Commission Joint Research Centre-Institute of Transuranium Elements, Hermann-von-Helmholtz-Platz 1, Eggenstein-Leopoldshafen, 76344, Germany
[3]ENRESA, Emilio Vargas 7, Madrid, Spain.
[4]Chemical Engineering Department, Universitat Politècnica de Catalunya UPC, Av Diagonal 647, 08028, Barcelona, Spain

ABSTRACT

The so-called Instant Release Fraction (IRF) is considered to govern the dose released from Spent Fuel repositories. Often, IRF calculations are based on estimations of fractions of inventory release based in fission gas release [1]. The IRF definition includes the inventory located within the Gap although a conservative approach also includes both the Grain Boundary (GB) and the pores of restructured HBS inventories.

A correction factor to estimate the fraction of Grain Boundary accessible for leaching has been determined and applied to spent fuel static leaching experiments carried out in the ITU Hot Cell facilities [2]. Experimental work focuses especially on the different properties of both the external rim area (containing the High Burn-up Structure (HBS)) and the internal area, to which we will refer as Out and Core sample, respectively. Maximal release will correspond to an extrapolation to simulate that all grain boundaries or pores are open and in contact with solution.

The correction factor has been determined from SEM studies taking into account the number of particles with HBS in Out sample, the porosity of HBS particles, and the amount of transgranular fractures during sample preparation.

INTRODUCTION

The data acquired during the previous European Project, Spent Fuel Stability (SFS) [1] points to a lack of experimental data regarding to IRF, which is considered to govern the dose released over a prolonged period of storage. For this reason, the Performance Assessment (PA) exercises are forced to use a conservative approach to integrate the IRF.

IRF is defined as the fraction of RN that can be released "Instantly" or, more accurately, faster than the matrix, when the water comes in contact with the SF after the failure of the canister in the repository.

Fuels with burnup (BU) higher than 40 MWd/kgU have an external layer called 'High Burn-Up Structure' (HBS) characterized by a higher BU than the core as well as higher porosity and plutonium content. The depth of the HBS increases with BU and depends on the fuel irradiation history [3,4]. The dissolution of this layer has not been studied in detail, yet.

There is agreement that radionuclides (RN) present at the gap should be considered as part of the IRF. As a conservative approach, the RN located in the GB and on the pores of the HBS are also included into the IRF. On the other hand, there is not a clear consensus on whether the RN allocated in the HBS region need to be considered as part of the IRF (due to its small grain size) or should be considered as part of the matrix release [1].

Even if the RN located at the GB or in the HBS pores are considered as part of the IRF, it is clear that these RN can only be released in case that all GB/pores can get in contact with water, which is not going to be the case in the repository.

With the aim to study the effect of HBS, experiments were carried out using two different powdered samples, one prepared from the centre of the pellet, free of HBS, and one from the outer part, containing HBS.

This paper focuses on the estimation of a correction factor to correlate the open Grain Boundary and HBS pores with the total GB/pores. These factors might allow the estimation of the maximal fractional releases from experimentally determined values.

EXPERIMENTAL DETAILS

Experiments were performed using a LWR Spent Nuclear Fuel (SF) irradiated in a commercial reactor with a mean Burn-Up of ~60 GWd/t_U.

Two SF samples were prepared from different radial positions of the pellet, one prepared from the centre, named Core and one from the outer part, enriched with particles from HBS region, labelled Out. The preparation procedure, in five steps, is described below.

Firstly, segments were cut from a pin. Secondly, the Core part was drilled out. In a third step the Outer part was mechanically separated from the Cladding applying pressure over the external layer of the cladding material. The fourth step was the grinding between 50 and 100 μm (with previous milling in the case of Out sample). Finally the powdered samples were cleaned with acetone to remove the fines attached to the surface produced during sample preparation.

SEM images obtained at 100x magnification were used to determine mean particle sizes of (68 ± 15) μm and (82 ± 8) μm for Core and Out sample respectively.

Static leaching experiments were carried out in glass centrifuge tubes as batch reactors under oxidising atmosphere. Partial replenishments were made after each sample. Two different leaching solutions were used, bicarbonate water (1mM $NaHCO_3$ + 19mM NaCl), and synthetic Bentonitic Granitic Water prepared according to the recipe from [5]. All reagents were supplied from Merck, Germany.

Solution analyses were performed by HR-ICP-MS of filtered (0.2 micron pore size) and unfiltered samples.

Each experiment was conducted in parallel for both SF samples using ~0.25 g of fuel with S/V ratio of 110-130 m^{-1}.

The solution was completely exchange after 1 and 3 days. Beyond this time partial replenishment with fresh solution was made after each sample to keep the volume constant.

RESULTS AND DISCUSSION

In these experiments, the RN determined in solution comprise the release from the oxidised layer, the GB, the pores HBS and also from the release of SF matrix.

In the present work an attempt to distinguish between the contribution of the oxidise layer, matrix and GB release is carried out, in order to determine the RN fast release contribution.

Not all grain boundaries or porosities in the HBS area will be accessible to the leaching solution, due to the particle size used in the experiments as mentioned in the introduction. Therefore the fast fractional release determined after matrix dissolution correction does not correspond to the maximal release expected for the case that all GB or HBS pores are open. Thus it is necessary to estimate the open GB and porous fraction in contact with leaching solution. Based on SEM analysis (at least on three independent images) correction factors had been calculated taking into account geometrical considerations between particle and grain size for both Core and Out sample.

Analyses of SEM images were carried out using both Corel Photo-Paint12 and UTHSCSA Image-Tool3 software, (IT3) [6].

To determine the above mentioned correction factors the following analysis were made:

- Determination of the ratio between non-HBS and HBS particles in Out sample.
- Determination of accessible pore volume in HBS particles.
- Determination of accessible GB for Core and Out particles.

Determination of the ratio between Core and HBS particles in Out sample.

Ceramographic investigations indicate that the HBS regions depth in this fuel is about 250 µm, which is on the upper limit at this burn-up based on published data.[7]

The Out sample corresponds to the external 600 µm (based on the nominal drilling tool diameter), and contains therefore not only fuel belonging to the HBS region. This observation made it necessary to calculate the fraction of fuel with HBS structure in the Out sample.

Based on the depth of the HBS region and nominal diameter of the drilling tool, the Out sample should contain 40 % of fuel from the HBS region. This value is a rough estimation and can only be considered as an upper limit of HBS fraction in the Out sample.

To obtain more accurate data, the total number of particles and the particles with HBS were counted in SEM images. A particle was considered to belong to the HBS region when the surface of the particle is clearly porous (figure 1b). With this method an average ratio of 19 % of particles with HBS in the Out sample was obtained. An example of particles counting with IT3 software is shown in figure 1, where HBS particles are marked with dots in the SEM image at 100x (a), detail of (a) at 1000x magnification (b).

Figure 1. a) SEM image (x100) of Out sample. HBS particles are marked with a dot. b) Centre of a) where porous particle is shown at 1000x.

Determination of accessible pore volume in HBS particles

A conservative approach considers that the RN located at the pores of HBS particles should be included as part of the IRF [5]. For this reason, it is interesting to determine the fraction of pores that is in contact with solution. Especially as the HBS particles have closed porosity.

To calculate this ratio it is necessary to determine first the porosity in a HBS particle.

The surface porosity in HBS particles was calculated from SEM images (4000x), determining the ratio between the total area of the image and the sum of the areas corresponding to the surface cross section of pores.

In good agreement with published results (15 %) an average porosity of 18 % [2] for HBS particles was determined with the IT3 program,. An example is shown in figure 2, were porous areas are marked in white.

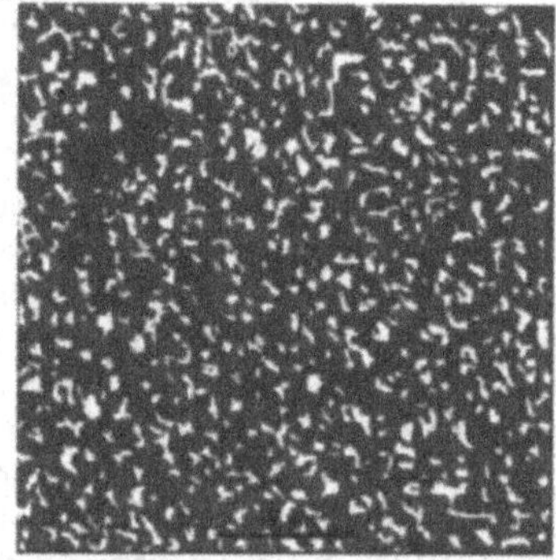

Figure 2. SEM image obtained at (x4000) of the HBS zone in the Out sample. White areas correspond to the considered cross-section of pores at the surface.

To calculate the fraction of open pores in contact with solution a pore size of 2 micron [3] and spherical geometry for both particles and pores had been considered. The total volume porosity had been calculated as the 18 % of the particle volume. The open pore porosity had been calculated estimating the number of pores bubbles that are broken at the surface and the half volume of a sphere with the diameter of the pore. Total pore number had been obtained from the ratio between the 18 % of the particle surface area and the cross-section area of one pore.

A ratio **pore volume/pore volume at the surface of 10** had been calculated.

Determination of leaching accessible GB

The GB area that are accessible for leaching corresponds to the GB area located at the faces of the particle added to the penetration area of the solution along the perimeter of the GB that cross the surface of the particle.

With these considerations, a geometrical factor (total GB/ available GB) will be calculated for both Core and Out sample.

Calculations were made with the assumption that the SF grains and particles have the geometrical shape of the tetrakaidecahedron (also named Kelvin Cell -KC) and that the perimeter of grains at the particle surface have the form of a pentagon. These geometric assumptions are based on the fact that the average number of faces that cut a surface is around 5 [8,9] and that the KC corresponds to a structure that represents better than sphere a 3-dimensional net of grains. [8-10].

The total area of GB in one particle was then calculated determining the number of SF grains (10 μm diameter) that fits on the volume of one particle and the surface area of one grain, corrected by the theoretical SF density (95 %).

To determine the GB at the particle surface is it necessary to consider that the central part of the fuel pellet breaks preferentially through to transgranular fractures [11]. SEM studies confirm that this is the main breaking process. However, regions where intergranular fractures dominate had been observed. The ratio between the area of intergranular regions and the particle area where calculated with the help of IT3 using a similar protocol to the determination of porosity in HBS particles.

As example, SEM image of the Core sample at 1000x magnification is shown in figure 3; where predominant intergranular fractures are marked in white. As average value, it was determined that 97 % of the surface in Core HBU had been broken through trans-granular process. Then, only 3% of the particle surface is estimated to contain open grain-boundaries.

The same determination was carried out for Out sample. In this case, large variability is obtained as particles coming from the pure HBS region over intermediate zones as found as well as pure Core like particles are found. As conservative approaches the same ratio as for the Core sample was selected (97 %) for all Out particles with non HBS, even in some particles belonging to the transition zone, only an area of 7 % had been broken through transgranular fracture.

GB area at the face of the non-HBS particle is then calculated as the 3 % of the surface area of the Out particle.

Due to the small grain size, 0.5 microns, HBS particles are assumed to be broken only through intergranular process.

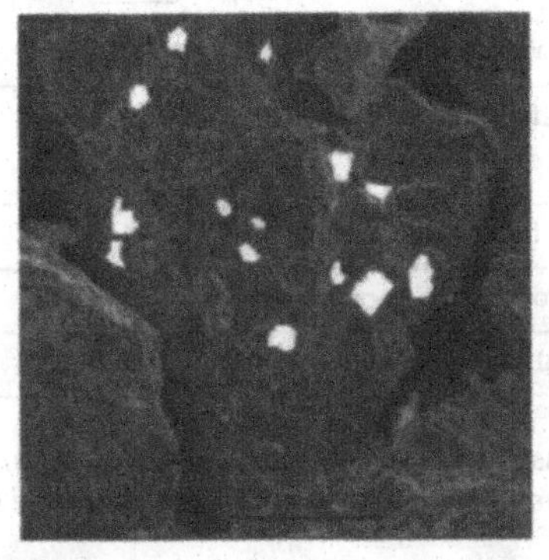

Figure 3. SEM images of Core sample at x1000 magnification factor. Zones considered to break through intergranular fractures are marked in white.

The GB area along the perimeter of broken grains is calculated determining the number of SF grains that fits into particle surface and the area of the sides of the pentagonal column. The edge of the side is related to grain size while the height is related to the penetration depth of water into the particle. In the present calculations is assumed a penetration depth of 1 micron for non-HBS particles and 0.5 micron for HBS particles.

The results from open GB are summarised in table II.

Table II. Results to estimate the open GB correlation factors.

	Core fraction	Out fraction (non-HBS particle)	Out fraction (HBS particle)
Total GB area (m^2)	$1.4 \cdot 10^{-7}$	$3.2 \cdot 10^{-7}$	$6.4 \cdot 10^{-6}$
Accessible GB area (m^2)	$1.2 \cdot 10^{-8}$	$2.2 \cdot 10^{-8}$	$2.0 \cdot 10^{-6}$
Correlation factor	**11**	**15**	**3**

The contribution of GB at the particle faces is one order of magnitude lower than the GB area exposed along the perimeter of broken grains. The difference between Core and Out sample is due to the different particle size of each sample.

Total correlation factors

The total factor to correlate the determined RN fractional release with the expected maximal release if all GB and pores can be leached for Out sample was calculated as a weighted value between the correlation factors obtained for particles with and without HBS taking into account the fraction of HBS in Out sample.

A summary of results to correlate open GB/pore with total GB/pore is shown in table III.

Table III. Summary of correlation factors between leaching available and total GB/pore.

	Core	Out
HBS fraction	_____	19 %
Surface porosity in HBS	_____	18 %
% of surface with transgranular fracture	97	Variable[1]
Factor for non HBS particles	11	15
Factor for HBS particles	_____	10 pore /3 GB
Total correlation factor	**11**	**13**

[1]Large variability as particles from like to HBS region to like to Core part had been observed. As a conservative approach, the same value as for the Core fraction was selected for the calculation.

Determination of maximal fast release

Table IV summarises the RN fractional release calculated from values collected from the values collected during the first week including the second washing with the factors shown in Table III.

Table IV. Corrected RN fast fraction release.

	Bic		BGW	
	Core	Out	Core	Out
Rb	5%	7%	9%	9%
Sr	7%	9%	16%	11%
Y	3%	1%	7%	5%
Zr	1%	< 0.01%	< 0.01%	< 0.01%
Mo	14%	10%	21%	8%
Tc	9%	4%	7%	5%
Ru	0.3%	0.3%	0.3%	0.4%
Rh	0.5%	0.5%	0.7%	0.8%
Te	3%	3%	3%	3%
Cs	7%	8%	8%	13%
La	2%	1%	4%	4%
Nd	4%	2%	6%	6%
U	1.6%	0.9%	0.7%	0.5%
Np	7%	3%	10%	4%
Pu	2.8%	1.3%	0.8%	1.4%
Am	2.6%	0.6%	0.3%	0.8%
Cm	2.9%	0.5%	0.5%	0.5%

Considering the uncertainty in SF leaching experiments (at least 10 %) and the different particle size used, no clear effect of HBS dissolution can be deducted from these results despite higher release of Mo and Np is observed for core part and Cs shows faster release in Out sample.

CONCLUSIONS

An estimation of the maximal release was carried out based on geometric relations between characteristics of SF and powder particles used in static experiments.

Release of some RN in core sample are higher than in Out sample. This difference can be due to the presence of HBS or to the different particle size used in the experiments. If the difference is due to HBS, this structure will act as a protective layer.

No clear difference between the two different waters has been observed.

ACKNOWLEDGMENTS

The authors would like to thank the staff of the EU Joint Research Centre Institute of Transuranium for their help on daily work and fruitfully discussions.

Authors wants to acknowledge the financial support through the NF-PRO project (FIW-CT-2003-02389) belonging to 6th Framework program of the European Commission.

REFERENCES

1. C Poinssot et al. Final Report of the European Project Spent Fuel Stability under Repository Conditions. CEA-R-6093. ISSN 0429-3460. 2005, pp 103.
2. Final Report of the European Project Near Field Processes (NF-PRO). In preparation.

3. J. Spino, K. Vennix and M. Coquerelle. Journal of Nuclear Materials. Vol. 231, 1996, pp 179-190.
4. T. Walker, D. Staicu, M. Sheindlin, D. Papaioannou, W. Goll, F. Sontheimer. Journal of Nuclear Materials, Vol. 350, 2006, pp 19-39.
5. A. M. Fernández. Preparación de una agua intersticial sintética bentonítica a diferentes salinidades (Preparation of a synthetic bentonitic interstitial water at different salinities). CIEMAT DMA/M214213/05, Madrid, 2005.
6. UTHSCSA Image Tool 3. University of Texas Health Science Center at San Antonio, Texas and available from the Internet by anonymous FTP from maxrad6.uthscsa.edu.
7. L. Johnson. Journal of Nuclear Materials, Vol. 346, 2005, pp 56-65.
8. L. Coble. Journal of Applied Physics, Vol. 32 (5), 1961, pp. 787-792.
9. K.M. Döbrich, C. Rau, C.E. Krill. Metallurgical and Materials Transaction A. Vol. 35A, 2004, pp. 1953-1961.
10. W.F. Hosford, www.cambridge.org. ISBN=13:9780511258336. 18/04/07.
11. Gray, W.J.; Wilson, C.N. Spent Fuel Dissolution Studies FY 1991 to 1994. PNL- 10540. 1995.

Mater. Res. Soc. Symp. Proc. Vol. 1107 © 2008 Materials Research Society

Radionuclides Release From the Spent Fuel Under Disposal Conditions: Re-evaluation of the Instant Release Fraction

Cécile Ferry[1], Jean-Paul Piron[2], Arnaud Poulesquen[1] and Christophe Poinssot[1]
[1] Department of Physico-chemistry, Commissariat à l'Energie Atomique, CEA-Saclay, Gif-sur-Yvette, 91191, France
[2] Department of Fuel Studies, Commissariat à l'Energie Atomique, CEA-Cadarache, Saint-Paul Lez Durance, 13108, France

ABSTRACT

The so-called Instantaneous Release Fraction (or IRF) corresponds to the rapid release of radioactivity by the spent fuel rod at the canister breaching time in repository and depends on the spent fuel evolution in a closed system. The effect of He accumulation on the spent fuel pellet microstructure and diffusion processes are here re-assessed in the light of the recent results issued from the projects Near-Field PROcesses and PRECCI. It allows diminishing the conservatism of the former IRF values of key safety relevant radionuclides for PWR UO_2 fuels.

INTRODUCTION

In repository, a fraction of the radionuclides (RN) inventory will be rapidly released from the spent fuel rod. In the framework of the European "*Spent Fuel Stability under Repository Conditions*" project [1] the so-called Instantaneous Release Fraction (or IRF) has been defined as the fraction of the inventory of safety-relevant radionuclides, which is located in fuel microstructures for which no confinement properties is anticipated or can be demonstrated at the time of canister breaching [2].

Hence, the IRF depends on (1) the intrinsic evolution of the spent fuel pellet microstructure before the canister breaching (i.e. during~10 ky.); (2) RN distribution within the spent fuel rod after discharge from the reactor and (3) its evolution with time before canister breaching.

IRF best and pessimistic estimates have been proposed by Johnson *et al.* [2] for PWR UO_2 fuels. These values take into account uncertainties on the spent nuclear fuel (SNF) evolution during a confinement phase of 10,000 years prior to the canister breaching. Hereafter the IRF values of the key-safety-relevant radionuclides are updated considering the recent results issued from the European NF-PRO and French PRECCI projects on (1) the diffusion processes in spent fuel, (2) evolution of the pellet microstructure with helium accumulation and (3) leaching data.

DIFFUSION PROCESSES IN SPENT FUEL

Due to diffusion processes, part of the inventory, which is initially located within grains after irradiation, may reach the fuel surface or grain-boundaries and contribute to the IRF. RN mobility processes in spent fuel have been detailed in a recent review [3]. Due to the low temperature expected during disposal, thermal diffusion is not relevant in repository conditions even in the long-term [4]. Regarding athermal diffusion, various modelling approaches have been proposed to estimate the α self-irradiation enhanced diffusion [5]. The upper estimate is based on extrapolation of the measured in-reactor athermal diffusion coefficient of U atoms under irradiation and yields a released fraction from 8 μm grains around 5% after 10,000 years of disposal. This value was taken into account in the first IRF quantitative assessment [2].

However, the effects of α decays on atom mobility in the spent fuel pellet have also been quantified through more physical modeling [5]. Taking into account the enhanced mobility of atoms induced by the α particles, due to electronic excitation, and by the recoil atoms, due to cascades of displacements, the model yields an enhanced diffusion for fission products atoms under α self-irradiation about three orders of magnitude lower than the previous upper estimate.

Furthermore, heavy ion bombardments of I-implanted UO_2 disks have been used to simulate the effects of α self irradiation on iodine mobility in UO_2. For the study of ballistic effects, I-implanted samples have been irradiated with Xe ions of 800 keV at fluences of 2×10^{16} and 5×10^{16} ions.cm^{-2}. Under these conditions the irradiation damage is similar to the ballistic damage accumulated in a MOX fuel of 60 GWd.t^{-1} after about 10,000 years of disposal and the upper estimate of the diffusion coefficient ($\sim 10^{-25}$ m^2.s^{-1} at the beginning of disposal) leads to a diffusion length of around 200 nm. The iodine profiles were measured by Secondary Ion Mass Spectrometry before and after irradiation. No measurable displacement of iodine could be detected (< 50 nm).These results are consistent with the lower value of the theoretical diffusion coefficient ($\sim 10^{-28}$ m^2.s^{-1}). The effect of the electronic stopping power of α particles electronic on I atom mobility was simulated by using ions of high energy and electronic power of 20 keV/nm. Some I-implanted UO_2 disks were irradiated with Br ions of energy varying between 170 and 220 MeV at a dose of 10^{15} ions per cm^2, and with Ag ions of 70 MeV at a dose of 5×10^{15} ions per cm^2. These last experiments are very similar to those performed by Hocking *et al.* [6], except that the authors used I ions for their irradiation. Contrary to Hocking *et al.* and although the conditions of irradiation were similar, no mobility of I atoms has been detected.

Models and studies of iodine mobility under heavy ion bombardments that simulate α self-irradiation effects converge to an α self-irradiation enhanced diffusion coefficient value which is at least two to three orders of magnitude lower than the current upper estimate. With this new value the contribution of α self-irradiation enhanced diffusion to IRF of fission products becomes negligible. As for the activation products, Pipon *et al.* [7] on the basis of experiments performed on implanted UO_2 have shown that thermal diffusion of chlorine in UO_2 is higher than for I and Cs. The same behaviour is observed under irradiation [8]. Due to the uncertainties concerning the behaviour of activation products, the contribution of diffusion process to the IRF of ^{36}Cl and ^{14}C remain an open question.

EVOLUTION OF THE SPENT FUEL PELLET MICROSTRUCTURE

In a closed system, the spent fuel pellet microstructure may evolve due to (i) chemical evolution of the spent fuel with radioactive decays, (ii) accumulated α-decay damage and (iii) production of He by α decays. Whereas chemical evolution and accumulated damage should not alter the spent fuel microstructure, the effect of helium accumulation is still an open issue.

After irradiation in reactor two main regions with different microstructures are observed: The rim zone at the periphery of the pellet, and the central and intermediate zones ($0 < r/Ro < 0.75$, with Ro the pellet radius) where the pristine grains are not restructured. Due to these major structural discrepancies these two regions of the pellet will present different mechanical properties and behaviour with helium accumulation.

Effect of helium in the central and intermediate zone of the spent fuel pellet

Considering the upper estimate of the diffusion coefficient, He release from 8 µm diameter grains to grain boundaries should be relatively low under disposal conditions (< 5%). Therefore in the not restructured zones, only grain micro-cracking due to over-pressurization or formation of gas bubbles with He accumulation should cause significant He release to grain boundaries, where helium can then accumulate.

The conditions of micro-cracking of grains due to the evolution of the intra-granular bubble population with helium production have been described in a previous paper [9]. The considered geometry corresponds to a homogeneous population of fission gases and He bubbles in grains, with a bubble size varying between 1 nm and 1 µm. The bubble pressure values at which tensile material failure is likely to occur have been discussed by Stout *et al.* in [10]. Considering a realistic density of intra-granular bubbles as a function of the bubble size at the end of irradiation and that all formed He atoms are trapped in pre-existing fission gas bubbles, the concentration of He to reach the critical bubble pressure is 4×10^{20} at.cm^{-3} (or ~ 0.6 at%) in 10 nm bubbles. For larger bubbles, it is around 10^{21} at.cm^{-3}. Considering the formation of new He bubbles, the minimal critical value is 3×10^{20} at.cm^{-3} and relates to the formation of 100 nm bubbles of helium. Table I shows that the mean pellet quantity of helium in UO_2 spent fuels is lower than the critical value even at the highest burnup. We conclude that the quantity of α–decay produced helium is not sufficient in UO_2 spent fuels to induce micro-cracking of grains of the pellet core, which is the first step before any significant He release to grain-boundaries occurs.

Literature contains some data on the transport of He in He implanted UO_2 disks. No detrimental effect of implantation has been evidenced at maximum implanted concentrations of 0.2 at% [11], 0.3 at% [12] and 0.6 at% [13]. At a maximum concentration of 1.1 at%, Guilbert *et al.* [14] observed flaking effects due to bubble precipitation. These various experimental results are consistent with the above proposed critical value for He concentration. However, only the sample surface (< 10 µm ~ grain size) including grains and grain boundaries is implanted with He in this type of experiments and extrapolation of these results to the grain stability is questionable.

From the above discussion we conclude that the spent fuel microstructure of the central and intermediate parts of the pellet should not significantly evolve with He accumulation. Therefore grain-boundaries should remain inaccessible to water and their inventories should not contribute to IRF.

Table I. Mean quantity of helium in the spent fuel pellet of UO_2 fuel with different burnup (CESAR code calculations)

	Mean pellet quantity of He (at.cm^{-3})			
Cooling time (years)	UO_2 (47.5 GWd.t^{-1})	UO_2 (60 GWd.t^{-1})	UO_2 (68 GWd.t^{-1})	UO_2 (75 GWd.t^{-1})
5000	1.2×10^{20}	1.3×10^{20}	1.4×10^{20}	1.5×10^{20}
10,000	***1.5×10^{20}***	***1.6×10^{20}***	***1.8×10^{20}***	***1.9×10^{20}***
50,000	2.6×10^{20}	2.7×10^{20}	2.9×10^{20}	3.1×10^{20}

Effect of helium on the rim microstructure

The rim (or High Burnup Structure, HBS) is characterized by sub-micrometer grains and micrometer pores with a total porosity which ranges between 10 and 15 % [15]. At the pellet edge, the fission rate is 2 to 3 times higher than the mean pellet value. Consequently the quantity of α-decay produced He is locally 2 to 3 times larger than the mean pellet quantities (Table I). Due to the small grain size (0.3 μm) and considering the upper value of diffusion for He, the fraction of α-decay produced He, which is released to pores, is estimated to be more than 50%.

In the rim geometry, large pores form sharp angles at the grain boundaries and constitute flaws from which fracture of the polycrystalline ceramic can initiate [16]. The underlying basis of the strength of brittle ceramics derives from the Griffith flaw concept [17]. This concept may be embodied in the simple expression:

$$K_I = A \times \sigma \times c^{1/2}, \quad (1)$$

Where, σ (MPa) is an applied tensile stress (assuming a uniform stress over the crack area), c (m) is a characteristic flaw dimension and K_I (MPa.m$^{1/2}$) the stress intensity factor. A is a constant depending on the crack geometry and type of loading. The most critical flaw in ceramics corresponds to $\sqrt{\pi}$ value for A.

Under equilibrium conditions of fracture (inert environment, high stress rates, low temperature), the flaw propagates spontaneously to failure when the stress intensity factor reaches the material toughness, K_{IC} (MPa.m$^{1/2}$), which defines the material toughness. An indication of the fracture toughness of the rim material can be inferred from the Vickers indentation tests [15, 18]. These results evidence an increase of the strength to fracture in the rim zone by a factor of two compared to un-irradiated UO_2 and the rest of the pellet. On the basis of Eq.1 and considering that pores constitute the limiting flaws in the rim, we can deduce the critical bubble pressure which leads to cracks propagation in the rim region as a function of the pore size. Hence, given characteristics of fission gas bubbles in the rim region at the end of irradiation and using the same approach as detailed in [9], we calculate the total number of gas (He + fission gas) atoms per bubble, n_{at}^{cr}, to reach the critical bubble pressure as a function of the bubble size. Table II gives the characteristics of fission gas bubbles (P_b^{eq}, bubble pressure; V_{at}^{eq}: gas atom volume; n_{at}^{eq} : number of gas atoms in bubble) at thermodynamic equilibrium at the end of irradiation in the rim region as a function of the bubble size.

Table II. Characteristics of fission gas bubbles in the rim region at the end of irradiation (T = 400°C and P_{hyd} = 16 MPa)

Bubble size (μm)	0.5	1	2	2.5
P_b^{eq} (MPa)	22	19	18	17
V_{at}^{eq} (10^{-28} m^3)	4	4.8	4.9	5.2
n_{at}^{eq} (at/bubble)	1.64×10^8	1.09×10^9	8.55×10^9	1.57×10^{10}

Table III reports the critical bubble pressure, P_b^{cr}, in the rim region. It corresponds to a fracture toughness of 2 MPa.m$^{-1/2}$. The total number of gas atoms (fission gas + He) to reach the critical bubble pressure in the pores, n_{at}^{cr}, has been calculated according to the Van der Waals equation. The corresponding concentration of He atoms assumes a uniform distribution of pore size and a total porosity of 15%. The mean distance between bubbles is also indicated for comparison to the He expected diffusion length after 10,000 years (about 0.15 μm). It is in good agreement with pore-to-pore surface distances as reported in [15].

Table III. Characteristics of (fission gas + He) bubbles at the critical bubble pressure in the rim region at T = 200°C

Bubble size (μm)	0.5	1	2	2.5
P_b^{cr} (MPa)	1600	1130	798	714
V_{at}^{cr} (10^{-29} m^3)	8.98	9.12	9.3	9.37
n_{at}^{cr} (at/bubble)	7.29×10^8	5.74×10^9	4.5×10^{10}	8.73×10^{10}
Density of bubbles (/m^3)	2.3×10^{18}	2.8×10^{17}	3.6×10^{16}	1.8×10^{16}
Mean Distance between bubbles (m)	7.6×10^{-7}	1.5×10^{-6}	3.0×10^{-6}	3.8×10^{-6}
Critical quantity of He (at.cm^{-3})	***1.3×10^{21}***	***1.3×10^{21}***	***1.3×10^{21}***	***1.3×10^{21}***

The critical He quantity is not sensitive to the relatively large bubble size of the rim region. If we take into account an initial quantity of fission gas in the bubble which is two times the quantity at equilibrium (Table II), the critical He quantity is around 20% lower than the value given in Table III. This value is about one order of magnitude higher than the mean pellet quantity of helium after 10,000 years of disposal (Table I). Considering that the maximal local concentration of helium in the rim region is 3 times larger than the mean pellet quantity, and that 100% is released by diffusion and grains micro-cracking to pores, it should not be sufficient to reach the conditions of crack propagation in the grain boundaries of the rim region. Therefore, rim inventories would not contribute to the IRF. Their release would be controlled by the aqueous dissolution kinetics of the rim.

ESTIMATED IRF VALUES

Leaching data of spent fuel have been detailed and discussed by Johnson *et al.* [2]. Grain boundary inventories of Cs and Sr are also given in [19] for a 60 GWd.t^{-1} UO_2 fuel. These leaching data are consistent with former data. Gap and grain boundaries inventories of the most relevant radionuclides as proposed in [2] are thus retained. The IRF values of key-safety radionuclides are revised in accordance with the new results, which essentially concern the evolution of the spent fuel during the confinement phase prior to the canister breaching.

Based on theoretical models comforted by experiments simulating α-decays effects on the atoms mobility in spent fuel, the release of fission products to grain-boundaries should not be significant even on the long-term. Thus, the contribution of the α self-irradiation enhanced diffusion to the IRF of fission products can be neglected. Due to the high mobility of chlorine that has been observed under heavy ions bombardment simulating fission effects, the contribution of diffusion to the IRF of activation products remains an open issue. The upper estimate of 5% is conservatively taken into account as a contribution to the IRF of ^{36}Cl and ^{14}C.

Furthermore, grain-boundaries and rim pores long-term stabilities with helium accumulation have been assessed by two different models depending on the microstructure properties. These models have shown that the α-decay produced helium quantity should not be sufficient to reach the critical bubble pressure in the rim region and pristine matrix of the SNF pellet. In conclusion, grain-boundaries and rim region inventories would not contribute to the IRF.

Conservative assumptions have been taken into account in the models of spent fuel evolution. Due to the lack of experimental validation on the long-term stability of grain boundaries and rim and modelling uncertainties (macroscopic approach), best and pessimistic estimates of *IRF(t = 0)* are proposed in Table IV. Best estimates correspond to the gap inventory fraction and are based on Fission Gas Release data as a function of burnup and leaching data [2]. Pessimistic estimates include inventories of gap, grain boundaries and all fission products in rim region (pores plus grains). It corresponds to the best evaluation of these inventories (BE values of table 7 in [2]). This approach leads to a large dispersion of the IRF of fission products for high burnup fuels between the best and pessimistic estimate values. This is due to the major contribution of the rim region to the pessimistic value of the IRF.

As for the activation products, due to the lack of data concerning their behaviour under irradiation and their location in the spent fuel rod after discharge from reactor, the IRF values as proposed in [2] are retained.

Table IV. IRF estimates (% of total inventory) for various radionuclides for PWR UO_2 fuel, Best estimate values, with Pessimistic estimate values in brackets

BURNUP (GWd/tU)	41	48	60	75
RN	IRF	IRF	IRF	IRF
fission gas	1 (2)	2 (4)	4 (8)	8 (16)
^{14}C	10	10	10	10
^{36}Cl	5	10	16	26
^{90}Sr	1 (2)	1 (3)	1 (5)	1 (9)
^{99}Tc, ^{107}Pd	0.1 (1)	0.1 (3)	0.1 (5)	0.1 (9)
^{129}I	1 (3)	2 (4)	4(8)	8 (16)
^{135}Cs, ^{137}Cs	1 (2)	2 (4)	4 (8)	8 (16)

DISCUSSION ABOUT UNCERTAINTIES AND CONCLUSION

The IRF of the major safety key-relevant radionuclides has been revised with regards to the most recent knowledge on the spent fuel long-term evolution prior to the canister breaching. IRF is the sum of RN inventories located in the zones of the spent fuel rod with low expected confinement properties at the time of canister breaching (i.e. after ~ 10 ky. in the French reference scenario). At the end of irradiation only free volumes of the spent fuel rod (gap, cracks, free surface) contribute to a rapid release of the activity during leaching tests. Given the above assessments, the spent fuel pellet microstructure is expected not to evolve before water arrival at the fuel surface under disposal conditions. The main question concerns the degree of confidence in the proposed models of evolution.

As for the contribution of the α self-irradiation enhanced diffusion, the available experimental data, which constitute indirect measurements of this process, support the theoretical approaches. Diffusion of fission products is not relevant even in the long-term. The degree of confidence on these results is relatively high.

The influence of helium accumulation on the spent fuel pellet microstructure has also been assessed by distinguishing the case of the rim from the rest of the pellet. Due to the relatively slow diffusion of helium atoms, the release to grain boundaries of the central and intermediate zones of the pellet should be limited unless important grain micro-cracking occurs. The conditions of grain micro-cracking due to accumulation of helium in pre-existing intra-granular bubbles or in new He bubbles have been assessed as a function of the bubble size assuming a homogeneous population of bubbles at equilibrium at the end of irradiation. Even if fission gas bubbles are over-pressurized at the end of irradiation, the contribution of initial fission gas pressure to the critical bubble pressure remains negligible. Taking into account a realistic density of fission gas bubbles for each bubble size, we conclude that the quantity of He in the pellet is not sufficient to reach the critical bubble pressure, even if all helium atoms are trapped in bubbles (pessimistic assumption). Furthermore, we assume that the gas characteristics of He and Xe are the same, whereas He atoms are smaller (pessimistic assumption). The main uncertainty of this model is the value of the critical bubble pressure, which is based on a fracture stress derived from three-point bending tests on poly-crystalline UO_2. We expect a larger value in mono-crystalline UO_2. Results are consistent with observations of He-implanted polycrystalline UO_2 at different concentrations but the implantation affects a very small volume (surface effects). Observations on α-doped UO_2 pellets with different accumulated damage should allow validation of this model. Since the proposed model is relatively conservative, the degree of confidence is good.

For the rim, due to the geometry of this region that is characterized by small grains and large pores, the conditions of fracture propagation in grain boundaries due to large fission gas bubbles have been determined based on Griffith theory of rupture. This approach allows determining the fracture strength (corresponding here to the critical pore pressure) versus the pore size and the fracture toughness. The distribution of pores size and the fracture toughness are derived from literature data. For the calculations, a homogeneous population of pores has been considered with a pore density leading to a fixed total porosity of 15%. However the results are not sensitive to the pore size. They show that the quantity of helium is not sufficient to lead to crack propagation in the rim region. The main assumption here concerns the stress intensity factor which depends on the applied stress and flaw size but also on the type of loading and crack geometry. A conservative expression has been used for the calculation. It would be necessary to validate this model by more microscopic approaches. The degree of confidence in this model is relatively poor due to the lack of other data to confirm its results.

In conclusion, best estimate values of IRF have been proposed. They correspond to the RN inventories located within free volume (gap, voids, cracks) of the spent fuel rod. These values are derived from leaching experiments of spent fuel fragments. Therefore proposed IRF values of safety relevant radionuclides are strongly dependent on the available leaching data. Data is currently lacking for high burnup UO_2 fuels. Furthermore, the locations of ^{14}C and ^{36}Cl inventories in the spent fuel rod are poorly known. More data are necessary to improve the IRF estimates for these RN. Due to the significant uncertainties associated with the modelling of the spent fuel long-term evolution, in particular for the rim evolution, pessimistic estimates of IRF have also been proposed. They include grain-boundaries and rim inventories. Since the rim inventory is the major contribution to IRF in high burnup fuels, a better understanding on the distribution of RN between rim pores and grains should allow decreasing the pessimistic estimates.

Finally, if the long-term stability of rim and grain-boundaries is confirmed, the surface area that will be accessible to water at the canister breaching, will be the same as observed after irradiation.

ACKNOWLEDGMENTS

This work is performed within the European NF-PRO project (Contract Number: FI6W-CT-2003-02389) and the co-financed CEA-EDF PRECCI project. Diffusion experiments have been provided by Dr. Ph. Garcia and collaborators. The authors thank L. Johnson for the review of this paper.

REFERENCES

1. C. Poinssot, C. Ferry, M. Kelm, B. Grambow, A. Martinez-Esparza, L. Johnson, Z. Andriambolona, J. Bruno, C. Cachoir, JM. Cavedon, H. Christensen, C. Corbel, C. Jegou, K. Lemmens, A. Loida, P. Lovera, F. Miserque, J. De Pablo, A. Poulesquen, J. Quinones, V. Rondinella, K. Spahiu, D. Wegen, "*Final report on the European project spent fuel stability under repository conditions*", CEA Report, CEA-R-6093 (2005), p. 104
2. L. Johnson, C. Ferry, C. Poinssot, P. Lovera, J. Nucl. Mat. 346, 56-65 (2005)
3. C. Ferry, C. Poinssot, V. Broudic, C. Cappelare, L. Desgranges, P. Garcia, C. Jégou, P. Lovera, P. Marimbeau, J.P. Piron, A. Poulesquen, D. Roudil, J.M. Gras, P. Bouffioux, "*Synthesis on the spent fuel long term evolution,*"CEA Report, CEA- R-6084, 2005, p.257.
4. C. Ferry, J.P. Piron, C. Poinssot, in Mat. Res. Soc. Symp. Proc., vol. 932, 513-520 (2006).
5. C. Ferry, P. Lovera, C. Poinssot, P. Garcia, J. Nucl. Mat. 346, 48 (2005).
6. W.H. Hocking, R.A. Verrall, and I.J. Muir, J. Nucl. Mat. 294, 45 (2001).
7. Y. Pipon, N. Toulhoat, N. Moncoffre, L.Raimbault, A.M. Scheidegger, F. Farges, G. Carlot , J. Nucl. Mat. 362, 416 (2007).
8. Y. Pipon, N. Toulhoat, N. Moncoffre, N. Bererd, H. Jaffrezic, M.F. Barthe, P. Desgardin, L. Raimbault, A.M. Scheidegger and G. Carlot, in Scientific Basis for Nuclear Waste Management, edited by D. Dunn (in press).
9. C. Ferry, J.P. Piron, R. Stout, in Scientific Basis for Nuclear Waste Management, edited by D. Dunn (in press).
10. R. B. Stout, C. Ferry, C. Poinssot, J.P. Piron, "Estimations of failure pressures in spent fuels from actinide alpha decay helium transported to fission gas bubbles,"10th Internat. Conf. on Environmental Remediation and Radioactive Waste Management, Sept2005, Glasgow, Scotland.
11. S. Guilbert, T. Sauvage, Ph. Garcia, G. Carlot, M.F. Barthe, P. Desgardin, G. Blondiaux, C. Corbel, J.P. Piron, J.M. Gras, J. Nucl. Mat. 327, 88 (2004).
12. G. Martin, P. Garcia, H. Labrim, T. Sauvage, G ; Carlot, P. Desgardin, M.F. Barthe et J.P. Piron, J. Nucl. Mat. 357, 198 (2006).
13. D. Roudil, X. Deschanels, P. Trocellier, C. Jegou, S. Peuget, J.M. Bart, J. Nucl. Mat. 325, 148-158 (2004).
14. S. Guilbert, T. Sauvage, H. Erralmi, M.F. Barthe, P. Desgardin, G. Blondiaux, C. Corbel, J.P. Piron, J.P., J. Nucl. Mat. 321, 121 (2003).
15. J. Spino, K.Vennix, M. Coquerelle, J. Nucl. Mat., 231 (1996).
16. A.G. Evans, R.W. Davidge, J. Nucl. Mat. 33, 249-260 (1969).
17. M.F. Ashby and D.R.H Jones, Matériaux 1. Propriétés et applications (Ed. Dunod, Paris, 1998), p.278
18. T.R.G. Kutty, K.N. Chandrasekharan, J.P. Panakkal, J.K. Ghosh, J. Mat. Science Lett. 6, 260 (1987).
19. D. Roudil, C. Jegou, V. Broudic, B. Muzeau, S. Peuget, X. Deschanels , J. Nucl. Mat. 362, 411 (2007)

Mater. Res. Soc. Symp. Proc. Vol. 1107 © 2008 Materials Research Society

Np-Incorporation Into K-boltwoodite

Lindsay C. Shuller[1], Rodney C. Ewing[1,2], and Udo Becker[2]
[1]Materials Science and Engineering, University of Michigan
[2]Geological Sciences, University of Michigan

ABSTRACT

Np-237 ($\tau_{1/2}$ = 2.1 million years) is a potentially important contributor to the total dose for a geologic repository under oxidizing conditions. Further, the Np^{5+}-complexes are mobile aqueous species. Several processes may limit the transport of Np, as well as other actinides: *i)* the precipitation of Np-solids, *ii)* the incorporation of Np into secondary uranium phases, and *iii)* the sorption and reduction of Np-complexes on Fe-oxide surfaces. This study utilizes quantum-mechanical calculations to determine the most energetically favorable Np^{5+}-incorporation mechanisms into uranyl phases, where Np^{5+}-substitution for U^{6+} requires a charge-balancing mechanism, such as the addition of H^+ into the structure. Experimental results suggest that uranyl structures with charged interlayer cations have a greater affinity for Np^{5+} than uranyl structures without interlayer cations. Therefore, the uranyl silicate phase boltwoodite ($KUO_2(SiO_3OH)(H_2O)_{1.5}$) is selected for this computational investigation. The charge-balancing mechanisms considered to occur with substitution include: *i)* addition of H^+, *ii)* substitution of Ca^{2+} for K^+, and *iii)* substitution of P^{5+} for Si^{4+}. While the incorporation energy results (1-3 eV) are higher than energies expected based on current experimental studies, solid-solution calculations are used to estimate the limit of Np incorporation for the P^{5+} substitution mechanism (10 ppm at ~100°C). The electronic structure of the boltwoodite structure provides insight into the electron density that may be involved in the incorporation of Np into the structure.

INTRODUCTION

Spent nuclear fuel (SNF) is primarily composed of uranium dioxide (UO_2) (95-99%), while the remaining 1-5% is composed of fission products (e.g., Cs, Sr, Tc) and transuranium elements (e.g., Pu, Np, Am). Over the very long term, the total radioactivity of SNF is dominated by the long-lived actinides [1], such as ^{237}Np ($\tau_{1/2}$ = 2.1 million years). Under oxidizing conditions, the UO_2 matrix of SNF alters, as the U^{4+} oxidizes to U^{6+}, and the uranyl molecule, UO_2^{2+}, forms complexes in solution, depending on the ground water chemistry. The oxidized uranium may precipitate as the bright yellow and orange U^{6+} phases that often form the corrosion rinds of altered uraninites [2,3]. The oxidation and dissolution of UO_2 is accompanied by the oxidation and release of the transuranium elements and fission products. Aqueous Np^{5+}-complexes are mobile in the environment; thus, research has focused on mechanisms that may reduce the mobility of the actinide complexes. Significant Np-immobilization mechanisms include precipitation as a solid Np-oxide phase, such as NpO_2 or Np_2O_5, incorporation of Np^{5+} into uranyl phases [4], and sorption and reduction of neptunyl (NpO_2^+)-complexes onto mineral surfaces.

Several experimental studies have examined Np-incorporation into a variety of uranyl phases including metaschoepite $[(UO_2)_4O(OH)_6](H_2O)_5$ [5,6], Na-schoepite $Na[(UO_2)_4O_2(OH)_5](H_2O)_5$ [7], studtite $[UO_2O_2(H_2O)_2](H_2O)_2$ [8], Na-compreignacite $Na_2[(UO_2)_3O_2(OH)_3]_2(H_2O)_7$ [6], and uranophane $Ca[(UO_2)(SiO_3OH)]_2(H_2O)_5$ [6]. Results suggest that structures with interlayer cations can incorporate more Np^{5+} than structures without interlayer cations. For example, Na-substituted schoepite incorporates more Np than regular schoepite [7]. The mechanism of this charged-balanced incorporation is not fully understood. This study uses quantum-mechanical density functional theory to investigate several charge-balanced Np-incorporation mechanisms in the uranyl silicate phase boltwoodite: $[K(UO_2)(SiO_3OH)(H_2O)_{1.5}]$.

Boltwoodite is a uranyl silicate (*$P2_1/m$*) composed of sheets of edge- and corner-sharing uranyl polyhedra and silicon tetrahedra, which follow the α-uranophane anion topology [9,10]. The interlayer is composed of water and either K^+ or a combination of K^+ and Na^+. For computational simplicity the interlayer was composed of water and K^+ (Figure 1). The charge-balanced incorporation mechanisms include the addition of an H^+ atom, the substitution of a divalent cation (Ca^{2+} or Mg^{2+}) for a monovalent cation (K^+) in the interlayer, and the substitution of P^{5+} for Si^{4+} in a tetrahedral site.

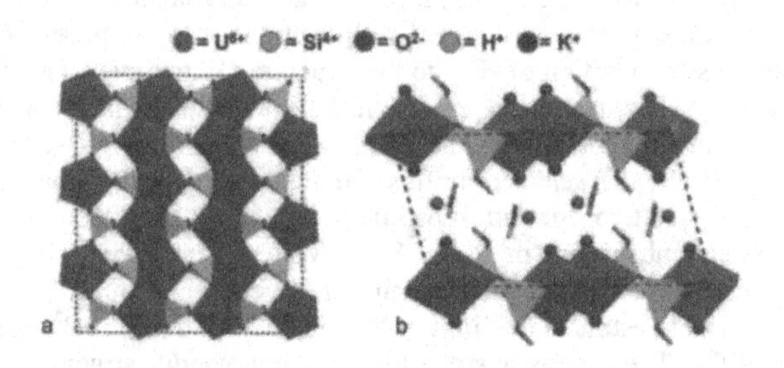

Figure 1. (a) The (100) plane of boltwoodite showing the edge- and corner-sharing U-polyhedra and Si-tetrahedra that comprise the boltwoodite sheets. (b) The (001) plane of boltwoodite showing the relation between the sheets and the interlayer species, which are K^+ and H_2O.

This study provides a detailed description of the charge-balanced mechanisms for Np-incorporation into boltwoodite, as well as a description of the electronic structure of boltwoodite, as derived using quantum mechanical calculations. Along with the theoretical incorporation energy for a (U, Np)-boltwoodite solid solution, the electronic structure comparison aids in the understanding of the boltwoodite system at the electronic level.

THEORY

Incorporation Energy Calculation

The incorporation energy associated with the substitution of Np^{5+} for U^{6+} was determined for a balanced reaction (Equation 1). When Np^{5+} substitutes for U^{6+}, the structure has an overall negative charge; therefore, a charge-balancing mechanism is necessary in order to retain charge-neutrality. The charge-balancing mechanisms investigated include the addition of H^+, the substitution of a divalent interlayer cation for a monovalent interlayer cation, and the substitution of P^{5+} for Si^{4+} in a tetrahedral site. The reaction changes depending on the charge-balancing mechanism, as well as the reference phases used to describe the Np-incorporation. Reaction 1 describes the substitution of Np^{5+} for U^{6+} in boltwoodite (bolt.) for the addition of an H^+ atom for charge balance. The source that we chose for Np is Np_2O_5 and the sink for U is UO_3. The source and sink in the reaction are considered as reference phases, as well as any other molecules or phases required to balance the reaction, in this case H_2O.

$$\text{bolt.} + \tfrac{1}{2}Np_2O_5 + \tfrac{1}{2}H_2O \rightarrow (\text{bolt-}U^{6+}\text{+}Np^{5+}\text{+}H^+) + UO_3 \quad (1)$$

The incorporation energy is the difference between the sum of the total energy of the products and that of the reactants. The total energy of a periodic solid calculation in CASTEP is the ground state energy for the system, where temperature is 0 K. Therefore, the entropy of the system is ignored, and the energy is equal to the enthalpy.

Computational Method

The total energies of the structures used in the incorporation energy calculation were determined using the density functional theory-based code CASTEP [11]. Ultrasoft pseudopotentials are used to describe the interactions between the core electrons and valence electrons. Some relativistic effects are included in the pseudopotentials. CASTEP uses a planewave approach with periodic boundary conditions to describe the behavior of the valence electrons. An energy cut-off for the planewaves of 800 eV was used to ensure that the final energy converged with respect to planewave energy cut-off. Due to the large number of electrons in the actinide phases, the computations are performed using the Γ k-point only. The electron exchange and correlation were approximated using the generalized gradient approximation (GGA) with the Perdew, Burke, and Ernzerhof (PBE) functional [12]. A spin polarized approach was used to account for the 2 unpaired 5f electrons associated with Np^{5+}. The BFGS algorithm was used for geometry optimizations.

RESULTS AND DISCUSSION

In the following, the energetics and structure of the different charge-balancing mechanisms are described. Due to the nature of the calculation, the charge-balancing mechanisms were not directly compared with each other; however, several issues about each mechanism are addressed.

H^+ addition

Four different H^+ locations were tested in the Np^{5+}-modified boltwoodite structure: on the neptunyl oxygen, on the water in the interlayer, on the apical [SiO_4] hydroxyl, and on a bridging [SiO_4] oxygen (Figure 2).

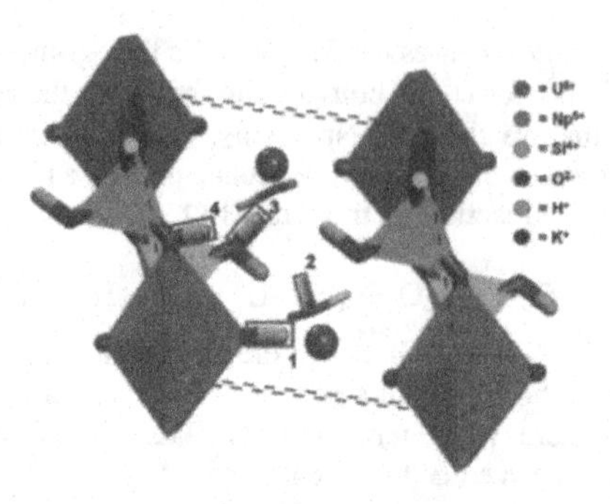

Figure 2. Boltwoodite unit cell showing Np^{5+}(green) substituted for U^{6+} (blue) and four possible H^+ positions - 1) neptunyl oxygen, 2) interlayer water, 3) apical [SiO_4] hydroxyl, and 4) bridging [SiO_4] oxygen.

The incorporation energy for each location was calculated using Reaction 1, where the energy of the modified boltwoodite structure on the right side of the reaction is unique to each H^+ location. The most energetically favorable H^+ addition location is on the neptunyl oxygen, which results in an incorporation energy of 1.4 eV (135 kJ/mol).

Interlayer cation substitution

The substitution of a divalent cation for a monovalent cation in the interlayer is another possible charge-balancing mechanism. Two possible substitution sites were investigated (Figure 3), as well as two possible divalent cations (Mg^{2+} and Ca^{2+}).

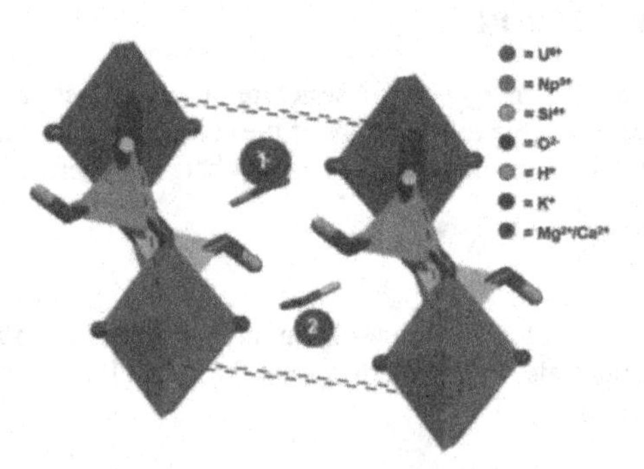

Figure 3. Boltwoodite unit cell showing Np^{5+}(green) substituted for U^{6+}(blue) and the possible interlayer sites for the divalent cation substitution (Site 1 and Site 2).

The distance between the substituted Np^{5+} and Site 1 is 4.34 Å, and the distance between the Np^{5+} and Site 2 is 4.07 Å. At Site 1, Mg^{2+} substitution is favored over Ca^{2+} by 1.3 eV (125 kJ/mol), while at Site 2, Ca^{2+} substitution is favored over Mg^{2+} by 2.1 eV (203 kJ/mol). Overall, the favored divalent substitution mechanism is the substitution of Ca^{2+} for K^{+} at Site 2.

Two reference phase cases were compared in order to determine the lowest incorporation energy for the divalent cation substitution mechanism – the oxide phases and the silicate phases (Reactions 2 and 3). The sources and sinks for Np/U are the same as in Reaction 1. The source/sink for Ca/K in the oxide reference phase case are CaO and K_2O; while, the source/sink for Ca/K in the silicate reference phase case are anorthite ($CaAl_2Si_2O_8$) and K-spar ($KAlSi_3O_8$).

$$\text{bolt.} + \tfrac{1}{2}Np_2O_5 + CaO \rightarrow (\text{bolt.-}U^{6+}\text{-}K^{+}\text{+}Np^{5+}\text{+}Ca^{2+}) + UO_3 + \tfrac{1}{2}K_2O \quad (2)$$

$$\text{bolt.} + \tfrac{1}{2}Np_2O_5 + CaAl_2Si_2O_8 + SiO_2 \rightarrow$$
$$(\text{bolt.-}U^{6+}\text{-}K^{+}\text{+}Np^{5+}\text{+}Ca^{2+}) + UO_3 + KAlSi_3O_8 + \tfrac{1}{2}Al_2O_3 \quad (3)$$

The structures for CaO and K_2O are computationally simple; however, silicate minerals are more common in nature, and, therefore, a more likely cation source in the environment. Reaction 4 describes the difference between the two reference phase cases.

$$CaO + KAlSi_3O_8 + \tfrac{1}{2}Al_2O_3 \rightarrow \tfrac{1}{2}K_2O + CaAl_2Si_2O_8 + SiO_2 \quad (4)$$

The silicate phases are too computationally intensive, such that the enthalpies of formation for these phases were taken from the literature [13]. The resulting change in enthalpy between the two reference phase cases is 1.2 eV (116 kJ/mol). The energy and enthalpy are used interchangeably in the analysis of quantum-mechanical results, as described above. The incorporation energy for the coupled substitution of Np^{5+} and Ca^{2+} for U^{6+} and K^{+} using the oxide reference phases is 2.6 eV (251 kJ/mol), while the incorporation energy using the silicate reference phases is only 1.4 eV (135 kJ/mol).

Layer cation substitution

The layer cation substitution mechanism involves the substitution of a P^{5+} for a Si^{4+} in a tetrahedral site. Several configurations for this mechanism were compared, and the most energetically favorable configuration is shown in Figure 4, where P^{5+} is located at the tetrahedron that is edge/corner-sharing with the Np^{5+} polyhedron. The shortest distance between the P^{5+} and the Np^{5+} is 3.168 Å.

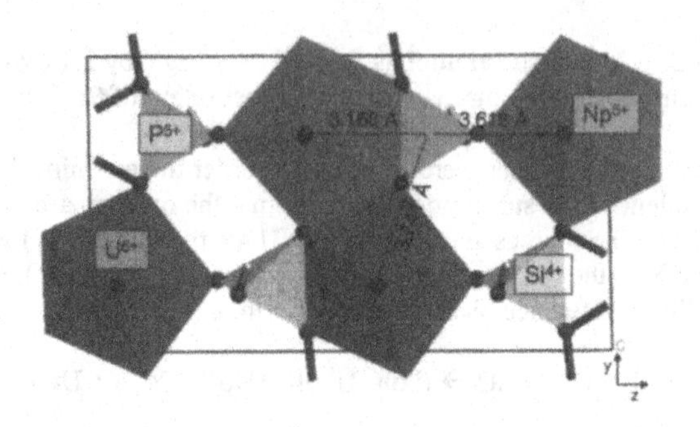

Figure 4. Modified boltwoodite structure, where Np^{5+} and P^{5+} replace half of the U^{6+} and Si^{4+}.

When we choose Np_2O_5 as the source for Np^{5+}, berlinite ($AlPO_4$) for P^{5+}, UO_3 as the sink for U^{6+}, α-quartz (SiO_2) as the sink for Si^{4+}, and corundum (Al_2O_3) to balance the reaction, the final incorporation energy is 1.24 eV (120 kJ/mol) or 0.62 eV/exchangeable cation (60 kJ/(mol exchangeable cations)). The balanced reaction for this reference phase case is:

$$(\text{bolt.}) + \tfrac{1}{2}Np_2O_5 + AlPO_4 \rightarrow (\text{bolt-}U^{6+}\text{-}Si^{4+}\text{+}Np^{5+}\text{+}P^{5+}) + UO_3 + SiO_2 + \tfrac{1}{2}Al_2O_3 \quad (5)$$

Other reference phases can be considered for the P^{5+} substitution mechanism, where the sources and sinks for the P^{5+} and Si^{4+} are uranyl phosphates and uranyl silicates. Due to the size of the systems, these reference phases are more computationally complex. Some of the reactions with different reference phases that will be explored are:

$$(\text{bolt.}) + \tfrac{1}{2}Np_2O_5\ \tfrac{1}{2}(\text{meta-autunite}) \rightarrow (\text{bolt-}U^{6+}\text{-}Si^{4+}\text{+}Np^{5+}\text{+}P^{5+}) + \tfrac{1}{2}(\text{uranophane}) + 1.5O_2 \quad (6)$$

$$(\text{bolt.}) + Ba(NpO_2)(PO_4)(H_2O) + O_2 \rightarrow (\text{bolt-}U^{6+}\text{-}Si^{4+}\text{+}Np^{5+}\text{+}P^{5+}) + (\text{sanbornite}) + UO_3 \quad (7)$$

$$(\text{bolt.}) + \tfrac{1}{2}Np_2O_5 + \tfrac{1}{2}(\text{parsonsite}) + H_2 + \tfrac{1}{2}UO_3 \rightarrow (\text{bolt-}U^{6+}\text{-}Si^{4+}\text{+}Np^{5+}\text{+}P^{5+}) + (\text{kasolite}) + O_2 \quad (8)$$

In Reactions 6-8, the sources for P^{5+} are meta-autunite ($Ca[(UO_2))PO_4)]_2(H_2O)_6$), a Np^{5+}-phase [14], and parsonsite ($Pb_2[(UO_2)(PO_4)]$), and the sinks for Si^{4+} are uranophane ($Ca[(UO_2)(SiO_3OH)]_2(H_2O)_5$), sanbornite ($BaSi_2O_5$), and kasolite ($Pb(UO_2)(SiO_4)H_2O$).

Np-incorporation limit

The thermodynamically stable limit for Np-incorporation into the uranyl phases is estimated from these quantum mechanical calculations by developing a theoretical solid solution series based on an individual charge balancing incorporation mechanisms. For the P^{5+} substitution mechanism, the boltwoodite unit cell has 2 unique U^{6+} sites, as well as 2 unique Si^{4+} sites. Therefore, a solid solution series was developed by generating optimized geometries for the following three structures: $K_2(UO_2)_2(SiO_3OH)_2(H_2O)_3$, $K_2(UO_2)(NpO_2)(SiO_3OH)(PO_3OH)(H_2O)_3$, and $K_2(NpO_2)_2(PO_3OH)_2(H_2O)_3$. The excess enthalpy of formation (ΔH), as calculated from the two end members and one intermediate member at x = 0.5, is fit to the Margules function (Equation 8). The excess free energy of formation (ΔG, Equation 9) is determined from the function for the excess enthalpy (Equation 8) and the excess configurational entropy without ordering (ΔS, Equation 10).

$$\Delta H(x) = A\,x\,(1-x) \quad (8)$$

$$\Delta G(x) = \Delta H(x) - T\Delta S(x) \quad (9)$$

$$\Delta S(x) = -R[x \ln x + (1-x)\ln(1-x)] \quad (10)$$

The A parameter in the Margules function is calculated using the excess enthalpy at 50% incorporation. Preliminary calculations indicate that A = 35.5 for Np-incorporation with the P^{5+} charge balancing mechanism, the most energetically favorable substitution mechanism that we found so far. The limit of Np-incorporation into boltwoodite with the P^{5+} charge balancing mechanism is then predicted at a range of repository temperatures (Table I).

Table I. Estimated limit of Np-incorporation with the P^{5+} charge-balancing mechanism at repository temperatures.

Temperature (°C)	estimated Np-incorporation limit (ppm)
100	11
150	42
200	120
300	580

Electronic structure

The band gap of boltwoodite is about 1.2 eV, which is within the range of a weak semiconductor. Because of these semiconducting properties, some delocalized valence electrons may be available for the incorporation of impurities, such as Np. The highest occupied molecular orbital (HOMO) level for boltwoodite is composed primarily of electron density from the U 5f orbitals, the edge-sharing O 2p orbitals, and the hydroxyl O 2p orbitals (Figure 5), while the lowest unoccupied molecular orbital (LUMO) level is composed primarily of electron density from U 5f orbitals and corner-sharing O 2p orbitals (Figure 6).

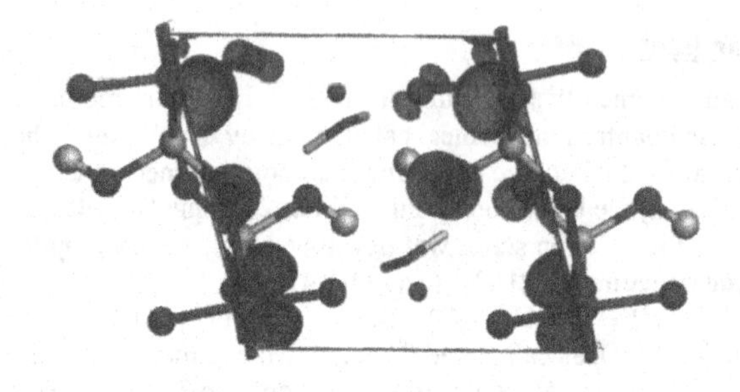

Figure 5. Boltwoodite unit cell where the electron density at the HOMO level is shown in blue.

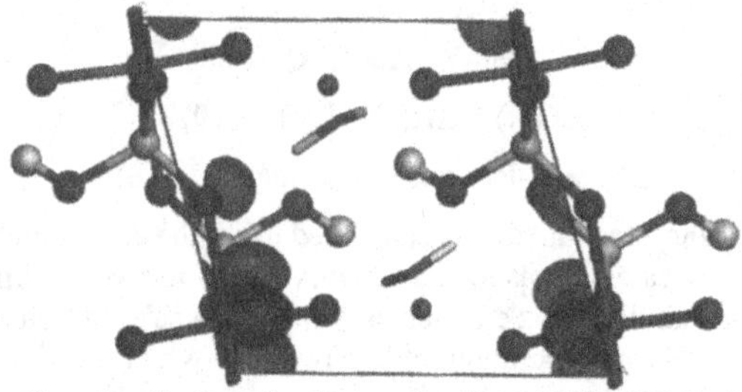

Figure 6. Boltwoodite unit cell where the electron density at the LUMO level is shown in blue.

SUMMARY

The purpose of this quantum-mechanical investigation was to gain a better understanding of the possible charge-balanced, Np^{5+}-incorporation mechanisms into the boltwoodite structure. The incorporation mechanisms included: the addition of H^+ bonded to the neptunyl oxygen, substitution of Ca^{2+} at the K^+ site closest to the incorporated Np^{5+}, and the substitution of P^{5+} at the Si^{4+} site closest to the incorporated Np^{5+}. For the P^{5+} substitution mechanism, the limit of Np^{5+}-incorporation is on the order of 10 to 100 ppm at 100°C to 300 °C. The calculated electronic structure of boltwoodite indicates significant electron density at the HOMO level contributed from the U 5f orbitals, the edge-sharing O 2p orbitals, and the hydroxyl O 2p orbitals. Future calculations comparing the Np^{5+}-modified boltwoodite electronic structure with the boltwoodite electronic structure presented here will provide insight into the nature of the neptunyl bonding in the boltwoodite structure.

ACKNOWLEDGEMENTS

This work was supported by the Office of Science and Technology and International of the Office of Civilian Radioactive Waste Management (DE-FE28-04RW12254). The views, opinions, findings, and conclusions or recommendations of the authors expressed herein do not necessarily state or reflect those of DOE/OCRWM/OSC/S&T.

REFERENCES

1. Hedin, A. (1997) Spent Nuclear Fuel – How dangerous is it? SKB technical Report 97-13, Swedish Nuclear Fuel and Waste Management Co.: 60.
2. Finch, R.J. and Ewing, R.C. (1992) The corrosion of uraninite under oxidizing conditions. Journal of Nuclear Materials, 190: 133-156.
3. Bruno, J. and Ewing, R.C. (2006) Spent nuclear fuel. Elements, 2: 343-349.
4. Burns, P.C., Ewing, R.C., and Miller, M.L. (1997) Incorporation mechanisms of actinide elements into the structures of U^{6+} phases formed during the oxidation of spent nuclear fuel. Journal of Nuclear Materials, 245: 1-9.
5. Buck, E.C., Finch, R.J., Finn, P.A., and Bates, J.K. (1998) Retention of neptunium in uranyl alteration phases formed during spent fuel corrosion. Materials Research Society Symposium Proceedings, 506: 87-94.
6. Burns, P.C., Deely, K.M., and Skanthakumar, S. (2004) Neptunium incorporation into uranyl compounds that form as alteration products of spent nuclear fuel: Implications for geologic repository performance. Radiochimica Acta, 92: 151-159.
7. Klingensmith, A.L., Deely, K.M., Kinman, W.S., Kelly, V., and Burns, P.C. (2007) Neptunium incorporation into sodium-substituted schoepite. American Mineralogist, 92: 662-669.
8. Douglas, M., Clark, S.B., Friese, J.I., Arey, B.W., Buck, E.C., and Hanson, B.D. (2005) Neptunium(V) partitioning to uranium(VI) oxide and peroxide solids. Environmental Science and Technology, 39: 4117-4124.
9. Burns, P.C., Miller, M.L., and Ewing, R.C. (1996) U^{6+}minerals and inorganic phases: A comparison and hierarchy of crystal structures. Canadian Mineralogist, 34: 845-880.
10. Burns, P.C. (2005) U^{6+} minerals and inorganic compounds: insights into an expanded structural hierarchy of crystal structures. Canadian Mineralogist, 43: 1839-1894.
11. Payne, M.C., Teter, M.P., Allan, D.C., Arias, T.A., and Joannopoulos, J.D. (1992) Iterative minimization techniques for abinitio total-energy calculations – molecular-dynamics and conjugate gradients. Reviews of Modern Physics, 64(4): 1045-1097.
12. Perdew, J.P., Burke, K., and Ernzerhof, M. (1996) Generalized gradient approximation made simple. Physical Review Letters, 77(18): 3865-3868.
13. Robie, R.A., and Hemingway, B.S. (1995) Thermodynamic properties of minerals and related substances at 298.15 K and 1 bar (10^5 pascals) pressure and at higher temperatures. U.S. Geological Survey Bulletin, 2131.
14. Forbes, T.Z., and Burns, P.C. (2006) Ba(NpO2)(PO4)(H2O), its relationship to the uranophane group, and implications for Np incorporation in uranyl minerals. American Mineralogist, 91(7): 1089-1093.

Engineered Barrier Systems I: Backfill and Buffer

Mater. Res. Soc. Symp. Proc. Vol. 1107 © 2008 Materials Research Society

Uranium(VI) Uptake by Synthetic Calcium Silicate Hydrates

J. Tits[1], T. Fujita[2], M. Tsukamoto[2] and E. Wieland[1]
[1]Paul Scherrer Institute, Villigen PSI, Switzerland
[2]Central Research Institute of the Electric Power Industry (CRIEPI), Tokyo, Japan

ABSTRACT

The immobilization of U(VI) by C-S-H phases under conditions relevant for the cementitious near field of a repository for radioactive waste has been investigated. C-S-H phases have been synthesized using two different procedures: the "direct reaction" method and the "solution reaction" method.

The stabilities of alkaline solutions of U(VI) (presence of precipitates or colloidal material) were studied prior to sorption and co-precipitation tests in order to determine the experimental U(VI) solubility limits. These U(VI) solubility limits were compared with the U(VI) solubilities obtained from thermodynamic speciation calculations assuming the presence of combinations of different solid U(VI) phases. The solid phase controlling U(VI) solubility in the present experiments was found to be $CaUO_4$(s).

The U(VI) uptake kinetics and sorption isotherms on C-S-H phases with different C:S ratios were determined under various chemical conditions; e.g., sorption and co-precipitation experiments and different pH's. U(VI) was found to sorb fast and very strongly on C-S-H phases with distribution ratios (R_d values) ranging in value between 10^3 L kg^{-1} and 10^6 L kg^{-1}. Both sorption and co-precipitation experiments resulted in R_d values which were very similar, thus indicating that no additional sorption sites for U(VI) were generated in the co-precipitation process. Furthermore, C-S-H synthesis procedures did not have a significant influence on U(VI) uptake. The U(VI) sorption isotherms were found to be non-linear, and further, increasing Ca concentrations resulted in increasing U(VI) uptake. The latter observation suggests that U(VI) uptake is controlled by a solubility-limiting process, while the former observation further indicates that pure Ca-uranate is not the solubility-limiting phase. It is proposed that a solid solution containing Ca and UO_2^{2+} could control U(VI) uptake by C-S-H phases.

INTRODUCTION

For disposal options involving the isolation of low- and intermediate-level radioactive waste in the cementitious near field of a deep underground geologic repository, uptake by calcium silicate hydrate (C-S-H) phases is expected to play an important role in retarding the migration of radionuclides in both the near field and the altered far field. In the near field, C-S-H phases are major components of hardened cement paste (e.g., [1]), and the interaction of hyperalkaline fluids from the repository with the mineral components of sedimentary rocks produces a range of C-S-H-type secondary minerals in the far field (pH plume) [2]. Studies concerning the retention of trace concentrations of radionuclides by C-S-H phases have mainly focused on adsorption as the dominant uptake mechanism. However, for some radionuclides other uptake mechanisms such as co-precipitation (solid-solution formation) may play an important role.

The objective of the present study was to investigate the interaction of U(VI) with C-S-H phases under conditions relevant to the cementitious near field of a repository for radioactive waste, to distinguish adsorption from co-precipitation (incorporation) processes on C-S-H phases, and to identify the type of processes (adsorption, surface precipitation, solid-solution formation...) controlling the U(VI) retention by cementitious materials. U(VI) sorption and co-precipitation studies were conducted in batch-type experiments on C-S-H phases with varying composition (CaO : SiO_2 (C:S) ratio), prepared using two different synthesis methods. The experiments were carried out in three different types of pore waters: i) alkali-free pore waters having Ca and Si concentrations in equilibrium with the respective C-S-H phases and a pH between 10.5 and 12.5 [3], (ii) low-alkali pore waters with a pH of ~12.0, and (iii) an artificial cement pore water (ACW) with pH 13.3, containing 0.112 M NaOH, and 0.14 M KOH. In low-alkali pore waters, variation in the Na_2SiO_3 concentration ($0 \leq [Na_2SiO_3] \leq 2 \cdot 10^{-3}$ M) originating from the synthesis procedure gave rise to small variations in pH ($11.96 < pH < 12.18$). In all pore water types, the Ca and Si concentrations are determined by the solubility of the C-S-H phases used, which depends on the C:S ratio.

MATERIALS AND METHODS

The solutions used throughout this study were prepared using Fluka or Merck "pro analysis" chemicals. Milli-Q water generated by a Millipore water purification system was used for the preparation of the solutions and for sample dilution. Sorption experiments were carried out in 40-mL polyallomere centrifuge tubes (Beckman Instruments, Inc.), which were washed, left overnight in a solution of 0.1 M HCl, and thoroughly rinsed with deionized water. The C-S-H phases and alkaline solutions were prepared and stored in perfluoralkoxy-copolymer (PFA) bottles. Furthermore, all experiments were carried out in a glove box under a nitrogen atmosphere (CO_2, O_2 $\leq$ 2 ppm) at room temperature (23±3)°C.

C-S-H phases were synthesized using two different preparative methods: (i) The first method is an adapted version of the so-called "direct reaction" method; i.e., silica fume (AEROSIL 300, Degussa-Huls AG, Baar, Switzerland) was mixed with CaO in Milli-Q water or ACW at ratios corresponding to the target C:S ratios of the C-S-H phases. Details of this preparative method as well as a detailed characterization of the C-S-H material are given elsewhere [3]. (ii) In the second method, denoted as "solution reaction" method, calcium hydroxide ($Ca(OH)_2$) solutions and sodium metasilicate ($Na_2SiO_3 \cdot 9H_2O$) solutions are used for the C-S-H synthesis with the aim of avoiding the presence of solid reactants in the starting mixtures. A detailed description is given elsewhere [4]. Briefly, 20 mL aliquots of a 0.015 M $Ca(OH)_2$ solution were mixed with $Na_2SiO_3 \cdot 9H_2O$ solutions of appropriate concentrations to give target C:S ratios between 0.66 and 1.8. For the preparation of the latter solutions appropriate amounts of $Na_2SiO_3 \cdot 9H_2O$ were dissolved in Milli-Q water or in a solution containing 0.228 M NaOH and 0.36 M KOH. The Na and K concentrations of the latter solution are twice the concentrations in ACW. Prior to use, the $Ca(OH)_2$ solutions were filtered using a membrane filter with a molecular weight cut-off of 30000 Dalton to remove any remaining calcium hydroxide colloids.

The solubility limits of U(VI) were tested in the different pore waters used in this study. The tests were carried out as follows: Small volumes (< 1 mL) of U(VI) stock solutions of known concentrations and prepared in 0.1 M HNO_3 were labelled with ^{233}U solutions and diluted with pore water solution. The total U(VI) concentration varied between $2.1 \cdot 10^{-8}$ M and 10^{-4} M. These solutions were shaken end-over-end for 1 day and 7 days. After equilibration, the ^{233}U activities in the homogeneous bulk solution and in the supernatant obtained after centrifugation (1 hour at 95'000 g (max)), were determined using liquid scintillation counting. Calculations using Stokes law suggested that any colloidal material with a diameter larger than ~ 20 nm should settle during centrifugation. With the above procedure it is possible to distinguish the radionuclide fraction in "true" solution from the radionuclide fraction present as colloidal material.

U(VI) sorption experiments were carried out using C-S-H phases with varying target C:S ratios synthesized following both the “solution reaction” procedure and the “direct reaction” procedure. The final solid to liquid (S:L) ratio of the C-S-H suspensions made following the former method varied between 0.8 g L^{-1} and 2.4 g L^{-1} and the S:L ratio of the suspensions made following the latter method was 5.0 g L^{-1}. The suspensions were shaken end-over-end for two weeks prior to the addition of U(VI) solution. The required U(VI) concentrations in all the suspensions were adjusted by adding appropriate amounts of a 10^{-2} M $UO_2(NO_3)_2$ solution and 0.1 mL $2.16 \cdot 10^{-4}$ M ^{233}U tracer solution (prepared in 0.1 M HNO_3). The centrifuge tubes were again shaken end-over-end in the glove box for time periods between 1 day and 120 days for the kinetic experiments and 30 days for the sorption isotherms. Co-precipitation kinetics tests with U(VI) were performed using procedures similar to those used in the sorption experiments. The main difference is related to tracer addition: In the co-precipitation tests, the $UO_2(NO_3)_2$ solution and the ^{233}U tracer were added immediately after mixing the chemicals used for the C-S-H synthesis instead of curing the C-S-H suspensions for 2 weeks prior to addition of U(VI) solution. The quantity of U(VI) sorbed was determined from the difference of total U(VI) added to the suspensions and the aqueous U(VI) concentration in the supernatant solution obtained after centrifugation for 1 hour at 95’000g (max). The aqueous U(VI) concentration was determined either by measuring the concentration of U(VI) in solution with inductively coupled plasma optical emission spectrometry (ICP-OES) or by measuring the aqueous ^{233}U activity by radioassay with a Canberra Packard Tri-carb 2250 CA liquid scintillation counter using an energy window between 100 keV and 200 keV. The ^{233}U sample solutions for analysis were prepared by mixing 5 mL aliquots with 15 mL scintillator (Ultima Gold XR, Packard Bioscience S.A.) prior to counting. Standards were prepared by mixing 0.01 ml ^{233}U tracer solution with 5 ml 10^{-3} M $Ca(OH)_2$ solution or 5 ml ACW and 15 ml scintillator. Samples and standards were measured together with blank samples containing 5 ml 10^{-3} M $Ca(OH)_2$ solution or 5 ml ACW and 15 ml scintillator.

RESULTS AND DISCUSSION

Experimental U(VI) solubility limits in alkaline pore waters

U(VI) solubility tests in four solutions representative of the different chemical conditions in the sorption tests were carried out [5]. The results of two of these tests in a solution containing 0.015 M $Ca(OH)_2$ and 10^{-5} M Si (pH=12.5) and in a solution containing $5 \cdot 10^{-5}$ M $Ca(OH)_2$, $2 \cdot 10^{-3}$ M Si and 0.02 M NaOH (pH=12.1), are presented in Figures 1a and b and a summary of all the results is given in Table 1. The figures show the solution concentrations of the radionuclides as a function of the added total U(VI) concentration in solution. The data refer to measured concentrations before and after centrifugation. Thus, the concentration determined after centrifugation represents the concentration of dissolved U(VI) plus the concentration of U(VI) associated with colloidal material smaller than ~20 nm. In the solution containing 0.015 M $Ca(OH)_2$, no reductions in the U(VI) solution concentrations were observed before and after centrifugation up to starting concentrations of about 10^{-5} M. The data fall on the line with slope = 1. Above this starting concentration threshold however, the U(VI) concentration in solution remains approximately constant at a concentration of 10^{-6} M to $7 \cdot 10^{-6}$ M, presumably due to the formation of an U(VI) solid phase. This decrease is more pronounced after centrifugation. In the solution containing $5 \cdot 10^{-5}$ M

$Ca(OH)_2$, $2 \cdot 10^{-3}$ M Si and 0.02 M Na, no solid phase formation was observed over the entire U(VI) concentration range investigated.

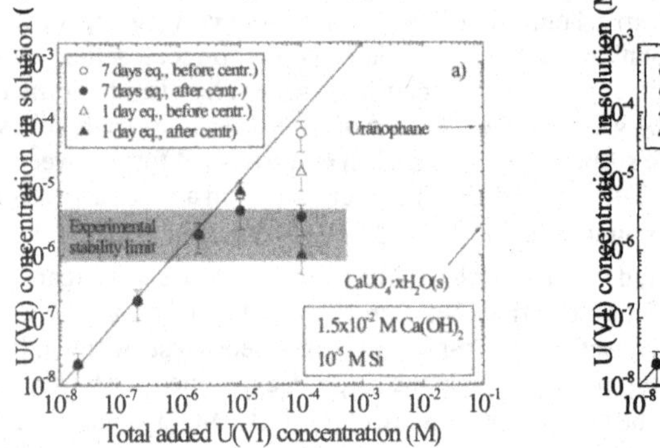

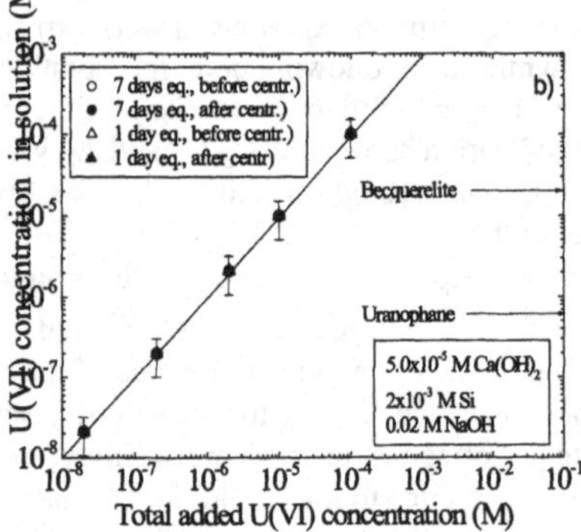

Figure 1: U(VI) concentration in a) a solution containing 0.015 M $Ca(OH)_2$ and 10^{-5} M Si, and in b) a solution containing $5 \cdot 10^{-5}$ M $Ca(OH)_2$, $2 \cdot 10^{-3}$ M Si and 0.02 M NaOH, before (open symbols) and after (closed symbols) centrifugation. Equilibration times were 1 day and 7 days. The shaded area represents the experimental solubility limit determined from the experimental data.

In general, all tests in ACW and in low-alkali solutions show a clear trend (Table 1): The solubility limit of U(VI) in the solutions increases with decreasing Ca concentration and increasing Si concentration. Furthermore, increasing pH (ACW at pH=13.3 compared to low-alkali solutions with 11.96<pH<12.18) result in higher solubility limits caused by higher concentrations of U(VI) hydroxy complexes in solution.

Table 1: Comparison of calculated $CaUO_4$(s) solubility with the experimentally determined U(VI) concentrations for the relevant solutions compositions used in this study.

Experiment	U(VI) solubility calculated with $\log_{10} {}^*K^0_{s,0}$=23.1 [7, 8]	Experimental U(VI) solubility limit
ACW	$7 \cdot 10^{-6}$ M	10^{-5} M
$1.5 \cdot 10^{-2}$ M $Ca(OH)_2$, 10^{-5}M Si, pH=12.18	$3 \cdot 10^{-6}$ M	$4 \cdot 10^{-6}$ M
$3 \cdot 10^{-3}$ M Ca, 10^{-2} M NaOH, 10^{-4} M Si, pH 12.09	$2 \cdot 10^{-5}$ M	$3 \cdot 10^{-5}$ M
$5 \cdot 10^{-5}$ M Ca, $2 \cdot 10^{-2}$ M NaOH, $2 \cdot 10^{-3}$ M Si, pH 11.96	No precipitation	No precipitation

Speciation calculations using the thermodynamic data reported in Hummel et al. [6] and Guillaumont et al. [7] were performed using schoepite ($(UO_2)_8O_2(OH)_{12} \cdot 12H_2O$), compreignacite ($K_2(UO_2)_6O_4(OH)_6 \cdot 8H_2O$(cr)), Na-uranate ($Na_2U_2O_7$(cr)), soddyite ($(UO_2)_2SiO_4 \cdot 2H_2O$(cr)), becquerelite ($Ca(UO_2)_6O_4(OH)_6 \cdot 8H_2O$(cr)), and uranophane ($Ca(UO_2)_2SiO_3(OH)_2 \cdot 5H_2O$(cr)) as possible candidates controlling the U(VI) solubility in the test solutions. None of these solids gave an acceptable agreement with all the experimentally determined U(VI) solubility limits [5]. Using schoepite, becquerelite, compreignacite or soddyite as solubility-limiting phases, the calculated solubility limits were found to be much higher than the experimentally observed [5]. On the other hand, when either Na-uranate or uranophane were assumed to control U(VI) solubility, the calculated solubility limits were found to be too low [5]. Nevertheless, excellent agreement between predicted solubility limits and experimental data was achieved for all relevant solution compositions by including an amorphous Ca-uranate ($CaUO_4(s)$) (solubility product, $\log_{10} K^0_{s,0} = 23.1$ [7,8]) in the calculations and excluding all the other U(VI) solids (Table 1, [5]). This suggests that $CaUO_4(s)$ controls the U(VI) concentrations in the given systems. Note, however, that this phase could be metastable and other U(VI) phases could control the U(VI) concentration in cementitious systems in the long run.

Effect of the synthesis procedure on the U(VI) sorption and co-precipitation kinetics

Figure 2 shows examples of the observed sorption and co-precipitation kinetics on C-S-H phases with a C:S ratio of 1.1 synthesized following the "direct reaction" method and the "solution reaction" method in alkali-free conditions and in ACW. These data show that the U(VI) sorption reaches a maximum after approximately 10 days equilibration. Subsequently, R_d values remain constant over the investigated time period of up to 120 days. The R_d values obtained were found to be independent of the synthesis method used to prepare the C-S-H phases. In all cases, both sorption and co-precipitation experiments resulted in similar R_d values, thus indicating that no additional sorption sites for U(VI) were generated in the co-precipitation process. Only in the experiments performed under ACW conditions (Figure 2b), R_d values tend to be slightly higher in the co-precipitation experiments. However, in most experiments, these differences are small and hardly significant taking into account the experimental uncertainties.

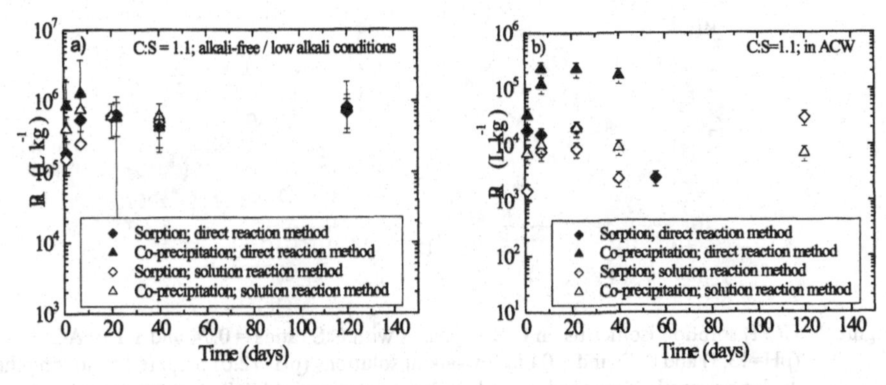

Figure 2: U(VI) sorption and co-precipitation kinetics on C-S-H phases with a C:S ratio of 1.1 a) under alkali-free and low-alkali conditions, and b) in ACW. Total U(VI) concentration = 10^{-6} M, S:L ratio = between 0.9 g L^{-1} and 5.0 g L^{-1}.

Effect of pH and Ca concentration on the U(VI) sorption isotherms

Sorption isotherms were determined for C-S-H phases synthesized in solutions at pH ~12 with C:S ratios of 0.73 and 1.04 and in ACW (pH = 13.3) with C:S ratios of 0.74 and 1.1 using the "solution reaction" method (Figures 3a and b). Similar curves were obtained for C-S-H phases prepared using the "direct reaction" method [5].

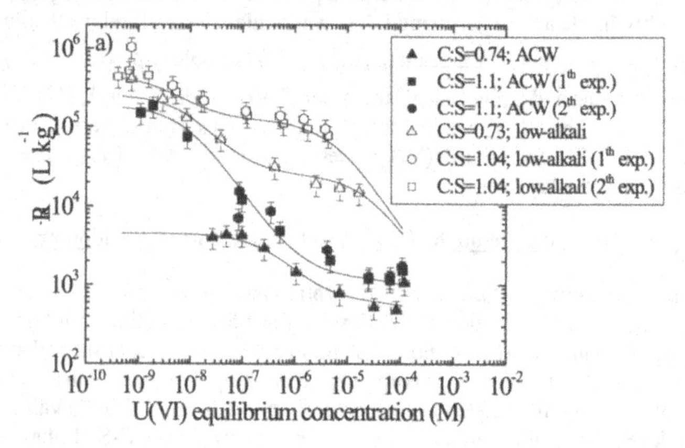

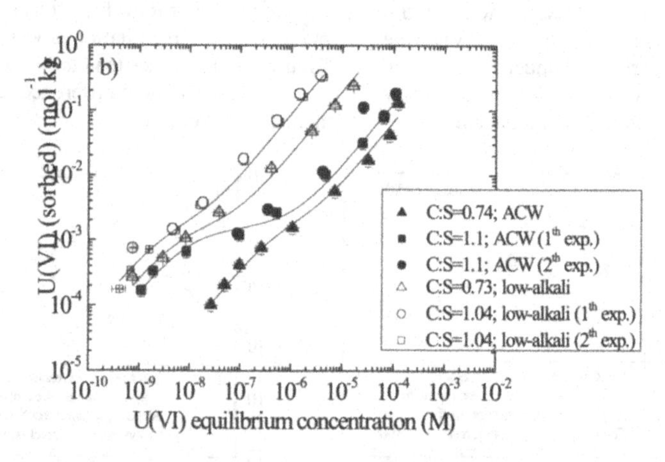

Figure 3: U(VI) sorption isotherms on C-S-H phases with C:S ratios = 0.74 and 1.1 in ACW (pH=13.3) and 0.73 and 1.04 in low-alkali solutions (pH~12.0) prepared following the "solution reaction" method. R_d values (a) and amount of U(VI) sorbed (b) versus the U(VI) equilibrium concentration. Solid lines are added to guide the eye.

The isotherms are non-linear with R_d values decreasing with increasing aqueous U(VI) concentrations. The R_d values for U(VI) uptake by C-S-H phases are lower in ACW (pH = 13.3) (R_d > 10^3 L kg^{-1}) than in the solutions of pH ~12 (R_d > 10^5 L kg^{-1}). The strong influence of pH on

U(VI) uptake might be attributed to changes in U(VI) speciation in solution. The trend towards a stronger uptake of U(VI) species in alkali-free or low-alkali systems (pH 10.5 to 12.5) can be explained by lower concentrations of aqueous uranyl-hydroxy complexes compared to ACW (pH=13.3). Another interesting finding concerns the solubility limit of U(VI) in these systems. Note that the experimental U(VI) solubility limit indicated for example by the data presented in Figure 1a (approximately 10^{-6} M to $7\cdot10^{-6}$M) was exceeded in the isotherm experiments. Nevertheless, the formation of a pure U(VI) bearing phase that would give rise to the suggested solubility limit, is not observed. The results from the isotherm measurements at pH ~12 and at pH = 13.3 further show a clear trend to higher R_d values with increasing C:S ratios of the C-S-H phases. These observations suggest that U(VI) uptake by C-S-H phases depends on the Ca concentration in solution, i.e. U(VI) uptake increases with increasing Ca concentration in solution. Note that the equilibrium Ca concentration of a C-S-H phase strongly depends on the C-S-H composition (Figure 4).

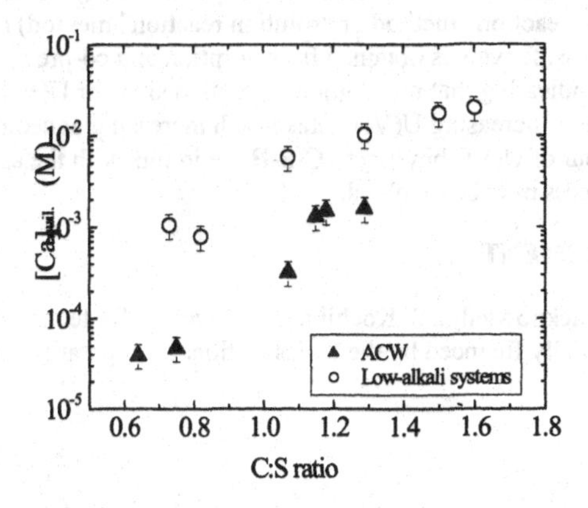

Figure 4: Ca solubility of C-S-H phases as a function of their C:S ratio under ACW conditions and in low-alkali systems [3,5].

With decreasing pH, the Ca concentration in equilibrium with a C-S-H phase having a specific C:S ratio increases due to the common ion effect. Interestingly, increasing U(VI) uptake by C-S-H with increasing Ca concentration is in contrast to previous observations on Sr and Ra binding to C-S-H phases [3,9] which showed a decreasing uptake with increasing aqueous Ca concentration. The latter data could be modelled in terms of a competition between Ca and Sr/Ra binding to C-S-H phases (ion exchange mechanism). Thus, in contrast to Sr and Ra, U(VI) appears to be bound to C-S-H phases in a "precipitation-type" mechanism in which Ca and U(VI) are simultaneously bound to the sorbent. This further implies the formation of a solid solution, in which U(VI) binding is controlled by the formation of species with variable stoichiometric composition incorporated in C-S-H phases.

Recent X-ray absorption fine structure (XAFS) investigations suggest the formation of an uranophane-like coordination environment of U(VI) upon uptake by C-S-H phases [10]. Identification of the solid solution involved, however, is not possible based on the presently available batch sorption data. The suggestion of solid solution formation with different endmember stoichiometries is also in line with the non-linear sorption behaviour of U(VI) in C-S-H systems as revealed from the sorption isotherm data. The formation of a pure U(VI) bearing solid phase having a fixed solubility is not indicated from these measurements. This finding is in contrast to the results from the stability tests with U(VI) solutions (concentration range 10^{-8} M - 10^{-3} M U(VI)) where it was observed that U(VI) solutions are not stable in the C-S-H equilibrium solutions at concentrations above about 10^{-5} M in ACW and 10^{-6} M at pH = 12.5.

CONCLUSIONS

In short term experiments, the solubility of U(VI) in cementitious pore waters appears to be controlled by Ca uranate ($\log_{10} K^0_{s,0} = 23.1$). This solid phase was identified and characterized earlier by Moroni and Glasser [8].

U(VI) shows a high affinity for C-S-H phases under highly alkaline conditions. The uptake of U(VI) by C-S-H phases was completed within approximately 10 days. The mode of C-S-H preparation ("direct reaction" method or "solution reaction" method) had no significant influence on the uptake process. R_d values obtained from sorption and co-precipitation experiments are very similar, thus indicating that no additional sorption sites for U(VI) were generated in the co-precipitation process. Increasing U(VI) uptake with increasing aqueous Ca concentrations and the non-linear behaviour of U(VI) binding to C-S-H are in line with the assumption that a solid solution formation process may be involved.

ACKNOWLEDGEMENT

We gratefully acknowledge S. Köchli and J.-P Dobler for technical and analytical assistance. This work was partially financed by the Swiss National Cooperative for the Disposal of Radioactive Waste (Nagra).

REFERENCES

1. B. Lothenbach and E. Wieland. Waste Manage. **26**, 706 (2006).
2. M.B. Crawford and D. Savage, Technical Report No. WE/93/20C, Britisch Geological Survey, Nothingham, UK. (1994).
3 J. Tits, E. Wieland, C.J. Müller, C. Landesmann and M.H. Bradbury. J. Coll. Interface Sci. **300**, 78-87 (2006).
4. D. Sugiyama and T. Fujita. Cem. Concr. Res. **36**, 227 (2006).
5. J. Tits, T. Fujita, E. Wieland, M. Harfouche, R. Dähn, S. Sugiyama. PSI report, Paul Scherrer Institut, Villigen, Switzerland. In press.
6. W. Hummel, U. Berner, E. Curti, F.J. Pearson and T. Thoenen. Nagra Technical Report NTB 02-06, Nagra, Wettingen, Switzerland and Universal Publishers/uPublish.com, Parkland, Florida (2002).
7. R. Guillaumont, T. Fanghänel, J. Fuger, I. Grenthe, V. Neck, D.A. Palmer and M.H. Rand. Update on the Chemical Thermodynamics of Uranium, Neptunium, Plutonium, Americium and Technetium. Elsevier, Issy-les-Moulineaux, France (2003).
8. L.P. Moroni, and F.P. Glasser. Waste Manage. **15**, 243-254 (1995).
9. J. Tits, K. Iijima, E. Wieland and G. Kamei. Radiochim. Acta **94**, 637-643 (2006).
10. M. Harfouche, E. Wieland, R. Dähn, T. Fujita, J. Tits, D. Kunz and M. Tsukamoto. J. Colloid Interface Sci. **303**, 195-204 (2006).

Mater. Res. Soc. Symp. Proc. Vol. 1107 © 2008 Materials Research Society

PHREEQC Modelling of Leaching of Major Elements and Heavy Metals From Cementitious Waste Forms

Evelien Martens[1], Diederik Jacques[1], Tom Van Gerven[2], Lian Wang[1] and Dirk Mallants[1]
[1]Belgian Nuclear Research Centre (SCK•CEN), Institute for Environment, Health and Safety, Boeretang 200, B-2400 Mol, Belgium
[2]Katholieke Universiteit Leuven, Faculty of Engineering, Department Of Chemical Engineering, de Croylaan 46, B-3001 Leuven, Belgium

ABSTRACT

In this study, Ca, Mg, Al, and Pb concentrations leached from uncarbonated and carbonated ordinary Portland cement – dried waste incinerator bottom ash samples during single extraction tests (EN12457 test) at a pH from 1 to 12, were modelled using the geochemical code PHREEQC. A good agreement was found between modelling results and experiments in terms of leached concentrations for Ca, Mg, and Al by defining a single set of pure mineralogical phases for both the uncarbonated and carbonated (three levels) samples. The model also predicted well the observed decrease in Ca leaching with increasing carbonation. Modelling results further revealed that leaching of Pb is not controlled by dissolution/precipitation of pure Pb containing minerals only (carbonates and (hydr)oxides). The addition of solid solutions (calcite-cerrusite and gibbsite-ferrihydrite-litharge solid solutions) and adsorption reactions on amorphous Fe- and Al-oxides improved the model representation of the experimentally observed amphoteric leaching profile of Pb from the cementitious material.

INTRODUCTION

Solidification/stabilization is a technique for immobilizing hazardous wastes in binding materials, mostly cement-based, to delay dissolution and release of toxic components to the environment. Large amounts of low-level radioactive wastes (LLW) are conditioned in this way to guarantee safe disposal and negligible radiological impact on humans and the environment for long time scales. Besides radioactive elements in LLW, a number of chemically toxic elements, such as Pb, are also present in those wastes. In this study, Ca, Mg, Al, and Pb concentrations leached from uncarbonated and carbonated ordinary Portland cement (OPC) – dried bottom ash from municipal solid waste incinerator (MSWI) samples during single extraction tests at different pH were modelled using the geochemical code PHREEQC. MSWI bottom ash is enriched in Pb with contents varying between 98 to 13700 mg/kg [1].

The objectives were (i) to define mineralogical phases that can be used in modelling to represent the leachate concentrations measured at different pH-values, (ii) to model the effect of carbonation on leaching behaviour in a single general model, and (iii) to assess the inclusion of mechanisms such as a formation of solid solutions and surface complexation on modelling of Pb leaching.

EXPERIMENTAL DATA

Mortars with a water-to-cement ratio of 0.5 were prepared by mixing 548 kg/m³ of OPC, 1096 kg/m³ dried MSWI bottom ash and 281 kg/m³ distilled water. The mixtures were poured in moulds of 150 x 150 x 150 mm and vibrated. After a 24 hour setting time, the samples were demoulded and cured for 28 days in a humid room (20°C, > 95% relative humidity (RH) and 0.035% CO_2). To have an uncarbonated sample, a layer of about 1.5 cm of material from the edges was removed. Sample cubes of 40 x 40 x 40 mm were prepared using a dry cutting technique. Uncarbonated samples ("B0") were dried in a vacuum oven at 40°C and stored at room temperature in a CO_2 free bag filled with nitrogen gas. The carbonated samples were prepared by conditioning fresh samples for 14 ("B14"), 30 ("B30") and 60 ("B60") days in a closed chamber at 37°C, > 90% RH and 20% CO_2.

A series of single extraction tests was conducted on each set of cubes using 10 g of particle-size reduced material in 100 ml distilled water acidified with different volumes of concentrated HNO_3 (analogous to the EN12457 test, [2]) resulting in a pH range from 1 to 12. The pH and composition of the leachate were analysed after 24 hours of contact time. Detailed experimental procedures and discussion of the experimental data are given in [3].

The amounts of hydrous ferric oxides (HFO) and amorphous aluminium minerals (AAM) were determined on B0 and B60 samples, to estimate the adsorption capacity for Pb by surface complexation. To determine the HFO content in MSWI bottom ash, the method of Ferdelman ([4], cited by [5]) by which HFO is selectively extracted by ascorbate, was used [6]. HFO contents for B0 and B60 samples are 1.31 g/l (one measurement) and 0.62 g/l (average of two repetitions), respectively. AAM was measured by oxalate extraction in the dark [7], giving values of 3.07 g/l and 3.25 g/l for the B0 and B60 sample, respectively (each an average of 3 repetitions). Additional details on the experimental procedure are given in [8].

MODEL BUILDING

Model description of the initial system

The four series of single extraction tests (one uncarbonated series and three carbonated series) were modelled with the geochemical code PHREEQC2.13 [9] to assess if the selected model minerals and processes are able to describe the release of constituents at different pH. The release of sodium and potassium was not modelled because the concentration of these two elements are unlikely controlled by solubility and sorption, the two mechanisms considered in this study. The focus of the comparison is on the major elements Ca, Mg, and Al and the trace element Pb, because the leachate concentrations of S and Fe were not measured and Si concentrations were only available for the uncarbonated series. The initial condition was defined by the total concentrations of the elements measured in the solid samples at the start of the extraction test (Table I), except for Si and C. Preliminary simulations indicated that leachate Ca concentrations are underpredicted by several orders of magnitude when the total Si concentration is used. The maximum measured leachate Si concentration (for uncarbonated samples, data from [10]) was only 106 mmol/l (liquid-to-solid ratio = 10, pH = 2) compared to a total measured concentration of 917 mmol/l. Simulation results for Ca became more acceptable when this maximum measured leachate Si concentration was used as input concentration. For the carbonated series, leachate Ca concentrations were underpredicted when the total C concentration is used (results not shown), which was also observed in other studies [11,12,13]

Table I. Measured total concentration and input concentrations used in the simulations.

Element	Sample	Measured total concentration (mmol/l)	Input concentration (mmol/l)
Si	B0, B14, B30 & B60	917	106
Ca	B0, B14, B30 & B60	476	476
Al	B0, B14, B30 & B60	128	128
Mg	B0, B14, B30 & B60	41	41
S(6)	B0, B14, B30 & B60	22	22
Fe(3)	B0, B14, B30 & B60	101	101
Na	B0, B14, B30 & B60	29	29
K	B0, B14, B30 & B60	17	17
Pb	B0, B14, B30 & B60	0.49	0.49
Cl	B0, B14, B30 & B60	45	45
C	B0	0	0
	B14	330	120
	B30	400	145
	B60	600	218

indicating either inappropriate thermodynamic data or slow reaction kinetics. Alternatively, a part of the carbonate is not available for reaction with Ca because it is possibly incorporated in stable cement phases [3,14]. The active carbonate concentration is therefore deduced from the difference between the free portlandite in uncarbonated and carbonated samples assuming that free portlandite is transformed into calcite. Reported ratios between total measured carbonate concentrations and calculated active ones ranged between 2.5 [14] and 3 [3]. A value of 2.75 is adapted in this study. The concentrations of the elements used in the simulations are also given in Table I. At each pH, equilibrium is calculated between the leachate and the mineralogical phases (Ca, Mg, Al and Pb), solid solutions and surface complexation (for Pb only).

Mineralogical phases controlling Ca, Mg, Al, and Si leaching

Mineral phases considered in the model to represent the key components of cement and MSWI ashes are: portlandite, CSH_0.8, CSH_1.1, CSH_1.8, hydrogarnet, brucite, gibbsite, amorphous $Al(OH)_3$, ferrihydrite, ettringite, gypsum, monosulfo-aluminate, calcite, magnesite, and monocarbo-aluminate. Dissolution reactions of these minerals and equilibrium constants are listed in Table II.

As the Ca/Si-ratio of CSH-gels continuously decreases with increasing carbonation (the decalcification process, [14, 15]), CSH-phases with different Ca/Si ratios may form and are included in the model. Although more sophisticated models such as a CSH solid solution model exist (e.g. [16]), a simpler approach is adopted with three separate mineralogical phases with a different ratio in this study. Other studies showed that this is sufficient to describe the Ca leaching during carbonation (e.g., [17]).

Brucite is included since many studies showed that this mineral controlled the solubility of Mg in bottom ash [13, 18] and cement [19]. Using brucite as a Mg controlling phase produced better results than using hydrotalcite ($Mg_4Al_2O_7.10H_2O$, results not shown) as the solubility controlling phase, which is consistent with the observations of Miyamoto [20] showing brucite rather than hydrotalcite formation in fresh cement samples. Magnesite possibly controls leaching in the carbonated samples [19].

Dijkstra et al. [18], Kirby and Rimstidt [21], and Meima and Comans [13] showed that the strongly pH-dependent leaching of Al from bottom ash is adequately described by the solubility

Table II. Dissolution reactions and thermodynamic data of the minerals used in the model.

Phase	reaction	log *K*	ref
Cement phases			
CSH_0.8	$Ca_{0.8}SiO_5H_{4.4} + 1.6\ H^+ = 0.8\ Ca^{2+} + H_4SiO_4 + H_2O$	11.1	[23]
CSH_1.1	$Ca_{1.1}SiO_7H_{7.8} + 2.2\ H^+ = 1.1\ Ca^{2+} + H_4SiO_4 + 3\ H_2O$	16.7	[23]
CSH_1.8	$Ca_{1.8}SiO_9H_{10.4} + 3.6\ H^+ = 1.8\ Ca^{2+} + H_4SiO_4 + 5\ H_2O$	32.6	[23]
Portlandite	$Ca(OH)_2 + 2\ H^+ = Ca^{2+} + 2\ H_2O$ [24],[22]	22.80	
Ettringite	$Ca_6Al_2(SO_4)_3(OH)_{12}{\cdot}26H_2O + 12\ H^+ = 2\ Al^{3+} + 3\ SO_4^{2-} + 6\ Ca^{2+} + 38\ H_2O$	56.7	[25]
Gypsum	$CaSO_4{\cdot}2H_2O = Ca^{2+} + SO_4^{2-} + 2\ H_2O$	-4.58	[9]
Monosulfo-aluminate	$Ca_4Al_2SO_{10}{\cdot}12H_2O + 12\ H^+ = 4\ Ca^{2+} + 2\ Al^{3+} + SO_4^{2-} + 18\ H_2O$	71.0	[24]
Hydrogarnet	$Ca_3Al_2O_{12}H_{12} + 12\ H^+ = 3\ Ca^{2+} + 2\ Al^{3+} + 12\ H_2O$	78	[24]
$Al(OH)_3$(am)	$Al(OH)_3 + OH^- = Al(OH)_4^-$	0.24	[22]
Gibbsite	$Al(OH)_3 + 3\ H^+ = Al^{3+} + 3\ H_2O$	7.76	[22]
Ferrihydrite	$Fe(OH)_3 + 3\ H^+ = Fe^{3+} + 3\ H_2O$	4.89	[26]
Brucite	$Mg(OH)_2 + 2\ H^+ = Mg^{2+} + 2\ H_2O$	16.84	[22]
Calcite	$CaCO_3 = Ca^{2+} + CO_3^{2-}$	-8.48	[9]
Magnesite	$MgCO_3 = Mg^{2+} - H^+ + HCO_3^-$	2.04	[22]
Monocarbo-aluminate	$Ca_4Al_2CO_9{\cdot}10H_2O + 13\ H^+ = 4\ Ca^{2+} + 2\ Al^{3+} + HCO_3^- + 16\ H_2O$	80.33	[24]
Pb phases			
Lead hydroxide	$Pb(OH)_2 + 2\ H^+ = Pb^{2+} + 2\ H_2O$	11	[27]
Alamosite	$PbSiO_3 + 2\ H^+ = H_2O + Pb^{2+} + SiO_2$	5.6733	[28]
Litharge	$PbO + 2\ H^+ = H_2O + Pb^{2+}$	12.6388	[28]
Pb_2SiO_4	$Pb_2SiO_4 + 4\ H^+ = SiO_2 + 2\ H_2O + 2\ Pb^{2+}$	18.0370	[28]
Laurionite	$PbOHCl + H^+ = Pb^{2+} + Cl^- + H_2O$	0.623	[26]
Blixite	$Pb_2(OH)_3Cl + 3\ H^+ = 2\ Pb^{2+} + 3\ H_2O + Cl^-$	8.793	[26]
Anglesite	$PbSO_4 = Pb^{2+} + SO_4^{2-}$	-7.79	[26]
Cerrusite	$PbCO_3 + H^+ = HCO3^- + Pb^{2+}$	-3.2091	[28]
Hydrocerrusite	$Pb_3(CO_3)_2(OH)_2 + 4\ H^+ = 2\ H_2O + 2\ HCO_3^- + 3\ Pb^{2+}$	1.8477	[28]
(Pb,Ca)CO₃ solid solution			
Cerrusite	$PbCO_3 + H^+ = HCO_3^- + Pb^{2+}$	-3.2091	[28]
Calcite	$CaCO_3 = Ca^{2+} + CO_3^{2-}$	-8.48	[9]
PbO,Al(OH)₃,Fe(OH)₃-solid solution			
Litharge	$PbO + 2\ H^+ = H_2O + Pb^{2+}$	12.64	[28]
Gibbsite	$Al(OH)_3 + 3\ H^+ = Al^{3+} + 3\ H_2O$	7.76	[28]
$Fe(OH)_3$	$Fe(OH)_3 + 3\ H^+ = Fe^{3+} + 3\ H_2O$	5.66	[28]

of Al(hydr)oxide phases as amorphous $Al(OH)_3$ and gibbsite in addition to hydrogarnet. In the presence of calcite, mono-carboaluminate is more stable than hydrogarnet [22].

Some minerals were included to control the leaching of Fe and S. Several authors [18, 13, 29] have shown that the amphoteric leaching behaviour of Fe in MSWI ash is fairly represented by the solubility of ferrihydrite. S leaching at high pH is possibly controlled by ettringite in MSWI bottom ash [30, 13], MSWI air-pollution-control residues [11] and cement [19]. At lower pH values, gypsum is likely the controlling phase for S and Ca in MSWI bottom ash [13, 18] and cement [19]. Also the cement mineral monosulfo-aluminate is included because it is more stable than ettringite at high pH [31].

Mineralogical phases controlling Pb leaching

Many authors [32, 33, amongst others] included only Pb hydroxide and Pb carbonates (cerrusite, hydrocerrusite) to model release of Pb by solubility. However, Johnson et al. [34] stated that leaching of Pb is not controlled by solubility of pure hydroxides and carbonates alone. An extensive set of Pb minerals is included as listed in Table II.

In addition to the pure minerals discussed above, two solid solutions were included to describe the Pb leaching: the ideal $(Ca,Pb)CO_3$ solid solution (as suggested by [21, 35, 36] and the ternary ideal solid solution with gibbsite, ferrihydrite and litharge as end-members [32].

Surface complexation reactions of Pb on HFO and AAM

Apart from solubility and solid solution mechanisms, Pb leaching may be controlled by sorption. Pb sorbs on HFO [37, 38, 33, 32, 39] and AAM [38, 39] and these two sorbing minerals were included in the model. The sorption sites on HFO are divided into low-capacity/high-affinity sites (HFO_s) and high-capacity/low-affinity sites (HFO_w) with 5.10^{-3} mol/mol Fe for HFO_s and 0.2 mol/mol Fe for HFO_w according to [37]. The number of reactive sites is obtained from the measured amount of HFO and the molecular weight of 89 g HFO/mol Fe [37]. Values are given in Table III. The reactive sites on AAM were calculated using the same specific surface area and concentration of binding sites as for HFO, using the latter as a surrogate for the former, because of lack of appropriate data, although AAM have a lower surface area than HFO [40]. Thermodynamic constants for the surface complexation reactions are from [37].

Table III. Total number of sorption sites on HFO and AAM.

Sample	HFO_w (mol/l)	HFO_s (mol/l)	AAM_w (mol/l)	AAM_s (mol/l)
B0	$2.95*10^{-3}$	$7.37*10^{-5}$	$6.89*10^{-3}$	$1.72*10^{-4}$
B60	$1.38*10^{-3}$	$3.46*10^{-5}$	$7.31*10^{-3}$	$1.83*10^{-4}$

RESULTS AND DISCUSSION

Leaching of major elements

Figure 1 compares the calculated leachate concentrations with the experimental data. Overall, predictions describe the observed trends relatively well. The model is able to predict the leached concentrations of the major elements within one or two orders of magnitude, for fresh, partially and fully carbonated cementitious waste samples. Since only equilibrium dissolution and precipitation reactions are included, it can be concluded that the leaching of the major elements is mainly solubility controlled.

The model reproduces the decreased Ca leaching with increasing degree of carbonation. However, this effect is more pronounced in the experimental data than in the model which may be due to a too high estimate of the carbonate input correction factor. The simulations show that a large set of minerals controlled the Ca leaching: portlandite and the three CSH-phases for the B0-sample, the three CSH-phases and calcite for the B14- and B30-samples, and CSH_0.8, CSH_1.1, and calcite for the B60-sample. In addition, the amount of CSH-phases with a higher Ca/Si-ratio decreases with increasing carbonation as the decalcification of the CSH-phases proceeds. Also monosulfo-aluminate and hydrogarnet dissolve to supply Ca-ions for the formation of calcite. The model fails to reproduce the decreasing leaching of Mg with increasing carbonation because it predicts only the formation of brucite for all samples (no Mg-carbonate phases were predicted). For Al, no real trend is visible in the experimental data. The Al-controlling phases in the model are also rather independent of carbonation level with the pH-dependency of the solubility of gibbsite as the most important factor. Due to the dissolution of monosulfo-aluminate and hydrogarnet at higher carbonation levels, gibbsite forms at higher pH with increasing carbonation.

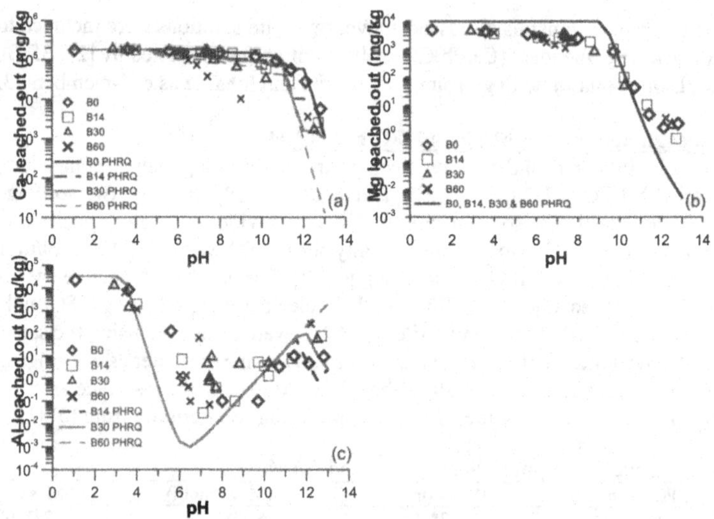

Figure 1. Comparison between calculated and measured leaching concentrations of (a) Ca, (b) Mg, and (c) Al for uncarbonated (B0) and three levels of carbonated (B14, B30, B60) OPC-MSWI bottom ash samples.

Leaching of Pb

Figure 2 shows predictions of Pb leaching using four alternative models: (1) model 1 based on [33] with cerrusite, hydrocerrusite and lead hydroxide, (2) model 2, the model with the complete list of Pb minerals (Table II), (3) model 3 as model 2 and two solid solutions , and (4) model 4 as model 3 with surface complexation on HFO and AAM. All models describe the amphoteric character of Pb leaching. Model 1 gave rather poor predictions. Predictions with model 2 indicate that the solubility controlling minerals are laurionite at pH 6, alamosite for $6.5 < pH < 9.5$, and blixite for $10 < pH < 13$. Model 2 did not predicted the effect of carbonation on Pb leaching (experimental data indicated that carbonation leads to lower Pb leaching for $pH < 7$ and $pH > 11$) because no Pb-containing carbonates (cerrusite, hydrocerrusite) precipitated.

When the ideal binary $(Ca,Pb)CO_3$ and ternary gibbsite-ferrihydrite-litharge solid solutions were included (model 3), the reduced Pb leaching in the carbonated samples is accurately predicted due to cerrusite precipitation in the $(Ca,Pb)CO_3$ solid solution.

For model 4, surface complexation constants were taken from the PHREEQC-database [9], but the log(K) for surface complexation on the weak sites was changed from 0.3 to 1.7 as suggested by Meima and Comans [38] following their recommendation of a general trend of 3 log units difference between complexation constants for high- and low-affinity sites. Including surface complexation reactions on HFO and AAM in the model (model 4) improved the description of the Pb leaching, especially for the uncarbonated sample. Note however that the number of reactive sites on AAM might be overestimated because the same number of reactive sites per gram material as for HFO was assumed. However, the total sorbent concentration measured in this study (4.38 g/l) corresponds well with the value obtained by Meima and

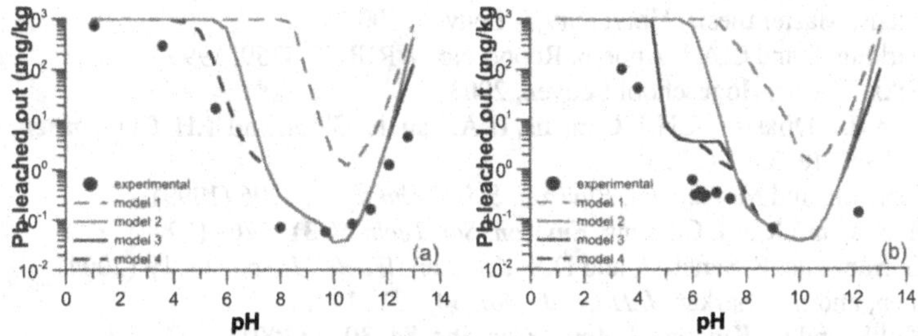

Figure 2. Comparison of the experimental and modelled Pb leaching curves for sample (a) B0 and (b) B60.

Comans [41] (5.06 g/l) for a fresh MSWI bottom ash sample. On the other hand, not all reactive sites are included in the model. For example, crystalline Fe-oxides (formed from HFO) have a smaller, but still significant specific surface area (e.g., 600 and 100 m^2/g for HFO and crystalline iron oxides, respectively, [18]). As the ascorbate extraction technique only measured HFO content, total adsorption capacity might be underestimated. An alternative is using dithionite extraction [42] measuring amorphous and crystalline iron (hydr)oxides.

CONCLUSIONS

A model was presented to simulate the leaching of major elements and the trace element Pb from uncarbonated and carbonated ordinary Portland cement – dried MSWI bottom ash samples as measured by series of single extraction tests at different pH between 1 and 12. Overall, model simulations and measurements agreed fairly well, mostly within one order of magnitude. The Ca leaching was controlled by portlandite, CSH-phases and calcite. Mg and Al were mainly controlled by solubility of brucite and gibbsite, respectively. The effect of carbonation on Ca leaching was fairly well reproduced. Since no formation of carbonate-containing Mg phases was predicted by the model, the effect of carbonation on Mg leaching was not reproduced.The amphoteric leaching behaviour was reproduced by three solubility-controlling minerals: laurionite, alamosite and blixite. The effect of carbonation on lead leaching (i.e., decreasing Pb leaching with increasing carbonation) is mimicked by introducing an ideal binary solid solution between calcite and cerrusite. Modelling results also indicated that surface complexation reactions on HFO and AAM are important mechanisms that control Pb leaching.

REFERENCES

1. IAWG, *Studies in Environmental Science* **67** (1997).
2. D.S. Kosson, H.A. van der Sloot, F. Sanchez, A.C. Garrabrants, *Environ. Eng. Sci.* **19 (3)**, 159 (2002).
3. T. Van Gerven, PhD. Thesis, University of Leuven, 2005
4. T.G. Ferdelman, Master thesis, University of Delaware, 1988.
5. J.E. Kostka, and G.W. Luther, *Geochim. Cosmochim. Acta* **58**, 1701 (1994).
6. G. Cornelis, T. Van Gerven, and C. Vandecasteele, *J. Hazard. Mater.* **A137**, 1284 (2006).
7. M.L. Jackson, C.H. Lim, and L.W. Zelazny, in *Methods of Soil Analysis Part I – Physical and Mineralogical Methods*, A. Klute (SSSA, 1996) pp.113-118.

8. E. Martens, Master thesis, University of Leuven, 2007.
9. D.L. Parkhurst, and C.A.J. Appelo, Report No. WRIR 99-4259, 1999.
10. L. Vinckx, Thesis, Hogeschool Leuven, 2003.
11. T. Astrup, J.J. Dijkstra, R.N.J. Comans, H.A. van der Sloot, and T.H. Christensen, *Environ. Sci. Technol.* **40**, 3551 (2006).
12. C.A. Johnson, and M. Kersten, *Environ. Sci. Technol.* **33**, 2296 (1999).
13. J.A. Meima, and R.N.J. Comans, *Environ. Sci. Technol.* **31**, 1269 (1997).
14. A.C. Garrabrants, F. Sanchez, and D.S. Knosson, *Waste Manag.* **24**, 19 (2004).
15. D. Bonen, and S.L. Sarkar, *J. Hazard. Mat.* **40**, 321 (1995).
16. D.A. Kulik, and M. Kersten, *J. Amer. Cer. Soc.* **84**, 3017 (2001).
17. L. De Windt, D. Pellegrini, and J. van der Lee, *J. Cont. Hydrol.* **68**, 165 (2004).
18. J.J. Dijkstra, H.A. van der Sloot,and R.N.J. Comans, *Appl. Geochem.* **21**, 335 (2006).
19. H.A. van der Sloot, *Cement and Concrete Research* **30**, 1079 (2000).
20. S. Miyamoto, S. Uehara, M. Sasoh, M. Sato, M. Toyohara, K. Idemitsu, and S. Matsumura, *J. Nuclear Sc. And Tech.* **43**, 1370 (2006).
21. C.S. Kirby, and J.D. Rimstidt, *Environ. Sci. Technol.* **28**, 443 (1994).
22. B. Lothenbach, and F. Winnefeld, *Cement and Concrete Research* **36**, 209 (2006).
29. T.T. Eighmy, J.D. Eusden, J.E. Krzanowski, D.S. Domingo, D. Stämpfli, J.R. Martin, and P.M. Erickson, *Environ. Sci. Technol.* **29**, 629 (1995).
30. A. Polettini, and R. Pomi, *J. Hazard. Mat.* **B113**, 209 (2004).
31. M. Chrysochoou, and D. Dermatas, *J. Hazard. Mat.* **136**, 20 (2006).
23. S.A. Stronach, and F.P.Glasser, *Adv. in Cem. Res.* **9**, 167 (1997).
24. X. Bourbon, ANDRA Report C.NT.ASCM.03.026.A. (2003).
25. M. Atkins, D. Macphee, A. Kindness, and F.P. Glasser, *Cement and Concrete Research* **21**, 991 (1991).
26. J.D. Allison, D.S. Brown, and K.J. Novo-Gradac, MINTEQA2/PRODEFA2, a geochemical assessment model for environmental systems: version 3.0. User's manual (1990).
27. L. De Windt, and R. Badreddine, *Waste Manag.* doi:10.1016/j.wasman.2006.07.019.
28. T.J. Wolery, EQ3/6, Technical report UCRL-MA-110662 PT I ed., Lawrence Livermore National Laboratory, Livermore, 1992.
32. C.E. Halim, S.A. Short, J.A. Scott, R. Amal, and G. Low, *J. Hazard. Mat.* **A125**, 45 (2005).
33. C. Jing, X. Meng, and G.P. Korfiatis, *J. Hazard. Mat.* **B114**, 101 (2004).
34. C.A. Johnson, M. Kersten, F. Ziegler, and H.C. Moor, *Waste Manag.* **16**, 129 (1996).
35. A. Godelitsas, J.M. Astilleros, K. Hallam, S. Harissopoulos, and A. Putnis, *Environ. Sci. Technol.* **37**, 3351 (2003).
36. A.A. Rouff, E.J. Elzinga, R.J. Reeder, and N.S. Fisher, *Geochim. Cosmochim. Acta* **69**, 5173 (2005).
37. D.A. Dzombak, and F.M.M. Morel, Surface Complexation Modeling – Hydrous Ferric Oxide (1990).
38. J.A. Meima, and R.N.J. Comans, *Environ. Sci. Technol.* **32**, 688 (1998).
39. Y. Xu, T. Boonfueng, L. Axe, S. Maeng, and T. Tyson, *J. Col. Interf. Sci.* **299**, 28 (2006).
40. M. Fan, T. Boonfueng, Y. Xu, L. Axe, and T.A. Tyson, *J. Col. Interf. Sci.* **281**, 39 (2005).
41. J.A. Meima, and R.N.J. Comans, *J. Geochem. Expl.* **62**, 299 (1998).
42. H.A. van der Sloot, A. van Zomeren, J.J. Dijkstra, J.C.L. Meeussen, R.N.J. Comans, and H. Scharff, ECN-RX—05-164 (2005).

Mater. Res. Soc. Symp. Proc. Vol. 1107 © 2008 Materials Research Society

Changes on the Mineralogical and Physical Properties of FEBEX Bentonite Due to Its Contact With Hyperalkaline Pore Fluids in Infiltration Tests

A. M. Fernández, A. Melón, D.M. Sánchez, M.P. Galán, R. Morante, L. Gutiérrez-Nebot, M.J. Turrero, A. Escribano
CIEMAT, Dpto. Medio Ambiente. Avda. Complutense 22, 28040 Madrid (Spain)

ABSTRACT

Two infiltration tests with a synthetic hyperalkaline solution at pH 13.5 were performed. The aim was to analyse the interaction of concrete solutions with compacted smectitic bentonites. The experiment was performed inside an anoxic glove box at 30-35 ºC with FEBEX bentonite compacted at a dry density of 1.65 g/cm^3 and a water content of 13.4%. Outflowing water coming from the bentonite was collected over time. A total of three times the bentonite pore volume was recovered and twelve chemical analyses of the resulting pore water were obtained. The ionic strength of the pore waters decreases with time from 0.20 M to 0.03 M and the pH increases from 7.9 to 9.0. After a test period of 595 days, the cells were dismantled, and the study of the state and alteration of the bentonite was undertaken. A modification in the adsorbed cation population, an increase in the Na- and K-smectites type, and the presence of a tri-octahedral smectite (saponite) and zeolites (phillipsite and chabazite) were observed.

INTRODUCTION

Concrete will be used as support of tunnels and galleries in the geological disposal of nuclear wastes in argillaceous host formations and also as waste containment material. The bentonite barrier will become saturated with the water resulting from the host-rock/concrete interaction. Different laboratory scale tests and geochemical modelling have been performed to study the interactions of clays with highly alkaline fluids [1-6]. The mineralogical alterations of smectites could lead to changes in different physico-chemical properties of the bentonite which could jeopardize its performance as a bentonite barrier. In this context, a concrete-bentonite interaction experiment has been performed at a high solid to liquid ratio with FEBEX bentonite. The aim of the experiment is to analyse the buffering capacity of the bentonite and the clay mineral stability in a high-pH environment over a long contact period.

EXPERIMENT

The tests were performed with FEBEX bentonite, extracted from the Cortijo de Archidona deposit (Almería, Spain). It consists of more than 90 percent montmorillonite, Ca and Mg being the main cations in the exchange complex, in which Na is also present. This bentonite originated from the alteration of pyroclastic volcanic rocks and contains numerous accessory and trace minerals [7,8,9].

Two infiltration tests in a small-scale hermetic cell- whose internal diameter is 50 mm and inner length 25 mm- have been running under anoxic conditions inside a glove box (<1 ppm O_2) for 595 days (1.65 years). 91.82 g of FEBEX bentonite were compacted at 1.65 g/cm^3 with a hygroscopic water content of 13.4% in each cell. The body of the cell is made of 10 mm thick methacrylate to prevent the deformation of the cell due to bentonite swelling, and the lids are made of 316L stainless steel (Figure 1).

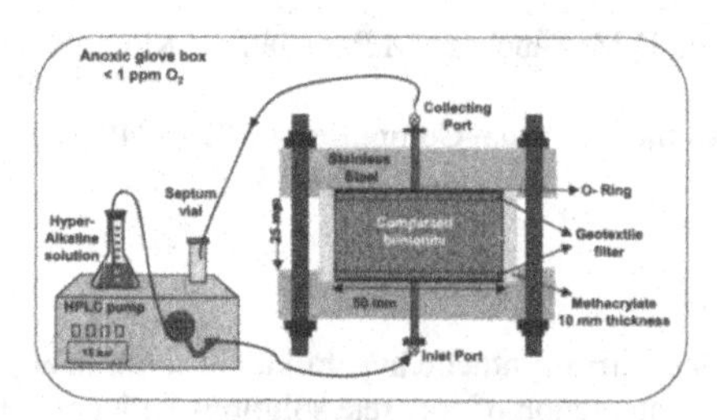

Figure 1. Experimental setup for the infiltration tests

Hydration of the bentonite took place at the bottom part of the cell through a geotextile filter. The infiltration water was a synthetic hyperalkaline water, which was injected with a pressure of 15 bars by means of a 307 Gilson® piston pump. At the top of the cell the outflowing pore water coming from the bentonite through another geotextile disk was collected inside a vacuum vial closed by a septum and analysed. The temperature inside the glove box was kept at 30-35 ºC.

The type of alkaline solution was a Na-K-OH water in equilibrium with portlandite, $Ca(OH)_2$, at pH 13.5 (Table I). This water is representative of an average pore water of a mortar made with CEM-I-SR type Porland cement (sulphate-resistant) at a 0.6 cement/water ratio and a 3:1sand/cement ratio.

Table I. Chemical composition of the alkaline concrete pore fluid solution used in the experiments

Reagent Grade Salts	**NaOH**	**$Ca(OH)_2$**	**K(OH)**	**Na_2SO_4**	**pH**
Concentration (g/L)	2.8744	0.0545	19.8742	2.1686	13.5

EXPERIMENTAL RESULTS

The cells (called CW-1 and CW-2) were dismantled after a test period of 595 days at laboratory conditions (22ºC). During their analysis, special care was taken to avoid any disturbance in the conditions of the clay, particularly its dry density and water content. Once the bentonite block was extracted from each cell, it was sawed in half in two cylindrical pieces, 1.25 thick and each piece into 1/8 sections (Figure 2).

Final physical state

The final appearance of the one of the bentonite blocks is shown in Figure 2. The average final water content is 29 percent and the dry density is 1.53 g/cm^3. A change of the initial dry density was detected due to the swelling of the bentonite (Table II).

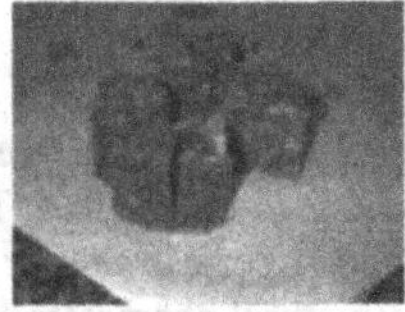

Figure 2. Appearance of the bentonite after dismantling and sampling

Table II. Characteristics of the infiltration tests

Initital conditions	CW-1	CW-2	Average	Final conditions	CW-1	CW-2	Average
Date	5/9/2005	5/9/2005	--	Dismantling date	23/04/2007	23/04/2007	--
Mass (g)	91.82	91.82	91.82	Mass (g)	106.02	105.36	105.69
Water content (%)	13.4	13.4	13.4	Water cont. (%)	29.8	28.1	29.0
Diameter (mm)	50.0	50.0	50.0	Diameter (mm)	50.71	50.74	50.73
Thickness (mm)	25.0	25.0	25.0	Thickness (mm)	27.38	26.86	27.12
Dry density (g/cm^3)	1.65	1.65	1.65	Dry density (g/cm^3)	1.51	1.54	1.53
Degree of Saturation	57 %	57%	57%	Degree of Saturation	101 %	101 %	101 %
Porosity	0.39	0.39	0.39	Porosity	0.44	0.43	0.44
Water volume, mL	19.08	19.08	19.08	Water intake (g)	24.34	23.11	23.73
				Outflowing water	60.0 mL	58.7 mL	59.4 mL

The evolution of the Darcy's velocity and hydraulic conductivity of the bentonite during the test are shown in Figure 3. An average of 59.5 mL of water was recovered at the end of the tests, which implies a replacement of around three porewater volumes of the bentonite after 595 days of the onset of the experiment. The final average value of the Darcy's velocity (Q) is $9.96 \cdot 10^{-5}$ dm^3/day and the hydraulic permeability (K) of the bentonite ranged from $3.7 \cdot 10^{-14}$ to $9.59 \cdot 10^{-14}$ m/s, being the last value similar to that obtained for the reference sample, $7.97 \cdot 10^{-14}$ m/s, at a dry density of 1.53 g/cm^3 in saturated conditions [7]. The variation of K observed is probably associated with changes in the density of the bentonite during the experiment due to the clay swelling caused by its gradual hydration up to final saturation.

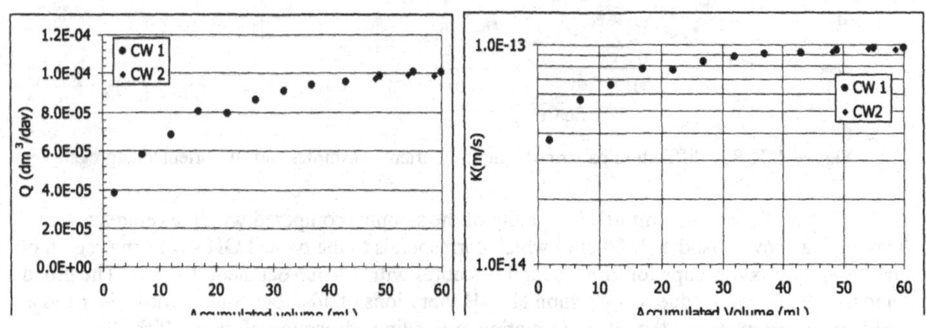

Figure 3. Darcy's velocity and hydraulic conductivity of the bentonite during the infiltration tests with hyperalkaline solutions

Mineralogy

The X-ray diffraction patterns of different samples taken at the top and bottom of each cell (CW-1 and CW-2) are shown in Figure 4. The smectite particles are not significantly altered, since the intensity of the 4.45 Å peak is similar to the reference FEBEX sample. However, there are clear changes in their structural characteristics as outlined by the peak at 1.52 Å, which indicates the presence of a tri-octahedral smectite (saponitic type). This peak mainly appears in the samples situated at the bottom of the cell (0-1.25 cm), close the hydration source (inlet point); although it seems that this peak also appears in the samples located at the top (1.25-2.5 cm). Furthermore, all the samples show a modification in the adsorbed cation population, mainly an increase in the Na-smectite type particles (peak at 12.8 Å), and also in the K-smectite type particles (peak at 11.5 Å), which implies an increase in illite/smectite mixed-layer formation (up to 40% according to the $°\Delta 2\theta$ (d(003)-d(002)) from EG solvated oriented aggregates). There are no significant changes in the peak area corresponding to calcite. Accessory minerals, such as feldspars (anorthite and albite) and quartz do not present signs of dissolution (which should enable the precipitation of CSH minerals or CASH-phases), as can be deduced from the intensity increase of the corresponding peaks. However, two types of zeolites were distinguished in the X-ray diffraction patterns from total fraction and oriented aggregates (<0.5 μm fraction): phillipsite and chabazite.

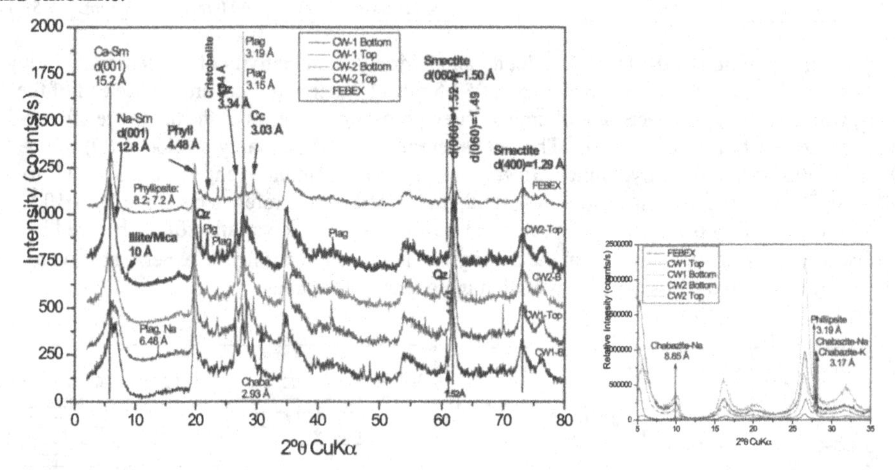

Figure 4. X-Ray diffraction pattern of randomly oriented samples and EG-oriented aggregate

Figure 5 shows the mid/FTIR spectra of the samples compared with the reference one. The spectra show a band at 3620 cm^{-1} which corresponds to the typical OH stretching region of structural hydroxyl groups for dioctahedral smectites with Al-rich octahedral sheets. The broad band near 3430 cm^{-1} is due to stretching H-O-H vibrations of adsorbed water, while the band at 1642 cm^{-1} corresponds to the OH deformation or bending adsorption of water. If the Si-O absorptions and OH bending bands in the 1300-400 cm^{-1} range are examined, only one broad, complex Si-O stretching vibration band at around 1030 cm^{-1} is seen, which is typical of a dioctahedral montmorillonite. The occupancy of the octahedral sheet strongly influences the position of the OH bending bands. In these samples, the presence of a peak at 915 cm^{-1} (δAlAlOH) and at 840 cm^{-1} (δAlMgOH), indicates a partial substitution of octahedral Al by Mg, typical of dioctahedral smectites. The bands at 524 cm^{-1} and 466 cm^{-1} correspond to Si-O-Al (octahedral Al) and Si-O-Si bending vibrations, respectively. The weak band at 790 cm^{-1} is caused by the Si-O stretching of quartz-CT. All these bands are similar to those found in the reference FEBEX sample. However, some modifications are also shown in some bands of the spectra indicating the development of a trioctahedral smectite, as in the upper and lower part of the bentonite sample. In the OH stretching region, a band characteristic of the trioctahedral smectites appears near 3668 cm^{-1} which corresponds to Mg_3OH units in the octahedral sheet. In the Si-O stretching-bending and OH bending region, the Mg_3OH bending vibration is observed at 690-670 cm^{-1}, while the δSi-O-Si band splits into two peaks at 466 cm^{-1} and 445 cm^{-1}, like saponite. The latter belongs to the Mg-O-H libration (bending) mode. Besides, a decrease of the Si-O-Al band at 520 cm^{-1} is observed in the upper part of the CW2 sample, which indicates a decrease in the octahedral cation content, and possible changes in the new tetrahedral sheets formed by octahedral cation leaching.

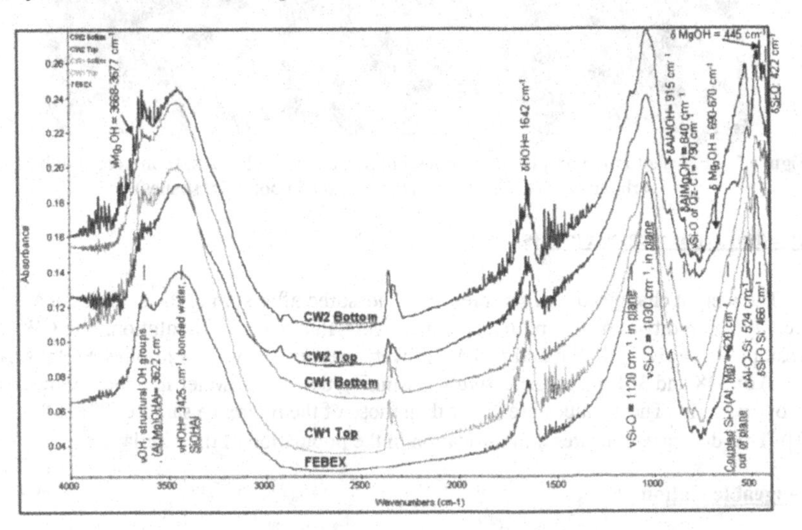

Figure 5. FTIR spectra of the bentonite (total fraction) after the infiltration test

The microstructure of the samples were analysed by SEM-EDX. The clay samples are mainly montmorillonite in appearance and composition. The Si/Al ratio is maintained with respect to the reference sample. However, an increase in the potassium content (from 1-2% to 4, 8 and 28% K_2O) was detected by EDX micro-analyses in several clay particles of the samples (Figure 6a,b), implying a decrease in the sodium and magnesium content. Nevertheless, no fibrous structures with a typical clay composition were found indicating illite formation. K-feldspars were found in the samples (Figure 7a) without neoformation-dissolution signs, although an increase of the intensity of the related peaks is observed in some sample by XRD. Other secondary minerals identified were calcite (Figure 7b) and spherical structures containing Cl and S. Besides, silica minerals with botroidal and spherical shapes were observed (Figure 7c).

Other types of alteration products of montmorillonite, such as CSH-gel phases were not detected in the samples probably because of the low temperature (~30-35°C) and/or the duration of the experiment. Zeolites were not observed by SEM, only by XRD analyses. It seems that the montmorillonite has undergone a first stage of alteration driven by its dissolution [5,6].

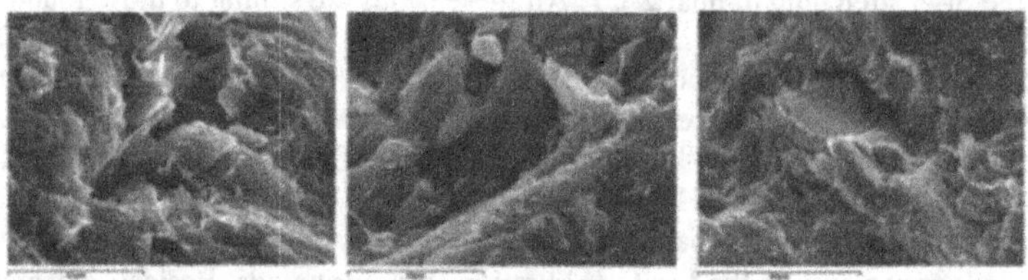

Figure 6. SEM photomicrographs of clay minerals: a) and b) K-rich clay minerals (K~4-28%), and c) clay mineral without calcium content (only K, Fe, Mg and Na)

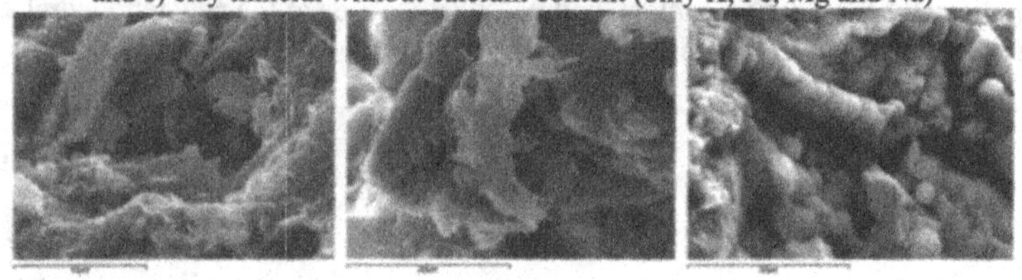

Figure 7. SEM photomicrographs of secondary minerals: a) K-feldspars; b) minerals with Si and carbonates; c) Silica crystals (spherical and botroidal shapes)

Water adsorption measurements

The weight changes of the samples were measured after storing the samples at a constant 85% relative humidity atmosphere in a chamber over-saturated in KCl-solution. The CW1 and CW2 samples adsorbed 148.9 g and 155.4 g of water, respectively. This implies a total surface area (SA) of 538 and 561 m^2/g, respectively; assuming that 1% of water in a monomolecular layer covers 35 m^2. These values are lower than those of the reference sample (725 ± 47 m^2/g, [7,8,9]). This decrease indicates a modification in the population of the interlayers.

Exchangeable Cations

The concentration of exchanged cations in two samples located at the top and bottom of the cell in the different tests (CW-1 and CW-2) is shown in Table III. These measurements were performed with a 0.5 N $CsNO_3$ solution at pH 8.2. The values are compared with those obtained from the reference FEBEX bentonite sample. A significant decrease can be observed in the total CEC obtained as a sum of the exchangeable cations with respect to the reference sample. A significant variation in the potassium content and a slight increase of sodium at the exchange positions is observed, resulting in a decrease of the calcium and magnesium contents. The increase of potassium implies the increase of illite layers in the smectitic phases of the FEBEX bentonite (up to 40%), which is originally composed of a smectite-illite mixed layer with ~11 % of illite layers. The decrease of magnesium at the interlayers may imply the migration of magnesium to octahedral sites forming Mg-rich smectites, such as saponite or stevensite.

Table III. Total exchange capacity and exchangeable cations after 595 days of bentonite interaction with hyperalkaline solution (in meq/100g)

Cell	Sample	Na	K	Mg	Ca	Sr	Ba	Σcations
–	FEBEX reference	27.12	2.5	30.96	34.77	0.38	--	95.79
CW1	At the top	32.19	24.55	1.32	17.35	0.18	0.02	76
CW1	At the bottom (close to the hydration source)	32.18	26.59	1.25	18.83	0.20	0.02	79
CW2	At the top	30.11	28.64	0.05	16.51	0.18	0.02	76
CW2	At the bottom (close to the hydration source)	30.45	26.60	0.16	18.79	0.18	0.02	76

Analysis of the outcoming porewaters

A total of 3.1 times the bentonite pore volume was recovered (~60 mL) during the test period, and twelve chemical analyses of the resulting pore water were carried out (Figure 8). The ionic strength of the chemical composition of the pore waters decreases with time from 0.20 M to 0.03 M. The pH increases from 7.9 to 9.0, but it is always lower than the infiltrating hyperalkaline water. Therefore, after an experiment of 595 days, the FEBEX bentonite maintains its buffer capacity, lowering the pH of the infiltrating water. The chemical composition of the different aliquots obtained was compared to that obtained in the same conditions (initial dry density and water content) but injecting distilled water, which was taken as reference pore water. This reference bentonite pore water contains Cl, Na, SO_4^{2-}, Ca and Mg as major ions [8,9].

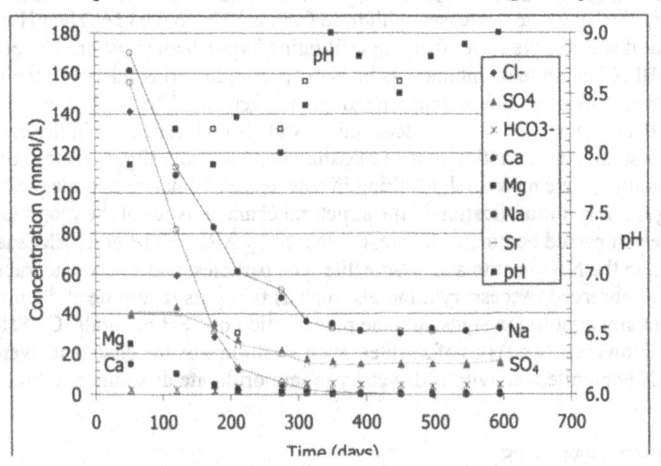

Figure 8. Chemical evolution of the bentonite pore water over time in the CW-1 (filled symbols) and CW-2 (open symbols) infiltration tests

The pH of the reference water is lower than that obtained from of the first aliquot sample extracted. Chloride concentration has a maximum value of 148 mmol/L after 52 days, and then decreases over time. It seems that the chloride salts were not totally removed from the bentonite after 595 days of experiment. The sulphate content maintained a constant value in the first two aliquots, it began to decrease after 176 days, and kept a constant value of 15 mmol/L after 312 days onwards, which is the inlet content of the infiltrating water. Calcium decreased with time, and reached the value of the synthetic hyperalkaline water, ~0.40 mmol/L, after 392 days, the same as that of sulphate. The bicarbonate concentration increased from an initial value of 2.72 mmol/L to 6.40 mmol/L after 312 days, when the sulphate and Ca contents reached a constant value. Afterwards, it decreased slightly, although the values were always higher than that obtained in the first aliquot. Magnesium concentration decreased with time; as well as, the sodium and potassium content. However, the concentration of Na and K was always lower than the concentration of the infiltrating water; i.e., the composition of the incoming water was never obtained at the end of the experiments. All the waters were oversaturated with respect to calcite and dolomite but subsaturated with respect to gypsum.

CONCLUSIONS

Two infiltration tests with hyperalkaline water (Na-K-OH fluid at pH 13.5) were performed with FEBEX compacted bentonite at anoxic conditions and at a temperature of 30-35°C during 595 days. At the end of the test, the saturated bentonite (29% of w.c.) had a final dry density of 1.53 g/cm^3. No significant changes in the hydraulic permeability of the bentonite was found, reaching a value of $9.59 \cdot 10^{-14}$ m/s, which was the expected value for the reference material at similar conditions of saturation. A total of 3.1 times the bentonite pore volume was recovered (~60 mL) during the test period. The ionic strength of the outflowing pore waters obtained from the bentonite decreased with time from 0.20 M to 0.03 M. The pH increased from 7.9 to 9.0, but it was always lower than the infiltrating hyperalkaline water. Therefore, after 506 days, the FEBEX bentonite maintained its buffer capacity, lowering the pH of the infiltrating water. However, a modification in the mineralogy and geochemistry of the bentonite was observed. A decrease of the CEC was detected, as well as a modification in the adsorbed cation population. A significant variation in the potassium content and a slight increase of sodium at the exchange positions were measured, resulting in a decrease of calcium and magnesium contents. These changes imply a modification in the structural characteristics of the montmorillonite in the 2.5 cm of the compacted bentonite sample, evidenced by XRD, FTIR and SEM analyses. There is an increase in the Na-smectite and K-smectite type particles, and a trioctahedral smectite (saponite) was observed. Accessory minerals, such as feldspars (anorthite and albite) and quartz, do not present any dissolution signs; and the precipitation of CSH minerals, CASH-phases were not detected. However, two types of zeolites, such as phillipsite and chabazite, were identified by XRD. FEBEX bentonite reactivity is driven by montmorillonite dissolution at high pH.

ACKNOWLEDGEMENTS

This test was performed in the context of the NF-PRO Project, financed by ENRESA (Spanish National Agency for Waste Management) and the European Commission (EC Contracts FI6W-CT-2003-02389). We thank A.E. González and R. Campos for their technical assistance during the laboratory work. E. Baldomero, from the "Luis Bru" Microscopy Electronic Centre of the Complutense University of Madrid, helped us with the SEM analyses.

REFERENCES

1. Karnland, O. Cement/bentonite interaction. Results from 16 month laboratory tests. SKB technical report 97-32 (1997).
2. Savage, D., Noy, D., Mihara, M. Modelling the interaction of bentonite with hyperalkaline fluids. Applied Geochemistry 17, 207-223 (2002).
3. Metcalfe, R., Walker, C. Proceedings of the International Workshop on bentonite-cement interaction in Repository environments. Posiva Working Report, 2004-25, 192 pp. (2004).
4. ANDRA. Ecoclay II: Effects of Cement on Clay Barrier Performance – Phase II. Final Report. EC project nº FIKW-CT-2000-00028. Andra Report. Nº CRPASCM04-0009, 381 pp (2005).
5. Gaucher, E., Blanc, P. Cement/Clay interactions. A review: Experiments, natural analogues and modeling. Waste Management 26, 776-788 (2006).
6. Cuevas, J., R., Vigil de la Villa, Ramírez, S., Sánchez, L, Fernández, R., Leguey, S. The alkaline reaction of FEBEX bentonite: a contribution to the study of the performance of bentonite/concrete engineered barrier systems. Journal of Iberian Geology 32(2), 151-171 (2006).
7. Huertas, F., P. Fariña, J. Farias, J.L. García-Siñeriz, M.V. Villar, A. M. Fernández, P.L. Martín, F.J. Elorza, A. Gens, M. Sánchez, A. Lloret, J. Samper, M.A. Martínez. Full-scale Engineered Barriers Experiment. Updated Final Report 1994-2004. *Publicación Técnica ENRESA* 05-0/2006. 590 pp. (2006).
8. Fernández, A.Mª. Caracterización y modelización del agua intersticial de materiales arcillosos. Estudio de la bentonita de Cortijo de Archidona. Ph.D. Thesis. Universidad Autónoma de Madrid. Editorial CIEMAT, 505 pp. Madrid. (2004).
9. Fernández, A.Mª., Baeyens, B., Bradbury, M., Rivas, P. Analysis of the pore water chemical composition of a Spanish compacted bentonite used in an engineered barrier. Physics and Chemistry of the Earth, Vol. 29/1, pp. 105-118 (2004).

Mater. Res. Soc. Symp. Proc. Vol. 1107 © 2008 Materials Research Society

A New Natural Analogue Study of the Interaction of Low-Alkali Cement Leachates and the Bentonite Buffer of a Radioactive Waste Repository

W. Russell Alexander[1], Carlo A. Arcilla[2], Ian G. McKinley[3], Hideki Kawamura[4], Yoshiaki Takahashi[5], Kaz Aoki[6] and Satoru Miyoshi[4]
[1]Bedrock Geosciences, Auenstein, Switzerland. [2]National Institute of Geological Sciences, University of the Philippines, Quezon City, Philippines. [3]McKinley Consulting, Dättwil, Switzerland. [4]Obayashi, Tokyo, Japan. [5]NUMO, Tokyo, Japan. [6]RWMC, Tokyo, Japan.

ABSTRACT

Bentonite plays a significant barrier role in many radioactive waste repository designs, where it has been chosen due to its favourable properties such as plasticity, swelling capacity, colloid filtration, low hydraulic conductivity and its stability in relevant geological environments. However, bentonite is unstable at high pH meaning that it could lose its favourable properties if interacted with hyperalkaline leachates from concrete construction materials (e.g. tunnel liners, grouts, etc.), seals and plugs and/or cementitious wastes in a repository. This fact has forced several national programmes to assess alternative construction and sealing materials such as low alkali cements. Recently, it has been assumed that the lower pH (typically pH 10-11) leachates of such cements will degrade bentonite to a much lesser degree than 'standard' OPC-based cement leachates (generally with an initial pH>13).

To date, few laboratory or *in situ* URL (underground rock laboratory) data are available to support the use of low alkali cements in conjunction with bentonites, partly because of the very slow kinetics involved. Consequently, a new project has focussed on finding an appropriate natural analogue site to provide long-term supporting data which will avoid the kinetic constraints of laboratory and URL experiments. Early results have identified an initial, very promising site at Mangatarem in the Philippines, where a quarry excavating bentonite and zeolites is found in the sedimentary carapace of the Zambales ophiolite. In the immediate vicinity of the quarry, ophiolite-derived hyperalkaline groundwaters are present and further field work (including geophysics surveys and borehole drilling) are now being planned to assess regional bentonite/hyperalkaline groundwater interaction. This paper presents an overview of the current status of the project and assesses the relevance of the study to improving understanding of low-alkali cement leachate/bentonite interaction.

INTRODUCTION

Bentonite is one of the most safety-critical components of the engineered barrier system for the disposal concepts developed for many types of radioactive waste (radwaste) [1]. The choice of bentonite results from its favourable properties – such as plasticity, swelling capacity, colloid filtration, low hydraulic conductivity, high retardation of key radionuclides – and its stability in relevant geological environments. However, bentonite – especially the swelling clay component that contributes to its essential barrier functions – is unstable at high pH. This led to some repository designs, especially for disposal of HLW (high-level waste) or SF (spent fuel) [e.g. 2], that specifically exclude the use of concrete from any sensitive areas containing bentonite, due to the fact that cementitious materials react with groundwater to produce initial leachates with pH >13, later falling to around pH 12.5 [3].

Such an option of avoiding the problem by constraining the design was considered acceptable during early, generic studies but, as national radwaste programmes move closer to implementation, it is increasingly recognised that constructing extensive facilities underground without concrete – a staple of the engineering community – would be difficult, expensive and potentially dangerous for workers [4]. This is especially the case in countries like Japan, where a volunteering approach to siting a HLW repository [5] means that repository construction could be in a technically challenging host rock.

A further area of concern involves ILW (long-lived intermediate level waste, also called TRU), particularly if this is co-disposed (or co-located) with HLW/SF. ILW contains large inventories of cementitious materials and hence, in principle, could pose a risk to the EBS (engineered barrier system) of HLW/SF, even if concrete was excluded from the HLW part of the repository (e.g. scenarios discussed in [2]). Indeed, some designs of the EBS for various kinds of ILW include a bentonite layer [e.g. 6], which is planned to act as an external barrier around concrete structures. To date, there has been no comprehensive demonstration that the performance of such a barrier can be assured for relevant periods of time [e.g. 7].

Recently, therefore, there have been extensive efforts to better understand the interactions of hyperalkaline fluids with bentonite [8,9], coupled with studies aimed at reducing the risk by development of low alkali cement formulations [e.g. 10]. The greatest challenge is bringing the information produced by laboratory (conventional and URL – underground rock laboratory) and modelling studies together to form a robust safety case. This is complicated by the inherently slow kinetics of such reactions and the commonly observed persistence of metastable phases for geological time periods (for a good overview of the issues involved, see [8]). Clearly, this is an area where natural analogues (see [11,12] for definition and examples) could play a valuable role – bridging the disparity in realism and timescales between laboratory studies and the systems represented in repository performance assessment [13].

NATURAL ANALOGUE CONCEPT

Although systems representative of leachates from both OPC (i.e. groundwaters with pH 12.5 and above) and low alkali cements (groundwater pH of 10 to 11) have previously been examined as natural analogues of cementitious repositories [9, 14, 15], neither produced data of relevance to the questions now being posed with regard to low alkali cements. The natural cements in Jordan [9, 14] are closely representative of the OPC (Ordinary Portland Cement) based materials traditionally used in repository designs and, although it is likely that a period of cement evolution similar to that of low alkali cements exists at the study sites, no data are currently available. An appropriate low alkali cement analogue was studied in Oman [15] but, as the focus of the work was on hydro- and biogeochemistry, no data of relevance to bentonite alteration were collected here either.

As such, an initial literature study was sponsored by RWMC of Japan with search parameters that include both aspects of the target geology and also themes of relevance to radwaste management programmes. Factors considered include:

- ophiolite terrains
- hyperalkaline groundwaters
- bentonites
- H_2 or CH_4 gas in groundwaters
- thermal groundwaters

- tuffaceous deposits
- coastal sites
- logistics (e.g. potential support from local mining operations, ease of transport etc)
- use in training junior staff

and indicated that no useful information on this topic could be "mined" from any past studies and hence the option of a new project was examined. The basic idea was to use a "top-down" approach to identify sites where bentonite deposits have been exposed to relevant hyperalkaline water for very long periods. Especially given the interest in low pH (or, more precisely, low alkali) cements, the focus was on sites that have natural waters with pH in the appropriate range (around 10-11). As indicated in [8], the cement leachate is simulated by natural hyperalkaline water, which, if the timescale of interaction can be determined, allows the models that are being developed to quantify the specific processes shown in the figure to be tested.

The challenge is to maximise the value of this test, by assuring that materials and boundary conditions are as similar as possible to those in a repository. Nevertheless, it must be emphasised that such sites are no more than an analogy of a repository, not a copy, and hence certain differences are inevitable (discussed below). Currently, the technical focus is on:

- long-term bentonite stability in analogue low alkali cement leachates
- if possible, same system as above interacting with seawater/brines for a coastal repository
- if possible, same system as above interacting with a range of leachate chemistries (cf. Table I) as the precise situation in a repository will depend both on the site conditions and the composition of the cementitious materials – neither of which have been fixed as yet
- low alkali cement leachate/host rock interaction – is there any?
- BPM (blind predictive modelling – see [16]) of the chemistry of safety relevant elements (eg Se), including *in situ* speciation
- microbiology of the system (cf. 15)
- staff training, including mentoring by experienced (in radwaste) international staff

In principle, there are a number of locations worldwide where such an analogue might be found, including Cyprus, Oman, California, Bosnia, Papua New Guinea, Japan and the Philippines. Based on a multi-attribute analysis, considering factors such as probability of finding suitable locations, relevance to Asian programmes, opportunity for training, low risk of disrupting calls for volunteer sites in Japan and cost-effectiveness, the Philippines is now under consideration as one of the preferred options for the Japanese programme and has recently been the focus of more detailed literature studies and a limited number of field investigations to confirm fundamental feasibility.

The hyperalkaline pH values (generally between pH 10 and 11, see [17] for examples) observed in the groundwaters are a product of the serpentinisation of ultramafic rock, in this case usually the ophiolites (Figure 1), a reaction which has several possible pathways (e.g. equation 1, from [18 after 19]) with the exact reaction pathway depending on Mg content of the precursor olivine/pyroxene or serpentine product, CO_2 fugacity, water-rock ratio, Ca^{2+} content of groundwater, etc. In simple terms:

$$15 \text{ olivine } (Fo_{90}) + 22.5\, H_2O + xC = \text{magnetite} + 7.5 \text{ serpentine} + 4.5 \text{ brucite} + xCH_4 + (4-2x)\, H_2 + 2OH^- \quad (1a)$$

$$15 \text{ olivine } (Fo_{90}) + 22.5\ H_2O = \text{magnetite} + 7.5 \text{ serpentine} + 4.5 \text{ brucite} + 4H_2 + 2OH^- \quad (1b)$$

Equation 1a is for systems open with respect to carbon input and equation 1b is for closed systems. The serpentinite mineral assemblages are very strongly reducing and the hyperalkaline waters are often effervescent with H_2 and/or CH_4 gas (see above) and geochemical modelling [e.g. 15, 19, 20] and experimental evidence [e.g. 21, 22] suggest abiogenic reduction during the serpentinisation process. Some of the reaction pathways are also strongly exothermic, frequently producing hydrothermal groundwaters which are often used as therapeutic springs.

RESULTS AND DISCUSSION

All hyperalkaline groundwaters studied so far in the Philippines originate from the 20 or so known ophiolite bodies which are widely scattered throughout the archipelago. Several ophiolites on the islands of Panay (around E121,000 N012,500), Luzon (around E120,500 N014,500) and Palawan (around E118,000 N010,000) have been examined as part of a preliminary assessment of suitable sites and, to date, the Mangatarem area (E120,180 N015,047) in the province of Pangasinan on the west central area of the island of Luzon in northern Philippines appears the most promising. This area hosts the largest known bentonite deposits in the country, with reserves of approximately four million tonnes. Here, bedded bentonites and zeolites belong to the Eocene Aksitero Formation (and most probably extend to the finer members of the overlying Moriones Formation), the sedimentary carapace of the Eocene Zambales ophiolite, and which (un)conformably overlie the cherts and upper pillow lavas of the ophiolite (cf. Figure 1). The well-bedded tuffaceous members of the formations have been authigenically transformed after deposition into bedded zeolite (mordenite [23]) and a bentonite dominated by montmorillonite and quartz which is similar in chemistry to the MX-80 bentonite which is under consideration as an EBS material in many national programmes (e.g. [1, 5, 6 and 8] - see also Table II). Individual beds range in thickness from a few centimetres to several metres and range in colour from light cream, beige, off-white, light to medium brown and greenish. Texture closely resembles a tuff, but sometimes exhibits finer grains.

In the immediate vicinity of the bentonite deposit, hyperalkaline groundwaters come to the surface as a hot spring and the chemistry is clearly that of an serpentinisation-derived (presumably from the nearby ophiolite) groundwater which has similarities to low alkali cement leachates (see Table I).

At both the Palawan and Panay sites, no exposed bentonite has yet been discovered and it seems likely that drilling would be required at these sites. However, it is not necessary to cover all technical questions at one site so it is likely that the focus at Palawan will be on coastal repository processes due to the close proximity (<1 km) of the site to the sea whereas a seemingly highly active microbial community at Panay (here microbial mats in and around the groundwater springs and seeps are the simplest way to find the alkaline groundwaters, backed up by establishing the groundwater pH) might prioritise this site for studies of microbial processes in a cementitious repository [cf. 24]. An intercomparison of the BPM work at all three sites would, based on the intrinsic differences in the groundwater chemistry, increase confidence in the results of the modelling.

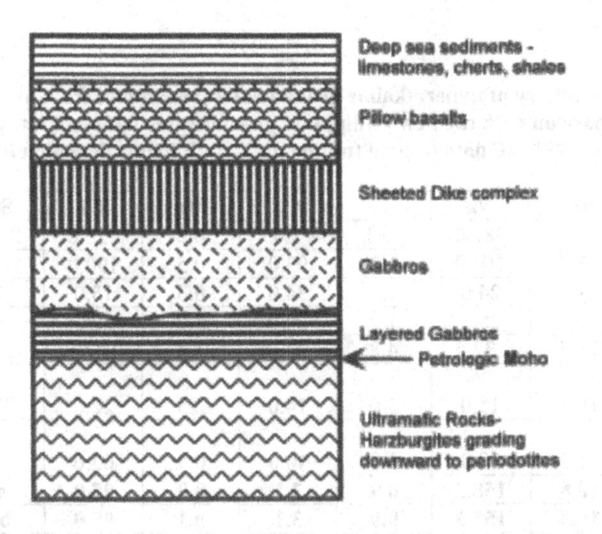

Figure 1. Cross section through an 'average' ophiolite. The serpentinisation reaction can take place in any of the igneous members of the entire sequence, hence leading to slightly different leachate chemistries. At Mangatarem, the bentonites lie on the pillow lavas and cherts at the top of the ophiolite sequence shown above.

CONCLUSIONS

Although still at a preliminary stage, the results of the project to date are very encouraging and indicate the potential of the sites in the Philippines to address the long-term alteration of bentonite (and possible repository host rocks) by low alkali cement leachates. Currently, the Mangatarem site looks the most promising for the study of bentonite/leachate interaction, but further work may be carried out at the Palawan and Panay sites, as these provide better examples of a coastal repository analogue and microbiological processes respectively. As each site will provide a different challenge in terms of site characterisation, logistics and interpretation, this project is particularly suitable for staff training.

Indeed, the proposed International Philippines Hyperalkaline Analogue Project (IPHAP) aims to maximise the extent of transfer of experience to younger staff [25], which is widely acknowledged to be a critical problem in many national waste management programmes [e.g. 26]. Participation in this project is open to radwaste implementing, regulating and R&D organisations and may be particularly suitable for the developing nuclear programmes in Asia.

ACKNOWLEDGEMENTS

This study was initiated within a project to develop an integrated natural analogue programme in Japan, which was funded and coordinated by RWMC. Developing the IPHAP proposal and initial fieldwork in the Philippines would have been impossible without logistical support by staff and students of the National Institute of Geological Sciences of the University of the Philippines. Thanks also to Profs J.H.Sharp and F.P.Glasser for their helpful reviews of the paper.

Table I: Hydrochemistry of hyperalkaline groundwaters: examples from around the world [27] for comparison with the new Philippines groundwater data and low pH cement leachates [28]. All data in ppm (nd: no data bdl: below detection limit).

Location	pH	Na	K	Ca	Mg	Cl	SO_4	SiO_2
Cyprus 3a	11.5	385.0	15.1	1.0	0.3	420.0	251.0	24.0
Cyprus 3b	11.2	163.0	1.2	93.0	0.5	190.0	207.0	3.4
Greece	11.3	24.0	1.0	34.0	0.3	15.0	3.0	2.0
Bosnia	11.7	35.0	1.5	29.0	7.0	20.0	2.0	0.9
Oman	11.5	132.5	4.8	34.0	1.3	127.5	22.5	3.0
New Caledonia	10.8	15.0	3.0	14.0	2.3	22.0	0.8	0.4
Western USA	11.5	19.0	1.0	40.0	0.3	63.0	0.4	0.4
Narra 1	**10.8**	**158.3**	**0.9**	**3.1**	**0.0**	**95.0**	**nd**	**nd**
Narra 2	**10.3**	**158.3**	**0.9**	**3.1**	**0.1**	**95.0**	**bdl**	**43.8**
Narra 3	**10.3**	**157.4**	**0.9**	**2.4**	**0.1**	**80.0**	**bdl**	**44.2**
Mangatarem	**11.1**	**28.0**	**0.5**	**18.6**	**0.2**	**17.4**	**5.1**	**nd**
Cement leachate ALL-MR f63	11.03	42	7.3	20	<0.5	52	12	49.2
3Cement leachate OL-SR f63	10.05	4400	150	4300	0.56	13000	247	32.1

Table II: Geochemical (XRD) analysis of the Mangatarem bentonite and MX-80 bentonite

Parameter (%)	Untreated Mangatarem	Average treated MX-80 (Wyoming) [29]
SiO_2	50.3 - 69.8	58.5
TiO_2	0.40 – 0.64	nd
Fe_2O_3	2.8 – 6.8	3.8
Al_2O_3	12.1 - 13.2	19.1
CaO	0.1 – 5.8	1.4
MgO	1.9 - 3.1	2.4
K_2O	0.4 - 1.2	0.5
Na_2O	1.2 - 1.3	2.1
CEC	55.2 – 102.5 meq/100 g	72.0 meq/100 g

DISCLAIMER

At the date this paper was written, web sites or links referenced herein were deemed to be useful supplementary material. Neither the authors nor the Materials Research Society warrants or assumes liability for the content or availability of web sites referenced in this paper.

REFERENCES

1. W.R.Alexander and I.G.McKinley (1999) The chemical basis of near-field containment in the Swiss high-level radioactive waste disposal concept. pp 47-69 *in* Chemical containment of wastes in the geosphere (eds. R.Metcalfe and C.A.Rochelle), Geol.Soc.Spec.Publ. No. 157, Geol Soc, London, UK.
2. Nagra (2002) Project Opalinus Clay – Safety Report. Nagra Technical Report NTB 02-05, Nagra, Wettingen, Switzerland.
3. A.Haworth, S.M.Sharland, P.W.Tasker and C.J.Tweed (1987) Evolution of the groundwater chemistry around a nuclear waste repository. *Sci. Basis Nucl. Waste Manag.*, **XI**, 425-434
4. W.R.Alexander and F.B.Neall (2007) Assessment of potential perturbations to Posiva's SF repository at Olkiluoto caused by construction and operation of the ONKALO facility. Posiva Working Report 2007-35, Posiva, Olkiluoto, Finland.
5. NUMO (2007) The NUMO structured approach to HLW disposal in Japan. NUMO Report TR-07-02, NUMO, Tokyo, Japan.
6. JAEA (2007) Second progress report on R&D for TRU waste disposal in Japan. JAEA Review 2007-010/FEPC TRU-TR2-2007-01, JAEA, Tokai, Japan.
7. H.Umeki (2007) Holistic assessment to put mobile radionuclides in perspective. Proc. MOFAP'07, January 16-19 2007, La Baule, France. (*in press*).
8. R.Metcalfe and C.Walker (2004) Proceedings of the International Workshop on Bentonite-Cement Interaction in Repository Environments 14–16 April 2004, Tokyo, Japan. NUMO Tech. Rep. NUMO-TR-04-05, NUMO, Tokyo, Japan.
9. W.R.Alexander, I.D. Clark, P.Degnan, M.Elie, G.Kamei, H.Khoury, U.Mäder, A.E.Milodowski, K.Pedersen, A.F.Pitty, E.Salameh, J.A.T.Smellie, I.Techer and L.Trotignon (2007) Cementitious natural analogues: safety assessment implications of the unique systems in Jordan. *Phys. Chem. Earth* (*submitted*).
10. M.N.Gray and B.S.Shenton (1998) For better concrete, take out some of the cement. *In* Proceedings of the 6th ACI/CANMET symposium on the durability of concrete, Bangkok, Thailand, 31st May – 5th June, 1998. CANMET Technical Report, CANMET, Ottawa, Canada.
11. W.M.Miller, W.R.Alexander, N.A.Chapman, I.G.McKinley, and J.A.T.Smellie (2000) Geological disposal of radioactive wastes and natural analogues. Waste management series, vol. 2, Pergamon, Amsterdam, The Netherlands.
12. www.natural-analogues.com (NAWG, the Natural Analogue Working Group, web site)
13. W.R.Alexander, A.Gautschi and P.Zuidema (1998) Thorough testing of performance assessment models: the necessary integration of *in situ* experiments, natural analogues and laboratory work. *Sci. Basis Nucl. Waste Manag.* **XXI**, 1013-1014
14. A.F.Pitty (*ed*), (2007) A natural analogue study of cement buffered, hyperalkaline groundwaters and their interaction with a repository host rock IV: an examination of the

Khushaym Matruk (central Jordan) and Maqarin (northern Jordan) sites. ANDRA Technical Report, ANDRA, Paris, France (*in press*).
15. I.G.McKinley, A.H.Bath, U.Berner, M.Cave and C.Neal (1988) Results of the Oman analogue study. *Radiochim Acta*, **44/45**, 311-316
16. S.M.Pate, I.G.McKinley and W.R.Alexander (1994) Use of natural analogue test cases to evaluate a new performance assessment TDB. CEC Report EUR15176EN, Brussels, Belgium.
17. I.Barnes and J.R.O'Neill (1969) The relationship between fluids in some fresh alpine-type ultramafics and possible modern serpentinisation, western United States. *Geol. Soc. Amer. Bull.*, **80**, 1947-1960
18. J.A.Sader, M.I.Leybourne, M.B.McClenaghan and S.M.Hamilton (2007) Low-temperature serpentinisation processes and kimberlite groundwater signatures in the Kirkland Lake and Lake Timiskiming kimberlite fields, Ontario, Canada: implications for diamond exploration. *Geochem.*, **7**, 3-21
19. T.A.Abrajano, N.C.Sturchio, J.K.Bohlke, G.L.Lyon, R.J.Poredar and C.M.Stevens (1988) Methane-hydrogen gas seeps, Zambales Ophiolite, Philippines: deep or shallow origin? *Chem. Geol.*, **71**, 211-222
20. C.Neal and G.Stanger (1983) Hydrogen generation from mantle source rocks in Oman. *Earth Planet. Sci. Lett.*, **66**, 315–320.
21. K.Wright and C.R.A.Catlow (1996) Calculations on the energetics of water dissolution in wadsleyite. *Phys. Chem. Mins* **23**, 38-41
22. H. Hosgörmez (2007) Origin of the natural gas seep of Çirali (Chimera), Turkey: Site of the first Olympic fire. *J. Asian Earth Sci.* **30**, 131-141
23. http://www.philzeolite.com/htms/saileDeposit.htm#body (the web site of the Saile bentonite/zeolite producing company)
24. J.M.West, P.Coombs, S.J.Gardner and C.A.Rochelle (1995) The microbiology of the Maqarin site, Jordan. A natural analogue for cementitious radioactive waste repositories. *Sci. Basis Nucl. Waste Manag.* **XVIII**, pp 181-189.
25. I.G.McKinley, W.R.Alexander, C.A.Arcilla, H.Kawamura and Y.Takahashi (2007) IPHAP: a new natural analogue of bentonite alteration by cement leachates. Proc ISRSM conference, Daejeon, ROK. KHMP report, Daejeon, ROK.
26. T.Kawata, H.Umeki and I.G.McKinley (2006) Knowledge Management: The Emperor's New Clothes? Proc. 11th International High-Level Radioactive Waste Management Conference 2006, Las Vegas, Nevada, April 30-May 4, pp.1236-1243
27. C.Neal and P.Shand (2002) Spring and surface water quality of the Cyprus Ophiolites. *Hydrol. Earth System Sci.*, **6**, 797-817
28. U.Vuorinen, J.Lehikoinen, H.Imoto, T.Yamamoto and M.Cruz Alonso (2005) Injection Grout for Deep Repositories, Subproject 1: Low-pH cementitious Grout for Larger Fractures, Leach Testing of Grout Mixes and Evaluation of the Long-Term Safety. Posiva Working Report 2004-46, Posiva, Olkiluoto, Finland.
29. O.Karnland, P.Sellin and S.Olsson (2006) Mineralogy and some physical properties of the San José bentonite – a natural analogue to buffer material exposed to saline groundwater. *Mat. Res. Soc. Symp. Proc*, **807**, 1-6

Mater. Res. Soc. Symp. Proc. Vol. 1107 © 2008 Materials Research Society

Migration Behaviour of Ferrous Ion in Compacted Bentonite Under Reducing Conditions Controlled With Potentiostat

Kazuya Idemitsu, Syeda Afsarun Nessa, Shigeru Yamazaki, Hirotomo Ikeuchi, Yaohiro Inagaki, Tatsumi Arima,
Dept. of Applied Quantum Physics and Nuclear Engineering, Kyushu Univ., Fukuoka, JAPAN

ABSTRACT

Carbon steel overpack is corroded by consuming oxygen introduced by repository construction after closure of the repository and then maintains the reducing environment in the vicinity of the repository. The migration of iron corrosion products through the buffer material will affect the migration of redox-sensitive radionuclides. Therefore, it is important to study the migration of iron corrosion products through the buffer material because it may affect the corrosion rate of overpack, and migration of redox-sensitive radionuclides. Electromigration experiments have been conducted with the source of iron ions supplied by anode corrosion of the iron coupon in compacted bentonite. The carbon steel coupon was connected as the working electrode to the potentiostat and was held at a constant supplied potential between - 650 to +300 mV vs. Ag/AgCl electrode for up to 168 hours. The amount of iron penetrated into a bentonite specimen was in good agreement with the calculated value from the corrosion current under the assumption that iron is dissolved as ferrous ions. A model using dispersion and electromigration could explain the measured iron profiles in the bentonite specimens. The fitted value of electromigration velocity depended on the potential supplied. On the other hand the fitted value of the dispersion coefficient did not depend on the potential supplied but a constant. This constant dispersion coefficient could be due to the much larger diffusion coefficient of ferrous ion in bentonite compared with the effect of mechanical dispersion. The experimental configurations used in this study are applicable to the examination of the migration behaviour of cations with the source of iron ions under a reducing condition controlled with a potentiostat.

INTRODUCTION

Carbon steel is one of the candidate overpack materials for high-level waste disposal [1]. Carbon steel overpack is corroded by consuming oxygen introduced by repository construction after closure of the repository. Corrosion products diffuse into buffer materials and then maintain the reducing environment in the vicinity of the repository [2]. The reducing condition is expected to retard the migration of redox-sensitive radionuclides. For example, a rare study on plutonium diffusion in compacted bentonite is available. It has been reported that no movement could be measured for plutonium diffusion in a concrete-bentonite system under an oxidizing condition in an experimental period as long as 5 years [3]. Our research group succeeded in migrating plutonium about 1mm in bentonite within a week under a reducing condition by using the electromigration technique [4]. The result shows that plutonium has a higher diffusion coefficient under the reducing condition in bentonite with iron corrosion products than under oxidizing conditions.

Our research group has developed and conducted electromigration experiments with the source of iron ions supplied by anode corrosion of iron coupons in compacted bentonite [4, 5, 6].

It is expected that iron ions could migrate as ferrous ions through the interlayer of montmorillonite replacing exchangeable sodium ions in the interlayer. This technique allows the maintaining and control of a reducing experimental condition and acceleration of the cation migration. However, the migration behavior of iron was complex and difficult to explain. This study aims to accumulate knowledge about the migration behavior of iron in compacted bentonite under the reducing conditions by this electromigration technique.

EXPERIMENTAL

Materials

A typical Japanese sodium bentonite, Kunipia-F, was used in this experiment. Kunipia-F contains approximately 95wt% of montmorillonite. The chemical composition of Kunipia-F is shown elsewhere [7]. Bentonite powder was compacted into cylinders with a diameter of 10 mm and a height of 10 mm with a dry density of around 1.4 Mg/m^3 (porosity 0.46). Each compacted bentonite was inserted in an acrylic resin column and saturated with water including 0.01 M of NaCl for one month.

Carbon steel, SM41B, was used in this study. Carbon steel was cut into cylindrical coupons with a diameter of 18 mm and thickness of 3 mm. The surface of the coupon was wet-polished with #1500 emery paper.

Electromigration tests

A carbon steel coupon was assembled with bentonite saturated by water including 0.01M of NaCl into an apparatus for electromigration as shown in Fig. 1. There was a reference electrode of Ag/AgCl and a counter electrode of platinum foil in the upper part of the apparatus with 0.01 M of NaCl solution. The carbon steel coupon was connected to a potentiostat as a work electrode and was supplied electrical potential in a range of - 650 to +300 mV vs. Ag/AgCl electrode at 25°C for up to 168 h. After supplying electrical potential, the bentonite specimen was pushed out from the column and was sliced in steps of 0.3 to 2mm. Each slice was submerged in 1N HCl solution to extract iron, and the liquid phase was separated by the centrifugal method. Then the supernatant was taken to be measured by Atomic Absorption Spectrometry.

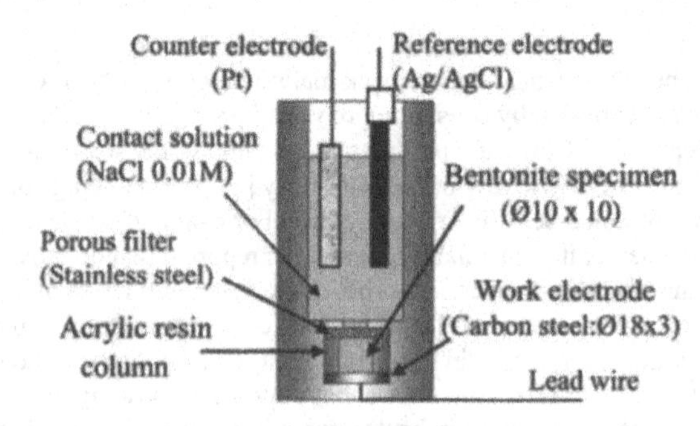

Figure 1. Experimental apparatus for electromigration.

RESULTS AND DISCUSSION

Profiles of iron in bentonite specimens

Typical profiles of iron infiltrated in bentonite specimens up to 168 h are shown in Fig. 2. The higher the potential supplied to a work electrode, the deeper the iron penetrated into the bentonite specimen. This means that cations, here iron ions, were accelerated toward the counter electrode.

Currents between carbon steel and counter electrode

The currents between carbon steel and the counter electrode, corrosion currents, are plotted as a function of time in Fig. 3. In most cases currents showed a peak at the beginning, then they dropped to a constant current. There is a tendency that the higher the potential supplied to a work electrode, the larger the constant currents.

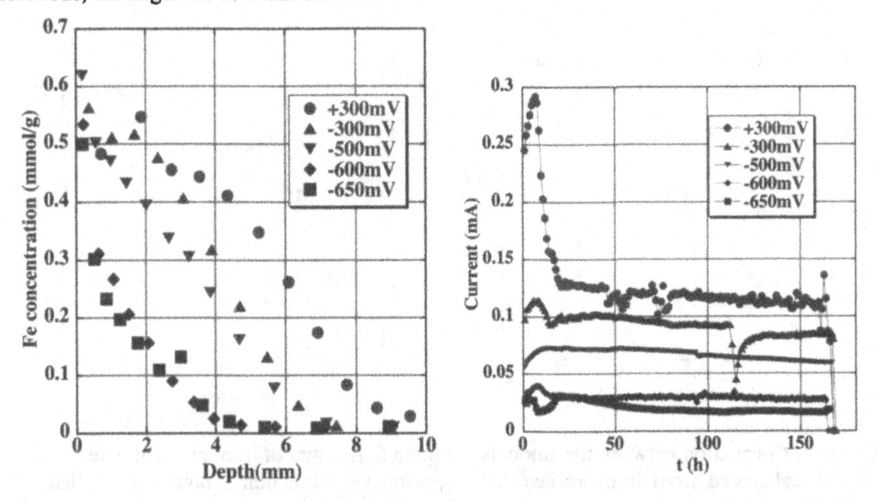

Figure 2. Iron profiles in compacted bentonites by electromigration up to 168 h.

Figure 3. Corrosion currents as a function of time up to 168 h.

Amount of iron in bentonite specimen

Both penetration depth and corrosion current depended on the potential supplied to the work electrode. There should be correlation between the penetration depth and corrosion current. Then we compared the amount of iron in the bentonite specimen calculated from an iron profile and the corresponding corrosion current. The amount of iron from an iron profile was calculated by summing the products of iron concentration and weight of each slice. On the other hand the amount of iron from the corrosion current was calculated as follows: (i) calculate the quantity of electricity by integrating the corrosion current with respect to time, (ii) convert the quantity of electricity to the amount of iron with the assumption that iron will corrode as the following reaction.

$$Fe \rightarrow Fe^{2+} + 2e^- \tag{1}$$

The amounts of iron calculated from corrosion currents are plotted as a function of the amount of iron calculated from the iron profiles in Fig. 4. Both values are in good agreement, so the assumption of the dissolution reaction (1) could be reasonable. Therefore, iron corroded to ferrous ions at the interface between carbon steel and the bentonite specimen.

The amounts of iron calculated from the iron profiles are also plotted as a function of the potential supplied to the work electrodes, carbon steels, in Fig. 5. The amount of iron increased with the potential; however it seemed to be saturated at higher potential. Little dependency of potential supplied to the work electrode on the iron profile was reported above +300 mV [6].

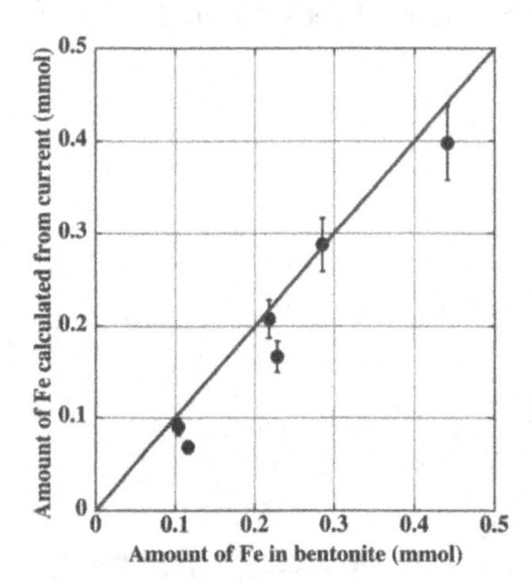

Figure 4. Correlation between the amounts of iron calculated from iron profiles and corrosion currents.

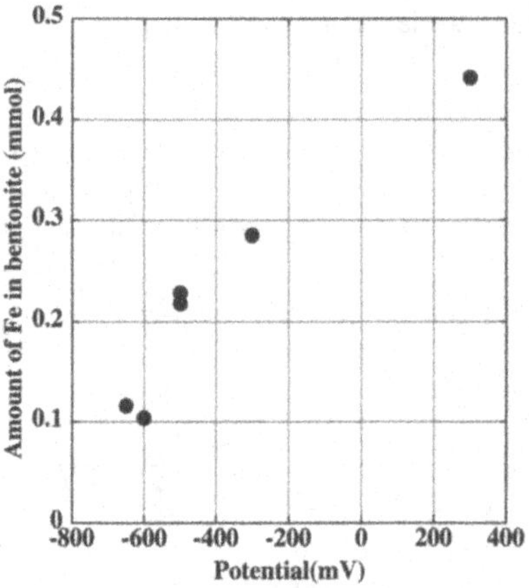

Figure 5. Amount of iron into bentonite specimen as a function of potential supplied to work electrode.

Model of electromigration

The movement of ions under the influence of the electric potential gradient, hence electromigration, can be described with the dispersion-convection equation,

$$\frac{\partial C}{\partial t} = D_a \frac{\partial^2 C}{\partial x^2} - V_a \frac{\partial C}{\partial x} \tag{2}$$

where D_a is an apparent dispersion coefficient, V_a is an apparent convection velocity including mostly electromigration of iron and negligible electro-osmotic flow of water. Just before solving this equation with the initial and boundary conditions in this experimental configuration, it is better to describe the solution with a plane source. Therefore, the initial condition is,

$$C = 0 \text{ at } x > 0, C = M\,\delta(0) \text{ at } x = 0, \tag{3}$$

where M is the total amount of migrating ions, and $\delta(0)$ is delta function. Because ions cannot migrate to the region $x < 0$, the boundary conditions are

$$\frac{\partial C}{\partial x} = 0 \text{ at } x = 0 \text{ and } C = 0 \text{ at } x = \infty. \tag{4}$$

An analytical solution for Eq.(2) is derived by introducing the initial condition and boundary conditions [8],

$$C(x,t) = \frac{M}{2\sqrt{\pi D_a t}}\left[\exp\left\{-\frac{(x-V_a t)^2}{4D_a t}\right\} + \exp\left\{-\frac{(x+V_a t)^2}{4D_a t}\right\}\right]. \tag{5}$$

To solve Eq.(2) with the boundary conditions in this experimental configuration, we considered successive spikes of ion in each time step, $M(t)$, which is a function of the corrosion current. Then we can obtain the solution by integrating Eq.(5) with respect to time known as Duhamel's principle [9]:

$$C(x,t) = \int_0^t \frac{M(t-\tau)}{2\sqrt{\pi D_a \tau}}\left[\exp\left\{-\frac{(x-V_a \tau)^2}{4D_a \tau}\right\} + \exp\left\{-\frac{(x+V_a \tau)^2}{4D_a \tau}\right\}\right] d\tau \tag{6}$$

where τ is the migration period of ions introduced into the specimen at 't - τ' migrate as shown in Fig. 6. We emphasize that this solution would be correct when the dispersion coefficient is independent of concentration.

Fitted curves are shown in Fig. 7.

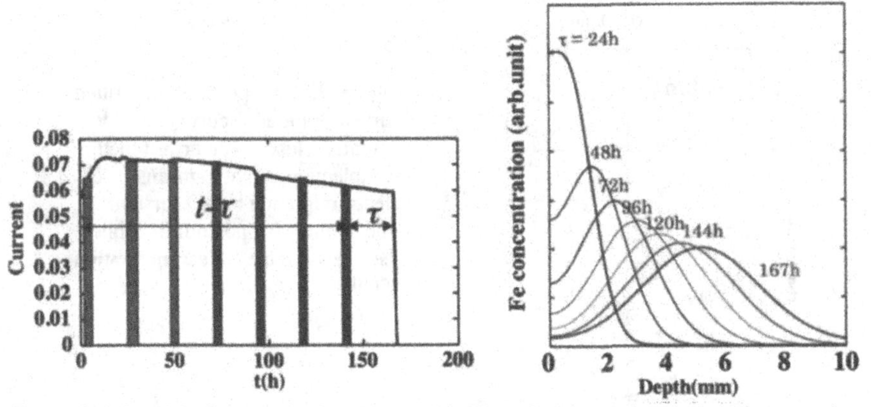

Figure 6. Schematic figures of migration model. The solution can be obtained by summing up profiles in the right graph.

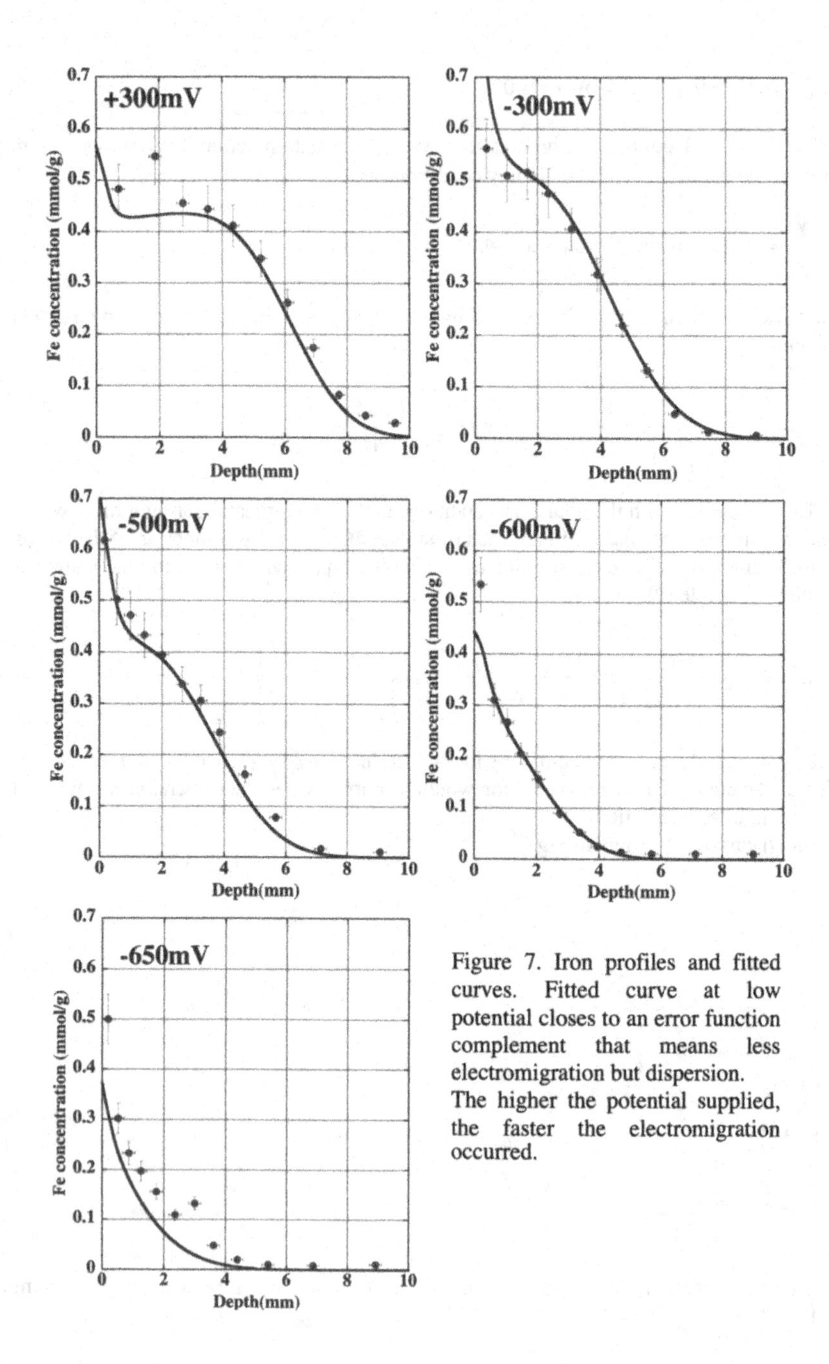

Figure 7. Iron profiles and fitted curves. Fitted curve at low potential closes to an error function complement that means less electromigration but dispersion. The higher the potential supplied, the faster the electromigration occurred.

There are only two fitting parameters such as apparent convection velocity and apparent dispersion coefficient. Fitting into measured iron profiles gives the apparent convection velocity and the apparent dispersion coefficient. Obtained values shown in Table I explain the profiles well as shown in Fig. 7.

Table I. Summary of results for iron.

Potential (mV)	Dry density (Mg/m^3)	V_a (m/s) by curve fitting	D_a (m^2/s) by curve fitting
+300	1.33	$(9.6\pm1.0)*10^{-9}$	$(2.0\pm1.0)*10^{-12}$
-300	1.45	$(7.2\pm0.5)*10^{-9}$	$(2.6\pm1.0)*10^{-12}$
-500	1.38	$(6.5\pm0.5)*10^{-9}$	$(2.1\pm1.0)*10^{-12}$
-600	1.51	$(3.8\pm0.5)*10^{-9}$	$(1.6\pm1.0)*10^{-12}$
-650	1.38	$(2.0\pm0.5)*10^{-9}$	$(2.4\pm1.0)*10^{-12}$

Obtained values are plotted as a function of potential supplied in Fig. 8. The apparent convection velocity that is mostly electromigration velocity depended on the potential supplied. This is reasonable because the electrical gradient as a driving force of cation migration increases with the potential. On the other hand the apparent dispersion coefficient was independent of the potential.

The relationship between an apparent molecular diffusion coefficient, $D_{a,m}$ and an apparent dispersion coefficient, D_a is described by the following equation:

$$D_a = D_{a,m} + \alpha V_a \qquad (7)$$

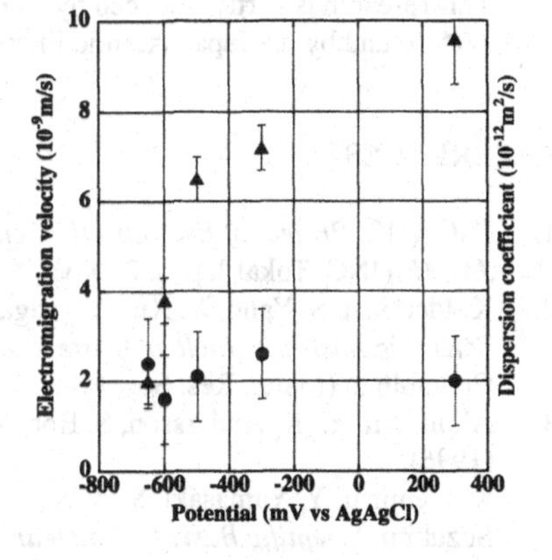

Figure 8. Fitted parameters as a function of the potential supplied.

where α is the dispersion length. It is necessary to know the values of the dispersion length to calculate apparent dispersion coefficients; however, they are generally small, e.g. for sodium ions, the dispersion length is 10^{-5} m at a dry density of 1.0 Mg/m^3 [10]. In addition to that, obtained apparent convection velocities are very small as shown in Table I. Then the second term of Eq.(7) is less than 10^{-13} m^2/s. The second term of Eq.(7), mechanical dispersion, was much smaller than the apparent dispersion coefficients obtained as shown in Table I. Therefore, the apparent dispersion coefficients obtained were mostly apparent diffusion coefficient of ferrous ion in bentonite, ca. 10^{-12} m^2/s [2, 6].

In this experiment the apparent dispersion coefficients obtained were mostly the apparent diffusion coefficients due to the large apparent diffusion coefficient of ferrous ion in bentonite. The second term of Eq.(7), however, might be comparable to the apparent dispersion coefficients for another ion that has small apparent diffusion coefficient in bentonite. When the apparent dispersion coefficients obtained depend on the potential supplied, we can estimate the apparent diffusion coefficient and dispersion length by using Eq.(7).

CONCLUSION

The following facts were obtained from the result of this electromigration experiment.
(1) This new technique allows the easy creation of a reducing condition, to accelerate cation migration and to introduce ferrous ions into bentonite.
(2) The constant dispersion coefficient obtained in this study could be due to the much larger diffusion coefficient of ferrous ions in bentonite compared with the effect of mechanical dispersion.
(3) We provide the method to obtain apparent diffusion coefficient and dispersion length when apparent dispersion coefficient depends on the potential supplied.

ACKNOWLEDGMENTS

This research is partly financed by a Grant-in-Aid for scientific research of Japan (contract B19360430) and by the Japan Atomic Energy Agency.

REFERENCES

1. JNC, H12: *Project of Establish the Scientific and Technical Basis for HLW Disposal in JAPAN* (JNC, Tokai Japan, 2000).
2. K. Idemitsu, S. Yano, X. Xia, Y. Inagaki, T. Arima, T. Mitsugashira, M. Hara, Y. Suzuki in *Scientific Basis for Nuclear Waste Management XXV*, edited by B.P. McGrail and G. A. Cragnolono (Mater. Res. Soc. Proc. **713**, Pittsburgh, PA, 2001) pp.113-120.
3. Albinsson, Y., K. Andersson, S. Börjesson, B. Allard, J. Contaminant Hydrology 12, 189 (1996).
4. K. Idemitsu, Y. Yamasaki, S. A. Nessa, Y. Inagaki, T. Arima, T. Mitsugashira, M. Hara, Y. Suzuki in *Scientific Basis for Nuclear Waste Management XXX*, edited by D.S. Dunn, C. Poinssot, B. Begg (Mater. Res. Soc. Proc. **985**, Pittsburgh, PA, 2007), NN11-7, pp.443-448.
5. K. Idemitsu, S. Yano, X. Xia, Y. Kikuchi, Y. Inagaki, T. Arima in *Scientific Basis for Nuclear Waste Management XXVI*, edited by R. J. Finch and D. B. Bullen (Mater. Res. Soc. Proc. **757**, Pittsburgh, PA, 2003) pp.657-664.
6. K. Idemitsu, X. Xia, Y. Kikuchi, Y. Inagaki, T. Arima in *Scientific Basis for Nuclear Waste Management XXVIII*, edited by John M. Hanchar, Simcha Stroes-Gascoyne, Lauren Browning (Mater. Res. Soc. Proc. **824**, Pittsburgh, PA, 2004) pp.491-496.
7. H. Sato, T. Ashida, Y. Kohara, M. Yui, and N. Sasaki. J. Nucl. Sci. Tech. **29**, 873 (1992).
8. K. Idemitsu, M. Yamamoto, Y. Yamasaki, Y. Inagaki and T. Arima in *Scientific Basis for Nuclear Waste Management XIX*, edited by Pierre Van Iseghem (Mater. Res. Soc. Proc. **932**, Pittsburgh, PA, 2006) pp.943-950.
9. Stanley J. Farlow. Partial Differential Equations for Scientists and Engineers, (John Wiley & Sons, Inc., New York, 1982), Part 2 section 14.
10. Higashihara, T., K. Kinoshita, S. Sato, and T. Kozaki. Appl. Clay Sci. **26**, 91 (2004).

Engineered Barrier Systems II: Containers

Mater. Res. Soc. Symp. Proc. Vol. 1107 © 2008 Materials Research Society

Effect of Weak Acid Additions on the General and Localized Corrosion Susceptibility of Alloy 22 in Chloride Solutions

Ricardo M. Carranza, C. Mabel Giordano, Martín A. Rodríguez and Raul B. Rebak[1]
Comisión Nacional de Energía Atómica, Av. Gral. Paz 1499
San Martín, 1650 Buenos Aires, Argentina
[1]Lawrence Livermore National Laboratory, PO Box 808 L-631, 7000 East Ave.
Livermore, CA, 94550, U.S.A.

ABSTRACT

Electrochemical studies such as cyclic potentiodynamic polarization (CPP) and electrochemical impedance spectroscopy (EIS) were performed to determine the corrosion behavior of Alloy 22 (N06022) in 1M NaCl solutions at various pH values from acidic to neutral at 90ºC. All the tested material was wrought Mill Annealed (MA). Tests were also performed in NaCl solutions containing weak organic acids such as oxalic, acetic, citric and picric acids.

Results show that the CR of Alloy 22 was significantly higher in solutions containing oxalic acid than in solutions of pure NaCl at the same pH. Citric and Picric acids showed a slightly higher CR, and Acetic acid maintained the CR of pure chloride solutions at the same pH. Organic acids revealed to be weak inhibitors for crevice corrosion. Higher concentration ratios, compared to nitrate ions, were needed to completely inhibit crevice corrosion in chloride solutions.

Results are discussed considering acid dissociation constants, buffer capacity and complex formation constants of the different weak acids.

INTRODUCTION

Alloy 22 (N06022) contains by weight approximately 22% chromium (Cr), 13% molybdenum (Mo), 3% tungsten (W) and approximately 3% iron (Fe). Alloy 22 was commercially designed to resist the most aggressive industrial applications, offering a low general CR (CR) both under oxidizing and reducing conditions.[1] Under oxidizing and acidic conditions Cr exerts its beneficial effect in the alloy. Chromium oxide is protective under acid conditions at moderate anodic potentials. However, if the environment is super oxidizing, Cr_2O_3 may dissolve into the electrolyte to form soluble chromate compounds. The super oxidizing conditions are not commonly found in the industry. They only exist in rare applications such as fuming nitric acid, oleum, etc. The environmental conditions or an underground repository will never evolve to become super oxidizing. Under reducing conditions the most beneficial alloying elements are Mo and W, which offer a low current for hydrogen discharge.[2] Due to its balanced content in Cr, Mo and W, Alloy 22 is used in hot chloride environments where austenitic stainless steels may fail by pitting corrosion and stress corrosion cracking (SCC).[1,2] Alloy 22 critical crevice temperatures (CCT) depend on the environment where the tests are performed. In the industry the CCT is generally measured using the ASTM G 48[3] technique which involves immersing creviced coupons in $FeCl_3$ solutions. Alloy 22 has some of the highest CCT among the nickel alloys.[4] However in other electrolytes the CCT will be different. For example, in 5 M $CaCl_2$ solution, Alloy 22 will suffer crevice corrosion (CC) under anodic polarization even at 30 ºC.[5] In general Alloy 22 is not highly susceptible to CC at temperatures below 60C.[6]

Alloy 22 is the material selected for the fabrication of the outer shell of the nuclear waste containers for the Yucca Mountain site.[7,8] Several papers have been published recently describing the general and localized corrosion behavior of Alloy 22 regarding its application for the nuclear waste containers.[9] It is also known that the addition of NO_3^- and other oxyanions to a Cl^--containing environment, decreases or eliminates the susceptibility of Alloy 22 to CC.[10,11] It has been recently reported that F^- may also act as an inhibitor to CC of Alloy 22.[12] Little is known on its corrosion behavior in organic acids.[13]

Oxalic acid ($H_2O_4C_2$) is one of the most aggressive alkane acids and is slightly oxidizing. Acetic acid ($H_4O_2C_2$) is a weak monocarboxylic acid and is classified as a weak acid, because it does not completely dissociate into its component ions when dissolved in aqueous solutions. At a concentration of 0.1 M, only about 1% of the molecules are ionized. Citric acid ($C_6H_8O_7$) is a weak tricarboxylic acid that shares the properties of other carboxylic acids. When heated above 175 °C, it decomposes through the loss of carbon dioxide and water. Citric acid is used in the biotechnology industry to passivate high purity process piping. Picric acid ($C_6H_3O_7N_3$) is an aromatic nitro compound ($ArNO_2$) which can be reduced to nitroso compounds ($ArNO$) or to hydroylamines ($ArNH_2$) by means of different reaction mechanisms.[14] The principal laboratory use of picric acid is in microscopy, where it is used as a reagent for staining samples.

Table I shows the dissociation K_a and metal complex formation K_{cpx} constants for the organic acids. It can be seen that oxalic acid is the strongest acid followed by picric, citric and acetic in decreasing acidity. The complexing character is more difficult to evaluate for it depends on the metal cation being complexed, the ionic force of the solution ant the temperature.

The objective of the current study was to use electrochemical methods and parameters to systematically assess the corrosion behavior of Alloy 22 (N06022) in sodium chloride solutions with additions of different organic acids as compared to the behavior in pure NaCl. General corrosion behavior was assessed by corrosion rate (CR) measurements using the Electrochemical Impedance Spectroscopy (EIS) technique, and localized corrosion behavior was evaluated using Cyclic Potentiodynamic Polarization (CPP) curves.

EXPERIMENTAL DETAILS

Specimens of Alloy 22 were prepared from wrought mill annealed plate (MA) stock. The chemical composition of the alloy in weight percent was 59.20% Ni, 20.62% Cr, 13.91% Mo, 2.68% W, 2.80% Fe, 0.01% Co, 0.14% Mn, 0.002% C, and 0.0001% S. Two types of specimens were used: (a) prismatic specimens: a variation of the ASTM G 5[15] specimen, and (b) prism crevice assemblies (PCA),[16] fabricated based on ASTM G 48[15] which contained 24 artificially creviced spots formed by a ceramic washer (crevice former) wrapped with a PTFE tape. The applied torque was 7.92 N-m. The tested surface areas were approximately 10 cm^2 for prismatic specimens and 14 cm^2 for PCA specimens. The specimens had a finished grinding of abrasive paper N° 600, and were degreased in acetone and washed in distilled water. Polishing was performed 1 hour prior to testing. Electrochemical measurements were conducted in a three-electrode, borosilicate glass cell (ASTM G 5)[15]. A water-cooled condenser combined with a water trap was used to avoid evaporation and the ingress of air. Solution temperature was controlled by immersing the cell in a thermostatically controlled water bath. The cell was equipped with both a water cooled Luggin and a saturated calomel reference electrode (SCE) which has a potential of 0.242 V more positive than the standard hydrogen electrode (SHE). A large area platinum foil was used as counter electrode. The electrochemical tests were carried out in 1M,

0.1M and 0.01M NaCl solutions, at different pH values between 1 and 6, with and without the addition of organic acids with concentrations from 0.001M to 2M. Small amounts of HCl were added in order to adjust solution pH only for pure NaCl solutions. Solution pH was systematically measured before and after the tests. The test temperature was 90.0 ± 0.1°C. Solutions were prepared with analytical grade chemicals and 18.2 MΩ resistivity water. Solutions for the cyclic potentiodynamic polarization curves were deaerated before (2 h) and during the experiments with high purity nitrogen. Solutions for EIS measurements were naturally aerated.

Table I. Dissociation [17] and metal complex formation [18,19] constants for organic acids. $M + n L \Leftrightarrow ML_n$, M: metal ion, L: organic ligand, ML_n: metal complex. Parentheses: T °C (25°C when not indicated) and ionic force of the solutions.

DISSOCIATION CONSTANTS		Acetic Acid Ethanoic CH_3CO_2H	Oxalic Acid $C_2H_2O_4$	Citric Acid $C_6H_8O_7$	Picric Acid $C_6H_3O_7N_3$
H^+	log K_{a1}	-4.757	-4.266	-6.396	-0.33
	log K_{a2}		-1.252	-4.761	
	log K_{a3}			-3.128	
METAL COMPLEX FORMATION CONSTANTS K_{cpx}					
Cr^{2+}	log(ML/M.L)	1.25 (0.3)	3.85 (0.1)		1.05 (18-25°C)
		1.80 (0)			
	log(ML_2/M.L^2)	2.15 (0.3)	6.81 (0.1)		
		2.92 (0)			
Cr^{3+}	log(ML/M.L)	4.63 (0.3)			
	log(ML_2/M.L^2)	7.08 (0.3)			
	log(ML_3/M.L^3)	9.6 (0.3)			3.2 (18-25°C)
Co^{2+}	log(ML/M.L)	1.10 (0.16)	3.84 (0.1)	5.00 (20°, 0.1)	
		0.71 (30°, 0.4)	3.25 (1.0)	4.83 (0.16)	
		0.81 (1.0)	4.72 (0)		
		1.40 (0)			
	log(ML_2/M.L^2)		5.60 (1.0)		2.85 (18-25°C)
			7.0 (0)		
	log(MHL/M.HL)		1.61 (0.1)	3.02 (20°, 0.1)	
				3.19 (0.16)	
	log($M(HL)_2$/M.$(HL)^2$)		2.89 (0.1)		
	log(M_2HL/M.H_2L)			1.25 (20°, 0.1)	
Ni^{2+}	log(ML/M.L)	0.74 (0.5)	5.16 (0)	5.40 (20°, 0.1)	2.89 (18-25°C)
		0.83-0.1 (1.0)		5.11 (0.16)	
		1.43 (0)			
	log(MHL/M.HL)			3.30 (20°, 0.1)	
				3.19 (0.16)	
	log(M_2HL/M.H_2L)			1.75 (20°, 0.1)	
Fe^{2+}	log(ML/M.L)	1.40 (0)	3.05 (1.0)	4.4 (20°, 0.1)	
	log(ML_2/M.L^2)		5.15 (1.0)		
	log(MHL/M.HL)			2.65 (20°, 0.1)	
Fe^{3+}	log(ML/M.L)	3.38 (20°, 0.1)	7.53 (0.5)	11.50 (20°, 0.1)	1.8 (18-25°C)
		3.2 (20°, 1.0)	0.1 (0.1)		
		3.23 (3.0)	7.59 (1.0)		
			7.54 (3.0)		
	log(ML_2/M.L^2)	6.5 (20°, 0.1)	13.64 (0.5)		
		6.22 (3.0)			
	log(ML_3/M.L^3)	8.3 (20°, 0.1)	18.49 (0.5)		3.1 (18-25°C)
	log(MHL/M.HL)		4.35 (0.5)		
	log($(ML)^2$/$M_2(H_{-1}L)_2$.H^2)			1.6 (20°, 0.1)	

CPP curves (ASTM G 61)[15] were performed using PCA specimens. The potential scan was started at the end potential of a 10 min. galvanostatic cathodic treatment of 50 $\mu A.cm^{-2}$ in the anodic direction at a scan rate of 0.167 mV/s. Usually CPP curves are started 150 mV more cathodic than E_{CORR}, but this treatment produces huge amounts of O_2 in the counter electrode for low pH solutions. The cathodic treatment produces negligible amounts of O_2 in the counter electrode independently of the pH of the solution. The scan direction was reversed when the current density reached 5 mA/cm^2 in the forward scan. Solutions tested were NaCl solutions (0.01M to 1M) with and without the addition of organic acids (0.001 to 2M). They were deaerated.

Electrochemical Impedance Spectroscopy (EIS) measurements were carried out at E_{CORR} in natural aerated solutions after 24h of immersion. The low frequency EIS polarization resistance (R_P) was used to calculate CR.[12]

RESULTS

Figure 1 shows the CR for pure 1M NaCl solutions from pH 1 to 6 and for 1M NaCl solutions with the addition of different concentrations of organic acids from 0.001M to 0.1M. For pure chloride solutions, CR was very low and independent of solution pH for pH values between 2 and 6 (CR ~ 0.1 $\mu m.yr^{-1}$). An increase of one order of magnitude was observed in the CR for pH 1 pure chloride solution (CR ~ 1 $\mu m.yr^{-1}$). Compared to pure chloride solutions of similar pH: a) CR was not modified by the addition of acetic acid for all the concentrations employed (0.1M, 0.01M and 0.001M), b)CR was slightly increased by the addition of the higher concentrations of citric acid (0.1 and 0.01M) and picric acid (0.05M* and 0.01M), and c)CR was significantly increased by the addition of oxalic acid for all the concentrations tested (0.1M, 0.01M and 0.001M) reaching the value of 24 $\mu m.yr^{-1}$ for the highest concentration.

Figure 2 shows CPP curves obtained using PCA specimens with crevice formers. All the curves presented a passive zone with low current densities from 1 to 10 $\mu A.cm^{-2}$. After the passive domain the increase of current densities with potential was due to CC initiation and/or to transpassive dissolution. Current hysteresis was always observed between the forward and the reverse scans. The repassivation potential chosen was the cross-over potential E_{CO}, i. e. the potential at which the reverse scan intersects the forward scan.

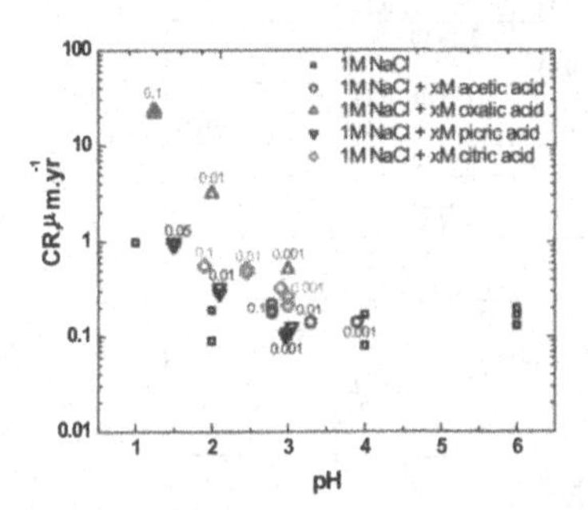

Figure 1. Corrosion rate, obtained from R_P EIS low frequency fitting parameter after 24h of immersion in naturally aerated solutions, as a function of pH and concentration of the added organic acid to 1M NaCl (figures next to symbols)

Abnormally high E_{CORR} values were obtained during the CPP curves in picric acid containing solutions (Figure 2). Consequently, the reverse scan intersected the forward scan in its cathodic domain. An extra cathodic reaction would produce misleading high E_{CO} values. Cyclic voltammetry (not shown) obtained at room temperature using a Pt electrode in deaerated 1M NaCl and 1M NaCl + 0.01M picric acid solutions both at pH 2, clearly demonstrate that the presence of

* Higher concentrations of picric acid are not possible in aqueous solution due to solubility limitations.

picric acid significantly increased the cathodic current at potentials more anodic than the corresponding ones to hydrogen evolution reaction at that pH. This current increase was attributed to picric acid reduction reaction.[14] Picric acid was then discarded for E_{CO} measurements. CC was observed in all the specimens in these solutions at the end of the CPP curves and the morphology of the attack can be observed in Figure 3. Typical shiny crystalline attack was observed under the crevice formers generally under all the 24 teeth of the crevice formers. Grain structure of the alloy was also lightly revealed under the crevice formers.

Figure 4 shows the values of E_{CO} as a function of solution pH and organic acid concentration for deaerated 1M NaCl with and without addition of organic acids. A large dispersion of the results was obtained. E_{CO} values between 0 mV_{SCE} and -200 mV_{SCE} were obtained independently of solution pH, composition and organic acid concentration. Neither inhibiting nor detrimental effect on localized corrosion could be attributed to organic acids in set A solutions.

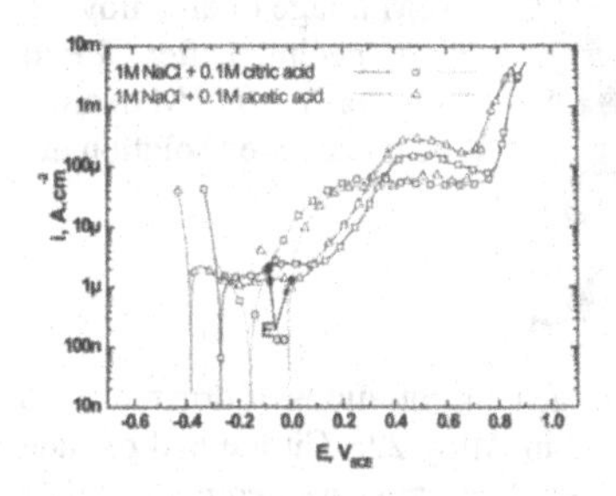

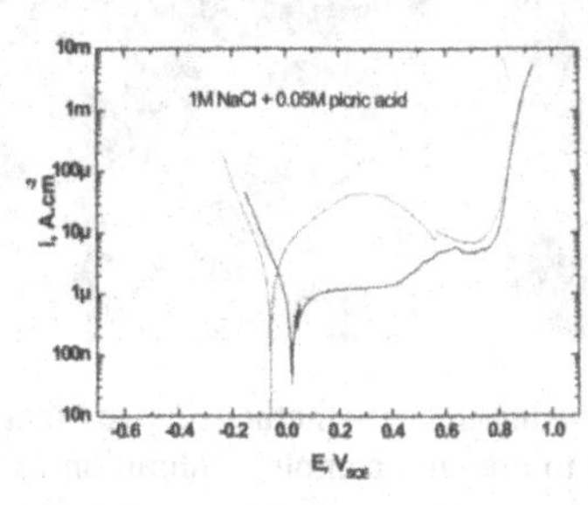

Figure 2. CPP curves for Alloy 22 at 90°C. Scan rate: 0.167 mV/s. Solid lines: forward scan. Dotted lines: reverse scan. E_{CO}: Cross-over repassivation potential.

Another set of CPP curves was run using higher ratios of organic acid concentration to chloride ion concentration $r_{org} = [Org]/[Cl^-]$ (set B solutions). Set A solutions shown in Figure 4 correspond to r_{org} 0.1 to 0.001 (r_{org} is zero for pure chloride solutions). In set B, r_{org} was increased to 1, 2, 5, 10 and 20. Solutions of set B were equimolar organic acid + sodium organic salt (weak acid + conjugate base) solutions in order to have the highest possible buffer capacity in the solutions at those concentrations (pH = pK_a). $[Cl^-]$ was decreased to 0.1 and 0.01M when needed due to precipitation. Figure 5 shows a CPP curve obtained for high r_{org} set B solutions. It can be seen that current hysteresis has disappeared in the tested specimens. No crevice attack was observed. Figure 6 shows E_{CO} as a function of r_{org} far all the solutions tested in the present work, including those already shown in Figure 4 (sets A and B). The average E_{CO} for pure chloride solutions was also included in the figure as horizontal dotted lines. A complete inhibition of CC was obtained for: a) acetic+acetate solutions with r_{org} higher than 10, b) citric+citrate solutions with r_{org} higher than 2, and c) oxalic+oxalate solutions with r_{org} higher than 2. Transpassivity potentials E_{20} were also included in Figure 6. E_{20} corresponds to the potential at which the current density attained 20 $\mu A.cm^2$ in the forward scan (Figure 5) when no CC was found.

DISCUSSION

As it is shown in Figure 1 only the addition of oxalic acid to chloride containing solutions significantly increased CR of Alloy 22 if compared to pure chloride solutions at the same pH. This fact implies that an additional effect on the CR, besides the low pH value, is produced by the presence of oxalic acid. From the K_{cpx} found in the bibliography (Table I) it can be seen that the highest values of K_{cpx} correspond to oxalic and citric acids. However, no K_{cpx} were found for complexes formed with Cr cations and citric acid while for picric and oxalic acids values were found only for Cr^{+2} cations complexes. The complexing power of citric and oxalic acids with Ni cations is similar, and it diminishes following: oxalic ~ citric > picric > acetic. No K_{cpx} values were found for these organic acids and Mo cations. If it is assumed that passive films on Alloy 22 are enriched in Cr and Mo[20], the hypothesis that a higher general CR is due to a higher complexing character of the solution components for Mo, Cr and Ni, can not be simply ruled out. Instead, additional quantitative information about K_{cpx} together with identification of the cations forming the passive film at the open circuit potential must be obtained.

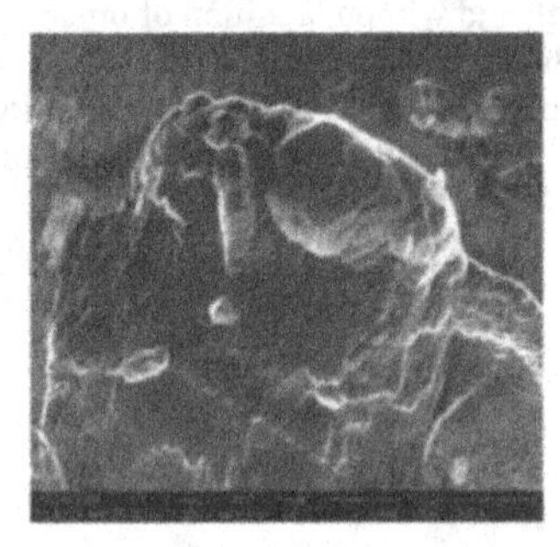

Figure 3. Photograph and SEM image of an Alloy 22 PCA specimen after CPP in 1M NaCl + 0.01M oxalic acid deaerated solution at 90°C.

Results obtained using the CPP method showed that large concentrations of organic acids were necessary in order to obtain a complete inhibition of CC in Alloy 22. Citrate and oxalate ions eliminated CC for concentration ratios r_{org} higher than 2 while acetates needed r_{org} equal or higher than 10. Much lower inhibitor to chloride ratios r were published for nitrate, carbonate, bicarbonate, and sulfate anions ($r = [Anion]/[Cl^-] \approx 0.1$) to completely inhibit chloride induced CC. [21,22,23] On the other hand, r higher than 2 are necessary to eliminate chloride induced CC in fluoride containing solutions with a chloride concentration of 0.01M while r higher than 5 are needed for chloride concentrations of 0.1M and 1M.[16] If the localized acidification model is valid for CC as was argued elsewhere,[24] one would expect that the weaker the organic acid, the stronger is the inhibition if only chemical reactions are involved in the elimination of free protons. One would expect for example that acetate ions ($\log K_a = -4.757$) would be a better inhibitor compared to fluoride ions ($\log K_a = -3.15$) in contradiction with which it was found in this work. It can then be argued that small anions as fluoride enter easily into the crevice and hence they are more able to inhibit CC. On the other hand, comparing the organic acids between them, the expected inhibiting strength according to K_a would be: acetic > citric > oxalic, which is not what was found in this work. The solutions used in the present work with r_{org} higher than 0.1 were prepared using equal molar concentrations of the organic acid and its corresponding sodium salt (conjugate base), in order to have the higher possible buffer capacity β. The higher the value of β the more difficult is to reduce the pH in the crevice to reach the low pH values needed to propagate it. The buffer capacity is independent of K_a and it is a function only of the total concentration of anions. It is then expected that different organic acids at $pH = pK_a$ (equal concentrations of the acid and its conjugate base) and the same total concentration have the same β and therefore, the same CC inhibiting effect. The observed differences could be attributed to differences in molecular size as it was previously invoked for fluoride ions. If we consider molecular sizes, the inhibiting power should vary following: acetic > oxalic > citric. Experimental results showed that larger citric and oxalic acids were better inhibitors than the smaller acetic acid, even though the total oxalate concentration was lower (lower β) than the total acetate concentration, under complete inhibition conditions (Figure 6).

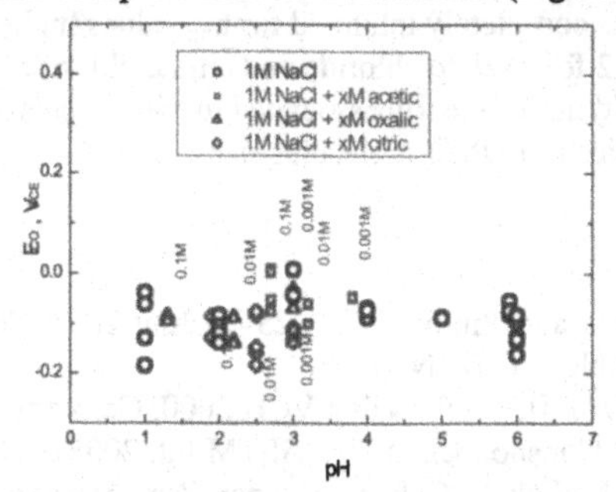

Figure 4. Cross-over Repasivation Potential as a function of solution pH. Deaerated solutions at 90 °C.

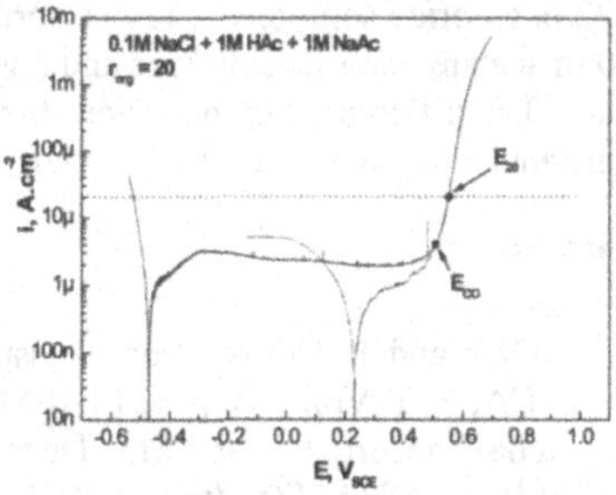

Figure 5. CPP curve for Alloy 22 at 90°C. Scan rate: 0.167 mV/s. Solid lines: forward scan. Dotted lines: reverse scan.

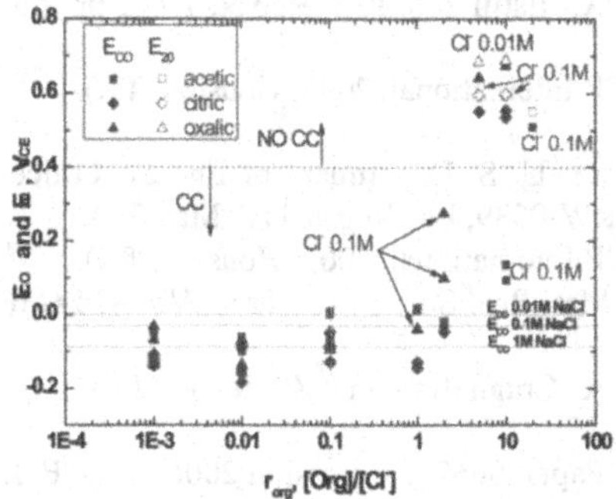

Figure 6. Cross-over Repassivation Potential and Transpassive Potential as a function of the ratio of organic inhibitor concentration to chloride ion concentration r_{org} = [Org]/[Cl⁻]. Full symbols: E_{CO}. Empty symbols: E_{20}. Dotted lines: mean E_{CO} for pure chloride solutions. Dashed line: E_{CO} limit for CC.

Consequently, differences in localized corrosion inhibition can not be easily attributed to differences in K_a, β or molecular sizes. A systematic study with other organic and inorganic compounds must be undertaken in order to elucidate the mechanism involved in CC inhibition in Alloy 22. This systematic study must include a modeling of the concentration profile inside the crevice for all the species present in solution. It must be taken into consideration that species concentration and ionic force inside de crevice are extremely high and commonly considerations about coupling effects between complexation, metal ion hydrolysis and acid dissociation in terms of the chemical equilibria are quite complex and different from those in diluted solutions.

CONCLUSIONS

Corrosion rates of Alloy 22 after 24h of immersion in 1M NaCl solutions with the addition of oxalic acid in concentrations ranging from 0.1M to 0.001M were higher than those obtained for pure 1M NaCl solutions at the same pH, while the addition of the same concentrations of acetic acid produced no CR increase. The addition of citric and picric acids produced a slight increase of the CR at the highest concentrations used. Additional tests are needed in order to determine with certainty whether or not an increase in general CR of Alloy 22 can be attributed to a high metal complexing strength of an organic ligand.

Large concentrations of organic anions were needed in order to eliminate crevice corrosion in Alloy 22 in chloride solutions. Crevice corrosion was completely inhibited for r_{org} values higher than 10 in acetate/chloride solutions and higher than 2 for oxalate/chloride and citrate/chloride solutions. The differences in inhibition strength could not be clearly associated to the dissociation constants of the acids, nor to their buffer capacities or to their molecular sizes.

REFERENCES

1. R.B. Rebak and P. Crook, Proc. Pressure Vessels and Piping Conf., 25-29 July 2004, San Diego, CA, PVP-Vol. 483, p. 131 (ASME, 2004: New York, NY).
2. R.B. Rebak in Corrosion and Env. Degradation, Vol. II, p. 69 (Wiley-VCH,2000, Germany).
3. Annual Book ASTM Standards, vol. 03.02 (West Conshohocken, PA: ASTM Int. 2005).
4. R. B. Rebak "Corrosion of Non-Ferrous Alloys. I. Nickel-, Cobalt-, Copper, Zirconium- and Titanium-Based Alloys" in Corrosion and Environmental Degradation, Volume II, p. 69 (Weinheim, Germany: Wiley-VCH, 2000).
5. K.J. Evans, A. Yilmaz, S.D. Day, L.L. Wong, J.C. Estill and, R.B. Rebak, JOM, pp. 56-61 (January 2005).
6. R.B. Rebak, Paper 05610, Corrosion/2005 (NACE International, 2005: Houston, TX).
7. G.M. Gordon, Corrosion, 58, 811 (2002).
8. Yucca Mountain Science and Engineering Report, U. S. Department of Energy, Office of Civilian Radioactive Waste Management, DOE/RW-0539, Las Vegas, NV, May 2001.
9. R.B. Rebak, Paper 05610, Corrosion/2005 (NACE International, 2005: Houston, TX).
10. D.S. Dunn, L. Yang, C. Wu, G.A. Cragnolino, Mat. Res. Soc. Symp. Proc. Vol 824 (MRS, 2004: Warrendale, PA)
11. D.S.Dunn, Y.-M. Pan, K. Chiang, L. Yang, G.A. Cragnolino and X. He, JOM, pp. 49-55 (January 2005).
12. R.M Carranza, M.A. Rodríguez, R.B. Rebak, Paper 06622, Corrosion/2006 (NACE Int., 2005: Houston, TX).
13 S.D. Day, M.T. Whalen, K.J. King, G.A. Hust, L.L. Wong, J.C. Estill, R.B. Rebak, Corrosion, 60, 804 (2004).
14. J. March, Adv. Organic Chemistry, Reactions, Mechanisms, and Strucutre. 3rd Ed., J. Wiley & Sons, p. 1103, N.Y., 1985.
15. Annual Book ASTM Standards, vol. 03.02 (West Conshohocken, PA: ASTM Int. 2005).
16 R.M. Carranza, M.A. Rodríguez, R.B. Rebak, Corrosion, 63, 480 (2007).
17. CRC Handbook of Chem. and Phys., David R. Lide, Ed., 85th Ed., CRC Press, NY, 2004.
18. Critical Stability Constants, R.M. Smith and A.E. Martell, eds., Plenum Press, NY, 1976.
19. Stability Constants, Part I: Organic Ligands, J. Bjerrum, G. Schwarzenbach, and L.G. Sillén, eds., The Chemical Society, London, 1957
20. A.C. Lloyd, J.J. Noël, S. McIntyre, D.W. Shoesmith, Electrochim. Acta 49, 3015 (2004).
21. D.S. Dunn, O. Pensado, and G.A. Cragnolino, Paper 05588, Corrosion/2005 (NACE Int., 2005: Houston, TX).
22. D.S. Dunn, Y.-M. Pan, L.Yang and G.A. Cragnolino, Corrosion, 61, 1078 (2005).
23. D.S. Dunn, Y.-M. Pan, L.Yang and G.A. Cragnolino, Corrosion, 62, 3 (2006).
24. R.M Carranza, M.A. Rodríguez, R.B. Rebak, Paper 07581, Corrosion/2007 (NACE Int., 2007: Houston, TX).

Mater. Res. Soc. Symp. Proc. Vol. 1107 © 2008 Materials Research Society

Waste Package Corrosion Studies Using Small Mockup Experiments

B.E. Anderson[1], K.B. Helean[2], C.R. Bryan[2], P.V. Brady[2] and R.C. Ewing[1]
[1]Department of Geological Sciences, The University of Michigan, Ann Arbor, MI, USA
[2]Sandia National Laboratories, Albuquerque, NM, USA

ABSTRACT

Understanding the corrosion of spent nuclear fuel (SNF) and the subsequent mobilization of released radionuclides, particularly under oxidizing conditions, is one of the key issues in evaluating the long-term performance of a nuclear waste repository. However, the large amounts of iron in the metal waste package may create locally reducing conditions that would lower corrosion rates for the SNF, as well as reduce the solubility of some key radionuclides, e.g., Tc and Np. In order to investigate the interactions among SNF-waste package-fluids, four small-scale (~1:40 by length) waste package mockups were constructed using metals similar to those proposed for use in waste packages at the proposed repository at Yucca Mountain. Each mockup experiment differed with respect to water input, exposure to the atmosphere, and temperature. Simulated Yucca Mountain process water (YMPW) was injected into three of the mockups at a rate of 200 μL per day for five days a week using a calibrated needle syringe. The YMPW was prepared by equilibrating 50 mg/L silica as sodium metasilicate with air, and adding enough HCl to lower the pH to 7.6 in contact with an excess of powdered calcite.

X-ray powder diffraction and scanning electron microscopy confirm that, where corrosion occurred, the dominant corrosion product in all cases was magnetite. In the high temperature (60°C) experiment, hematite and a fibrous, Fe-O-Cl phase were also identified. The Fe(II)/Fe(III) ratios measured in the corrosion products using a wet chemistry technique indicate extremely low oxygen fugacities (10^{-36} bar). Experiments are in progress in which 0.1g powdered UO_2 was included in the mock-up in order to investigate the relative kinetics of Fe and U oxidation and to identify the U corrosion products formed under these conditions.

INTRODUCTION

Ferrous iron, Fe^{2+}, has been shown to reduce UO_2^{2+} to $UO_{2(s)}$ [1], and some ferrous iron-bearing ion-exchange materials adsorb radionuclides and heavy metals [2]. High electron availability leads to the reduction and subsequent immobilization of problematic dissolved species such as TcO_4^-, NpO_2^+, and UO_2^{2+} and may also inhibit corrosion of spent nuclear fuel.

To test for the presence of locally reducing conditions within a waste package, four small-scale models or "mockups" of a carbon-steel based waste package proposed for use in YMR were manufactured and regularly injected with YMPW, a simplified pore water composition described below. If Fe^{2+}-bearing phases, rather than fully oxidized phases such as goethite, are produced during corrosion, then locally reducing conditions may be present. Furthermore, assuming that the oxidation of the zero valent Fe in steel is approximately at equilibrium, the ratio of Fe(II) to Fe(III) in the corrosion products may be used in conjunction with pH data and known thermodynamic constants to directly calculate an oxygen fugacity or pe for the system. The focus of this paper is on the nature of Yucca Mountain waste package steel corrosion products and the local redox conditions inside the waste package as indicated by Fe(II) to Fe(III) ratios.

EXPERIMENT

Description of mockups

Four small-scale (~1:40 by length) waste package mockups were constructed using 316 stainless steel (composition: $Fe_{62.075}\,C_{0.02}\,Mn_{2.0}\,P_{0.045}\,S_{0.03}\,Si_{0.75}\,Ni_{14.0}\,Cr_{18.0}\,Mo_{3.0}\,N_{0.08}$ [3]), the same material as the proposed Yucca Mountain waste packages, for the body, end-caps, and fittings (Figure 1). This steel corrodes less rapidly under most conditions than the A-516 carbon steel (composition: $Fe_{97.87}\,C_{0.31}\,Mn_{1.3}\,P_{0.035}\,S_{0.035}\,Si_{0.45}$ [4]) proposed for use as guides for spent nuclear fuel inside the waste packages. In order to maintain the same ratio of body-interior surface area to guide surface area, twenty-five 1 x 10 x 0.2 cm strips of A-516 carbon steel were inserted into each waste package, and inert polytetrafluoroethylene (PTFE) balls with diameter 9.53 mm were used to separate the steel strips and fill in the excess void space. The A-516 steel and 316 stainless steel were obtained from Laboratory Testing, Inc., a DOE-approved supplier, and are certified to meet the ASTM standards for those materials. Each mockup had an internal radius of 38.1 mm, an internal length of 12.3 mm, two upper ports, and one lower port with an Ultratorr fitting and a heavy-gauge 1.6 cm diameter glass test tube for effluent sample collection. The caps on both ends of each mockup were sealed using viton O-rings and parafilm to prevent higher corrosion rates on the straight-threaded end-caps.

The four mockups differed with respect to water input, exposure to the atmosphere, and temperature. In most cases, the upper port on the opposite side from the lower port was covered by a rubber septum for the introduction of the aqueous phase. The exception was mockup C, which was not injected with water but left entirely open to an atmosphere with greater than 90% humidity and partial pressures of oxygen and carbon dioxide of $10^{-0.7}$ bar and $10^{-3.5}$ bar, respectively. The second upper ports of mockups A and D were sealed using a Swagelok snubber, while that of mockup B was left open to the same atmospheric conditions as that of mockup C. Because of their increased exposure to the atmosphere, the "open" mockups were more likely to accurately simulate conditions at Yucca Mountain. In all cases, gases were able to escape through the upper ports, thus preventing internal pressure buildup. Humid conditions were maintained by enclosing the mockup setup in a sealed plastic bag connected by tygon tubing with a dewer of CO_2-equilibrated deionized water constantly agitated by an aquarium pump. A Traceable® digital hygrometer was used to check these conditions. Mockups A, B, and C were allowed to corrode at room temperature, and mockup D was maintained at 60°C using a hot plate. Mockup D is significantly different from the other three mockups because it started out as only a scoping study. Ideally other mockups would be run with, for instance, elevated temperature and high relative humidity, in order to better gauge the effect of changing variables. Differences among the four waste packages are summarized in Table I.

Yucca Mountain process water

A simulated Yucca Mountain process water (YMPW) was injected into mockups A and B at a rate of 200 μL per day five days a week using a calibrated needle syringe. Scaling by volume, this rate is equivalent to the introduction of 1.3 mL of water per minute in a full-size waste package. YMPW consists of 50 mg/L silica, 38.3 mg/L Na, enough hydrochloric acid to lower the pH to 7.6, and an excess of powdered calcite. The solution was equilibrated with the atmosphere for 5 days, filtered, and allowed to requilibrate with the atmosphere for an additional 5 days. The final pH stabilized at 7.5. While historically J-13 well water [5, 6], pore waters from the unsaturated zone near Yucca Mountain [7],and various calculated waters have been used to approximate the composition of fluids entering a breached waste package, there is no universally accepted optimal fluid composition for source term work because of large uncertainties in, for example, breach time, fluid sources, and the extent of prior fluid-rock interactions. As shown in Table II, the major element chemistry of YMPW is similar to that of waters currently found at the mountain. Minor compositional differences are unlikely to be significant for this study given the high influence of interactions with the waste package on the evolution of water chemistry. A similar fluid, YMPW-2, a pH 7.9 dilute Na-Ca-HCO_3- silicate water with approximately 0.8 ppm chloride and 1.0 ppm fluoride, was injected into mockup D at a rate of 1 mL per week.

Characterization

Mockups A and B were allowed to corrode at room temperature and >90% relative humidity until the test tubes in the lower port were nearly full of effluent. Mockup C was sampled at the same time as mockups A and B, and mockup D was sampled at 30 and 90 days. During sampling, the mockups were disassembled and the effluent analyzed for pH using a Ross electrode with a Symphony SB70P meter. Characterization of corrosion products included X-ray powder diffraction (XRD), scanning electron microscopy (SEM) with energy dispersive spectroscopy (EDS), and transmission electron microscopy (TEM).
Fe(II)/Fe(III) ratios were measured in both the corrosion products and the effluents of mockups A and B using a method combining the standard Pratt and ferrozine methods [8, 9]. In this method, known as the ferrozine micro-method for solids [10], samples are dissolved in H_2SO_4 and HF, and the resulting solution is analyzed using the spectrophotometric agent ferrozine . As in the Pratt method, oxidation is prevented by immediate and continuous boiling of the acid mixture during digestion. In mockup D, these ratios were measured using a standard $K_2Cr_2O_7$ method. One potential weakness with nearly all wet chemical methods for determining Fe(II)/Fe(III) ratios is that the Fe(II) can be easily oxidized during sample digestion. Given the likelihood of this occurring in our analyses, the numbers reported for Fe(II)/Fe(III) ratios are *minimum* values.

The software packages EQ3 and EQ6 were used to calculate speciation, equilibrium, and oxygen fugacities using the measured effluent pH and Fe(II)/Fe(III) ratios of both the effluent and solids. The major assumptions that went into these calculations were: 1) the system is saturated with $CaCO_3$, 2) the system is saturated with amorphous silica, 3) Na^+, K^+, and Al^{3+} are present, having been eroded from the surrounding tuff at Yucca Mountain, and 4) oxygen fugacity will be at approximately ambient levels.

RESULTS AND DISCUSSION

Very little corrosion was observed in mockup C even after 1.5 years exposure to humid air, but in mockups A, B, and D there was sufficient corrosion to analyze the samples by x-ray diffraction (XRD) (Figure 2). In all three cases, the major phase identified was magnetite, Fe_3O_4, or maghemite, which is structurally identical to magnetite but fully-oxidized. Fe(II) was observed in all samples, suggesting the presence of magnetite. In mockup D at 90 days, this ratio suggests that 12% of the Fe was present as magnetite. (Because of a delay in transferring the sample during preparations for the titration, some of the Fe^{2+} originally present may have been oxidized before it could be measured.) In mockups A and B the Fe(II)/Fe(III) ratio is so high that magnetite alone cannot account for all of the Fe(II), suggesting the presence of poorly crystalline phases. Hematite, Fe_2O_3, and a phase with a (001) *d*-spacing of 12.87 angstroms were also identified in mockup D, the high temperature study (see Table III). The high d-spacing phase, seen only in mockup D after 90 days, was characterized using SEM with EDS and found to preferentially incorporate Cl, most likely as Fe-O-Cl. This "fibrous" phase located near an oxidizing zone also appears to have a layered structure (Figure 3).

These results suggest that local conditions inside of the waste package will be at least somewhat reducing, and magnetite will be an important corrosion product. Hematite is expected to be present in significant amounts at elevated temperatures, which will likely be the case at YMR. If the fibrous Cl-rich phase seen in mockup D also forms in a real waste package, it could be important for the uptake of radionuclides and in maintaining reducing conditions because the observed morphology, chemistry, and interlayer *d*-spacing suggest that this phase may be related to green rust, a layered Fe-oxyhydrate that, when sulfate or carbonate is present in the interlayer, has been shown to remove over 99.8% of pertechnetate from solution [11].

CONCLUSIONS

The presence of reduced iron in magnetite and high aqueous Fe(II)/Fe(III) ratios in the waste package mockups suggests not only that these radionuclide-adsorbing minerals will be present in Yucca Mountain waste packages, but also that the abundance of reduced iron will cause locally reducing conditions. However, there are a number of important questions that this study does not address, such as: 1) the accuracy of EQ3/6 and similar programs in estimating oxygen fugacity in these situations, 2) the effect of switching from carbon steel, examined in this study, to the more corrosion-resistant stainless steel, 3) the effect of aluminum used to join the steel basket in the waste package, 4) the degree of porosity reduction and any effect that might have in preventing corrosion in the interior of the waste package, and, most importantly, 5) the effect that these locally reducing conditions will have on spent nuclear fuel corrosion and corrosion products.

Thermodynamic calculations that set ferrous iron concentrations in heterogeneous equilibrium with magnetite and ferric iron concentrations in heterogeneous equilibrium with goethite indicate that oxygen fugacity within a corroding waste package may be as low as 10^{-62} at a pH of 7, and 10^{-54} at a pH of 9. In this paper these fugacities are calculated based on thermodynamic relationships, although to take non-equilibrium into account, in the future they may be interpolated from direct measurements of Fe(II)/Fe(III) ratios as a function of known oxygen fugacity. These low oxygen fugacities also depend on having a steel type that corrodes relatively easily, providing electrons from the zero valent Fe. Stainless steel, which has been proposed to replace carbon steel in the waste packages, is less likely to corrode and therefore less likely to contribute to either a reducing environment of the generation of minerals that inhibit radionuclide transport. More mockups tests may be performed to investigate these issues. The mockups used in these studies may be re-packed with stainless steel instead of carbon steel and tested in a manner similar to mockups A and B.

Of greater concern than the exact oxygen fugacity is the effect these conditions will have on the degradation of spent nuclear fuel and the retention of released radionuclides. For instance, if the volume increase resulting from corrosion results in a significant porosity decrease, there may be a self-sealing effect that prevents significant oxygen from approaching spent nuclear fuel in the center of the waste package. While this is unlikely to happen without a very large amount of corrosion taking place, this large amount of corroding Fe may still have a significant effect on spent nuclear fuel corrosion processes. Under reducing and Si-rich conditions, UO_2 is known to alter to coffinite [12], a notably different structure than uranophane and other products of oxidative corrosion and alteration. For this reason, two additional mockups will be run in similar fashion to mockups A and B, but these will contain 0.1g UO_2. These studies will provide further insight into the effects of A516 steel corrosion products on radionuclide release and transport.

ACKNOWLEDGMENTS

B. E. Anderson is thankful for fellowships from the Office of Civilian Radioactive Waste Management and the National Science Foundation. This work was supported by the Office of Science and Technology and International (OST&I) of the Office of Civilian Radioactive Waste Management (DE-FE28-04RW12254). The views, opinions, findings and conclusions or recommendations of the authors expressed herein do not necessarily state or reflect those of DOE/OCRWM/OSTI. Sandia is a multiprogram laboratory operated by Sandia Corporation, a Lockheed Martin Company, for US DOE's NNSA under contract DE-AC04-94AL85000.

Table I. Test matrix for waste package mockup experiments.

Mockup	Atmosphere	YMPW Volume	Relative Humidity	Temperature (°C)	Sampling
A	Closed	200 µl/day	100%	25	~ 1 year
B	Open	200 µl/day	100%	25	~ 1 year
C	Open	-	100%	25	~ 1 year
D	Closed	1 mL/week YMPW-2	environmental	60	30 and 90 days

Table II. Saturated (J-13) and unsaturated (UZ) zone compositions (mM) from Yucca Mountain [5, 7] compared to YMPW and YMPW-2.

Component	J-13	UZ4-TP-7	UZ calc'd	YMPW[a]	YMPW-2
Na^+	1.96	2.09	1.70-5.22	0.83	2.00
Ca^{++}	0.29	2.00	2.02-2.35	0.45	0.87
SiO_2^{aq}	1.07	1.48	0.70	0.83	0.90
HCO_3^-	2.34	2.82	5.93-6.51	1.01	1.95
pH	7	7.4	7.4-7.6	8.3	7.9

[a] Calculated values given known Na and Si inputs, atmospheric carbon dioxide levels and calcite equilibria. Actual values produced may vary slightly.

Table III: Major and minor corrosion product phases, pH, Fe(II)/Fe(III) ratios, and calculated oxygen fugacities present in the mockups based on XRD, SEM with EDS, chemical analyses, and using the geochemical modeling tool EQ3/6. When enough material for multiple samples was available, error estimates for the Fe(II)/Fe(III) ratios were made.

Mockup	Sample Time	Major Phases	Minor Phases	pH	Effluent Fe(II)/FeIII) ratio	Solids Fe(II)/Fe(III) ratio	oxygen fugacity
A	~ 1.5 years	magnetite	-	8.94	3.57	0.72	10^{-53}
B	~ 1.5 years	magnetite	-	8.74	0.11 +/- 0.06	0.60 +/- 0.05	10^{-46}
B	~ 0.5 years	magnetite	-	6.64	0.66 +/- 0.04	n/a	10^{-36}
C	~ 1.5 years	n/a	n/a	n/a	n/a	n/a	n/a
D	30 days	magnetite	hematite	n/a	n/a	n/a	n/a
D	90 days	magnetite	hematite, Fe-O-Cl	n/a	n/a	0.04	n/a

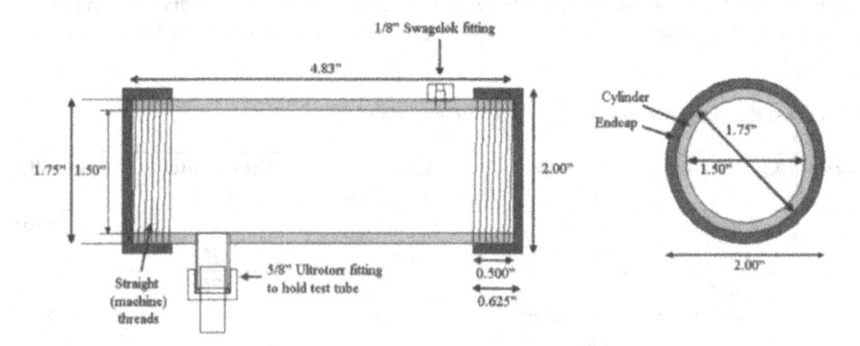

Figure 1. Schematic of a mockup viewed from two angles.

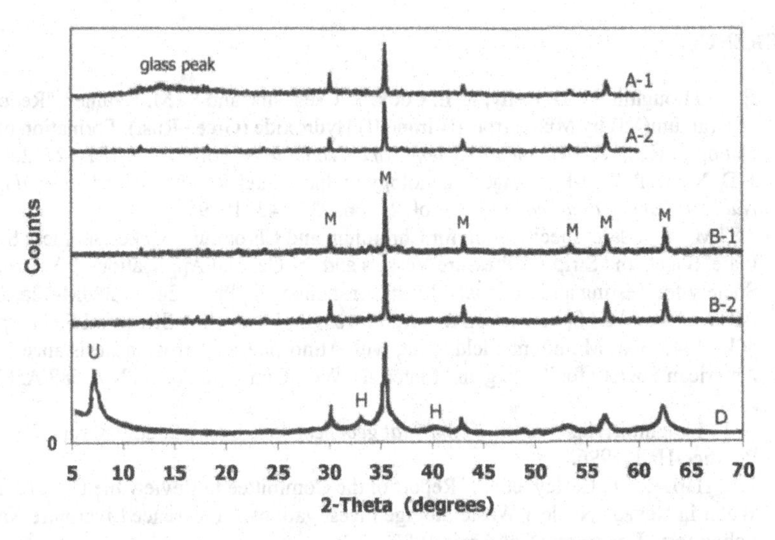

Figure 2. X-ray diffraction spectra showing of corrosion products showing magnetite (M), hematite (H), and an unidentified mineral (U) with large d-spacing in mockup D.

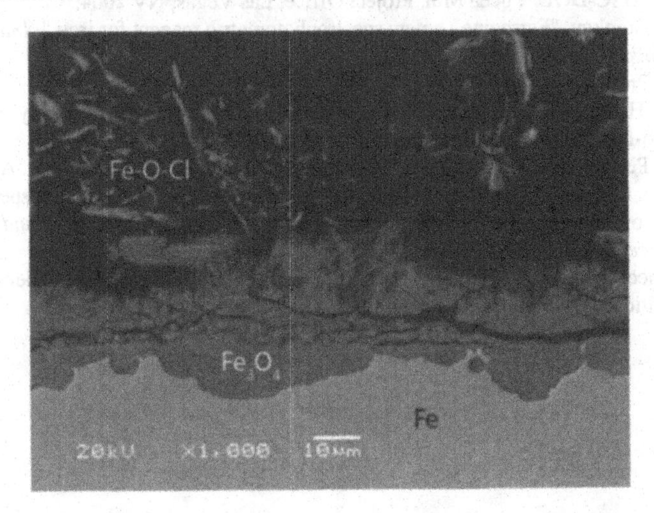

Figure 3. Backscattered electron micrographs of Mockup D corroded steel of at 90 days. The polished cross-section shows oxidation "blisters" pock-marking the steel surface and loosely consolidated fibers of a Cl-rich phase.

REFERENCES

[1] E. J. O'Loughlin, S. D. Kelly, R. E. Cook, R. Csencsits, and K. M. Kemner, "Reduction of Uranium(VI) by Mixed Iron(II)/Iron(III) Hydroxide (Green Rust): Formation of UO_2 Nanoparticles," *Environmental Science and Technology*, vol. 37, pp. 721-727, 2003.

[2] J. D. Navratil, "Ion-Exchange Technology in Spent Fuel-Reprocessing," *Journal of Nuclear Science and Technology*, vol. 26, pp. 735-743, 1989.

[3] ASTM, "Standard Specification for Chromium and Chromium-Nickel Stainless Steel Plate, Sheet, and Strip for Pressure Vessels and for General Applications," American Society for Testing and Materials, West Conshohocken, PN A 240/A 240M-02a, 2002.

[4] ASTM, "Standard Specification for High-Strength, Low-Alloy Structural Steel, up to 50ksi [345Mpa] Minimum Yield Point, with Atmospheric Corrosion Resistance," American Society for Testing and Materials, West Conshohocken, PN A 588/A588M-05, 2005.

[5] D. L. Langmuir, *Aqueous environmental geochemistry*. Upper Saddle River, NJ: Prentice-Hall, 1996.

[6] J. E. Harrar, J. F. Carley, et al., "Report of the Committee to Review the Use of J-13 Well Water in Nevada Nuclear Waste Storage Investigations.," Lawrence Livermore National Laboratory, Livermore, California 1990.

[7] P. S. Domski, "In-package chemistry abstraction for wasteforms at Yucca Mountain, Rev 03B," BSC-DOE, Yucca Mtn. Project Office, Las Vegas, NV 2004.

[8] L. L. Stookey, "Ferrozine - a new spectrophotometric reagent for iron," *Analytical Chemistry*, vol. 42, pp. 779, 1970.

[9] J. H. Pratt, *American Journal of Science*, vol. 48, pp. 149, 1984.

[10] J. W. Husler, B. E. Anderson, K. B. Helean, C. R. Bryan, and P. V. Brady, "Ferrozine Micro-method for Determination of Iron in Solids," 2007.

[11] S. E. Pepper, D. J. Bunker, N. D. Bryan, F. R. Livens, J. M. Charnock, R. A. D. Pattrick, and D. Collison, "Treatment of radioactive wastes: An X-ray absorption spectroscopy study of the reaction of technetium with green rust," *Journal of Colloid and Interface Science*, vol. 268, pp. 408-412, 2003.

[12] J. Janeczek and R. C. Ewing, "Dissolution and alteration of uraninite under reducing conditions," *Jounal of Nuclear Materials*, vol. 190, pp. 157-173, 1992.

Mater. Res. Soc. Symp. Proc. Vol. 1107 © 2008 Materials Research Society

Repassivation Potential of Alloy 22 in Sodium and Calcium Chloride Brines

Raul B. Rebak,[1] Gabriel O. Ilevbare[2] and Ricardo M. Carranza[3]
[1]Lawrence Livermore National Laboratory, Livermore, CA 94550, USA
[2]Electric Power Research Institute, Palo Alto, CA, 94304, USA
[3]Atomic Energy Commission of Argentina, 1650 San Martin, Argentina

ABSTRACT

A comprehensive matrix of 60 tests was designed to explore the effect of calcium chloride vs. sodium chloride and the ratio R of nitrate concentration over chloride concentration on the repassivation potential of Alloy 22. Tests were conducted using the cyclic potentiodynamic polarization (CPP) technique at 75°C and at 90°C. Results show that at a ratio R of 0.18 and higher nitrate was able to inhibit the crevice corrosion in Alloy 22 induced by chloride. Current results fail to show in a consistent way a different effect on the repassivation potential of Alloy 22 for calcium chloride solutions than for sodium chloride solutions.

INTRODUCTION

Alloy 22 (N06022) is a nickel base alloy especially designed to be resistant to all forms of corrosion. Alloy 22 contains approximately 56% nickel (Ni), 22% chromium (Cr), 13% molybdenum (Mo), 3% tungsten (W) and 3% iron (Fe) (ASTM B 575). [1] Because of its high level of Cr, Alloy 22 remains passive in most industrial environments and therefore has an exceptionally low general corrosion rate. [2-6] Because this Alloy 22 is Ni based, it does not suffer environmentally assisted cracking in hot chloride solutions. [3] The resistance of Alloy 22 to localized corrosion in chloride solutions is given by the combined presence of Cr, Mo and W. [7-12] However, Alloy 22 may suffer crevice corrosion when it is anodically polarized in chloride-containing solutions. [8-10,13-15] The presence of nitrate (NO_3^-) in the solution minimizes or eliminates the susceptibility of Alloy 22 to crevice corrosion. [8-10,16-23] The value of the ratio R = $[NO_3^-]/[Cl^-]$ has a strong effect on the susceptibility of Alloy 22 to crevice corrosion. [16-25] The higher the nitrate to chloride ratio R, the stronger is the inhibition by nitrate. The minimum required R value for inhibition may depend on other experimental variables such as total concentration of chloride and temperature. Other anions in solution were also reported to inhibit crevice corrosion in Alloy 22. [19-20, 26-28]

The objective of this work was to examine the susceptibility of Alloy 22 in several electrolyte solutions containing sodium chloride (NaCl), calcium chloride ($CaCl_2$) and sodium nitrate ($NaNO_3$) using the cyclic potentiodynamic polarization (CPP) technique. [29] Ratios R of nitrate over chloride from 0.0086 to 0.25 were investigated.

EXPERIMENTAL TECHNIQUE

Alloy 22 specimens were prepared from 1-inch thick plate. The specimens were creviced using a ceramic washer and PTFE tape. [15,30,31] The specimens were multiple crevice assemblies (MCA) [23] or lollipops. All the tested specimens had a finished grinding of abrasive paper number 600 and were degreased in acetone and treated ultrasonically for 5 minutes in de-ionized

(DI) water 1 hour prior to the start of testing. Specimens were cut from as-welded (ASW) plates. The weld in the plates was produced with matching filler metal using Gas Tungsten Arc Welding (GTAW). The welded specimens were not all weld metal but contained a weld seam band across the center of the specimen, varying in width from approximately 8 to 15 mm.

The electrochemical tests were carried out in ten different NaCl, $CaCl_2$ and $NaNO_3$ electrolytes (Table 1). All the solutions were rather concentrated, from a little over 2 molar (M) (Electrolyte 1) to more than 6 molar (Electrolyte 10). The ratio R of nitrate over chloride varied from 0.0086 (Electrolyte 9) to 0.25 (Electrolyte 4). The addition of either NaCl or $CaCl_2$ was used as a testing variable. For example, Electrolytes 2 and 3 have the same total chloride concentration and same ratio R; however, Electrolyte 2 was rich in NaCl and Electrolyte 3 was rich in $CaCl_2$. The same is applicable for Electrolytes 5 and 6. The pH of the solutions was not adjusted, and was near neutral. The testing temperatures were 75°C and 90°C.

Table 1 – Matrix of tested electrolyte solutions

Electro-lyte	Composition	$[Cl^-]$ NaCl	$[Cl^-]$ $CaCl_2$	Total $[Cl^-]$	$[NO_3^-]$	
1	1 M NaCl + 0.5 M $CaCl_2$ + 0.05 M $NaNO_3$	1	1	2	0.05	0.025
2	4 M NaCl + 0.5 M $CaCl_2$ + 0.05 M $NaNO_3$	4	1	5	0.05	0.01
3	1 M NaCl + 2 M $CaCl_2$ + 0.05 M $NaNO_3$	1	4	5	0.05	0.01
4	1 M NaCl + 0.5 M $CaCl_2$ + 0.5 M $NaNO_3$	1	1	2	0.5	0.25
5	1 M NaCl + 2 M $CaCl_2$ + 0.5 M $NaNO_3$	1	4	5	0.5	0.1
6	4 M NaCl + 0.5 M $CaCl_2$ + 0.5 M $NaNO_3$	4	1	5	0.5	0.1
7	1.8 M NaCl + 0.5 M $CaCl_2$ + 0.05 M $NaNO_3$	1.8	1	2.8	0.05	0.018
8	1.8 M NaCl + 0.5 M $CaCl_2$ + 0.5 M $NaNO_3$	1.8	1	2.8	0.5	0.18
9	1.8 M NaCl + 2 M $CaCl_2$ + 0.05 M $NaNO_3$	1.8	4	5.8	0.05	0.0086
10	1.8 M NaCl + 2 M $CaCl_2$ + 0.5 M $NaNO_3$	1.8	4	5.8	0.5	0.086

The electrochemical tests were conducted in a one-liter, three-electrode, borosilicate glass flask (ASTM G5). [29] A water-cooled condenser combined with a water trap was used to avoid evaporation of the solution and to prevent the ingress of air (oxygen). All the tests were carried out at ambient pressure. The reference electrode was saturated silver chloride (SSC), which at ambient temperature has a potential of 199 mV more positive than the standard hydrogen electrode (SHE). The reference electrode was connected to the solution through a water-jacketed Luggin probe so that the electrode was maintained at near ambient temperature. The counter electrode was a flag (36 cm^2) of platinum foil spot-welded to a platinum wire. All the potentials in this paper are reported in the SSC scale. Nitrogen (N_2) was purged through the solution at a flow rate of 100cc/min for 24 hours while the corrosion potential (E_{corr}) was monitored. Nitrogen bubbling was continued throughout all the electrochemical tests. The specimens were immersed for 24 hours in the deaerated electrolytes at temperature while nitrogen gas was purged through the solution. The open circuit potential of the working electrodes were recorded during the 24-hr immersion and the value at the end of the 24-hr immersion was called the corrosion potential (E_{corr}-24hr). After the 24-hour immersion, a cyclic potentiodynamic polarization (CPP) was performed. In the CPP tests, the potential scan was started approximately 100 mV below E_{corr} at a set scan rate of 0.167 mV/s. The scan direction was generally reversed when the current density reached 30 mA/cm^2 in the forward scan. The total applied current density was higher than recommended by the ASTM standard G 61 of 5 mA/cm^2. The CPP test is a fast and efficient method to determine crevice corrosion resistance of commercial alloys. In the forward scan of the CPP, the potentials for which the current density is 20 and 200 $\mu A/cm^2$ are called E20 and E200.[9,18,21,31] These parameters represent values of breakdown potentials. In the reverse scan of the CPP, the values of potentials for which the current density is 10 and 1 $\mu A/cm^2$ are called ER10 and ER1. The potential at which the reverse scan intersects the forward scan is called repassivation potential cross over (ERCO). ER10, ER1 and ERCO represent values of repassivation potentials.

After the tests, the specimens were examined in an optical stereomicroscope at a magnification of 20 times to establish the mode and location of the attack.

RESULTS

Cyclic potentiodynamic polarization (CPP)

Figure 1 shows the CPP curves for specimens JE1733 and JE1707 tested in Electrolyte 2 (NaCl dominated) at 75°C and 90°C. The passive current density for both specimens was practically the same. The range of passivity was wide (higher than 700 mV). The breakdown potential at 75°C was slightly higher than at 90°C. Table 2 shows that E20 for JE1733 was +488 mV SSC while the E20 for JE1707 was +304 mV SSC. Both specimens suffered a significant hysteresis in the return scan suggesting the presence of crevice corrosion. The repassivation potential ER1 was practically the same for both specimens (ER1 was –111 mV SSC for JE1733 and –100 mV SSC for JE1707) (Table 2 and Figure 1). The repassivation potential was even slightly lower in the lower temperature solution. This could be an artifact of the test method and related to diffusion processes. If the scan rate is decreased ten times or the Tsujikawa-Hisamatsu method is used, it is likely that the trend between ER1 vs. temperature may disappear or even reverse itself. Electrolyte 2 is a high chloride low nitrate solution (R = 0.01) and therefore it is assumed that Alloy 22 would suffer crevice corrosion for the high applied potentials at the tested temperatures (Figure 1). Table 2 shows that both of these specimens suffered crevice corrosion as well as abundant transpassivity due to the high final applied current density of 30 mA/cm^2.

Figure 2 shows information similar to Figure 1 but for a $CaCl_2$ dominated solution (Electrolyte 3). The same discussion above for Figure 1 can be applied to discuss Figure 2. The repassivation potential ER1 was lower for JE1720 tested at 75°C (ER1 = -105 mV SSC) than for JE1722 tested at 90°C (ER1 = -80 mV SSC). Analyses of Figures 1 and 2 seem to suggest that NaCl rich solutions yielded slightly lower repassivation potentials than the $CaCl_2$ rich solutions.

Figure 3 shows comparatively the behavior of Alloy 22 in the NaCl rich and $CaCl_2$ rich electrolytes (Electrolytes 2 and 3) for solutions with R = 0.01 at 90°C. Figure 3 shows that under the tested conditions Alloy 22 had practically the same behavior in both electrolytes, showing little or no influence of the cations in the solution. Figure 4 shows the effect of nitrate in the solution (comparing electrolytes 7 and 8). For the lower R value of 0.018 there was a noticeable hysteresis in the reverse scan suggesting the presence of crevice corrosion; however, when R = 0.18, there was no reverse scan hysteresis. Table 2 shows that crevice corrosion was observed in specimen JE1758 (R = 0.018); but in specimen JE1766 only transpassive dissolution was present after the CPP tests. Results from Figure 4 show that for a ratio R = 0.18 at 75°C crevice corrosion was fully inhibited by nitrate. Similar findings have been reported before.[8-10,16-23]

Crevice Repassivation Potential

Figure 5 shows the crevice repassivation potential (ERCO) of Alloy 22 as a function of the temperature in NaCl and $CaCl_2$ rich electrolytes (Electrolytes 5 and 6) for solutions with R = 0.1. For both solutions the ERCO decreased as the temperature increased, and, at each temperature the ERCO was lower in the $CaCl_2$ rich solution than in the NaCl rich solution. An analysis of the data in Table 2 shows that this trend on the effect of the cation is not clear or even opposite in Electrolytes 2 and 3 where R = 0.01 (Figure 3). More systematic studies are needed to explore the effect of the cation on the repassivation potential of Alloy 22.

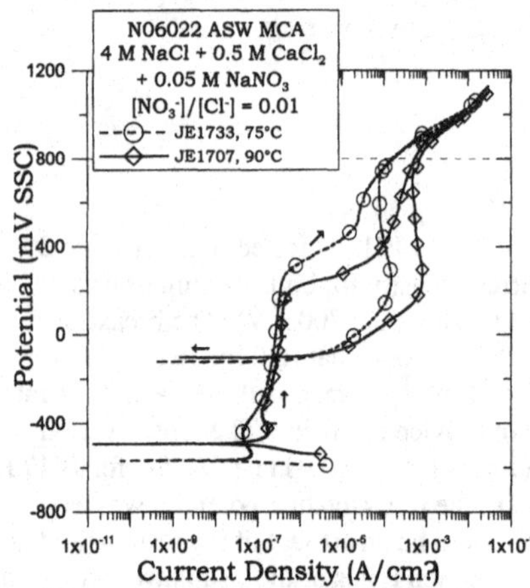

Figure 1 – Cyclic potentiodynamic polarization (CPP) in Electrolyte 2, R = 0.01

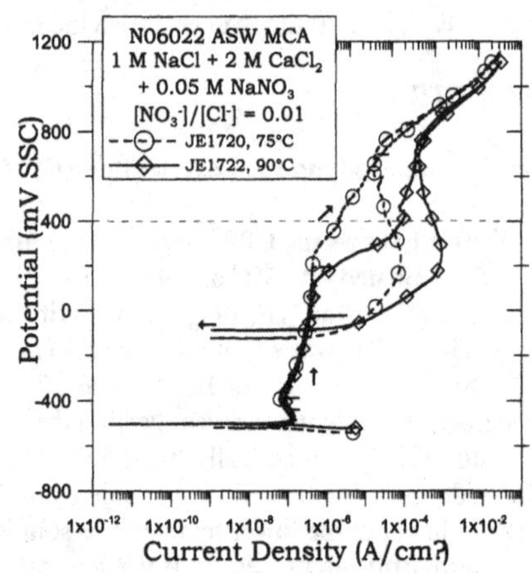

Figure 2 – Cyclic potentiodynamic polarization (CPP) in Electrolyte 3, R = 0.01

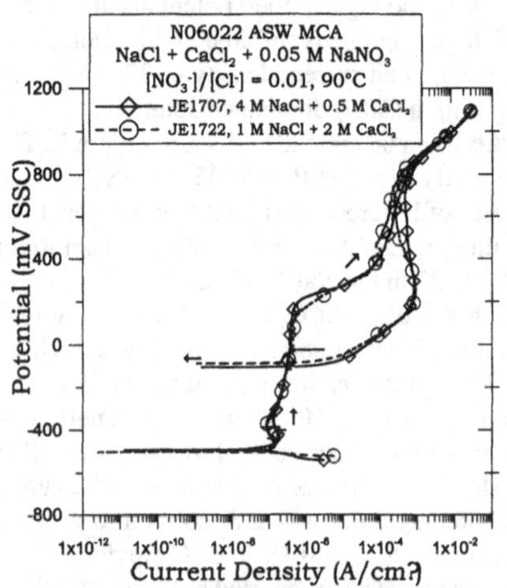

Figure 3 – CPP comparison between Electrolytes 2 and 3 at 90°C, R = 0.01

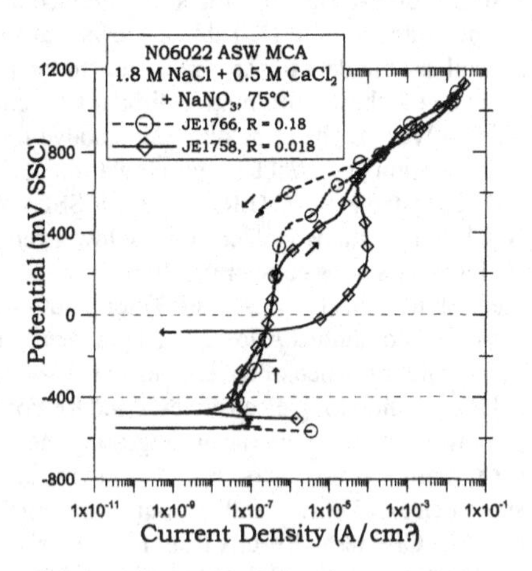

Figure 4 – CPP in Electrolytes 7 and 8 at 90°C. Effect of nitrate

Table 2 – Parameters from the CPP Tests. All potentials are in mV SSC
All the specimens suffered transpassive dissolution after the CPP tests
CC= Crevice Corrosion, CC-II = Type II Crevice Corrosion [Ref. 21].

Electro-lyte	T, °C	Specimen	E_{corr}	E20	E200	ER10	ER1	ERCO	Attack
1	75	JE1723	-478	540	791	33	-69	-82	CC
1	75	JE1721	-486	723	785	24	-60	-71	CC
1	75	JE1725	-354	493	739	25	-54	-62	CC
1	90	JE1711	-464	373	609	-21	-69	-72	CC
1	90	JE1702	-478	299	565	-34	-79	-83	CC
1	90	JE1728	-456	467	682	-20	-66	-68	CC
2	75	JE1730	-490	609	836	-38	-111	-123	CC
2	75	JE1705	-461	282	620	-34	-80	-84	CC
2	75	JE1733	-489	488	815	-42	-111	-121	CC
2	90	JE1718	-487	199	296	-50	-73	-75	CC
2	90	JE1707	-447	304	540	-67	-100	-103	CC
2	90	JE1729	-489	200	293	-58	-100	-104	CC
3	75	JE1727	-402	680	833	-22	-97	-106	CC
3	75	JE1717	-430	337	801	-50	-104	-110	CC
3	75	JE1720	-442	674	838	-27	-105	-117	CC
3	90	JE1722	-440	294	716	-45	-80	-84	CC
3	90	JE1712	-462	303	379	0	-15	-15	CC
3	90	JE1714	-438	235	301	-18	-33	-35	CC
4	75	JE1706	-411	689	766	672	586	643	No CC
4	75	JE1735	-446	646	776	684	609	805	No CC
4	75	JE1724	-438	620	758	686	611	831	No CC
4	90	JE1704	-449	670	757	602	480	592	No CC
4	90	JE1726	-386	512	712	620	517	819	No CC
4	90	JE1734	-475	613	722	621	444	218	No CC
5	75	JE1709	-398	714	840	145	-41	-62	CC-II
5	75	JE1732	-435	686	831	120	-51	-70	CC-II
5	75	JE1719	-430	701	845	738	-43	-66	CC-II
5	90	JE1715	-430	516	793	-55	-101	-104	CC
5	90	JE1713	-434	606	790	-40	-88	-93	CC-II
5	90	JE1701	-394	572	797	-43	-95	-101	CC
6	75	JE1736	-474	651	821	689	-46	-70	CC-II
6	75	JE1703	-417	709	833	131	-40	-60	CC-II
6	75	JE1716	-488	676	830	734	54	-2	CC-II
6	90	JE1731	-477	541	781	-26	-86	-94	CC
6	90	JE1710	-489	674	782	55	-68	-81	CC-II
6	90	JE1708	-470	561	784	-14	-79	-87	CC

Table 2 - Continued									
7	75	JE1740	-483						N/A
7	75	JE1758	-406	511	775	5	-66	-76	CC
7	75	JE1762	-514	577	767	78	-35	-51	CC
7	75	JE1767	-359	532	787	22	-70	-87	CC
7	90	JE1742	-405						N/A
7	90	JE1750	-505	418	683	-29	-94	-101	CC
7	90	JE1755	-509	416	659	-46	-92	-96	CC
7	90	JE1768	-514	441	705	-25	-91	-97	CC
8	75	JE1739	-447						N/A
8	75	JE1754	-491	584	772	682	600	825	No CC
8	75	JE1764	-471	637	795	694	620	813	No CC
8	75	JE1766	-480	644	787	690	612	818	No CC
8	90	JE1746	-471	476	735	613	-16	-33	CC-II
8	90	JE1748	-469	477	727	620	87	47	No CC
8	90	JE1759	-482	555	745	635	526	192	No CC
9	75	JE1743	-434	157	737	-41	-91	-94	CC
9	75	JE1751	-454	683	844	-37	-113	-125	CC
9	75	JE1760	-444	673	840	10	-83	-99	CC
9	90	JE1737	-432						N/A
9	90	JE1749	-494	481	781	-72	-107	-110	CC
9	90	JE1761	-462	238	357	-53	-76	-78	CC
9	90	JE1765	-476	256	336	-37	-62	-64	CC
10	75	JE1747	-426	526	840	29	-67	-78	CC-II
10	75	JE1756	-398	714	859	89	-61	-78	CC-II
10	75	JE1757	-445	703	845	738	41	-12	No CC
10	90	JE1752	-444	604	803	-15	-89	-99	CC-II
10	90	JE1753	-452	615	818	-7	-88	-98	CC
10	90	JE1763	-420	424	769	-28	-73	-78	CC

Figure 6 shows the repassivation potential ER1 as a function of R for all the solutions in Table 2. For R between 0.0086 and 0.1, the values of ER1 were low, near 0 mV or below. All the specimens tested under these conditions suffered crevice corrosion (Table 2). For the specimens tested in the solutions with R = 0.086 and 0.1, it appears that the ER1 values at 75°C were slightly higher than at 90°C, confirming that a higher temperature generally produces lower repassivation potentials for some electrolyte solutions. For the electrolyte with R = 0.18, there was a separation of the behavior of the alloy at 75°C and 90°C. The alloy had higher values of ER1 at 75°C, while there was a large scattering for the values of ER1 measured at 90°C. Finally for R = 0.25, ER1 was higher than 400 mV at both tested temperatures.

Corrosion Mode of the Tested Specimens

Figure 7 shows crevice corrosion and transpassivity in specimen JE1714 tested in the low R Electrolyte 3 solution at 90°C. Figure 8 shows only transpassivity in specimen JE1734 tested in

Electrolyte 4. Because of the value R of 0.25, specimen JE1734 suffered only transpassive dissolution in spite of the high anodic potentials applied during the test.

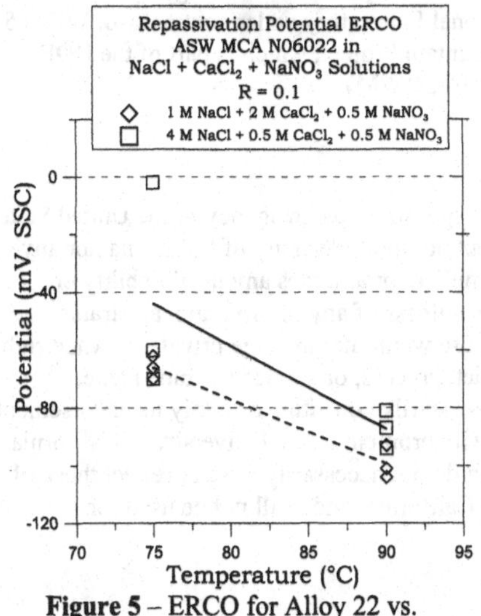

Figure 5 – ERCO for Alloy 22 vs. temperature. Effect of NaCl vs. $CaCl_2$

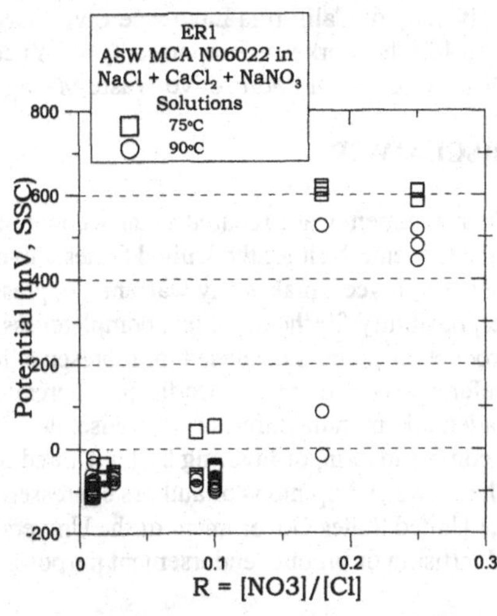

Figure 6 – ER1 for all the tests vs. ratio R

Figure 7– Specimen JE1714 after CPP in Electrolyte 3, R = 0.01 at 90°C. Crevice corrosion and transpassivity are observed.

Figure 8– Specimen JE1734 after CPP in Electrolyte 4, R = 0.25 at 90°C. Only transpassivity is observed.

SUMMARY

1. It was confirmed that Alloy 22 suffers crevice corrosion when polarized in solutions that have a low ratio R of nitrate concentration over chloride concentration
2. When the ratio R was higher than 0.18, inhibition of crevice corrosion was observed in spite of the high base concentration of chloride.
3. Studies of the effect of Ca vs. Na are not definitive. While in the R = 0.1 electrolytes Ca seemed more detrimental than Na, the same trend was not evident for R = 0.01 solutions.

ACKNOWLEDGMENTS

This work was performed under the auspices of the U. S. Department of Energy by the University of California Lawrence Livermore National Laboratory under contract No. W-7405-Eng-48. The work was supported by the Yucca Mountain Project, which is part of the DOE Office of Civilian Radioactive Waste Management (OCRWM)

DISCLAIMER

REFERENCES

1 ASTM International, Standard B575, Vol. 02.04 (ASTM, 2002: West Conshohocken, PA).
2 Haynes International, "Hastelloy C-22 Alloy", Brochure H-2019E (Haynes International, 1997: Kokomo, IN).
3 R. B. Rebak in Corrosion and Environmental Degradation, Volume II, p. 69, Wiley-VCH, Weinheim, Germany (2000).
4 R. B. Rebak and P. Crook, "Nickel Alloys for Corrosive Environments," *Advanced Mater. & Proc.*, **157**, 37, 2000.
5 R. B. Rebak and P. Crook, "Influence of the Environment on the General Corrosion Rate of Alloy 22," PVP-Vol. 483 pp. 131-136 (ASME, 2004: New York, NY).
6 R. B. Rebak and Joe H. Payer, "Passive Corrosion Behavior of Alloy 22," ANS Conf. International High Level Radioactive Waste Management, Las Vegas 30Apr-04May 2006.
7 R. B. Rebak and P. Crook, "Improved Pitting and Crevice Corrosion Resistance of Nickel and Cobalt Based Alloys," ECPV 98-17, pp. 289-302 (The Electrochemical Society, 1999: Pennington York, NJ).
8 B. A. Kehler, G. O. Ilevbare and J. R. Scully, *Corrosion*, 1042 (2001).
9 K. J. Evans and R. B. Rebak in Corrosion Science – A Retrospective and Current Status in Honor of Robert P. Frankenthal, PV 2002-13, p. 344-354 (The Electrochemical Society, 2002: Pennington, NJ).
10 K. J. Evans, S. D. Day, G. O. Ilevbare, M. T. Whalen, K. J. King, G. A. Hust, L. L. Wong, J. C. Estill and R. B. Rebak, PVP-Vol. 467, Transportation, Storage and Disposal of Radioactive Materials – 2003, p. 55 (ASME, 2003: New York, NY).
11 Y-M. Pan, D. S. Dunn and G. A. Cragnolino in Environmentally Assisted Cracking: Predictive Methods for Risk Assessment and Evaluation of Materials, Equipment and Structures, STP 1401, pp. 273-288 (West Conshohocken, PA: ASTM 2000).

12 R. B. Rebak in Environmentally Assisted Cracking: Predictive Methods for Risk Assessment and Evaluation of Materials, Equipment and Structures, STP 1401, pp. 289-300 (West Conshohocken, PA: ASTM 2000).
13 C. S. Brossia, L. Browning, D. S. Dunn, O. C. Moghissi, O. Pensado and L. Yang, "Effect of Environment on the Corrosion of Waste Package and Drip Shield Materials," Publication of the Center for Nuclear Waste Regulatory Analyses (CNWRA 2001-03), September 2001.
14 D. S. Dunn, L. Yang, Y.-M. Pan and G. A. Cragnolino, "Localized Corrosion Susceptibility of Alloy 22," Paper 03697 (NACE International, 2003: Houston, TX).
15 K. J. Evans, A. Yilmaz, S. D. Day, L. L. Wong, J. C. Estill and R. B. Rebak, "Comparison of Electrochemical Methods to Determine Crevice Corrosion Repassivation Potential of Alloy 22 in Chloride Solutions," *JOM*, p. 56, January 2005.
16 G. A. Cragnolino, D. S. Dunn and Y.-M. Pan, "Localized Corrosion Susceptibility of Alloy 22 as a Waste Package Container Material," Scientific Basis for Nuclear Waste Management XXV, Vol. 713 (Materials Research Society 2002: Warrendale, PA).
17 D. S. Dunn and C. S. Brossia, "Assessment of Passive and Localized Corrosion Processes for Alloy 22 as a High-Level Nuclear Waste Container Material," Paper 02548 (NACE International, 2002: Houston, TX).
18 J. H. Lee, T. Summers and R. B. Rebak, "A Performance Assessment Model for Localized Corrosion Susceptibility of Alloy 22 in Chloride Containing Brines for High Level Nuclear Waste Disposal Container," Paper 04692 (NACE International, 2004: Houston, TX).
19 D. S. Dunn, L. Yang, C. Wu and G. A. Cragnolino, Material Research Society Symposium, Spring 2004, San Francisco, Proc. Vol. 824 (MRS, 2004: Warrendale, PA).
20 D. S. Dunn, Y.-M. Pan, L. Yang and G. A Cragnolino and X. He, "Localized Corrosion Resistance and Mechanical Properties of Alloy 22 Waste Package Outer Containers" *JOM*, January 2005, pp 49-55.
21 R. B. Rebak, "Factors Affecting the Crevice Corrosion Susceptibility of Alloy 22," Paper 05610, Corrosion/2005 (NACE International, 2005: Houston, TX).
22 D. S. Dunn, Y.-M. Pan, L. Yang and G. A Cragnolino, Corrosion, 61, 11, 1076, 2005.
23 G. O. Ilevbare, K. J. King, S. R. Gordon, H. A. Elayat, G. E. Gdowski and T. S. E. Gdowski, Journal of The Electrochemical Society, 152, 12, B547-B554, 2005.
24 D. S. Dunn, Y.-M. Pan, L. Yang,, and G. A. Cragnolino, *Corrosion*, **61**, 1078 (2005).
25 D. S. Dunn, Y.-M. Pan, L. Yang, and G. A. Cragnolino, *Corrosion*, **62**, 3 (2006).
26 G. O. Ilevbare, *Corrosion*, **62**, 340 (2006).
27 R. M. Carranza, M. A. Rodríguez, and R. B. Rebak, *Corrosion*, **63**, 480 (2007).
28 R. B. Rebak, "Mechanisms of Inhibition of Crevice Corrosion in Alloy 22," in proceedings of Scientific Basis for Nuclear Waste Management XXX, (MRS, 2006: Warrendale, PA).
29 ASTM International, Volume 03.02 "Wear and Erosion; Metal Corrosion" (ASTM International, 2003: West Conshohocken, PA).
30 K. J. Evans, L. L. Wong and R. B. Rebak "Determination of the Crevice Repassivation Potential of Alloy 22 by a Potentiodynamic-Galvanostatic-Potentiostatic Method," PVP-ASME Vol. 483, pp. 137-149 (American Society of Mechanical Engineers, 2004: New York, NY).
31 K. J. Evans and R. B. Rebak "Determination of the Crevice Repassivation Potential of Alloy 22 by a Potentiodynamic-Galvanostatic-Potentiostatic Method," (to be published in JAI, the journal or ASTM International).

Mater. Res. Soc. Symp. Proc. Vol. 1107 © 2008 Materials Research Society

Apparent Inversion of the Effect of Alloyed Molybdenum for Corrosion of Ordinary and Enhanced 316L Stainless Steel in Sulfuric Acid

Gloria Kwong[1], Anatolie Carcea and Roger C. Newman[2]

[1]Ontario Power Generation (OPG)
22 St. Clair Avenue East,
Toronto, Ontario, Canada
M4T 2S3

[2]University of Toronto
Dept of Chemical Engineering & Applied Chemistry
200 College Street,
Toronto, Ontario, Canada
M5S 3E5

ABSTRACT

An aging assessment of the OPG waste resin storage system predicted the potential for premature failure of the carbon steel resin liners. Consequently, resin liners made of 316L stainless steel with a minimum content of 2.5% molybdenum were selected to replace the carbon steel liners. The 2.5% Mo 316L stainless steel was specified to enhance pitting resistance in the spent resin environment. With the additional Mo, one would expect that a brief electrochemical corrosion test will reveal the superiority of such alloy over conventional 316L steel. This study reports a contrary experience.

Introduction

Type 316L stainless steel is a workhorse material, but some believe that it does not meet modern requirements for applications with sulfuric acids, as in OPG spent ion exchange resins. This is due to the low Mo content dictated by the economics of production: typically 2.1%. So manufacturers have introduced special grades containing no less than 2.5 or 2.7% Mo, usually with nitrogen addition. The nitrogen improves pitting resistance and stabilizes the austenite phase. One naturally expects that a brief electrochemical corrosion test will show the superiority of such a material over ordinary 316L steel. This study observed the opposite effect of Mo alloyed in ordinary and enhanced 316L stainless steel in sulfuric acid.

The anodic dissolution of stainless steels in dilute acids has been studied using voltammetry and surface analysis [1-5]. Both Ni and Mo enrich on the dissolving surface, and this is the fundamental reason for the reduction in critical current density for passivation (i_{crit}) by these alloying elements. The effect of Mo is particularly strong if expressed as the reduction in i_{crit} per atom-percent of Mo addition. This inhibiting effect on dissolution is believed – at least by the present authors – to be the underlying reason for the beneficial effect of alloyed Mo on localized corrosion resistance in salt water [6]. Wanklyn [7] associated this with the stability of MoO_2 in acid. Probably the authors of Ref 2 would not agree with such interpretations – they used to draw a line between ordinary acid corrosion and pitting, preferring to associate the latter with subtle effects on ion transport in the passive film.

Experimental Procedures

Two 316L steels with different Mo contents were obtained as ¼" thick plate. Their compositions are shown in Table 1. Specimens for electrochemical study were cut as small blocks and mounted in resin to expose one face to various concentrations of sulphuric acid. Lacquer was used to mask the edges, leaving an exposed area of 0.1- 0.2 cm^2. The surface was abraded to a 1200 grit finish just before each experiment. The electrochemical cell was a 1 liter glass vessel containing 250 ml of solution, deoxygenated by high-purity nitrogen. The counter electrode was platinum and the reference electrode mercury/mercurous sulfate (MSE: 680 mV vs SHE). The solution was heated to 60°C and maintained at that temperature during the measurements. Solutions containing sulfuric acid and sodium chloride at various concentrations were prepared using analytical grade reagents and de-ionized water. Limiting current measurements showed an oxygen content in the range of 50-100 ppb. A PARSTAT 2263 software-controlled electrochemical system was used to perform dc and ac electrochemistry; since corrosion rates needed to be estimated at quite short intervals, the impedance tests generally used the 'multisine' mode with a pseudo-random noise input. This gave slightly noisy but interpretable results. Every experiment started by polarizing the specimen at -1.25 V (MSE) for 10 seconds, a procedure that was known to give sensible and reproducible anodic behavior, provided that only a short time had elapsed since abrasion.

Table 1 Compositions in weight % of the two austenitic stainless steels.

316L	Cr	Ni	Mo	C	Si	Mn	P	S	N	Cu	Co	Fe
Low Mo	16.92	10.43	2.19	0.018	0.37	1.46	0.034	0.001	0.038	0.49	NA	Bal
High Mo	16.67	10.61	2.60	0.02	0.48	1.64	0.028	0.001	0.048	0.38	0.17	Bal

Results and Discussion

The first set of solutions selected (0.1M H_2SO_4 + 0.025M NaCl) showed a consistent reversal in the corrosion behaviour of the steels compared with expectations. Bode plots for the 2 steels after various immersion periods are shown in Figures 1a and 1b. The low Mo alloy consistently shows a higher polarization resistance than the high Mo alloy after different immersion periods. Figure 1c shows the corrosion potentials versus time for the 2 steels.

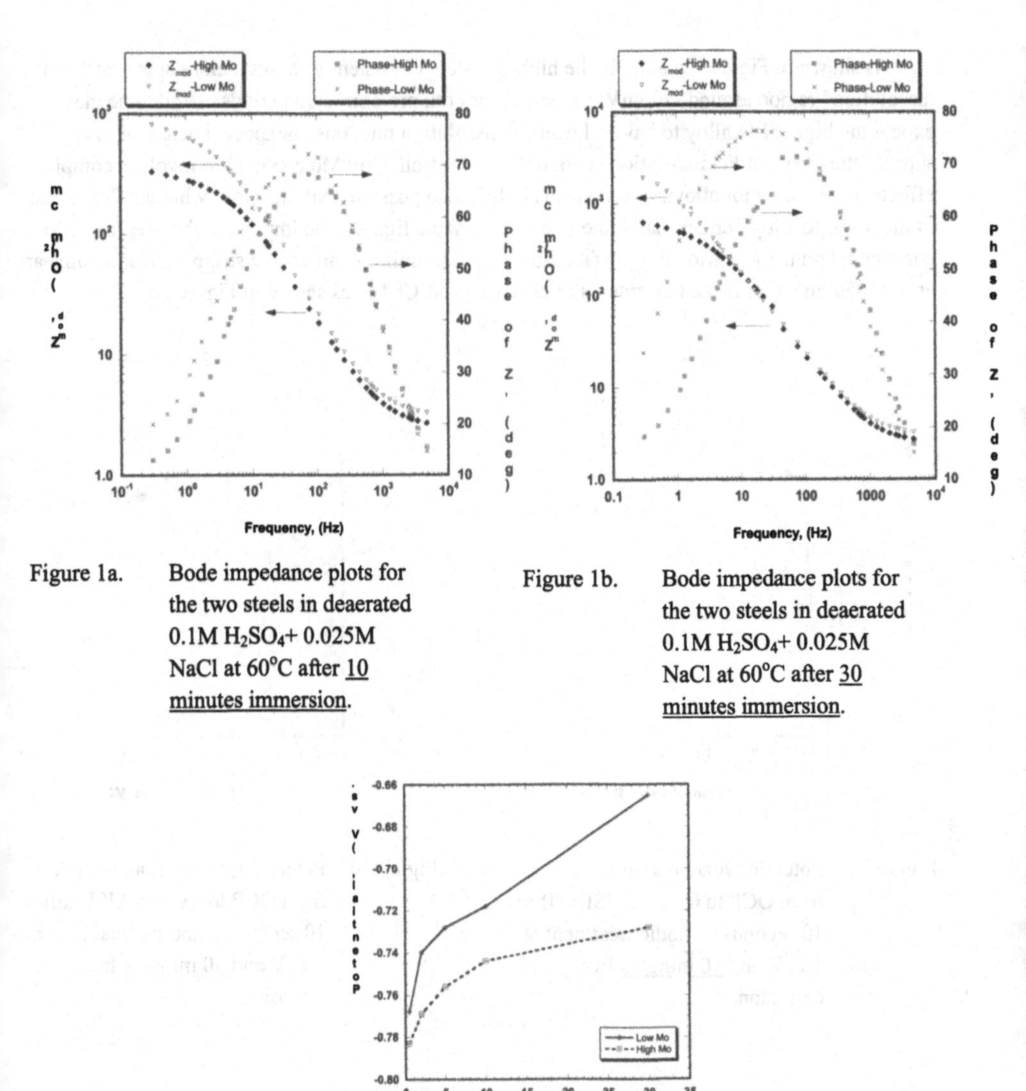

Figure 1a. Bode impedance plots for the two steels in deaerated 0.1M H_2SO_4+ 0.025M NaCl at 60°C after <u>10 minutes immersion</u>.

Figure 1b. Bode impedance plots for the two steels in deaerated 0.1M H_2SO_4+ 0.025M NaCl at 60°C after <u>30 minutes immersion</u>.

Figure 1c. Corrosion potential versus time of the two steels in 0.1M H_2SO_4 + 0.025M NaCl.

Many identical experiments were run to confirm this behavior. To aid interpretation, polarization techniques were also used. The specimens were allowed to corrode freely for different lengths of time, and the potential was scanned in the positive direction. This hardly original procedure turned out to give novel results.

As shown in Figures 2a and 2b, the high-Mo steel consistently shows a more apparent "iron dissolution" region around -700mV (vs. MSE) for both pre-corrosion periods. While one may expect the higher Mo alloy to have a lower Fe dissolution rate, this unexpected behavior may support the idea that Fe dissolution is not entirely controlled by Mo content but involves complex effects of all the major alloying elements [1]. It is also possible that the slightly higher Cr content in the low-Mo alloy perhaps has some effect. The same figures also imply a higher degree of Ni enrichment on the low-Mo alloy surface, since a peak similar to an anodic stripping feature appears around -500mV (confirmed by studies of Ni-Cr and Ni-Cr-Mo as shown in Figure 2c).

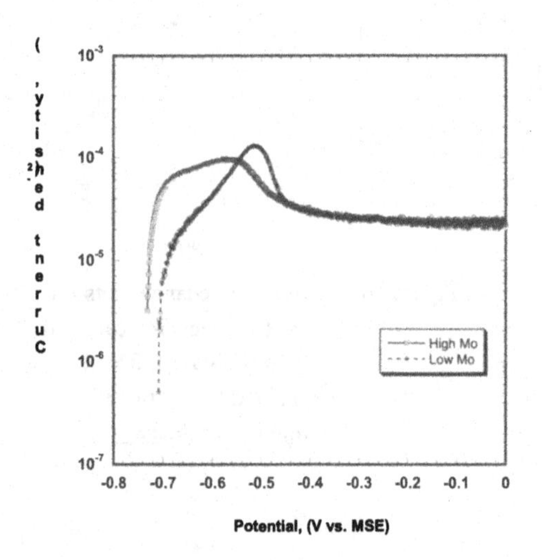

Figure 2a Potentiodynamic scan at 3mV/s, from OCP to 0V (vs. MSE) after 10 seconds cathodic treatment at -1.25V and <u>10 minutes</u> free corrosion.

10^{-3}
10^{-4}
10^{-5}
10^{-6}
10^{-7}
-0.8 -0.7 -0.6 -0.5 -0.4 -0.3 -0.2 -0.1
High Mo
Low Mo
Potential, (V vs. MSE)

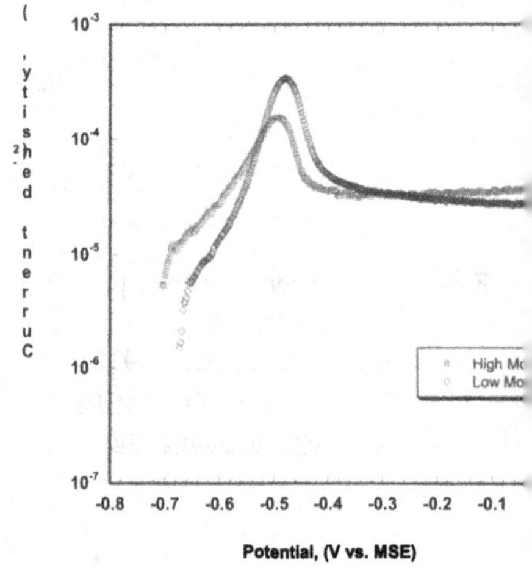

Figure 2b Potentiodynamic scan at 3mV/s, from OCP to 0V (vs. MSE) after 10 seconds cathodic treatment at -1.25V and <u>30 minutes</u> free corrosion.

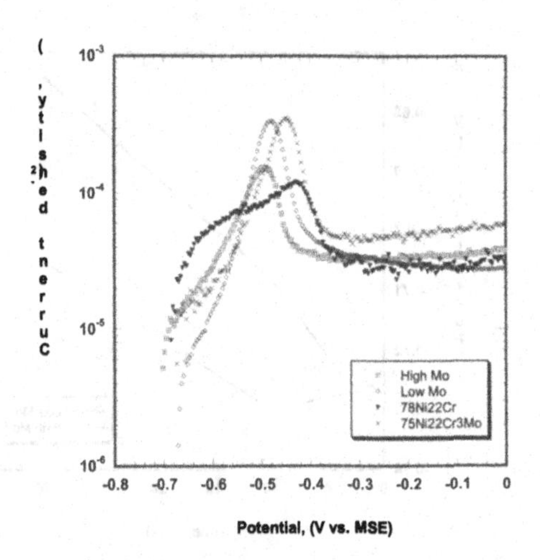

Figure 2c Comparison of Ni "stripping" peak after <u>30 minutes</u> free corrosion with 78Ni22Cr and 75Ni22Cr3Mo, in 0.1M H_2SO_4 + 0.025M NaCl.

This observation suggests that while element enrichment is initiated by selective dissolution of iron, the degree of enrichment may not be linearly proportional to the dissolution rate. Figure 2d illustrates the difference in Ni enrichment between 3 minute and 10 minute immersion periods. Corrosion potentials for the 2 steels after various free corrosion periods are shown in Figure 2e.

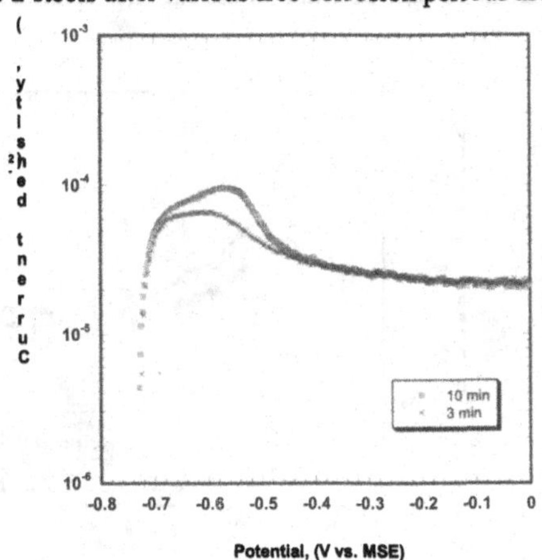

Figure 2d Current density difference for the high Mo steel between 3 minutes and 10 minutes corrosion in 0.1M H_2SO_4 + 0.025M NaCl

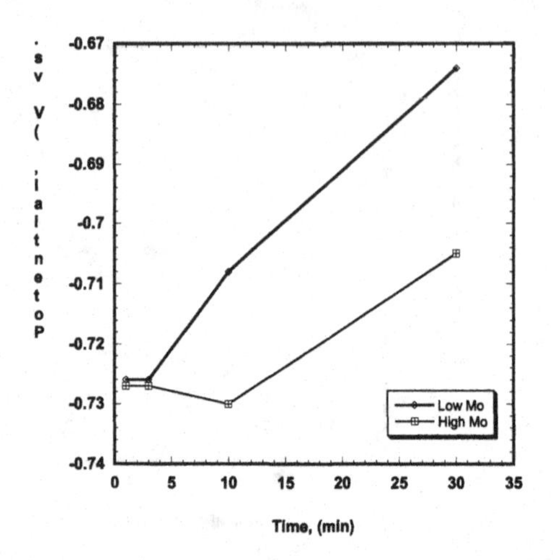

Figure 2e Corrosion potential for the two steels in 0.1M H_2SO_4 + 0.025M NaCl after 10 seconds cathodic treatment at -1.25V and free corrosion for various periods

In stronger acid (1M H_2SO_4 + 0.025M), the materials behaved almost equally, as shown in Figure 3a. Again, the double peak behavior in these curves seems to be a novel observation despite the simple procedure. The low Mo alloy experiences more apparent iron dissolution in stronger acid but continues to show greater evidence of Ni enrichment than the high Mo steel. Note that the Ni "stripping" peak occurs at a lower potential in the stronger acid, in line with the behavior of pure Ni.

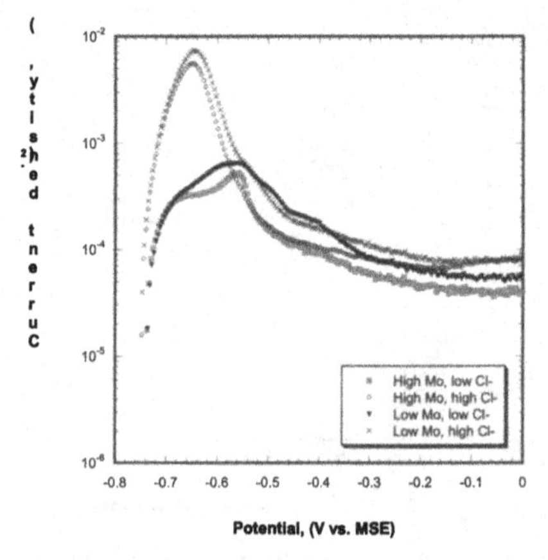

Figure 3a Potentiodynamic scan at 3mV/s, from OCP to 0V (vs. MSE) after 10 seconds cathodic treatment at -1.25V and <u>30 minutes free corrosion</u> in 1M H_2SO_4 + 0.025M Cl^- (low) to 0.25M Cl^-.(high).

In strong acid, more significant Ni dissolution reduces the blocking effect of Ni enrichment. On a linear scale, Figure 3b, we can see a slight superiority of the high-Mo steel in the Fe dissolution region when corroded freely in strong solution (1M H_2SO_4 + 0.25M NaCl). The low Mo steel continues to exhibit more evidence of Ni surface enrichment than the high Mo steel (Figure 3c) although the Ni peak is significantly less when exposed to weak acid.

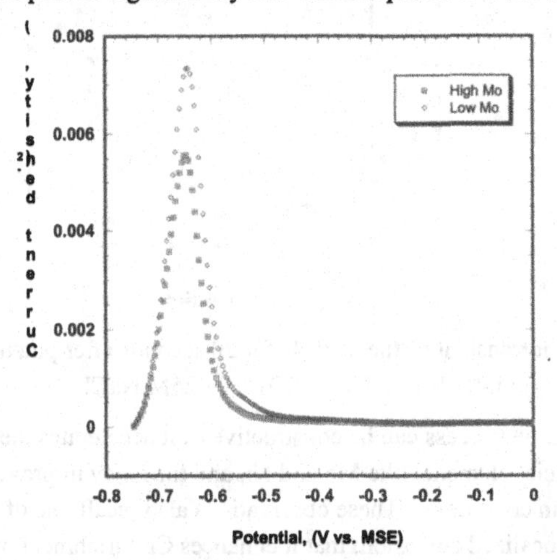

Figure 3b Enlarged details of the two steels showing current density behavior between -0.8 to -0.5V (vs. MSE) after 10 seconds cathodic treatment (-1.25V) and 30 minutes free corrosion in 1M H_2SO_4 + 0.25M NaCl.

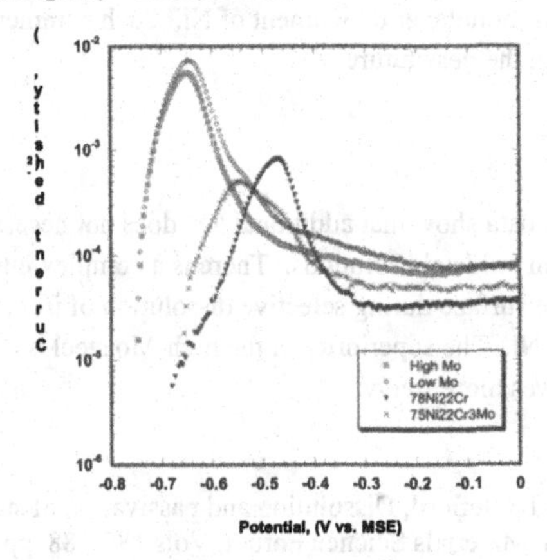

Figure 3c Potentiodynamic scan at 3mV/s, from OCP to 0V (vs. MSE) after 10 seconds cathodic treatment at -1.25V and 30 minutes free corrosion, in 1M H_2SO_4 + 0.25M NaCl.

Potential stepping experiments from -1.25 V to -0.7 V showed that indeed the low-Mo alloy had a lower current density at the onset of the corrosion process, as shown in Figure 4.

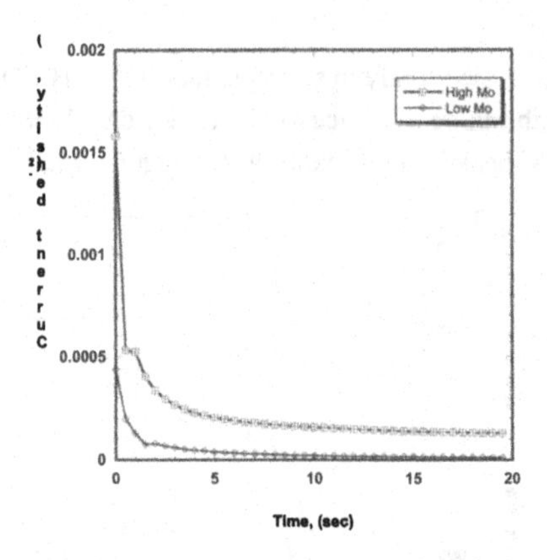

Figure 4 Current transient of the 2 steels for 20 seconds after potential step from -1.25V (10 seconds) to -0.7V in 0.1M H_2SO_4 + 0.025M NaCl

The iron dissolution process can be constructive as it accelerates the onset of elemental enrichment of beneficial elements like Mo and Ni, and thus may improve the corrosion resistance of the alloy under certain conditions. These observations also recall one of the numerous theories for the effect of Mo on localized corrosion: that it enhances Cr enrichment by accelerating initial dissolution of iron [8]. Since alloying elements like Mo and Ni move around by surface diffusion during Fe dissolution, it appears that there can be constructive or destructive mutual effects on the corrosion resistance. Possibly the presence of more than a critical amount of Mo prevents the formation of a continuous monolayer enrichment of Ni. Such comments will form the basis of a surface analytical study in the near future.

Conclusion

The electrochemical data show that additional Mo does not necessary improve the corrosion resistance of 316L steel in mild acid solutions. There is a complex interaction between alloying elements enriching on the surface during selective dissolution of iron, most likely involving the redistribution of Mo and Ni. The superiority of the high-Mo steel is only seen in strong acid solution where Ni dissolves more freely.

References

1. L. Wegrelius and I. Olefjord, Dissolution and passivation of stainless steels exposed to hydrochloric acid, Materials Science Forum, Vols 185-188, pp 347-356 (1995).
2. I. Olefjord and C.R. Clayton, ISIJ International, 31, 134-141 (1991).
3. H. Knote, S. Hofmann and H. Fischmeister, Fresenius' Z. Anal. Chem., 329, 292-297 (1987).
4. I. Olefjord, B. Brox and U. Jelvestam, J. Electrochem. Soc., 132, 2854-2861 (1985).
5. I. Olefjord, Mater. Sci. Eng., 42, 161-171 (1980).
6. N. J. Laycock and R.C. Newman, Corros. Sci., 39, 1771 (1997).
7. J. N. Wanklyn, Corros. Sci., 21, 211-225 (1981).
8. P.I. Marshall and G.T. Burstein, Corros. Sci., 24, 363-478 (1984).

Mobilization, Migration and Retention

Mater. Res. Soc. Symp. Proc. Vol. 1107 © 2008 Materials Research Society

Grimsel Test Site - Phase VI: Review of Accomplishments and Next Generation of *In-Situ* Experiments Under Repository Relevant Boundary Conditions

Ingo Blechschmidt, Stratis Vomvoris, Joerg Rueedi and Andrew James Martin
Nagra (National Cooperative for the Disposal of Radioactive Waste)
5430 Wettingen, Switzerland

ABSTRACT

The Grimsel Test Site owned and operated by Nagra is located in the Swiss Alps (www.grimsel.com). The Sixth Phase of investigations was started in 2003 with a ten-year planning horizon. With the investigations and projects of Phase VI the focus has shifted more towards projects assessing perturbation effects of repository implementation and projects evaluating and demonstrating engineering and operational aspects of the repository system. More than 17 international partners participate in the various projects, which form the basic organisational "elements" of Phase VI. Scientific and engineering interaction among the different projects is ensured via an annual meeting and several experimental team meetings throughout the year. On-going projects include: evaluation of full-scale engineered systems under simulated heat production and long-term natural saturation (NF-Pro/FEBEX), gas migration through engineered barrier systems (GMT, finished this year), emplacement of a shotcrete low-pH plug (ESDRED/Module IV), testing and evaluation of standard monitoring techniques (TEM).

Numerous *in-situ* experiments with inactive tracers and radionuclides were successfully carried out over the past few years at the Grimsel Test Site (GTS). For the GTS Phase VI, three major projects have been initiated to simulate the long-term behaviour of contamination plumes in the repository near-field and the surrounding host rock:

- The CFM (Colloid Formation and Migration) project, which focuses on colloid generation and migration from a bentonite source doped with radionuclides;
- The LCS (Long-Term Cement Studies) project, which aims at improving the understanding of low-pH cement interaction effects in water conducting features;
- The LTD (Long-Term Diffusion) project, which aims at *in-situ* verification of long-term diffusion concepts for radionuclides.

As Phase VI approaches its mid-term point, what are the next steps planned? The accomplishments assessed to date and the opportunities with the on-going projects as well as new projects – currently under discussion – are presented herein.

INTRODUCTION

In 1983, Nagra (the Swiss National Cooperative for the Disposal of Radioactive Waste; www.nagra.ch) initiated a wide range of in-situ experiments in its underground rock laboratory, the Grimsel Test Site (GTS), located in the crystalline rocks of the Aare Massif of the Swiss Alps [1, 2, 3]. This facility, which is not considered for waste disposal, provides convenient horizontal access to the rock 450 m below the flanks of the Juchlistock Mountain and allows experiments to be carried out within a network of tunnels and caverns, with infrastructure and services found in a normal surface laboratory. In particular, the GTS includes an IAEA level B radiation-controlled zone, which allows *in-situ* experiments with radionuclides – including actinides such as thorium, uranium, neptunium, plutonium and americium [4, 5, 6].

Based on the wide range of experiences gained since 1983 and considering the input received and Nagra's own needs, in 2003 Nagra initiated GTS Phase VI with a 10 year planning horizon. GTS Phase VI objectives have also developed from the ones set in 1983 [7, 8]:

- Develop further and maintain know-how for key engineering issues such as handling, emplacement, monitoring and retrieval of high-level waste.
- Apply state-of-the-art science to validate key models over long periods (all waste types) by longer-term radionuclide retardation projects.
- Raise confidence and acceptance in key concepts prior to the repository licensing/construction by full-scale engineering projects.
- Act as a platform for scientific collaboration in the waste management community by providing access to a facility with flexible, open boundary conditions.
- Provide a centre for training future generations of "nuclear waste"-experts (considering the needs of implementers, regulators and research organisations).
- Provide an infrastructure for technical PR (Public Relation).

The participation of several partner organizations provided the expert resources needed for complex, multi-disciplinary experiments and projects and also allowed costs to be shared. GTS projects are currently run by international teams drawn from 17 partner organizations from the Czech Republic, Finland, France, Germany, Japan, United Kingdom, Spain, Sweden and Switzerland, as well as numerous universities, institutes and companies from around the world. The European Union (with the Swiss State Secretariat for Education and Research; SER) provides financial support to some projects.

GRIMSEL PHASE VI – ACCOMPLISHMENTS TO DATE

The Full-scale Engineered Barrier Experiment (FEBEX), Gas Migration Test in the EBS and Geosphere (GMT) and the High pH-Plume in Fractured Rocks (HPF) are long-term projects that were initiated in Phase V already and were integrated in Phase VI in 2003. Their accomplishments are discussed in [9].

Colloid Formation and Migration (CFM)

CFM (Colloid Formation and Migration) is dedicated to study the generation of colloids from a bentonite-based engineered barrier system (EBS) and to investigate the influence of such colloids on radionuclide migration in a fractured host rock under advective flow conditions. The CFM project is the current project in a series of experiments conducted in the radionuclide retardation program of the GTS since 1984 and is subdivided into 2 Phases. Phase I began in 2004 and will be completed in 2008. The particular characteristic of CFM is that the flow field conditions will be closer to repository relevant conditions than the preceding experiments at the GTS. The field experiments are supported by:

- Laboratory studies to increase process understanding and to support the *in-situ* experiment as well as the modelling task (batch, core and column experiments), and
- Numerical modelling (hydraulic modelling and colloid generation/transport modelling) for site characterisation and experimental design.

Geochemical issues such as actinide-colloids interaction as a function of radionuclide speciation will also be investigated both *in-situ* and in the laboratory. The principle layout of the *in-situ* experiments in the tunnel is shown in Figure 1 and consists of an emplacement borehole (grey), a bentonite source (blue) and a sealed tunnel surface around the shear zone (orange). Additional installations such as observation boreholes (green), extraction borehole (red) and sealing boreholes (brown) will be added, as required, later in the experiment.

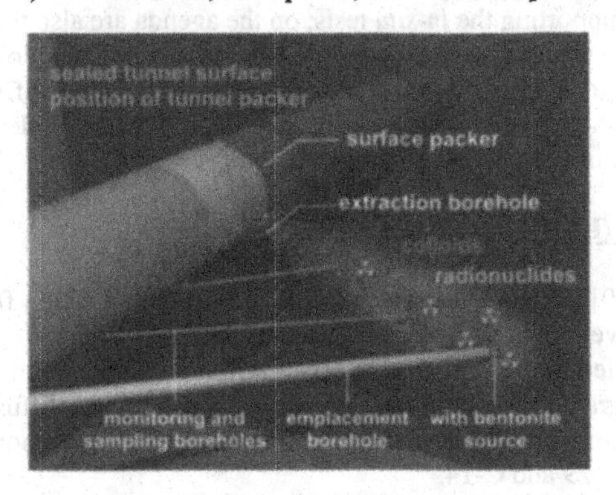

Figure 1. Simplified layout of the *in-situ* experiment.

The sealing of the tunnel surface was one of the major experimental tasks in 2006/2007 and it was accomplished through the installation of a steel tunnel-packer (6 m long and 3.5 m in diameter), fondly referred to as the CFM "yellow submarine" (Photo 1).

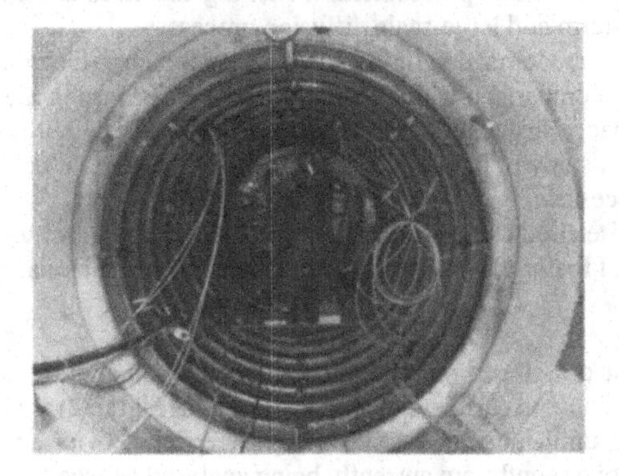

Photo 1. CFM tunnel-packer CFM "yellow submarine" installed along the experimental site.

To date a series of preparation tracer tests have been performed within the target shear zone to evaluate suitable flow fields for the colloid migration experiment (assessment of advective travel times and recovered tracer mass and estimation of dispersion parameters in the shear zone flow fields). The laboratory group has initiated detailed studies of the following issues:

- Colloid generation (physical erosion, geochemical alteration etc.).

- Colloid transport/retardation and stability (filtration effects, sedimentation, groundwater chemistry).
- Radionuclide association (colloid-RN binding, reversibility, radio-colloids).
- Bentonite intercomparison (MX-80, Febex, Kunigel).

The international modelling group has also been focussing its activities with first priority to the short-term aim of supporting the *in-situ* tests; on the agenda are also the medium-term aims, namely, supporting PA relevant studies on colloid generation and on colloid-facilitated radionuclide transport (including irreversibility of sorption and matrix diffusion/filtration effects). The next step is the detailed planning of the bentonite/tracers/radionuclides emplacement, expected to be initiated in Phase II of the project in 2008.

Long Term Diffusion (LTD)

Matrix diffusion [10] is the process by which solutes in groundwater, flowing in discrete hydraulically-conductive fractures, penetrate into the adjacent rock mass and are transported through an interconnected micro-porous network of micro fractures, grain boundary pores, intragranular and intergranular pores in the rock matrix. Rock matrix diffusion is particularly important when determining dose and risk calculations for weakly- and non-sorbing radionuclides such as I-129 and C-14.

The Long Term Diffusion (LTD) project is a series of experiments which aims to obtain quantitative information on matrix diffusion under *in-situ* conditions. The project is divided into four work packages in Phase I (2005-2008) of the LTD project which are:

- Work-package 1: An *in-situ* monopole diffusion experiment where radionuclide tracers diffuse into undisturbed rock matrix with subsequent geochemical analysis of matrix samples combined with predictive and post mortem modelling exercises to increase confidence in the modelling of long-term and large-scale diffusion processes.
- Work-package 2: Characterisation of the pore space geometry (including determination of *in-situ* porosity for comparison with laboratory-derived data) using the C-14 doped PMMA (polymerpolymethacrylate) resin injection and NHC-9 chemical porosimetry techniques.
- Work-package 3: A study of natural tracers in the rock matrix to elucidate evidence for long-term diffusion processes.
- Work-package 4: Detailed characterisation of the flow paths in a water-conducting fracture and investigation of the *in-situ* matrix diffusion paths in core material from earlier GTS experiments.

Circulation of a cocktail of sorbing, weakly sorbing and non-sorbing radionuclides (^{3}H, ^{22}Na, ^{131}I, ^{134}Cs) in the monopole experiment of work-package 1 (Figure 2) was started on June 6, 2007 and will continue until June 2008. Water samples retrieved twice a month and more frequently in the first two months are currently being analysed to ascertain the amount of sorption and diffusion.

Field activities in other work packages are now complete. Laboratory analyses of C-14 and NHC-9 resin impregnated overcores and other rock samples are being carried out in order to improve knowledge of matrix porosity and to refine predictive diffusion models carried out before the start of each *in-situ* experiment.

The main aim of the next phase (Phase II) is to investigate the migration of radionuclides from a fracture into the rock matrix. This is because most performance assessment (PA) models use a parameter that drives the transport from the fracture into the pore matrix. A common approach is to assume that only a limited part of the contact surface between fracture and matrix, the so called flow wetted surface (FWS), can be used for radionuclide transport. In order to examine the effect of FWS on the extent of diffusion, dipole experiments are planned for Phase II (from 2009) where sorbing and safety relevant radionuclides will be circulated in a shear zone.

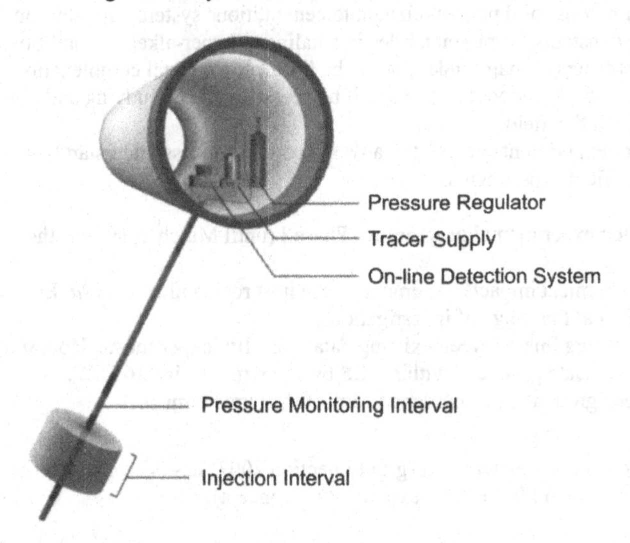

Figure 2. Simplified layout of *in-situ* monopole experiment (LTD, work-package 1) with injection interval for the tracer cocktail between 6.90 and 7.60 m. The total length of the borehole is 8.02 m, with a diameter of 56 mm.

Long Term Cement Studies (LCS)

Based on the results of the HPF project [11], the basic need for further studies related to cement leaching/hyper-alkaline plume production related issues was formulated with an emphasis on *in-situ* field experiments with more repository realistic boundary conditions (longer time frame, lower water fluxes etc) and focussed on low alkali cements. A modelling module was also defined with the emphasis on understanding processes and mechanisms to better facilitate use of the experimental results in national waste programmes worldwide. Thus, the overall aim of the LCS project is to increase the understanding of OPC (Ordinary Portland Cement) and low pH cement interaction effects with the host rock in the repository near field so that modelling can make confident, robust and safety-relevant assessments of future system behaviour, irrespective of repository host rock, EBS and waste type. The following main tasks have been identified and partly initiated:

- Field experiment(s) at the Grimsel Test Site: The main thrust of the LCS *in-situ* experiment will be the utilisation of more relevant hydraulic gradients, different kinds of solid sources

(OPC and low-pH cement) and a significantly longer time frame. In addition, the use of bentonite for sealing boreholes will be investigated, offering the opportunity to examine the different bentonite – cement interactions.

- Thermodynamic modelling: A significant modelling module to bridge the time scales and different repository designs and materials and to enable robust and defensible predictions of the effects of the different cement leachates on repository performance.
- Database development to compare and select basic thermodynamic data (NEA, JAEA, etc.) for the numerous solid phases relevant to cementitious systems. In addition, as many simple solubility constants for radionuclides in alkaline to hyper-alkaline conditions are highly uncertain (orders of magnitude) due to the little known metal complexation constants beyond the first hydroxide association, this will be investigated by studying and comparing existing work done in this field.
- Laboratory experiments geared towards supporting process understanding, modelling efforts and *in-situ* field experiments.

The detailed experimental program for Phase I (until March 2009) has the following major aims:

- Focus on chemical impacts of cement on the host rock and *vice versa*. Engineering issues as a by-product at this stage of investigations.
- Provide a strong link between existing data (e.g. HPF experiments, laboratory experiments, etc.) and new data produced within LCS by using micro fine 16 OPC.
- Use of fresh grout and pre-hardened cement as source terms.

The site for the first experiment (grout injection/2007) has been characterised and instrumented (see also Figure 3 for experimental concept).

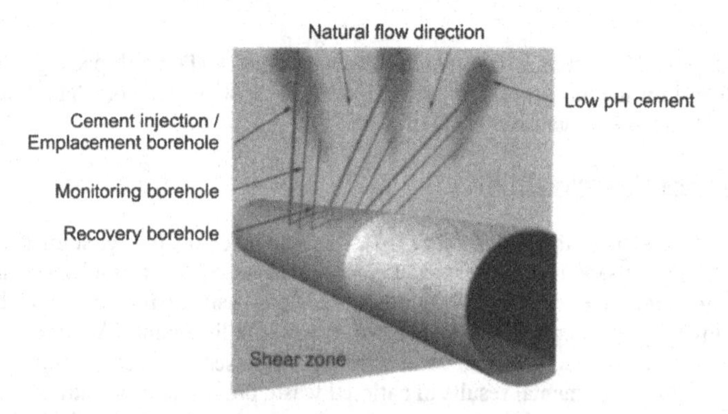

Figure 3. Experimental concept for the first experiments planned.

A review of available thermodynamic datasets currently used, with a special emphasis on solid phases (an interim Project report is in progress) is on-going. For their critical evaluation, a number of benchmark experiments is being defined and will be modelled.

Test and Evaluation of standard Monitoring techniques – TEM

In many countries, the permanent closure of a geological repository will be guided by observations in the vicinity of the disposal site or pilot facilities. In addition to traditional monitoring techniques with hard-wired connections, novel wireless and non-intrusive techniques are being considered, the goal of TEM is to test and evaluate existing monitoring techniques under realistic conditions. The novelty of TEM is that these techniques are being evaluated in a common full-scale realistic setup provided by the ESDRED (Engineered Studies and Demonstration of Repository Designs) shotcrete plug experiment installed at the GTS at the end of 2006 and beginning of 2007 (Figure 4). The concrete plug provides a confining force for a 1 m thick bentonite buffer made up of 5 sections of highly compacted bentonite blocks. Specific tasks within TEM are:

i) Investigate the efficiency of existing *wireless magneto-inductive (MI) transmission techniques*
ii) Evaluate the seismic tomography as a *non-intrusive monitoring technique*
iii) Provide additional data for the hydration process within the ESDRED shotcrete plug.

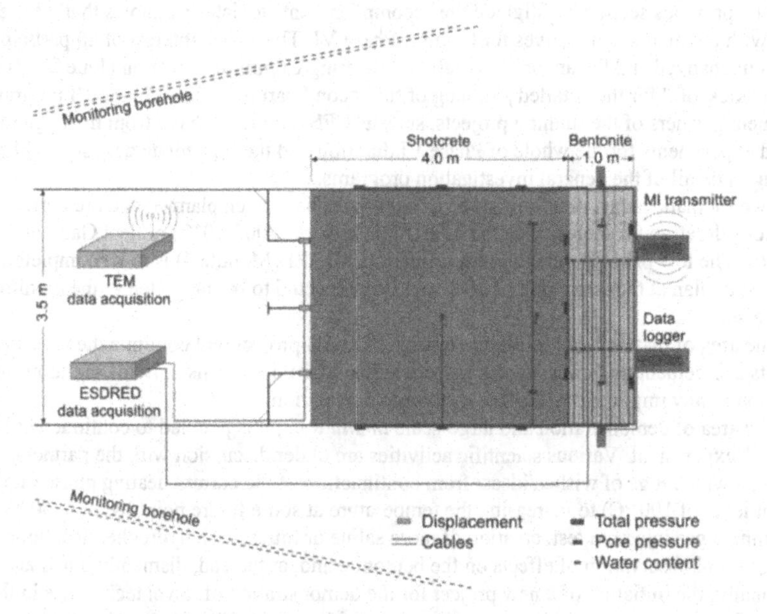

Figure 4. Layout of the ESDRED shotcrete plug and TEM experiments with conventionally wired (green), wireless (red) sensors, monitoring boreholes for the seismic surveys (grey) and data acquisition.

Since March 2007, the bentonite sections are being artificially saturated by injecting water between each section with the aim of developing a swelling pressure of circa 5 MPa when fully saturated. This ESDRED plug experiment also provides the opportunity to carry out a direct evaluation of different 'standard' monitoring techniques that could be considered in future repository monitoring programs, namely, non-intrusive monitoring and conventional monitoring – with hard wired sensors to the data acquisition system and wireless systems. A unique opportunity emerges to test and evaluate the three monitoring methods in a situation which would have similarities with a repository vault-end seal.

With respect to non-intrusive techniques, initial tomographic surveys began prior to any constructions in the tunnel. Repeat surveys followed after completion of the plug (March 2007), during initial saturation of the bentonite (April 2007) and in July 2007. The overall quality of the seismic recordings is judged to be very good, with signal frequencies up to 5 kHz, being observed. However, no significant effects of the saturating bentonite or the concrete plug could be seen in the first arrival waves. It is expected that full-waveform inversions, which will be performed in the future, will help to detect the small changes also at this initial stage of the experiment.

GRIMSEL PHASE VI – FINAL REMARKS AND THE NEXT STEPS

A planned milestone for Grimsel Phase VI was set at the mid-point of its duration, namely 2008. The previous section highlighted the accomplishments to date and shows that we are on course with the planned objectives for Grimsel Phase VI. The strong interest of all participants' remains unchanged and the annual International Steering Committee meeting (June 2007) was also the "kick-off" for the detailed planning of the second part of Grimsel Phase VI program. The experiment partners of the running projects, such as CFM and LTD, have from the beginning planned experiments for the whole of Phase VI duration and the immediate next step will be the planning in detail of the general investigation programs.

In two thematic areas, monitoring and gas activities have been planned and are currently under consideration for inclusion in the 7th EU Framework (2008-2012) project GasMig & MoDeRn. The low pH shotcrete plug experiment (ESDRED/Module 4) is to be completed according to plan at the beginning of 2008 and it is expected to be integrated to the monitoring project also.

In the area of material and technique testing, the LCS project will continue the investigation of grouts and cement materials, their long-term behaviour under realistic field conditions and evaluation of any implications for repository implementation.

In the area of demonstration and large-scale *in-situ* tests, it is planned to continue with the "FEBEX" experiment. Various scientific activities are under discussion with the partners, covering a wide range of wishes/ideas: from continuation of the current heating phase (at the constant level of 100 °C) to increasing the temperature at some future point by up to 50 %, performing a gas injection test, creation of more saline or nitrate-rich hydrochemical boundary conditions and observation of effects on the bentonite and, at the end, dismantling and analysis. Additionally, the initiation of a new project for the demonstration of novel techniques in the context of repository programs is under discussion with partners. Focus of such a project may be the emplacement techniques and in particular Tele-Handling (TH).

Finally, increased activities are expected with respect to training and technique testing. In fall 2007 a new 3 year project under the leadership of CRIEPI (Japan) on advanced techniques for fractured rock is to be initiated. Training, in general, will continue and expand, through (a) a stronger and direct cooperation with the IAEA, where Grimsel is one of the founding members of the IAEA Network of Centres of Excellence; (b) the consolidation of the current ad-hoc inclusion of Grimsel in the Master and PhD Program of University of Berne and of the ETH

Zurich; (c) the continuous support of the ITC (International Training Centre – School of Underground Waste Storage and Disposal) program and (d) the coordination with other underground facilities for complementary training activities.

AKNOWLEDGEMENTS

Nagra would like to acknowledge the support of the many partner organisations and a large number of contractors who have contributed to the success of the last almost two and a half decades of GTS projects and who are taking an active role in approaching the challenges of future decades.

REFERENCES

1. C. McCombie, I.G. McKinley and S. Vomvoris, "Contributions of the Grimsel Test Site (GTS) to Swiss site characterisation programmes – High level radioactive waste management: Proceedings of the sixth annual international conference": Las Vegas, Nevada, April 30-May 5, 1995. American Nuclear Society, La Grange Park (1995).
2. S. Vomvoris, W. Kickmaier and I.G. McKinley, "Grimsel Test Site – the next decades". Sci. Basis Nucl. Waste Manag. XXVI, 13-21 (2003).
3. W. Kickmaier, S. Vomvoris and I. G. McKinley, "Brothers Grimsel". Nucl. Engin. Internat. 50, 10-13 (2005a).
4. I.G. McKinley, W.R. Alexander, C. Bajo, U. Frick, J. Hadermann, F.A. Herzog and E. Hoehn, "The radionuclide migration experiment at the Grimsel rock laboratory, Switzerland". Sci. Basis Nucl. Waste Manag. XI, 179-187 (1988).
5. P.A. Smith, W. R. Alexander, W. Kickmaier, K. Ota, B. Frieg and I.G. McKinley, "Development and Testing of Radionuclide Transport Models for Fractured Rock: Examples from the Nagra/JNC Radionuclide Migration Programme in the Grimsel Test Site, Switzerland". J. Contam. Hydrol. 47, 335-348 (2001).
6. Möri, W.R. Alexander, H. Geckeis, W. Hauser, T. Schaefer, J. Eikenberg, T. Fierz, C. Degueldre and T. Missana, "The Colloid and Radionuclide Retardation experiment (CRR) at the Grimsel Test Site (GTS): influence of bentonite colloids on radionuclide migration in a fractured rock". Colloids and Surfaces A, 217, 33-47 (2003).
7. W. Kickmaier, S. Vomvoris; I.G. McKinley and W.R. Alexander: Grimsel Test Site: challenges for the 21st century and beyond. European Geologist / 19, pp. 19-22 (2005b).
8. S. Vomvoris, W. Kickmaier, I. McKinley, "Grimsel Test Site: 20 years of research in fractured crystalline rocks – Experience gained and future needs", in Proceedings of the Second International Symposium on Dynamics of Fluid in Fractured Rock, Report No. LBNL-54275, Berkeley National Laboratory, Feb. 10-12, 2004.
9. S. Vomvoris and W. Kickmaier, Grimsel Test Site – Phase VI: Review of accomplishments and next steps, Proc. 11th ICEM2007, ICEM07-7239 (to be published).
10. Neretnieks, I., 1980. Diffusion in the rock matrix: an important factor in radionuclide retardation? J. Geophys. Res. 85, 4379-4397.
11. J.M. Soler, W. Pfingsten, B. Paris, U.K. Mäder, B. Frieg, F. Neall, G. Källvenius, M. Yui, Y. Yoshida, P. Shi, Ch. A. Rochelle and D.J. Noy, Grimsel Test Site – Investigation Phase V, HPF-Experiment: Modelling Report, June 2006, Nagra Technical Report 05-01, Nagra, Wettingen, Switzerland.

Mater. Res. Soc. Symp. Proc. Vol. 1107 © 2008 Materials Research Society

Molecular Characterisation of Dissolved Organic Matter (DOM) in Groundwaters from the Äspö Underground Research Laboratory, Sweden: A Novel "Finger Printing" Tool for Palaeohydrological Assessment

Christopher H. Vane*[1], Alexander W. Kim[1], Antoni E. Milodowski[1], John Smellie[2], Eva-Lena Tullborg[3] and Julia M. West[1]

[1]British Geological Survey, Kingsley Dunham Centre, Keyworth, Nottingham, NG12 5GG, United Kingdom
[2]Conterra AB, Box 8180, 10420 Stockholm, Sweden
[3]Terralogica AB, Östra Annekärrsvägen 17443 72, Gråbo, Sweden

ABSTRACT

The molecular signature of dissolved organic matter (DOM) in groundwaters can be used as a tool when investigating the palaeohydrological response of groundwater systems in relation to changes in recharge environment, and also for examining groundwater compartmentalisation, mixing and transport at underground repositories for radioactive waste. The DOM in groundwaters from two compartmentalised bodies of groundwater of distinctly different origin within the Äspö Underground Research Laboratory (URL), Sweden and in Baltic seawater has been isolated using tangential flow ultrafiltration (TUF) and diafiltration. Recoveries of DOM ranged from 34.7 to 0.1 mg/L with substantial differences in the concentrations of the groundwaters collected only 120 m apart. Analysis by infrared spectroscopy (IR) and pyrolysis-gas chromatography-mass spectrometry (Py-GC-MS) of the isolated DOM revealed that the groundwaters contained abundant alkylphenols which may represent heavily decomposed proteins or lignins originating from biopolymers contained within soils. The difference in the distribution and relative abundance of major pyrolysis products groups such as alkyphenols confirmed that the groundwater and Baltic seawater DOM samples were chemically distinct indicating minimal infiltration of marine groundwater derived by recharge from the Baltic or earlier Littorina Sea within the two compartmentalised groundwater bodies.

INTRODUCTION

Safety considerations for the disposal of radioactive waste and spent fuel in a deep geological repository must take into account the hydrogeological evolution of the site over a timescale of the order of 10^5 to 10^6 years. An important issue considered is the effect of climate change on the deep groundwater system. During the last million years or so (the Quaternary Period), European climate has alternated between extremes of ice ages and conditions much warmer than today. Large areas of northern Europe were covered by ice sheets and experienced extensive permafrost, whilst southern Europe was more arid. The present climate state is not representative of the climate that existed for much of the Quaternary, and it could be argued that the present-day groundwater conditions are not an adequate basis for assessing the long-term safety of a repository. The stability of groundwater conditions is one of the most important safety requirements, because the chemical composition of the water and its flow are key factors that will influence the transport of radionuclides in the geosphere to the surface. The impacts of past Quaternary climate changes on the stability of deep groundwater systems (palaeohydrogeology) are therefore of significant interest to predicting long-term repository safety assessment [1,2].

Understanding of the palaeohydrogeological evolution of groundwater systems is normally elicited from the interpretation of the inorganic chemistry, stable isotopes, fluid inclusion and noble gas composition characteristics of groundwaters and/or mineralogical features formed within groundwater systems [1,3,4]. However, a complementary approach to evaluate flow paths, mixing and compartmentalisation of groundwater masses is to characterise the dissolved organic matter (DOM) contained in the groundwaters. DOM in groundwaters may originate from several sources including: surface soils during infiltration in the recharge area; organic matter contained in the host rocks; or *in situ* production from natural microbial biomass sources. Although the mineralogical record may sometimes preserve a more permanent palaeohydrogeological record compared to the more transient record provided by groundwater chemistry, the present study sought to look for the presence of organic molecules directly in the groundwaters rather than in the minerals precipitated from the groundwater. This has a number of potential advantages:

- It is easier to collect large volume samples of water;
- Water samples can be relatively easily processed to concentrate biomarkers for analysis, and;
- Water samples are easier to analyse, compared to the careful separation and analysis of individual mineral generations from complex fracture mineralisation.

Operationally defined DOM (0.2 μm- 1000 Da) in fresh and marine waters is generated from the secretion and transformation of biomolecules and is primarily comprised of non-living polymeric materials, which have been selectively preserved as well as viruses and some colloids [6]. The DOM in aquatic environments such as Pacific and Atlantic oceans, rivers, estuaries and freshwater wetlands have been isolated using tangential-flow ultrafiltration (TUF), and characterised at the molecular level using a variety of analytical techniques in order to better understand the global carbon cycle as well as improve knowledge of DOM source, transport and environmental fate [7,8,9]. These studies have shown that DOM from marine, estuarine and river environments can be chemically differentiated since terrestrial biomolecules such as lignin are abundant in stream and river waters but absent or extremely low in open marine waters [10,11]. This study sought to investigate whether the molecular "fingerprints" of DOM from groundwaters could provide complementary information to more traditional geochemical and mineralogical palaeohydrogeological methods.

EXPERIMENT

Two ground waters were collected from different boreholes in the tunnel of the Äspö underground research laboratory (URL), namely; KR0012B (redox alcove) representing relatively recent shallow meteoric recharged groundwater, and; KA1755A, which based on its highly depleted ^{18}O isotopic signature is interpreted to represent older groundwater with a possible large component of glacial melt water. An additional sample of modern Baltic seawater from the Northern Misterhult Archipelago Nature Reserve was also collected. A procedural blank (28 L of 18 MΩ distilled water) was treated in an identical manner to the other samples in order to differentiate artefacts from sampling, prefiltration, ultrafiltration, freeze drying and analysis.

Each sample was filtered and collected in two 28 L polypropylene storage barrels (Nalgene). Ultrafiltration was performed using a Millipore TUF system fitted with a 1000 Da cut-off cassette filter. Each sample yielded 1-1.2 L of retentate after 18.5-20 h of processing. The salts were removed from the retentates by diafiltration and the DOM freeze-dried for 48 h.

Infrared Spectra of KR0012B and Baltic Seawater were obtained using a Bio-Rad FTX3000MX series IR. Pyrolysis-gas chromatography-mass spectrometry (Py-GC-MS) was performed using a platinum resistance heated Chemical Data Systems (CDS) AS2500plus, connected to a Carlo Erba Mega 500 series gas chromatograph (GC). The platinum coil was heated at 610°C for 10 s. Products were separated using a fused silica Varian Factor 4 VF-1MS column (60 m length × 0.32 mm i.d. × 0.25 μm film thickness). The flow rate of helium carrier gas was 1 mL/min. The oven temperature was programmed from 30°C to 300 °C at 4 °C min^{-1} and held isothermally at 300 °C for 15 min. The GC was directly coupled to a Varian 1200L triple quadropole GC/MS/MS system operated in EI mode at 70 eV with a mass range 30-550. Products were identified by comparison of their mass spectra and relative retention times with compounds reported in the literature [12, 13,14] and National Bureau of Standards library.

DISCUSSION

Isolated ultrafiltered dissolved organic matter (DOM)

A comparison of the dry weights of DOM samples showed that KR0012B and Baltic seawater had the highest DOM contents at 970.8 mg and 81.8 mg respectively (Table I). In contrast KA1755A had the lowest DOM value of 2.8 mg and the dH_2O water yielded 1.4 mg (Table I). The presence of DOM in the procedural blank maybe explained by leaching of DOM from pre-filters, TUF filters, tubing or storage barrels. Waters isolated by TUF from the North Sea and Ems-Dollart Estuary have reported concentrations in the range of 6-14 mg/L. In contrast, in this current study, the Baltic seawater had a lower DOM concentration at 2.9 mg/L as compared to the reported North Sea values [15]. One explanation is that the Baltic seawater DOM is diluted with riverine runoff. The large difference in the amounts of DOM isolated from KR0012B and KA1755A is surprising given that the instrumented sites were only around 120 m apart and is entirely consistent with the notion that the two groundwaters have different origins and thus potentially different amounts and sources of organic matter.

Table I. Dissolved organic matter (>1000 Da) isolated from Äspö URL

Sample	Non-purgable organic carbon (mg/L)	Weight DOM powder (mg)	Weight DOM (mg/L)
KR0012B	17.0	970.8	34.7
Misterhult (Baltic seawater)	4.9	81.8	2.9
KA1755A	2.9	2.8	0.1
Procedural blank (distilled H_2O)	<0.5	1.4	0.1

Infrared (IR) spectroscopy

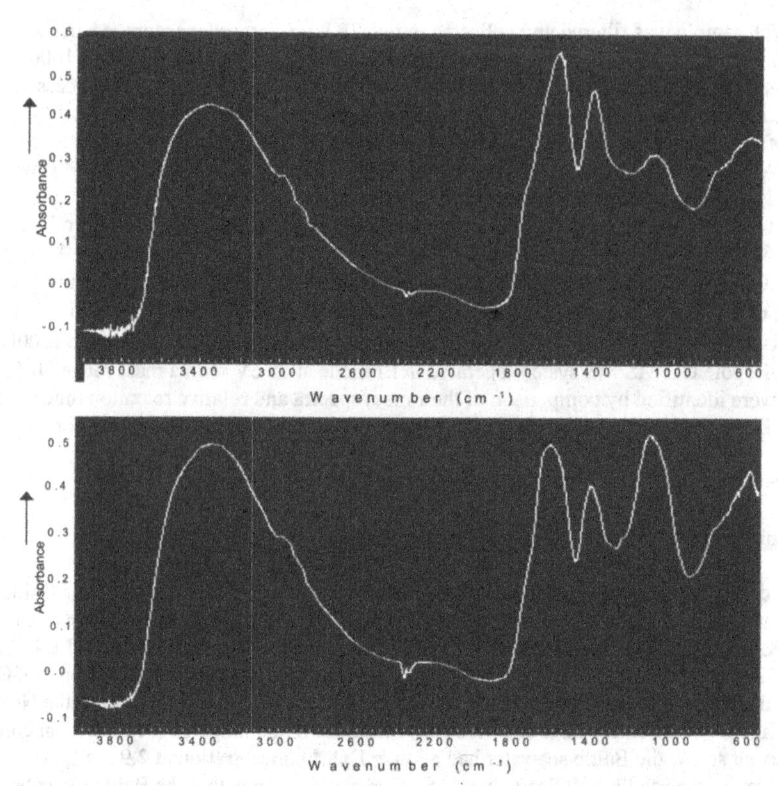

Figure 1. Infrared spectra of dissolved organic matter isolated from groundwater and seawater.

Both IR spectra of DOM from KR0012B and Baltic seawater showed an intense, broad signal centred at 3440 cm^{-1}, the absorption band can be assigned to a number of localised vibrations including O-H stretching from alcohols and phenols as well as N-H stretching from amines and or amide groups (Figure 1). However, the relative abundance of moieties with O-H as compared to those with stretching N-H bonds cannot be determined due to the weaker absorption of the latter [16]. The appearance of a shoulder in both spectra at 2957 cm^{-1} is probably due to CH stretching in aliphatic moieties. In the case of KR0012B a slight shoulder centered at 1740 cm^{-1} was observed, this could be C=O of H-bonded carboxyl groups and C=O of ketonic carbonyl groups which are not conjugated to the aromatic ring [17]. Previous IR studies of peat, river and soil humic acids have reported that the most prominent feature of the spectra was a peak at ~1725 cm^{-1} thus the functional group chemistry of DOM in this current work does not correspond to that of humic acids [18]. Comparison of the spectra in the region of 1600 cm^{-1} revealed intense signals indicative of C=C in plane aromatic and/or asymmetric –COO^- stretch in both KR0012B and Baltic seawater which suggested a significant aromatic component (Figure 1). IR spectra for KA17755A were not obtained due to the low recovery of DOM.

The slightly stronger and sharper form of the peak at ~1600 cm^{-1} in KR0012B may therefore indicate that that the groundwater sample has a higher aromatic content than the Baltic seawater. The KR0012B and Baltic seawater DOM exhibited strong signals centered at 1413 and 1406 cm^{-1} respectively, which can be assigned in part to aromatic ring stretching vibrations and associated C-H in plane deformation (Figure 1). The appearance of the peak at ~1090 cm^{-1} could be due to C-O deformation in secondary alcohols and aliphatic ethers. Previous studies of polysaccharides from wood and microbial cell walls have reported intense absorptions in the range 1050-1170 $cm-^{1}$ thus the peak at ~1090 cm^{-1} may be due to cellulose or xylans. The lower intensity of the peak at ~1090 cm^{-1} in DOM from KR0012B as compared to Baltic seawater suggests that the groundwater could have a lower polysaccharide content than the seawater. Overall, analysis by IR confirmed that the groundwaters and Baltic seawater DOM samples were chemically distinct thus supporting the notion that DOM of KR0012B had not been significantly modified by infiltration of DOM sourced from the Baltic Sea which lies in close proximity to the Äspö URL.

Pyrolysis-gas chromatography-mass spectrometry

Analytical pyrolysis (Py-GC-MS) was performed on DOM preparations of Baltic seawater and Äspö groundwaters from boreholes KR0012B and KA1755A (Figure 2). A comparison of the distribution of DOM products showed that both KR0012B groundwater and Baltic seawater contained a mixture of monomers, which contrasts with the limited distribution of monomers from KA1755A (Figure 2). The simple aromatic compound toluene was encountered in all three waters. However, this product has a number of possible origins including phenyalaline-containing proteins, protein derivatives, or can be produced as a secondary reaction product during pyrolysis of the polysaccharide cellulose [14]. Similarly, N-containing molecules with multiple origins including proteinaceous organic matter such as pyrrole and methylpyrrole were observed in KR0012B and Baltic seawater. Other compounds common to all DOM preparations included the N-containing compounds indole and 3-methylindole (*m/z* 117+131) the latter of which originates in part from the amino acid-moiety tryptophan, a major component of algal proteins (Figure 2) [19].

A variety of different polysaccharide products were observed in relatively high abundance as compared to other protein or phenolic products in the two Äspö URL groundwaters and Baltic seawater (Figure 2). Pyrolysis of KR0012B gave polysaccharides tentatively identified as cyclohexy-1,3diene, 2-methylcyclopenten-1-one and 3-methylcyclopenten-1-one, as well as 2,3-propylfuran (Figure 2). Pyrolysis of cellulose and xylans (e.g. plant derived polysaccharides) also yield similar products, however thermal depolymerisation of cellulose generally yields large amounts of 1,6-anhydro-β-D-glucopyranose [20]. The absence of 1,6-anhydro-β-D-glucopyranose in KR0012B does not necessarily exclude a plant origin for the polysaccharide products since other studies have shown that 1,6-anhydro-β-D-glucopyranose can decompose at elevated pyrolysis temperatures [14].

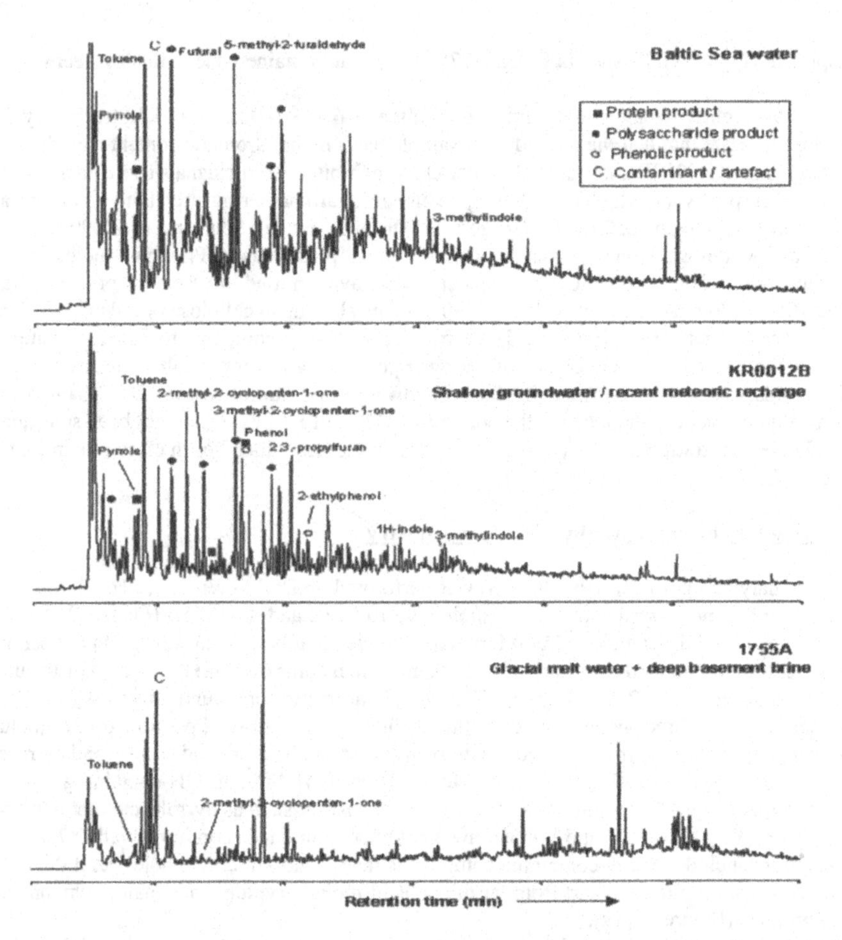

Figure 2. Chromatogram of the total ion current of the pyrolysis products of ultrafiltered DOM from Baltic seawater and groundwaters from boreholes KR0012B and KA1755A.

The polysaccharides from Baltic seawater DOM included those tentatively identified as furfural, 2-methyl-5-ethylfuran, 5-methylfuraldehyde, which probably originate from multiple sources such as plants (cellulose), marine humic substances and algae. Aromatic products including methylbenzenes, xylenes and a variety of phenolic molecules were encountered in KR0012B, in contrast only phenolic products were observed in KA1755A and Baltic seawater (Figures 2). Mass chromatograms reflecting alkylphenols are shown for the DOM preparations in Figure 3. Alkylphenols originate from a number of biopolymers including decayed lignins, polyphenols and proteins as well as being pyrolysis products of green, red and brown algae [13]. One plausible explanation for the broader distribution of alkylphenols in KR0012B as compared to the other DOM samples could be that KR0012B contains alkylphenols from multiple sources including those derived on land from woody plant matter in soils and peat. Such an assumption appears reasonable given that combined the inorganic hydrogeochemical and stable isotope data suggested relatively recent shallow meteoric recharge. However, it must be noted that the absence of methoxyphenolic products - the main marker for lignin - either excludes a land plant origin or alternatively indicates that the original lignin monomers have undergone extensive degradation [12] in the soil.

The pattern of pyrolysis products from all three waters was different confirming the notion that the three selected end-member waters (Baltic seawater, old groundwater with a possible high glacial melt water content and relatively recent shallow meteoric recharged groundwater) can be differentiated using DOM signatures. Furthermore, the clear difference between KA1755A and KR0012B suggests that the two groundwaters are isolated and compartmentalised. Overall the simple assemblage of pyrolysis products in KA1755A is considered to be consistent with this groundwater being sourced from deep basement brine mixed with a large component of glacial meltwater in that the biological productivity, and thus generation of DOM precursor molecules, would be restricted in these environments.

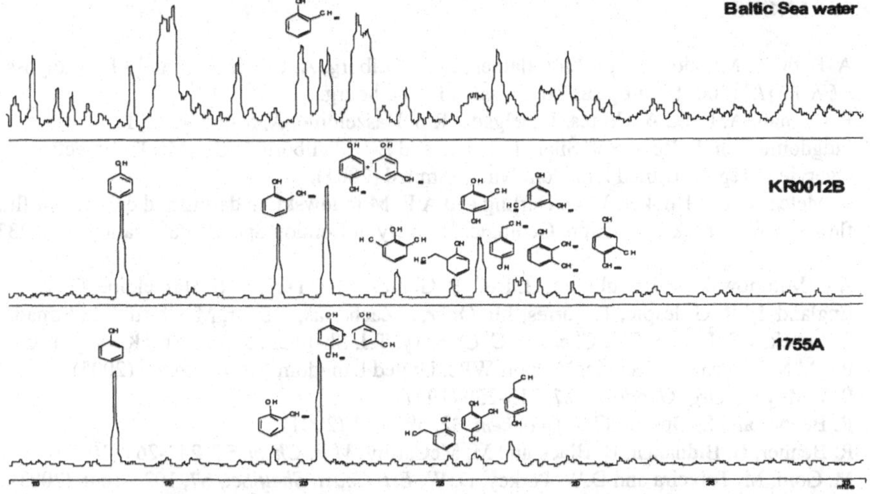

Figure 3. Partial mass chromatograms (*m/z* 94, 107, 108, 121, 122) of C_0-C_2 alkylphenols from Baltic seawater, and groundwaters from boreholes KR0012B and KA1755A. Note: only peaks with chemical structures are alkylphenols.

CONCLUSIONS

This pilot study has demonstrated that the chemical composition of DOM can potentially provide a new tool to evaluate palaeohydrological evolution of groundwater systems in relation to changes in recharge environment, mixing and transport of groundwaters at underground repositories for radioactive waste. Molecular level characterisation of DOM extracts from two Äspö groundwaters and modern Baltic seawater revealed different distributions of polysaccharide, protein and alkylphenols as well as other unidentified moieties reflecting the palaeoenvironment and transport history. DOM in both Äspö groundwater samples was distinct from that in modern Baltic seawater indicating little evidence of mixing with Baltic or earlier Littorina Sea waters. The extraction and analysis of DOM in groundwater provides palaeohydrogeological information that complements and supplements traditional hydrogeochemical and mineralogical approaches, in that it appears to have some utility in identifying compartmentalised waters and may potentially be used to infer the palaeoenvironment at recharge and the extent of biological activity.

ACKNOWLEDGMENTS

This pilot study was funded under the British Geological Survey's (BGS) project BioTran. SKB and staff at the Äspö URL are thanked for granting access to collect groundwater samples for this study, and for providing logistic support during the sampling programme. This paper is published by permission of both the Executive Director, BGS (NERC) and SKB.

REFERENCES

1. A. Bath, A. Milodowski, P. Ruotsalainen, E-T. Tullborg, A. Cortés Ruiz and J.F. Aranyossy, *EUR 19613* (2000), European Commission Luxembourg.
2. P. Degnan, A. Bath, A. Cortés, J. Delgado, R.S. Haszeldine, A. Milodowski, I. Puigdomenech, F. Recreo, J. Šilar, T. Torres and E-L. Tullborg. PADAMOT: Project Overview Report. United Kingdom Nirex Limited (2005).
3. R. Metcalfe, P.J. Hooker, W.G. Darling and A.E. Milodowski, in dating and duration of fluid flow and fluid-rock interaction. Geological Society of London Special Publication, 144, 233-260.
4. A.E. Milodowski, E-L. Tullborg, B. Buil, P. Gómez, M-J. Turrero, S. Haszeldine, G England, M.R. Gillespie, T. Torres, J.E. Ortiz, J. Zachariáš, J. Silar, M. Chvátal, L. Strnad, O. Šebek, J.E. Bouch, S.R. Chenery, C. Chenery, T.J. Shepherd and J.A. McKervey, J.A. PADAMOT Project Technical Report WP2. United Kingdom Nirex Limited (2005).
5. P.A. Meyers, *Org. Geochem*, **27**, 213-250 (1997).
6. R. Benner and S. Opsahl, *Org. Geochem.* **32**, 597-611 (2001).
7. R. Benner, B. Biddanda, B. Black and M, McCarthy, *Mar. Chem*, **57**, 243-263 (1997).
8. M. Goni, M. Teixeira and D.W. Perkey, D. W. *Est. Coast. Shelf Sci.* **57**, 1023-1048 (2003).
9. X.Q. Lu, N. Maie, J.V. Hanna, D.L. Childers and R. Jaffe, R. *Water Res.* **37**, 2599-2606 (2003).
10. S.W. Frazier, K.O, Nowack, K.M, Goins, F.S. Cannon, L.A. Kaplan and P.G. Hatcher, *J. Anal. Appl. Pyrolysis*, **70**, 99-128 (2003).
11. S. Opsahl and R. Benner, Nature, **386**, 480-482 (1987).
12. C.H. Vane *Int. Biodet. Biodeg*, **51**, 67-75 (2003).
13. J.D.H. van Heemst, P.F. van Bergen, B.A. Stankiewicz and J.W. de Leeuw, *J. Anal. Appl. Pyrolysis,* **52**, 239-256 (1999).
14. J. Templier, S. Derenne, J-P. Croue and C. Largeau, *Org. Geochem.* **36**, 1418-1442 (2005).
15. J.D.H. van Heemst, L. Megens, P.G. Hatcher and J.W. de Leeuw, *Org. Geochem*, **31**, 847-857 (2000).
16. D.H. Williams and I. Fleming, *Spectroscopic methods in organic chemistry*, edited by P. Sykes (McGraw-Hill, Maidenhead, 1966), p. 55.
17. C.H. Vane, C. H. *Appl. Spectroscopy*, **57,** 514-517 (2003).

18. Y. Inbar, Y. Chen and Y Hadar, *Soil Sci. Soc. of Amer. J.* **54**, 1316-1323 (1990).
19. S. Tsuge and H. Matsubara, *J. Anal. Appl. Pyrolysis,* **8**, 49-64 (1985).
20. J. Ralph and R. Hatfield, *J. Agric.Food Chem.* **39**, 1426-1437 (1991).

Mater. Res. Soc. Symp. Proc. Vol. 1107 © 2008 Materials Research Society

Modeling of Radionuclide Migration Through Fractured Rock in a HLW Repository With Multiple Canisters

Doo-Hyun Lim[1], Masahiro Uchida[2], Koichiro Hatanaka[3], and Atsushi Sawada[4]
[1] *Golder Associates Inc., 18300 NE Union Hill Road, Suite 200, Redmond WA 98052, USA*
[2] *Japan Atomic Energy Agency*, 959-31 Jorinji, Toki-shi, Gifu 509-5102, Japan
[3] *Japan Atomic Energy Agency*, Horonobe-Cho, Teshio-Gun, Hokkaido 098-3224, Japan
[4] *Japan Atomic Energy Agency*, 4-33 Muramatsu, Tokai-mura, Ibaraki 319-1194, Japan

ABSTRACT

An integrated numerical model for groundwater flow and radionuclide migration analyses in a water-saturated HLW repository with a multiple-canister configuration is developed by incorporating the heterogeneity of fractured host rock based on the previous multiple-canister model (MCFT2D [1, 2]). The current model incorporates i) heterogeneity of the fractured host rock represented stochastically by discrete fractures, ii) disposal-pit vertical emplacement concept, iii) representation of the waste package consisting of a waste canister and a bentonite-filled buffer, and iv) a user-determined repository configuration of multiple canisters using the repository parameters such as disposal tunnel spacing, waste package pitch, tunnel diameter, the number of tunnels in a repository, and the number of canisters in a tunnel. The current model can facilitate investigations into the effects of heterogeneous fractured host rock on water flow and nuclide migration for the different configurations of multiple canisters, as well as optimization of the repository design parameters in terms of release of nuclides from the repository.

INTRODUCTION

Migration of radionuclide (Cs-135) in a water-saturated high-level radioactive waste (HLW) repository was analyzed numerically by a two-dimensional numerical model (MCFT2D) incorporating both a multiple-canister configuration and a non-uniform horizontal flow field of the host rock [1, 2]. The previous studies [1, 2] showed that the migration of Cs-135 in a repository with multiple canisters is significantly influenced not only by the configuration of canisters but also by the groundwater flow conditions. For more robust analysis, heterogeneity of the fractured host rock should be taken into account explicitly in the flow and transport analyses.

The objective of this study is to develop a reliable numerical flow and transport model incorporating multiple-canister configurations and heterogeneity of the fractured rock by discrete-fracture network (DFN) in two-dimensional space.

MODEL DEVELOPMENT

Conceptual Model

A hypothetical water-saturated HLW repository is modeled based on the disposal-pit-vertical-emplacement method [3] in two-dimensional space. As shown in figure 1(a), a hypothetical repository consists of N disposal tunnels containing N_y packages in each tunnel. The total number of canisters is $N{\cdot}N_y$ for the repository with an area of $L_x{\times}L_y$, where $L_x=(N{\cdot}x_D)$, $L_y=(N_y{\cdot}y_D+2a_D)$. x_D is the disposal tunnel spacing. y_D is the waste package pitch in the disposal tunnel. a_D is assume to be $(x_D\text{-}t_D)/2$. t_D is the diameter of the disposal tunnel. Because the rock zone around the tunnel could be altered by the excavation, the area of $(N_y{\cdot}y_D){\cdot}t_D$ for each tunnel is considered to be an excavation disturbed zone (EDZ).

A waste package includes an overpack and a bentonite-filled buffer surrounding the overpack. The bentonite-filled buffer is considered as a mixture of 70wt% bentonite and 30wt% sand [3]. The overpack, which has a minimum design lifetime of 1000 years for the reference case of the Japanese concept [3], contains the vitrified waste matrix confining radionuclides.

Heterogeneity of the host rock, which plays an important role for groundwater flow and radionuclide transport [4, 5], is represented by a discrete fracture network (DFN) (see figure 1(b)) . Two types of fractures are considered in the current model. First, "background fractures" are generated in the host rock based on distribution functions of the fracture geometry parameters. Second, secondary fractures created in EDZ due to the excavation of tunnel are called "EDZ fractures".

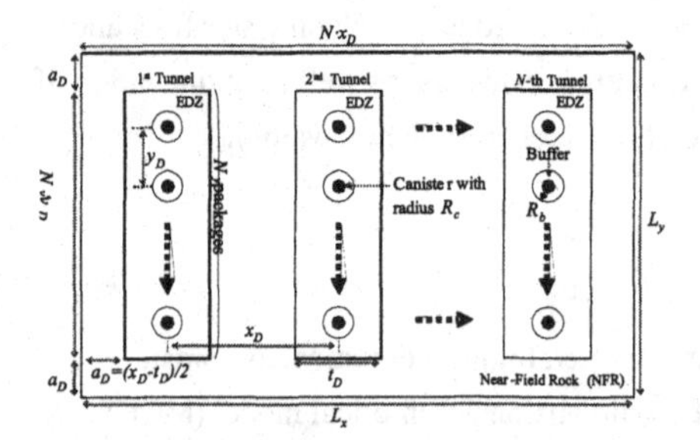

(a) Model Domain for MCFT2D [1]

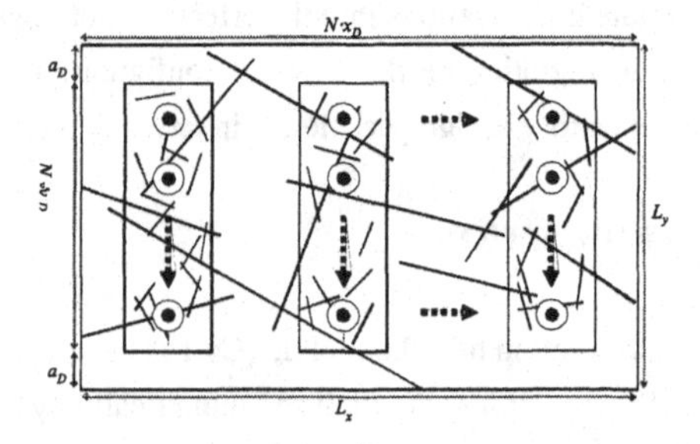

(b) Model Domain for MCFT2D-DFN

Figure 1. A concept with multiple canisters without DFN (a) and with DFN (b) based on the disposal-pit-vertical-emplacement concept in a two-dimensional space.

Discrete Fracture Network

A fracture segment is generated in the two-dimensional model space based on the statistics of the fracture geometry parameters [6, 7]. For the spatial distribution of background fractures, it is assumed that the center point of the fracture is generated randomly inside the domain. For the background fractures in the current mode, the power law distribution is adopted for the fracture size (or length), and it ensures a high frequency of large fractures that are considered to be important in the nuclide transport analysis [3, 8]. A Fisher distribution is implemented for the orientation of two orthogonal groups of vertical fractures. A lognormal distribution is considered for fracture aperture.

The excavation of underground tunnels might cause hydrological changes in the vicinity of the tunnels [3,9]. For spatial distribution of EDZ fractures, it is assumed that the center point of the EDZ fractures is generated randomly inside each tunnel area $((N_y{\cdot}y_D){\cdot}t_D)$. It implies that EDZ fractures are mainly due to the excavation of tunnels by neglecting the excavation of pits for emplacement of waste packages. Length of EDZ fractures is assumed to be constant. EDZ fractures are assumed to be distributed uniformly to maximize the connectivity with background fractures. For aperture of EDZ fractures, it is considered to be identical to the background fractures. The number of EDZ fractures is calculated by a percentage of the total number of background fractures in the current model.

Groundwater Flow and Radionuclide Transport

The domain for groundwater flow and radionuclide transport can be sub-divided into three different regions, i.e., buffer region, fractured region, and rock matrix region. The bentonite-filled buffer region is considered to be a homogeneous isotropic water-conducting medium, while the rock matrix is treated as a water-stagnant region (i.e., no flow and diffusion dominant region). The fractured region is identified based on the fractures forming the flow and transport paths. Each triangular region (or element) is homogenized [7, 10] using the Kozeny-Carman equation based on identified intersecting fractures. The hydraulic properties (i.e., fracture porosity, hydraulic conductivity) were calculated [7, 10] for each triangular region representing the fractured region. Each of the canisters is assumed to be impermeable.

The governing equation [7] for the steady-state space-dependent hydraulic head $h(x,y)$ is given in two-dimensional Cartesian coordinates for time-independent, incompressible Darcy flow with space-dependent hydraulic conductivity $K(x,y)$ as,

$$\frac{\partial}{\partial x}\left[K(x,y)\frac{\partial h(x,y)}{\partial x}\right]+\frac{\partial}{\partial y}\left[K(x,y)\frac{\partial h(x,y)}{\partial y}\right]=0. \qquad (1)$$

With assumptions that hydraulic properties of buffer is homogeneous isotropic, $K(x,y)$ is constant for the entire buffer region, such as $K(x,y)=K_B$ for buffer, while the homogenized hydraulic conductivity for each element is $K(x,y)=K_e$, where e represents the triangular element including fractures forming the flow and transport paths. The left and right boundaries of the $L_x \times L_y$ domain are prescribed as constant, h_H and h_L ($h_H > h_L$), while the top and bottom of the domain are prescribed as the no flow boundary. Darcy velocity normal to the surface of the canister is prescribed as zero for each of the canisters.

The governing equation (1) is solved numerically using the finite element method for the given boundary conditions [7]. Triangular finite-elements are used to discretize the domain. Hydraulic head at the internal nodes and Darcy velocity at the center of weight for each of the elements are obtained numerically. As a result, a heterogeneous flow field of the repository with multiple canisters is established over the entire model.

Radionuclide migration is simulated based on the established groundwater flow field. For radionuclide transport analysis, all $N \cdot N_y$ canisters are assumed to fail simultaneously 1000 years after emplacement. Release of radionuclides starts instantaneously at the surface of each of the canisters by assuming that mass transport inside the canister and mass transfer between the canister and the surrounding buffer are neglected. Radionuclides released from each of the canisters migrate through the pore water in the buffer by molecular diffusion, and then move by advection and/or diffusion in the fractured zone and the rock matrix. Some of the radionuclides released from the upstream canisters could migrate into the buffer regions surrounding the downstream canisters [1, 2]. Temperature effects are neglected in this study.

The transport equation for advection, molecular diffusion, sorption, and radioactive decay in homogeneous isotropic medium j (i.e., buffer, rock matrix, and homogenized fractured regions) for a single radionuclide [1] is simulated using the random-walk method taking into account the local mass conservation (LMC) error [1, 11] with a transfer probability [1], P, at the buffer-host rock interface, which is a diffusion dominant region, as

$$P = \frac{\varepsilon_B \sqrt{\dfrac{D_{p,B}}{R_B}}}{\varepsilon_B \sqrt{\dfrac{D_{p,B}}{R_B}} + \varepsilon_R \sqrt{\dfrac{D_{p,R}}{R_R}}}. \tag{2}$$

Note that the description of the parameters in (2) is given in Table I.

NUMERICAL SIMULATION

Numerical code and Input Parameters

A two-dimensional numerical code, MCFT2D-DFN (Multiple-Canister Flow and Transport code in 2-Dimensional space with Discrete-Fracture Network), is developed based on MCFT2D [1] code for the disposal-pit-vertical-emplacement method in a water-saturated HLW repository. The finite-element method is used for analysis of time-independent and incompressible Darcy flow. A random-walk reflection method [1] taking into account the local mass conservation at the discontinuous interface is incorporated in MCFT2D-DFN for analysis of nuclide migration by advection, diffusion, sorption, and radioactive decay. Using MCFT2D-DFN, a hypothetical repository in the heterogeneous rock can be modeled for any combination of the number of disposal tunnels (N) and the number of waste packages for each tunnel (N_y) depending on the computing resources available, i.e., $\{N \times N_y\}$ canister configuration.

Specifications and input parameters for the repository system are determined mainly based on the Japanese HLW repository concept [3]. The radius of the canister is prescribed as R_c=0.41m. The outer radius of the buffer is R_b=1.11m. For the hard rock system, the disposal tunnel spacing, x_D, is 10m, the waste package pitch in the disposal tunnel, y_D, is 4.44m, and the diameter of the disposal tunnel, t_D, is 5m. For numerical experiments in the current study, Cs-135, a half-life of 2.30×10^6 yr, is selected since it is considered to be a major contributor to the peak exposure dose rate [1, 3]. Input parameters for hydraulic and transport properties of Cs-135 are given in Table I.

Table I. Input Parameters of Hydraulic and Transport (Cs-135) Properties [1, 3]

Symbol	Description	Value
ε_B	Effective porosity in the buffer region	0.41
ε_R	Effective porosity in the host rock region	0.02
K_B	Hydraulic conductivity in the buffer region	4.50×10^{-13} m/s
K_R	Hydraulic conductivity in the host rock region	10^{-6} m/s
$D_{p,B}$	Pore diffusion coefficient of Cs-135 in the buffer region	4.62×10^{-2} m^2/yr
$D_{p,R}$	Pore diffusion coefficient of Cs-135 in the host rock region	4.62×10^{-2} m^2/yr [a]
$K_{d,B}$	Distribution coefficient of Cs-135 in the buffer region	0.01 m^3/kg
$K_{d,R}$	Distribution coefficient of Cs-135 in the host rock region	0.05 m^3/kg
$\rho_{s,B}$	Density of the solid material in the buffer region	1600 kg/m^3
$\rho_{s,R}$	Density of the solid material in the host rock region	2640 kg/m^3
R_B	Retardation factor of Cs-135 in the buffer region	24 [b]
R_R	Retardation factor of Cs-135 in the host rock region	6469 [b]

[a] Assumed parameter (e.g., $D_{p,R}$=4.73×10^{-3} m^2/yr [3])

[b] Calculated values based on given ε, ρ, and K for each region [1].

For the fracture size represented by the power law distribution in the 2D space, exponent of 2.0 and minimum length of 1m are used for the current simulation by assuming a perfectly planar surface [8,12],while exponent of 3.0 was used for the 3D space [12]. The Fisher constant κ=10, is used [3], and arithmetic mean and standard deviation of the aperture are 4.3×10^{-5} m and

8.1×10^{-5} m, respectively [3]. The linear frequency of the background fracture is prescribed as 0.5 in this simulation. The EDZ fracture is not considered in this simulation, while one can specify the intensity of the EDZ fractures as a ratio to that of the background fractures in MCFT2D-DFN.

The hydraulic gradient is 0.01 [3]. To investigate solely the effects of a heterogeneous flow field on transport of long-lived nuclides, such as Cs-135, mass reduction by radioactive decay process is not taken into account in the current study because the radioactive decay process is independent of groundwater flow conditions.

Preliminary Simulation using MCFT2D-DFN

For a given sets of input parameters, simulation result of a single realization is shown in figure 2 for a 50-canister case (5×10 configuration). Figure 2(a) shows a generated DFN for a domain of 5×10 canister configuration. Because not all fractures contribute to the flow and transport, connected fractures with the left boundary, right boundary, and outer surface of each of canister packages are identified by investigating intersections for all fractures as shown in figure 2(b). Figure 2(c) shows the finite-element triangular region intersected by connected fractures. Figures 2(d) and 2(e) show the results of contour of hydraulic head and Darcy velocity vectors, respectively.

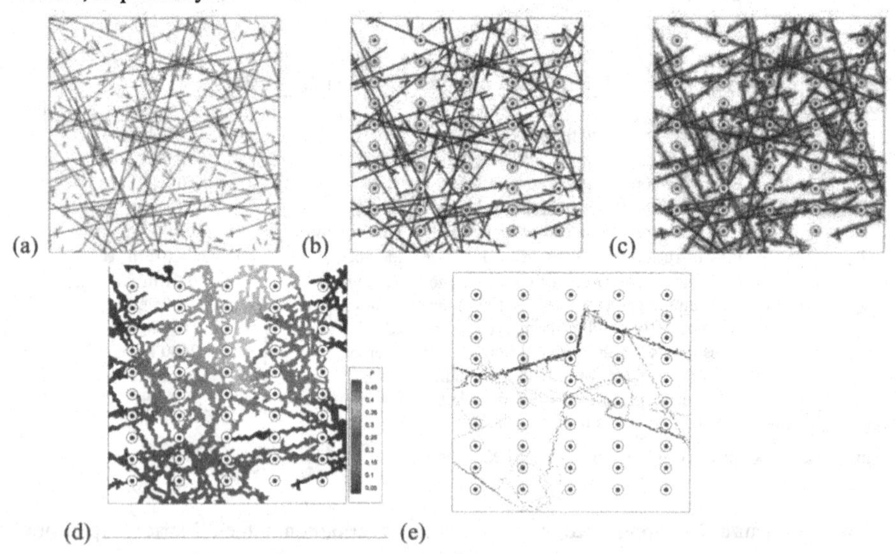

Figure 2. 5x10 canister configuration. (a) DFN generation, (b) fractures forming flow and transport paths, (c) finite-elements covering only fractures for flow and transport, (d) contour plot of hydraulic head, and (e) Darcy velocity vector.

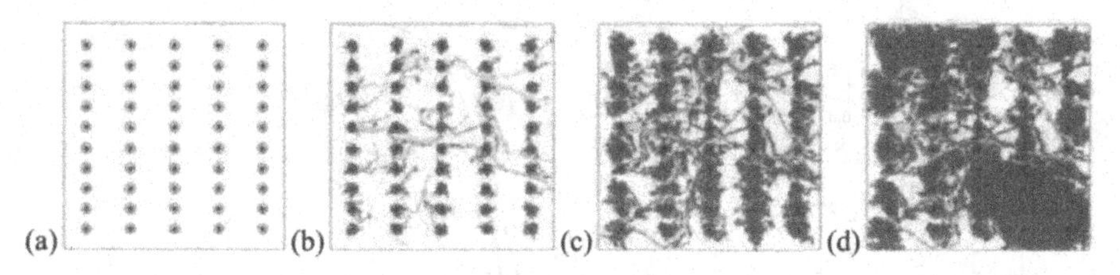

Figure 3. Particle paths. (a) Step=300, (b) Step=4960, (c) Step=29708, and (d) Step=3487472

Cs-135 migration is simulated based on the established groundwater flow velocity field shown in figure 2(e). In the current preliminary simulation, the total mass of Cs was prescribed as a constant, and the mass of Cs released from each canister was assumed to be identical from canister to canister. If particles reach the edge of the model domain boundaries, migration is no longer considered in the current model. Figures 3(a)~3(d) show the particle travel paths for different time steps. For the particular simulation steps shown in Figures 3(a)~3(d), 0%, 2.3%, 29%, and 100% of released particles from each of 50 canisters are escaped from the modeling domain with 5×10 configuration. Results in Figures 3(a)~3(d) confirmed that the radionuclide transport paths are highly dependent on heterogeneous flow, which is caused by the heterogeneity of a discrete-fracture network.

Furthermore, the relatively small flow rates on the upper left and lower right regions in the domain (see figure 2(e)), suggested that the molecular diffusion was the dominant transport mechanism in those region showing the random movement as shown in figure 3(d), bearing in mind the caveat that LMC treatment is not considered for the graphical results in figure 3.

Figure 4 shows the cumulative distribution function (CDF) of Cs-135 transport (10^3 particles) considering the LMC (local mass conservation) treatment for different canister configuration. The median residence time for 1x1, 4x4, and 5x10 configurations are 1.117E+03yr, 1.585E+05yr, and 2.668E+05yr, respectively. Because the residence time for multiple canisters with different configurations is highly dependent on the heterogeneous flow and transport paths as shown in figure 4, it is difficult to estimate the mass release rate (or residence time) from the multiple canisters by the ratio of the number of canisters (or number of tunnels) to the result obtained for the single canister concept.

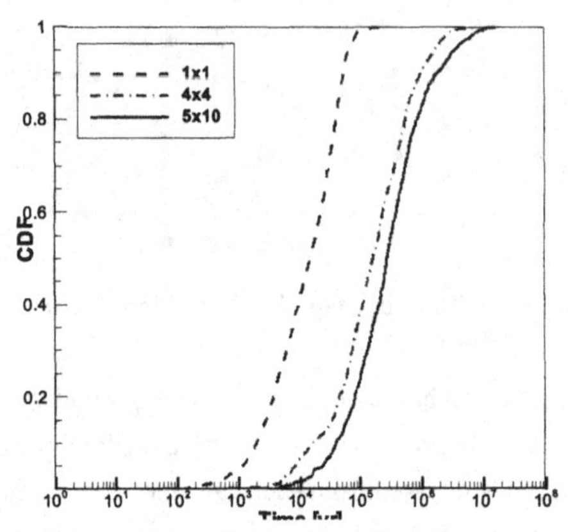

Figure 4. CDF of residence time of Cs-135. Single canister by 1×1 configuration. 16 canisters by 4×4 configuration. 50 canisters by 5×10 configuration.

DISCUSSION AND CONCLUSIONS

An integrated model for groundwater flow and radionuclide migration analyses in a water-saturated HLW repository with a multiple-canister configuration is developed in the two-dimensional space by incorporating the heterogeneity of fractured host rock. Preliminary simulations show that the radionuclide migration is highly dependent on the heterogeneous flow and transport paths. More rigorous simulations using MCFT2D_DFN are needed for more reliable analysis with respect to site-specific DFN parameters, and various configurations of the multiple canisters ($N \times N_y$). As a future study, a model can be extended to the three-dimensional space.

ACKNOWLEDGMENTS

This work was partially supported by the International Fellowship Program at Japan Atomic Energy Agency (formerly Japan Nuclear Cycle Development Institute). The first author would like to thank Prof. Joonhong Ahn and Prof. Paul L. Chambré of University of California, Berkeley for their guidance of the conceptual model development, Dr. Kaname Miyahara of JAEA for his helpful comment and discussion, and Dr. William S. Dershowitz of Golder Associates for his support. For the development of the current code (MCFT2D-DFN), some parts of FFDF code [7, 10] are used with some modifications.

REFERENCES

1. D. Lim, *Nuclear Technology*, **156**, 222-245 (2006).
2. D. Lim, Mater. Res. Soc. Proc. **932**, 243-250 (2006).
3. Japan Nuclear Cycle Development Institute, JNC TN 1410 2000, **001~005** (2000).
4. National Research Council, Rock Fractures and Fluid Flow: Contemporary Understanding and Applications, National Academy Press, Washington, D.C. (1996).
5. M. Sahimi, Flow and Transport in Porous Media and Fractured Rock, VCH (1995).
6. Long, J. C. S., J. S. Remer, C. R. Wilson, and P. A. Witherspoon, *Water Resources Research*, *18*(3), 645-658 (1982).
7. D. Lim, "Mass Transport Analysis in the Near Field of Geologic Repository", Ph.D. Thesis, University of California, Berkeley (2002).
8. P. R. La Pointe, *Int. J. Rock Mech. Min. Sci.*, **39**, 381-388 (2002).
9. A. Poteri and M. Laitinen, POSIVA 99-15 (1999).
10. D. Lim, J. Ahn, and P. L. Chambré, "Modeling of Contaminant Transport in Fractured Near-Field Rock", *Proc. 10th Int. HLW Management Conf.*, ANS (2003).
11. D. Lim, *Water Resour. Res.*, ***42***, W02601, (2006).
12. A. Sawada, Y. Ohnishi, H. Ohtsu, Y. Ijiri, and S. Nishiyama, Rock Engineering (Problems and Approaches in Underground Construction),**1**, 203-210, KSRM (2002).

Mater. Res. Soc. Symp. Proc. Vol. 1107 © 2008 Materials Research Society

A New Method for Real-Time Monitoring of Grout Spread Through Fractured Rocks

Alasdair E. Henderson[1], Iain A. Robertson[1], John M Whitfield[2], Graham F.G. Garrard[3], Nicholas G. Swannell[3], Hansruedi Fisch[4]
[1]Ritchies Division of Edmund Nuttall Ltd, Glasgow Road, Kilsyth, Glasgow G65 9BL, U.K.
[2]UKAEA Dounreay Division, Dounreay, Thurso, Caithness KW14 7TZ, U.K.
[3]Halcrow Group, Burderop Park, Swindon, Wiltshire SN4 0QD, U.K.
[4]Solexperts AG, Mettlenbachstr. 25, Postfach 122, CH-8617 Mönchaltorf, Switzerland.

ABSTRACT

Reducing water ingress into the Shaft at Dounreay is essential for the success of future intermediate level waste (ILW) recovery using the dry retrieval method. The reduction is being realised by forming an engineered barrier of ultrafine cementitious grout injected into the fractured rock surrounding the Shaft. Grout penetration of 6m in <50μm fractures is being reliably achieved, with a pattern of repeated injections ultimately reducing rock mass permeability by up to three orders of magnitude.

An extensive field trials period, involving over 200 grout mix designs and the construction of a full scale demonstration barrier, has yielded several new field techniques that improve the quality and reliability of cementitious grout injection for engineered barriers.

In particular, a new method has been developed for tracking in real-time the spread of ultrafine cementitious grout through fractured rock and relating the injection characteristics to barrier design. Fieldwork by the multi-disciplinary international team included developing the injection and real-time monitoring techniques, pre- and post injection hydro-geological testing to quantify the magnitude and extent of changes in rock mass permeability, and correlation of grout spread with injection parameters to inform the main works grouting programme.

INTRODUCTION

The D1225 Shaft at Dounreay Nuclear Establishment on the north coast of Scotland was an authorised disposal facility for ILW from 1959 until the last deposition in 1971. The shaft was constructed entirely in rock with a nominal diameter of 4.6m and is lined only over the upper 8.0m of its 65m depth. In addition to the 620m^3 of recorded ILW disposals, the shaft space is flooded by groundwater. As part of the ongoing decommissioning programme at Dounreay, the contents of the shaft are to be retrieved, sorted and consigned to alternative storage and the contaminated groundwater removed and treated.

The shaft contents will be recovered using the dry retrieval method: the shaft water level will be reduced until an item of solid waste is visible, then remote handling equipment will retrieve and sort the solid waste and the process will repeat until the shaft has been emptied. For this gradual dewatering method to be successful the groundwater ingress rate must be limited to a value capable of being dealt with by the existing site liquid effluent treatment plant.

To accomplish this, an engineered barrier is being constructed around the Shaft by the controlled injection of stable ultrafine cementitious grouts into the bedded and jointed rock surrounding the shaft. The grout is injected via drilled boreholes using an ascending stage sequence. The barrier design relies on grout spread from each stage treating a volume of surrounding rock. Using a split-spacing borehole technique with several series of injections, a multiple overlap between treated volumes is achieved and beneficial redundancy introduced.

Injections are controlled using the Grout Intensity Number (GIN) method proposed by Lombardi and Deere [1, 2] where the stop criterion is described by a pressure/volume curve of equal energy (a 'GIN' in units of bar.litres/metre) truncated by conventional pressure and volume limits. The method allows all parts of a given volume of homogenous rock to be grouted at the same intensity i.e. the same amount of work is done whether an injection terminates at high pressure/low volume, low pressure/high volume or at a point on the GIN curve between these extremes. The GIN method allows appropriate pressure to be applied to increase grout penetration, but prevents the combination of high pressure and high volume injections, thereby limiting the risk of ground heave.

Understanding the nature and range of the grout spread in the discontinuities within the rock mass is therefore a fundamental aspect of the design of the injection point layout and definition of the controlling GIN for each homogenous zone and thus the grout penetration experiment described here formed an important element of the early grouting site trials phase of the D1225 Shaft Isolation Project.

SITE GEOLOGY

The pre-works site trials, of which the P1 grout penetration experiment formed an early part, were undertaken in an area lying 65m to the north east of the D1225 Shaft and approximately along geological strike. Stratigraphically, the sequence is generally at the same level as the Shaft and as both the D1225 Shaft and the trials area lie on the foreshore line, they share a topographic similarity. Faulting is present at both areas and is characterised by steep dips (75° to 82°) and throws of between 3 and 20m.

The Caithness Flags at Dounreay generally comprise a cyclic sequence of silty limestones (A horizons), bituminous siltstones (B horizons), siltstones (C horizons) and sandstones (D horizons) with a typical upward sequence of A, B, C, D, C, B, A. The cycles repeat at 6 to 10m intervals numerous times, allowing successive A or A/B units to be numbered as marker beds up the sequence. The C and D horizons demonstrate more developed jointing, predominantly along bedding planes and are associated with the flowing features identified during hydrogeological site characterisation work. The beds dip at approximately 10° to the north west.

Extensive site characterisation work [3, 4] was undertaken in advance of the grouting trials to locate accurately known and suspected faults, confirm the geological model and to establish a hydrogeological baseline.

PENETRATION FIELD EXPERIMENT

Concept

Lombardi [2] describes an experimental method where a trial borehole is injected with grout while the injection pressure, injected volume and grout penetration are continuously monitored. Assuming reasonably uniform radial spread from the injection point, a 'test GIN' can be determined for a given penetration. The GIN for the spacing required by the design can then be estimated using the expression:

$$R = R_t \sqrt[3]{(GIN/GIN_t)} \quad (1)$$

where:

R	- penetration required in the prototype (design)
R_t	- penetration estimated from the test
GIN	- the GIN to be applied to the prototype
GIN_t	- the test GIN corresponding to the test penetration

Capturing injection pressure and volume data is straightforward with modern grouting equipment and is discussed briefly later. However, the practicality of determining the penetration distance and corresponding grout injection volume is a fundamental difficulty and explains in part why the experimental derivation of the GIN is rarely used in conventional grouting work.

An option study undertaken to look at the practical aspects of detecting grout concluded that the most cost effective and practicable solution to assessing the form of the grout flow would be to use sensors to detect grout arrival in real time at a number of observation holes located around the injection borehole. For this to work, it was recognised that fluid grout detection must be performed in a sealed section of observation hole to prevent the hole becoming a preferential sink for the injection grout and adversely affecting the development of the grout front from the injection point. Laboratory trials of various forms of detection probe (pressure, temperature, conductivity and pH) were undertaken and showed that the down-hole pH transducer was the most effective at detecting grout and that a change in pH could detect and distinguish grout both at dilute concentrations (imminent arrival) and at full concentration (fluid grout arrival). The site characterisation of the field trials area showed that in situ measurement of pH would be unaffected by the local rock and groundwater chemistry.

Experimental Layout

The field experiment was arranged as a 75m deep central injection borehole (P1) with three 75m observation bores (PO11 – PO13) arrayed equally on a 4m radius and a further three observation holes (PO14 – PO16) arrayed equally on a 6m radius with a 60° offset from the inner observation bores – Figure 1. Boreholes P1 and PO11 – PO13 were geophysically logged using wireline tools (televiewer, gamma log, flow log) and based on the stratigraphy and hydrogeology inferred from the logs the strata was subdivided into a number of contiguous 3m sections that were accurately referenced to the stratigraphical position within the sequence. These 3m sections were then hydrogeologically tested to estimate the hydraulic conductivity and character of the ground around the boreholes prior to grout injection. This characterisation was expressed in terms of borehole 'zones' (Table 1), analogous to Lombardi's 'homogeneous zones', with an experimental GIN being sought for each.

Borehole P1 was then grouted in ascending stages that matched the stratigraphic positions of the hydrotesting. After the grout penetration experiment, a further two boreholes (PO17, PO18) were drilled 1.5m and 4.5m respectively from P1 and hydrogeologically tested, again in continuous 3m intervals located at the same stratigraphic position within the sequence as the pre-grouting tests. The purpose of the post-grouting hydrotests was to quantify the change in rock mass transmissivity caused by a single grout injection. As noted previously, the completed barrier used a split-spacing technique and therefore each piece of ground will experience multiple grout injections

All boreholes were of a similar depth to those anticipated for the main Shaft Isolation works and passed through several cycles of the stratigraphical sequence ensuring that penetration trials were undertaken in all types of ground. In total, the borehole depth allowed 25 stages to be injected and monitored.

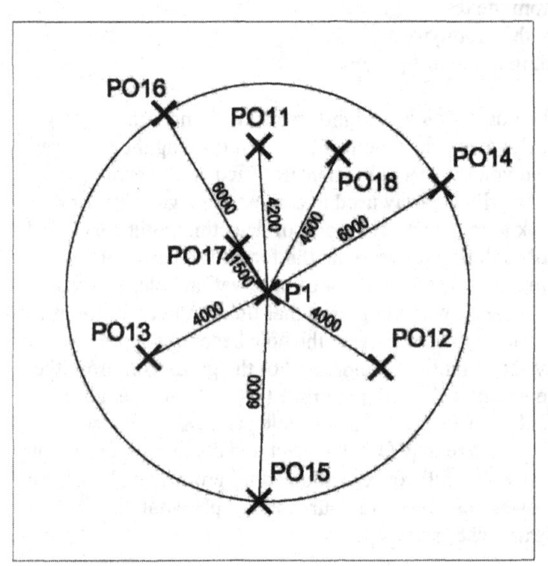

Figure 1 – Plan View on Field Experiment (dimensions in mm)

Table I – Borehole Zoning Definitions

Zone	Name	Description	Anticipated Grouting Properties
W	Weathered Zone	Closely spaced discontinuities, High or very high permeability.	Low pressure/high take. Likely to stop on volume limit.
F	Fault or Joint Zone	Clusters of SV joints or zone of broken ground. Moderate or high permeability	Low pressure/high take. Likely to stop on volume limit.
1	Major Bedding Planes	Clusters of SH discontinuities, identified as continuous across the site. Moderate or high permeability.	Low to medium pressure/high take. Require high GIN to ensure saturation.
2	Minor Bedding Planes	Multiple SH fissures with range of low permeabilities, low or moderate permeability.	High pressure/low take. High GIN and split spacing required to ensure full saturation.
3	Intact Rock	Very few fractures (0 or 1 per 5m) Very low or low permeability.	High pressure/low take. High GIN and multiple split spacing required to ensure full saturation.

Drilling

The boreholes for the field experiment were drilled using Boart Longyear DB520 hydraulic rotary drilling rigs employing an HQ wireline coring system (producing a 96mm bore diameter) and a clean water flushing medium. Wireline coring was selected as the preferred drilling method for reasons of radiological waste minimisation and, whilst relatively slow compared with percussive methods, provided a stiff drill string and good bore verticality.

During earlier drilling trials, various polymeric flushing media were assessed. However, these were not used during the penetration experiment nor in subsequent works, primarily due to concerns about clogging of discontinuity apertures and the development of borehole wall 'skin' effects, but also because clean water worked adequately as a flushing medium and was found to make solids control more manageable.

Hydrogeological Testing

Hydrotesting in each of the boreholes was undertaken by the constant head injection method. The test interval was 3m, with each stage being isolated using a double packer straddle with a down-hole shut-in tool. Test pressures of 1 – 3bar were used from 5m to 20mbgl while differential pressures of 3 – 6bar were used for intervals deeper than 20mbgl. The decline of the flow rate was recorded as a function of time, with the test duration being determined by the time necessary to record sufficient data of the transient formation response to be analysed in a semi-log plot (typically 20 – 30minutes per interval). Analyses of the test results were based on the conventional steady-state approximation equation or straight line analysis as appropriate. The results provided estimates of hydraulic conductivity (transmissivity/interval length, reported as 'Pregrout k' in the results) for ungrouted ground ranging over 6 orders of magnitude between $3x10^{-4}$m/s and $2x10^{-10}$m/s.

Grouting

The primary grout used in the penetration field trials was developed during an earlier phase of grout material trials and comprised an ultrafine (d_{95}=16μm) cement colloidally mixed with water, a superplasticiser and a silica fume stabilising agent.

Down-hole equipment comprised air-inflated double packers mounted on a steel mandrel with a variable (1 – 3m) straddle. Grout was delivered through Kevlar grout lines, with injection commencing near the base of the borehole and progressing in 3m ascending stages.

The grout mixing and injection plant comprised a pair of Colcrete SD200 high shear colloidal mixers fitted with automatic batching control. Batched grout was held in an agitator tank prior to being injected using paired opposing phase single acting 0-100bar piston pumps. Each pumpset was equipped with an in-line electro-magnetic flow meter and 0-100bar pressure transducer with signals being returned to the control system in a separate grouting control module.

The grout injection was controlled using bespoke software operating on a desktop computer, capable of controlling and datalogging up to six simultaneous injections with different parameters for each. For a given injection, stage information (borehole number, depth) was entered along with the defining injection parameters (GIN, maximum pressure, maximum volume, pressure corrections for stage depth). Flow rate for each pump was selected using ABB controllers and was generally held constant for each injection at between 3 to 5 litres/minute. For the purposes of the penetration experiment, the GIN parameter and volume limit were set at an artificially high level to allow injection to continue until grout detection occurred.

Down-Hole Monitoring

Each of the observation bores PO11 – PO16 was fitted with a double packer set with a 5.5m straddle. Hach Lange pH and pressure/temperature transducers were fitted to the centre of the straddles and hard-wired back to a datalogger at the surface.

Water flushing lines were included in the observation bore packer sets to counteract the effect of pH buffering from earlier injections. By this method, the pH in the test interval could be reduced sufficiently between vertically adjacent injections to allow fresh grout arrival to be detected.

Experimental Method

Each injection commenced with seating of the injection and observation packer sets at the test stage, with the 5.5m observation interval being centred vertically opposite the shorter 3m grout injection interval. Datalogging of the down-hole transducers in bores PO11 – PO16 was commenced some time in advance of grout injection to establish a pre-injection baseline. Where necessary, water flushing of the observation bores to re-establish neutral pH was undertaken until the packer sets where inflated. The injection lines and interval were filled with primary grout and injection commenced at a constant flow. Logged data from the grout control system and observation transducers was graphed and displayed in real-time to alert the operators to grout arrivals. Post analysis of the combined data allowed each fresh grout arrival, signified by a pH of 12.7, to be correlated with an injection pressure and volume, thereby providing the experimental GIN for that penetration distance from the following expression:

$$GIN = p_t . V_t / L \qquad (2)$$

where,

p_t - the injection pressure at time of grout arrival (bar)
V_t - the injected volume at time of grout arrival (litres)
L - the injection interval length (m)

The pattern of grout arrival over the six observation bores together with known arrival times permitted assessment of the penetration form and directional bias where present, and for these data to be related to the stratigraphy and joint orientation using the results from the pre-grouting televiewer and gamma logs.

RESULTS

25 grout injections were undertaken during the experiment, comprising one zone F, one zone W, eleven zone 1, eight zone 2 and four zone 3 intervals. Practical difficulties during grouting near the base of the borehole and packer bypass through subvertical joints is thought to have led to some fissure clogging that may have affected grout acceptance in lower stages (48-51m, 51-53.5m, 56-59m)

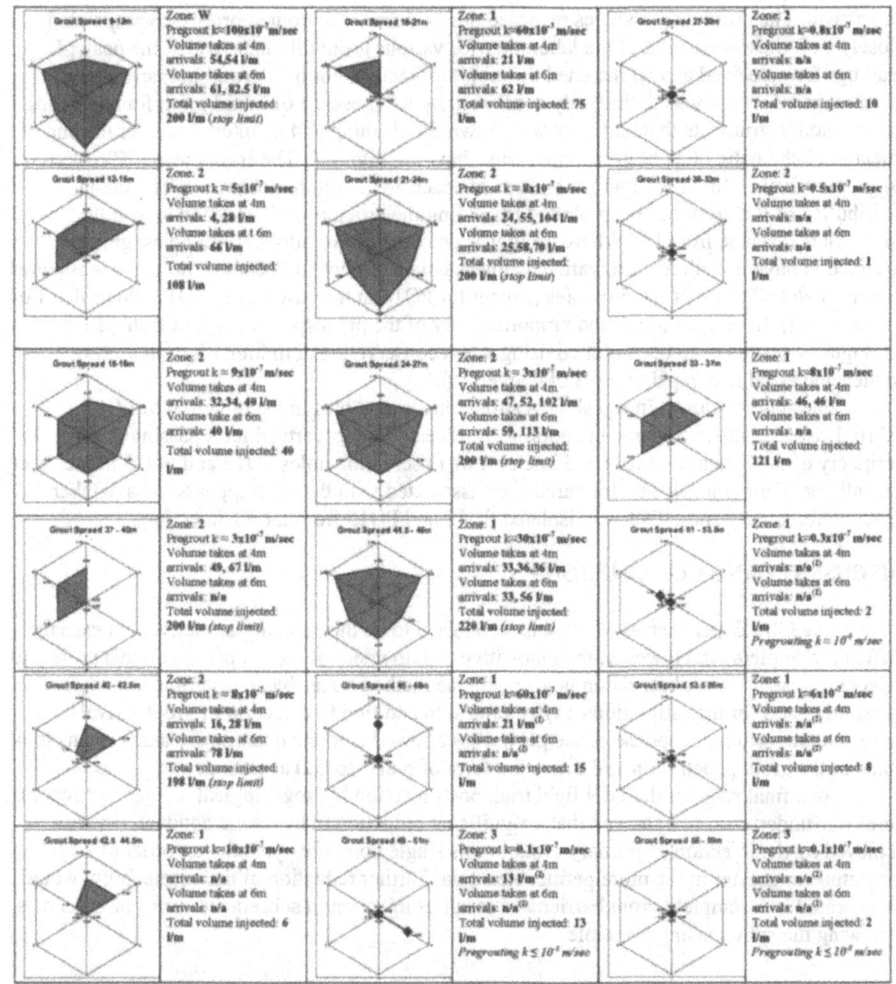

Figure 2 – Schematic Grout Injection Results (9 to 59mbgl)

The temperature probes in the observation bores generally showed no response at any point during grout injection. Given the extensive clean water flushing that was necessary to reduce pH after each stage, it is likely than any grout-driven temperature response was masked by the water/ground temperature gradient. The pressure transducers often showed some response, both to start of injection and grout arrival. However, the responses were highly variable in form and scale and, coupled with their unreliable nature, were not a credible indicator of grout arrival. The pH transducers were found to be unaffected by the start of injection or any resulting groundwater motion, but showed a distinct and pronounced response to grout approach and arrival. The absolute pH values recorded during perceived grout approach corresponded closely with those measured in the laboratory for various grout dilutions, whilst the peak pH readings for perceived arrival matched the laboratory values for neat grout. In many cases, physical grout arrival was definitively confirmed by the presence of liquid grout found coating observation instruments that had been withdrawn for cleaning and maintenance after the end of injection. Schematic plots of grout spread are shown in Figure 2. The stages below 59m recorded no grout arrival at 4m or 6m and for reasons of space are therefore not shown. The radial distribution of the grout has been plotted by joining the grout arrivals for each hole at the appropriate distance from P1. Where grout was detected in two adjacent 4m holes, grout is assumed to have travelled 4m towards the intermediate 6m hole. Otherwise, no grout is assumed to have travelled towards the 6m holes (except for PO16 in the case where PO17 shows that flow has occurred). In recognition of the proportionality of the pH response to grout front proximity, the Figure 2 results have been plotted using a less conservative definition of grout arrival: a sudden and rapid rise in pH or a pH of above 11.5.

The stages below 37mbgl show a strong directional bias in the direction of PO11 and PO16. However, the site characterisation work revealed a sub-vertical fault passing through the periphery of the penetration trial and intersecting observation holes PO12 and PO15 immediately beneath the 37mbgl level. The fractured rock associated with the fault appears to have taken grout preferentially and effectively isolated PO12 and PO15 from the P1 injection.

DISCUSSION AND CONCLUSIONS

The GIN Experimental Method is recognised to be theoretically attractive, but extremely difficult to implement. However, the innovative development of the pH probes on this project to detect grout arrivals in observation holes made the method a feasible option for the study both of penetration and for the estimation of GIN values. In addition to detection of grout arrivals, the clear and proportional response of the pH probes allowed credible quantitative assessments to be made of the grout penetration and hence the form of radial spread at each stage.

As a final stage of the GIN field trial, post-injection hydrogeological testing (not reported here) was undertaken and showed that a significant reduction in hydraulic conductivity was achieved by the injection of primary grout in this single borehole, typically an order of magnitude or greater in the more permeable strata. Further reduction of the permeability would be expected in a complete grout barrier as a result of injection in subsequent series of boreholes following the split-spacing principle.

ACKNOWLEDGEMENTS

The authors gratefully acknowledge the support of UKAEA Dounreay Division and thank them for granting permission to publish this paper.

REFERENCES

1. G. Lombardi and D. Deere, J. Water Power and Dam Constr., 15-22 (June 1993)
2. G. Lombardi, Int. J. Hydropower and Dams, 4, 62-66 (1996).
3. Halcrow Group, UKAEA Report 54324/3/REP/HAL/1361/IS/01, 2005.
4. Halcrow Group, UKAEA Report 54324/3/REP/HAL/1362/IS/01, 2006.

Mater. Res. Soc. Symp. Proc. Vol. 1107 © 2008 Materials Research Society

Observation of Diffusion Behavior of Trace Elements in Hardened Cement Pastes by LA-ICP-MS

Taiji Chida and Daisuke Sugiyama
Central Research Institute of Electric Power Industry,
2-11-1, Iwado-Kita, Komae, Tokyo 201-8511, Japan

Abstract

Diffusion coefficients of iodine (I), cesium (Cs) and strontium (Sr) in solid cement blocks of ordinary portland cement (OPC) and low-heat portland cement containing 30 wt% fly ash (FAC) are measured by in-diffusion experiment. By using "laser ablation inductively coupled plasma mass spectrometry", the penetration profiles of trace elements by diffusion were obtained quantitatively in cement solids. The apparent diffusion coefficients of the near-surface of the cement samples were estimated to be about 10^{-13} m^2 $s^{-1}$for I, Cs and Sr. Lower diffusion coefficients were observed for FAC than those obtained for OPC.

Introduction

For the performance assessment of radioactive waste disposal, the diffusion of radionuclides in cementitious materials, which are used as an engineered barrier [1], is an important factor. However, there is much to clarify in the diffusion of radionuclides because it is difficult to observe trace-element distribution in solid materials in detail. In particular, because elements like iodine (I) hardly sorb on cementitious materials, it is important to clarify the diffusion of these elements in an engineered barrier. Recently, laser ablation inductively coupled plasma mass spectrometry (LA-ICP-MS) is used to analyze trace elements in rocks, metals, and other materials [2-4]. The notable feature of LA-ICP-MS is to determine trace elements in solid quantitatively without special preparation.

In this study, the diffusion behavior of cesium (Cs), strontium (Sr) and I in hardened cement pastes was examined by an in-diffusion technique. The detailed penetration profiles of trace elements in hardened cement pastes were successively observed by LA-ICP-MS, and diffusion coefficients were estimated from the penetration profiles.

Experimental

Materials

Solid block samples of ordinary portland cement (OPC) and low-heat portland cement containing 30 wt% fly ash (FAC) were prepared for the diffusion experiments. The chemical compositions of OPC and FAC used in this study are shown in Table 1. The cements were mixed with deionized water in water/cement clinkers (w/c) of different mixing ratios of 0.35, 0.70 and 1.0, in order to give different diffusivities in solid. To prevent bleeding (separation of water at the top of the paste) at high w/c ratios of 0.7 and 1.0, the cement pastes were repeatedly mixed before setting. The cement pastes were allowed to set for 24 h at 30°C, and cured in water for 91 days at 50°C. The porosity of the hardened cement pastes was measured by mercury intrusion porosimetry, and these data are shown in Table 2. Hardened OPC and FAC were cut in the form of 20 × 20 × 20 mm using a diamond cutter, and were covered with epoxy resin except on one surface, for the block samples in the in-diffusion experiments.

Cement-equilibrated solutions were prepared by placing crushed OPC or FAC (< 250 μm) in contact with deionized water at a solid/liquid ratio of 1/10. After 28 days of equilibration, each solution was filtered through a membrane filter of 0.45 μm pore size.

In-diffusion experiment

Figure 1 shows an illustration of the in-diffusion experiment setup. The cement block samples coated with epoxy resin were immersed in a cement-equilibrated solution at reduced pressure to fill their pores with the solution. The samples were then placed in polyethylene vessels and left in contact with the cement-equilibrated solution for one week. Then, to set the

Table 1 Chemical compositions of OPC, LPC and fly ash.

	Chemical Composition (%)									
	ig. Loss	insol	SiO_2	$A_{12}O_3$	Fe_2O_3	CaO	MgO	SO_3	Na_2O	K_2O
OPC	1.5	0.2	21.2	5.2	2.8	64.2	1.5	2	0.3	0.5
LPC	8.7	0.1	25.4	3.5	3.5	62.5	1.1	2.2	0.2	0.4
Fly Ash	-	-	56.2	29.9	3.8	3.5	1.4	-	-	-

Table 2 Porosities of OPC and FAC used in experiment.

Sample	OPC35	OPC70	OPC100	FAC35	FAC70
w/c	0.35	0.70	1.00	0.35	0.70
Porosity	15%	40%	52%	23%	48%

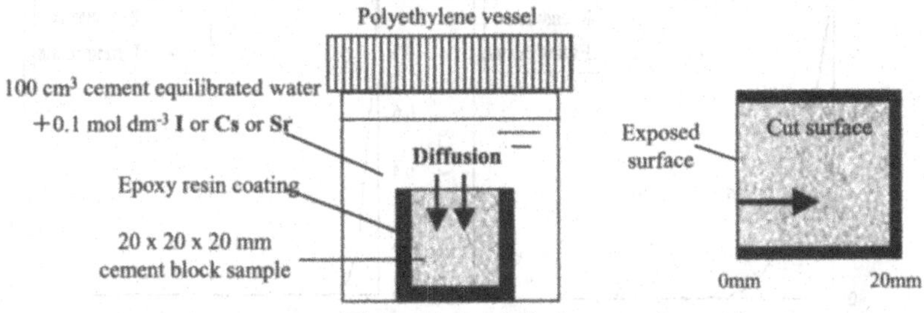

Figure 1 In-diffusion experiment.

concentration of Cs, Sr or I ion in the reservoir to 0.1 mol dm^{-3}, CsCl, $SrCl_2$ or KI solution was added. The total amount of solution in the reservoir was set to 100 cm^3. After 5 days, the cement block samples were removed from the reservoir solution, and allowed to dry in an argon atmosphere for 5, 20 and 180 days. The pH remained constant at 12.2 ± 0.2, and the concentrations of Cs, Sr and I also remained constant at 0.10 ± 0.01 mol dm^{-3} during the experiments. The preparation of the solutions and the in-diffusion experiments were carried out in an argon-filled glove box.

LA-ICP-MS

In this study, LA-ICP-MS was used to determine the penetration profiles of Cs, Sr and I in hardened cement paste. The block samples were cut in the vertical direction for the exposed surface using a diamond drill saw and the cross section was polished with abrasive paper (#1500). To determine the penetration profiles of trace elements, the polished surfaces were line-scanned in the range of 0 to 10 mm from the exposed surface. As operation conditions for the Nd:YAG laser (λ = 213 nm), following were used: scan speed, 80 μm; laser rate, 10 Hz; the laser spot size range, 30 - 80 μm and laser energy range, 10 - 40 J cm^{-2}. The mass of the material ablated with the laser was normalized by the mass of aluminum in the samples, which was generally distributed in the cement paste samples and was expected to be relatively immobile during the in-diffusion experiments. This internal standard method has been used in previous studies [e.g., 5-7].

Result

The penetration profiles of Sr in OPC35 and FAC35 block samples are shown in Figure 2.

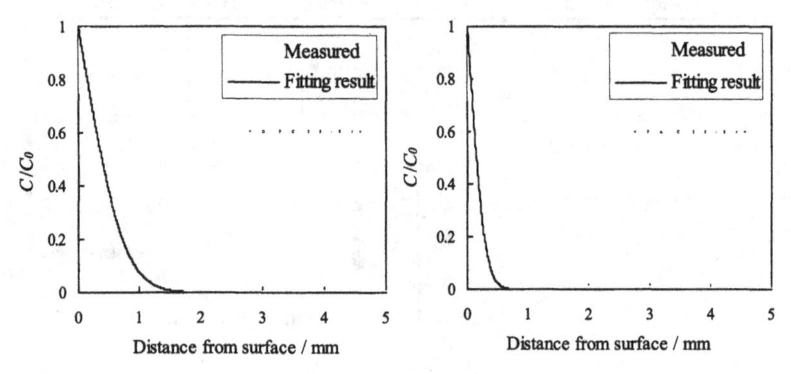

Figure 2 Penetration profiles of strontium after diffusion into OPC and FAC block samples for 5 days.

C is the concentration of diffused ions in porous media (mol dm^{-3}) and C_0 is the concentration of diffused ions in the exposed surface of the block samples (mol dm^{-3}). In this study, we assumed that C/C_0 is nearly equal to (counts of measured point by LA-ICP-MS)/(counts of the exposed surface by LA-ICP-MS). The lines of fitting result are described below. As shown in Fig. 2, the penetration profiles for Sr in the FAC samples were sharper than those in the OPC samples. The profiles of Cs and I were also similar to that of Sr.

The apparent diffusion coefficients were evaluated from the penetration profiles. The diffusion experiments were in a one-dimensional nonsteady state; therefore, the concentration of diffused ions can be written by Fick's second law

$$\frac{\partial C}{\partial t} = D_a \frac{\partial^2 C}{\partial x^2} , \quad (1)$$

where C is the concentration of diffused ions in solid phase (mol dm^{-3}), t the time (second), x the depth into the solid matrix (m) and D_a the apparent diffusion coefficient (m^2 s^{-1}). The relation of D_a to the effective diffusion coefficient (D_e) is

$$D_a = D_e/a, \quad (2)$$

$$a = e + \rho R_d, \quad (3)$$

where a is the capacity factor [-], e the total pore volume [-], ρ the density of matrix [g/m^3] and R_d the number of adsorbed ions per unit mass of the porous medium divided by the number of ions per unit volume of liquid [m^3/g].Then, the initial condition is

$$C = 0, \quad \text{at } x = 0, \quad t = 0, \tag{4}$$

and the boundary condition is

$$C = C_0, \quad \text{at } x = 0, \quad t > 0, \tag{5}$$

$$C = 0, \quad \text{at } x = \infty, \quad t > 0, \tag{6}$$

where C_0 is the concentration of diffused ions at the exposed surface of the solid (mol dm^{-3}). Then the theoretical expression [8] for Eq. (1) is

$$\frac{C}{C_0} = \text{erfc}\left(\frac{x}{2\sqrt{D_a t}}\right), \tag{7}$$

where "erfc" is the error function complement. The fitting results are shown in Figure 2, and the obtained apparent diffusion coefficients (D_a) of Cs, Sr and I in the cement block samples are shown in Table 3.

Discussion

Generally, the larger the porosity of the medium is, the larger the diffusion coefficient becomes [9, 10]. As shown in Table 2, OPC35 has the smallest porosity among all the samples (smaller than those of FAC35 and FAC70). However, the apparent diffusion coefficients of Cs, Sr and I for the FAC samples are almost the same as those for the OPC samples, or are smaller than those for the OPC samples. Figure 3 shows the pore size distributions of the OPC35 and

Table 3 Apparent diffusion coefficients (5 days)

[$m^2\ s^{-1}$]

	OPC35	OPC70	OPC100	FAC35	FAC70
Cs	4.1? 10^{-13}	7.3? 10^{-13}	3.2? 10^{-13}	3.8? 10^{-13}	8.1? 10^{-13}
Sr	3.7? 10^{-13}	1.1? 10^{-12}	6.0? 10^{-13}	5.6? 10^{-14}	3.8? 10^{-13}
I	5.5? 10^{-13}	9.5? 10^{-13}	1.0? 10^{-12}	4.3? 10^{-13}	3.5? 10^{-13}

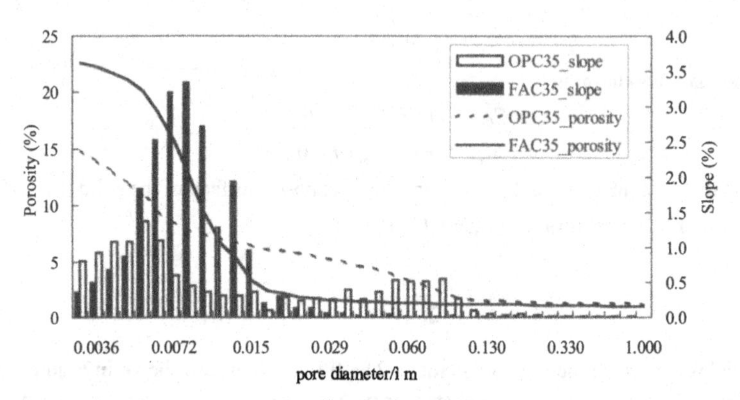

Figure 3 Pore size distributions of OPC35 and FAC35.

FAC35 samples determined by mercury intrusion porosimetry. The pore size of the FAC35 sample is smaller than that of the OPC35 sample. This difference in pore size is due to the pozzolanic reaction induced by mixing fly ash, and is reflected on the diffusion coefficients of the samples. In addition, we should consider that capillary pores are discontinuous in cements made with the addition of pozzolanic materials such as fly ash [11].

In this study, the penetration profiles and diffusion coefficients of the samples were also measured by in-diffusion experiment for 20 days and 180 days. Figure 4 shows the penetration profiles of Cs in the OPC35 samples after 5, 20 and 180 days. With time, the diffusion of Cs increasingly progresses. The penetration profiles in Fig. 4 show evidence of slow diffusion near the surface and fast diffusion in deeper parts. This diffusion behavior has been reported in previous studies [e.g., 12, 13]. The apparent diffusion coefficients were estimated by the application of Eq. (7) to the near-surface of the block samples, and are shown in Table 4 and 5. As shown in Figure 5, boundary condition (6) is not satisfied because the diffused ions reach deep into the samples, as shown in the penetration profiles for 20 days and 180 days. However, in this discussion, we focused on only the near-surface where the slope of the diffused ion concentration is high (at a depth less than 2 mm). Then, we assumed that the boundary condition is satisfied in this range, and Eq. (7) was applied to the experimental results for the evaluation of the apparent diffusion coefficients. As a feature, apparent diffusion coefficient decreases with increasing immersion time. It is possible that the structure of the cement samples changes in a more complicated manner with long-time immersion. We need more experimental data and more extensive discussion of this point.

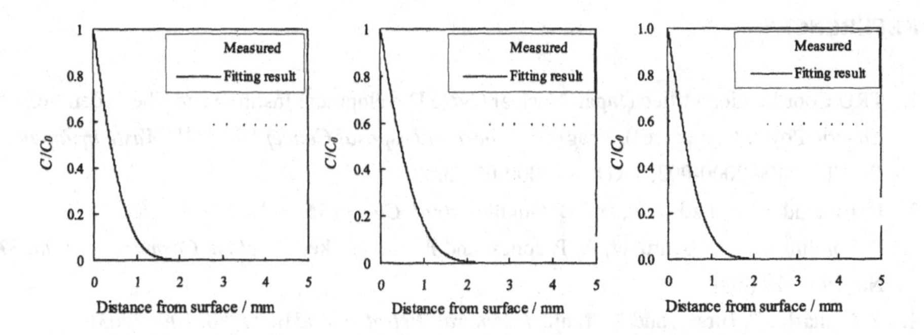

Figure 4 Penetration profiles of cesium after diffusion into OPC35 block samples for 5, 20 and 180 days.

Table 4 Apparent diffusion coefficients (20 days)

[$m^2 s^{-1}$]

	OPC35	OPC70	OPC100	FAC35	FAC70
Cs	1.5?10^{-13}	1.8?10^{-13}	2.1?10^{-13}	7.7?10^{-14}	3.9?10^{-13}
Sr	2.2?10^{-13}	-	1.9?10^{-13}	1.2?10^{-14}	5.4?10^{-14}
I	1.5?10^{-13}	8.3?10^{-13}	5.3?10^{-13}	2.9?10^{-14}	1.8?10^{-12}

Table 5 Apparent diffusion coefficients (180 days)

[$m^2 s^{-1}$]

	OPC35	OPC70	OPC100	FAC35	FAC70
Cs	1.3?10^{-14}	1.5?10^{-14}	3.5?10^{-14}	9.9?10^{-15}	5.2?10^{-14}
Sr	7.7?10^{-14}	1.7?10^{-14}	1.7?10^{-14}	4.2?10^{-15}	1.5?10^{-14}
I	7.7?10^{-13}	5.2?10^{-14}	1.6?10^{-13}	1.5?10^{-14}	4.2?10^{-13}

Conclusion

Detailed penetration profiles of Cs, Sr and I in hardened cement pastes were successively obtained by LA-ICP-MS, and the apparent diffusion coefficients of Cs, Sr, and I in cement samples for a 5-day immersion were estimated to be 10^{-12} - 10^{-13} $m^2 s^{-1}$ for OPC, and 10^{-13} - 10^{-14} $m^2 s^{-1}$ for FAC. The lower diffusion coefficients observed for FAC than for OPC could be explained by the smaller pore size distribution in the FAC solid matrix.

REFERENCES

1. TRU Coordination Office (Japan Nuclear Cycle Development Institute and The Federation of Electric Power Companies), *Progress Report on Disposal Concept for TRU Waste in Japan*, JNC TY 1400 2000-002, TRU TR-2000-02 (2000).
2. B. Hattendorf, C. Latkoczy, and D. Günther, *Anal. Chem.*, **75**, 341A (2003).
3. E. Tomlinson, I. D. Schruver, A. P. Jones, and F. Vanhaecke, *Geochim. Cosmochim. Acta*, **69**, No. 19, 4719 (2005).
4. Y. Orihashi, T. Hirata, and K. Tanji, *J. Mineral. Petrol. Sci. (JMPS)*, **98**, 109 (2003).
5. D. M. Wayne, T. A. Diaz, R. J. Fairhurst, R. L. Orndorff, and D. V. Pete, *Appl. Geochim.*, **21**, 1410 (2006).
6. C. A. Heinrich, T. Pettke, W. E. Halter, M. Aigner-Torres, A.Audétat, D. Günther, and B. Hattendorf, *Geochim. Cosmochim. Acta*, **67**, No. 18, 3473 (2003).
7. D. A. Zedgenizov, S. Rege, W. L. Griffin, H. Kagi, and V. S. Shatsky, *Chem. Geol.*, **240**, 151 (2007).
8. J. Crank, *Mathematics of Diffusion 2nd ed.* (Oxford University Press, London, 1975), p. 20.
9. A. Atkinson and A. K. Nickerson, *J. Mat. Sci.*, **19**, 3068 (1984).
10. K. Haga, S. Sutou, M. Hironaga, S. Tanaka, and S. Nagasaki, *Cement Concr. Res.*, **35**, 1764 (2005).
11. H. F. W. Taylor, *Cement Chemistry 2nd ed.* (Thomas Telford Services Ltd., London, 1997), p. 291.
12. A. Atkinson and A. K. Nickerson, *Nucl. Technol.*, **81**, 100 (1988).
13. K. Idemitsu, H. Furuya, R. Tatsumi, s. Yonezawa, Y. Inagaki, and S. Sato in *Scientific Basis for Nuclear Waste Management XIV*, edited by T. Abrajano, Jr. and L. H. Johnson, (Mater. Res. Soc. Proc. **212**, Boston, MA, 1991) pp. 427-432.

Mater. Res. Soc. Symp. Proc. Vol. 1107 © 2008 Materials Research Society

Magnetite Sorption Capacity for Strontium as a Function of pH

Joan de Pablo [1,2], Miquel Rovira [1,2], Javier Giménez [1,2], Ignasi Casas[1] and Frederic Clarens[2]
[1] Dept. Chemical Engineering, Universitat Politècnica de Catalunya (UPC), Avda. Diagonal 647, 08028 Barcelona, Spain
[2] CTM Centre Tecnològic, Avda. Bases Manresa 1, 08240 Manresa, Spain

ABSTRACT

The ubiquity of iron oxide minerals and their ability to retain metals on their surface can represent an important retardation factor to the mobility of radionuclides. In a deep repository for the spent nuclear fuel, the intrusion of the groundwater might produce the anoxic corrosion of the iron, with magnetite as one of the end-products. In this study, as expected considering the strontium speciation in solution, strontium is sorbed onto magnetite at alkaline pH values while at acidic pH the sorption is negligible. Magnetite is able to sorb more than the 50% of the strontium from a $8\cdot10^{-6}$ $mol\cdot dm^{-3}$ solution at the pH range representative of most groundwater (7-9). A surface complexation model has been applied to the experimental data, allowing to explain the results using the Diffuse Layer Model (DLM) and considering the formation of the inner-sphere complex $>FeOHSr^{2+}$ (with a calculated logK=2.7±0.3). Considering these data, the magnetite capacity to retain strontium and other radionuclides is discussed.

INTRODUCTION

In the high-level radioactive waste repository concept, spent nuclear fuel is designed to be encapsulated in steel canisters which represent the first physical barrier to radionuclide migration. Besides constituting a physical barrier for an eventual groundwater intrusion to the repository, the steel canister provides an effective chemical trap for the release of several radionuclides present in the fuel. Magnetite (Fe_3O_4) has been identified as a corrosion product of carbon steel canisters intended to be used as containers in a HLW final deep repository [1]. For this reason, the study of radionuclide retention role of magnetite is an issue of crucial importance for the performance assessment of spent nuclear fuel iron-containing canisters. In previous works [2-4] we have studied the magnetite sorption capacity for different radionuclides while in this work we present the results obtained with the sorption of strontium. In this work, the study of the interaction between Sr(II) and commercial magnetite is reported.

In the literature, several works have been focused on the study of Sr(II) interaction with hydrous ferric oxides. Todorovic et al. [5] studied the sorption of Sr(II) at trace level concentrations onto magnetite and hematite. For magnetite, sorption extent increased with pH, being almost negligible at pH 4, and for hematite, sorption was found to be very small at the pH values studied (4 and 9.4). Ebner et al. [6] determined sorption isotherms for Sr onto magnetite as well as in magnetite (80%)-silica composite at the pH range 5-9, and they found that solid saturation increased with pH and with the presence of silica.

Sorption processes are incorporated into performance assessment models through the use of empirical sorption isotherms, which assume that the sorption of a given radionuclide is instantaneous, reversible, and increases linearly with the concentration in solution. In general such an approach is not able to account for changes in the chemistry of the system or variations in the mineral-water interface. On the other hand, Surface Complexation Models have been used to model contaminant sorption on oxide surfaces over a wide range of chemical conditions. Dzombak and Morel [7] collected and treated experimental data of Sr(II) and hydrous ferric oxides interaction and considered the surface complexation reactions detailed in Table I, in which they defined strong and weak sites to describe the sorption mechanism.

Table I. Complexation constants for Sr(II) onto iron oxhidroxides from Dzomback and Morel (1990)

	log K
$>Fe^{s}OH + Sr^{+2} \Leftrightarrow >Fe^{s}OHSr^{+2}$	5.01
$>Fe^{w}OH + Sr^{+2} \Leftrightarrow >Fe^{w}OHSr^{+2}$	-6.58
$>Fe^{w}OH + Sr^{+2} + H_2O \Leftrightarrow >Fe^{w}OHSr^{+2} + 2H^{+}$	-17.60

s: strong, w: weak

The objective of this work was to study the role of magnetite on Sr immobilization, considering both the effect of pH and solid/liquid ratio. Surface Complexation Models have been used in order to obtain a mechanism able to explain the experimental data.

EXPERIMENTAL DETAILS

The commercial magnetite (Aldrich) presents the following characteristics: purity of 98%, particle size <5 μm and specific surface area 1.58±0.01 $m^2.g^{-1}$, determined by the BET methodology. X-Ray diffractogram showed a small percentage of Fe(III) oxide.

All the experiments were performed at room temperature. Sr solutions were prepared by dissolving $Sr(NO_3)_2$ in ultra pure water and 0.1 $mol.kg^{-1}$ NaCl was used as ionic media. Sorption tests were conducted in 20 cm^3 polystyrene tubes. Tubes containing weight amounts of magnetite and 10 cm^3 of aqueous phase with $8 \cdot 10^{-6}$ $mol.dm^{-3}$ Sr, were shaken vigorously in a rotatory mixer, the pH was adjusted by using either HCl or NaOH. The experiments were performed open to the atmosphere. After phase separation by centrifugation and filtration (0.22 μm pore size filter), Sr was determined in the liquid phase by means of ICP-MS and the final pH was measured. From preliminary tests, we selected a stirring time of 24 h, well above the time experimentally determined to reach equilibrium (5 h).

RESULTS AND DISCUSSION

Sr sorption is low at acidic pH and increases at alkaline pH, as it can be seen in figure 1, where the sorption extent S(%) is represented as a function of pH. S(%) was calculated according to equation 1.

$$S(\%) = 100 \frac{C_O - C}{C_O} \quad (1)$$

where C_O and C denote initial and equilibrium Sr concentrations, respectively. The sorption dependency with pH is in agreement with the trends reported in the literature and previously described.

The modelling of the experimental results was performed by means of the FITEQL code version 4.0 [8]. The best fits of the data, which are presented as lines in figure 1, were achieved by using the Diffuse Layer Model (DLM), considering the reactions presented in table II, the site density for magnetite (2.31 sites.nm^{-2} [9]), the system equilibrated with atmospheric $CO_2(g)$ as well as the formation of the complex $>FeOHSr^{+2}$ as stated in equation 2.

$$>FeOH + Sr^{+2} \Leftrightarrow >FeOHSr^{+2} \quad (2)$$

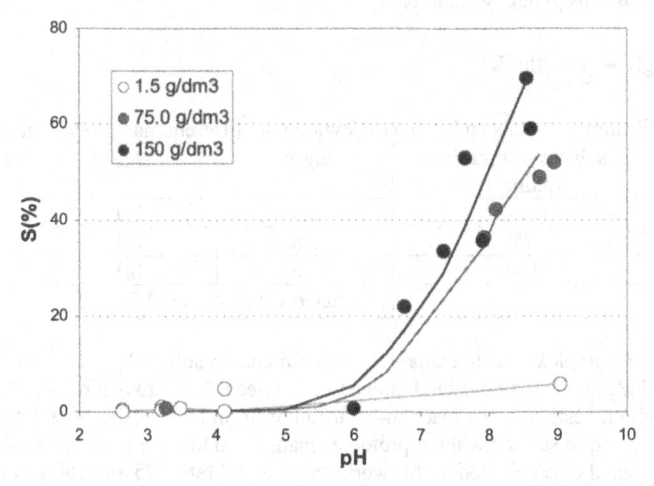

Figure 1. Influence of pH on Sr sorption at different solid/liquid ratios

Table II. Equilibrium constants used for modelling experimental data

	log K	References
Surface reactions		
$>FeOH_2^+ \Leftrightarrow >FeOH + H^+$	-5.1	[10]
$>FeOH \Leftrightarrow >FeO^- + H^+$	-9.1	
Aqueous reactions		
$H_2CO_3 \Leftrightarrow 2H^+ + CO_3^{-2}$	-17.34	[11]
$HCO_3^- \Leftrightarrow H^+ + CO_3^{-2}$	-10.88	
$H_2CO_3 \Leftrightarrow H_2O + CO_2(g)$	1.472	
$SrCO_3 \Leftrightarrow Sr^{+2} + CO_3^{-2}$	-3.69	
$SrHCO_3^- \Leftrightarrow Sr^{+2} + CO_3^{-2} + H^+$	-10.63	
$SrOH^+ + H^+ \Leftrightarrow Sr^{2+}$	13.51	
$NaCO_3^- \Leftrightarrow Na^+ + CO_3^{-2}$	-2.00	
$NaHCO_3 \Leftrightarrow Na^+ + H^+ + CO_3^{-2}$	-12.36	
$SrCO_3(c) \Leftrightarrow Sr^{+2} + CO_3^{-2}$	8.40	

Constants corresponding to aqueous reactions were corrected to 0.1 mol.kg^{-1} ionic strength by using the Davis equation

The DLM assumes that protonation/deprotonation and sorption occur in one plane at the surface-solution interface, and that background electrolytes are inert and not sorbed at the mineral surface. Data fitting allowed determining log K (see table III) for the different solid/liquid ratios. Equilibrium constants were averaged using the method proposed by Dzomback and Morel [7], in which the weighting factor is defined according to equation 3.

$$w_i = \frac{(1/\sigma_{\log K})_i}{\sum (1/\sigma_{\log K})_i} \tag{3}$$

where $\sigma_{\log K}$ is the standard deviation determined by FITEQL for the *i*th data set. The best estimate for logK is thus given by equation 4.

$$\log K = \sum w_i (\log K) \tag{4}$$

Table III. Equilibrium constants and standard deviation at different mass/volume ratios

solid/liquid ratio ($g.dm^{-3}$)	**log K**	$\boldsymbol{\sigma_{\log k}}$
1.5	3.7	2.1
75	2.8	0.1
150	2.6	0.1
	Estimated value: $\log K = 2.7 \pm 0.3$	

The surface complex used to explain the experimental results ($>FeOHSr^{+2}$) was also used by Dzombak and Morel [7] to understand interaction between Sr-hydrous ferric oxides. Axe and Anderson [12] who investigated the reaction-diffusion of Sr on ferrihydrite, postulated that Sr(II) was adsorbed in the solid surface without proton exchange and that the complex $>FeOHSr^{+2}$. In Figure 2, experimental data obtained in this work (mass/liquid ratios 75 and 150 g.dm-3) are compared with data reported in the literature for Sr(II)-magnetite interaction by using the distribution coefficient defined in equation 5 which has been calculated for all data.

$$K_d = q/c \tag{5}$$

where c is the equilibrium Sr concentration in the aqueous phase, and q is equilibrium Sr concentration in the solid phase calculated by mass balance.

In Figure 2, it can be observed the fairly good agreement between data obtained in this work and data obtained by Todorovic et al. [5] where the sorption of Sr(II) at trace level concentrations onto magnetite was studied. On the other hand, in case of Ebner et al. [6], there is an important deviation which can be explained by the fact that these authors performed their experiments under $N_2(g)$ atmosphere, avoiding the formation of carbonate-strontium complexes that inhibit the sorption.

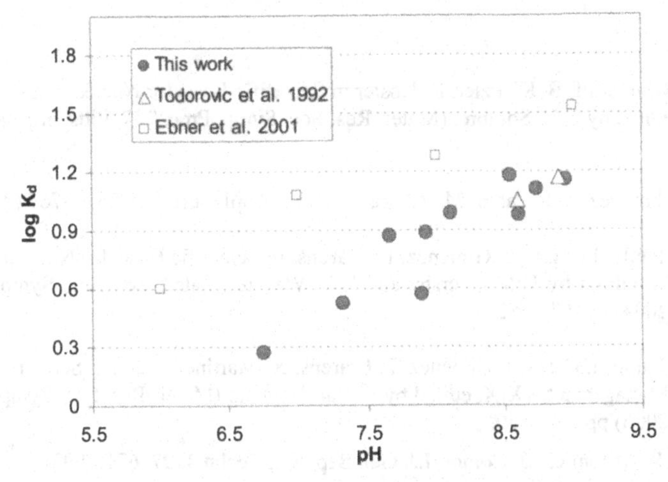

Figure 2. Comparison of the pH influence on the distribution coefficient between data obtained in this work (mass/liquid ratio 75 and 150 $g.dm^{-3}$) and data reported by Todorovic et al. [5] and Ebner et al. [6]

CONCLUSIONS

The main conclusions withdrawn for this work are:

- Sr-magnetite interaction has been interpreted by Surface Complexation Modelling assuming the formation of the inner-sphere complex $>FeOHSr^{+2}$ and by using the Diffuse Layer Model (DLM). The model obtained, allow incorporating the sorption mechanism into performance assessment of a HLW, covering a wide range of chemical conditions and improving the traditional use of empirical equations.
- The sorption of Sr onto magnetite is significant at basic pH, while sorption is inhibited at low pH in good agreement with data reported in the literature.
- The results obtained indicate that Sr sorption extent may be significant at pH values representatives of natural groundwater, as well as under the conditions expected to prevail in a HLW.
- Data obtained in this work at open atmosphere show an important decrease on sorption extent compared to works performed under $N_2(g)$ atmosphere, behaviour that can be explained by the absence of carbonate-strontium complexes that inhibit Sr sorption in presence of $N_2(g)$.

ACKNOWLEDGEMENTS

Thanks are also due to Aurora Martínez-Esparza for their interest in our work. This work was funded by ENRESA.

REFERENCES

1. E. Smailos, W. Swarzkopf, B. Kienzler, R. Köster in Scientific Basis for Nuclear Waste Management XV, edited by C.G. Sombret (Mater. Res. Soc. Symp. Proc. 257, Pittsburg PA, 1992) pp. 399-406.

2. M. Martínez, J. Giménez, J. de Pablo, M. Rovira, L. Duro, Appl. Surf. Sci. 252, 3767 (2006).

3. M. Rovira, J. de Pablo, I. Casas, J. Giménez, F. Clarens, in Scientific Basis for Nuclear Waste Management XXVII, edited by V:M Oversby and L.O. Werme (Mater. Res. Soc. Symp. Proc. 807, Pittsburg PA, 2004) pp. 677-682.

4. M. Rovira, J. de Pablo, I. Casas, J. Giménez, F. Clarens, X. Martínez-Lladó in Scientific Basis for Nuclear Waste Management XXIX, edited by P. Van Iseghem (Mater. Res. Soc. Symp. Proc. 932, Pittsburg PA, 2006) pp. 143-150.

5. M. Todorovic, S.K. Milonjic, J.J. Comor, I.J. Gal, Sep. Sci. Technol. 27, 671(1992).

6. A.D. Ebner, J.A. Ritter, J.D. Navratil, Ind. Eng. Chem. Res. 40, 1615 (2001).

7. D.A. Dzombak, F.M. Morel in Surface Complexation Modeling. Hydrous Ferric Oxide. (Wiley-Interscience, New York, 1990).

8. A.L. Herbelin, J.C. Westall, FITEQL 4.0: a Computer Program for Determination of Chemical Equilibrium Constants from Experimental Data. (Department of Chemistry, Oregon State University, Corvallis, 1999).

9. J.A. Davies, D.B. Kent, Rev. Mineralogy 23, 117 (1990).

10. T. Missana, M. García, C. Maffiotte, Uranium(VI) sorption on goethite: Experimental study and surface complexation modelling. ENRESA Report 02/2003 (Madrid, Spain, 2003).

11. J.W. Ball, D. K. Nordstrom WATEQ4F. User's manual with revised thermodynamic data base and test cases for calculating speciation of major, trace and redox elements in natural waters: U.S. Geological Survey Open-File Report 90-129, (1991) .

12. L. Axe, P.R. Anderson, J. Colloid Interf. Sci. 175, 157 (1995).

Mater. Res. Soc. Symp. Proc. Vol. 1107 © 2008 Materials Research Society

Kinetics of UO_2(s) Dissolution in the Presence of Hypochlorite, Chlorite, and Chlorate Solutions

Rosa Sureda[1], Ignasi Casas[1], Javier Giménez[1], Joan de Pablo[1,2]
[1]Dept. Chemical Engineering, Universitat Politècnica de Catalunya, Avda. Diagonal 647, 08028 Barcelona, Spain.
[2]Environmental Technology Area-CTM Centre Tecnològic, Avda. Bases de Manresa 1, 08242 Manresa, Spain.

ABSTRACT

The influence of hypochlorite, chlorite and chlorate in the UO_2 dissolution rate has been studied experimentally using a continuous flow-through reactor. Uranium concentration in each outflow solution was measured as a function of time and dissolution rates were determined once the steady-state was reached. The results obtained show that the influence of the hypochlorite anion concentration on the UO_2 dissolution rate can be expressed by the following empirical equation:

$r_{diss} = 10^{-8.7\pm0.1} \cdot [ClO^-]^{0.28\pm0.04}$

The dissolution rates obtained in this work were higher than those previously determined in presence of either oxygen or hydrogen peroxide using the same experimental methodology.

In contrast, neither chlorate nor chlorite had any significant effect on the UO_2 dissolution rates under the experimental conditions of this work.

INTRODUCTION

Different studies have been carried out concerning the UO_2 or spent nuclear fuel dissolution in the presence of the main molecular oxidants produced in the radiolysis of water: O_2 [1-6], and H_2O_2 [7-11]. However the effect of the oxidants produced in the radiolysis of groundwaters with a relatively high chloride concentration and brines has been much less studied, in spite of the fact that these ground waters (brines) are expected in saline geological environments. Radiation can alter the chemistry of the brine, producing many radiolytic products such as Cl_2^- and some oxychlorides [12,13], in particular ClO_2^-, ClO_3^-, and ClO^-.

The hypochlorite ion (ClO^-) was found to be formed after the α irradiation of a NaCl-brine due to the chloride ion oxidation [14,15] and the formation of ClO^- as a function of initial chloride concentration and alpha activity concentration was also studied [16,17]. It has been demonstrated that the formation of ClO^- in solution has an important influence on the chemistry of the actinides, for example, it was observed that the ClO^- formed was able to oxidize Am(III) and Pu(IV) in solution [14]. Kim et al. [18] studied the solubility of amorphous schoepite in solutions with and without ClO^-, obtaining solubility values two orders of magnitude higher in the presence of the hypochlorite ion with a concentration of 10^{-3} mol·dm^{-3}.

On the other hand, the formation of chlorate by gamma radiolysis from NaCl brines (6 mol·dm^{-3}) at ambient temperature and dose rates between 0.1 and 1 kGy/h has been demonstrated by Kelm and Bohnert [19-21], ClO_3^- being formed proportionally to the dose. The authors could measure ClO_3^- concentrations between $8.1 \cdot 10^{-5}$mol·dm^{-3} and $1.89 \cdot 10^{-4}$ mol·dm^{-3} at 35°C and doses between 3335 kGy (403 days) and 9094 kGy (415 days), when the temperature was increased from 35°C to 90°C, the chlorate ion concentration obtained decreased more than one order of magnitude.

Although some studies have shown the influence of some of these oxychlorides on the chemistry of actinides in the near-field, the kinetics of dissolution of UO_2 or spent fuel in the presence of those oxidants has not to the best of our knowledge been reported.

In this sense, we can cite the study on the influence of the hypochlorite ion concentration on the non irradiated UO_2 dissolution, carried out in batch experiments at three different ClO^- concentrations [9]. The authors obtained the following relationship between dissolution rate and ClO^- concentration in solution $r=k\cdot[ClO^-]^{1.00\pm0.04}$ in the range of ClO^- concentrations $2\cdot10^{-5}$-0.01 $mol\cdot dm^{-3}$. We could not find more kinetic data on the influence of ClO^-, ClO_2^-, or ClO_3^- on the UO_2 dissolution.

The aim of this work was to determine the UO_2 dissolution rate in the presence of varying concentrations of hypochlorite, chlorite or chlorate by using a flow through reactor [5].

EXPERIMENTAL

UO_2 dissolution rates were determined by using the same continuous flow-through reactor as that previously used for UO_2 dissolution rate determinations under different conditions [5,22].

1g of UO_2 with a particle size of 100-320 μm was introduced in a KONTES column. The solid was previously washed with $HClO_4$. The surface area of the solid, determined by the BET method (Flowsorb II 2300, Micrometrics), gave a value of $0.010\pm0.001\ m^2\cdot g^{-1}$. Experiments were carried out with a flow rate of 0.18-0.21 mL/min at pH=8-9 in N_2, and in $NaClO_4$ 0,1M ionic medium. Three different concentrations of chlorate, chlorite, and hypochlorite were used: 10^{-3}, 10^{-4}, and 10^{-5} $mol\cdot dm^{-3}$.

Samples of the outflow solution were taken periodically (see Table I), filtered through 0.22 μm pore size filters and acidified with concentrated HNO_3. During the collection of each sample the flow rate was measured by weight. Uranium concentration in solution was determined by ICP-MS.

UO_2 dissolution rates were determined as follows: the uranium concentration in solution was measured at given time intervals. Once the steady-state was reached, the dissolution rate was calculated according to the following equation:

$$r = q\cdot[U] / m\cdot S \qquad \text{(eq.1)}$$

where r ($mol\cdot m^{-2}\cdot s^{-1}$) is the dissolution rate, q ($dm^3\cdot s^{-1}$) is the flow rate, [U] ($mol\cdot dm^{-3}$) is the steady-state uranium concentration of the output solution, m (g) is the mass of UO_2, and S ($m^2\cdot g^{-1}$) is the specific surface area of the solid.

RESULTS AND DISCUSSION

Effect of hypochlorite ion

Table I shows the uranium concentrations measured as a function of time for the three experiments performed with hypochlorite ion. From the uranium concentrations at the steady state and using eq. 1, the dissolution rates were determined (Figure 1). The UO_2 dissolution rate values are $(9.7\pm1,8)\cdot10^{-11}$, $(1.5\pm0,3)\cdot10^{-10}$, and $(3.4\pm0.1)\cdot10^{-10}$ $mol\cdot m^{-2}\cdot s^{-1}$ at hypochlorite ion concentrations of 10^{-5}, 10^{-4}, and 10^{-3} $mol\cdot dm^{-3}$, respectively.

Table I. Uranium concentrations at the steady-state in the experiments in the presence of ClO^-. Concentrations given in $mol\cdot dm^{-3}$.

$[ClO^-]=10^{-5}$		$[ClO^-]=10^{-4}$		$[ClO^-]=10^{-3}$	
Time (d)	$[U]_t$	Time (d)	$[U]_t$	Time (d)	$[U]_t$
0.6	7.38E-07	1	3.09E-07	1	7.56E-07
0.7	6.22E-07	2	4.82E-07	2	1.02E-06
1.5	2.53E-07	3	5.30E-07	3	1.08E-06
1.7	2.85E-07	4	4.93E-07	5	8.89E-07
2.5	2.10E-07	5	6.01E-07	6	8.87E-07
2.7	2.97E-07	6	3.13E-07	6.5	8.96E-07
3.5	2.81E-07	7	4.18E-07	7	8.02E-07
3.7	3.83E-07	8	4.07E-07	8	7.51E-07
4.7	3.92E-07			9	8.24E-07
5.6	2.39E-07			10	8.04E-07
5.8	2.79E-07			11	8.01E-07
6.7	4.11E-07			12	7.72E-07
8.7	2.86E-07			13	1.04E-06
8.8	3.16E-07			14	1.06E-06
9.7	2.44E-07			15	9.68E-07
10.7	3.21E-07			16	1.03E-06
10.4	3.03E-07				
11.7	2.50E-07				
12.4	2.41E-07				

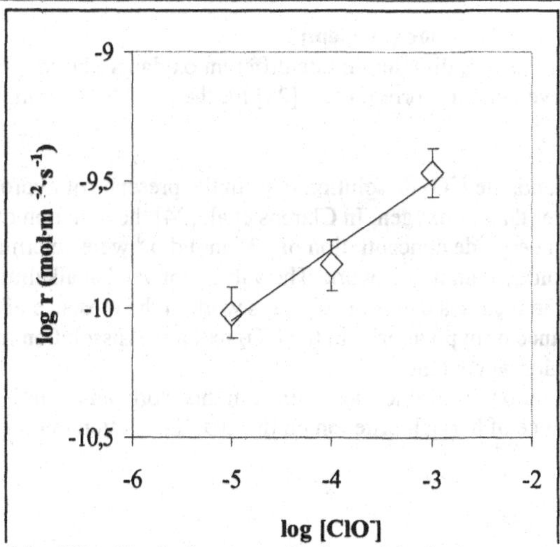

Figure 1. Variation of the UO_2 dissolution rate with hypochlorite ion concentration in solution.

In order to determine the oxidation capacity of the hypochlorite ion against UO_2, the dissolution rates were compared with the dissolution rates obtained in the presence of other oxidants (see Figure 2). Previous studies have mainly used the oxidants oxygen and hydrogen peroxide, due to their formation through the radiolysis of water. UO_2 dissolution rates in the presence of oxygen in free-carbonate solutions have been found to be relatively low at neutral to alkaline pH values. Torrero et al. [23] deduced an empirical relationship between UO_2 dissolution rate and both proton and oxygen concentrations in solution. According to this equation, UO_2 dissolution rates at neutral to alkaline pH were lower than 10^{-11} $mol \cdot m^{-2} \cdot s^{-1}$, which is lower than the dissolution rates in the presence of hypochlorite ion determined in the current work.

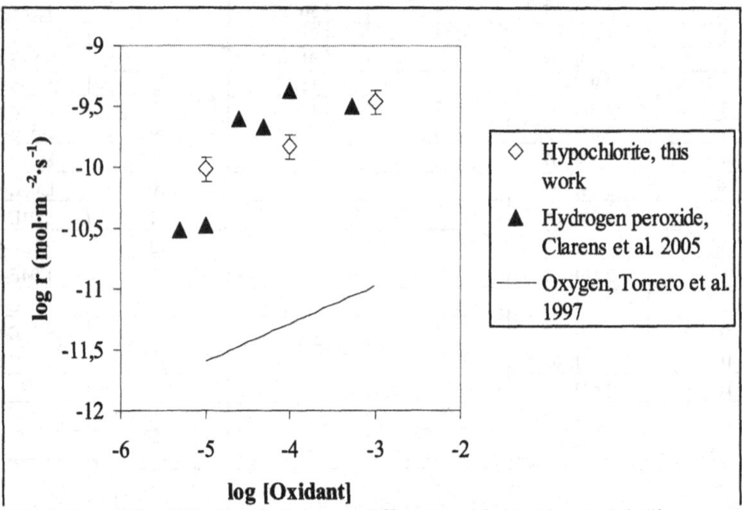

Figure 2. Variation of the UO_2 dissolution rate different oxidants (the straight line was obtained by using the model developed in Torrero et al. [23] for the UO_2 dissolution in the presence of oxygen).

On the other hand, the UO_2 dissolution rates in the presence of hydrogen peroxide have been found to be higher than in oxygen. In Clarens et al. [24] the variation of the dissolution rate with pH at a hydrogen peroxide concentration of 10^{-5} $mol \cdot dm^{-3}$ were determined using the same experimental methodology than in this work. The values obtained at alkaline pH were of the same order of magnitude than the rates determined in this work in the presence of the hypochlorite ion, indicating the importance of hypochlorite in the UO_2 oxidative dissolution mechanism in solutions with high chloride content.

From the dissolution rate values determined in this work, an empirical equation might be deduced for the influence of hypochlorite ion on the UO_2 dissolution rate:

$$r_{diss} = 10^{-8.7 \pm 0.1} \cdot [ClO^-]^{0.28 \pm 0.04}$$

Effect of chlorite and chlorate ions on the UO_2 dissolution

From the dissolution experiments carried out in the presence of variable concentrations of chlorite and chlorate the main result obtained was that uranium concentrations at the steady-state were not different from the ones determined in the 'blank' experiment (UO_2 dissolution in a solution containing only the ionic medium). This result implies that neither chlorite nor chlorate seem to significantly affect the UO_2 dissolution in the experimental conditions of this work.

CONCLUSIONS

Hypochlorite ion influences the UO_2 dissolution according to the following empirical equation:

$$r_{diss} = 10^{-8.7\pm0.1} \cdot [ClO^-]^{0.28\pm0.04}$$

The dissolution rates determined in the presence of hypochlorite ion are higher than the ones determined in oxygen and in hydrogen peroxide in similar experimental conditions. These relatively high dissolution rates point to the fact that hypochlorite ion should be considered when establishing the uranium and other radionuclides release from spent nuclear fuel in high-chloride concentration groundwaters. In contrast, neither chlorite nor chlorate caused any effect on the UO_2 dissolution rates in the experimental conditions of this work.

ACKNOWLEDGMENTS

Thanks are due to Aurora Martínez-Esparza for her valuable comments and suggestions. This work was financially supported by ENRESA (Spanish Radioactive Waste Management Co). Rosa Sureda thanks and the 'Cátedra ARGOS' (CSN-UPC collaboration) for financial support

REFERENCES

1. W.J. Gray and C.N. Wilson, Spent fuel dissolution studies: FY 1991 to 1994. Report PNL-10540, USA (1995).
2. D.W. Shoesmith, Fuel corrosion processes under waste disposal conditions, J. Nucl. Mater. 282 (2000) 1-31.
3. M.E. Torrero, E. Baraj, J. de Pablo, J. Giménez and I. Casas, Kinetics of corrosion and dissolution of uranium dioxide as a function of pH, Int. J. Chem. Kinet. 29 (1997) 261-267.
4. J. de Pablo, I. Casas, J. Giménez, M. Molera, M. Rovira, L. Duro and J. Bruno, The oxidative dissolution mechanism of uranium dioxide. I. The effect of temperature in hydrogen carbonate medium, Geochim. et Cosmochim. Acta 63 (1999) 3097-3103.
5. J. de Pablo, I. Casas, J. Giménez, F. Clarens, L. Duro and J. Bruno, The oxidative dissolution mechanism of uranium dioxide. The effect of pH and oxygen partial pressure, Mater. Res. Soc. Symp. Proc. 807 (2004) 83-88.
6. J. Giménez, F. Clarens, I. Casas, M. Rovira, J. de Pablo, J. Bruno, Oxidation and dissolution of UO_2 in bicarbonate media: Implications for the spent nuclear fuel oxidative dissolution mechanism, J. Nucl. Mater. 345 (2005) 232-238.

7. J.B. Hiskey, Kinetics of uranium dioxide dissolution in carbonate solutions, Trans. Inst. Min. Metall. Sect. C 89 (1980) 145-171.
8. D.W. Shoesmith and S. Sunder, The prediction of nuclear fuel (UO_2) dissolution rates under waste disposal conditions, J. Nucl. Mater. 190 (1992) 20-35.
9. J. Giménez, E. Baraj, M.E. Torrero, I. Casas and J. de Pablo, Effect of H_2O_2, NaClO and Fe on the dissolution of unirradiated UO_2 in NaCl 5 mol kg^{-1}. Comparison with spent fuel dissolution experiments, J. Nucl. Mater. 238 (1996) 64-69.
10. E. Ekeroth and M. Jonsson, Dissolution of UO_2 by radiolytic oxidants, J. Nucl. Mater. 322 (2003) 242-248.
11. F. Clarens, J. de Pablo, I. Casas, J. Giménez, M. Rovira, J. Merino, E. Cera, J. Bruno, J. Quiñones, A. Martínez-Esparza, The oxidative dissolution of unirradiated UO_2 by hydrogen peroxide as a function of pH, J. Nucl. Mater. 345 (2005) 225-231.
12. W.J. Gray, Effects of radiation on the oxidation potential of salt brine, Mater. Res. Soc. Symp. Proc. 112 (1988) 405-413.
13. M. Kelm, E. Bohnert, Gamma radiolysis of NaCl brine: Effect of dissolved radiolysis gases on the radiolytic yield of long-lived products, J. Nucl. Mater. 346 (2005) 1-4.
14. K. Büppelmann, S. Magirius, Ch. Lierse, J.I. Kim, Radiolytic oxidation of americium(III) to americium(V) and plutonium(IV) to plutonium(VI) in saline solution, J. Less-Common Metals 122 (1986) 329-336.
15. K. Büppelmann, J.I. Kim, Ch. Lierse, The redox-behaviour of plutonium in saline solutions under radiolysis effects, Radiochim. Acta 44/45 (1988) 65-70.
16. M. Kelm, I. Pashalidis, J.I. Kim, Spectroscopic investigation on the formation of hypochlorite by alpha radiolysis in concentrated NaCl solutions, Appl. Radiat. Isotopes 51 (1999) 637-642.
17. Th. Hartmann, P. Paviet-Hartmann, Ch. Wetteland, N. Lu, Spectroscopic determination of hypochlorous acid, in chloride brine solutions, featuring 5 MeV proton beam line experiments, Radiat. Phys. Chem. 66 (2003) 335-341.
18. W.H. Kim, K.C. Choi, K.K. Park, T.Y. Eom, Effects of hypochlorite ion on the solubility of amorphous schoepite at 25°C in neutral to alkaline aqueous solutions, Radiochim. Acta 66/67 (1994) 45-49.
19. B.G. Ershow, M. Kelm, E. Janata, A.V. Gordeev, E. Bohnert, Radiation-chemical effects in the near-field of a final disposal site: role of bromine on the radiolytic processes in NaCl-solutions, Radiochim. Acta 90 (2002) 617-622.
20. M. Kelm, E. Bohnert, Radiation chemical effects in the near field of a final disposal site – I: Radiolytic products formed in concentrated NaCl solutions, Nucl. Technol. 129 (2000) 119-122.
21. M. Kelm, E. Bohnert, Radiation chemical effects in the near field of a final disposal site – II: Simulation of the radiolytic processes in concentrated NaCl solutions, Nucl. Technol. 129 (2000) 123-130.
22. J. Giménez, F. Clarens, I. Casas, M. Rovira, J. de Pablo, J. Bruno, Oxidation and dissolution of UO_2 in bicarbonate media: Implications for the spent nuclear fuel oxidative dissolution mechanism. J. Nucl. Mater. 345 (2005) 232-238.
23. M.E. Torrero, E. Baraj, J. de Pablo, J. Giménez, I. Casas, Kinetics of corrosion and dissolution of uranium dioxide as a function of pH. Int. J. Chem. Kinet. 29 (1997) 261-267.
24. F. Clarens, J. de Pablo, I. Casas, J. Giménez, M. Rovira, J. Merino, E. Cera, J. Bruno, J. Quiñones, A. Martínez-Esparza, The oxidative dissolution of unirradiated UO_2 by hydrogen peroxide as a function of pH. J. Nucl. Mater. 345 (2005) 225-231.

Mater. Res. Soc. Symp. Proc. Vol. 1107 © 2008 Materials Research Society

Long-Term Predictions of the Concentration of α-isosaccharinic Acid in Cement Pore Water

Martin A. Glaus[1], Luc R. Van Loon[1], Bernhard Schwyn[2], Sarah Vines[3], Steve J. Williams[3], Peter Larsson[4], Ignasi Puigdomenech[4]
[1]Laboratory for Waste Management, Paul Scherrer Institut (PSI), CH-5232 Villigen, Switzerland
[2]National Cooperative for the Disposal of Radioactive Waste (Nagra), Wettingen, Switzerland
[3]Nuclear Decommissioning Authority (NDA), Radioactive Waste Management Directorate, Curie Avenue, Harwell Didcot, Oxon, OX11 0RH, UK
[4]Swedish Nuclear Fuel and Waste Management Co (SKB), Box 5864, SE-102 40 Stockholm, Sweden

ABSTRACT

The long-term prediction of the equilibrium concentration of α-isosaccharinic acid (α-ISA) in cement pore water is a crucial step in the assessment of the role of cellulose in the safety of a cementitious repository. The aim of the present contribution is to summarise recent efforts in identifying the most important processes leading to the formation or degradation of α-ISA and in predicting its most likely concentrations in cement pore water. The issues considered are the kinetics involved in the formation of α-ISA, reactions of α-ISA with dissolved or solid compounds that may lead to limitations of its pore water concentrations and the chemical stability of α-ISA in a heterogeneous alkaline environment. Some new results are presented showing that α-ISA is degraded to low-molecular weight organic compounds in the presence of oxygen, whereas such processes occur only to a minor extent under anaerobic conditions. It is concluded that the processes involved in the degradation of cellulose under alkaline conditions are not sufficiently understood to explain fully the observed concentrations of α-ISA in long-term experiments.

INTRODUCTION

Cellulose is a main constituent of the organic matter in low- and intermediate level radioactive wastes. The degradation of cellulosic material in the cementitious alkaline environment of the repositories for these wastes results in the formation of water-soluble species. Among the various reaction products of the alkaline degradation of cellulose, α-isosaccharinic acid (α-ISA) is one of the most important complexants formed [1-3]. Cellulose degradation products may thus have a detrimental effect on the barrier function of cement because they enhance the solubility and decrease the sorption of many cations of the transition metal, lanthanide and actinide series. The long-term estimation of the equilibrium concentration of α-ISA in cement pore water is consequently an important step in the assessment of the role of cellulose in the safety of such a repository.

Van Loon and Glaus [3] proposed a reaction scheme which was used to predict quantitatively the equilibrium concentration of α-ISA in cement pore water. The scheme comprises the degradation reactions of cellulose that lead to the formation of α-ISA and reactions that lead to the disappearance of α-ISA from the solution phase by precipitation or sorption reactions [4,5]. For a typical reference case of cellulose containing waste in a repository for low and intermediate level waste in Switzerland, the extent of cellulose degradation turned out to be a critical parameter in this scheme. In a rather conservative approach, where complete conversion of cellulose to α-ISA is assumed, the produced amount of α-ISA exceeds the sorption capacity of cement. Large concentrations of α-ISA in the order of $>10^{-2}$ M would result, which would lead to a significant alteration of radionuclide speciation and mobility [6-9]. In a more realistic scenario, where the slow kinetics of cellulose degradation are taken into account, the amount of α-ISA formed would be less than the sorption capacity of cement, resulting in α-ISA concentrations between 10^{-5} and 10^{-4} M. Negligible effects on the speciation of lanthanide or actinide ions are expected in this concentration range. Increased efforts were therefore undertaken in projects co-funded between Nagra, Nirex (now incorporated in NDA), SKB and PSI to consolidate the knowledge on formation and degradation reactions of α-ISA under the conditions of a cement pore water. The present contribution gives an overview of the most important results of these experiments and the conclusions drawn.

FORMATION OF α-ISA FROM THE ALKALINE DEGRADATION OF CELLULOSE

The degradation of cellulose under alkaline conditions has been reviewed elsewhere [e.g. 10-12]. α-ISA is formed basically from two different types of reaction (cf. Fig. 1).

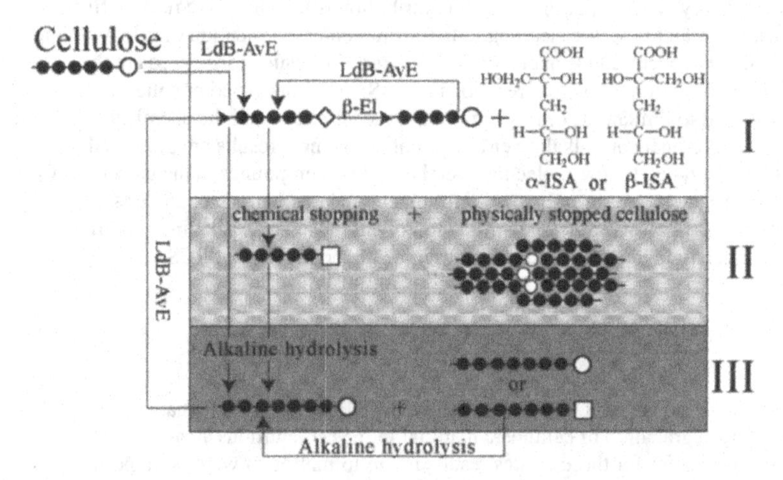

Figure 1. Simplified reaction scheme for the processes involved in the alkaline degradation of cellulose. The chain lengths are not represented to scale and do not indicate changes in chain length due to progression of the reactions. Side reactions (cf. the text) are not shown. The symbols used are: glucose monomeric units (●); reducing end-groups (O, aldehyde functional group); chemically stopped end-groups (□, metasaccharinic acid); activated end-groups (◇, enolate functional group); intermediate product (◆, 4-deoxy-2,3-hexodiulose). The reactions referred to are the Lobry de Bruin – Alberda van Ekenstein transformation (LdB-AvE) and a β-elimination (β-El).

The first type is a conversion of the reducing end-group of the polysaccharide chain into a reactive interim product, which is then clipped off (referred to as the 'propagation reaction', cf. box I of Fig. 1) to form α-ISA and β-ISA (a diastereomeric form of α-ISA with less complexing strength [3]) in approximately equal amounts. Depending on the cations present, parallel pathways may lead to the formation of short-chain carboxylic acids other than ISA [13]. Cellulose degrades only partly by the propagation reaction type, because either the reactive interim product is converted to a non-reactive metasaccharinic end-group (chemical stopping), or because the degradation reaches crystalline regions of the polysaccharide chain, where the peeling-off reaction cannot further proceed owing to steric reasons (physical stopping), cf. box II in Fig. 1. Propagation and stopping reaction taken together are frequently referred to in the literature as the peeling-off reaction. Kinetic data for the peeling-off process are available in temperature ranges between 25 and 130 °C [3,14]. The second type, frequently referred to as the mid-chain scission or alkaline hydrolysis, is a random cleavage of glycosidic bonds within the polysaccharide chain. A reducing end-group is recreated on one of the two chain fragments, whereupon the peeling-off reaction is reinitiated, cf. box III in Fig. 1. Cellulose may thus be completely degraded to monomeric products through the concerted action of peeling-off reaction and mid-chain scission. However, owing to the very slow reaction rate of mid-chain scission, reliable kinetic data have only been obtained in a high-temperature range between 140 °C and 180 °C [15,16].

Predictions of the long-term fate of cellulose in an alkaline repository environment are contradictory. Pavasars et al. [17] adopted a kinetic model proposed by Van Loon and Glaus [3,12] to obtain best-fit parameter values for the different reaction rate constants involved. Using a value of $3 \cdot 10^{-6}$ h^{-1} for the mid-chain scission occurring in a typical cement pore water at room temperature (~0.3 M OH^-) they concluded that cellulose is completely degraded within ~100 to 500 years. In contrast, Van Loon and Glaus obtained a reaction rate constant of ~10^{-10} h^{-1} [3,12] when extrapolating a value for the same conditions from experiments carried out at high temperatures by an Arrhenius type of relation. These authors came to the conclusion the stopped part of cellulose would be stable for ~10^4 years and would degrade completely over a period of 10^5 to 10^6 years.

EXPERIMENTAL DETAILS

The materials and experimental procedures have been described in previous publications [18-21]. The lactone form of α-ISA was prepared according to [22] using a modification given in [18]. Briefly, four different cellulosic materials (Aldrich purified cellulose, cellulose tissues, cotton and recycling paper) were contacted in closed PTFE containers with $Ca(OH)_2$ and an artificial cement pore water (ACW-I) simulating the chemical conditions of the first stage of cement degradation. The composition of ACW-I was: 114 mmol dm^{-3} Na, 180 mmol dm^{-3} K, ~2 mmol dm^{-3} Ca and a pH of 13.3. The reaction mixtures were kept at 25 ±2 °C under an inert-gas atmosphere (O_2, CO_2 < 5 ppm) and under exclusion of light. After various times of reaction the samples were centrifuged, filtered and analysed for pH, total alkalinity and concentration of non-purgeable dissolved organic carbon (DOC) by a TOC-V WP® device (Shimadzu, Reinach, Switzerland) using UV-promoted persulfate wet oxidation. α-ISA and β-ISA and other carbohydrates were measured by high-performance anion exchange chromatography with pulsed amperometric detection (HPAEC-PAD, Dionex DX-600), low-molecular weight carboxylic acids by high-performance ion exclusion chromatography (HPIEC, Dionex DX-600).

The experiments at elevated temperatures were kept in an oven at 60 and 90 °C under laboratory atmosphere. In order to minimise the influence of oxygen these suspensions were flushed with argon during preparation and each sampling.

The reactivity of α-ISA in heterogeneous alkaline systems was investigated under similar experimental conditions. Solutions of the lactone form of α-ISA were prepared in ACW-I at various concentrations and with various amounts of $Ca(OH)_2$ and kept at 25 ±2 °C or 90 ±2 °C either under exclusion of air in an inert-gas glove box, under argon protective gas or under laboratory air. Aliquots of these samples were analysed using the same procedures as in the cellulose degradation experiments.

RESULTS AND DISCUSSION

Degradation of cellulose at room temperature

Fig. 2 shows the evolution of the sum concentration of α-ISA + β-ISA (denoted to as ISA) and DOC for the four cellulosic materials as a function of reaction time. The data between 0 and ~2 y have been published earlier [19]. Evaporative loss of water was small and could be neglected.

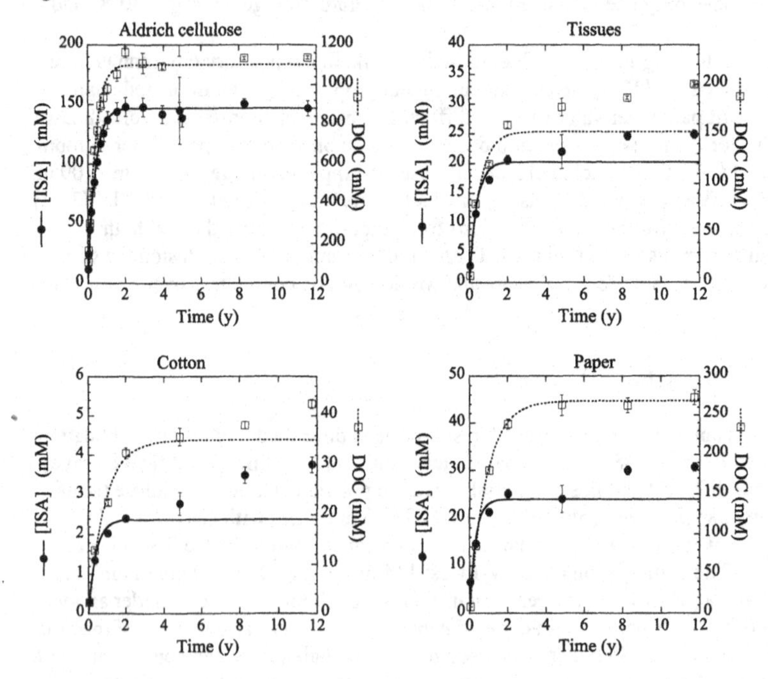

Figure 2. Evolution of sum concentrations of ISA and DOC (corrected for extractable compounds, cf. the text) as a function of time during degradation of various cellulose types in ACW-I in experiments carried out at room temperature. The model (lines) applied to the data comprises the peeling-off process solely and is based on the reaction rate constants given in [19].

No correction of the concentrations of the organic compounds for comparison with the earlier measurements was therefore necessary. DOC concentrations were corrected for organic compounds extracted within the first day of contact of the cellulose with ACW I according to [19] (16 mM for Aldrich cellulose, 70 mM for tissues, 4 mM for cotton, 200 mM for paper). A 1:6 proportion of the ordinates is chosen to represent directly the amount of ISA as compared to the total amount of organic carbon present in solution.

Supporting experiments [21] demonstrated that α-ISA remained stable in the homogeneous aqueous ACW-I phase and that no α-ISA could be detected in the solids, suggesting that the concentrations of reaction products in the solution phase are relevant for the assessment of the extent of cellulose degradation as a function of time. The close position of the ISA and DOC data confirms the earlier conclusion that the two stereoisomers of ISA are the main products of the alkaline degradation of cellulose. Also included in the plots are the model predictions based on the data measured between 0 and 3 years [19]. It can be seen that experimental data up to three years are in good agreement with the model, whereas the data between three and 12 years are underestimated by the model throughout. This suggests that the overall process of cellulose degradation is dominated by the peeling-off process in the first three years, while another slow process becomes rate-limiting in the second phase of reaction. The increase of ISA concentration parallels the increase of DOC in general. This is an indication that the peeling-off process is ultimately responsible for the further degradation of cellulose. In view of this observation two processes explaining possibly the slower second phase of cellulose degradation may be considered, (i) the mid-chain scission of the polysaccharide chain combined with subsequent peeling-off of reconstituted reducing end-groups, or (ii) the strongly decelerated reaction of reducing end-groups becoming temporarily accessible for nucleophilic attack in a dynamic equilibrium between crystalline and amorphous structures of cellulose. Whereas the former process is well known from experiments at high temperature [15,16], no quantitative information on the second process is available from literature. Based on the experimental data available it is not possible to discriminate between these two processes. However, fitting the data of the slow reaction phase tentatively to the mid-chain scission process results in reaction rate constants between 10^{-8} and 10^{-7} h^{-1} [21], which is still significantly different from the value obtained by extrapolation from high temperature data (~10^{-10} h^{-1}, cf. [12]). It can also be concluded from the results that the reaction rate constant proposed by Pavasars et al. [17] is much too large. Almost complete degradation of tissues, cotton and paper would be expected for such a case [21].

Degradation of cellulose at elevated temperature

Results for the degradation of cellulose at 90 °C are shown in Fig. 3. Similar results ending up with similar extents of cellulose degradation were obtained for 60 °C (details in [20]). The experiments at elevated temperatures show that a degradation model comprising the peeling-off reaction and mid-chain scission is not appropriate to describe the observed course of reaction. An almost instantaneous phase (< 1d) of peeling-off would be expected for 60 °C and 90 °C. The steady increase of cellulose degradation observed for reaction times between 1 and ~200 d cannot be explained by mid-chain scission. Complete disintegration of the polysaccharide chain with formation of short-chain carboxylic acids would be expected in such a case. It can thus be concluded that a reaction model consisting of peeling-off process and mid-chain scission is not sufficient to explain fully the experimental observations. With such a model, it would also not be expected that a pure cellulose with low degree of polymerisation (*DP*, ~120 for Aldrich cellulose) behaves the same way as cotton, a cellulose molecule with a *DP* of ~1800. The extent of cellulose degradation by the peeling-off process scales directly to the *DP* of the cellulose molecule. This corroborates the conclusion that cellulose degradation between 1 and ~200 d is not caused by the peeling-off process. No indication of the nature of the processes involved during this phase of cellulose degradation was found in the literature.

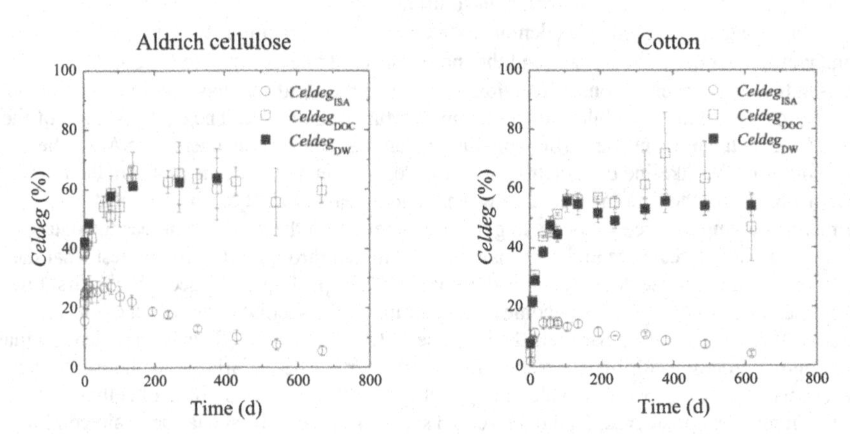

Figure 3. Extent of cellulose degradation (*Celdeg*) of purified cellulose (left) and cotton (right) in the presence of $Ca(OH)_2$ at pH 13.3 and 90 °C. The initial ratio of weight of cellulose to volume of liquid was ~10 g dm^{-3}. The data are average values from three independent experiments. *Celdeg* was either calculated from ISA, DOC or DW (dry weight) data.

Chemical reactivity of α-ISA in heterogeneous alkaline systems

The results of a typical experiment carried out under exclusion of air in an inert-gas glove box are shown in Figure 4 (left-hand plot).

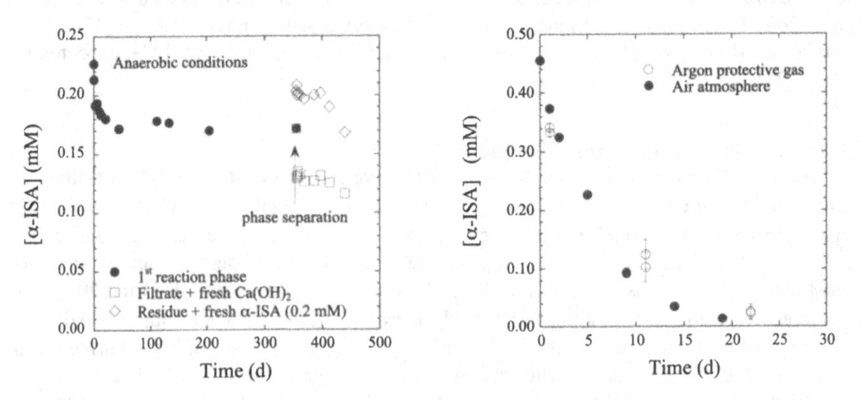

Figure 4. Reactivity of α-ISA in suspensions of $Ca(OH)_2$ in ACW-I under strict exclusion of air (left-hand figure, inert-gas glove box) and under no or partial exclusion of air (right-hand figure).

After an rapid initial decrease of concentration of α-ISA with a concomitant production of short-chain carboxylic acids (mainly glycolate, formate, lactate, acetate) a constant phase was observed, during which the concentration of α-ISA and the reaction products remained constant. In a second phase of the experiment the liquid phase was separated from $Ca(OH)_2$, and a fresh portion of $Ca(OH)_2$ was added to the liquid phase, whereas a fresh solution of α-ISA in ACW-I was added to the $Ca(OH)_2$. As can be seen from Figure 4 (left-hand plot) α-ISA was converted only upon addition of fresh $Ca(OH)_2$. The right-hand plot of Figure 4 shows the reactivity of α-ISA in the presence of traces or full amounts of air. Note that α-ISA remained stable for the time periods investigated in blank experiments without $Ca(OH)_2$ (not shown). It can be concluded from these experiments that stoichiometric amounts of an oxidising agent (O_2, or unknown impurities in $Ca(OH)_2$) were required to promote the conversion of α-ISA to short-chain carboxylic acids.

CONCLUSIONS

Quantitative long-term calculations of the concentration of α-ISA in cement pore water require extrapolations from the duration of laboratory scale experiments to geological time spans. Such a step can only be done if the chemical processes involved are thoroughly understood. The long-term degradation experiments of cellulose at room temperature and the experiments at elevated temperature clearly show that such a condition is not fulfilled. The current understanding of cellulose degradation under alkaline conditions goes mainly back to investigations carried out within the scope of paper production, where cellulose has to be separated from other wood components such as lignin. In view of a minimisation of the loss of cellulose by unwanted side reactions, such as the peeling-off process, the chemical mechanisms of such processes were investigated on time scales relevant for the industrial processes applied. The combination of peeling-off process and mid-chain scission turned out to be sufficient to explain the chemical behaviour of cellulose in a temperature range between room temperature and 200 °C for time scales of several hours. The experiments carried out in the frame of the present work demonstrate that other reactions, unimportant at short time scales, become dominant at extended time scales. Inconsistencies with respect to a simple model comprising peeling-off reaction and mid-chain scission were particularly found in experiments at moderate temperatures (60 and 90 °C, [20]).

It can be concluded that both estimates previously made for the long-term stability of cellulose under alkaline conditions are not appropriate. The prediction that cellulose may be completely degraded within ~100 to 500 years [17] underestimates the stability of cellulose, because the reaction rate constant for mid-chain scission was largely overestimated. The prediction that large parts of cellulose are stable for ~10^4 years and degrade completely over a period of 10^5 to 10^6 years [12] overestimates the stability of cellulose, because either the reaction rate constant for mid-chain scission was underestimated, or — more probably — because additional processes leading ultimately to a degradation of cellulose via the peeling-off process are involved in the alkaline degradation of cellulose.

In view of the lack of a process based understanding for the long-term behaviour of cellulose under alkaline conditions and the evidenced chemical stability of α-ISA under repository conditions it is recommended to assume the formation of α-ISA being scaled stoichiometrically to the amount of cellulose present in the wastes in a conservative performance assessment scenario. For a more realistic scenario, however, arguments may be found from the degradation experiments at room temperature that cellulose degradation in the first few hundreds up to thousands of years proceeds mainly via the peeling-off process, which may lead to reduced amounts of α-ISA formed [21]. With a given amount of α-ISA produced as a function of time, its pore water concentration can readily be calculated using thermodynamic data for the formation of sparingly soluble salts with Ca^{2+} [4] and for sorption of α-ISA to cement [5].

ACKNOWLEDGMENTS

The authors would like to thank Andreas Laube, Werner Müller and Roger Rossé for technical assistance. The support of R. Krebser, U. Seifried (Atisholz, AG, Switzerland) and R. Keil in analytical work is greatly acknowledged.

REFERENCES

1 B.F. Greenfield, G.J. Holtom, M.H. Hurdus, N. O'Kelly, N.J. Pilkington, A. Rosevear, M.W. Spindler, S.J. Williams, *Mat. Res. Soc. Symp. Proc.* **353**, 1151 (1995).
2 X. Bourbon, P. Toulhoat, *Radiochim. Acta* **74**, 315 (1996).
3 L.R. Van Loon, M.A. Glaus, *PSI Bericht* 98-07, Paul Scherrer Institut, Villigen, Switzerland. Also published as *Nagra Technical Report*, NTB 97-04, Nagra, Wettingen, Switzerland (1998).
4 K. Vercammen, M.A. Glaus, L.R. Van Loon, *Acta Chem. Scand.* **53**, 241 (1999).
5 L.R. Van Loon, M.A. Glaus, S. Stallone, A. Laube, *Environ. Sci. Technol.* **31**, 1243 (1997).
6 L.R. Van Loon, M.A. Glaus, A. Laube, S. Stallone, *Radiochim. Acta* **86**, 183 (1999).
7 K. Vercammen, M.A. Glaus, L.R. Van Loon, *Radiochim. Acta* 89, 393 (1999).
8 *Nirex Report* N/031, (2001).
9 P. Warwick, N. Evans, S. Vines, *Radiochim. Acta* **94**, 363 (2006).
10 R.L. Whistler, J.N. BeMiller, *Adv. Carbohydr. Chem. Biochem.* **13**, 289 (1958).
11 C.J. Knill, J.F. Kennedy, *Carbohydr. Polym.* **51**, 281–300 (2003).
12 L.R. Van Loon, M.A. Glaus, *J. Environ. Polym. Degrad.* **5**, 97 (1997).
13 G. Machell, G.N. Richards, *J. Chem. Soc.*, 1924 (1960).
14 D.W. Haas, B.F. Hrutfiord, K.V. Sarkanen, *J. Appl. Polym. Sci.* **11**, 587 (1967).
15 O. Franzon, O. Samuelson, *Svensk Paperstidning* **23**, 872 (1957).
16 Y.Z. Lai, K.V. Sarkanen, *Cellul. Chem. Technol.* **1**, 517 (1967).
17 I. Pavasars, J. Hagberg, H. Borén, B. Allard, *J. Polym. Environ.* **11**, 39 (2003).
18 M.A. Glaus, L.R. Van Loon, S. Achatz, A. Chodura, K. Fischer, *Anal. Chim. Acta* **398**, 111 (1999).
19 L.R. Van Loon, M.A. Glaus, A. Laube, S. Stallone, S., *J. Environ. Polym. Degrad.* **7**, 41 (1999).
20 M.A. Glaus, L.R. Van Loon, *PSI Bericht* 04-01, Paul Scherrer Institut, Villigen, Switzerland. Also published as *Nagra Technical Report*, NTB 03-08, Nagra, Wettingen, Switzerland (2004).
21 M.A. Glaus, L.R. Van Loon, submitted to *Environ. Sci. Technol.*
22 R.L. Whistler, J.N. BeMiller, *In:* Methods in carbohydrate chemistry (M.L. Wolfrom, J.N. BeMiller, eds.). Vol. **2**, pp. 477–479 (1963).

Mater. Res. Soc. Symp. Proc. Vol. 1107 © 2008 Materials Research Society

Release Behaviour of Sr-90 From Hydraulically Retrieved ILW Sludge

J J Hastings
Nexia Solutions Ltd, Sellafield, Seascale, Cumbria, CA20 1PG, UK.

ABSTRACT

Globally, the nuclear industry has a large number of legacy wastes that are stored in ponds, silos and tanks that are nearing the end of their design lifetime and hence said wastes need processing. In the UK there are significant quantities of radioactive sludge that have arisen from the corrosion of early Magnox fuel cans which have been stored underwater. As part of the present aggressive clean-up programme these materials will be retrieved, separated, processed and immobilised as dry waste forms for long-term storage. It is envisaged that hydraulic retrieval will be used for these ILW sludges resulting in some activity being released from the sludge phase to the process liquors challenging downstream ion exchange effluent treatment plants.

In order to understand this challenge, experiments have been conducted on sludge in ILW storage ponds and during sludge transfer operations to study the activity released from said sludges. In particular the solubility, adsorption behaviour of Sr-90 is discussed and how this and other aspects of the sludge chemistry impact upon the ion exchange effluent treatment process. The novel methodologies employed to obtain this data is also discussed.

INTRODUCTION

There are a variety of legacy sludge/slurry wastes that currently reside in ponds, silos and tanks on nuclear sites across the world. At Sellafield they largely arise from Britain's early Magnox and AGR nuclear generation programmes. Fuel from the reactors was shipped in water-filled transport flasks and stored in ponds at Sellafield. Some of the fuel remained in the ponds for many years and the Magnox cans corroded to form significant quantities of sludge now classified as Intermediate Level Waste. These pond facilities are now nearing the end of their service life and thus there is requirement to retrieve the wastes and decommission them. Figure 1 illustrates the stages of fuel can corrosion and Figure 2 shows a typical pond facility.

Corrosion of the Magnox cans produces a sludge that is predominately Magnesium hydroxide or Brucite $Mg(OH)_2$ but also contains other magnesium compounds, such as Artenites ($Mg_2CO_3(OH)_2.3H_2O$) [1]. The sludge is formed as the oxidation layer on the Magnox metal surface spalls off into the surrounding liquor.

Where sludge has been left undisturbed for prolonged periods of time there is a build up of activity within the sludge pore liquor. When the sludge is disturbed this pore liquor and its associated activity is released into the surrounding liquor and there are significant uncertainties in quantifying the rate and amounts of activity that will be released into the process liquor from the sludges as they are disturbed during retrieval operations. It is important that this activity release behaviour is understood so that the challenge to downstream effluent treatment plants can be estimated and the appropriate activity abatement technology can be installed.

In the case of the Sellafield Magnox sludge wastes and the installed Site ion exchange abatement technology it is particularly important to understand the behaviour of strontium as its solution/sorption chemistry is rather complex and strontium abatement is significantly hindered by the presence of magnesium competitor ions.

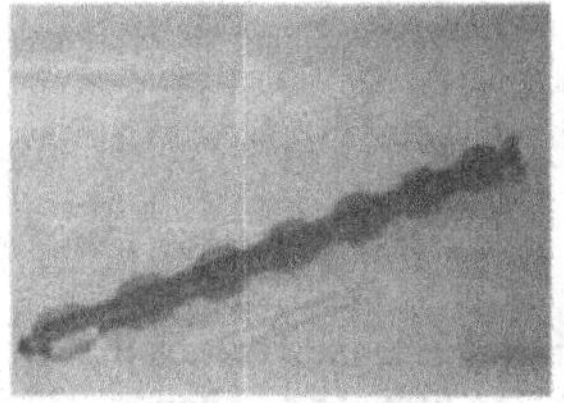

A - Magnox Fuel Can

B - Magnox - *Partially Corroded Swarf and Magnesium Hydroxide Sludge*

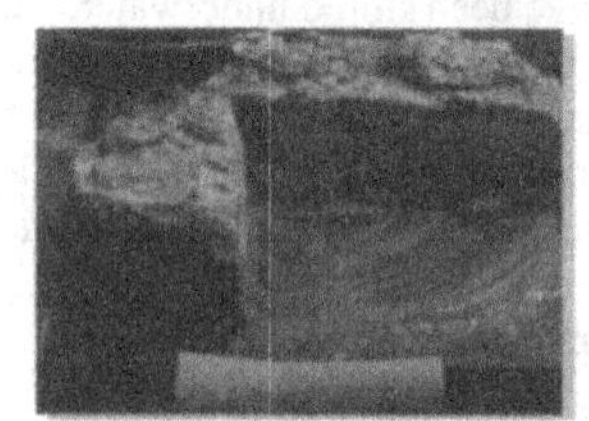

C - Magnox Sludge - *Active Magnesium Hydroxide Sludge Sample*

Figure 1: Corrosion of Magnox cans.

Figure 2: A Sellafield pond.

THEORY

In the reactor fission products form within the uranium bar with the amount produced being dependent on factors such as fuel burn-up and operating temperature. Upon storage under water the uranium corrodes and these fission products are released [2, 3].

Uranium corrodes to form uranium dioxide, hydrogen and uranium hydride, the uranium hydride is short lived and reacts further to form uranium dioxide and hydrogen. As the uranium fuel corrodes, a thin passive film of corrosion product is formed which adheres strongly to the uranium surface [4]. Once this film has been formed it is stable. However, above a certain thickness, this passive layer becomes unstable due to internal strains resulting in the outermost surface of the film spalling off leading to uranium corrosion products, incorporating fission products, being present in the corroded Magnox sludge.

The subsequent behaviour of the fission products is controlled by their solution chemistry which, in turn, is influenced by a number of factors. An investigation into the dissolution kinetics of magnesium hydroxide and the sorption, and subsequent release, of caesium, strontium and plutonium has been carried out [5]. This showed that both plutonium and strontium exhibited near complete sorption on to magnesium hydroxide with simulated pond liquor. The primary mechanism for the sorption of plutonium was thought to be an electrostatic one although the larger hydration shell of strontium cations was considered to have precluded this as the primary mechanism for strontium. However, there was evidence to show that the strontium incorporation into the $Mg(OH)_2$ lattice probably occurs via dehydration of the adsorbed $Sr(OH_2)_6$ unit and subsequent diffusion into the lattice.

Solubility and release of strontium-90 from legacy waste pond liquor.

To further the understanding of the release of strontium-90 from legacy waste sludges modelling studies and active characterisation work have been performed. The modelling work has been used to understand how the solubility of strontium-90 is controlled under pond storage conditions and active characterisation work to provide underpinning activity release data.

To inhibit corrosion of the Magnox fuel cans the storage ponds are caustic dosed to pH 10.5 to 11. Typically the Sr-90 in the pond liquor is approximately 2E-10moles/litre and approximately 7.5E-06 moles/litre in the sludge. The Sr-90 that is associated with the sludge is either physically trapped within voids in the sludge phase or it is adsorbed on the surfaces of the various insoluble magnesium phases (See Figure 3) that can form within the pond. Upon disturbance or retrieval of this sludge some of the Sr-90 is released to the bulk liquor phase, however predicting the amount transferred is complicated by how strongly the strontium is adsorbed to the insoluble magnesium phases and the amount of disruption of the sludge structure leading to the loss of trapped pore liquor containing soluble Sr-90.

The modelled solubility of pure $SrCO_3$ is not considered to control Sr-90 activity in the storage ponds as the solubilities predicted by the model are much higher than that observed. A more likely case is that Sr activity is limited by the co-precipitation (adsorption) of Sr-90 into the lattice of Mg solids (and possibly any Ca solids) that forms in the Magnox sludge.

Sr-90 has been shown to co-precipitate with Mg [6]. In an attempt to assess this, a co-precipitation model was developed to describe the trends in Mg, Cs-137 and Sr-90 behaviour. This was applied to the scenario where the pH of a pond liquor sample is adjusted from pH 9.8 to pH 12 by addition of NaOH.

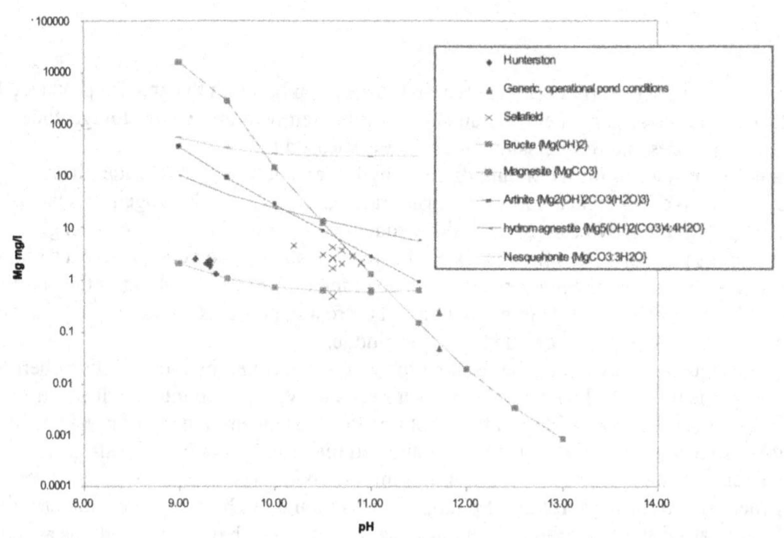

Figure 3: Modelled solubility of the magnesium phase as a function of pH and comparison to measured data.

EXPERIMENTAL

Active waste characterisation.

Two approaches have been taken to measure the release behaviour of strontium-90 from the sludge solids to the process liquor. The first involves taking of small sludge samples and performing experiments in the laboratory, the second is to perform in-situ experiments in the storage pond itself. It is considered that, although it is harder to control all the experimental parameters (e.g. the exact amount of sludge that is disturbed), the latter offers several advantages over the ex-situ use of small sludge samples;

- In-situ experiments allow much greater quantities of sludge to be used in the experiment and therefore results are less susceptible to sludge heterogeneity.
- In the in-situ experiment real pond liquor is used as opposed to a simulated liquor.
- The measurements are derived from liquor samples and this negates the need for extracting and transporting of more active sludge samples from the plant.
- With in-situ experiments it is possible to isolate undisturbed sludge, whereas the action of taking a sludge sample can result in some disturbance and loss of entrained activity.

Ex-situ sludge sample experiments

0.5g quantities (limited by ALARP constraints) of wet sludge were placed into three containers containing 30, 100 and 1000 mls of a saturated $Mg(OH)_2$ liquor that had been pH corrected to 11.4 using NaOH. These were shaken vigorously and the resulting contacted liquors analysed. The strontium release fraction (RF) was observed to increase with dilution from 1.2 to 3%. It is probable that the strontium solubility limit is not reached in these solutions but the dilution must aid the disruption of the sludge phase and thereby in turn increase the release of strontium into the bulk liquor. The release fraction is defined as:

$$\text{Release Fraction (RF)} = \frac{\text{total activity transferred to the contact liquor}}{\text{total activity content of the sludge.}}$$

It is also important to note that RFs are not fundamental measurements but are reliant on the conditions under which they are measured. The RFs are sensitive to such parameters as applied shear, dilution, pH and water chemistry.

At much higher sludge solids concentrations, 20-30% $^{w}/_{w}$, mixing resulted in significant changes in the particle size distribution of the sample. However, below 15% $^{w}/_{w}$ very little change in the particle size distribution is observed. The fact that comparable release fractions have been observed in these systems suggests that the majority of the activity released was from trapped pore liquor within the sludge rather than from the degradation of the sludge particulates.

In-situ monitoring of sludge transfer experiments

Liquor samples have been used to monitor sludge settling following sludge transfer operations. Figure 4 illustrates the different behaviour of Cs-137, Sr-90 and alpha material in this settling experiment. So that the settling behaviour of the above activities could be assessed the relative activities measured with time were normalised with the settling decay curve measured using a turbidity probe. Here the Cs-137 is clearly soluble and is independent of the settling solids whereas nearly half of the strontium-90 present is associated with particulate and nearly all the alpha material is associated with fine particulate.

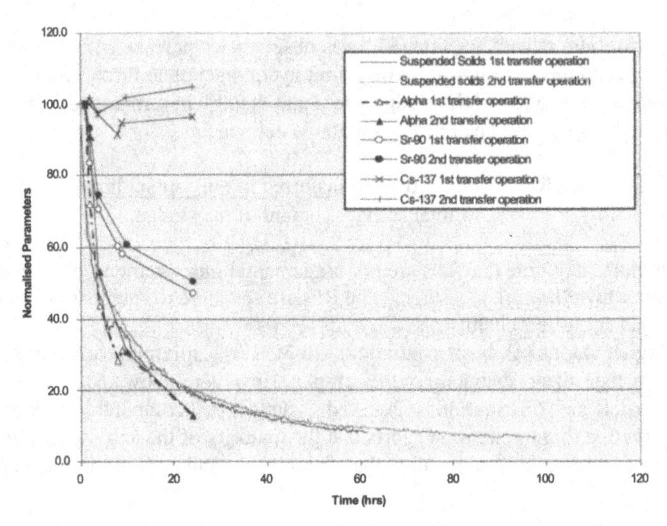

Figure 4: Normalised settling characteristics for alpha activity, Cs-137 and Sr-90 and suspended solids following a Magnox slurry transfer operation.

In-situ bell jar experiments.

In pursuit of further in-situ experiments on plant, Nexia Solutions has developed a 'bell jar' that can be deployed into the sludge in a waste storage pond. This equipment isolates a known amount of sludge and overstanding liquor from the bulk of the pond and a rotor then mobilises the isolated sludge and the activity released is monitored by taking liquor samples before and after the disturbance.

The bell jar consists of a stainless steel cylinder approximately 200 mm in diameter and 550 mm tall with a submersible motor unit and impeller at the top. It has externally mounted cameras to monitor its deployment and the open end of the mixing chamber sinks into the sludge to a depth of approximately 15 cm. The internal volume of the chamber is approximately 17 litres and it is designed to isolate approximately 4.5 litres of sludge.

Assuming the sludge is ≈30% solids ($^{V}/_{V}$), this gives an approximate 7% solids slurry on mixing. The volume of sludge disturbed is monitored via a visual depth gauge. The bell jar is fitted with four sludge samplers around the periphery of the mixing chamber so that the chemistry and radiochemistry of the sludge may also be assessed ex-situ.

The apparatus described above is still undergoing inactive trials. However, an earlier version of the bell jar system has been used to monitor both activity release from disturbed sludges and settling characteristics in an adjacent, near neutral pH, storage pond. This pond has a more complex chemistry than the caustic dosed pond with a greater quantity of organic material in the sludge. The sludges from this pond tended to either settle with a distinct settling front or as a clearing "fog". This was achieved by video analysis and monitoring a series of visual markers within the bell jar versus time. Figure 5 shows a distinct settling interface observed (just above marking 12) in one of the near neutral pH ponds.

Figure 5: Bell Jar used to monitor settling characteristics and activity release from the sludge to the bulk liquor phase.

The activity in the bell jar was assessed by a taking baseline liquor sample before any disturbance and samples following a sludge disturbance versus time. In contrast to other observations made in the alkaline dosed ponds, the Sr-90 activity in the liquor decreased slightly after mixing and settling the sludge. It is thought that the decrease may be the result of some component of the sludge acting as a mild ion-exchange medium and active suspended solids being removed from the liquor by flocculation of particles during the settling process.

CONCLUSIONS

Where sludge has been left undisturbed for prolonged periods of time there is a build up of activity within the sludge pore liquor. Upon the disturbance of this sludge, activity is released to the mobilisation liquor. This has been shown to be largely, from the pore liquor rather than from degraded particulate. Even in slurry streams containing 20-30% $^{W}/_{W}$ solids the strontium activity is mainly released from trapped pore liquor within the sludge rather than from the degradation of sludge particulate. Strontium has been shown to co-precipitate into the lattice of Mg solids (and possibly any Ca solids) that forms in the Magnox sludge. However, it is the detailed local chemistry (eg pH and relative ion concentrations) that determines the balance of activity between the solid and liquid phases. This is still not fully understood and further inactive work is required to consider the key mechanisms that might control strontium solubility and for further active measurements to underpin these.

Two approaches have been taken to measure the release behaviour of strontium-90 (and other species) from the sludge solids to the process liquor. The in-situ bell jar technique has been demonstrated to have a number of advantages over the traditional extraction of small sludge samples, these being,

- In-situ experiments allow much greater quantities of sludge to be used in the experiment and therefore are less susceptible to sludge heterogeneity.
- In the in-situ experiment real pond liquor is used as opposed to simulated liquor.
- The measurements are derived from liquor samples and this negates the need for extracting and transporting of more active sludge samples from the plant.
- With in-situ experiments it is possible to isolate undisturbed sludge, whereas the action of taking a sludge sample can result in some disturbance and loss of activity.

It should be noted that ex-situ samples will be required for radiochemical measurements for inventory purposes.

RECOMMENDATIONS

It is recommended that further bell jar experiments are performed and that this is complemented with inactive strontium studies to assess release behaviour from sludge phases.

ACKNOWLEDGMENTS

The author would like to thank Sellafield Ltd for discussions and funding this work.

REFERENCES

1. M E Dunbabin, NSTS(04)5238, (2004).
2. D A Hilton & B L Bullock, CEGB Berkeley Nuclear Laboratories Report RD/B/N1159, (1968).
3. D A Hilton, I H Robins & I R Brookes, CEGB Berkeley Nuclear Laboratories Report RD/B/N4106, (1978).
4. S Maguire and R W Burrows, R&T/NG/GEN/REP/0013/03, (2003).
5. A G Gault and A B Eilbeck, RDR 1399, (1998).
6. B J Hands, Nexia Solutions 5899, (2005).

Mater. Res. Soc. Symp. Proc. Vol. 1107 © 2008 Materials Research Society

Radionuclide Retardation in Granitic Rocks by Matrix Diffusion and Sorption

P. Hölttä[1], M. Siitari-Kauppi[1], M. Kelokaski[1] and V. Tukiainen[2]
[1]Laboratory of Radiochemistry, Department of Chemistry, University of Helsinki, P.O. Box 55, FIN-00014 University of Helsinki, Finland
[2]VTT Processes, Nuclear Energy, P.O.Box 1608, FIN-02044 VTT, Finland

ABSTRACT

Radionuclide retardation in mica gneiss, unaltered, moderately and strongly altered tonalite was studied by a thin section, batch, in-diffusion and column methods. Objectives were to examine retention processes in different scales and understand the influence of the rock matrix heterogeneity. Attempts were made for a more detailed interpretation of experiments using migration models used in performance assessments adapted for interpreting the laboratory scale experiments. Batch experiments were explained adequately using matrix diffusion-sorption model, instantaneous kinetic sorption model or model in which both mechanisms were taken into account. A numerical code FTRANS was able to interpret in-diffusion of calcium into the saturated porous matrix. Elution curves of calcium for the moderately and strongly altered tonalite fracture columns were explained adequately using FTRANS code and parameters obtained from in-diffusion calculations. K_d-values for intact rock obtained from fracture column experiments were lower than K_d-values for crushed rock indicating that batch experiments overestimate the retardation of radionuclides. Higher sorption and fair dependence on fraction size was obtained for altered tonalites due to the composition of alteration minerals and large specific surface areas.

INTRODUCTION

Granitic rock would act as the ultimate barrier, retarding the migration of radionuclides to the biosphere if radionuclides are released through engineered barriers in a repository of highly radioactive spent nuclear fuel. In transport models, radionuclide retardation has been usually taken into account by the K_d-concept, in which a retardation factor is used to apply the distribution ratio to radionuclide transport [1, 2]. Determination of the retardation factor by fracture flow experiments is a direct approach to application of the effects of sorption in radionuclide transport models. K_d values conventionally used in transport models have been determined for crushed rock with different specific surface areas and values may not be valid for modelling radionuclide transport in fractures. The effect of mineral composition and specific surface area on radionuclide sorption has been discussed earlier [3-5]. In this work retardation of radionuclides in granitic rock was studied by means of the different methods. Objective was to examine the processes causing retention in solute transport through rock fractures, especially focused on the matrix diffusion and sorption.

ROCK MATRIX CHARACTERIZATION

The rock samples representing different rock features and porosities were obtained from the hole SY-KR7 drilled in the Syyry area in Sievi, Western Finland. Drill hole SY-KR7 [6] intersects a medium-grained slightly schistose tonalite zone containing a few fine-grained and homogeneous mica gneiss inclusions. A quantitative mineral composition for the rock samples was obtained by a point counting method from the thin sections [5].

Table I. Particle diameter (mm), specific surface area ($m^2 \cdot g^{-1}$), porosity and density for fractions of mica gneiss and tonalites.

	Mica gneiss	Unaltered tonalite	Moderately altered tonalite	Strongly altered tonalite
Surface area ($m^2 \cdot g^{-1}$)				
0 – 0.1	0.40	0.40	3.35	1.95
0.1 – 0.3	0.30	0.22	1.73	2.17
0.3 – 0.5	0.17	0.11	1.26	1.37
0.5 – 0.8	0.17	0.19	0.89	1.23
0.8 – 1.0	0.13	0.10	0.90	0.66
1.0 – 2.0	0.12	0.09	0.55	1.35
2.0 – 3.15	0.08	0.11	0.61	1.24
Intact rock	0.07	0.05	0.40	0.50
Porosity ε (%)	0.2 - 0.8	0.16	2.0 – 7.5	2.0 -7.5
Density ρ ($g \cdot cm^{-3}$)	2.79	2.80	2.4	2.4

The total porosity and the surface areas of mineral grains available for sorption and migration of species were determined by the ^{14}C-PMMA method [7, 8]. Pore apertures and geometry in the mineral phases were analyzed by scanning electron microscopy (SEM), and the minerals and sorbed tracer were quantified by energy dispersive X-ray microanalysis (EDX). The crushed rock samples were sieved into the seven fractions from 0.01 mm to 3.15 mm. The specific surface areas for each fraction were determined by the BET nitrogen adsorption method. Characterization of rock samples is given in Table I.

EXPERIMENTS

The mass distribution ratios, K_d–values, were determined using thin section, batch, in-diffusion, crushed rock and fracture column methods. Degassed synthetic granitic groundwater of low salinity was used in all experiments. The sorption of tracers onto different minerals was estimated from autoradiograph taken from the thin sections of thickness 30 μm. The surface based distribution coefficient, R_a, was calculated using a geometric sorption area determined by digital image processing of autoradiograph. K_d–values were calculated from R_a–values using a thin section thickness and rock density. Spiked solution, containing ^{22}Na, ^{45}Ca, ^{85}Sr or ^{134}Cs, came into contact with equilibrated crushed rock in polypropylene centrifuge tubes which were shaken gently. The mass distribution ratio, K_d, was determined as the ratio of the radionuclide concentration in the solid phase to the aqueous concentration. In-diffusion of calcium into mica gneiss, unaltered, moderately and strongly altered tonalite was determined using 2 x 2 x 2 cm rock cubes. The rock samples for in-diffusion experiments were evacuated and equilibrated with the synthetic granitic groundwater for two weeks. Equilibrated rock cubes were immersed in spiked solution, containing ^{45}Ca.

The preparation of crushed rock and fracture columns and the apparatus used in the column experiments have been described in more detail in ref. [5]. A short tracer pulse (5 μL) was injected into the water flow using an injection loop (Rheodyne) and the out flowing tracer was collected. The volumetric flow rate was determined before and after each experiment by weighing the water collected over known time periods. The flow conditions in the columns were determined using tritiated water or chloride (HTO, $^{36}Cl^-$) which were injected simultaneously with a sorbing tracer. The retardation of sodium (^{22}Na), calcium (^{45}Ca) and strontium (^{85}Sr) was studied using water flow rates of 5–20 μL·min^{-1}. The beta activities of HTO, ^{36}Cl and ^{45}Ca were determined by liquid scintillation counting and gamma activities of ^{22}Na and ^{85}Sr were detected using a Wizard gamma counter.

MODELLING APPROACH

A model based on a numerical code FTRANS [9] was developed for a more detailed interpretation of static batch and in-diffusion experiments. In the model, crushed rock grains are assumed to be spheres of equal radius. When modelling concept was applied to in-diffusion of calcium rock cube was approximated to be sphere having a diameter of 2.6 cm. The grains are assumed to be porous which means that there exists a connected network of water filled micro fractures through the whole grain. The sorption into the pore is quantified by the mass based distribution coefficient K_d (m^3·kg^{-1}). In this approach matrix diffusion from water into the rock grains is taken into account. The equations describe the dynamic behaviour of the concentration of nuclides in the water surrounding the cube and in the water of the pores inside the matrix:

$$\frac{\partial C}{\partial t} = -AD_e \left.\frac{\partial C'}{\partial r}\right|_{r=a} \quad (1)$$

$$\frac{1}{r^2}\frac{\partial}{\partial r}\left(r^2 D_e \frac{\partial C'}{\partial r}\right) = \frac{\partial}{\partial t}(\varepsilon' R' C') \quad (2)$$

where C is tracer concentration in the solution surrounding the crushed rock (mol/m^3), C' is tracer concentration in the pore water inside the rock grains (mol/m^3), D_e is the effective diffusion coefficient (m^2/s), ε' is the porosity of crushed rock grains, R' is retardation coefficient in matrix, r is the distance from the center of the sphere (m), a is the radius of the sphere (m) and A is the surface area of the spheres divided by the volume of the surrounding water (1/m). When we know the total volume of rock used in the experiment and the volume of the surrounding water the quantity A in equation (1) can be written as:

$$A = \frac{V_{rock}}{V_{water}}\frac{A_{rock}}{V_{rock}} = \frac{V_{rock}}{V_{water}}\frac{3}{a}, \quad (3)$$

where V_{rock} is the total volume of rock used in the experiment (m^3), V_{water} is the volume of the surrounding water (m^3), A_{rock} is the total surface area of spherical particles (m^2). The initial conditions for the concentrations are

$$C(0) = C_0, \quad (4)$$

$$C_p(r,0) = 0, \quad (5)$$

where C_0 is the concentration in the surrounding water after sorption on the surfaces of the particles (mol/m^3). The concentration is assumed to be continuous in the interface of pores and surrounding water. The radial concentration distribution is also symmetric. The boundary conditions for the concentration are therefore

$$C_p(a,t) = C(t);\ t > 0 \quad (6)$$

$$\lim r \longrightarrow \infty (rC_p(r,t)) = 0 \quad (7)$$

The model for crushed rock column includes 1-dimensional advection, dispersion, and diffusion from the flowing water into the crushed rock grains. Tracer transport in flowing water is described:

$$\frac{\partial C}{\partial t} = \frac{\alpha_L v}{\varepsilon}\frac{\partial^2 C}{\partial x^2} - \frac{v}{\varepsilon}\frac{\partial C}{\partial x} + \Gamma \qquad (8)$$

and in the pore water inside crushed rock grains by:

$$\frac{D_e}{r^2}\frac{\partial}{\partial r}\left(r^2\frac{\partial C'}{\partial r}\right) = \frac{\partial}{\partial t}(\varepsilon' R' C') \qquad (9)$$

Tracer flux from water to matrix (or vice versa) is given by:

$$\Gamma = -AD_e \left.\frac{\partial C'}{\partial r}\right|_{r=a} \qquad (10)$$

where a is the radius of the spheres.

The retardation factor expresses the ratio of water velocity to the radionuclide velocity. The inverse of retardation factor represents the fraction of the total radionuclide inventory that is dissolved in the water and thus considered mobile. The retardation factors in a fracture ($R_{f,i}$) and in the rock matrix ($R_{p,i}$) are given by:

$$R_{f,i} = 1 + \frac{2}{2b}K_{a,i} \qquad (11)$$

$$R_{p,i} = 1 + \frac{\rho_s(1-\varepsilon_p)K_{d,i}}{\varepsilon_p} \qquad (12)$$

where 2b is fracture aperture (m), $K_{a,i}$ is the area-based distribution coefficient of nuclide i (m^3/m^2), $K_{d,i}$ is mass-based distribution coefficient of nuclide i (m^3/kg), and ε_p is porosity of the rock matrix. K_d can be derived from K_a by multiplying by the specific surface area of the rock:

$$K_{d,i} = K_{a,i}a_f \qquad (13)$$

where a_f (m^2/kg) is the specific surface area of the rock.

RESULTS AND DISCUSSION

Attempts for a more detailed interpretation of static batch experiments have been conducted using different modelling approaches. In the first approach matrix diffusion from the water into the rock grains was taken into account. The model explained the recurrent observation in batch experiments that after an initial rapid decrease in tracer concentration of solution after that there is a slower trend of decrease. An example of a model fit for unaltered mica gneiss is shown in Figure 1. At the beginning of the experiment the nuclide sorbs only to the outer surfaces of the rock particles. After this the nuclide starts to diffuse into the porous rock matrix. During the diffusion in the water filled pores the nuclide also sorbs onto the mineral surfaces of the pores. The sorption into the pore surfaces is assumed to be instantaneous equilibrium reaction, which can be quantified by the mass based distribution coefficient K_d. In the second approach instantaneous kinetic sorption onto the outer surfaces of grains was taken into account. Rapid decrease of the tracer concentration in the solution occurs at the beginning of the experiment. After that the measured concentrations were on the same level or decreased very slowly compared to the instantaneous transition. This kinetic sorption explained retardation on the small grains of 0-100 μm, 100-300 μm and 300-500 μm. An example of a model fit for moderately altered tonalite is shown in Figure 1. In the third approach the instantaneous kinetic sorption and matrix diffusion onto the outer surfaces of the grains were taken into account. Matrix diffusion alone was not able to explain extensively retardation on unaltered tonalite. Especially retardation on the grain size of 500-800 μm showed first a rapid instantaneous decrease of the tracer concentration in the solution indicating dominating kinetic sorption. Then a slow decrease in tracer concentration was explained by matrix diffusion into the rock pores. An example of a model fit for unaltered tonalite is shown in Figure 1.

Decrease in calcium concentration in the spiked solution of rock cubes as a function of time is shown in Figure 2. A numerical code FTRANS was able to interprete in-diffusion of calcium into moderately and strongly altered tonalite in which rock matrix was assumed to be saturated. Experimental and calculated in-diffusion of calcium into moderately altered tonalite rock cube is shown in Figure 2. The model curve was obtained using fitting parameters: R=250, $D_e=3.9\cdot10^{-11}m^2\cdot s^{-1}$, $K_d=0.002\ m^3\cdot kg^{-1}$. Experimental and calculated in-diffusion of calcium into strongly altered tonalite rock cube is shown in Figure 2. The model curve was obtained using parameters: R=600, $D_e=8.3\cdot10^{-11}m^2\cdot s^{-1}$, $K_d=0.008\ m^3\cdot kg^{-1}$. Calcium concentration profiles calculated using obtained D_e-values indicated that after 1850 hour a total saturation in highly porous rock cubes was reached. Owing to insignificant decrease in calcium concentration in the solution after 2000 hour following time, FTRANS code was not able to interprete in-diffusion of calcium into low porous mica gneiss and unaltered tonalite rock cubes. The extent of diffusion of the tracer into the rock cubes was supported by earlier through-diffusion experiments. The mass distribution ratio, K_d, was determined in in-diffusion experiments also as the ratio of the radionuclide concentration in the solid phase to the aqueous concentration.

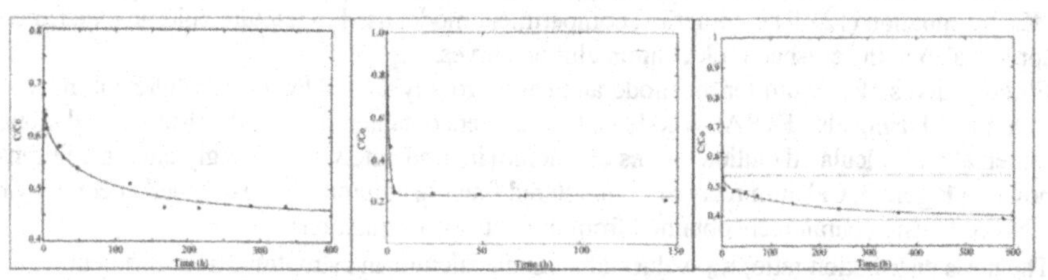

Figure 1. Experimental and calculated sorption of calcium into rock grains. Matrix diffusion model fit to mica gneiss (left). Kinetic sorption model fit to moderately altered tonalite (middle). Kinetic sorption and matrix diffusion model fit to unaltered tonalite (right).

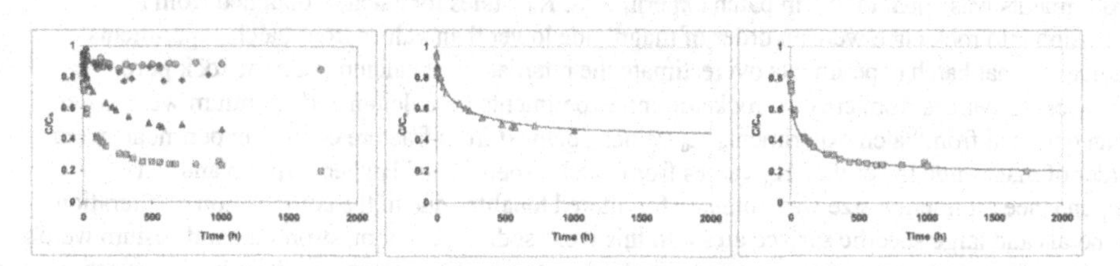

Figure 2. *Experimental in-diffusion of calcium into rock cubes (left). Mica gneiss (o), unaltered tonalite (◊), moderately altered tonalite (Δ) and strongly altered tonalite (□). Experimental and calculated in-diffusion of calcium into rock cubes, moderately altered tonalite (middle) and strongly altered tonalite (right).*

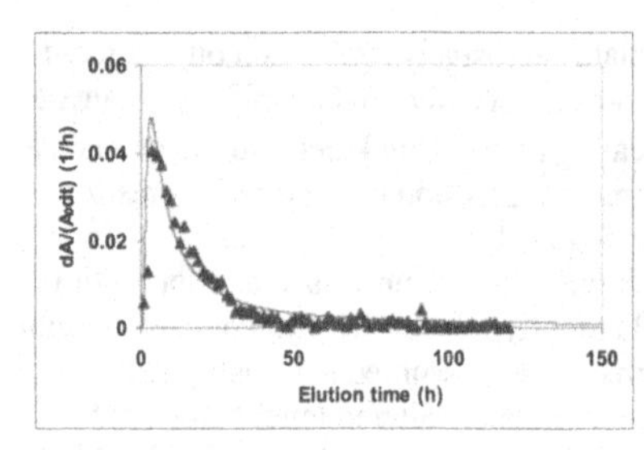

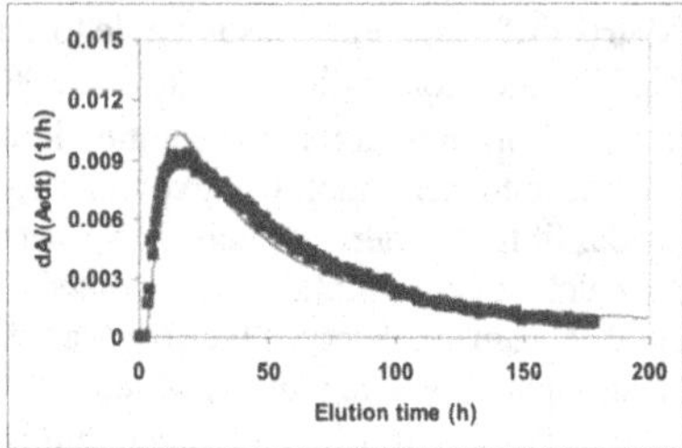

Figure 3. Experimental and calculated elution curves of calcium for moderately altered (left) and strongly altered (right) tonalite fracture columns.

The flow conditions and transport behaviour of HTO, chloride, sodium, calcium and strontium in fracture columns were interpreted using a applied numerical compartment model [10] which calculates the advection and hydrodynamic dispersion in the columns. Each experiment was modelled using the same parameters throughout the flow rate range. The effect of matrix diffusion was calculated using an analytic solution to the advection-matrix diffusion problem with semi-infinite boundary conditions in the matrix and a sudden release of the tracer at the inlet [5]. In the calculations for sorbing radionuclides retardation in the rock pore volumes is taken account. In matrix diffusion calculations retardation in rock pore volumes, R_p, is related to the mass distribution ratio, K_d, by equation (12). The numerical compartment model used in fracture column experiments was applied also to the crushed rock column elution curves.

Elution curves of calcium for the moderately and strongly altered tonalite fracture columns were interpreted using also FTRANS code and parameters obtained from in-diffusion calculations. Experimental and calculated elution curves of calcium in moderately and strongly altered columns are shown in Figure 3. Calcium retardation in altered fracture columns was explained adequately by FTRANS code using parameters obtained from in-diffusion calculations.

The mass distribution ratio, K_d -values of sodium, calcium and strontium from different approaches are compared in Table II. Cesium values are available only from thin section and batch experiments. Values from batch and crushed rock column experiments are given for the fraction size 1.25 mm. K_d-values calculated from thin section R_a-values were in agreement with K_d-values obtained from batch experiments indicating that the specific surface area usable in thin section experiments was equal to that in batch experiments. K_d values for calcium obtained from in-diffusion into rock cube were an order of magnitude lower than values from batch experiments indicating that batch experiments overestimate the retardation of calcium onto the rock pore surfaces. K_d-values from crushed rock column experiments for calcium and strontium were lower than obtained from batch experiments K_d-values obtained from fracture column experiments were order of magnitude lower than K_d-values from batch experiments. Higher sorption and fair dependence on fraction size was obtained for altered tonalites due to the composition of alteration minerals and large specific surface areas. In this work sodium, calcium, strontium and cesium were used as tracers to compare various experimental techniques. Experiments with relevant nuclear waste nuclides are needed for more detailed discussion on the differences in K_d values and the implication for performance assessment calculations.

Table II. Comparison of K_d ($m^3 \cdot kg^{-1}$) values of ^{22}Na, ^{45}Ca, ^{85}Sr and ^{134}Cs.

	Thin section	Batch	Rock cube	Crushed rock column	Fracture column
Mica gneiss					
Na	0.006	0.003		0.001	0.0006
Ca	0.002	0.008	0.0004	0.003	0.0008
Sr	0.003	0.035		0.003	0.004
Cs	0.06	1.1			
Unaltered tonalite					
Na	0.006	0.003		0.001	0.0003
Ca	0.004	0.009	0.0004	0.001	0.001
Sr	0.007	0.036		0.002	0.004
Cs	0.09	0.66			
Moderately altered tonalite					
Na	0.33	0.39		0.020	0.05
Ca	0.02	0.02	0.003	0.015	0.002
Sr	0.02	0.06		0.023	0.0015
Cs	0.57	3.4			
Strongly altered tonalite					
Na	0.15	0.15		0.01	0.02
Ca	0.03	0.02	0.01	0.01	0.008
Sr	0.03	0.04		0.02	0.015
Cs	0.54	1.7			

CONCLUSIONS

In this work retardation of sodium, calcium, strontium and cesium was studied on mica gneiss, unaltered, moderately and strongly altered tonalite using different static and dynamic methods. Attempts were made for a more detailed interpretation of batch and column experiments using different modelling approaches. Instantaneous kinetic sorption model was able to interpret calcium sorption onto outer surfaces of small grains. Matrix diffusion-sorption model fit was able to explain calcium sorption into mica gneiss and large altered tonalite grains due to high intragranular porosity. Calcium retardation into unaltered tonalites was explained adequately using combination of instantaneous kinetic sorption and matrix diffusion.

A numerical code FTRANS was able to interpret in-diffusion of weakly sorbing calcium into the saturated porous matrix. The code was not able to interpret in-diffusion into low porous mica gneiss and unaltered tonalite owing to too short following time and thus insignificant decrease in calcium concentration. Elution curves of calcium for the moderately and strongly altered tonalite fracture columns were explained adequately using FTRANS code and parameters obtained from in-diffusion calculations. Higher sorption and fair dependence on fraction size was obtained for altered tonalites due to the composition of alteration minerals and large specific surface areas. K_d-values for intact rock obtained from fracture column experiments were lower than K_d-values for crushed rock indicating that batch experiments overestimate the retardation of radionuclides. The experimental results presented here cannot be transferred directly to the spatial and temporal scales that prevail in the underground repository. However, this knowledge and understanding of the retention processes is transferable to different scales from laboratory to in–situ conditions.

ACKNOWLEDGMENTS

This work was part of the Finnish Research Programme on Nuclear Waste Management (JYT2001) funded by the Ministry of Trade and Industry.

REFERENCES

1. R. S. Rundberg, In: Radionuclide sorption from the safety evaluation perspective, Proceedings of a NEA Workshop, OECD (1992).
2. H. Johansson, M. Siitari-Kauppi, M. Skålberg, E. L. Tullborg, J. Contam. Hydrol. 35 41-53 (1998).
3. P. Hölttä, M. Siitari-Kauppi, P. Huikuri, A. Lindberg, A. Hautojärvi, In Scientific Basis for Nuclear Waste Management XX, Mater. Res. Soc. Proc. 465, 789-796 (1997).
4. M. Siitari-Kauppi, P. Hölttä, S. Pinnioja, A. Lindberg, In Scientific Basis for Nuclear Waste Management XXII, Mater. Res. Soc. Proc. 556, 1099-1106 (1999).
5. P. Hölttä, "Radionuclie migration in crystalline rock fractures–Laboratory study of matrix diffusion," Doctoral Thesis, University of Helsinki. Report Series in Radiochemistry 20/2002, 55 p. + Appendices (2002).
6. A. Lindberg, M. Paananen, Teollisuuden Voima Oy, Site Investigations Work Report 92-34, Helsinki, (1992) (in Finnish).
7. K-H. Hellmuth, M. Siitari-Kauppi and A. Lindberg, *Journal of Contaminant Hydrology* **13**, 403 (1993).
8. M. Siitari-Kauppi, "Development of ^{14}C-polymethylmethacrylate method for the characterisation of low porosity media," Doctoral Thesis, University of Helsinki. Report Series in Radiochemistry 17/2002, 156 p (2002).
9. INTERA Environmental Consultants, Inc. FTRANS: A two dimensional code for simulating fluid flow and transport of radioactive nuclides in fractured rock for repository performance assessment. Houston, Technical report ONWI-426 (1983).
10. H. Nordman and T. Vieno, Near-field model Repcom. Rep. YJT-94-14, Nuclear waste Commission of Finnish Power Companies (1994).

Mater. Res. Soc. Symp. Proc. Vol. 1107 © 2008 Materials Research Society

Modelling Transport of ^{14}C-Labelled Natural Organic Matter (NOM) in Boom Clay

Alice Ionescu[1], Norbert Maes[2], and Dirk Mallants[2]
[1]Agency for Radwaste Management (ANDRAD), Str. Campului 1, Mioveni 115400, Romania
[2]Belgian Nuclear Research Centre (SCK•CEN), Boeretang 200, 2400 Mol, Belgium

ABSTRACT

In Belgium, the Boom Clay formation is considered to be the reference formation for HLW disposal R&D. Assessments to date have shown that the host clay layer is a very efficient barrier for the containment of the disposed radionuclides. Due to absence of significant water movement), diffusion - the dominant transport mechanism, combined with generally high retardation of radionuclides, leads to extremely slow radionuclide migration. However, trivalent lanthanides and actinides form easily complexes with the fulvic and humic acids which occur in Boom Clay and in its interstitial water. Colloidal transport may possibly result in enhanced radionuclide mobility, therefore the mechanisms of colloidal transport must be better understood. Numerical modeling of colloidal facilitated radionuclide transport is regarded an important means for evaluating its importance for long-term safety.

The paper presents results from modeling experimental data obtained in the framework of the EC TRANCOM-II project, and addresses the migration behavior of relevant radionuclides in a reducing clay environment, with special emphasis on the role of the Natural Organic Matter (NOM) [1]. Percolation type experiments, using stable ^{14}C-labelled NOM, have been interpreted by means of the numerical code HYDRUS-1D [2]. Tracer solution collected at regular intervals was used for inverse modeling with the HYDRUS-1D numerical code to identify the most likely migration processes and the associated parameters. Typical colloid transport submodels tested included kinetically controlled attachment/detachment and kinetically controlled straining and liberation.

INTRODUCTION

Normally, in reducing clay environments, low values of mobile radionuclide concentrations in the clay pore-water are due mainly to solubility limitation. However, the presence of NOM may enhance the solubility due to complexation/colloid formation and/or may influence the sorption behaviour of radionuclides. Transport processes characteristics of Boom Clay were investigated by means of laboratory and in-situ experiments, taking account of NOM. In this paper, a set of laboratory migration experiments was carried out to clarify the possible role of mobile NOM as radionuclide carrier. For this purpose, radionuclide sources (^{241}Am) were prepared with concentrations as close as possible to their expected equilibrium concentration under in-situ Boom Clay conditions and in contact with ^{14}C-labeled BC Organic Matter (^{14}COM), so-called "double-labeled" migration experiments (details can be found in [1,3]). This paper focuses on the migration of the ^{14}COM. Breakthrough data of the ^{14}C labeled OM from these percolation experiments were used to inversely estimate meaningful values for migration parameters capable of describing colloid transport (sorption, attachment/detachment rates, straining).

MATERIALS AND METHODS

In so-called percolation type migration experiments, the radionuclide source is brought on a filter paper and confined between two clay cores. At one end, real Boom Clay water (RBCW) is forced into the clay core under constant pressure while at the other end percolating fluid is collected for monitoring. Characteristics of the experimental set-up are given in Table I. Further note that the percolation experiments were done with a fixed total plug length but with varying end-plug lengths (i.e., distance of the source to the outlet of the clay core), and varying Darcy velocities (Table II). The ^{14}C-labeled OM solution is characterized by large molecular weight molecules (30 000 MWCO), and the OM particle sizes range between 2.1 and 5 nm [1].

Table I. Experimental details for percolation tests.

Parameter	**Data**
Length of the clay core (cm)	7.2
Cross-sectional area, S (cm^2)	11.34
Length of the filter (cm)	0.2
Filter porosity, η (-)	0.4
Filter bulk density, ρ_b (g cm^{-3})	1.7
Used amount of source solution (cm^3)	0.2
Recovered activity at the outlet (%)	See Table II
Average flow rate, Q (cm day^{-1})	
Pore-water diffusion coefficient D_p (cm^2 day^{-1})	0.0432
Tracer half life, $T_{1/2}$ (days)	2.094×10^6
Source activity concentration, corrected for losses (Bq cm^{-3}), $C_0 = A_0/VOL$	18.1
Saturated water flux (Darcy velocity), $q_D = Q/S$ (cm day^{-1})	See Table II

Table II: Overview of Am-OM double labeled data. Travel length is distance between source (M2) and outlet (more details in [1, 3]).

Experiment code	**Travel length, *TL* (cm)**	**Experiment duration, *T* (d)**	**Average flow rate, *Q* (cm^3/d)**	**Hydraulic conductivity, K_h (10^{-12} m/s) [P (MPa)]**	**Recovered ^{14}C activity (Bq)**
ShortPAmCOM	1.2	620	0.164	1.3 [0.94]	2 720 (60.1 %)
LongPAmCOM	6.0	620	0.185	1.5[0.91]	2147 (59.2 %)
HiPAmCOM	3.6	620	0.173	0.9 [1.29]	2170 (59.8 %)
MePAmCOM	3.6	621	0.393	2.9[0.94]	2610 (72 %)
LowPAmCOM	3.6	621	0.304	2.5[0.85]	2650 (73.1 %)

THEORY AND MODELS

Colloid transport may be described by a combination of the convection-dispersion-retardation equation and colloid attachment theory [2]:

$$\frac{\partial(\eta_e c)}{\partial t} + \rho\frac{\partial(s_e)}{\partial t} + \rho\frac{\partial(s_1)}{\partial t} + \rho\frac{\partial(s_2)}{\partial t} = \frac{\partial}{\partial x}\left(\eta_e D\frac{\partial c}{\partial x}\right) - \frac{\partial(q_D c)}{\partial x} - \mu_w \eta c - \mu_s \rho(s_e + s_1 + s_2) \quad (1)$$

where c is the colloid concentration in the aqueous phase [N_c cm^{-3}], ρ is the bulk density (g cm^{-3}), s is the colloid solid phase concentration [N_c g^{-1}], subscripts e, 1 and 2 represent equilibrium and two kinetic sorption sites, respectively, N_c is the number of colloids, μ_w and μ_s represent decay rates in the liquid and solid phases (d^{-1}), respectively, t is time (d), $D = D_P + \alpha \times v$ is hydrodynamic dispersion coefficient (cm^2 d^{-1}), D_p is diffusion coefficient (cm^2 d^{-1}), α is dispersivity (cm), q_D is Darcy flux (cm d^{-1}), η_e is effective porosity (-), $v = q_D/\eta_e$ is pore-water velocity (cm d^{-1}) λ is the decay constant (d^{-1}), and x is distance (cm).

Considering the flexibility in the modeling by using a non-linear sorption model, the Freundlich sorption isotherm relating dissolved (c) and adsorbed (s) concentrations has been used in the present model:

$$s = k_F \cdot c^b \tag{2}$$

where b is the Freundlich exponent (-) and k_F is the Freundlich sorption coefficient (cm^3 g^{-1}).

Mass transfer between the aqueous and solid kinetic phases can be described as (the indexes 1, respectively 2, have been dropped in the following equation)

$$\underbrace{\rho \frac{\partial s}{\partial t}}_{\textit{accumulation on solid phase}} = \underbrace{\eta_e k_a \Psi c}_{\textit{attachment / straining}} - \underbrace{k_d \rho s}_{\textit{detachment / liberation}} \tag{3}$$

where k_a is the first-order deposition (attachment) coefficient [d^{-1}], k_d is the first-order entrainment (detachment) coefficient [d^{-1}], and Ψ is the colloid retention function [-] (see further). Attachment is the removal of colloids from solution via collision with and fixation to the solid phase, and it is dependent on colloid-colloid, colloid-solvent, and colloid-porous media interactions [4, 5]. To simulate reductions in the attachment coefficient k_a as a result of filling of favorable attachment sites, a Langmuirian dynamics equation may be used to describe the decrease of Ψ with increasing colloid mass retention (~blocking mechanism) [2]

$$\psi = 1 - S / S_{max} \tag{4}$$

where S_{max} is the maximum solid phase concentration [N_c g^{-1}]). Alternatively, a depth-depending blocking coefficient may be invoked to characterize the so-called straining process, where straining means the entrapment of colloids in down gradient pores and at grain junctions that are too small to allow particle passage [2]:

$$\psi = \left(\frac{d_c + x - x_0}{d_c} \right)^{-\beta} \tag{5}$$

where d_c is colloid diameter (cm), β is a fitting parameter (-) that controls the shape of the colloid spatial distribution, x is depth (cm) and x_0 is depth of the column inlet or textural interface (cm).

RESULTS AND DISCUSSIONS

Breakthrough curves and application of the attachment-detachment model to experimental data

Observed breakthrough curves for the ^{14}C-labeled OM reveal that peak concentrations and time of the peak (Figure 1) are influenced by 1) the position of the source within the clay core, and 2) by the flow rate, which is highest for MePAmCOM (highest K_h×pressure value) and lowest for HiPAmCOM (lowest K_h×pressure value). (Table II). For similar water fluxes (0.015 – 0.017 cm/d) ^{14}C recovery is independent of travel length and on average about 60%. For the higher flow rates (0.028 – 0.036 cm/d) recovery is higher too (~72-73 %).

For modeling purposes, three sections (materials) have been considered (see Table II for further details): the inlet domain M1, a very thin source layer M2 (0.6 mm long) and the outlet domain M3. The source layer M2 is modeled as a medium with a higher effective porosity (η_e=0.3), in such a way as to accommodate the 200 mm^3 of applied tracer solution containing dissolved ^{14}C-labelled OM.

Size and/or charge exclusion have been taken into account by using an effective porosity, η_e, of 0.13 [1, annex 16], instead of the effective porosity of 0.37 obtained for the conservative tracer HTO [6]. For the dispersivity α, a value of 0.1 mm has been used, based on initial sensitivity analysis (results not shown) [6]. Due to the small duration of the experiments compared to the ^{14}C half life, decay is neglected in the simulation.

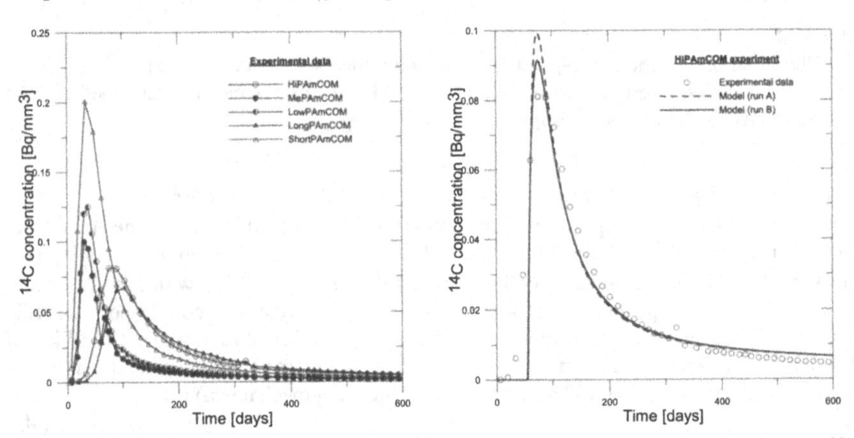

Figure 1: Experimental ^{14}C labelled OM breakthrough curves (left); fitted and observed ^{14}C breakthrough curve for HiPAmCOM (right).

The delay of the peak arrival time T_{max}, compared to the advective water travel time through the clay core T_w, in the experimental BTCs suggests the occurrence of a sorption type mechanism for OM transport. The calculated retardation coefficient R varies between 2.4 and 3.1, which suggests mildly retarded OM transport. Based on these retardation values R, Freundlich k_F values were calculated assuming linear reversible sorption. The calculated k_F values were taken as starting values for the inverse calculation with the model based on Eq. 1.

Colloid parameter starting values have been taken from previous calculations, based on simulations with the POPCORN code [1, annex 17], except for the values for the straining and liberation rates, processes that are not modeled within the POPCORN code. Initial sensitivity analysis without considering the detachment (k_{det} = 0, for the attachment-detachment model) and liberation terms (k_{lib} = 0, for the straining model) failed in predicting the observed breakthrough curve (results not shown). Therefore, further calculations always included mechanisms of detachment and liberation.

Modeling ^{14}COM migration

Transport of ^{14}COM is described with an advection–dispersion transport model that accounted for sorption, first-order kinetic attachment–detachment and straining-liberation.

Boundary conditions for ^{14}COM migration consider a zero concentration flux at the inlet and a zero-gradient condition at the outlet. Initial condition is given by the concentration in the source layer M2.

Modeling of the migration of OM was done by considering three different sorption sites: site (1) equilibrium with non-linear Freundlich sorption (submodel 1), site (2) kinetically controlled attachment/detachment (submodel 2) and site (3) kinetically controlled straining and liberation (submodel 3). Part of the ^{14}COM colloids are presumed to be sorbed on the solid surface (the equilibrium site – submodel 1). Sorption is considered to take place both at the inlet and the outlet layers of the clay column, but it is neglected in the thin source layer M2. Colloid attachment (submodel 2) was taken into account in all the three layers (i.e., M1, M2, and M3).

Presence of the Ψ-parameter in the attachment/straining component of the mass-transfer equation (3) allows one to account for different straining/blocking mechanisms. The simulation of the BTC curves for ^{14}COM colloids taking into account advection, dispersion and 'clean-bed' attachment-detachment model ($\Psi = 1$ and $k_d = 0$) for both kinetic sites failed in reproducing both the peak (including peak arrival time) and the tailings of the BTC curves. Introducing nonlinear sorption (Freundlich isotherm with $b = 1.1$) for layer M3 (outlet), resulted in an improvement in matching the peak concentration (including the peak arrival time), but the slow residual release at the tailings was still not met. Therefore, for the first kinetic site, colloid mass removal from solution is considered to take place by Langmuir blocking (Eq. 4). The second kinetic site considers straining (see further). Although the starting values of the attachment/detachment coefficients are the same for both kinetic sites, this may not be necessarily true since attachment depends, among other factors, on colloid-solid and colloid-colloid interactions. Due to the heterogeneity of the solid phase and of the ^{14}COM colloids, the use of different values for the attachment and detachment coefficients may be required. When depth-dependent straining was invoked (submodel 3, Eq. 5) in addition to attachment and detachment, the liberation rate was first set to zero. Taking the straining process into account in the column inlet part is justified by the fact that back-diffusion can divert the ^{14}C-labelled OM from the source zone M2 into the inlet component of the clay column.

Simulations started with the experiment HiPAmCOM. Inverse calculations resulted in a set of parameter values descriptive of colloid transport of OM. The optimized parameters for the HiPAmCOM experiment are then used to simulate the other four experiments (forward calculations).

Inverse calculations for HiPAmCOM experiment (^{14}COM)

Inverse calculations for determination of the advection dispersion transport and colloid model parameters have been performed with the HYDRUS-1D code for the HiPAmCOM experiment. The fitting had to deal with a very large number of parameters for three materials. In a first instance, following parameters were fitted in a staged manner: equilibrium sorption parameters k_F (Freundlich distribution coefficient) and b (Freundlich exponent), effective porosity η_e and dispersivity α in the inlet and outlet layers, site-1 attachment (k_{att1}) and detachment (k_{det1}) rates for the inlet (M1) and outlet (M3) layers of the clay core, attachment ($k_{att1} = k_{att2}$) and detachment ($k_{det1} = k_{det2}$) rates for the source layer M2 (site-1 identical to site-2), site-2 straining (k_{str}) and liberation (k_{lib}) rates for the inlet (M1) and outlet (M3) layers, the medium grain diameter of the clay d_c and empirical factor β in the depth-dependent straining function (Eq. 5). The values that had thus been obtained were kept constant while continuing the fitting with the next set of two to three parameters, until all parameters were optimized. Final results are referred-to as run A (r^2=0.926). At this stage, no parameter constraints have been imposed during optimization. A second global optimization (run B) has been undertaken where all parameters simultaneously were allowed to vary in a 10% interval from the optimal values obtained in run A. The fitted parameter values for run A and B are shown in Table III, while observed and fitted BTC are displayed in Figure 1. Including the diffusion coefficient D_p in the optimization run B reduces the calculated peak by about 10% and brings it in much better agreement with the data (r^2=0.947), while the tailing of the simulated BTC is now in good agreement with the observed BTC.

Table III: Optimized parameter values in the first (diffusion fixed) and second fit (diffusion optimized). Here NF means 'parameter not fitted', and f stands for 'fixed parameter value'. (units in mm, g and d).

Param.	Layer M1 & M3		Layer M2		Param.	Layer M1 & M3		Layer M2	
	Run A	Run B	Run A	Run B		Run A	Run B	Run A	Run B
η_e	0.14	NF	0.3 (f)	0.3 (f)	k_{det1}	0.0014	0.0012	0.0017	(0.0015)
D_p	4.32 (f)	6.43	4.32 (f)	3.94	k_{att2} (k_{str})	0.1011	0.090	0.017	(0.015)
k_F	83.86	92.01	0 (f)	0 (f)	k_{det2} (k_{lib})	0.010	0.0095	0.0017	(0.0015)
b	0.39	0.43	1 (f)	1 (f)	S_{max2} (β)	0.418	0.45	10^{25} (f)	10^{25} (f)
k_{att1}	0.017	0.015	0.017	(0.016)	d_c	0.0010	0.00095	0.0010	0.00091

Direct calculations for MePAmCOM, LowPAmCOM, ShortPAmCOM and LongPAmCOM experiments (^{14}COM)

The optimized parameters resulting from run A and B were then used to simulate in a direct way the remaining BTCs. It can be seen (from Figure 2 and Table IV) that the forward simulations are in good agreement with the experimental data when the source is located in the middle of the clay column (MePAmCOM, LowPAmCOM), but for experiments with a different position of the source (i.e., LongPAmCOM and ShortPAmCOM), the simulations fail in representing the data for both run 1 (using run A parameters) and run 2 (using run B parameters). Run 2 gave a better agreement with the experimental data for all the simulated experiments, except for the LongAmCOM data for which the model underestimates the peak concentration by more than a factor of 2. The difference is in agreement with the hypothesis of the different structure of the source layer compared to the rest of the column (by assigning a higher porosity). Also, the heterogeneity of the clay and of the organic matter may be accounted for by using different values for the attachment and detachment coefficients for the two kinetic sites in the source layer. For both parameter sets considered, there is a discrepancy between the magnitude and arrival of the first measured and simulated concentration breakthrough (Table IV). For experiment ShortPAmCOM, travel length is 12 mm compared to 36 for HighPAmCOM, while for experiment LongPAmCOM a longer travel length (60 mm) is used. Owing to the resulting difference in travel time, several processes (for example, dispersion, kinetically controlled attachment/detachment, etc) that are distance and/or time dependent may not be well described on the basis of parameter values derived from tests with different space and time scales.

Experiment type	Run no.	Peak ^{14}C concentration [Bq mm^{-3}]		Time of peak [days]		Recovered activity [Bq]		Coefficient of regression
		EXP	CALC	EXP	CALC	EXP	CALC	
HiPAmCOM	1.	0.081	0.098	75.4	74.9	2170	2149	0.926
	2.		0.089		78.2		2207	0.947
MePAmCOM	1.	0.1	0.22	30.7	29.2	2610	2982	0.768
	2.		0.19		32.0		3035	0.825
LowPAmCOM	1.	0.125	0.18	38.3	39.6	2650	2753	0.605
	2.		0.15		43.1		2810	0.485
LongPAmCOM	1.	0.067	0.04	103.7	180.4	2146	1480	0.0006
	2.		0.03		195.7		1360	0.014
ShortPAmCOM	1.	0.20	0.53	33.3	14.4	2720	4923	0.460
	2.		0.45		14.1		5121	0.587

Table IV: Optimization results for HiPAmCOM and the effect of using the fitted parameters for simulating other experiments. Experimental (EXP) and calculated (CALC) results.

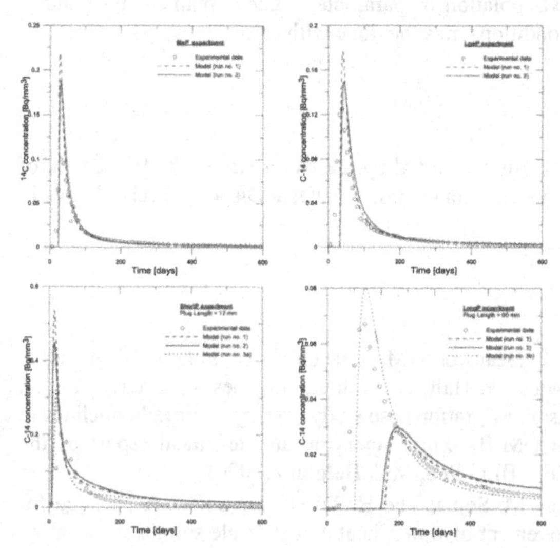

Figure 2: Simulated BTC for MePAmCOM, LowPAmCOM, ShortPAmCOM, and LongPAmCOM experiments, using fitted parameters from HiPAmCOM experiment.

Additional sensitivity analyses have shown that the most influential parameters for the ShortP- and LongPAmCOM experiments are: the attachment k_{att}, Freundlich distribution k_F, and dispersion coefficients D_p. Straining did not seem to influence the peak and time of peak, but it has an important role in explaining the tailing of the BTC. When parameters for ShortPAmCOM were fitted, best-fit values (run 3a, $r^2 = 0.62$) were as follows: 6.45 $mm^2\ d^{-1}(D_p)$, 0.026 d^{-1} (k_{att}), and 125,8 $mm^3\ g^{-1}$ (k_F). Best-fit parameter values for LongPAmCOM were (run 3b, r^2 =0.77): 8.36 mm^2/d (D_p), 0.0088 d^{-1} (k_{att}), and 60 $mm^3\ g^{-1}$ (k_F).

CONCLUSIONS

Migration of colloidal OM in undisturbed clay cores was described by means of combining the advective-dispersive-retardation equation and attachment-detachment theory. By inverse modeling, best-fit parameter values were derived for the experimental setup with the OM source spiked in the middle. Forward modeling for the same setup, but with different flow rates, gave calculated breakthrough curves (BTC) in good agreement with the experimental data. The forward modeling failed, however, when applied to a different setup (tracer source at different location). For the latter case, additional inverse modeling produced parameter values quite different compared to the initial runs. Sensitivity analysis revealed the most influential parameters were the attachment, Freundlich sorption and dispersion coefficients. Straining had most influence on the tailing of the BTC. Results further suggest that colloid migration parameters are sensitive to the experimental setup, more specific to the spatial scale used in deriving parameter values. Therefore, extrapolation of parameter values obtained from small-scale core samples to large scale in-situ conditions must be done with care.

ACKNOWLEDGEMENTS

The work of A. Ionescu was financially supported by the IAEA within the framework of URL project TC INT/9/173 "Training in and Demonstration of Waste Disposal Technologies in Underground Research Facilities".

REFERENCES

[1] N. Maes, L. Wang, G. Delécaut, T. Beauwens, M. Van Geet, M. Put, E. Weetjens, J. Marivoet, J. Van der Lee, P. Warwick, A. Hall, G. Walker, A. Maes, C. Bruggeman, D. Bennett, T. Hicks, J. Higgo, D. Galson, Migration case study: Transport of radionuclides in a reducing clay sediment (TRANCOM-II). Final scientific and technical report of the TRANCOM-II EC project, SCK·CEN-BLG-988, Mol, Belgium (2004).

[2] J. Simunek, M. Th. Van Genuchten, M. Sejna, The HYDRUS-1D Software package for simulating the One-dimensional movement of water, heat and multiple solutes in variably-saturated media, version 3.0 (2005).

[3] N. Maes, L. Wang, T. Hicks, D. Bennett, P. Warwick, T. Hall, G. Walker, A. Dierckx, The role of natural organic matter in the migration behaviour of americium in the Boom Clay – Part I: Migration experiments. Physics and Chemistry of the Earth 31, 541-547 (2006).

[4] S. A. Bradford, M. Bettahar, J. Simunek, M. Th. Van Genuchten, Straining and attachment of colloids in physically heterogeneous porous media, Vadose Zone J. 3:384-389 (2004)

[5] S. A. Bradford, J. Simunek, M. Bettahar, Y. F. Tadassa, M. Th. Van Genuchten, S.R. Yates, Straining of colloids at textural interfaces, Water Res. Research, vol. 41, W10404, (2005).

[6] A. Ionescu, Progress report no. 2, Internal report SCK-CEN, (2006).

Mater. Res. Soc. Symp. Proc. Vol. 1107 © 2008 Materials Research Society

Release of U, REE and Th From Palmottu Granite

Mira K. Markovaara-Koivisto[1], Nuria Marcos[2], David Read[3], Antero Lindberg[4], Marja Siitari-Kauppi[5], Kirsti P. Loukola-Ruskeeniemi[1]
[1] *Helsinki University of Technology, PO Box 6200, FIN-02015 TKK, Finland*
[2] *Saanio & Riekkola Oy, Laulukuja 4, FIN-00420 HELSINKI, Finland*
[3] *Enterpris, University of Reading, Whiteknights, PO Box 227, Reading, RG6 6AB, UK*
[4] *Geological Survey of Finland, Betonimiehenkuja 4, FIN-02150 ESPOO, Finland*
[5] *University of Helsinki, PO Box 55, FIN-00014 HELSINKI UNIVERSITY, Finland*

Keywords: granitic rock, release of U, REE and Th, Palmottu

ABSTRACT

Interpretation of trace metal mobility in geological environments is often hampered by conflicting data from alternative experimental protocols and the lack of detailed mineralogical characterization of the host medium. To illustrate this issue, the release of uranium, thorium and the rare earth elements (REE) was investigated in polished rock slab samples from the U-Th deposit at the Palmottu Natural Analogue study site (SW Finland) by means of leaching experiments. The samples were sequentially leached with artificial groundwater of moderately high carbonate content at pH8, and nitric acid solutions at pH5 and pH3. The mineralogy and composition of the U, Th and REE mineral phases was studied using SEM-WDS and EDAX methods before and after each leaching step. In parallel, leaching was carried out on crushed material of the same samples and the leachates analysed by ICP-MS.

The most notable U minerals are uraninite, uranophane and two secondary U-Pb phases. Thorium occurs predominantly in monazite and at lower concentrations in uraninite. Accessory thorite is also present, which together with monazite contains most of the REE. Differential leaching of the elements was noted across all phases on the timescale of the experiments. Uraninite is partly dissolved at pH3. The main secondary uranium phase, uranophane, was stable in moderately acidic solution, but easily dissolved in the artificial groundwater and at pH3. Some release of REE was observed although the main REE-bearing phase, monazite, showed no evidence of degradation.

This study provides insights in the preferential release of radionuclides in granitic bedrock. An understanding of these processes is essential when assessing the safety of a spent fuel repository. Once released from the primary waste form U is expected to precipitate as secondary phases within micro fractures, as observed at Palmottu and numerous other deposits.

INTRODUCTION

Spent nuclear fuel from Finnish power plants is planned to be disposed of deep in the bedrock at Olkiluoto (western Finland), and comprehensive assessment of the future performance of the repository is required. Early failure of the engineered barrier system, in particular defects in the copper–iron canisters containing the fuel could lead to interactions with the local groundwater. On the longer term, glacial melt waters, which may intrude into the bedrock due to the high hydrostatic pressure under an ice sheet also have to be taken into account. The effects of melt waters have already been observed at several natural uranium deposits in Finland [1, 2] and Sweden [3] following the Weichselian episode.

In this study, the behaviour of uranium (U), thorium (Th) and the rare earth elements (REE) was investigated in samples from the Palmottu uranium deposit. The study provides an indication of how uranium and associated elements could migrate from the repository to the surrounding bedrock in the event the canisters were breached.

The aim of this study is to clarify how U, Th and the REE minerals behave during leaching and to extrapolate this information to predict the likely behaviour of analogous elements in the spent fuel should this come into contact with different groundwater types. Other than uranium, the most important components of the spent fuel are the transuranics and certain long-lived fission products. Hence, the focus of this study was the mode of occurrence of uranium, thorium and the REE, the latter taken as surrogates of the trivalent actinides.

EXPERIMENTAL

Samples

The granitic rock samples are taken from a boulder (De) and drill cores (R389 and R390) at Palmottu (Figure 1); details of the locations are given in Blomqvist et al., 2002 [2].

Uranophane on fresh fracture surfaces through the boulder can be seen with the naked eye. The samples comprise mostly of quartz and K-feldspars; plagioclase is almost totally sericitised. Fe-oxyhydroxide staining is common.

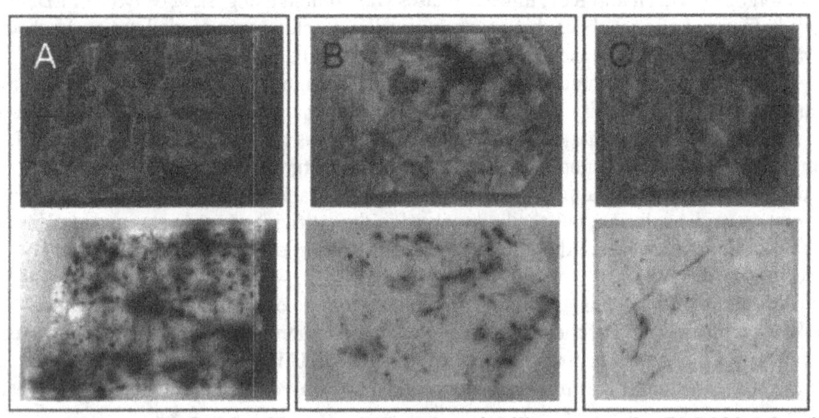

Figure 1. Photograph of A) boulder sample De B3, and drill core samples B) R390 A6 and C) R389 A3 with corresponding autoradiographs. Dark areas correspond to radioactive mineral phases. Width of samples: A 3.5 cm, B 4.1 cm and C 3.4 cm.

Distributions of the radioactive minerals in the selected rock slab samples were visualised on roentgen film (Figure 1). The sawn and slightly polished rock surface was exposed on Kodak X-omat MA roentgen film for 32 days. Well-defined dark spots, typically with a surrounding halo, indicate primary U phases. The film is also darkened where secondary U phases occur along inter- and intragranular fissures [4].

Leaching experiments

The release of the above elements was investigated by sequential leaching experiments on rock slabs and crushed rock samples. Figure 2 illustrates the leaching procedure.

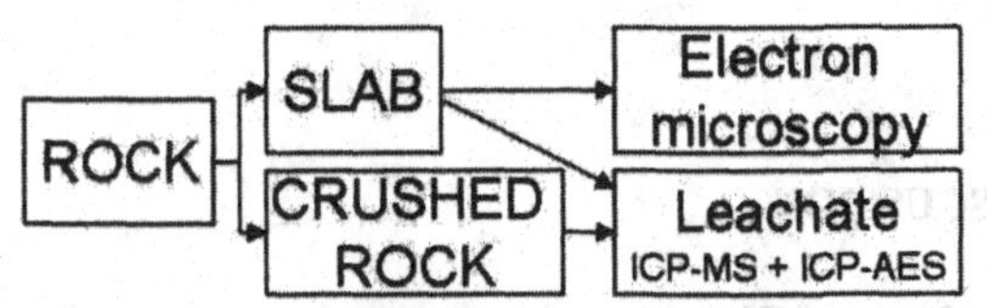

Figure 2. Leaching procedure and microscopy studies on samples.

In the first leaching test, the rock slabs were placed in acid-cleaned plastic containers and 10–15 ml of HNO_3 solution (pH5) added. The samples were left to leach for one week. After examination by electron microscopy, the same rock slabs were removed to fresh acid-cleaned containers, which were filled with 10–15 ml of HNO_3 solution at pH3, and leached for a further week. The leachates obtained were then analysed.

In the second leaching test, 2 g of crushed rock was weighed into acid-cleaned bottles and 10 ml of a synthetic solution ("Allard water" [6]) was added to each sample bottle. The mixtures were constantly agitated for 20 minutes and then centrifuged for 2 min at 2000 r.p.m. The resultant supernatants were then analysed.

In the third leaching test, 2 g of crushed rock was weighed into acid-cleaned bottles and about 10 ml of HNO_3 solution (pH5) added to each sample bottle. The bottles were sealed and the rock–acid mixtures were allowed to react for 20 minutes while under constant mechanical agitation. The mixtures were then centrifuged for 2 min at 2000 r.p.m., and the resultant supernatants analysed. The solid residues were dried at room temperature for at least 50 hours and leached again with a HNO_3 solution at pH3 using the procedure described above. The leachate obtained was then analysed.

Analytical methods

Microscopy

Elemental distributions on the surface of the rock slabs were analysed by SEM-WDS at the Geological Survey of Finland (GSF) using a Cameca SX100-type microprobe. Back-scattered electron (BSE) images were recorded by scanning electron microscopy from the areas of interest. All the microprobe analyses were carried out with 15 keV acceleration voltage and 10-25 nA beam current.

The mineral phases in the rock slabs were studied by field-emission scanning electron microscopy and energy dispersive X-Ray microanalysis (EDAX) after each leaching treatment. FESEM imaging and EDAX analyses were carried out at the University of Helsinki (HITACHI S-4800). Secondary electron (SE) and back-scattered (BSE) images were recorded from areas containing the mineral phases of interest. The chemical conmposition of phases was determined with EDAX (Oxford instruments + INCA program for analyses). The beam resolution was 10 µm. Voltage of the electron beam was set at 20 keV to ensure good quality analyses. In a SE+YAGBSE image the minerals are visualised according their chemical content, with different minerals appearing in varying shades of gray.

Mass-spectrometry

The trace elements in the leachates obtained from the rock slabs and crushed rocks were determined by Inductively Coupled Plasma–Mass Spectrometry (ICP-MS, Perkin Elmer PE SCIEX model ELAN 5000). The major elements in the leachates were determined by Inductively Coupled Plasma–Atomic Emission Spectrometry (ICP-AES, Thermo Jarrell Ash Corp. model Iris Advantage, Duo).

RESULTS AND DISCUSSION

Uranium, thorium and REE mineralogy

Most of the secondary U phases are found within 0.5 mm of primary U minerals, taken to be the source. The secondary U phases present are uranophane, which is dominant, and two U-Pb silicate phases yet to be fully characterised. One of the latter has been tentatively identified as kasolite. The primary U phase, uraninite, is invariably altered and known to be very old, around 1678-1741 Ma [2,5]. Th and REEs are present in monazite, uraninite and thorite with relatively minor contributions from inclusions (zircon etc.) in biotite and plagioclase. Those uraninite grains which are Th-rich, also tend to be enriched in heavy REE (HREE. Gd-Lu) and Y [1].

The uraninite grains at shallow depths display poorer crystal form and contain less uranium and lead than those at depth, presumably due to groundwater leaching. The released uranium occurs in secondary uranophane, which fills micro fissures in quartz and the pores of altered feldspar. Acicular uranophane is been observed in open fractures [4].

Figure 3 shows a uraninite grain (about 200 µm) in K-feldspar, and a U-silicate filling in a fracture extending to the uraninite grain. There is an alteration rim around the uraninite and a zircon grain next to it. The U-silicate filling in the fracture was totally dissolved at pH3. The uraninite grain was also partially dissolved at pH3, which can be seen as small cavities on the grain surface.

Secondary uranium phases were soluble in artificial groundwater and the strongly acidic solution (pH3), but they were not significantly dissolved in mildly acidic solution (pH5).

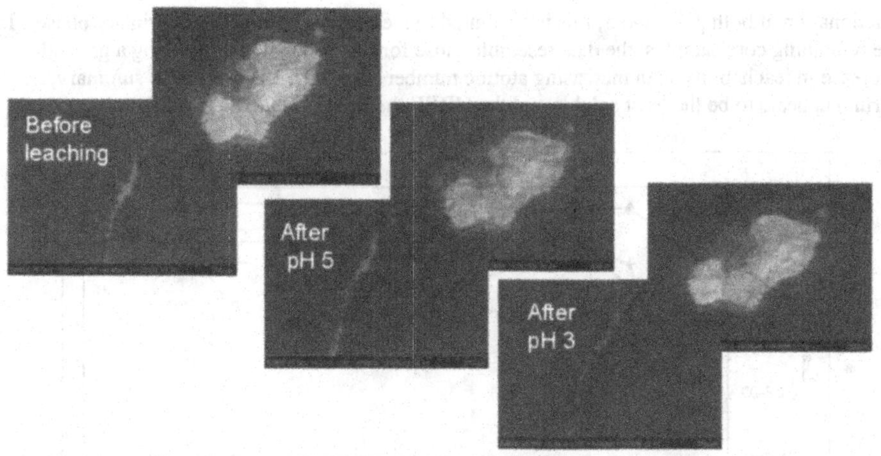

Figure 3. Large uraninite grain and U-silicate filling a fracture in K-feldspar before leaching, after leaching at pH5, and after leaching at pH3 (x250).

Release of U, Th, and REEs

Table I shows the release of Th and U in the different solutions.

Table I. Release of uranium and thorium in different solutions.

	Artificial groundwater	Solution at pH 5	Solution at pH 3
U / crushed samples	1% of total content	0.1% of total content	0.5-2% of total U content
Th / crushed samples	No release observed	1 ppm Th	traces
U / rock slabs	No data	0.3-0.7% of total U content	1-4% of total U content
Th / rock slabs	No data	traces	Below detection limits

Figure 4 shows rock normalised REE distributions for the leachates of crushed rocks in artificial groundwater and solutions at pH 3 and 5. All show evidence of fractionation with the exception of the boulder (De) at pH 3, where weathering is most apparent. Here, the REE are released congruently and the pattern would seem to represent complete dissolution of REE-bearing phase(s) in the strongly acidic solution. In all other cases, although the data are incomplete, there is a general increase in leachability across the series with the heavier REE released preferentially. The two core samples (R389, R390) give similar patterns and significantly lower concentrations in solution than the boulder for the same leachate. Leachates from the boulder sample also contained the highest concentrations of U and Th.

As expected, leaching was much reduced in the rock slabs (Figure 5). The boulder sample (De) again gave the highest concentrations in solution and showed correspondingly less evidence of HREE fractionation. In contrast to the crushed material there is clear evidence of cerium fractionation at both pH 3 and 5; this is attributed to the presence of Ce(IV) in primary phases. In the remaining core samples, the data resemble those for crushed material, showing a general decrease in leachability with increasing atomic number across the REE series. In summary, cerium appears to be the least soluble and the HREE the least soluble of the REE.

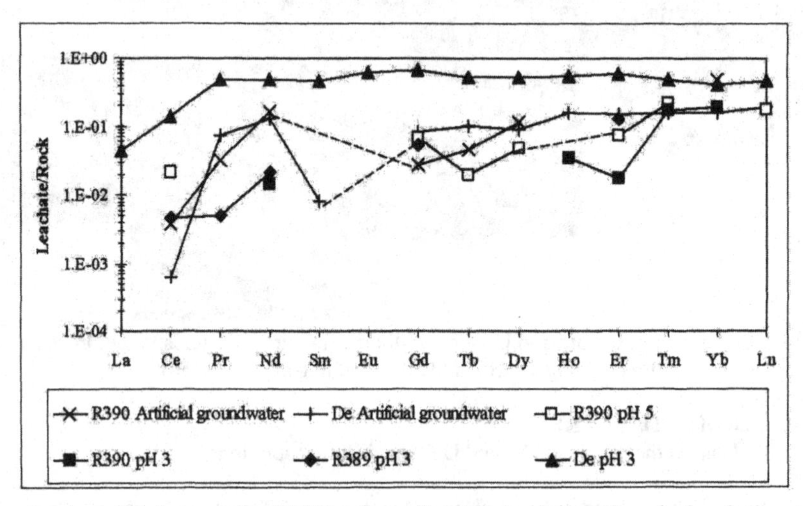

Figure 4. Selected rock normalized REE patterns from the leachates of crushed rocks in artificial groundwater and in solution at pH5 and pH3.

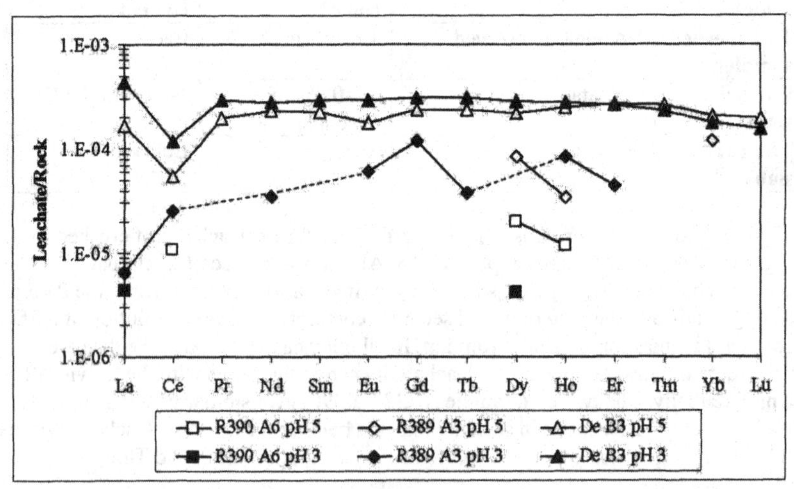

Figure 5. Rock normalized REE patterns from the leachates of rock slabs in solution at pH5 and pH3.

The composition of the artificial groundwater used in the experiments is close to that of distilled water equilibrated with crushed granite [6]. It easily leached U from uranophane owing to the presence of ligands, such as HCO_3^-, which enhance dissolution. In the absence of such ligands, the solubility of U in secondary phases is limited at pH5. Similarly, primary uraninite did not dissolve significantly at pH5 but at pH3 the grains showed signs of weathering.

CONCLUSIONS

The secondary U phases studies here dissolve to an extent in artificial groundwater and totally in strongly acidic solution (pH3). Primary U and Th, the latter taken as indicative of Pu(IV) in the waste, dissolves only in strongly acidic solution. REE as analogues for trivalent actinides (e.g. Am) in the nuclear waste, are soluble in both artificial groundwater and moderately acidic solution, being most soluble in strongly acidic solution.

Once released from a breached nuclear waste canister U is expected to precipitate as secondary minerals along microfractures in the rock. According to these experiments, groundwater containing ligands, especially carbonate, may cause subsequent remobilisation.

These results highlight significant discrepancies between data obtained using different leach protocols for reasons that are readily understood. It is recommended that all such data are treated with caution and viewed in the context of individual mineral phase stability.

ACKNOWLEDGEMENTS

Financial support for this study was provided by the Finnish Research Programme on Nuclear Waste Management (KYT).

REFERENCES

1. D. Read, S. Black, T. Buckby, K-H. Hellmuth, N. Marcos and M. Siitari-Kauppi, 2007. Secondary uranium mineralization in southern Finland and its relationship to recent glacial events. Global and Planetary Change, doi: 10.1016/j.gloplacha.2007.02.006.
2. R. Blomqvist, T. Ruskeeniemi, J. Kaija, L. Ahonen, M. Paananen, J. Smellie, B. Grundfelt, K. Pedersen, J. Bruno, L. Pérez del Villar, E. Cera, K. Rasilainen, P. Pitkänen, J. Suksi, J. Casanova, D. Read and S. Frape, 2000. The Palmottu Analogue Project – Phase II: Transport of radionuclides in a natural flow system at Palmottu. CEC Report EUR 19611.
3. R. Löfvendahl and E. Holm, 1981. Radioactive disequilibria and apparent ages of secondary uranium minerals from Sweden. Lithos 14: 182-201.
4. D. Read, M. Siitari-Kauppi, M. Kelokaski, S. Black, T. Buckby, N. Marcos, J. Kaija and K-H. Hellmuth, 2004. Natural geochemical fluxes in Finland as indicators of nuclear repository safety. Helsinki University of Technology, Laboratory of Rock Engineering, Research Report 34, Espoo. 58 p.
5. T. Ruskeeniemi, A. Lindberg, L. Pérez del Villar, R. Blomqvist, J. Suksi, A. Blyth and E. Cera, 2002. Uranium mineralogy with implications from mobilisation of uranium at

Palmottu. In. "Proc. 8th CEC Natural Analogue Working Group Meeting", Strasbourg. CEC Report EUR 19118.
6. B. Allard, F. Karlsson and I. Neretnieks, 1991. SKB Technical Report 91-50.

Mater. Res. Soc. Symp. Proc. Vol. 1107 © 2008 Materials Research Society

Comparative Analysis of Results From Deterministic Calculations of the Release of Radioactivity From Geological Disposal of Spent Fuel in Crystalline Rocks

Slavka Prvakova, Pavel V. Amosov[1] and Karl-Fredrik Nilsson
EC-JRC, Institute for Energy,
P.O. Box 2, 1755 ZGPetten, Netherlands.
[1]Mining Institute, MI KSC,
Fersman Street, 24, Apatity, Murmansk Region, 184200, Russia.

ABSTRACT

This paper presents results from deterministic calculations of radionuclide migration in a deep geological environment. The concept assumes single canister with spent nuclear fuel situated in bentonite (near-field) and surrounded by crystalline host rock (far-field). The results are presented in the form of release rates from the near-field and far-field, which contains also a part of advective release from the system of two single fractures. Additionally, radiological risk is evaluated in the form of dose rates for a water drinking scenario for an exposed population group. The analyses have been done by using the methodologies and computer codes used at the Institute for Energy EC-JRC in the Netherlands and the Mining Institute KSC RAS in Russia.

INTRODUCTION

Development of a safety case for disposal of radioactive waste involves consideration of the evolution of the waste and engineered barrier systems, and the interactions between these complex systems, which are also evolving. The presented paper compares results from deterministic calculations of the radionuclide releases from the hypothetical geological repository situated in crystalline host rock. Radionuclide migration in the deep geological environment was simulated as a one-dimensional model for the timescale of one million years over which the repository system has to provide well-functioning barriers against radionuclide transport. There are future plans to carry out uncertainty analyses of the stochastically varied input parameters carried out by probabilistic techniques, taking into account simultaneous variation of all input data.

CONCEPTUAL MODEL

The conceptual model (see figure 1) assumes a single canister with spent nuclear fuel placed in bentonite which is surrounded by crystalline host rock (granite). The repository is expected to be situated below the water table and so saturated conditions are assumed to prevail for the whole time of disposal. The source term consists of the radionuclides C-14, Cl-36, Ni-59, Se-79, Nb-94, Tc-99, I-129, Cs-135, Pu-239 and Am-243. The canister is assumed to fail completely 1000 years after disposal. No sorption is modelled in the region of the source term.

Within both barriers, bentonite and part of the host rock, transport is assumed to be purely diffusive. Radionuclides can sorb onto these materials according to the material specific distribution coefficients and may be subject to solubility constraints. Few of them, namely C-14, Cl-36 and I-129, can be subject to anion exclusion processes which reduces their effective porosity. The input data regarding the half-lives, solubility limits and distribution coefficients together were obtained from the Swedish SR-Can Report [1].

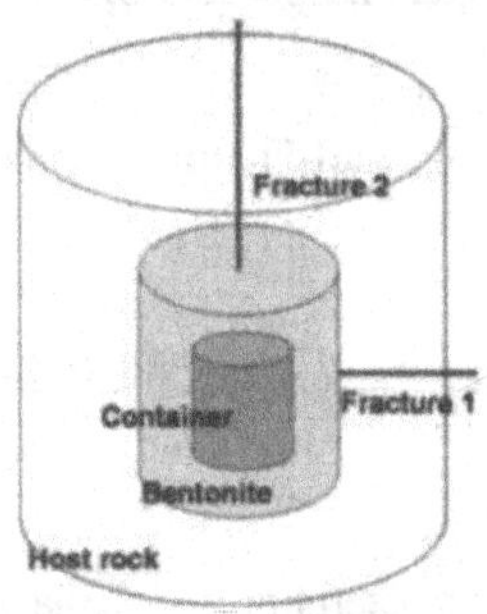

Figure 1. Conceptual model.

Two single fractures are considered to be emplaced at the outer face of the host rock (see Figure 1). The transport in the flowing fractures is advectively-dominated. The end points of fractures are assumed to be in direct connection with the source of drinking water used by the habitants living in the vicinity of the repository. This determines the necessity to assess the expected dose rates. The excavated disturbed zone is not included in the concept.

The barriers have symmetrical cylindrical character. The height of the canister is 4.83 m, outer diameter of the canister is 1.05 m, outer bentonite diameter 2.30 m, thickness of the bentonite below the canister is 0.50 m and thickness of the bentonite above canister is 1.50 m. The geometry of both fractures is the same but flow rates differ by one order of magnitude. The aperture is 5.10^{-4} m, with length 500 m and width 1 m. The Darcy velocity in the horizontal fracture is q_l= 0.02 $m^3/(m^2year)$, and in the vertical fracture q_l=0.2 $m^3/(m^2year)$.

METHODOLOGY AND MODELLING TOOLS

The conceptual model described above consists of two parts. The first one solves the diffusive dominated transport of radionuclides with radioactive decay, dissolution/precipitation and linear equilibrium sorption. The calculation gives the release rates of the radionuclides from the bentonite as well as the from porous host rock. Furthermore, the obtained level of activity at the entrance of fractures is used for the second part of the model which calculates the advective transport through the system of two single fractures. The resulting release rates are then transformed into dose rates which show the level of foreseen doses to an exposed population in the water drinking scenario.

Methodology presented by Mining Institute KSC RAS

The first part of the model with a diffusion dominant transport process was simulated using the code PORFLOW [2], developed by Analytic & Computational Research, Inc., ACRi. The governing equations of the code are presented in [2].

Due to the assumption of diffusion equation written in Cartesian coordinates, it is necessary to transform the cylindrical system (shown in Figure 1) into corresponding Cartesian coordinates. The transformation applies the condition of constant surface area of the canister. Corresponding quadratic shape was approximated with a rectangular parallelepiped shape.

In general, PORFLOW is not very suitable for modelling the canister containing spent fuel. In the area of source term (canister), PORFLOW is restricted to simulate two types of conditions. The first condition allows immediate dissolution of radionuclides which are very soluble. Also Nb-94 is included in this group due to its sufficiently high distribution coefficient as well as high value of water in the canister. The second condition presents the solubility limit approach which has been applied for the rest of radionuclides. The advective transport of dissolved nuclides is calculated on the basis of a single fracture model with the boundary condition of constant concentration at the fracture entrance [3].

In our case, it is necessary to define the constant concentration arising from the diffusive equations. Knowing the concentration of dissolved nuclide *n* in water at the entrance of fracture $C_{fn}(0, t)$, it is possible to make the interpolation $C_{fn} = \frac{1}{T}\int_0^T C_{fn}(0,t)dt$, where T is time (one million years in our case). The concentration of dissolved nuclide *n* in water at the entrance of fracture can be used to calculate the following parameters:

- release of nuclide *n* from the rock by advective transport through the fracture, given by

$$F_n(t) = C_{fn}(L_f, t) \cdot Q, \quad (1)$$

where Q presents a flow rate from the fracture, given by $Q = V_f \cdot 2b \cdot W$,

where L_f is the length, 2b is aperture, W is the width and V_f is flow rate in the fracture.

- dose rate of nuclide *n* for the water drinking scenario, given by

$$E(t) = \sum_{n=1}^{10} F_n(t) \cdot EDF_n, \quad (2)$$

where EDF_n is the dose conversion factor for ingestion for nuclide *n*.

Methodology presented by Institute for Energy EC-JRC

The modelling of radionuclide migration is performed with the code PROPER [4], developed by Swedish SKB together with Serco Assurance. The near-field transport is handled by sub-module COMP23 and the far-field by sub-module FARF31. Both of them allow simplified 1D calculations of radionuclide migration in an integrated way. The equations upon which are the codes based are presented in [5, 6].

RESULTS

Release of radionuclides from bentonite

The modelled release rates from bentonite as a function of time are shown in Figure 2a (Institute for Energy) and Figure 2b (Mining Institute). The first impression from visual comparison of the figures shows good qualitative agreement between the curves. Both figures show the clear dominance of radionuclide C-14 up to 10^4 years. Detailed description of the modelled release rates is shown in Table II. The nuclides are ordered according to their radiological relevance.

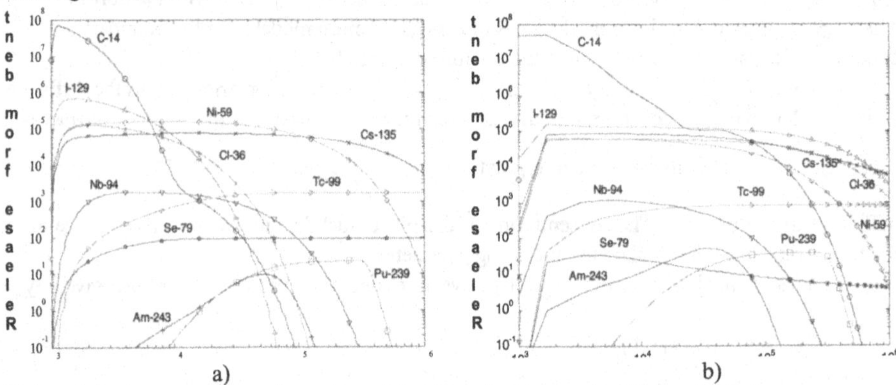

Figure 2. Release rates from bentonite, a) Institute for Energy, b) Mining Institute.

Table I. Detailed description of the modelled release rates of all considered radionuclides.

Radionuclide	Qualitative assessment	Quantitative assessment	Special features
C-14	agreement in the early stages up to 4 000 years	agreement in the early stages up to 4 000 years	difference appears after 4 000 years, curve descends faster in calculations performed by PROPER
I-129	agreement in the early stages	Curves are sufficiently close in the early stages	difference appears after 5 000 years, curve descends faster in calculations performed by PROPER
Cs-135, Tc-99, Nb-94	agreement	agreement	significantly successful agreement between curves
Ni-59	correspondence between curves over the entire timescale	deviation of the maximum release rate of an order of magnitude	higher values in calculations performed by PROPER
Cl-36	agreement in the early stages up to 10 000 years	sufficiently close agreement between curves up to 10 000 years	difference appears after 10 000 years, curve descends faster in calculations performed by PROPER
Se-79	significant difference	release rates are sufficiently close	curve is slightly increasing with time in calculations from PROPER and is

			decreasing in calculations from PORFLOW
Am-243, Pu-239	correspondence between curves	deviation in one order of magnitude	almost no time delay from the point the curves start to decrease

The search for the reason of the different results should be concentrated on the source term and consistency in the assignment of the migration parameters. Regarding description of the source term, one should note that PORFLOW requires to keep condition of constant surface area of the canister, which reduces in volume by 18% due to the transfer of cylindrical system into the corresponding Cartesian coordinates. Consequently, it causes modification of the initial concentration level in the model of immediate dissolution. Furthermore, the radionuclides drawn by the model of immediate dissolution (C-14, Cl-36, I-129) and which are non-sorbing on bentonite material, have a difference in maximum release rates as well as deviation in curves after 5000 years. However, there is significantly good agreement between curves of the nuclides Nb-94 and Cs-135, which are also drawn by the model of immediate dissolution but are low-sorbing on bentonite. The best qualitative and quantitative agreement is reached for nuclide Tc-99. The highest difference in the release rates is for to the high sorbing nuclides Pu-239 and Am-243.

It can be concluded that the current version of PORFLOW is not suitable for modelling such source term and should be used with caution in similar investigation.

Release of radionuclides from host rock

The calculated release rates from the host rock as a function of time are presented in Figure 3a (Institute for Energy) and Figure 3b (Mining Institute). Visual comparison of the figures shows certain qualitative but not quantitative agreement.

Results from both PROPER and PORFLOW show that I-129 and Cl-36 are the dominant radionuclides with maximum dose much higher than the other radionuclides (in some cases 6-7 orders of magnitudes).

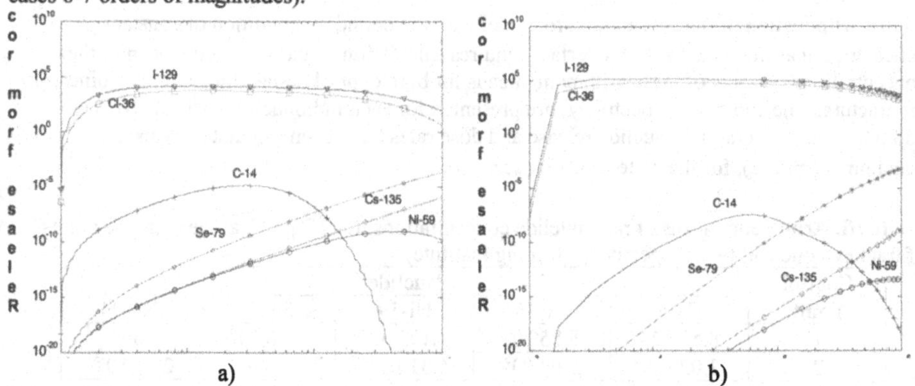

Figure 3. Release rates from rock, a) Institute for Energy, b) Mining Institute.

Detailed description of release rates is presented in the Table II. Radionuclides with release rates lower than 10^{-20} Bq/yr are not presented.

Table II. Detailed description of the modelled release rates of all considered radionuclides.

Radionuclide	Qualitative assessment	Quantitative assessment	Special features
I-129	good agreement	deviation of the maximum release rate of an order of magnitude	faster rise of curve up to the maximum release rate in calculations performed by PORFLOW
Cl-36	sufficiently good agreement	deviation of the maximum release rate of an order of magnitude	faster rise of curve up to the maximum release rate in calculations performed by PORFLOW
C-14	agreement in shape, shift in time	difference over the entire timescale	shift in time after the maximum is reached in results from PORFLOW
Se-79	correspondence between curves	deviation of the maximum release rate of an order of magnitude	one order of magnitude higher values of release rate from calculations performed by PORFLOW
Cs-135	agreement between the curves	deviation in numerical values	higher values in calculations performed by PROPER
Ni-59	correspondence between curves in ascending mode	deviation in numerical values	higher values in calculations performed by PROPER, small part of descending curve occurs in calculations from PROPER unlike PORFLOW

The differences between the obtained results cannot be explained by the source term (geometry, solubility models). There are also other factors, for example the exact assignment of the migration parameters of the radionuclides in the host rock as well as the algorithm for calculation of the release rate, which can have an influence on the results.

Release from the fractures and dose rates

Application of the single fracture model with the boundary condition of constant concentration at the rock-fracture interface requires calculation of the arithmetic mean of these concentrations. Results of these arithmetic means for both control points (the entrance points into the fractures one and two, respectively) are presented for each radionuclide in Table III. Two additional parameters, radionuclide release and dose rates have been calculated by using equation (1) and (2), for the water drinking scenario.

Table III. Arithmetic means of radionuclide concentrations, C_{sn} [Bq/m^3], at the entrance points of fractures one and two, respectively (Mining Institute).

Control point	Nuclide				
	C-14	Cl-36	Ni-59	Se-79	Nb-94
1	$8.53 \cdot 10^{-7}$	$8.85 \cdot 10^{5}$	$9.12 \cdot 10^{-12}$	$5.10 \cdot 10^{-2}$	$1.76 \cdot 10^{-32}$
2	$3.02 \cdot 10^{-7}$	$5.03 \cdot 10^{5}$	$5.31 \cdot 10^{-12}$	$2.04 \cdot 10^{-2}$	$2.61 \cdot 10^{-31}$

Control point	Nuclide				
	Tc-99	I-129	Cs-135	Pu-239	Am-243
1	$3.48 \cdot 10^{-25}$	$1.97 \cdot 10^{6}$	$3.97 \cdot 10^{-8}$	$1.81 \cdot 10^{-38}$	$1.09 \cdot 10^{-41}$
2	$4.27 \cdot 10^{-24}$	$1.23 \cdot 10^{6}$	$1.82 \cdot 10^{-8}$	$7.01 \cdot 10^{-38}$	$2.96 \cdot 10^{-41}$

Analysis of the resulting concentrations at the fracture ends shows that there are only two nuclides I-129 and Cl-36 with release rates from the fractures above 10^{-20} Bq/yr (see Figure 4a from the Institute for Energy and Figure 4b from the Mining Institute), which are considered radiologically relevant. The concentration level of the remaining radionuclides is so low that there is no meaning to consider them in our analyses.

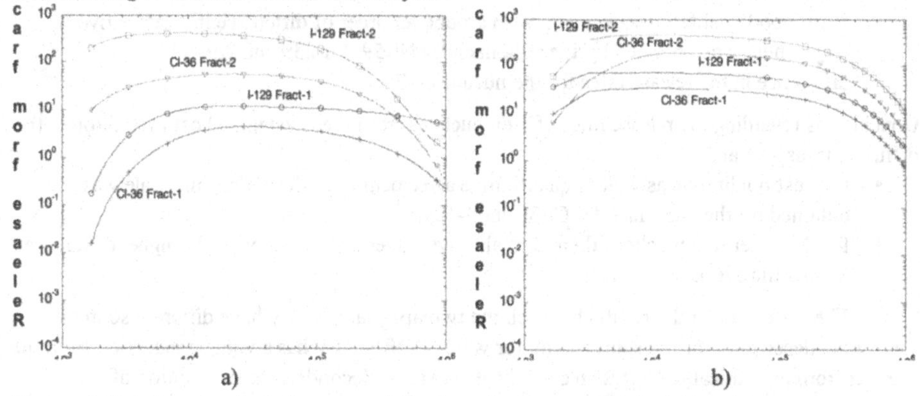

Figure 4. Release rates from fractures, a) Institute for Energy, b) Mining Institute.

The release rates from fracture 2 (presented in Figure 4) are higher than the corresponding curves for fracture 1, which to a certain extent is conditioned by higher water flow in fracture 2. The time behaviour of curves from both fractures has the same character. Results obtained from the PORFLOW indicate that the contribution to the resulting release rate is lower from the advective portion than from the diffusion. PROPER does not have any option for such separation.

The calculated dose rates for the water drinking scenario are not presented in graphical form because the shape of curves is the same as in the case of release rates from the fractures, presented in Figure 4. The comparison of results obtained from both institutes clearly shows good qualitative agreement between the release rates of dominant radionuclides. However, the differences in numerical values are higher. The results from both methodological approaches confirmed that the highest contribution comes from the radionuclide I-129. The obtained maximum dose rates are $5 \cdot 10^{-8}$ Sv/yr and 10^{-7} Sv/yr for the calculations of Mining Institute and Institute for Energy, respectively, which are both well bellow the limits, recommended by IAEA and the Russian regulatory offices.

CONCLUSIONS

The results present comparative analysis study of the release rates from the deep geological repository situated in a crystalline host rock (granite), solved within the frame of a single 1D problem but with using different methodological approaches.
Conclusions regarding the release rates of radionuclides from bentonite are:

- the best qualitative as well as quantitative agreement over the entire time scale was obtained for the radionuclides Cs-135, Nb-94 and Tc-99,
- good qualitative as well as quantitative agreement on certain time range was obtained by the radionuclides C-14, Cl-36 and I-129,
- fairly good qualitative agreement, with around an order of difference in quantitative assessment was obtained by the radionuclides Ni-59, Pu-239 and Am-243,
- difference in the release rates for the nuclide Se-79.

Conclusions regarding the release rates of radionuclides from the crystalline host rock through the diffusive transport are:

- the best qualitative as well as quantitative agreement over the entire time scale was obtained by the radionuclides Cl-36 and I-129,
- for the other radionuclides there is qualitative agreement but apparently higher difference in quantitative assessment.

The difference in the results between the two approaches may have different sources. Firstly the description of the source term, for which PORFLOW has a volume reduction from the transfer from cylindrical to Cartesian co-ordinate system. Secondly, the description of radionuclide migration differs as regards assignment of parameters and algorithms for release rate computation. The difference in the geometrical description and discretization may also affect the results.

The computed dose rates from both codes are well below 1 mSv/y, which is recommended limit by IAEA [7] and the Russian Regulatory Offices [8].

REFERENCES

1. Interim data report for the safety assessment SR-Can. – Swedish Nuclear Fuel and Waste Management Co. R-04-34 (2004), 76 p.
2. ACRi. PORFLOW: A Software Tool for Multiphase Fluid Flow, Heat and Mass Transport in Fractured Porous Media. User's manual. Version 3.07. Analytic and Computational Research, Inc. (1996), 326 p.
3. D.H. Tang, O.E. Frind, E.A. Sudicky, *Water Resources Res.* **3**, 555-564 (1981).
4. PROPER Monitor user's manual, Version 3.2. Swedish Nuclear Fuel and Waste Management Co., Stockholm (2000), 68 p.
5. K.A. Cilffe, M. Kelly, COMP23 version 1.2.2 user's manual. Swedish Nuclear Fuel and Waste Management Co. R-04-64 (2004), 76 p.
6. M. Elert, B. Gylling, M. Lindgren, Assessment model validity document FARF31. Swedish Nuclear Fuel and Waste Management Co. R-04-51 (2004), 39 p.

7. International basic safety standards for protection against ionizing radiation and for the safety of radiation sources. IAEA Safety Fundamentals, Safety Series No. 115. Vienna, Austria (1996), 328 p.
8. Normy radiacionnoj bezopasnosti (NRB-99).-M.:Minzdrav, Russian Federation (1999), 115p.

Performance Assessment

Mater. Res. Soc. Symp. Proc. Vol. 1107 © 2008 Materials Research Society

Development and Validation of a Model of Uranium Release to Groundwater From Legacy Disposals at the UK Low Level Waste Repository

J.S. Small[1], C. Lennon[1], S. Kwong[1] and R.J. Scott[2]
[1] Nexia Solutions Ltd, Warrington, UK
[2] LLWR, Holmrook, Cumbria, UK

ABSTRACT

A previous radiological assessment of the UK Low Level Waste Repository (LLWR) has considered how the prevailing reducing chemical conditions in disposal trenches, may limit uranium release through the extreme low solubility of U(IV) solids. This study considers the additional effects that the physical and chemical nature of the uranium wastes may have on the release of uranium. Fluoride process residues produced by refining of uranium metal comprise the majority of the legacy inventory. Based on historic records and descriptions of the uranium wastes a conceptual model has been developed which bounds the release rate of uranium present as inclusions and dissolved in the solid residues by the dissolution rate of a magnesium fluoride matrix. The model is represented in a 3-dimensional groundwater flow and geochemical model. Initial findings indicate that the model correctly represents the range of fluoride and uranium concentrations that are measured in leachate from the LLWR trenches. Incorporation of this model in future safety assessments, together with a reduction in the derived inventory of uranium, is likely to result in a significant lowering of the peak groundwater dose to acceptable levels, even in the case that the site re-oxidizes. The study builds confidence in the inherent safety features that are provided by the sparingly soluble uranium waste residues and the reducing chemical conditions of the LLWR trenches.

INTRODUCTION

The UK Low Level Waste Repository (LLWR) located close to the village of Drigg, Cumbria has served as a national repository for LLW since 1959. Until 1988 LLW was backfilled into trenches excavated into glacial clays and covered with an interim cap. The current disposal practice comprises emplacement of compacted waste in steel ISO-freight containers, with void space filled with a cement grout, with the containers stored in an engineered Vault. A post-closure radiological assessment was undertaken in 2002 [1], which highlighted potential doses that may arise from uranium disposals to the trenches. The near-field conceptual model that formed the basis of these calculations [1,2] considered that the release of uranium would be limited by the low solubility of UO_2 stable under the prevailing reducing chemical conditions. A biogeochemical reactive transport model of the site [2] indicated that after periods of around 3,000 to 4,000 years the trenches may re-oxidize to the conditions in the surrounding geosphere and uranium may potentially become more mobile. The radiological assessment of the groundwater pathway considering this scenario [1] determined that after 10,000 years Ra-226 and Pb-210 daughters of U-234 would arrive in the biosphere, leading to doses that may exceed the 1e-6 yr^{-1} risk target.

These previous assessments were based on descriptive information in the UK Radioactive Waste Inventory and the assumption that the dissolution of the uranium wastes would be instantaneous up to the solubility limit. This paper presents the findings of an approach using inventory information from archived disposal records which contain further descriptions of the physical and chemical properties of the uranium wastes. Further information and literature regarding fluoride residues resulting from uranium metal refining in which form the majority of the trench uranium is present has also been considered. This inventory information has been used to develop a model that represents the behaviour of these materials and the release of uranium. LLWR site monitoring data include analyses of major chemical elements, uranium and other radionuclides in leachate samples recovered from locations within the trenches. These data provide a means to independently validate the conceptual and computational model.

THE NATURE OF LEGACY URANIUM DISPOSALS

Information concerning the nature of the legacy disposals of uranium to the LLWR trenches from 1959 to 1988 has been obtained by a review of archived disposal records from UK waste consigners. The study has re-examined the inventory of uranium, Th-232 and Ra-226, since these radionuclides are important to the groundwater, gas and human intrusion dose pathways. The inventory study has mapped the spatial distribution of waste consignments and waste types and other structural features of the trenches, such as soil firebreaks. This information is required to develop the 3-dimensional (3d) hydrogeochemical model.

Inventory mapping has shown that 95% of the trench uranium represents disposals from the refining of uranium metal for Magnox fuel fabrication [3]. The bulk of the waste comprises process residues and scrap materials. The process residues comprise filter cakes from refining of uranium ore and yellowcake (impure UO_3). Ore refining and waste recycling processes were designed and optimised to economically produce a residue with <0.1 wt% uranium, although in practice some higher uranium concentration residues (<0.9 wt% uranium) could arise. Fluoride residues result from slags formed by the reduction of UF_4 to uranium metal by Mg or Ca metal. Descriptions of the processes [3] indicate that the fluoride slags were ground and leached in 60% HNO_3, at 95 °C for 8 hours to extract uranium, prior to consignment to the LLWR. Thus, remaining traces of uranium will be strongly bound and included in the fluoride matrix.

Amongst the uranium refining wastestreams, fluoride reduction slags comprise a significant proportion (> 50 %) of the trench uranium inventory. MgF_2 is the principal form, although some CaF_2 is recorded. Information concerning the physical and chemical nature of these materials is recorded in past research papers dating from the 1950s from the Atomic Energy Research Establishment. These include detailed, grain size, compositional and X-ray diffraction (XRD) studies of MgF_2 and CaF_2 slags after leaching by HNO_3. Magnesium slags are notable in that they contain particles ("shots") of metallic uranium that comprise more than 85% of the uranium. Some shots were encrusted in a black crystalline phase that appeared to inhibit leaching. Particle size distribution curves are provided which indicate that leached material in the Tyler sieve, mesh size 8- 28 (~5mm to 0.6mm) contain the highest uranium contents and that finer grained particles, (down to micron scale) contain less uranium. This is consistent with the larger particles containing discrete metallic uranium shots. CaF_2 has been described as being present as five distinct coloured phases, which have varying uranium contents. Red and smoky varieties contain relatively high uranium contents (7-15%) but yellow and white varieties have lower U contents (0.19, 0.68%). XRD analysis shows changes in crystal lattice spacings amongst these samples, compared to pure CaF_2 which suggests that the uranium is present in a 'solid-solution' within the CaF_2 rather than as discrete metallic shots as in the case of MgF_2. A discrete uranium-rich phase appears to be present in red coloured CaF_2 which has the highest uranium contents.

The process of UF_4 reduction by Mg or Ca metal is a technology used in several countries and additional understanding is provided by recent literature from these programmes e.g. [4,5].

Chaudhury *et al.* [4] present an XRD analysis of MgF_2 slag which indicates the presence of both MgF_2 and UO_2, which contrasts with the UK materials, which indicated that uranium was present mainly as metallic inclusions. Avvaru *et al.* [5] present a model of metal leaching from fluorides that considers the development of a porous alteration layer observed to be developed under strongly acidic conditions.

Zhang *et al.* [6] examine the dissolution of fluorite (CaF_2) under less aggressive pH conditions between pH 2.6 and 7 at 25°C and 100°C. These studies are relevant to the dissolution behaviour of CaF_2 under groundwater conditions and in particular in the LLWR trenches where leachates with pH 6 have been sampled [2]. Zhang *et al.* [6] conclude that the rate of fluorite dissolution is controlled by H^+ and Ca^{2+} ion activities (a) such that the dissolution rate (r) is expressed as:

$$-r = k\left(\left(a_H^+\right)^2 \Big/ \left(a_{Ca}^{2+}\right)\right)^{\alpha} \qquad \text{Equation (1)}$$

Where k is the rate constant and α is the order with respect to the H^+ activity versus the Ca^{2+} activity. Zhang *et al.*. [6] present X-ray photoelectron spectroscopy (XPS) and Secondary Ionisation Mass Spectroscopy (SIMS) investigations, which show that the fluorite surface becomes depleted in Ca^{2+} and that H^+ and Cl^- ions diffuse into the solid. This effect may be considered to be generally consistent, but at a much smaller scale, with a porous layer developed under strongly acidic leaching of MgF_2 [5].

A MODEL OF URANIUM RELEASE FROM FLUORIDE RESIDUES

Based on the review of literature and information concerning the acid recycling processes operating to recover easily leached uranium from fluoride reduction slag it is evident that any freely dissolvable uranium metal or UO_2 is unlikely to be present in these specific wastes. Consequently, further leaching of uranium from these materials in the disposal environment is likely to be slow and limited by the dissolution of the poorly soluble fluoride residue under groundwater conditions. Figure 1 provides a schematic representation of a fluoride residue particle. As the fluoride matrix dissolves uranium present as a trace "solid solution" within the fluoride matrix will be released. In addition, dissolution of the matrix will lead to the exposure of metallic uranium and UO_2 inclusions to groundwater and dissolution under the prevailing chemical conditions.

Most near-surface groundwaters are undersaturated with respect to CaF_2 (fluorite), and in particular groundwaters in the vicinity of the LLWR have quite low fluoride concentrations, typically around 0.2 mg l^{-1}. Dissolution of the predominantly MgF_2 residues will depend on the product of Mg and F concentrations. For the MgF_2 mineral sellaite the solubility product (K_{sp}) is:

$$K_{sp} = \left[Mg^{2+}\right]\left[F^-\right] = 10^{-9.3843} \qquad \text{Equation (2)}$$

Similarly for fluorite:

$$K_{sp} = \left[Ca^{2+}\right]\left[F^-\right] = 10^{-10.037} \qquad \text{Equation (3)}$$

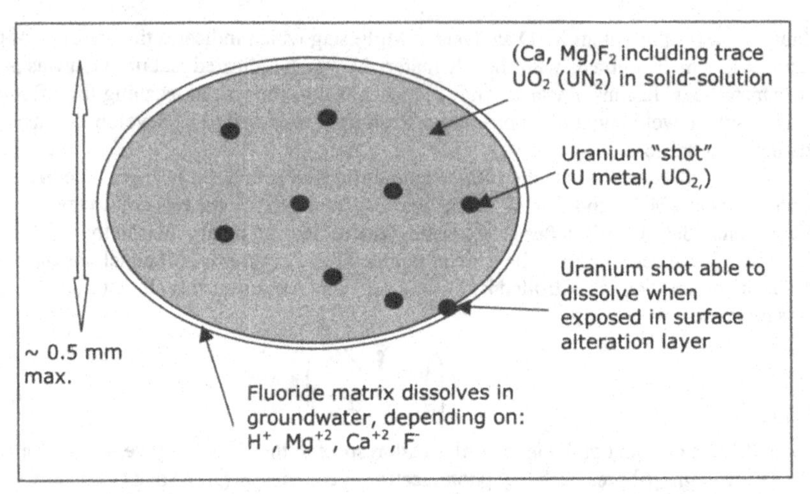

Figure 1. Conceptual model of uranium release controlled by the dissolution of fluoride residues.

Sellaite has a larger solubility product than fluorite. Thus in groundwaters with a supply of Ca^{2+}, MgF_2 will be unstable with respect to CaF_2. The Mg rich fluoride residues are therefore likely to re-crystallize to form fluorite. Mg fluoride slag formed in the uranium reduction process may be more soluble than the mineral phase sellaite, e.g. it may be comprised of amorphous or vitreous material and this would provide a further driving force for the re-crystallization process. It is also possible that other more soluble metal fluorides (e.g, Fe, Zn) may be present.

While the relative solubilities of fluoride may drive the re-crystallisation process the reaction and hence the release of uranium may be limited by kinetics of the dissolution process. The study by Zhang *et al.* [6], described above provides a basis to examine kinetic effects of fluoride dissolution making the assumption that MgF_2 behaves similarly to fluorite.

Computational representation

The model of fluorite dissolution [6] (Equation 1) is readily represented in the PHREEQC geochemical modelling code [7] and can be incorporated into the 3d groundwater flow, transport and reactive code PHAST [8]. The rate constant (*r*) defined by Equation 1 is used to define rates of MgF_2 dissolution at far from equilibrium conditions. Close to equilibrium with MgF_2 the model reduces the rate by:

$$rate = -r \times (1 - Q/K) \qquad \text{Equation (4)}$$

where Q is the ion activity product and K is the equilibrium constant for sellaite. The PHREEQC model considers that the kinetic reactant has a formula MgF_2U_n, where (n) is a fraction representing the mass of uranium included in the fluoride matrix (n=2.6e-4 represents 0.1wt% U). The rate constant k and α parameter (Equation 1) used in the model were based on data in [6]. A rate constant of $\log k = -1.0$ mol m^{-2} min^{-1} was used in a base case run. Compared to the rates measured by Zhang *et al.* [6] for 20-40 mesh size fluorite (surface area 0.67 m^2 g^{-1}) this represents relatively rapid dissolution, which should be conservative with respect to uranium release. Further work is required to quantify these kinetic parameters for MgF_2 residues, and these values are a starting point for sensitivity analyses.

The model allows fluorite to precipitate to equilibrium when it is oversaturated. Other equilibrium phase constraints in the model were the presence of calcite, siderite and $Fe(OH)_3$ minerals at a log partial pressure of CO_2 of -2.0. These constraints represent the pH and Eh conditions in the geosphere at the LLWR [1,2]. The thermodynamic data base file *llnl.dat* containing data from the file thermo.com.V8.R6.230 distributed with PHREEQC was used for the model as at includes data for sellaite.

This PHREEQC model has been included in a 3d groundwater flow and transport of a generic non specific region of the LLWR trenches to examine how, under hydrogeological conditions typical of the trenches, the dissolution behaviour of the MgF_2 residues can be examined and concentrations of fluoride and uranium determined. Figure 2 presents output showing in plan view the development of a plume of uranium developed from regions containing the uranium containing MgF_2 residue which are interspersed amongst other wastes and firebreaks in the trench that is excavated in glacial drift. Groundwater flow is calculated by the model driven by a hydraulic gradient of 7.0e-3 and considering hydraulic conductivities of 1.0e-4, 1.15e-5 and 4.0e-5 m s^{-1} for the trench waste, firebreaks and drift sediments respectively, which are based on hydrogeological studies at the LLWR [1]. The model simulates that maximum uranium concentrations developed in the vicinity of the fluoride residues attain 5.0e-7 mol kgw^{-1} (molality in water) In other downstream regions of the trenches the plume has a concentration of around 8.e-8 mol kgw^{-1}.

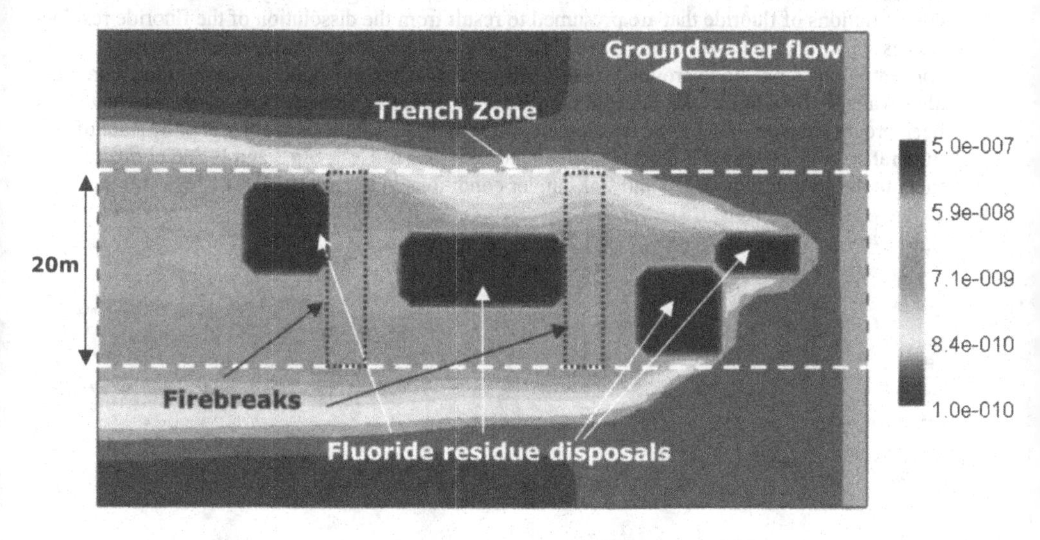

Figure 2. Generic trench PHAST model output of U concentration (mol kgw^{-1}) for leaching of fluoride residues containing 1 wt% U (monochrome print of colour output).

VALIDATION WITH SITE DATA

Site monitoring data are available from the LLWR trenches which can be used to independently validate the conceptual model of uranium release from fluoride residues and its computational representation. Since 1989 the area of the LLWR trenches containing fluoride residues have been covered by an interim cap. More than 80 vent probes comprising perforated, galvanised steel pipes are installed through the cap and into the disposed wastes primarily in order to monitor the generation of gases from the wastes. The vent probes have been used to sample leachate from the wastes and these have been analysed for major elements, radionuclides and other potential contaminants such as heavy metals and organic compounds. Previously these site data, in particular concentrations of redox sensitive species have been used to build confidence in the biogeochemical model of the LLWR trenches [1,2]. The vent probe data include values for F, U, Mg and Ca concentrations in the leachate present from around 50 of the vent probes which have contacted leachate. Data sets are available from an initial sampling during 1989-1990 shortly after installation of the trench cap. More recent samplings during 1998, 2002/2003 have recovered fewer samples as the trenches are now drier due to the effect of the cap.

Fluoride concentration and mineral equilibria

Concentrations of fluoride in natural groundwaters are normally limited (<1 mg l^{-1}) by the solubility of CaF_2 (fluorite). The dilute groundwaters at the LLWR have an even lower background of around 0.2 mg l^{-1}. LLWR trench leachate has significantly elevated concentrations of fluoride that are presumed to result from the dissolution of the fluoride residue wastes. The trench dataset has a mode around 3 mg l^{-1} fluoride and some notably higher concentrations range up to around 30 mg l^{-1} fluoride. Magnesium and Ca concentration data are also available for some of the leachate samples with measured fluoride. These analyses have been processed with PHREEQC to calculate the activities of aqueous species and the state of mineral saturation. Figure 3 plots the calculated activities of Ca^{2+}, Mg^{2+} and F^- and compares them to the solubility of MgF_2 and CaF_2 under conditions in the trenches.

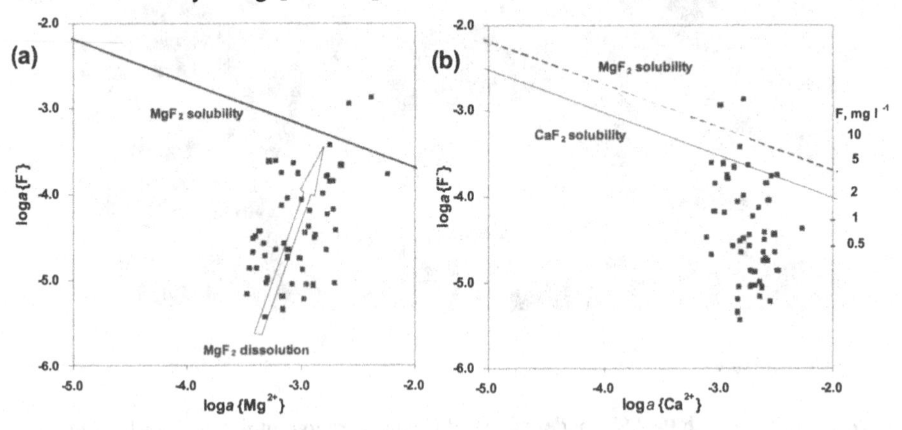

Figure 3. Phase diagrams showing the solubility of (a) MgF_2 and (b) CaF_2 and data from LLWR trench leachate samples, the slope of the arrow (a) represents congruent MgF_2 dissolution.

For Mg (Figure 3a) it is noted that the leachate data define a trend of increasing Mg^{2+} and F^- activity which is in the mole ratio 1:2, consistent with the dissolution of MgF_2. It is also noted that, apart from two samples from one specific location, all the leachate analyses are undersaturated with respect to MgF_2. CaF_2 has a lower solubility than MgF_2 (Figure 3b) and the fluoride concentration data closely approach the CaF_2 solubility line. These relationships are consistent with the conceptual model of dissolution of the more soluble MgF_2 residue and the reaction of released F^- with Ca^{2+} present in groundwater resulting in the precipitation of CaF_2. The data suggest that the fluoride residues approach chemical equilibrium in the trench leachate.

Model Comparison

Figure 4 compares frequency distributions of the concentration of fluoride and uranium measured in trench leachate with modelled concentrations from the PHAST model of fluoride residue dissolution and uranium release. For both fluoride and uranium the frequency distribution of the model output is for the trench zone (highlighted in Figure 2) where the vent probes are located. The majority of the modelled fluoride concentrations (Figure 4a) range between $10^{-4.9}$ and $10^{-4.2}$ mol kgw^{-1}, which is within the measured range of the majority of the leachate data. Higher modelled fluoride concentrations ($10^{-3.1}$ mol kgw^{-1}) represent the regions of the model where the fluoride residue is present, these concentrations are more comparable to anomalous measured fluoride concentrations (30 mg l^{-1}, $10^{-2.8}$ mol kgw^{-1}) that occur in one area.

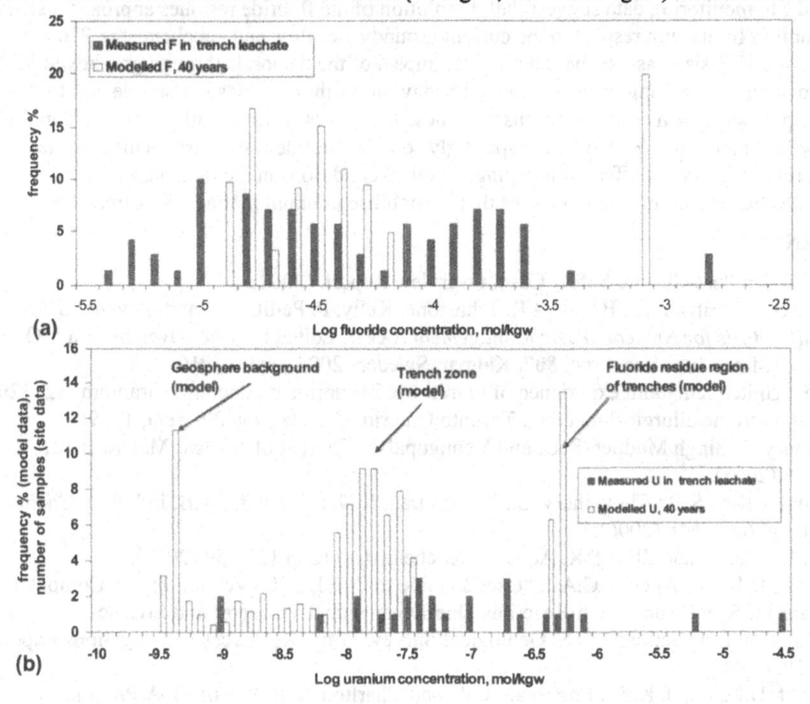

Figure 4. Comparison of modelled fluoride (a) uranium (b) concentrations and leachate concentrations.

The model also represents quite well the range of measured uranium concentrations in trench leachate (Figure 4b). The model distribution reflects regions where the solid fluoride is present, which have highest uranium, concentrations ($10^{-6.4}$ mol kgw^{-1}) and the downstream plume in the trench zone($10^{-7.9}$ mol kgw^{-1}). These uranium concentrations are comparable with the measured concentrations. Two leachate samples are noted to have significantly higher uranium concentrations ($> 10^{-6}$ mol kgw^{-1}) and these also have high fluoride concentrations. Such U and F concentrations could result from a more soluble fluoride phase, such as influenced by the presence of other metal fluorides (e.g. Fe, Zn) or a higher uranium content of the residue. Overall the model is considered to produce a good representation of the behaviour of the uranium-bearing fluoride residues, which explains the measured concentrations of uranium in groundwater.

CONCLUSIONS

The conceptual model of uranium behaviour in the LLWR trenches has been updated to consider specific information concerning the chemical and physical form of fluoride uranium refining residues that comprise the majority of this legacy disposal inventory. Historic information regarding these residues indicates that they are resistant to strong acid leaching and thus will be quite insoluble in groundwater in the LLWR trenches. The fluoride dissolution model represents well the measured uranium and fluoride concentrations in trench leachate. The model and site monitoring data suggest that dissolution of the fluoride residues approach a steady state (solubility limit) with respect to the current groundwater flow and geochemistry. This provides a sound basis to assess the radiological impact of the disposals through groundwater. The improved understanding of the nature and behaviour of these wastes is also relevant to the other dose pathways such as human intrusion. These new investigations build confidence in the inherent safety function provided by the sparingly soluble fluoride waste form. This together with the previously known effect of reducing chemical conditions in the trenches [1,2] demonstrates the important safety features that immobilise uranium in the LLWR trenches.

REFERENCES

1. BNFL. Drigg Post-Closure Safety Case: Overview Report. (2002).
2. Small. J.S., Abraitis, P.K., Beadle, I.R. Johnstone, Kelly, P. Pettit, C.L. and Stevens, G.A. in *Scientific Basis for Nuclear Waste Management XXVII.* Edited by V.M. Oversby and L.O. Werme, (Mater. Res. Soc. Proc. **807,** Kalmar, Sweden, 2004), pp905-910.
3. Page, H. United Kingdom experience of uranium tetrafluoride production. Uranium' 82. 12th Annual hydrometallurgical meeting, Toronto Ontario, Canada Aug. 29- Sept, 1, 1982.
4. Chaudhury, S. Singh Mudher, K.D. and Venugopal, V. Journal of Nuclear Materials 322, p 119-125 (2003).
5. Avvaru, B., Roy, S.B., Chowdhury, S., Hareendran, K.N. and Pandit, A.B. Ind. Eng. Chem. Res. 45, p7639-7648 (2006).
6. Zhang, R., Hu. A. and Zhang, X. Aquatic Geochemistry, 12, p 123-159 (2006).
7. Parkhurst, D.L. and Appelo, C.A.J. "User's Guide to PHREEQC (Version 2) – A Computer Program for Speciation, Batch-Reaction, One-Dimensional Transport and Inverse Geochemical Calculations". U.S. Geological Survey, Water Resources Investigations Report 99-4259. (1999).
8. Parkhurst, D.L., Kipp, K. L., Engesgaard, P., and Charlton, S. R. " PHAST-A Program for Simulating Ground-Water Flow, Solute Transport, and multicomponent Geochemical Reactions", U.S. Geological Survey, Techniques and Methods 6-A8, 154p. (2004).

Mater. Res. Soc. Symp. Proc. Vol. 1107 © 2008 Materials Research Society

Near-Field Modelling in the Safety Assessment SR-Can

C. Fredrik Vahlund
Swedish Nuclear Fuel and Waste Management Co, SKB
Box 250
SE-101 24 Stockholm, Sweden

ABSTRACT

Spent nuclear fuel from the Swedish energy programme will be stored in an underground repository situated in saturated fractured rock at a depth of approximately 500 m. This paper describes numerical simulations of radionuclide migration in the near-field (consisting of a canister filled with spent fuel and an engineered system backfilled with swelling clays) for the recently completed safety assessment SR-Can [1] using a Matlab / Simulink code. Handling of input data for the models from the site descriptive programme from on-going investigations at two candidate sites and the numerical modelling concept are discussed.

INTRODUCTION

The Swedish Nuclear Fuel and Waste Management Co. (SKB) is jointly owned by the operators of the Swedish nuclear power plants and is responsible for interim storage and final disposal of the spent radioactive fuel produced within the Swedish nuclear energy programme. A research programme for developing a repository system has been ongoing since the 1970s and has resulted in SKB suggesting a KBS-3 type of repository for final storage of the spent fuel. The KBS-3 method is based on storing the spent fuel in corrosion resistant copper canisters with a cast iron insert to provide mechanical stability. The canisters are to be placed in an underground repository constructed in saturated granitic rock at a depth of around 500 m. Deposition holes where the canisters are placed and tunnels connecting the holes will be backfilled with swelling clay in order to provide a suitable environment for the canisters. For the deposition holes, a bentonite clay is planned to be used, while the deposition tunnels, based on the present design, will be backfilled with some other sort of swelling clay, for instance Friedland clay with rather different properties to the bentonite buffer clay. At emplacement, the clay will be comparatively dry, but as water from the fractured rock surrounding the repository enters the clay system, the clay becomes more saturated and swells to ideally seal off fractures and limit the inflow of water. The low hydraulic conductivity of the clay in the deposition hole (the buffer) ensures diffusion to be the dominating mechanism for radionuclide migration and that the transport time is relatively long. In the deposition tunnel a larger conductivity is allowed and advective transport must be considered in those parts of the system. Modelled point releases (particles) that are assumed to be transported through the system are either corrodants migrating towards the canister and corrosion products and possibly radionuclides in the case of a damaged canister (either initially or later in time, through corrosion). Based on knowledge gained in the current development program, the likelihood for having an initially damaged canister is so low that it was ruled out in SKB's most recent safety assessment for the KBS-3 system [1]. Performing studies on this case is however still relevant as it is suitable for demonstrating the capabilities for system understanding. In the present paper, the migration calculations performed in order to investigate the effect of this are presented. In addition to this canister failure mode (the pinhole case using the terminology of the SR-Can assessment), a case where a low ionic strength glacial groundwater erodes the buffer which results in advective conditions in the deposition hole and earlier failure of the canister, as well as a much faster migration of radionuclides from the canister and canister failures due to earthquake induced secondary shear movements of fractures intersecting deposition holes, were also considered in the SR-Can assessment.

Apart from the safety assessment programme, much of the research activities are targeted at site investigations for a future site for the repository. Two sites are being investigated, one at Forsmark north of Stockholm, and one at Oskarshamn in the south. At the selected site, SKB will apply to build a repository able to store up to approximately 10 000 tonnes of spent fuel. An application, which includes the presently ongoing safety assessment SR-Site, to construct a deep repository will be presented to the regulators in 2009. The SR-Can assessment, where the methodology to be used in SR-Site was demonstrated is part of the preparatory work for SR-Site. The current paper reports the near-field simulations conducted within the SR-Can assessment.

METHODOLOGY

In order to perform the near-field migration calculations a number of supplementary calculations were required to determine conditions in the system calculations that were performed:

Hydrogeological simulations

The different flow properties needed for the migration calculations were obtained through hydrogeological simulations which were mainly performed using the ConnectFlow suite of modelling software [2]. The ConnectFlow suite includes the NAMMU [3, 4] continuum porous medium, CPM, module (where the properties of the fractured media are distributed over porous elements) and the NAPSAC [5] discrete fracture network, DFN, module (where flow in a fracture network surrounded by an in-permeable rock matrix is considered) Output from the hydrogeological simulations provides transport times and path length of advectively transported particles through the different near-field components and flow rates in and location of fractures intersecting the near-field. These calculations are also used to provide flow properties needed for radionuclide migration calculations through the geosphere and discharge points in the biosphere.

Groundwater chemistry simulations

Groundwater compositions are modelled through advection, mixing and chemical reactions with fracture filling minerals and are needed when assigning migration properties of the near-field materials and for determining the solubility limits inside the canister [6]. These calculations produce the spatial and temporal distribution of salt and reference groundwaters, which are then used as input to calculations that include reactions with some fracture filling minerals such as calcite, silica, iron(III) hydroxides, etc, which are abundant and equilibrate relatively quickly with circulating groundwaters. The result of this modelling is an estimate of the detailed groundwater composition. Chemical reactions have been modelled with the PHREEQC code [7].

Solubilities inside the canister

Inside the canister, nuclides are dissolved in water filling the canister void (after the integrity of the canister has been breached). Some nuclides are solubility limited, i.e. the concentration has a maximum level above which the species will precipitate. Radionuclides of a particular element are assumed to share the solubility in proportions to the amount available for each isotope. Given groundwater speciations at different times (obtained from groundwater simulations), the solubilities of different species were calculated using the computer codes PHREEQC and the thermodynamic data base SKB-TDB (Duro *et al.* [8] which is based on Nagra/PSI TDB 01/01 [9]). The codes are able, based on thermodynamic considerations, to predict the solubilities for different groundwater speciations. However, in addition to the calculations, a certain amount of expert judgement is required to evaluate whether the favourable phases calculated from the thermodynamic database are likely to form or not; see Duro *et al.* [10].

Inventory calculations

The initial radionuclide inventory of the spent fuel is taken as that 40 years after operation, reflecting the planned interim storage time of the Swedish nuclear fuel, calculated using the CASMO-4 code. Inventories at later times are then calculated by the radionuclide migration code.

Input data used in the simulations

For performing the simulations, a vast amount of input data is required (both due to the many components in the system but mainly because a large number or radionuclides are simulated). The full input data set is given in the SR-Can Data report [11] and in the SR-Can main report [1]. The majority of the input data were given as distribution functions and the full set of probabilistic input data were generated by the commercial code @Risk.

Migration calculations

Migration calculations were performed using a Matlab implementation of the COMP23 code, a Fortran 77 code that has been used for near-field migration calculations in several previous safety assessments at SKB [12, 13]. Using Matlab together with Simulink, which provides a graphical user-interface to Matlab, is a suitable platform-independent way to perform these types of calculations. The model considers instant release of radionuclides in the fuel at contact with water, release congruently with fuel dissolution as well as solubility limiting phases inside the canister. Outside the canister, transport through diffusion and advection is modelled together with sorption on the different near-field materials. Radionuclide decay with branching chains is considered throughout the system.
Figure 1a shows the near-field part of the KBS-3 system consisting of the canister, buffer material in the deposition hole and the backfilled deposition tunnels, surrounded by rock. The near-field is connected to the far-field through a number of sinks (corresponding to water bearing fractures intersecting the different near-field components) through which the radionuclides are able to leave the near-field and enter the far-field. In the figure these sinks are denoted Q1-Q3.

These three sinks are based on models in previous safety assessments i.e. [14-16], and represent:

Q1 a fracture intersecting the deposition hole. In the discrete fracture model, DFN, used for hydrogeological modelling, several fractures may intersect the deposition hole and these could be located anywhere along its longitudinal axis. However, to simplify the near-field migration model, the flow rates of all fractures intersecting the deposition hole is added and assigned to one single fracture. This fracture is placed on the opposite side of the buffer to the canister defect, at the same vertical position as the defect, hence minimising the distance and the diffusional transport resistance.

Q2 the excavation damaged zone, EDZ, is treated in the hydrogeology model as a thin conductive layer located at the bottom of the deposition tunnel. As further explained [11] the extension of the EDZ in the longitudinal direction is dependant on the excavation method (tunnel boring machine or drill and blast).

Q3 a larger fracture zone intersecting the deposition tunnel. As the deposition tunnel in the hydrogeological model is intersected by several fractures and fracture zones all with different properties, the location of Q3 is obtained from the hydrogeological model by tracking advecting particles released in the centre of the deposition tunnel over each deposition hole. The fracture zone these particles enter is considered to be the Q3 fracture zone. As the distance between the deposition hole and this fracture zone differs, the longitudinal dimensions of the modelled deposition tunnel may be different for different deposition holes.

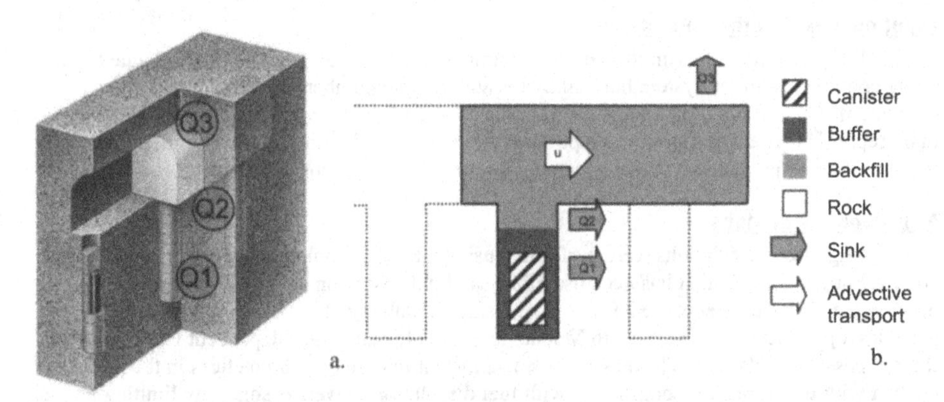

Figure 1. a. The near-field model (canister, buffer in the deposition hole and back-filled deposition tunnels) and sinks, numbered Q1-Q3, connecting the near- and the far-field. **b.** Conceptual model of the near-field. Advective transport with velocity *u* occurs only in the tunnel part of the model (light grey). Neighbouring canister positions not included in the near-field model are indicated by dotted lines. The extent of the model in the downstream direction is given by the hydrogeological simulation and is recalculated for each single canister position.

In Figure 1b, except for the Q1-Q3 sinks, no transport of radionuclides is allowed over the boundaries of the model. Advective transport with a transport velocity *u* obtained from the hydrogeological simulation is considered in the deposition tunnel. At the upstream boundary of the deposition tunnel (the left-hand boundary in Figure 1b) a zero flux boundary condition is implemented and no radionuclides can be transported over that boundary. At the downstream boundary of the tunnel (the right-hand boundary in Figure 1b), particles may leave the model with an advective transport velocity *u*. The symmetric properties of the deposition tunnel with deposition holes up- and downstream are indicated by dotted lines. As radionuclide migration calculations are performed for one canister position at a time these are not considered in the simulations. As all advectively transported material is assumed to leave the model through the Q3 sink, no advective transport is assumed to occur at the downstream side of the Q3 fracture zone. The downstream boundary was therefore placed close to the Q3 fracture zone. Simulating a much larger portion of the tunnel would clearly be possible. It was, however, decided that this was not necessary from a computational and an assessment point of view.

Boundary conditions and sinks in the model

The canister is modelled to have a defect through which radionuclides leave the canister and enter the system of buffer and backfill compartments. Transport out of the canister is diffusional and, in order to reduce the necessary number of elements required, an additional diffusional resistance is added when going from the smaller defect to the larger neighbouring compartment [17]. After entering the buffer-backfill system of compartments, the radionuclides will (either decay or) be transported out of the system through the Q1-Q3 sinks. No diffusion or advective transport is allowed over the other boundaries of the model. In the case of having a small advective transport velocity, diffusional transport may dominate and particles would migrate over the upstream boundary. This is, however, not allowed in the model, both in order to simplify the calculations by keeping the model smaller and to avoid having to specify a diffusional boundary condition. The transport rate out of the near-field is obtained from the product of the concentration and the transport rate, Q. For advectively transported particles the rate is:

$$Q_{advective} = \frac{L}{t}\varepsilon A \qquad (1)$$

Where L and t are the length and advective travel time in the tunnel obtained by tracking particles released from the top of the deposition hole (in the hydrogeological model) to the fracture intersecting the tunnel through which the particles leave the near-field model. ε and A represent the porosity and the cross-sectional area of the tunnel, respectively. Diffusive transport out of the tunnel into the intersecting fracture zone must also be considered. The transport rate for this was assumed to be correlated to the flow in the fracture intersecting the tunnel (obtained from the hydrogeological model).
The equivalent flow-rate, $Q_{diffusional}$, can be written [11]:

$$Q_{diffusional} = 2\sqrt{\frac{4D_w Le\left(Q_f/\sqrt{a}\right)}{\pi}} \qquad U = \frac{Q_f}{w\sqrt{a}} \qquad (2)$$

where L is the length of the fracture zone intersection with the tunnel, U is the Darcy velocity in the fracture zone averaged over the fracture cross-sectional area, e is the transport aperture of the fracture intersecting the tunnel, D_w the diffusivity of water, w is the fracture zone thickness and a is the area of the fracture plane intersecting the tunnel.

Cases simulated

For the safety assessment, hydrogeological simulations for the two sites Forsmark and Laxemar (Oskarshamn) were conducted for a number of different cases investigated for instance different modelling concepts. Of these, four cases, Table 1, were used for migration calculations.

Table 1, Hydrogeological simulation cases progressed to migration calculations.

Case	Modification compared to base case
Forsmark, Base case	-
Forsmark, Conductive tunnel	Increased tunnel conductivity 100 times
Forsmark, Conductive EDZ	Increased EDZ conductivity 100 times
Laxemar, Base case	-

Advective conditions in the deposition hole

In addition to the case described in the current paper, which is denoted in SR-Can as the pin-hole case (actually regarded to be the less probable case) which is based on the assumption that the canister has a defect through the copper shell, simulations were also conducted for a case where high flow rates in fractures intersecting the deposition would result in having advective conditions in the deposition hole and hence a more rapid corrosion than in the previous case. The advective (conditions in the deposition hole) case were simulated in the SR-Can assessment both using the code described in the previous paper and using an analytical approach. For both cases, the geometry is far less complex than in the pin-hole case. This case, and results from the simulations, are described in detail in the SR-Can report [1].

RESULTS

Figure 3 shows the mean annual dose (based on dose conversion factors presented in [1]) for the base case and the variants of the base case for a highly conductive tunnel and a highly conductive EDZ. It can be seen in the figure that the dose from the Q1 transport path is relatively unchanged between the base case and the variants for the two cases. The main reason is that the distance to the Q1 fracture, which is located close to the canister defect, is short and is not influenced by changed properties in other parts of the model. For the Q2 transport path, it is possible to see an increased dose for the case where the EDZ conductivity has been increased (the left hand figure). A corresponding increase in the release through Q3 may also be seen in the case where the backfilled deposition tunnels are given a higher conductivity.

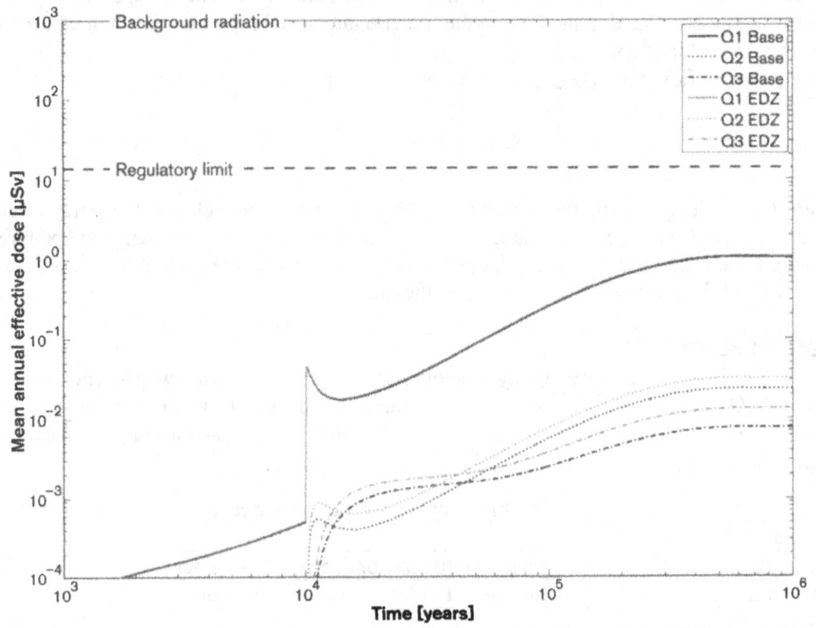

Figure 2. Mean annual effective dose for the base case (black lines) and the highly conductive EDZ case (grey lines) for Forsmark.

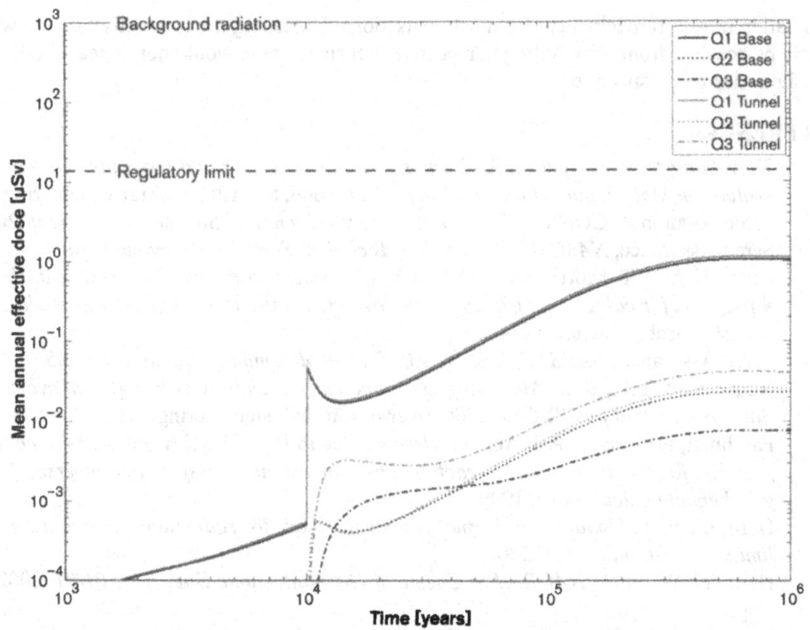

Figure 3, Mean annual effective dose for the base case (black lines) the highly conductive tunnel case (grey lines) for Forsmark.

DISCUSSION

The Matlab/Simulink implementation of the previously used COMP23 code was found to be a suitable code for safety assessments. On a normal desk top PC, the total simulation time for the SR-Can base case (7000-8000 realisations) was a few days. For the case where the conductivity of the backfilled tunnel was increased, the transport length in the tunnel (to the Q3 fracture zone) increased and consequently also the number of compartments required. This increase in number of compartments resulted in longer computational times for that case.
In the assessment calculations, where a total of 37 radionuclides were simulated, the importance of an increase in advective transport in the deposition tunnel varied for different elements. For strongly sorbing radionuclides, the Q1 exit path is the predominating source of release from the near-field, and many of these will not appear in the part of the system where the diffusion length is long. One of the most important radionuclides in the analysis, Ra-226, belongs to that category of nuclides. For elements that do not sorb on the near-field material, the effect of increased advective transport in the deposition tunnel is more important. The most important radionuclide in that group is I-129, which will appear everywhere in the near-field system. I-129 also dominates the release at earlier time, due both to its mobility and also to the fact that part of the I-129 inventory is instantly released upon contact with water (the rapid increase in dose at 10^4 years visible in Figure 2 and Figure 3 is related to I-129 release). It can be seen in Figure 3 that the effect of an increased conductivity in the deposition tunnel is of limited importance, not only to the total release (as Q1 is the dominating exit path), but also to the release through the Q3 sink. It must be noted that, although the conductivity was increased a hundredfold, diffusion is still the

dominating transport mechanism for most realisations. Increasing the conductivity more would clearly be possible from a modelling perspective, but such a case would not be realistic from a repository design perspective.

REFERENCES

1. SKB, *Long-term safety for KBS-3 repositories at Forsmark and Laxemar – a first evaluation Main report of the SR-Can project*. 2006, Svensk Kärnbränslehantering AB.
2. Serco Assurance, *CONNECTFLOW Release 9.0 Technical Summary Document*. 2005.
3. Serco Assurance, *NAMMU, Release 9.0, Technical Summary Document*. 2005.
4. Cliffe, K.A., S.T. Morris, and J.D. Porter, *Assessment model validity document. NAMMU: A program for calculating groundwater flow and transport through porous media*. 1998, Svensk Kärnbränslehantering AB.
5. Serco Assurance, *NAPSAC, Release 9.0, Technical Summary Document*. 2005.
6. Auqué Sanz, L.F., et al., *Modelling of groundwater chemistry over a glacial cycle. Background data for SR-Can*. 2006, Svensk Kärnbränslehantering AB.
7. Parkhurst, D.L. and C.A.J. Appelo, *User's guide to PHREEQC (version 2) – a computer program for speciation, batch-reaction, one-dimensional transport and inverse geochemical calculations*. 1999.
8. Duro, L., et al., *Update of a thermodynamic database for radionuclides to assist solubility limits calculation for PA*. 2005.
9. Hummel, W., et al., *NAGRA/PSI Chemical Thermodynamic Data Base 01/01*. 2002, Nagra.
10. Duro, L., et al., *Determination and assessment of the concentration limits to be used in SR-Can*. 2006, Svensk Kärnbränslehantering AB.
11. SKB, *Data Report for the Safety Assessment SR-Can*. 2006, Svensk Kärnbränslehantering AB.
12. Vahlund, F. and H. Hermansson, *Compulink – Implementing the COMP23 conceptualisation in Simulink*. 2006.
13. Romero, A.L., *The near-field transport in a repository for high-level nuclear waste*. 1995, Department of Chemical Engineering and Technology, Royal Institute of Technology Stockholm Sweden: Department of Chemical Engineering and Technology, Royal Institute of Technology, Stockholm, Sweden.
14. SKB, *Deep repository for spent nuclear fuel. SR 97 - Post-closure safety. Main report - Volume I, Volume II and Summary*. 1999, Svensk Kärnbränslehantering AB.
15. Lindgren, M. and F. Lindström, *SR 97 - Radionuclide transport calculations*. 1999, Svensk Kärnbränslehantering AB.
16. SKB, *Interim main report of the safety assessment SR-Can*. 2004, Svensk Kärnbränslehantering AB.
17. Kelly, M. and K.A. Cliffe, *Validity Document for COMP23*. 2006, Svensk Kärnbränslehantering AB.

Mater. Res. Soc. Symp. Proc. Vol. 1107 © 2008 Materials Research Society

Illustration of HLW Repository Performance: Using Alternative Yardsticks to Assess Modeled Radionuclide Fluxes

Kaname Miyahara and Tomoko Kato
Geological Isolation Research and Development Directorate,
Japan Atomic Energy Agency (JAEA),
4-33 Muramatsu, Tokai, Ibaraki, 319-1194 Japan

ABSTRACT

Complementary indicators have been used in developing a safety case in order to avoid uncertainties in the biosphere modeling used to estimate conventional dose or risk. For example, radionuclide fluxes can be used to evaluate the effectiveness of barrier performance. However, it is difficult to define relevant yardsticks for comparison, because the fluxes of naturally occurring radionuclides due to geological processes vary considerably depending on time and location. This paper discusses the relevance of alternative yardsticks for assessing modeled radionuclide fluxes by selecting yardsticks calculated from fluxes of natural radionuclides at the groundwater discharge point from the geosphere to an aquifer; these are then compared with fluxes of repository-derived radionuclides at the same point. Such yardsticks avoid surface geological processes that may also contribute to natural fluxes, allowing comparison at a suitable, common evaluation point that avoids dependence on site-specific conditions. The effectiveness and robustness of barrier performance is demonstrated using the developed yardsticks and the sensitivity of the analysis to groundwater flux is illustrated.

INTRODUCTION

Complementary safety indicators, such as radionuclide concentration and flux in groundwaters, have been used in developing safety cases for deep geological disposal of radioactive waste based on multiple lines of reasoning [1-3]. These indicators can be used to avoid uncertainties in the biosphere modeling used to estimate conventional dose or risk, which inevitably arise from the assumptions made regarding future human behavior. These indicators can also demonstrate the effectiveness of barrier performance, which is useful for showing redundancy within the natural and engineered barrier system, i.e. that safety is not dependent on the performance of a single component of the disposal system. Such presentations can provide a more transparent means of communicating a safety concept, particularly to non-technical stakeholders, even though alone they are inadequate indicators of total system safety.

For example, the effectiveness of barrier performance can be evaluated based simply on calculated fluxes of radionuclides to the accessible environment, if relevant yardsticks for comparison can be defined. Natural radionuclide fluxes (natural fluxes) due to geological processes (e.g. erosion and river flow) can potentially be used for this purpose [4,5]. However, these fluxes vary considerably depending on location and time. The question is, therefore, whether it is possible to define alternative yardsticks that may be more appropriate to a generic model.

This paper discusses relevant yardsticks based on natural fluxes for illustrating the effectiveness and robustness of the H12 natural and engineered barrier system performance [6,7], currently assessed by a generic model.

THEORY

Fluxes of repository-derived radionuclides (repository-derived fluxes) are estimated on the basis of models of near-field and far-field radionuclide migration. Generally, the following two types of data are required for quantifying natural fluxes: concentrations of the radionuclides of interest in the rock, water, etc. and the rates of the relevant geological processes [2]. An approach to showing the effectiveness of barrier performance is to compare the repository-derived fluxes crossing the geosphere–biosphere interface with their natural equivalents in an appropriate context (yardsticks). When used as yardsticks, natural fluxes due to surface geological processes such as erosion and river flow depend on site-specific conditions and vary considerably with time. Moreover, the evaluation point for such surface geological processes does not necessarily coincide with that for repository-derived fluxes, which can be estimated without considering such surface geological processes. Therefore, when selecting yardsticks to assess repository-derived radionuclide fluxes, the following problems should be solved:

- Avoiding surface geological processes that may contribute to natural fluxes
- Comparison at a suitable, common evaluation point
- Avoiding dependence on site-specific conditions.

Conceptual model

One possibility for circumventing this difficulty is to select yardsticks calculated from fluxes of natural radionuclides at the groundwater discharge point from the geosphere to the aquifer, which are then compared with repository-derived fluxes at the same point. The advantage of this is that it does not include any of the complex, variable, time-dependent and highly site-specific processes in the surface environment.

For the present purpose, the H12 generic conceptual model has been selected as shown in Figure 1. Each waste package is assumed to be located 100 m from a single downstream major water-conducting fault (MWCF) [6,8]. All radionuclides released from the repository (40,000 waste packages) are then assumed to migrate upwards through the MWCF, traveling the shortest distance to the aquifer. At the discharge point from the MWCF to the aquifer, these fluxes are compared with natural fluxes in such groundwater.

It is expected that a structured siting and design process will ensure favorable geological conditions for a repository. This allows the assumption to be made that the groundwater flux at repository depth (e.g. 300 m), which is one of the most sensitive parameters for barrier performance, is not more than the average value in the host rock above the repository. In this study, groundwater flux through the repository is therefore conservatively assumed to be the same as that through the overlying host rock.

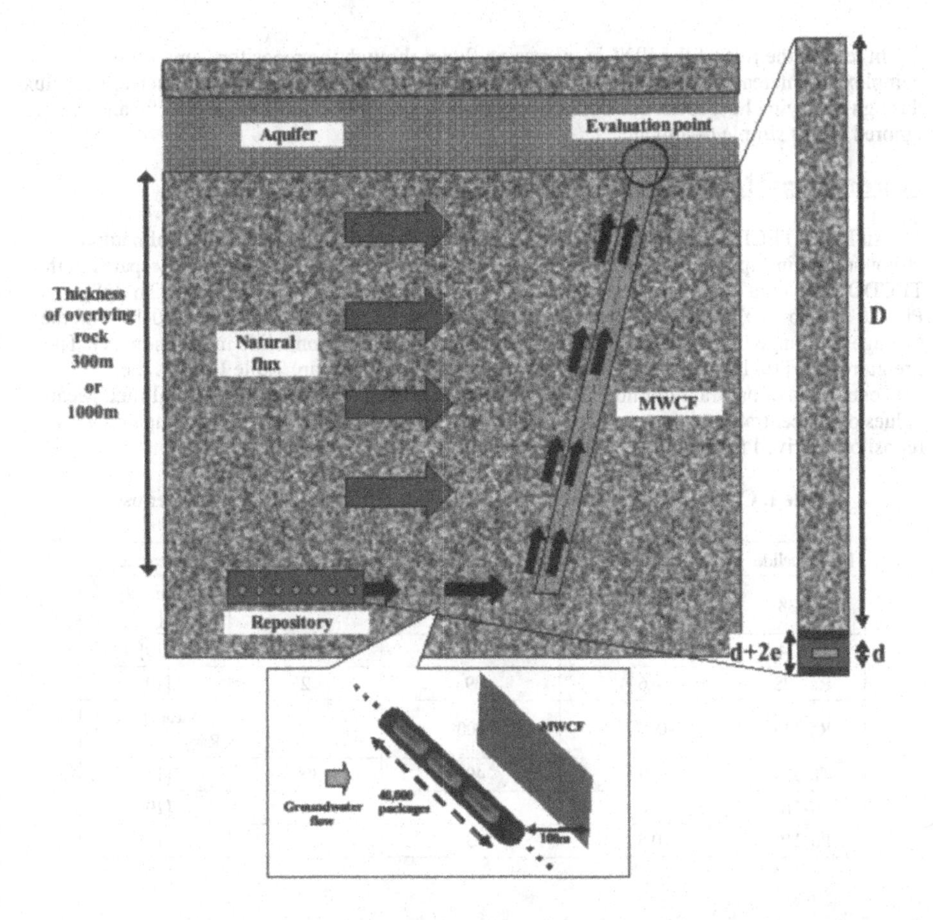

Figure 1. System for calculating radionuclide fluxes and geometry for calculating total groundwater flux

Calculation of natural fluxes

Natural fluxes can be calculated by the following equation based on the geometry in Figure 1:

[Natural flux (Bq/y)] = [Total groundwater flux (m^3/y)] x [Concentrations of naturally occurring radionuclides (Bq/m^3)]

The total groundwater flux is taken to be proportional to the thickness of the overlying rock (D); therefore equals the groundwater flux associated with one waste package x D/(d+2e), where

- Diameter of repository drift (d) = 2.2 m for horizontal emplacement
- Excavation disturbed zone (EDZ) thickness (e) = 0.5 m.

In reality, the role of the EDZ in increasing fluxes through the repository area is rather complex and increases of both permeability and gradient at lesser depths would increase the flux through overlying host rock. Such effects are, however, very site and design specific and are ignored in this simple treatment.

Concentrations of naturally occurring radionuclides

An IAEA TECDOC [9] documents the average and range for concentrations of natural radionuclides in Japan. As the data for concentration in groundwater are relatively sparse in the TECDOC [9], data for several radionuclides (i.e. U-238, U-234, Ra-228, Ra-226, Th-228, Pb-210 and Po-210) were extended by information derived from the literatures [10-13] and the average and range of concentration were set for each natural radionuclide in groundwater. The literature data [10-13] were collected from several regions in Japan. Table I shows the concentrations of natural radionuclides in groundwater used for calculating natural flux. Mean values of concentration data were used for calculating the natural fluxes to be compared with repository-derived fluxes.

Table I. Concentrations of natural radionuclides in Japanese groundwater used for calculating natural fluxes

Radionuclide	Minimum [$Bq\ m^{-3}$]	Maximum [$Bq\ m^{-3}$]	Mean [$Bq\ m^{-3}$]	Reference
U-238	0.2	4	2	Mean: [9] Range: [10],[11]
U-234	2	7	3	[10]
Ra-228	0.6	9	2	[10]
Ra-226	0.07	800	2	Mean: [10] Range :[11],[12]
Th-228	90	400	300	[13]
Pb-210	3	8	6	[10]
Po-210	0.6	3	2	[10]

Comparison with repository-derived fluxes

The repository concept, models and parameters used for calculating radionuclide flux from a repository and natural flux were based on those of the H12 reference case. The following are the main features of the H12 generic model that are relevant for this study. The lifetime of the overpack is assumed to be 1,000 years. The radionuclides released from the vitrified waste diffuse into the buffer and, at the outer boundary of the buffer, are assumed to be instantaneously mixed within the EDZ. All radionuclides released from the engineered barrier system (EBS) are assumed to migrate through the host rock surrounding the repository along a set of representative channels, taking account of the heterogeneity of real fractures and channels with respect to transmissivity. Parameters related to groundwater are important as they affect the outer boundary condition of the EBS and transport times in the geosphere [6,8].

The following flux ratio is used to represent barrier performance:
[Flux ratio] = [Repository-derived flux**] / [Natural yardstick fluxes]

**fluxes from each component of the multibarrier system are considered

Fluxes of radionuclides that do not occur naturally were converted, based on their radiotoxicity for ingestion [14], to an "equivalent flux" of U-238.

A sensitivity analysis with regard to groundwater flux was carried out. Groundwater flux through the EDZ ranges from 1×10^{-5} to 1×10^{2} m^3/y for repository-derived fluxes, which was arbitrarily chosen for the sensitivity analysis based on the value for the H12 Reference Case (1×10^{-3} m^3/y). The values for representative minimum and maximum thicknesses of the overlying rock are selected as 300 and 1000 m.

RESULTS AND DISCUSSION

Sensitivity analysis (1): EBS performance

Figure 2 shows the flux ratio used to compare summed peak fluxes of all radionuclides and radionuclides which have natural counterpart (e. g. U-238, U-234, Ra-226, Pb-210, Po-210, Ra-228, Th-228) from the EBS with natural fluxes. The radionuclide fluxes from the EBS are a maximum of around 2 or 3 orders of magnitude higher than the natural fluxes. Release from the EBS increases almost linearly with groundwater flux up to 10^{-1} m^3/y; above this value, the release becomes insensitive to flux. Therefore, the flux ratio is almost constant up to 10^{-1} m^3/y and decreases above this value.

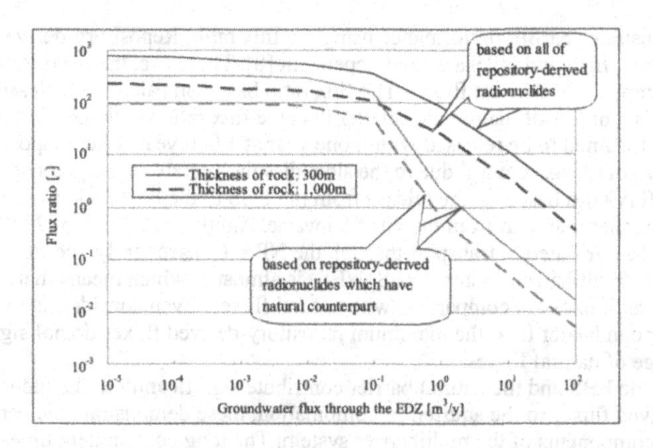

Figure 2. Comparison between radionuclide peak fluxes from EBS and natural fluxes

Sensitivity analysis (2): EBS + geosphere performance

Figure 3 shows the flux ratio used to compare radionuclide fluxes from the natural barrier system (NBS, i.e. the host rock and the MWCF) and natural fluxes. The summed peak fluxes of all radionuclides and radionuclides which have natural counterpart (e. g. U-238, U-234, Ra-226, Pb-210, Po-210, Ra-228, Th-228) from the NBS are comparable with natural fluxes. Performance is dominated by the NBS at low water flow and the EBS at higher flow, resulting in the observed peak. The net effect is a relatively low sensitivity to groundwater flux in the range of about 10^{-3} to 10^{2} m^3/y.

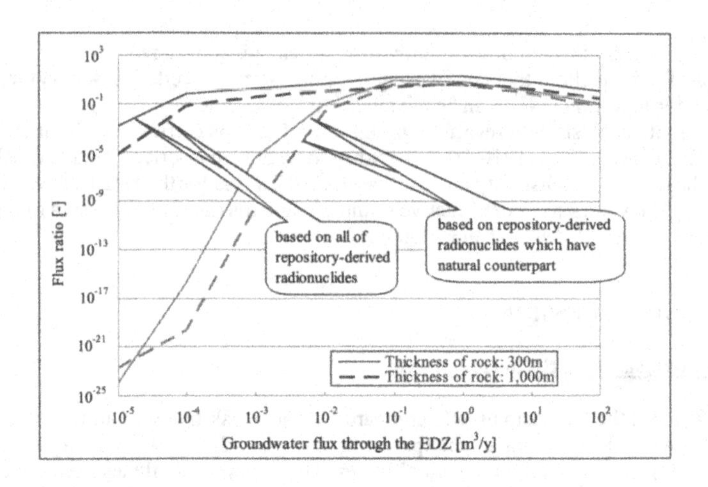

Figure 3. Comparison between radionuclide peak fluxes from NBS and natural fluxes

Illustration of barrier performance

Figure 4 illustrates barrier performance using the flux ratio. Repository-derived fluxes from the vitrified waste, EBS and NBS are time-dependent [8]. Therefore, the maximum fluxes are chosen for the repository-derived fluxes. The flux ratio based on calculated release from the vitrified waste is 5 orders of magnitude smaller than the flux ratio would be if all nuclides in the waste form are assumed to be released within one year at 1,000 years after disposal, indicating the effectiveness of release control due to the slow dissolution rate of the glass waste form. In particular, the flux ratio based on the release from the EBS to the host rock is a further 4 orders of magnitude smaller than that from the vitrified waste. Another reduction of 2 to 3 orders of magnitude can be attributed to transport through the NBS. Consequently, the flux ratio for releases from the multibarrier system (EBS+NBS) is almost 1, which means that repository-derived fluxes are comparable with natural fluxes. Even considering uncertainty or variability of groundwater flux, the maximum repository-derived fluxes do not significantly exceed the range of natural fluxes.

Thus, both the EBS and the natural barrier contribute significantly to the reduction of repository-derived fluxes to the aquifer. In combination, these demonstrate the performance of the individual components of the multibarrier system. The long containment times within the overpack, slow release from the glass matrix and transport through the bentonite and the rock mean that many radionuclides decay to insignificant levels within these barriers, before reaching the aquifer.

At the discharge point, radionuclide releases from the repository will be diluted in the aquifer and the surface environment. This means that natural fluxes are conservatively estimated in this study, without taking into consideration the large volume of these dilution media (e.g. $10^8 m^3/y$ for river flow in the H12 reference case, which would actually result in several orders of magnitude further decrease in the flux ratios shown in Figure 4).

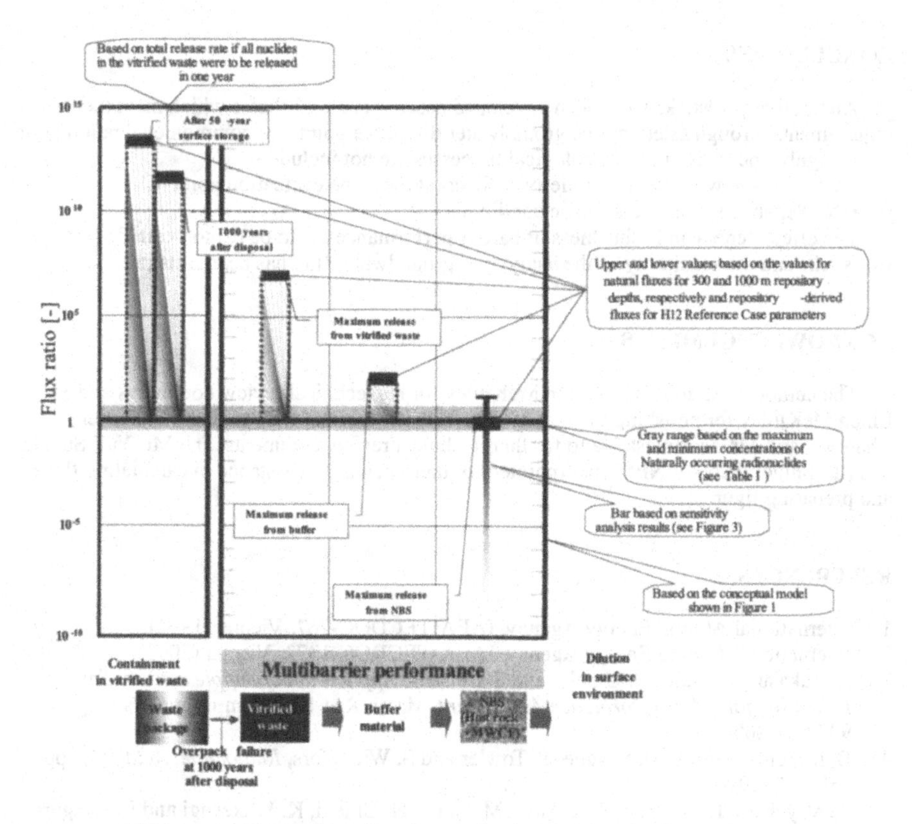

Figure 4. Illustration of barrier performance using the flux ratio

Further considerations for improving the relevance of this approach

This approach is applied to the H12 generic model for performance assessment, in which each waste package is assumed to be located 100 m from the MWCF. To improve realism, it is necessary to evaluate the geometric effect of the layout on repository-derived fluxes, as radionuclides released from many waste packages will take the same transport pathways. Considering overlapping releases from 100 packages results only in a one order of magnitude higher flux than the single package case illustrated - i.e. concentrations are higher but the integrated release from the repository is an order of magnitude lower [6].

Even if site-specific conditions are considered, it is possible to use groundwater discharge points as the evaluation points for repository-derived and natural fluxes. Regarding geological disposal of other wastes such as TRU wastes, it is possible to apply this approach, taking account of the relevant diameter of repository drifts when deriving natural flux ratios.

CONCLUSIONS

Alternative yardsticks have been developed based on natural fluxes which meet the following requirements through selecting the groundwater discharge point as a common evaluation point:

- Highly uncertain surface geological processes are not included
- Comparison with repository-derived fluxes at the same evaluation point
- No dependence on site-specific conditions.

The effectiveness and robustness of barrier performance could be demonstrated using the yardsticks and the sensitivity of the analysis to groundwater flux has been illustrated.

ACKNOWLEDGEMENTS

The authors wish to thank Dr. Ian McKinley for his technical review comments and Ms. Linda McKinley for polishing the manuscript. The authors also wish to thank Prof. Neil Chapman and Mr. Morimasa Naito for their technical review comments, and Mr. Yuji Suzuki and Mr. Shigeru Koo of NESI Incorporated for their technical assistance in calculating fluxes and preparing figures.

REFERENCES

1. International Atomic Energy Agency, IAEA-TECDOC-767, Vienna (1994).
2. International Atomic Energy Agency, IAEA-TECDOC-1372, Vienna (2003).
3. A. Takasu, M. Naito, H. Umeki, and S. Masuda. *Application of Supplementary Safety Indicators for H12 Performance Assessment*, Mater. Res. Soc. Symp. Proc. **663**, pp. 907-917(2000).
4. B. Miller, G. Smith, D. Savage, P. Towler and S. Wingefors, *Radiochim. Acta*, **74**, pp. 289-295 (1996).
5. K. Miyahara, H. Makino, A. Takasu, M. Naito, H. Umeki, K. Wakasugi and K. Ishiguro. *Application of Non-dose Risk Indicators for Confidence-Building in the H12 Safety Assessment*, International Atomic Energy Agency, IAEA TECDOC-1282, pp.113-125, Vienna (2002).
6. Japan Nuclear Cycle Development Institute, JNC TN1410 2000-001; JNC TN1410 2000-004 (2000).
7. H. Umeki. *Key Aspects of the H12 Safety Case*, Mater. Res. Soc. Symp. Proc. **663**, pp. 701-711 (2000).
8. K. Miyahara, H. Makino, T. Kato, K. Wakasugi, A. Sawada, Y. Ijiri, A. Takasu, M. Naito and H. Umeki. *An Overview of the H12 Performance Assessment in Perspective*, Mater. Res. Soc. Symp. Proc. **713**, pp. 177-187 (2001).
9. International Atomic Energy Agency, IAEA-TECDOC-1464, Vienna (2005).
10. N. Tomoyose, A. Tanahara, M. Takemura, A.Toguchi and H. Taira, *Geochem. J.*, **31**, pp.227-233 (in Japanese) (1997).
11. K. Kametani and T. Matsumura, *RADIOISOTOPES*, **32**, pp.18-21 (in Japanese) (1983).
12. K. Kametani, K. Tomura, *RADIOISOTOPES*, **25**, pp.38-40 (in Japanese) (1976).
13. T. Hashimoto and T. Kubota, *RADIOISOTOPES*, **38**, pp.415-420 (in Japanese) (1989).
14. Science and Technology Agency: Notification No.5 (in Japanese) (2000).

Mater. Res. Soc. Symp. Proc. Vol. 1107 © 2008 Materials Research Society

Characterisation of Radionuclide Migration and Plant Uptake for Performance Assessment

Simon A. Mathias, Andrew M. Ireson, Adrian P. Butler, Bethanna M. Jackson,
Howard S. Wheater
Department of Civil and Environmental Engineering, Imperial College London,
London SW7 2AZ, U.K.

ABSTRACT

Unsaturated vegetated soils are an important component in performance assessment models used to quantify risks associated with deep engineered repositories for underground radioactive waste disposal. Therefore, experimental studies, funded by Nirex over nearly 20 years, have been undertaken at Imperial College to study the transfer of radionuclides (Cl-36, I-129, Tc-99) from contaminated groundwater into crops. In parallel to this has been a modelling programme to aid interpretation of the experimental data, obtain parameter values characterising transport in soil and plant uptake and provide new representations of near-surface processes for performance assessment. A particular challenge in achieving these objectives is that the scale of the experimental work (typically $\leq$ 1m) is much smaller than that required in performance assessment. In this paper, a new methodology is developed for upscaling model results obtained at the experimental scale for use in catchment scale models. The method is based on characterising soil heterogeneity using soil texture. This has the advantage of allowing hydrological and radionuclide transport parameters to be correlated in a consistent manner. An initial investigation into the calculation of effective (i.e. upscaled) hydrological and transport parameters has been undertaken and shows the results to be potentially highly (and non-linearly) sensitive to soil properties. Consequently, they have important implications for future site characterisation programmes supporting a proposed underground waste repository.

INTRODUCTION

To date, experimental studies at Imperial College have focussed on understanding the processes that control upward migration and uptake of radionuclides in vegetated soils. Associated with this work has been a modelling programme concerned with analysis of the experimental data and the representation and parameterisation of those processes in physically-based models [2]. This work has now reached an important juncture, as several of the key elements of interest in radioactive waste disposal (i.e. chlorine, iodine, technetium, and selenium) have been studied using laboratory soil columns and (in the case of chlorine) field lysimeters. However, in order for this work to feed into and inform a future safety assessment, a clear, consistent and defensible methodology is needed for producing model-grid scale parameters from laboratory and lysimeter scale data.

There is a vast array of different upscaling techniques in the literature reflected by the correspondingly large number of review papers on the subject (see [7] and references therein). In this paper, we adopt the stream-tube approach to explore the upscaled response of a horizontally heterogeneous field of lysimeter experiments similar to those carried out by [2]. In the stream-tube approach, horizontal heterogeneity in the unsaturated zone is conceptualised as a bundle of independent parallel soil columns, each with different but internally homogeneous soil hydraulic properties [7]. The essential assumption is that the stream tubes are large enough to minimise the effects of local inter-stream-tube interactions. Additionally, it is necessary that the upscaled unit is large enough to include several variogram ranges of local variability [1].

A model was developed for a hypothetical scenario, comprising a one hectare field of 1 m^2 columns of a sandy loam soil (based on experimental lysimeter studies at Imperial's Field Research Station at Silwood Park, Ascot, UK) cropped with a sward of perennial ryegrass (*Lolium perenne*), forced by measured daily potential evapotranspiration and precipitation data at Silwood, from June 1989 to June 1994 with a fixed water table boundary. For the first six months, the water table is assumed to be uncontaminated. After six months, the water table acts as a constant source of Cl-36. Each column is assumed to be internally homogeneous. Two water table depths are considered, 70 cm (consistent with the lysimeter experiments carried out at the field sites [2]) and 200 cm (considered more typical of a well drained agricultural soil). The field is assumed to be vertically homogeneous but horizontally heterogeneous, and flow is assumed to be vertical only. Therefore, the upscaled transport of water and solute can be obtained from the linear sum of all the local columns. The same approach was adopted for steady unsaturated flow studies by [8] and for transient solute transport with steady state flow by [1]. For simplicity, spatial variability is assumed to be caused by variations in the soil texture amongst the individual columns. Therefore, there are only two variable parameters, % clay and % sand. Corresponding hydraulic parameters are obtained using the ROSETTA database [5]. Upscaled soil texture parameter sensitivity is examined using a set of flow and solute transport objective functions.

DEVELOPMENT OF THE NUMERICAL MODEL

The first step is to develop a coupled flow and transport model suitable for simulating field-scale plant uptake of Cl-36 from a fixed water table. In this section, the necessary modelling equations and solution procedure are described for the individual soil columns. A summary diagram of the model setup is shown in Figure 1.

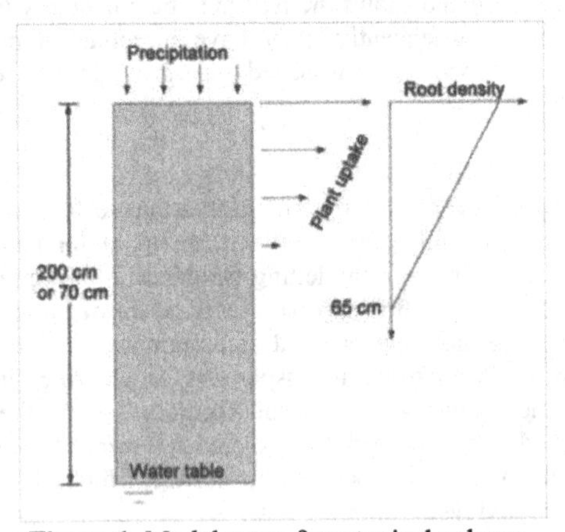

Figure 1. Model setup for a typical column.

One-dimensional flow of water through a partially saturated, single porous medium is described by Richards' equation

$$\frac{\partial\theta}{\partial t} = -\frac{\partial q}{\partial z} - q_s \quad (1)$$

subject to initial and boundary conditions:

$$\begin{aligned} &\psi = z - z_{wt}, && 0 \le z \le z_{wt}, && t = 0 \\ &q = P(t), && z = 0, && t > 0 \\ &\psi = 0, && z = z_{wt}, && t > 0 \end{aligned} \quad (2)$$

where θ [-] is moisture content, t [T] is time, z [L] is depth, q [LT^{-1}] is the flow rate of water per unit area (hereafter referred to as the water flux), $q_s(z,t)$ [T^{-1}] is a sink term due to water uptake by plant roots in response to evapotranspiration, ψ [L] is pressure head, $P(t)$ [LT^{-1}] is the rate of precipitation and z_{wt} [L] is the water table depth (set to 70 cm or 200 cm).

Solution of equation (1) requires the auxiliary functions

$$\frac{\partial\theta}{\partial t} = \left(S_e S_s + \frac{d\theta}{d\psi}\right)\frac{\partial\psi}{\partial t} \quad \text{and} \quad q = -K(\psi)\left(\frac{\partial\psi}{\partial z} - 1\right) \quad (3)$$

where S_s [L^{-1}] is a specific storage coefficient whilst S_e [-] and K [LT^{-1}] are the effective saturation and hydraulic conductivity respectively, found using the van Genuchten functions [6]

$$S_e = \frac{\theta - \theta_r}{\theta_s - \theta_r} = \left[\frac{1}{1 + |\alpha\psi|^n}\right]^m \quad \text{and} \quad K = K_s S_e^L\left[1 - \left(1 - S_e^{1/m}\right)^m\right]^2 \quad (4)$$

where θ_r [-] and θ_s [-] are the residual and saturated moisture contents, K_s is the saturated hydraulic conductivity, $m = 1 - 1/n$ and α [L^{-1}], n [-] and L [-] are empirical parameters.

The sink term due to evapotranspiration inducing water uptake by plant roots, q_s is obtained using a Feddes root-extraction function [4]

$$q_s = PE(t)\frac{\rho_r(z)}{\int_0^{z_r}\rho_r(z)dz} \times \begin{cases} 0, & \psi > \psi_{an} \\ 1, & \psi_{an} \ge \psi > \psi_d \\ 1 - \dfrac{\psi - \psi_d}{\psi_w - \psi_d}, & \psi_d \ge \psi > \psi_w \\ 0, & \psi_w \ge \psi \end{cases} \quad (5)$$

where $PE(t)$ [LT^{-1}] is the rate of potential evapotranspiration, ψ_{an} [L], ψ_d [L] and ψ_w [L] are the anaerobiosis point, the point below which water stress commences and wilting point (set to -5, -400 and -15,000 cm, respectively, see [4]) and ρ_r [L.L^{-3}] is the root density, which in this case is assumed to decrease linearly with depth such that

$$\rho_r(z) = \begin{cases} \rho_{r,\max}\left(1 - z/z_r\right), & z \le z_r \\ 0, & z > z_r \end{cases} \quad (6)$$

where z_r [L] and $\rho_{r,max}$ [L.L^{-3}] are the maximum root depth and the maximum root density (set to 61.9 cm and 7.28×10^{-4} cm/cm^3 respectively, see [3]).

The mass conservation relationship for decay corrected soil water Cl-36 concentration c [ML^{-3}] can be written as

$$c\frac{\partial\theta}{\partial t}+\theta\frac{\partial c}{\partial t}=\frac{\partial}{\partial z}\left[\left(\alpha_L|q|+\theta D_P\right)\frac{\partial c}{\partial z}-qc\right]-2\pi\alpha'\rho_r c \tag{7}$$

subject to the initial and boundary conditions:

$$\begin{aligned}
&c=0, && 0\le z\le z_{wt}, && t=0\\
&qc-\left(\alpha_L|q|+\theta D_P\right)\frac{\partial c}{\partial z}=0, && z=0, && t>0\\
&qc-\left(\alpha_L|q|+\theta D_P\right)\frac{\partial c}{\partial z}=0, && z=z_{wt}, && 0<t\le 6\text{ months}\\
&qc-\left(\alpha_L|q|+\theta D_P\right)\frac{\partial c}{\partial z}=qc_0 && z=z_{wt}. && t>6\text{ months}
\end{aligned} \tag{8}$$

where α_L [L] is the longitudinal dispersivity associated with hydrodynamic dispersion (set to $z_{wt}/10$), D_P [L^{2}T^{-1}] is the pore diffusion coefficient of Cl-36 in the soil (set to 8.6×10^{-3} cm^2/day, see [3]), α' [L^{2}T^{-1}] is the solute root uptake coefficient (set to 7.78 cm^2/day, see [2]) and c_0 [ML^{-3}] is a constant boundary concentration. In all plots in this paper, solute concentrations are normalised by dividing by c_0. The 6 month delay time was adopted to allow a 'hydrological warm up' before solute injection.

The above set of partial differential equations was discretised in space using finite differences. The resulting set of ordinary differential equations (with respect to time) was then integrated using the stiff integrator ODE15s available in MATLAB. From a sensitivity study, it was found that a grid space-step of 5 cm was sufficient for solution convergence.

GENERATION OF SPATIALLY RANDOM SOIL TEXTURE

As stated earlier, for the representation of spatial heterogeneity, only parameters associated with soil texture are assumed to vary. In this paper we look at 10% and 20% textural sub-triangles that encompass soils similar in texture to those found at Silwood (see Figure 2a). By 10% textural sub-triangle, we mean an area on the textural triangle enclosed by an equilateral triangle of 10% side-length. The triangles were discretised into 105 equally spaced soil texture nodes (STN). Each STN represents a possible soil texture. Van Genuchten parameters (θ_r, θ_s, α, n, K_s and L) were then obtained for the 105 STNs using the ROSETTA database [5], which is based on a neural network analysis of 2134 soil samples. Figure 3 shows plots of $\theta(\psi)$ and $K(\psi)$ for the 105 textures of both the 10% and 20% triangles. Also shown are the corresponding Silwood soil curves obtained from lysimeter experiments [2]. Note that the Silwood soil curves were based on an empirical function that truncates the hydraulic conductivity at low matric potentials. It was previously found that without this truncation, when the soil dries out the hydraulic conductivity becomes too low to allow subsequent infiltration. ROSETTA only gives parameters for the van Genuchten functions. Therefore, to replicate this residual behaviour, the van Genuchten hydraulic conductivity function was truncated at ψ = -3000 cm. A sensitivity analysis showed that this had a negligible effect on total water uptake and total radionuclide uptake, which are the main variables of interest. Also note that although the Silwood soil has a much higher porosity than the ROSETTA soils, the effective porosity (i.e. θ_s - θ_r) is comparable.

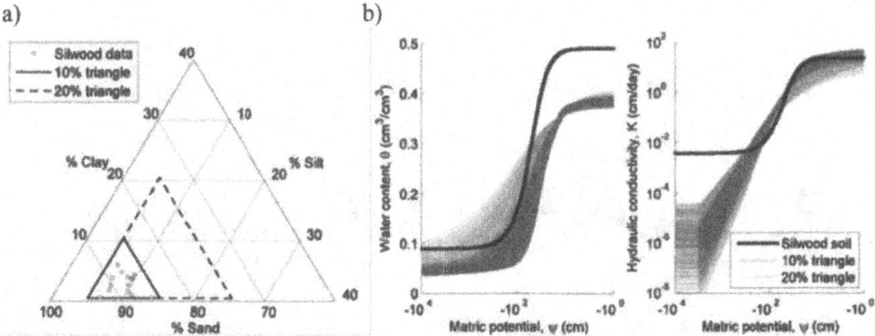

Figure 2. a) Part of textural triangle showing Silwood soil texture data and the 10% and 20% sub-triangles. b) Moisture content and hydraulic conductivity plots against matric potential for Silwood soil (studied by [2]) and the 105 STNs taken from the 10% and 20% textural sub-triangles.

UPSCALED RESPONSE

The upscaled response for the 10% and 20% sub-triangles was obtained by averaging model output from all the individual STNs. Figure 3a summarises the modelling output from the 70 cm 20% triangle simulations. It can be seen in Figure 3a(ii) that all the potential evapotranspiration is satisfied (i.e. actual = potential). This is not surprising, as the water table is just 10 cm below the maximum rooting depth. Figure 3a(iii) shows the lower boundary water flux. A positive flux is drainage, whereas a negative flux is inflow arising from evapotranspiration losses. Significant inflow only occurs in the summer. Figure 3a(iv) shows the total normalised root solute uptake. Corresponding to the water flux, significant uptake only occurs during the summer. In the first year there is no uptake as the water table is largely uncontaminated. Substantially more uptake occurs in 1990 compared to other years as this was a particularly dry year. The main quantities of interest are the cumulative water and solute uptake, for which the mean, the mean ± one standard deviation, σ, the minimum and the maximum are also plotted (see Figures 3a(v) and 3a(vi)). The relative ranges (between the maximum and minimum) are very small. Therefore, it can be said that soil texture has little influence in this scenario.

Figure 3b summarises the modelling output from the 200 cm 20% triangle simulation. In this case, potential evapotranspiration is not satisfied (i.e. actual < potential) during 1990 and 1994. The difference from the 70 cm simulation is due to the additional depth of water table. It can also be seen that the lower boundary water flux is much smoother and of lower amplitude (compare Figures 3a(iii) and 3b(iii)). This is due to the column being sufficiently deep to accommodate the precipitation long enough that it is used up by plant uptake (with the 70 cm column most of it leaches to the water table). Accordingly, root solute uptake is much smaller (compare Figures 3a(iv) and 3b(iv)). Consequently, the relative range is significantly larger than with the 70 cm columns because plant uptake is much more dependant on soil moisture status, and hence soil texture.

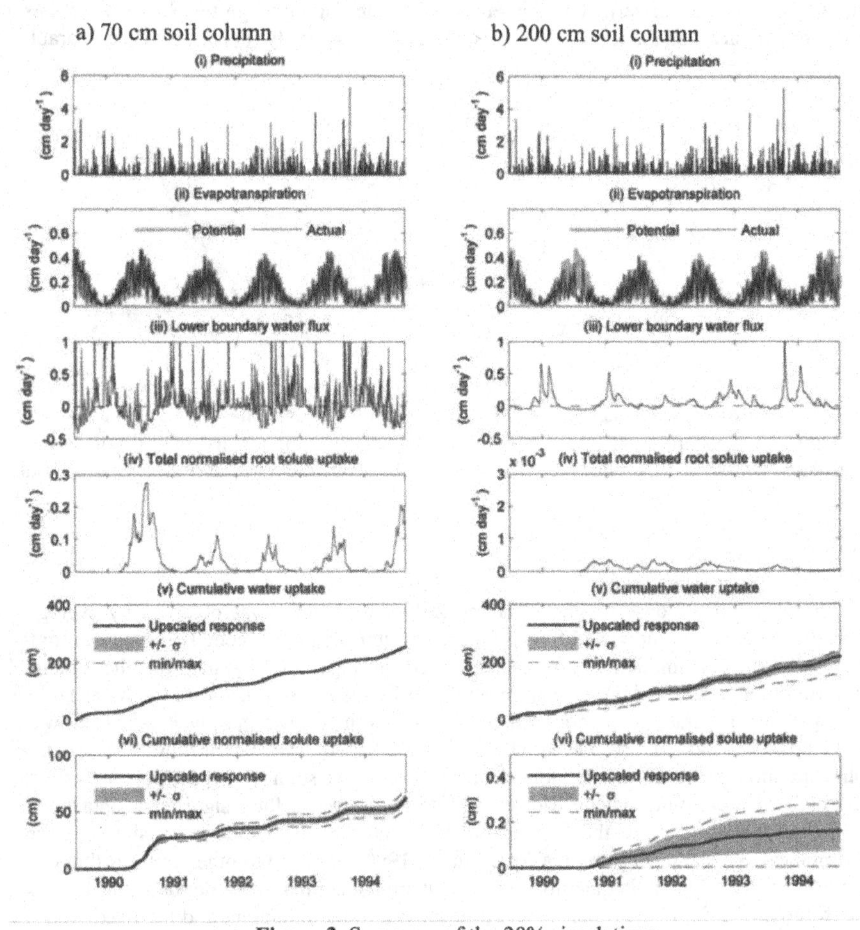

Figure 3. Summary of the 20% simulations.

Recall that the upscaled response is given by the sum of the outputs from the individual columns. To explore the possibility of representing the upscaled response using a single homogenous column, we compare the output of each individual column with the upscaled output, using the following objective functions:

- *NSE*1 - compares moisture content for each depth and time;
- *NSE*2 - compares solute concentration for each depth and time;
- *NSE*3 - compares total water uptake for each time;
- *NSE*4 - compares total solute uptake for each time;

where *NSE* is the Nash Sutcliffe efficiency criterion

$$NSE = 1 - \sum_{n=1}^{N}(u_n - m_n)^2 \Big/ \sum_{n=1}^{N}(u_n - \bar{u})^2 \quad (9)$$

and u_n is an upscaled result, m_n is the result for an individual STN, $\bar{u}$ is the mean upscaled result (i.e. for *NSE*1, $\bar{u}$ is the upscaled moisture content averaged over time and space). The advantage of using the *NSE* as opposed to other error measures (e.g. root mean squared error) is that *NSE* is normalised to the mean response, which allows values of *NSE*1, *NSE*2, *NSE*3 and *NSE*4 to be compared.

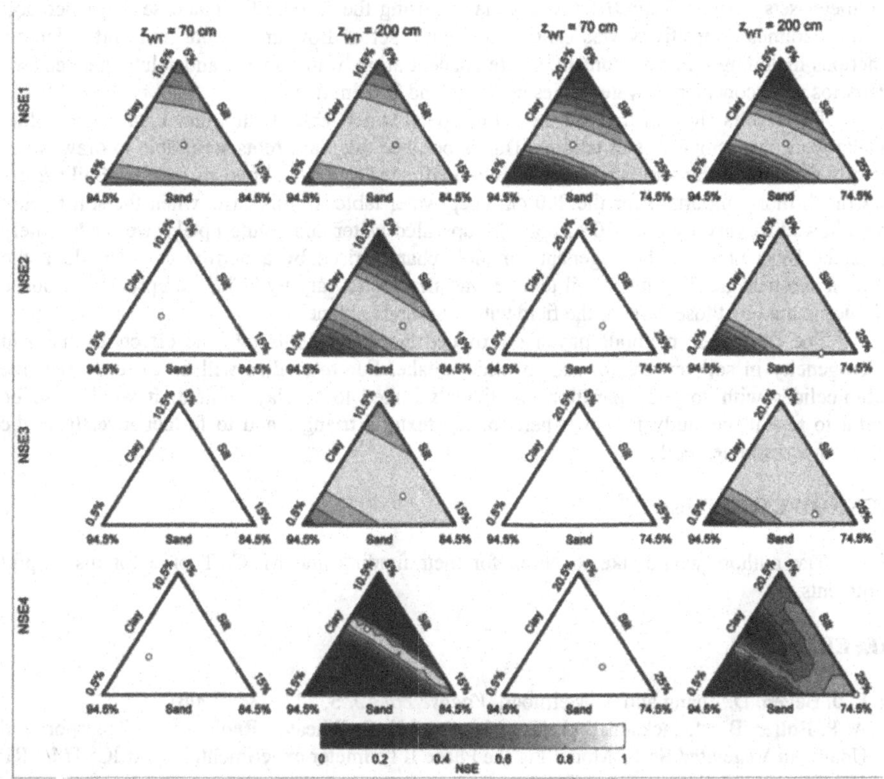

Figure 4. Comparison of NSE1, NSE2, NSE3 and NSE4. The circular markers show the locations of STNs with the highest NSE values.

Figure 4 shows contour plots of NSE1, NSE2, NSE3 and NSE4. It is clear that for the 70 cm soil column, soil texture has no effect on uptake. This is because the water table is so close to the ground surface that the plant roots are able to draw water directly from the water table and are largely unaffected by soil moisture status. For the 200 cm 10% triangle column there is a corridor of good model performance. However, when the triangle is extended to 20% this pattern is lost, and the column that gives the best upscaled representation is the one with the lowest sand content, demonstrating the highly non-linear response to soil texture.

DISCUSSION AND CONCLUSIONS

In this paper, a stream tube methodology has been developed and applied for upscaling heterogeneity in soil texture in the context of Cl-36 uptake by a perennial ryegrass crop from a shallow water table. This involved the development and application of a transient coupled unsaturated flow and solute transport model. Simulations were undertaken for hydraulic parameter-sets derived from 105 textural classes using the ROSETTA database. Upscaled soil texture parameter sensitivity was examined using a set of flow and solute transport objective functions including moisture content, solute concentration, water uptake and solute uptake. Two scenarios were considered, water tables of 70 cm and 200 cm depth.

The results showed that for the 70 cm deep water table, both water uptake and solute uptake were insensitive to soil texture. This is because the plant roots were able to draw water directly from the water table and were largely unaffected by the soil moisture status. Soil texture was much more important for the 200 cm deep water table. In this case, when the soil texture was allowed to vary by only 10%, both the upscaled water and solute uptake were adequately predicted by a range of homogeneous models characterised by a narrow corridor along the textural sub-triangle. When the soil texture was allowed to vary by 20%, the upscaled response was dominated by those parts of the field with low sand content.

The developed methodology has provided helpful insights into the effects of sub-grid heterogeneity in soil texture for radionuclide uptake. Future studies will be to relate reactive radionuclides with sorption partition coefficients linked to % clay content. It would also be useful to repeat the study for other parts of the textural triangle and to further investigate the effect of water table depth.

ACKNOWLEDGEMENTS

The authors would like to Nirex for their funding and M. C. Thorne for his helpful comments.

REFERENCES

1. K. J. Beven, D. Zhang and A. Mermoud, *Vadose Zone J.* **5**, 222–233 (2006).
2. A. P. Butler, B. M. Jackson, J. A. Tompkins and H. S. Wheater, Radionuclide Transport and Uptake in Vegetated Soils: Modelling the Phase II lysimeter experiment, Report ICSTM/NRP 002, Imperial College London (2002).
3. A. M. Ireson, S. A. Mathias and A. P. Butler, A Preliminary Investigation into Upscaling Radionuclide Migration and Uptake in Vegetated Soils, Report ICSTM/NRP 020, Imperial College London (2007).
4. R. A. Feddes, P. Kowalik, K. Kolinska-Malinka and H. Zaradny, *J. Hydrol.* **31**, 13–26 (1976).
5. M. G. Schaap, F. J. Leij and T. van Genuchten, *J. Hydrol.* **251**, 163–176 (2001).
6. M. T. van Genuchten, *Soils, Soil Sci. Soc. America J.* **44**, 892–898 (1980).
7. H. Vereecken, R. Kasteel, J. Vanderborght and T. Harter, *Vadose Zone J.* **6**, 1–28 (2007).
8. J. Zhu and B. P. Mohanty, *Vadose Zone J.* **3**, 1464–1470 (2004).

Mater. Res. Soc. Symp. Proc. Vol. 1107 © 2008 Materials Research Society

Evaluating Chemical Toxicity of Surface Disposal of LILW-SL in Belgium

D. Mallants[1], L. Wang[1], E. Weetjens[1], W. Cool[2]
[1]Belgian Nuclear Research Centre (SCK•CEN), Boeretang 200, 2400 Mol, Belgium
[2]Belgian Agency for Radioactive Waste and Enriched Fissile Materials (ONDRAF/NIRAS), Kunstlaan 14, 1210, Brussel, Belgium

ABSTRACT

ONDRAF/NIRAS is developing and evaluating a surface disposal concept for low and intermediate level short-lived radioactive waste (LILW-SL) at Dessel, Belgium. In support of ONDRAF/NIRAS's assignment, SCK•CEN carried out long-term performance assessment calculations for the inorganic non-radioactive components that are present in LILW-SL. This paper summarizes the results obtained from calculations that were done for a heavily engineered surface disposal facility at the nuclear zone of Mol/Dessel. The calculations address the migration of chemotoxic elements from the disposed waste to groundwater.

Screening calculations were performed first to decide which non-radioactive components could potentially increase concentrations in groundwater to levels above the groundwater standards. On the basis of very conservative calculations, only 6 out of 41 chemical elements could not be classified as having a negligible impact on man and environment. For each of these six elements (B, Be, Cd, Pb, Sb, and Zn), the source term was characterized in terms of its chemical form (i.e., metal, oxide, or salt), and a macroscopic transport model built that would capture the small-scale dissolution processes relevant to element release from a cementitious waste container. Furthermore, reliable transport parameters in support of the convection-dispersion-retardation (CDR) transport calculations were determined. This included derivation of (1) solubility for a cementitious near field environment based on thermodynamic equilibrium calculations with The Geochemist's Workbench, and (2) distribution coefficients based on a compilation of literature values. Scoping calculations illustrated the effects of transport parameter uncertainty on the rates at which inorganic components in LILW-SL leach to groundwater.

INTRODUCTION

In the period 1998 to 2006, termed the *preliminary project phase*, ONDRAF/NIRAS worked in formal partnerships with representatives of four municipalities that hosted nuclear facilities to jointly develop disposal concepts for implementation at a site chosen within their municipal boundaries. Following technical studies and social and political dialogue in Belgium, the government has decided in 2006 that plans should be developed for a near surface disposal facility to accept LILW-SL (category A waste) at a site within the Mol/Dessel nuclear zone. In the *project phase*, now ongoing, detailed design and safety assessment studies are being carried out starting from the surface facility design developed within the *preliminary project phase* by the STOLA-Dessel partnership. This design will be the basis for the evaluation of the chemical toxicity of LILW-SL.

LILW-SL is known to contain a large fraction of inorganic non-radiological components [1]. Long-term management of such waste is done by isolation from man and environment. This isolation is based on physical and chemical encapsulation of the waste in a disposal facility. In the present study long-term impact from non-radiological components present in category A waste is being investigated for a near-surface disposal facility in the Mol/Dessel nuclear zone. The assessment is done by verifying whether or not the very slow release of such components will lead to a significant increase of their concentration in shallow groundwater, and comparison is made between estimated groundwater concentrations versus groundwater quality standards.

INVENTORY OF NON-RADIOLOGICAL COMPONENTS

Periodically, ONDRAF/NIRAS is updating its inventory of existing waste and expected future productions (until 2070). Such an inventory has been established in 1998 [2] and 2003 [3] and will be further updated in 2008-2009. Based on the 2003 inventory a volume of 70500 m^3 category A waste is expected. The main waste streams are as follows: 50% from the dismantling of the nuclear power plants (NPPs), 23% from dismantling other installations such as research installations, 20% operational NPP waste, and 7% generated by operating other installations of the nuclear fuel cycle.

The chemical composition (i.e., organic and inorganic non-radiological components) of category A waste corresponds to nearly 166000 tonne, of which concrete, sand and cement account for the bulk of total mass [3]. These components are used in immobilization matrices and will be generated during decommissioning of nuclear facilities. Based on the 1998 category A inventory, a total of 41 non-radiological elements has been identified, belonging to one or more of the following five categories:

- Oxides: Nearly 73% of total mass belongs to this category. Main components are sand and cement (decommissioning waste). Incinerator ashes from CILVA and EC installations of BELGOPROCESS also belong to this category (for a typical composition, see Table I).
- Metals: Total mass represented by metals amounts to 21%, mainly carbon steel.
- Salts: Despite their small mass (0.08%), they are still important in the safety evaluations because they usually have a high solubility and low sorption owing to their ionic chemical form.
- Cellulose: Mainly super-compacted and cemented paper, wood, and cotton (~ 0.2% of total mass). This category has not been considered further in this study, because this category is not considered to contain chemotoxic components. Note, however, that degradation products of cellulose may affect the sorption and mobility of certain elements.
- Plastics: To this category belong the organic mono- and polymers such as polyvinylchloride, latex, etc. (~2.5 % of total mass). They are considered not to be potentially chemotoxic, but their degradation products likely have similar effects as those from cellulose.

Table I Composition of CILVA and Evence-Coppée ashes.

Element	Ashes CILVA	Ashes Evence-Coppée	Element	Ashes CILVA	Ashes Evence-Coppée
Al	1050%	695%	Mg	121%	80%
B	30%	20%	Mn	28%	19%
Ba	69%	46%	Na	131%	87%
C	500%	3690%	Ni	34%	22%
Ca	693%	461%	O	3820%	2540%
Cd	22%	14%	Pb	55%	36%
Cl	159%	106%	S	79%	52%
Cr	26%	18%	Sb	15%	10%
Cu	281%	187%	Si	1530%	1010%
F	18%	12%	Sr	17%	11%
Fe	905%	602%	Ti	183%	121%
Ga	18%	12%	Zn	120%	80%
K	102%	68%			

SCREENING CALCULATIONS

Inorganic inventories are known to be very heterogeneous with a large number of elements present in different chemical forms [1]. Many of these elements, however, are present in very small quantities, have a low solubility, and/or are already present in groundwater and/or surface water, soils and sediments at relatively large concentrations. To identify elements that could possibly lead to an important concentration increase, a number of simplified screening analyses was carried out. Elements that comply with one of five screening criteria are considered as unimportant and hence screened out for further analysis. Five screening criteria were applied, three quantitative and two qualitative [4]:

- Comparison between solubility limit in the concrete conditioned waste and quality standards for groundwater (VLAREM II, [5]). Elements whose solubility limit in a high pH environment is smaller than their maximum tolerable concentration in groundwater are screened out.
- Comparison between a very conservative estimate of element concentration in groundwater and quality standards for groundwater. The conservative estimate is obtained by assuming a complete and instantaneous dissolution of the entire chemical inventory in a relatively small volume of a sandy aquifer (270000 m^3), with dissolved elements being partitioned between solid and water phase using a conservative distribution coefficient K_d.
- When both stable and radioactive isotopes for one and the same element were present, radiotoxicity was compared with chemotoxicity. If the element was more radiotoxic than chemotoxic, it was screened out here, because it is then more relevant to evaluate its radiotoxicity in a radiological assessment.
- A considerable number of elements are ubiquitous in nature, such as calcium, carbon, chlorine, magnesium, aluminium, silicon, etc. They are thus naturally present in soil and groundwater at relatively high concentrations, while in the disposal installation most of these elements mainly occur in packaging (concrete) and conditioning (grout) materials

(e.g., S is present for 72% of its mass in the conditioning matrix). As a result, their release from the engineered barriers will be extremely slow, such that natural levels likely will not be exceeded. In other words, their presence in category A waste will not result in deterioration of groundwater quality.

- Qualitative criteria (i.e., aesthetic) were used to screen out elements such as iron and manganese: water quality guidelines for these elements are based more on aesthetic (taste, smell, or colour of water) and economic considerations than on toxicity considerations.

When these five criteria were applied to each of the 41 elements using the 1998 chemical inventory [2], most elements were screened out (Figure 1). For the few elements that did not comply with any of the criteria (B, Be, Cd, Pb, Sb, Zn), detailed calculations by means of a numerical transport model were carried out to assess the impact on groundwater quality.

DETAILED ASSESSMENTS

Waste forms considered

The inorganic non-radiological elements considered in this paper may be present in different chemical forms, but mainly as metals, oxides, and salts (Table II). Most important waste streams that contribute to the inorganic components inventory are described below:

- Incinerator ashes: Metals such as Cd, Pb, and Zn are mainly present in incinerator ashes as oxides. Also the element B is present as oxide, most probably as B_2O_3. These oxides are likely to reside in Al-Ca-Fe silicates [6]. Within the CILVA installation at BELGOPROCESS, incinerator ashes are put in 200 litre C-steel drums that are subsequently super-compacted. The compacted drums are collected in a 400 litre drum and grouted.
- Metals: Most important chemotoxic elements such as Be, Cd, Pb, and Sb are present as metallic pieces. Zinc originates for one third of its total mass from galvanised carbon steel drums (200 and 400 litre drums). Zinc is not considered in the Flemish environmental legislation as being chemotoxic [5], it is rather considered as an unwanted chemical at elevated concentrations. Lead originates mainly from components used in shielding radioactive sources. It is usually present as a lead plate or brick. Metallic beryllium waste will be produced after dismantling of SCK•CEN's BR2 reactor, cadmium is used in neutron shielding, and antimony is alloyed with lead as a material in shielding applications.
- Glass: Glass and borosilicate glass are important sources of boron (boron in glass exists as B_2O_3). Glass and borosilicate glass contain, respectively, 3.3 and 3.1 weight % of B [7]. Both glass types differ mainly in their Na, K, and Ba composition. To simplify calculations, dissolution of glass (and thus B) is assumed to occur in the same way as dissolution of oxides from incinerator ashes.

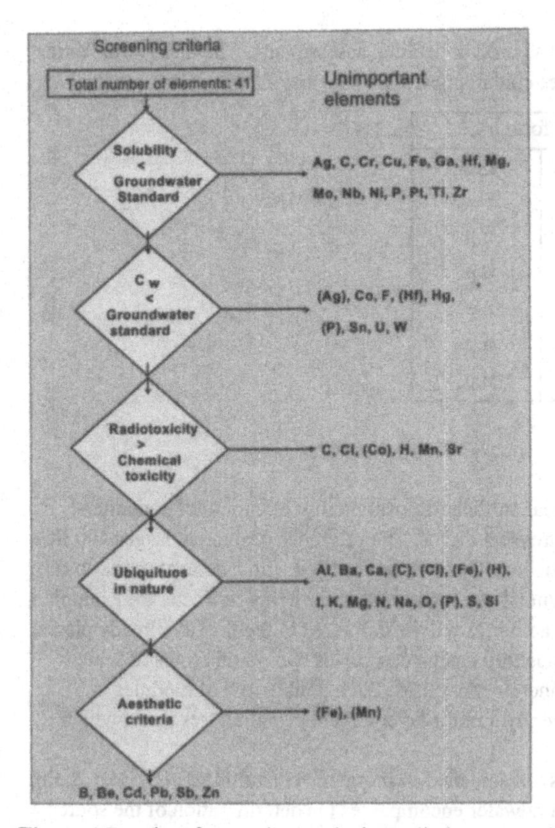

Figure 1 Results of screening analysis applied to non-radioactive elements present in LILW-SL.

- Evaporator concentrates: Boric acid (H_3BO_3) is used in the primary cooling circuit of nuclear power reactors to control the reactivity by absorption of excess neutrons [8]. Upon treatment of the water of the primary cooling circuit, boric acid together with radioactive substances is separated via an evaporation and condensation step. The resulting concentrate contains mainly sodium borate ($Na_2B_4O_7.10H_2O$). When the pH is above 10, boron is mainly present as dissociated boric acid. In Belgian NPPs this concentrate is mixed with grout in 400 litre drums. Evaporator concentrates produced at Belgian NPP's Doel and Tihange have a mean boron concentration of respectively 55000 and 65000 mg litre^{-1} [9]. Addition of lime ($Ca(OH)_2$) to boron containing waste is done to enhance hardening of the cement, by generation of insoluble or slightly soluble compounds [10].

Table II Volume conditioned waste (m^3) as used in detailed assessments. Chemical form "oxide" with 2300 m^3 represents incinerator ashes containing Cd, Pb, Sb, and Zn (n.p. = not present).

Element	Chemical form		
	Metal	Oxide	Salt
Boron (B)	n.p.	12242	2720
Beryllium (Be)	84	n.p.	n.p.
Cadmium (Cd)	629	2300	n.p.
Lead (Pb)	9738	2300	n.p.
Antimony (Sb)	185	2300	n.p.
Zinc (Zn)	58244	2300	n.p.

Disposal facility and models

Wastes will be delivered to the near surface disposal facility in variety of containers, although most will be in standard 400 litre steel drums. The reference design envisages 400 litre drums to be placed in a standard concrete box. This will be in-filled with a cement grout, to form a monolith for emplacement in the repository, see Figure 2. Larger items of waste, for example from decommissioning and dismantling activities will be delivered to the facility already placed into concrete boxes. The preliminary repository design comprises two double rows of sealed concrete disposal modules filled with concrete monoliths. Each double row of modules is protected against penetration by rainwater by a multi-layer cover several metres thick, which forms a tumulus.

Detailed calculation of the release of screened-in inorganic contaminants (B, Be, Cd, Pb, Sb, Zn) from the disposal facility to groundwater encompasses (1) determination of the source term and (2) subsequent application of the CDR-model. The source term determines the initial pore-water concentration of the chemical species in the internal volume of the monolith. The chemical species will subsequently migrate through the concrete walls of the monolith and through the engineered structure of the module. The unit of calculation is a simplified 2D-model for the monolith, which was tested for simplification versus more complex 3D models (Figure 2). Total fluxes to groundwater representative of the entire inventory were obtained by integrating fluxes over all monoliths. Contributions of the module floor and filled inspection rooms to containment were disregarded because, for non-decaying contaminants with a very long leaching behaviour, peak fluxes are relatively insensitive to such additional engineered barriers. Furthermore, the model thus invoked is conservative.

Source term calculations

Among the processes considered in the release modelling, the source term calculations are a key component for they determine the maximum dissolved concentration of contaminants in the monolith pore-water. Since the maximum contaminant flux from each monolith is proportional to the maximum dissolved concentration in the source zone, it is vital to have accurate values of such concentrations. While developing the source term models for the different chemical forms, care was taken to use conservative hypothesis and/or parameter values whenever insufficient information was available to do otherwise. Dissolved concentration in the source zone is in many cases governed by the solubility limit, which was obtained by means of geochemical calculations. Solubility calculations are strongly dependent on the imposed chemical environment (e.g., pH, Eh, pore-water composition). Also, because the calculations are done typically for thousands to tens of thousands of years, assumptions have to be made about the stability of the chemical environment providing for low solubility and high sorption. Although within this period chemical degradation of concrete will have resulted in a considerable decrease in pH and Ca/Si ratio, sorption and solubility are often not negatively impacted (see further).

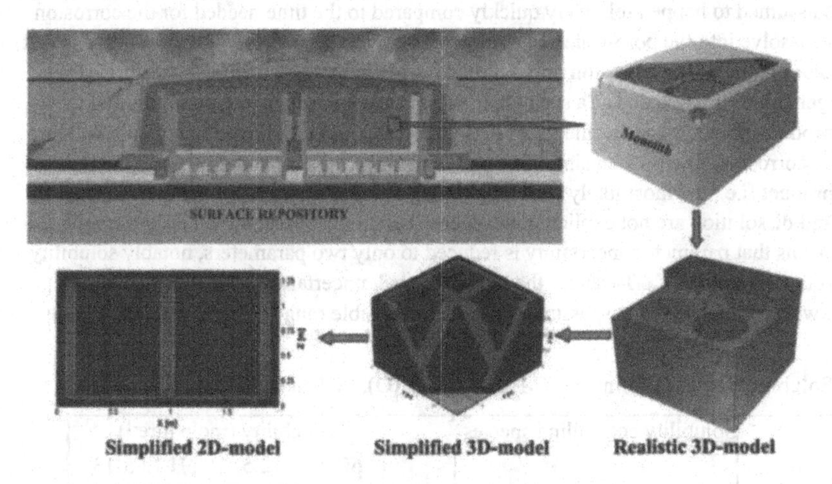

Figure 2 Cross-section of near-surface disposal facility and disposal containers (monoliths) (top), and derived numerical models of varying complexity used for leaching calculations (bottom).

The smallest unit for the overall source term calculation is the monolith internal useful volume, where waste and backfill material have been considered as a homogeneously mixed entity, resulting in a uniform concentration distribution inside the monolith. The physical and chemical properties of the concrete backfill and matrix were assumed identical. Regarding the dissolution processes relevant for most waste forms, it is conservatively assumed that all metals are completely and instantaneously dissolved (unless the metal is solubility limited, see further). This simplified source term model does not explicitly account for a realistic corrosion/dissolution of metals from solidified waste components. However, due to the slow release from the monolith owing to slow diffusion inside the concrete walls and backfill, and due to the strong retardation on concrete, the release from the monoliths will mimic the slow leaching process. As will be further demonstrated in detail, the key parameters used in determining the dissolution and release from the conditioned waste are the solubility and the sorption onto the solid concrete phases.

Source term models were developed for the three chemical forms identified, i.e. salts, metals, and oxides. Although in reality the physico-chemical processes involved in the gradual dissolution of these three components may be quite different, long-term modelling of the such processes requires a simplified approach using relatively well defined parameters.

The major simplification in the source term model concerns the dissolution of metals and oxides. For metallic waste, the corrosion process that would normally lead to a gradual release of ionic species from the metallic form is simplified into a solubility controlled dissolution of corrosion products that would exist on the surface of the metallic pieces. The corrosion process itself and the associated generation of corrosion products (usually hydroxides, carbonates, or silicates) is assumed to happen relatively quickly compared to the time needed for the corrosion products to dissolve into the pore-water. In other words, when the corrosion rate is at least as fast as the dissolution rate of the corrosion products, the corrosion process itself can be skipped (i.e., it is no longer explicitly included as a separate process) and replaced by the dissolution of the corrosion products. Since also the kinetics of dissolution was not considered in our calculations, the complex corrosion process then simplifies into one where the only unknowns are the type of corrosion product (i.e., the most likely mineral phase) and its solubility. As both the kinetics of corrosion and dissolution are not explicitly considered here, the approach used is conservative. This also means that parameter uncertainty is reduced to only two parameters, notably solubility and the sorption parameter K_d. To assess the impact of such uncertainty, scoping calculations were made with contrasting values, assumed to cover a sensible range of parameter uncertainty.

Table III Solubility limits C_s for metals (M) and oxides (O), * taken from [12].

Element	Solubility controlling species		Solubility (mole litre^{-1})	
			pH 11 – 12.5	pH 12.5-13
Beryllium (Be)	BeO	M	3×10^{-6}	3.3×10^{-5}
Lead (Pb)	$Pb_3(CO_3)_2(OH)_2$	M, O	10^{-3}	10^{-2}
	$PbSiO_3$	M, O	10^{-3}	5×10^{-3}
Cadmium (Cd)	$Cd(OH)_2$	M, O	3.8×10^{-7}	7.3×10^{-7}
	CdO	M	8.7×10^{-6}	1.7×10^{-5}
Antimony (Sb)	Sb_4O_6	M, O	3×10^{-4}	1.1×10^{-3}
	$Sb(OH)_3$	M, O	4.8×10^{-7}	1.4×10^{-6}
Zinc (Zn)	Zn_2SiO_4	M	2×10^{-5}	1.2×10^{-4}
	ZnO	M	2.3×10^{-5}	2×10^{-4}
	$Zn(OH)_2$	M	1×10^{-4}	8.3×10^{-4}
	$CaZn_2(OH)_6.2H_2O$	O*	3.3×10^{-4}	8.7×10^{-4}

For oxides it is assumed that dissolution is governed by pure mineral phases. Source term modelling is restricted to the determination of the most likely mineral phase present and its solubility. The geochemical calculations done by means of the Geochemist's Workbench [11] and the LLNL database v8r6 for the oxides are basically the same as those for the metals. From these calculations always two or more possible mineral phases were retained for transport modeling (Table III). In such a way the uncertainty about the type of solubility controlling mineral and its solubility was included: a reference calculation was made with the mineral having the highest solubility, and alternative calculations were made with the other less soluble mineral(s). In the same way effects of pH on solubility were included, i.e. by carrying out one CDR-model based leaching calculation with a solubility representative for the pH range 12.5-13 (i.e. the period of little or no chemical concrete degradation), and a second calculation with a lower pH representative for a period characterized by more pronounced chemical degradation (pH range 11-12.5).

For the only salt present (B), the current geochemical calculations could not identify a mineral phase that would impose a solubility limit. In other words, all boron salts were assumed to be completely and instantaneously dissolved (conservative assumption in view of the poorly soluble calcium compounds present). Furthermore, because the geochemical calculations could not be carried out using the appropriate geochemical boundary conditions (i.e. the ionic strength imposed during the modelling was lower than the presumed real one, since the activity coefficient corrections were based on the Debye-Hückel method), considerable uncertainty exists as to whether boron salts are solubility limited or not.

Finally, the same approach is followed to define the source term models for salts, metals, and oxides, using following 3 steps (step 1 and 2 can be combined into one step because it concerns geochemical calculations with the same code):

1. Determine whether solubility controlling mineral phases (including salts) are present in alkaline and slightly reducing conditions.
2. Calculate the solubility for the chosen solubility controlling mineral phase including salts and determine a best estimate value for the K_d.
3. Calculate the initial dissolved concentration in the pore-water of the hypothetical homogeneous waste zone using the previously derived best estimate K_d and solubility.

Solubility limits calculated using The Geochemist's Workbench ([11]) and the LLNL database (v8r6) are considered as best estimate. Table III shows the best estimate solubility values for each element and for each solubility controlling species identified by the code. For most elements two or more species have been given, because scientific evidence was lacking to exclude one or two of the other. Note that solubility in the pH range 12.5-13 is always higher than the values for the pH range 11-12.5. This is due to the amphoteric character of the metals (see Figure 3). At high pH, speciation of such metals is determined by a number of hydrolysed species. Because of the large stability of such species, and because of the large concentration of the ligand OH^-, solubility will be high at high pH, but lower at lower pH. Long-term chemical degradation of concrete characterized by pH decrease (or decrease in Ca/Si ratio) will thus not immediately result in a reduced performance of concrete engineered barriers. This is also true for the sorption process, where K_d values typical of pH 12.5 may be used up to several tens of thousands of years to represent degraded concrete up to pH of ~10 (Ca/Si ratio of 0.7) [13].

Distribution coefficients K_d were compiled from the literature, and best estimate values together with the best-fit probability density function are given in Table IV.

Table IV Distribution coefficient K_d (litre/kg) used in leaching modeling [4].

Element	Best estimate (distribution)	Minimum	Maximum
B	1 (log-uniform)	0.01	100
Be	0.032 (log-uniform)	0.001	1
Cd	10 (log-uniform)	1	100
Pb	500 (log-uniform)	50	5000
Sb	10 (log-uniform)	1	100
Zn	100 (log-uniform)	10	1000

Calculated fluxes for zinc

The approach for calculating releases of inorganic contaminants is illustrated for the element zinc (i.e., zinc oxides and metallic zinc). Fluxes from a single monolith for best estimate C_s and K_d (or the derived retardation factor $R = 2751$) values are shown in Figure 4. Effects of uncertainties in K_d (or R) and C_s have been included, by using, respectively, minimum and maximum K_d's (R's) from Table IV and a lower solubility (2.5×10^{-5} mol litre^{-1}) for the solubility controlling mineral calciumzincate $CaZn_2(OH)_6.2H_2O$, taken from [14]. Source term calculations revealed that zinc is solubility limited only for the lowest solubility in combination with $R = 276$ and $R = 2751$. For all other parameter combinations, sorption onto concrete is high enough to reduce zinc pore-water concentration below the solubility limit. As a result of parameter uncertainty, differences in maximum flux to groundwater may be easily two orders of magnitude. Figure 4 further shows effects of concrete degradation on leaching of metallic zinc for two potential corrosion products (minerals $Zn(OH)_2$ and Zn_2SiO_4).

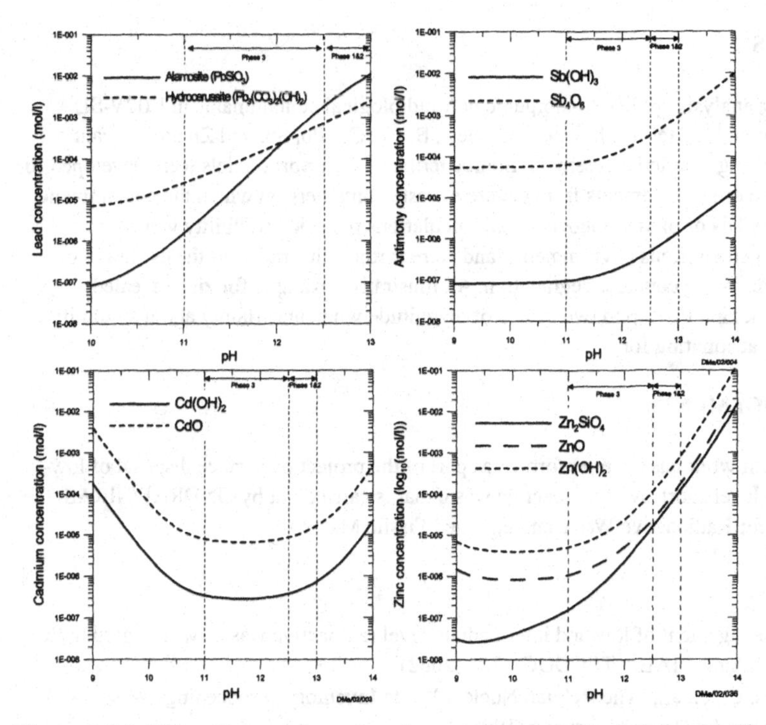

Figure 3 Solubility-pH dependency based on geochemical calculations with The Geochemist's Workbench [11].

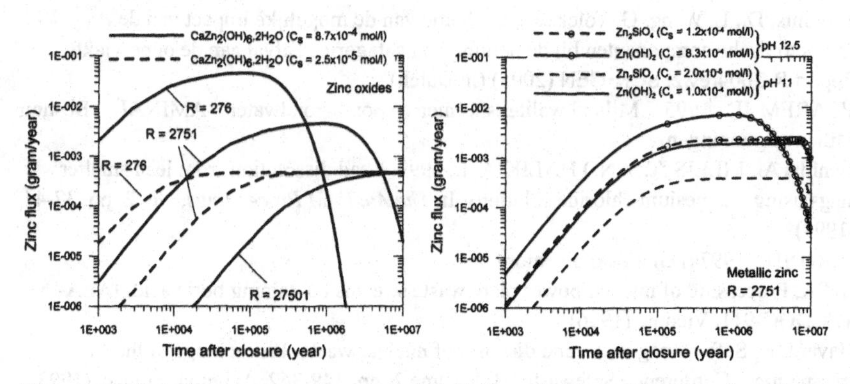

Figure 4 Effect of uncertainty about R and C_s (left), and effect of concrete degradation (pH 11 versus 12.5) for two different mineral phase (right) on zinc leaching for a single monolith.

CONCLUSIONS

Screening analysis applied to inorganic non-radiological contaminants in LILW-SL indicated that from a long term safety point of view, B, Be, Cd, Pb, Sb, and Zn are relevant components requiring detailed assessment. Source term and transport models were developed for calculating release of such elements from concrete waste containers, in which the contaminants were present as metals or oxides. Geochemical calculations provided solubility values representative of cementitious environments, and were further illustrative of the evolution of solubility with increasing concrete degradation. An illustrative example for zinc revealed differences in peak zinc flux up to two orders of magnitude when uncertainty about solubility and sorption was accounting for.

ACKNOWLEDGEMENTS

The present work has been performed as part of the project on surface disposal of low- and intermediate level short-lived radioactive waste that is carried out by ONDRAF/NIRAS, the Belgian Agency for Radioactive Waste and enriched Fissile Materials.

REFERENCES

1. IAEA, Management of low and intermediate level radioactive wastes with regard to their chemical toxicity, IAEA-TECDOC-1325 (2002).
2. Cosemans, C., et. al., 'The Belgian Nuclear Waste Inventory', Proceedings Waste Management '98, Tucson, Arizona (1998).
3. Cosemans, C., Braeckeveldt, M., De Preter, P., 'The inventory of radioactive waste as an integrated part of low-level radioactive waste management system, ICEM05, (2005).
4. Mallants, D., L. Wang, G. Volckaert, Evaluatie van de mogelijke impact van de chemotoxische componenten bij de berging van categorie A afval aan de oppervlakte, Report R-3604rev.2, SCK•CEN (2005) (in Dutch).
5. VLAREM-II, 1995. Milieukwaliteitsnormen voor grondwater. AMINAL, Bestuur Milieuvergunningen.
6. Xenidis A., LILLIS, T., AND HALIKIA, I., 1999. Lead dissolution from lead smelter slags using magnesium chloride solutions. In *The AusIMM Proceedings*, No. 2, pp. 37-44, (1999).
7. NIROND, (1997a) chemical inventory
8. IAEA, Processing of nuclear power plant waste streams containing boric acid. IAEA-TECDOC-911, Vienna, (1996)
9. Havard, P., Safe management and disposal of nuclear waste. Proceedings of the International Conference Safewaste 93, Volume 2, pp. 349-362, Avignon, France, (1993).
10. IAEA, Improved cement solidification of low and intermediate level radioactive wastes. IAEA-Technical Reports Series-350, Vienna, (1993).
11. Bethke, C.M., *The Geochemist's Workbench* ®, A User's Guide to Rxn, Act2, Tact, React, and GtPlot, Release 3.1, Hydrogeology Program, University of Illinois, (2000).

12. Ziegler, F., A.C., Johnson, The solubility of calcium zinkcate ($CaZn_2(OH)_6.2H_2O$). Cement and concrete research, 31: 1327-1332 (2001).
13. Richet, C., C. Galle, P. Le Bescop, H. Peycelon, S. Bejaoui, I. Tovena, I. Pointeau, V. L'Hostis, P. Lovera, Synthèse des connaissances sur le comportement à long terme des bétons applications aux colis cimentés, Rapport CEA-R-6050 (2004).
14. Moulin, I., W.E.E., Stone, J., Sanz, J.-Y., Bottero, F., Mosnier, and C., Haehnel, Lead and zinc retention during hydration of tri-calcium silicate: a study by sorption isotherms and ^{29}Si nuclear magnetic resonance spectroscopy. Langmuir, 15: 2829-2835 (1999).

AUTHOR INDEX

SUBJECT INDEX

Printed in the United States
By Bookmasters